Quantum Transport in Mesoscopic Systems

Quantum Transport in Mesoscopic Systems

Editors

David Sánchez
Michael Moskalets

MDPI • Basel • Beijing • Wuhan • Barcelona • Belgrade • Manchester • Tokyo • Cluj • Tianjin

Editors

David Sánchez
Institute for Cross-Disciplinary
Physics and Complex Systems
IFISC (UIB-CSIC)
Spain

Michael Moskalets
NTU "Kharkiv Polytechnic Institute"
Ukraine

Editorial Office
MDPI
St. Alban-Anlage 66
4052 Basel, Switzerland

This is a reprint of articles from the Special Issue published online in the open access journal *Entropy* (ISSN 1099-4300) (available at: https://www.mdpi.com/journal/entropy/special_issues/Quantum_Transport).

For citation purposes, cite each article independently as indicated on the article page online and as indicated below:

LastName, A.A.; LastName, B.B.; LastName, C.C. Article Title. *Journal Name* **Year**, *Article Number*, Page Range.

ISBN 978-3-03943-366-7 (Hbk)
ISBN 978-3-03943-367-4 (PDF)

Cover image courtesy of Michele Filippone and Gwendal Fève.

Contents

About the Editors

David Sánchez (Senior Lecturer) gained his Ph.D. in Physics at the Autonomous University of Madrid (2002), and his MA in Hispanic Philology at UNED (2014). He was also a postdoctoral researcher at the University of Geneva (2002–2004). Later he was a Ramon y Cajal fellow (2005–2008). Since 2011, he has been an associate professor at the University of the Balearic Islands, and today he is also a faculty member of the Institute for Cross-Disciplinary Physics and Complex Systems IFISC (UIB-CSIC). He has been a visiting scholar at the universities of Indiana, Texas, California, Stanford and ETH Zürich, has more than a hundred published papers in scientific journals, and has spoken at more than sixty conferences. His main research areas include nanophysics, quantum thermodynamics and language variation.

Michael Moskalets (Leading Research Fellow) gained his Dr. of Science in Theoretical Physics at the Institute for Single Crystals National Academy of Science of Ukraine (2008). He is a leading research fellow at the Department of Metal and Semiconductor Physics, at the National Technical University "Kharkiv Polytechnic Institute", Kharkiv, Ukraine. He has been a visiting scholar at the Geneva University, Aalto University, ENS Lyon, IFISC (UIB-CSIC) Palma de Mallorca, RWTH Aachen University, as well as the National Chiao Tung University of Taiwan. More than 100 of his papers have been published in scientific journals. He is the author of the book *"Scattering Matrix Approach to Non-Stationary Quantum Transport"*, Imperial College Press, London, 2011. His main research areas include quantum coherent single-electronics, nanophysics, quantum thermodynamics

Editorial

Quantum Transport in Mesoscopic Systems

David Sánchez [1],* and Michael Moskalets [2],*

[1] Institute for Cross-Disciplinary Physics and Complex Systems IFISC (UIB-CSIC), E-07122 Palma de Mallorca, Spain

[2] Department of Metal and Semiconductor Physics, National Technical University "Kharkiv Polytechnic Institute", 61002 Kharkiv, Ukraine

* Correspondence: david.sanchez@uib.es (D.S.); michael.moskalets@icloud.com (M.M.)

Received: 20 August 2020; Accepted: 26 August 2020; Published: 1 September 2020

Keywords: quantum transport; mesoscopic systems; nanophysics; quantum thermodynamics; quantum noise; quantum pumping; Kondo effect; thermoelectrics; heat transport

Mesoscopic physics has become a mature field. Its theoretical foundations and main models were established in the last two decades of the past century [1,2]. Ever since, quantum transport techniques have served as an excellent tool to understand the intriguing properties of charge carriers in nanoscale conductors [3,4]. However, in the last few years, the number of applications has grown so quickly that even experts find it difficult to stay updated with the recent advancements. The goal of the present special issue is to give a current snapshot of the field by means of a collection of review papers and research works that discuss the hottest theoretical questions and experimental results.

While the average current was the focus of early studies, the interest has gradually shifted to time-resolved transport. The motivation is partly due to new devices, such as single-electron emitters, which are able to inject quantized current pulses onto a Fermi sea for the investigation of inelastic and interaction effects upon electronic collisions. This is the subject of the review paper by Filippone et al. [5], in which Fermi liquid theories are employed to analyze strong correlations (Coulomb interactions) in the out-of-equilibrium dynamics of mesoscopic capacitors, a type of mesoscopic system whose response is purely dynamical. Here, dynamics is enforced via a time-dependent potential applied to a nearby gate. Under certain circumstances, the interplay of this potential and Coulomb interactions can lead to fractionalization effects in a single-electron transfer (quantized pumping). Chen and Zhu [6] find quantum pumping for a double-barrier system in the adiabatic limit. The novelty lies in their consideration of Dirac–Weyl quasiparticles. Tokura [7] also consider slow potentials, but the system is now an interferometer that allows not only for Aharanov–Bohm phases but also for spin-dependent shifts, due to both Rashba and Dresselhaus spin-orbit couplings. Meanwhile, Hashimoto and Uchiyama [8] tackle the nonadiabatic regime and present a complete analysis of the pumped charge, spin, and energy induced by temperature modulations in the attached reservoirs. A particularly useful approach that deals with this kind of problems is based on generalized master rate equations. Moldoveanu, Manolescu, and Gumundsson [9] illustrate the power of this method for a hybrid quantum-dot system that hosts both electronic and bosonic degrees of freedom. Among other things, they solve the master equations, including many-body effects in the transient response to time-dependent signals applied at the contact regions. Dynamically driven quantum devices are also suitable systems for testing alternative theoretical formulations. An example is the work of Pandey et al. [10], in which the Bohmian quantum theory is utilized to elucidate the role of non-Markovian conditions in graphene probed at very high frequencies.

In the recent cross-fertilization between thermodynamics and quantum physics, mesoscopic systems play a pivotal contribution. In their review article, Ansari, van Steensel, and Nazarov [11] connect information-theoretic concepts with the evaluation of entropy in quantum systems. They illustrate their discussion by calculating the entropy of various quantum heat engines. Quantum point contacts are prototypical mesoscopic devices that can precisely work as heat engines. It is, therefore, of utmost importance to understand their maximum generated power, as discussed by Kheradsoud et al. [12]. Interestingly, they find that power, efficiency, and fluctuations are bounded by thermodynamic uncertainty relations. Additionally, Bustos, Marún, and Calvo [13] introduce mechanical degrees of freedom to analyze the dynamics of quantum motors built, e.g., from double quantum dots coupled to rotors. A Langevin approach allows them to generically describe both motors and pumps out of equilibrium, which are relevant for quantum refrigeration setups. Remarkably, some of the well-established results in linear response (Onsager reciprocity, fluctuation-dissipation relations) also hold far from equilibrium. Maisel and López [14] demonstrate, with a capacitively coupled doubled quantum dot system, that it is possible to find bias configurations that lead to stalling currents, around which the above results were verified.

Quantum conductors constitute excellent platforms for the measurement and manipulation of thermal gradients and currents while keeping the quantum character of energy carriers. Biele and D'Agosta [15] review the standard theoretical approaches to quantum thermal transport (Landauer–Büttiker formalism and Boltzmann equation), pointing to their strengths and limitations. To overcome the latter, they discuss advanced methods, such as time-dependent density functional theory, the nonequilibrium Green's functions approach, and density-matrix formulations. Atomistic computations are reviewed by Medrano Sandonas et al. [16] in the context of nanophononics. Clearly, phonons should be taken into account in any general description of heat transport in nanodevices, especially in molecular junctions. A density-functional tight-binding module specifically designed to deal with phonon transport is able to compute the phonon conductivity of molecules sandwiched between thermally biased metallic contacts. Perroni and Cataudella [17] consider the case of a fullerene and study the combined influence of vibrations and Coulomb interactions in the thermoelectric transport through the molecule.

In confined mesoscopic systems, electron–electron interactions can lead to strong correlations visible in transport measurements. A celebrated phenomenon is the Kondo effect, where the unpaired spin of an electron localized inside a quantum dot forms, at a low temperature, a many-body singlet with the spin density arising from conduction electrons that propagate in the leads attached to the dot. Tettamanzi [18] review the Kondo and the Kondo–Fano effects in silicon nanostructures, taking into account correlations between pseudospins belonging to different degeneracy points in the conduction bands. Simultaneous fluctuations in both the spin and pseudospin degrees of freedom give rise to higher symmetry Kondo states. Lee, Dong, and Lei [19] propose a multiterminal setup comprising a quantum dot attached to two ferromagnetic contacts and one superconducting lead, with the aim of assessing both local and nonlocal conductances within a slave-boson mean-field approximation. Their main finding is a competition between the superconductivity proximity effect, Kondo correlations, and spin polarizations that could be analyzed with a careful study of the conductance.

We began this Editorial with an emphasis on time-dependent currents. We would like to finish our presentation with a somewhat related quantity, namely, the noise, since current fluctuations are defined from time correlators. Bulka and Luczak [20] analyze the electric current noise in a ring structure that supports persistent currents. When the ring is pierced by an external magnetic field, the interference pattern is affected by the Aharonov–Bohm effect and this is reflected in the noise as a function of the flux. In mesoscopic conductors, not only charge, but also heat fluctuations, are significant. This leads to heat and mixed charge-heat correlators, as illustrated by Ronetti et al. [21] for a harmonically driven quantum Hall bar. The system shows quasiparticle excitations of fractional charge (Laughlin states), and it is demonstrated that the mixed noise differs for integer and fractional filling factors. Finally, current–current correlations can provide us with valuable information about the electronic traversal time through mesoscopic constrictions. Ridley, Sentef, and Tuovinen [22] calculate the cross-correlations of graphene nanoribbons and find that the sample disorder increased the traversal time.

Overall, these papers represent an outstanding perspective of current research in nanophysics. They show that the field is actively developing and alive with problems that are interesting to a great variety of physicists, whose concerns range from condensed matter to quantum information and thermodynamics. There is still plenty of room at the bottom, which implies fruitful opportunities in the near future.

Funding: This work was funded by AEI grant numbers MAT2017-82639 and MDM2017-0711.

Acknowledgments: We express our thanks to the authors of the above contributions and to the journal Entropy and MDPI for their support during the preparation of the special issue.

Conflicts of Interest: The authors declare no conflict of interest.

References

1. Büttiker, M. Four-terminal Phase-Coherent Conductance. *Phys. Rev. Lett.* **1986**, *57*, 1761.
2. Imry, Y. *Introduction to Mesoscopic Physics*; Oxford University Press: Oxford, UK, 1997.
3. Nazarov, Y.V.; Blanter, Y.M. *Quantum Transport: Introduction to Nanoscience*; Cambridge University Press: Cambridge, UK, 2009.
4. Ihn, T. *Semiconductor Nanostructures: Quantum States and Electronic Transport*; Oxford University Press: Oxford, UK, 2009.
5. Filippone, M.; Marguerite, A.; Le Hur, K.; Fève, G.; Mora, C. Phase-Coherent Dynamics of Quantum Devices with Local Interactions. *Entropy* **2020**, *22*, 847.
6. Chen, X.; Zhu, R. Quantum Pumping with Adiabatically Modulated Barriers in Three-Band Pseudospin-1 Dirac–Weyl Systems. *Entropy* **2019**, *21*, 209.
7. Tokura, Y. Quantum Adiabatic Pumping in Rashba-Dresselhaus-Aharonov-Bohm Interferometer. *Entropy* **2019**, *21*, 828.
8. Hashimoto, K.; Uchiyama, C. Nonadiabaticity in Quantum Pumping Phenomena under Relaxation. *Entropy* **2019**, *21*, 842.
9. Moldoveanu, V.; Manolescu, A.; Gudmundsson, V. Generalized Master Equation Approach to Time-Dependent Many-Body Transport. *Entropy* **2019**, *21*, 731.
10. Pandey, D.; Colomés, E.; Albareda, G.; Oriols, X. Stochastic Schrödinger Equations and Conditional States: A General Non-Markovian Quantum Electron Transport Simulator for THz Electronics. *Entropy* **2019**, *21*, 1148.
11. Ansari, M.H.; van Steensel, A.; Nazarov, Y.V. Entropy Production in Quantum is Different. *Entropy* **2019**, *21*, 854.
12. Kheradsoud, S.; Dashti, N.; Misiorny, M.; Potts, P.P.; Splettstoesser, J.; Samuelsson, P. Power, Efficiency and Fluctuations in a Quantum Point Contact as Steady-State Thermoelectric Heat Engine. *Entropy* **2019**, *21*, 777.
13. Bustos-Marún, R.A.; Calvo, H.L. Thermodynamics and Steady State of Quantum Motors and Pumps Far from Equilibrium. *Entropy* **2019**, *21*, 824.
14. Maisel, L.; López, R. Effective Equilibrium in Out-of-Equilibrium Interacting Coupled Nanoconductors. *Entropy* **2020**, *22*, 8.
15. Biele, R.; D'Agosta, R. Beyond the State of the Art: Novel Approaches for Thermal and Electrical Transport in Nanoscale Devices. *Entropy* **2019**, *21*, 752.
16. Medrano Sandonas, L.; Gutierrez, R.; Pecchia, A.; Croy, A.; Cuniberti, G. Quantum Phonon Transport in Nanomaterials: Combining Atomistic with Non-Equilibrium Green's Function Techniques. *Entropy* **2019**, *21*, 735.
17. Perroni, C.A.; Cataudella, V. On the Role of Local Many-Body Interactions on the Thermoelectric Properties of Fullerene Junctions. *Entropy* **2019**, *21*, 754.
18. Tettamanzi, G.C. Unusual Quantum Transport Mechanisms in Silicon Nano-Devices. *Entropy* **2019**, *21*, 676.
19. Lee, C.; Dong, B.; Lei, X.-L. Enhanced Negative Nonlocal Conductance in an Interacting Quantum Dot Connected to Two Ferromagnetic Leads and One Superconducting Lead. *Entropy* **2019**, *21*, 1003.
20. Bułka, B.R.; Łuczak, J. Current Correlations in a Quantum Dot Ring: A Role of Quantum Interference. *Entropy* **2019**, *21*, 527.

21. Ronetti, F.; Acciai, M.; Ferraro, D.; Rech, J.; Jonckheere, T.; Martin, T.; Sassetti, M. Symmetry Properties of Mixed and Heat Photo-Assisted Noise in the Quantum Hall Regime. *Entropy* **2019**, *21*, 730.

22. Ridley, M.; Sentef, M.A.; Tuovinen, R. Electron Traversal Times in Disordered Graphene Nanoribbons. *Entropy* **2019**, *21*, 737.

Review

Phase-Coherent Dynamics of Quantum Devices with Local Interactions

Michele Filippone [1],*, Arthur Marguerite [2], Karyn Le Hur [3], Gwendal Fève [4] and Christophe Mora [5]

[1] Department of Quantum Matter Physics, University of Geneva 24 Quai Ernest-Ansermet, CH-1211 Geneva, Switzerland
[2] Department of Condensed Matter Physics, Weizmann Institute of Science, Rehovot 7610001, Israel; arthur.marguerite@weizmann.ac.il
[3] CPHT, CNRS, Institut Polytechnique de Paris, Route de Saclay, 91128 Palaiseau, France; karyn.le-hur@polytechnique.edu
[4] Laboratoire de Physique de l'Ecole Normale Supérieure, ENS, Université PSL, CNRS, Sorbonne Université, Université de Paris, F-75005 Paris, France; gwendal.feve@lpa.ens.fr
[5] Laboratoire Matériaux et Phénomènes Quantiques, CNRS, Université de Paris, F-75013 Paris, France; christophe.mora@u-paris.fr
* Correspondence: michele.filippone@unige.ch

Received: 22 April 2020; Accepted: 2 July 2020; Published: 31 July 2020

Abstract: This review illustrates how Local Fermi Liquid (LFL) theories describe the strongly correlated and coherent low-energy dynamics of quantum dot devices. This approach consists in an effective elastic scattering theory, accounting exactly for strong correlations. Here, we focus on the mesoscopic capacitor and recent experiments achieving a Coulomb-induced quantum state transfer. Extending to out-of-equilibrium regimes, aimed at triggered single electron emission, we illustrate how inelastic effects become crucial, requiring approaches beyond LFLs, shedding new light on past experimental data by showing clear interaction effects in the dynamics of mesoscopic capacitors.

Keywords: dynamics of strongly correlated quantum systems; quantum transport; mesoscopic physics; quantum dots; quantum capacitor; local fermi liquids; kondo effect; coulomb blockade

---

1. Introduction

The manipulation of local electrostatic potentials and electron Coulomb interactions has been pivotal to control quantized charges in solid state devices. Coulomb blockade [1–3] has revealed to be a formidable tool to trapping and manipulating single electrons in localized regions behaving as highly tunable artificial impurities, so called quantum dots. Beyond a clear practical interest, which make quantum dots promising candidates to become the building block of a quantum processor [4–6], hybrid [7] quantum dot systems also became a formidable platform to address the dynamics of many-body systems in a controlled fashion, and a comprehensive theory, which could establish the role of Coulomb interactions when these systems are strongly driven out of equilibrium, is still under construction.

Beyond theoretical interest, this question is important for ongoing experiments with mesoscopic devices aimed towards the full control of single electrons out of equilibrium. Figure 1 reports some of these experiments [8–19], in addition to the mesoscopic capacitor [20–26], which will be extensively discussed in this review. These experiments and significant others [27–33] have a common working principle: A fast [34] time-dependent voltage drive $V(t)$, applied either on metallic or gating contacts, triggers emission of well defined electronic excitations. Remarkably, these experiments achieved to generate, manipulate, and detect single electrons on top of a complex many-body state such as the Fermi sea. A comprehensive review of these experiments can be found in Ref. [35].

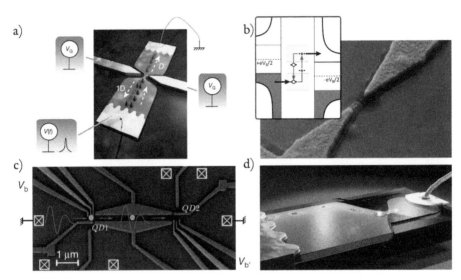

Figure 1. Some recent experiments achieving real-time control of single electrons. (**a**) Leviton generation by a Lorentzian voltage pulse in metallic contacts, generating a noiseless wave-packet carrying the electron charge e [8–13]. This wave-packet is partitioned on a Quantum Point Contact (QPC), whose transmission D is controlled by the split-gate voltage V_G. (**b**) Single quantum level electron turnstile [18,19]. Two superconductors, biased by a voltage V_B, are connected by a single-level quantum dot. Inset—Working principle of the device: A gate voltage controls the orbital energy of the quantum dot, which is filled by the left superconductor and emptied in the right one. (**c**) Long-range single-electron transfer via a radio-frequency pulse between two distant quantum dots QD1 and QD2 [14–17]. The electron "surfs" along the moving potential generated by the radio-frequency source and is transferred along a one-dimensional channel from QD1 to QD2. (**d**) The mesoscopic capacitor [20–26], in which a gate-driven quantum dot emits single electrons through a QPC in a two-dimensional electron gas. This platform will be extensively discussed in this review.

In this context, interactions are usually considered detrimental, as they are responsible for inelastic effects leading to diffusion and dephasing [36]. Interaction screening or, alternatively, the disappearance of such inelastic effects at low driving energies or temperatures [37–42] is thus crucial to identify single-electron long-lived excitations (quasi-particles) close to the Fermi surface. The possibility of identifying such excitations, even in the presence of strong Coulomb interactions, is the core of the Fermi liquid theory of electron gases in solids [43,44], usually identified with the $\propto T^2$ suppression of resistivities in bulk metals. It is the validity of this theory for conventional metals that actually underpins the success of Landauer–Büttiker elastic scattering theory [45–47] to describe coherent transport in mesoscopic devices.

The aim of this review is to show how a similar approach can also be devised to describe transport in mesoscopic conductors involving the interaction of artificial quantum impurities. In these systems, electron-electron interactions are only significant in the confined and local quantum dot regions, and not in the leads for instance, therefore we use the terminology of a Local Fermi Liquid theory (LFL) in contrast to the conventional Fermi liquid approach for bulk interactions. Originally, the first LFL approach [48] was introduced to derive the low energy thermodynamic and transport properties of Kondo local scatterers in materials doped with magnetic impurities [49]. In this review, we will show how LFLs provide the unifying framework to describe both elastic scattering and strong correlation phenomena in the out-of-equilibrium dynamics of mesoscopic devices. This approach makes also clear how inelastic effects, induced by Coulomb interactions, become visible and unavoidable as soon as such systems are strongly driven out of equilibrium. We will discuss how extensions of LFLs and related approaches describe such regimes as well.

As a paradigmatic example, we will focus on recent experiments showing the electron transfer with Coulomb interactions [50], (see Figure 2), and, in more detail, on the mesoscopic capacitor [20–26], (see Figure 6). The mesoscopic capacitor does not support the DC transport, and it makes possible the direct investigation and control of the coherent dynamics of charge carriers. The LFL description of such devices entails the seminal results relying on self-consistent elastic scattering approaches by Büttiker and collaborators [51–56], but it also allows one to describe effects induced by strong Coulomb correlations, which remain nevertheless elastic and coherent. The intuition provided by the LFL approach is a powerful lens through which it is possible to explore various out-of-equilibrium phenomena, which are coherent in nature but are governed by Coulomb interactions. As an example, we will show how a bold treatment of Coulomb interaction unveils originally overlooked strong dynamical effects, triggered by interactions, in past experimental measurements showing fractionalization effects in out-of-equilibrium charge emission from a driven mesoscopic capacitor [25].

This review is structured as follows. In Section 2, we give a simple example showing how Coulomb interactions trigger phase-coherent electron state transfer in experiments as those reported in Ref. [50], Figure 2. Section 3 discusses how the effective LFL approach [57–64] provides the unified framework describing such coherent phenomena. In Section 4, we consider the study of the low-energy dynamics of the mesoscopic capacitor, in which the LFL approach has been fruitfully applied [65–69], showing novel quantum coherent effects. Section 5 extends the LFL approach out of equilibrium and describes signatures of interactions in measurements of strongly driven mesoscopic capacitors [25].

2. Phase-Coherence in Quantum Devices with Local Interactions

To illustrate the restoration of phase coherence at low temperatures in the presence of interactions, we consider two counter-propagating edge states entering a metallic quantum dot, or cavity. Such a system was recently realized as a constitutive element of the Mach–Zender interferometer of Ref. [50], reported in Figure 2. In that experiment, the observation of fully preserved Mach–Zehnder oscillations, in a system in which a quantum Hall edge state penetrates a metallic floating island demonstrates an interaction-induced, restored phase coherence [70,71].

The dominant electron-electron interactions in the cavity have the form of a charging energy [1–3]

$$\mathcal{H}_c = E_c[N - \mathcal{N}_g(t)]^2, \tag{1}$$

in which N is the number of electrons in the island, C_g is the geometric capacitance, and $\mathcal{N}_g = C_g V_g(t)/e$ is the dimensionless gate voltage, which corresponds to the number of charges that would set in the cavity if N was a classical, non-quantized, quantity. We also define the charging energy $E_c = e^2/2C_g$: The energy cost required the addition of one electron in the isolated cavity. For the present discussion, we neglect the time-dependence of the gate-potential V_g, which will be reintroduced to describe driven settings. In the linear-dispersion approximation, the right/left-moving fermions $\Psi_{R,L}$ in Figure 2, moving with Fermi velocity v_F, are described by the Hamiltonian:

$$\mathcal{H}_{kin} = v_F \hbar \sum_{\alpha=R/L} \int_{-\infty}^{\infty} dx\, \Psi_\alpha^\dagger(x)(-i\alpha\partial_x)\Psi_\alpha(x), \tag{2}$$

with the sign $\alpha = +/-$ multiplying the ∂_x operator for right- and left-movers respectively. The floating island occupies the semi-infinite one-dimensional space located at $x > 0$ with the corresponding charge $N = \sum_\alpha \int_0^\infty dx \Psi_\alpha^\dagger(x)\Psi_\alpha(x)$. It is important to stress that the model (1)–(2) is general and effective in describing different quantum dot devices. It was originally suggested by Matveev to describe quantum dots connected to leads through a single conduction channel [72] and it equally describes the mesoscopic capacitor, see Sections 4 and 5.

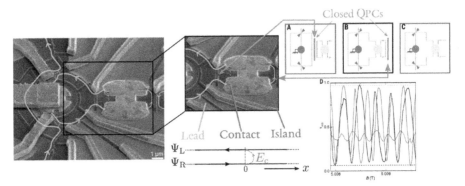

Figure 2. Left—Mach–Zehnder interferometer with a floating metallic island (colored in yellow) [50]. The green lines denote chiral quantum Hall edge states, which can enter the floating island passing through a gate-tunable QPC (in blue). An additional QPC separates the floating island from an additional reservoir on its right. Center—The floating island is described by two infinite counter-propagating edges, exchanging electrons coherently thanks to the charging energy E_c of the island (red arrow). Right—Mach–Zehnder visibility of the device as a function of magnetic field B. Oscillation of this quantity as function of B signals quantum coherent interference between two paths encircling an Aharonov–Bohm flux. In the situation sketched in box A, the first QPC is closed and the interferometer is disconnected from the island and visibility oscillations are observed, as expected (red line). Remarkably, the oscillations persist (black line) in the situation sketched in box B, where the leftmost QPC is open and one edge channel enters the floating island. The visibility oscillations are only suppressed in the situation sketched in box C, where the rightmost QPC is also open and the island is connected to a further reservoir (blue line).

The model (1)–(2) characterizes an open-dot limit in the sense that it does not contain an explicit backscattering term coupling the L and R channels. It can be solved exactly by relying on the bosonization formalism [73–76], which, in this specific case, maps interacting fermions onto non-interacting bosons [72,77,78]. Using this mapping, one can show that the charging energy E_c perfectly converts, far from the contact, right-movers into left-movers. This fact is made apparent by the "reflection" Green function G_{LR} [77]:

$$G_{LR} = \left\langle \mathcal{T}_\tau \Psi_L^\dagger(x,\tau) \Psi_R(x',0) \right\rangle \simeq e^{-i2\pi\mathcal{N}_g} \frac{T/2v_F}{\sin\left[\frac{\pi T}{\hbar}\left(\tau + i(x+x') - i\frac{\pi\hbar}{E_c e^C}\right)\right]}, \tag{3}$$

which we consider at finite temperature T. In Equation (3), $\mathcal{T}_\tau$ is the usual time-ordering operator defined as $\mathcal{T}_\tau A(\tau)B(\tau') = \theta(\tau - \tau')A(\tau)B(\tau') \pm \theta(\tau' - \tau)B(\tau')A(\tau)$, in which the sign $+/-$ is chosen depending on the bosonic/fermionic statistics of the operators A and B [79] and $\theta(\tau)$ is the Heaviside step function. As first noted by Aleiner and Glazman [77], the form of G_{LR} at large (imaginary) time τ corresponds to the elastic reflection of the electrons incident on the dot, with a well-defined scattering phase $\pi\mathcal{N}_g$. The correlation function (3) would be identical if the interacting dot was replaced with a non-interacting wire of length $v_F\pi\hbar/E_c\gamma$ (with $\ln\gamma = C \simeq 0.5772$ being Euler's constant), imprinting a phase $\pi\mathcal{N}_g$ when electrons are back-reflected at the end of the wire [80].

The physical picture behind Equation (3) is that an electron entering and thereby charging the island violates energy conservation at low temperature and must escape on a time scale $\hbar/E_c$ fixed by the uncertainty principle. The release of this incoming electron can happen either elastically, in which case the electron keeps its energy, or inelastically via the excitation of electron-hole pairs. As we discuss in Section 3.1, inelastic processes are suppressed by the phase space factor $(\varepsilon/E_c)^2$, ε being the energy of the incoming electron, and they die out at low energy or large distance (time), reestablishing purely elastic scattering despite a nominally strong interaction.

Equation (3) is thus a remarkable example of how interactions trigger coherent effects in mesoscopic devices. It has been derived here for an open dot, a specific limit in which the charge quantization of the island is fully suppressed. However, the restoration of phase coherence at low energy is more general and applies for an arbitrary lead-island transmission, in particular in the tunneling limit where the charge states of the quantum dot are well quantized [1–3]. This quantization is known to induce Coulomb blockade in the conductance of the device, see Figure 3. Nevertheless, a Coulomb blockaded dot acts at low energy as an elastic scatterer imprinting a phase δ [81,82] related to its average occupation $\langle N \rangle$ via the Friedel sum rule, see Section 3.2. For weak transmissions, $\langle N \rangle$ strongly deviates from the classical value $\mathcal{N}_g$. These features constitute the main characteristics of the local Fermi liquid picture detailed in the forthcoming sections.

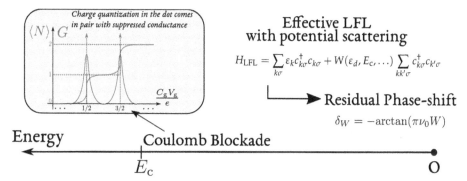

Figure 3. Coulomb blockade and emergent LFL behavior. When the typical energy of the system (temperature, bias-voltage, ...) is smaller than the charging energy E_c, charge quantization $Q = e \langle N \rangle$ in the dot suppresses the conductance G of the system. Degeneracy between different charge occupations lead to conductance peaks, which become larger the stronger the tunnel exchange of electrons with the leads. Conductance peaks and charge quantization disappear in the open-dot limit. For any tunneling strength, the dot behaves as an elastic scatterer described by the LFL theory (8), with potential scattering of strength W, inducing a phase-shift δ_W on lead electrons set by the dot occupation $\langle N \rangle$.

3. What Are Local Fermi Liquids and Why Are They Important to Understand Quantum-Dot Devices?

In this Section, we introduce the local Fermi liquid theory and discuss its application to quantum transport devices. The general system considered in this paper is a central interacting region, such as a quantum dot, connected to leads described as non-interacting electronic reservoirs. The Hamiltonian takes the general form:

$$\mathcal{H} = \mathcal{H}_{\text{res}} + \mathcal{H}_{\text{res-dot}} + \mathcal{H}_{\text{dot}} + \mathcal{H}_{\text{c}}. \tag{4}$$

The first term describes the lead reservoir, which could be either a normal metal [14], a chiral edge state in the quantum Hall regime [22,29], or a superconductor [18]. In the case of a normal metal, it is given by:

$$\mathcal{H}_{\text{res}} = \sum_k \varepsilon_k c_k^\dagger c_k , \tag{5}$$

in which c_k annihilates a fermion in the eigenstate state k of energy ε_k in the reservoir. For instance, in Figure 2, the reservoir modes correspond to the $x < 0$ components of the operators $\Psi_{\text{R/L}}$. The field $\Psi_{\text{res}}(x) = \theta(-x)\Psi_{\text{R}}(x) + \theta(x)\Psi_{\text{L}}(-x)$, with $\theta(x)$ the Heaviside step function, unfolds the chiral field onto the interval $x \in [-\infty, \infty]$ and its Fourier transform $c_k = \int_{-\infty}^{\infty} dx e^{-ikx}\Psi_{\text{res}}(x)$ recovers Equation (5) from Equation (2), with $\varepsilon_k = \hbar v_F k$.

The single particle physics of the quantum dot is described instead by:

$$\mathcal{H}_{\text{dot}} = \sum_l (\varepsilon_d + \varepsilon_l) n_l \tag{6}$$

in which $n_l = d_l^\dagger d_l$ counts the occupation of the orbital level l and d_l annihilates fermions in that state. The spectrum can be either discrete for a finite size quantum dot or dense for a metallic dot as in the case of Figure 2. We also introduced the orbital energy ε_d as a reference. $\mathcal{H}_{\text{res-dot}}$ describes the exchange of electrons between dot and reservoir. It has generally the form of a tunneling Hamiltonian:

$$\mathcal{H}_{\text{res-dot}} = t \sum_{k,l} \left[c_k^\dagger d_l + d_l^\dagger c_k \right], \tag{7}$$

in which we neglect, for simplicity, any k dependence of the tunneling amplitude t. The charging energy $\mathcal{H}_c$ is given in Equation (1) with the dot occupation operator $N = \sum_l n_l$.

Without any approximation, deriving the out-of-equilibrium dynamics of interacting models such as Equation (4) is a formidable task. The presence of local interactions leads to inelastic scattering events, creating particle-hole pairs when electron scatter on the dot (see Figure 4). From a technical point of view, such processes are difficult to handle and, even if these difficulties are overcome, one has to identify the dominant physical mechanisms governing the charge dynamics. In our discussion, interactions are usually controlled by the charging energy E_c, which cannot be treated perturbatively in Coulomb blockade regimes. The possibility to rely on Wick's theorem [83], when performing perturbative calculations in the exchange term $\mathcal{H}_{\text{res-dot}}$, is also denied. Thus, one has to look for a more efficient theoretical approach.

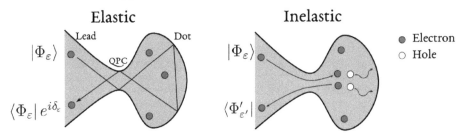

Figure 4. Difference between elastic (**left**) and inelastic (**right**) events for electrons scattering on a quantum dot. In the elastic case, electrons do not change energy ε. The wave function is preserved and the only residual effect of scattering is a phase-shift δ_ε. In the inelastic case, many-body interactions trigger the creation of particle-hole pairs. Outgoing electrons are then emitted in a state $|\Phi'_{\varepsilon'}\rangle$ of energy ε' different from the initial ε and phase coherence is gradually lost.

3.1. The Local Fermi Liquid

The local Fermi liquid approach is justified by the physical picture already presented in Section 2, namely that an incoming reservoir electron with an energy much smaller than the charging energy of the quantum dot is effectively scattered in a purely elastic way [77]. At temperatures well below the charging energy E_c, energy conservation prevents any permanent change in the charge of the quantum dot and each electron entering the dot must be compensated by an electron leaving it within the (short) time $\hbar/E_c$ fixed by the uncertainty principle. The electron escape can occur via elastic or inelastic processes, sketched in Figure 4, depending on whether the electron energy is preserved or not. Inelastic processes cause decoherence and call for a many-body approach to be properly evaluated.

At low energy ε of the incoming electron, the inelastic processes are typically suppressed by the ratio $(\varepsilon/E_{\text{FL}})^2$ [43,44]. E_{FL} is a Fermi liquid energy scale, typically of the order of the charging energy E_c. Nevertheless, in the presence of spin-fluctuations, the emergence of strong Kondo correlations, to be discussed in Section 3.3.2, can sensibly reduce E_{FL} down to the Kondo energy scale T_K, (see Equation (26)). Therefore, in the $\varepsilon \to 0$ limit, inelastic processes are ignored and the scattering is purely elastic. It is described within a single-particle formalism where the scattering by the quantum dot imprints a phase shift δ_W to the outgoing electronic wave functions. This phase shift alone incorporates all interaction and correlation effects.

The simplest model entailing these features is a free Fermi gas in which a delta barrier located at $x = 0$ (the entrance of the dot) scatters elastically quasi-particles. In the language of second quantization, the delta barrier is described by the electron operator $\Psi_{res}^\dagger(x = 0)\Psi_{res}(x = 0)$, and its strength W has to depend on the parameters of the parent model (such as the orbital energy ε_d, the charging energy E_c, etc.). Switching to momentum space, such model is a free Fermi gas with a potential scattering term:

$$\mathcal{H}_{LFL} = \sum_{k\sigma} \varepsilon_k c_{k\sigma}^\dagger c_{k\sigma} + W(\varepsilon_d, E_c, \ldots) \sum_{kk'\sigma} c_{k\sigma}^\dagger c_{k'\sigma} + \mathcal{O}\left(\frac{\varepsilon}{E_{FL}}\right)^2 \tag{8}$$

where the scattering potential leads, as shown in Appendix A, to the quasi-particle phase shift:

$$\delta_W = -\arctan\left(\pi\nu_0 W\right), \tag{9}$$

in which ν_0 is the density of states of the lead electrons at the Fermi energy, see also Equation (A21) in Appendix A for its rigorous definition. In this *Local Fermi Liquid* Hamiltonian, σ labels either a spin polarization or a channel. The number of channels in the lead can be controlled by the opening of a quantum point contact [84]. The potential strength W can be cumbersome to compute, but it is nevertheless related to the occupancy of the quantum dot via the Friedel sum rule, as explained in Section 3.2.

The simplicity of the local Fermi liquid Hamiltonian (8) makes it powerful to evaluate low energy properties. Being non-interacting, it also includes the restoration of phase coherence in the scattering of electrons seen in Section 2. An important assumption that we made is that the system exhibits a Fermi liquid ground state, or Fermi liquid fixed point in the language of the renormalization group. Non-Fermi liquid fixed points exist and cannot be described by such Hamiltonian [85], but they are generally fine-tuned and unstable with respect to perturbations. Furthermore, Equation (8) is not applicable to genuine out-of-equilibrium regimes, when the perturbations are too strong or vary too fast with respect to the Fermi liquid energy scale E_{FL} (typically of the order of the charging energy E_c).

3.2. The Role of the Friedel Sum Rule in Local Fermi Liquid Theories

In the process of electron backscattering by the quantum dot, or by the interacting central region, the phase shift relates the incoming and outgoing electronic wavefunctions $\Psi_L(0^-) = e^{2i\delta_\varepsilon}\Psi_R(0^-)$, see Appendix A for an explicit illustration on the resonant level non-interacting model. The Friedel sum rule [86] establishes the relation between the average charge occupation of the dot $\langle N \rangle$ and this phase-shift δ. Its form in the case of M conducting channels reads:

$$\langle N \rangle = \frac{1}{\pi} \sum_{\sigma=1}^{M} \delta_\sigma. \tag{10}$$

The Friedel sum rule has been proven rigorously for interacting models [87,88]. It is valid as long the ground state has a Fermi liquid character. Physically, it can be understood in the following way: The derivative of the phase shift δ_ε with respect to energy defines (up to $\hbar$) the Wigner-Smith scattering time [89], see Equation (A38), that is the time delay experienced by a scattered electron. In the presence of a continuous flow of electrons, a time delay implies that some fraction of the electronic charge has been (pumped) deposited in the quantum dot [51,90]. Therefore the phase shift amounts to a left-over charge and it does not matter that electrons are interacting on the quantum dot as long as they are not in the leads, which is the essence of the local Fermi liquid approach.

The Friedel sum rule (10) combined with Equation (9) relates the dot occupancy to the potential scattering strength. For the single-channel case ($M = 1$), one finds:

$$\langle N \rangle = -\frac{1}{\pi} \arctan \left(\pi \nu_0 W \right) . \tag{11}$$

This is an important result because the dot occupation $\langle N \rangle$ is a thermodynamic quantity, which can be also accessed in interacting models, allowing us to address quantitatively the close-to-equilibrium dynamics of driven settings, as we will discuss in Section 4.5.

We emphasize that the local Fermi liquid approach of Equation (8) can be extended to perturbatively include inelastic scattering and higher-order energy corrections and relate these terms to thermodynamic observables. This program has been realized in detail for the Anderson and Kondo models [57–64,91].

3.3. Derivation of the LFL Theory in the Coulomb Blockade and Anderson Model

We show now how the effective theory (8) can be explicitly derived from realistic models describing Coulomb blockaded quantum dot devices [66]. We focus on the Coulomb Blockade Model (CBM) [2,3]:

$$\mathcal{H}_{\mathrm{CBM}} = \sum_k \varepsilon_k c_k^\dagger c_k + t \sum_{k,l} \left[c_k^\dagger d_l + d_l^\dagger c_k \right] + \sum_l (\varepsilon_d + \varepsilon_l) d_l^\dagger d_l + E_{\mathrm{c}} \left(N - \mathcal{N}_{\mathrm{g}} \right)^2 \tag{12}$$

and the Anderson Impurity Model (AIM), which, in its standard form, reads [92,93]:

$$\mathcal{H}_{\mathrm{AIM}} = \sum_{k,\sigma} \varepsilon_{k,\sigma} c_{k,\sigma}^\dagger c_{k,\sigma} + t \sum_{k,\sigma} \left[c_{k,\sigma}^\dagger d_\sigma + d_\sigma^\dagger c_{k,\sigma} \right] + \varepsilon_d \sum_\sigma d_\sigma^\dagger d_\sigma + U n_\uparrow n_\downarrow . \tag{13}$$

Adding $-eV_{\mathrm{g}}N + E_{\mathrm{c}}\mathcal{N}_{\mathrm{g}}^2$ to the AIM and for $U = E_{\mathrm{c}}$, the charging energy (1) becomes apparent in Equation (13), as in Equations (4)–(12). The CBM coincides with the Hamiltonian (4) and describes the mesoscopic capacitor in the Quantum Hall regime: A reservoir of spinless fermions c_k of momentum k is tunnel coupled to an island with discrete spectrum ε_l. The AIM includes the spin degree of freedom and considers a single interacting level in the quantum dot. This model encompasses Kondo correlated regimes [49,94] and describes well the experiments [95,96].

To derive the LFL Hamiltonian (8), we rely on the Schrieffer–Wolff (SW) transformation [44,97], first devised to map the AIM [94] onto the Coqblin–Schrieffer model [98], and that we extend here to the CBM. Far from the charge degeneracy points, in the $t \ll E_{\mathrm{c}}$ limit, the ground-state charge configuration $n = \langle N \rangle$ is fixed by the gate potential V_{g} and fluctuations to $n \pm 1$ require energies of order E_{c}. For temperatures much lower than E_{c}, the charge degree of freedom of the quantum dot is frozen, acting but virtually on the low energy behavior of the system. The SW transformation is a controlled procedure to diagonalize perturbatively in t the Hamiltonian. The Hamiltonian is separated in two parts $\mathcal{H} = \mathcal{H}_0 + \mathcal{H}_{\mathrm{res-dot}}$, in which $\mathcal{H}_0$ is diagonal in the charge sectors labeled by the eigenvalues n of the dot occupation N, which are mixed by the tunneling Hamiltonian $\mathcal{H}_{\mathrm{res-dot}}$, involving the tunneling amplitude t. The perturbative diagonalization consists of finding the Hermitian operator S (of order t) generating the unitary $U = e^{iS}$ rotating the Hamiltonian in the diagonal form $\mathcal{H}' = U^\dagger \mathcal{H} U$. To leading order in S we find:

$$\mathcal{H}' = \mathcal{H}_0 + \mathcal{H}_{\mathrm{res-dot}} + i \left[S, \mathcal{H}_0 \right] + O \left(t^2 \right) . \tag{14}$$

This Hamiltonian is block diagonal if the condition:

$$i\mathcal{H}_{\text{res}-\text{dot}} = [S, \mathcal{H}_0] \tag{15}$$

is fulfilled and Equation (14) becomes:

$$\mathcal{H}' = \mathcal{H}_0 + \frac{i}{2}[S, \mathcal{H}_{\text{res}-\text{dot}}], \tag{16}$$

which is then projected on separated charge sectors.

3.3.1. Coulomb Blockade Model

To derive the effective low-energy form of the CBM model, it is useful, following Grabert [99,100], to decouple the charge occupancy of the dot from the fermionic degree of freedom of the electrons. This is achieved by adding the operator $\hat{n} = \sum_n |n\rangle\langle n|$, measuring to the dot occupation number. The fermionic operators d_l in Equation (12) are replaced by new operators describing a non-interacting electron gas in the dot. The Hamiltonian (12) acquires then the form:

$$\mathcal{H}_{\text{CBM}} = \sum_k \varepsilon_k c_k^\dagger c_k + t \sum_{n,k,l} \left[d_l^\dagger c_k |n+1\rangle\langle n| + \text{h.c.} \right] + \sum_l \varepsilon_l d_l^\dagger d_l + \varepsilon_d \hat{n} + E_c \left(\hat{n} - \mathcal{N}_g \right)^2. \tag{17}$$

The operator $S = s + s^\dagger$ fulfilling the condition (15) reads:

$$s = it \sum_{k,l,n} s_{kln} c_k^\dagger d_l |n-1\rangle\langle n|, \qquad s_{kln} = \frac{1}{\varepsilon_l - \varepsilon_k + E_c(2n-1) + \varepsilon_d}. \tag{18}$$

This operator, when inserted into Equation (16), also generates higher order couplings between sectors of charge n and $n \pm 2$, which we neglect in the present discussion. The Hamiltonian becomes then block diagonal in the sectors given by different values of n. For $(\mathcal{N}_g - \varepsilon_d C_g/e) \in [-1/2, 1/2]$, the lowest energy sector corresponds to $n = 0$ and the effective Hamiltonian reads $\mathcal{H}'_{\text{CBM}} = \mathcal{H}_0 + \mathcal{H}_B$, with:

$$\mathcal{H}_B = \frac{t^2}{2} \sum_{kk'll'} \left(s_{kl0} d_{l'}^\dagger c_{k'} c_k^\dagger d_l - s_{kl1} c_k^\dagger d_l d_{l'}^\dagger c_{k'} + \text{h.c.} \right). \tag{19}$$

This interaction can be simplified by a mean-field treatment:

$$d_l^\dagger c_k c_{k'}^\dagger d_{l'} = \left\langle d_l^\dagger d_{l'} \right\rangle c_k c_{k'}^\dagger + \left\langle c_k c_{k'}^\dagger \right\rangle d_l^\dagger d_{l'} = \delta_{ll'} \theta(-\varepsilon_l) c_k c_{k'}^\dagger + \delta_{kk'} \theta(\varepsilon_k) d_l^\dagger d_{l'}, \tag{20}$$

allowing to carry out part of the sums in Equation (19). Notice that the orbital energy ε_d does not appear in Equation (20) as it is now only associated to the charge degree of freedom n, while the Fermi gases corresponding to c_k and d_l have the same Fermi energy $E_F = 0$. One thus finds the effective low energy model, which, to leading order, reads [66]:

$$\mathcal{H}'_{\text{CBM}} = \mathcal{H}_0 + \frac{g}{\nu_0} \ln \left(\frac{E_c - \varepsilon_d}{E_c + \varepsilon_d} \right) \left[\sum_{ll'} d_l^\dagger d_{l'} - \sum_{kk'} c_k^\dagger c_{k'} \right] \tag{21}$$

in which we have introduced the dimensionless conductance $g = (\nu_0 t)^2$, corresponding to the conductance of the Quantum Point Contact (QPC) connecting dot and lead in units of e^2/h. This Hamiltonian describes two decoupled Fermi gases, but affected by potential scattering with opposite amplitudes. Equation (21) coincides with the LFL Hamiltonian (8) for the lead electrons.

The phase-shift δ_W (9) allows for the calculation of the charge occupation of the dot to leading order by applying the Friedel sum rule (10):

$$\langle N \rangle = \frac{\delta_W}{\pi} = g \ln \left(\frac{E_c - \varepsilon_d}{E_c + \varepsilon_d} \right). \tag{22}$$

This result reproduces the direct calculation of the dot occupation [99–101], showing the validity of the LFL model (8), with Friedel sum rule for the CBM. The extension to M channels is obtained by replacing $g \rightarrow M(\nu_0 t)^2$ in Equation (22). The extended proof to next-to-leading order in g is given in Ref. [66].

3.3.2. Anderson Impurity Model

Considering the internal spin degree of freedom in the AIM (13), it does not fundamentally affect the effective LFL behavior. Nevertheless, in the case where a single electron is trapped in the quantum dot $(-U \ll \varepsilon_d \ll 0)$, the derivation of Equation (8) is more involved, and sketched in Figure 5. The SW transformation maps the AIM onto a Kondo Hamiltonian including a potential scattering term [94]:

$$\mathcal{H}'_{AM} = \mathcal{H}_0 + J\mathbf{S} \cdot \mathbf{s} + W \sum_{kk'\sigma} c^\dagger_{k\sigma} c_{k'\sigma} . \tag{23}$$

The spin of the electron in the quantum dot $\mathbf{S}$ is coupled anti-ferromagnetically to the local spin of the lead electrons $\mathbf{s} = \sum_{kk'\tau\tau'} c^\dagger_{k\tau} \frac{\sigma_{\tau\tau'}}{2} c_{k'\tau'}$, with $\sigma_{\tau\tau'}$ the vector composed of the Pauli matrices, and

$$J = \frac{2\Gamma}{\pi\nu_0} \left(\frac{1}{\varepsilon_d + U} - \frac{1}{\varepsilon_d} \right), \qquad W = -\frac{\Gamma}{2\pi\nu_0} \left(\frac{1}{\varepsilon_d + U} + \frac{1}{\varepsilon_d} \right), \tag{24}$$

in which we introduced the hybridization energy $\Gamma = \pi\nu_0 t^2$, corresponding to the width acquired by the orbital level when coupled to the lead and which depends on the density of states of the lead electrons at the Fermi energy ν_0, see also Equation (A21) in Appendix A for its rigorous definition. Neglecting for the moment the Kondo anti-ferromagnetic coupling controlled by J, the LFL Hamiltonian (8) is directly recovered. Nevertheless, the potential scattering term is absent $(W = 0)$ at the particle-hole symmetric point $\varepsilon_d = -U/2$. At this point, the charge on the dot is fixed to one by symmetry, and the absence of potential scattering allows to derive various rigorous results, for instance concerning the ground state properties relying on Bethe ansatz [102,103]. It is a well-established fact that the system described by Equation (23) behaves as a LFL at low energies [48,104,105] and that the Friedel sum rule applies [87]. As a consequence, the Kondo coupling is responsible for the phase-shift of the low energy quasi-particles. Particle-hole symmetry, spin degeneracy, and Friedel sum rule fix the Kondo phase-shift to $\delta_K = \pi/2$. The Friedel sum rule states that:

$$\langle N \rangle = 2\frac{\delta_K}{\pi}, \tag{25}$$

$\langle N \rangle = 1$ because of particle-hole symmetry and the factor 2 signals spin degeneracy, fixing $\delta_K = \pi/2$. The detailed description of the Kondo effect is far beyond the scope of this review and we direct the interested reader to Ref. [49] for a comprehensive review and to Refs. [106–111] for the description of the low energy fixed point relying on boundary conformal field theory. For the scopes of this review it is enough to mention that below the Kondo temperature: [102,103,112,113]

$$T_K = \frac{e^{\frac{1}{4}} \gamma}{2\pi} \sqrt{\frac{2U\Gamma}{\pi}} e^{\frac{\pi\varepsilon_d(\varepsilon_d+U)}{2U\Gamma}}, \tag{26}$$

the spin-exchange coupling J in Equation (23) flows to infinity in the renormalization group sense. The relevance of this interaction brings the itinerant electrons to screen the local spin-degree of freedom of the quantum dot and phase-shifts the resulting quasi-particles by δ_K, see Figure 5. The phase-shift $\pi/2$ acquires thus a simple interpretation in one dimension [44,106]: Writing $\Psi_R = e^{2i\delta}\Psi_L$ for a given spin channel at the impurity site, $\delta = \delta_K = \pi/2$ leads to $\Psi_R + \Psi_L = 0$. The fact that the wave-function is zero at the impurity site, corresponds to the situation in which an electron screens the impurity spin, leading to Pauli blockade (due to Pauli principle), thus preventing other electrons with the same spin to access the impurity site. This dynamical screening of the impurity spin forms the so called "Kondo cloud" [114–117], see Figure 5. It is responsible for increasing the local density of states and leads to the Abrikosov–Suhl resonance [118], which causes the increase, below T_K, of the dot conductance in Coulomb blockaded regimes [93,119,120]. Remarkably, the Kondo phase-shift $\delta_K = \pi/2$ and the Kondo screening cloud have been also directly observed in two recent distinct experiments [121,122]. In Section 4.7, we illustrate how such phenomena also affect the dynamical properties of the mesoscopic capacitor in a non-trivial way.

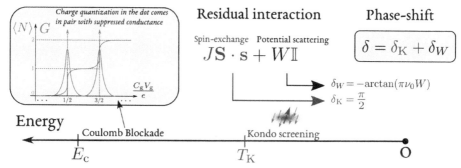

Figure 5. Modification of the physical scenario of Figure 3 in the presence of spin-exchange interactions between the dot and lead electrons in the Anderson Impurity Model (AIM). Spin-exchange interactions trigger the formation of the Kondo singlet below the Kondo temperature T_K, which is responsible for an additional elastic $\delta_K = \pi/2$ phase-shift of lead electrons in the effective Local Fermi Liquid (LFL) theory.

It remains to establish the combined role of spin-exchange and potential scattering on the low-energy quasi-particles. Remarkably, the phase-shift δ_W, caused by potential scattering, is additive to δ_K [123–125]:

$$\delta = \delta_K + \delta_W, \tag{27}$$

and can thus be calculated independently. The validity of the above expression is demonstrated by comparison with the exact Bethe ansatz solution of the AIM [102,103]. Inserting the expression (9) for the phase-shift caused by the potential scattering in the Friedel sum rule one finds:

$$\langle N \rangle = \frac{2}{\pi}[\delta_K - \arctan(\pi\nu_0 W)] = 1 + \frac{\Gamma}{\pi}\left(\frac{1}{\varepsilon_d + U} + \frac{1}{\varepsilon_d}\right). \tag{28}$$

This expression is consistent with the condition $\langle N \rangle = 1$, imposed by particle-hole symmetry, but also with the static charge susceptibility χ_c, which was derived with the Bethe ansatz [126]:

$$\chi_c = -\left.\frac{\partial \langle N \rangle}{\partial \varepsilon_d}\right|_{\varepsilon_d = -\frac{U}{2}} = \frac{8\Gamma}{\pi U^2}\left(1 + \frac{12\Gamma}{\pi U} + \dots\right). \tag{29}$$

The extension of this proof to next-to-leading order in t, is given in Refs. [66,68] and it shows how LFL approaches are effective in providing analytic predictions out of particle-hole symmetry as well, extending Bethe ansatz results.

Entropy **2020**, *22*, 847

This discussion concludes our demonstration of the persistence of elastic and coherent effects triggered by interactions at equilibrium. Local Fermi liquids provide a general framework to describe interacting and non-interacting systems at low energy, within an effective elastic scattering theory. Nevertheless, it is important to stress that LFL theories can fail in specific cases, such as overscreened Kondo impurities [127,128], and that their validity is limited to close-to-equilibrium/low-energy limits. It is thus expected that interactions become crucial as soon as such systems are driven out of equilibrium. We will illustrate now how the LFL theory allows to describe exotic, but still coherent in nature, dynamical effects in a paradigmatic setup such as the mesoscopic capacitor.

4. The Mesoscopic Capacitor

The mesoscopic capacitor in Figure 6 plays a central role in the quest to achieve full control of scalable coherent quantum systems [4,129,130]. A mesoscopic capacitor is an electron cavity coupled to a lead via a QPC and capacitively coupled to a metallic gate [51–53]. The interest in this device stems from the absence of DC transport, making possible the investigation and control of the coherent dynamics of single electrons. The first experimental realization of this system was a two-dimensional cavity in the quantum Hall regime [20,21], exchanging electrons with the edge of a bulk two-dimensional electron gas (2DEG). Operated out of equilibrium and in the weak tunneling limit, this system allows the triggered emission of single electrons [22–24], and paved the way to the realization of single-electron quantum optics experiments [131–134], as well as probing electron fractionalization [25,135], accounted by the scattering of charge density waves (plasmons) in the conductor [136–143], and their relaxation [26]. On-demand single-electron sources were also recently realized with real-time switching of tunnel-barriers [27–32], electron sound-wave surfing [14,16,144], generation of levitons [8–11,13], and superconducting turnstiles [18,19]. We direct again the interested reader to Ref. [35] for a comprehensive review of these experiments.

The key question concerning the dynamics of a mesoscopic capacitor is which electronic state, carrying a current $\mathcal{I}$, is emitted from the cavity following a change in the gate voltage V_g. The linear response is characterized by the admittance $\mathcal{A}(\omega)$,

$$\mathcal{I}(\omega) = \mathcal{A}(\omega)V_g(\omega) + \mathcal{O}(V_g^2). \tag{30}$$

In their seminal work, Büttiker and coworkers showed that the low-frequency admittance of a mesoscopic capacitor reproduces the one of a classical RC circuit [51–53],

$$\mathcal{A}(\omega) = -i\omega C(1 + i\omega R_q C) + \mathcal{O}(\omega^3), \tag{31}$$

in which both the capacitance C and the *charge relaxation resistance* R_q probe novel coherent dynamical quantum effects. The capacitance C was originally interpreted as an *electro-chemical capacitance* $1/C = 1/C_g + 1/C_q$, series of a *geometric* (C_g) and a *quantum* (C_q) contribution [21,51–53]. The geometric contribution is classical and depends on the shape of the capacitive contact between gate and quantum dot. The quantum contribution is a manifestation of the Pauli exclusion principle and was found proportional to the local density of states in the cavity, see Figure 7. Remarkably, the charge relaxation resistance $R_q = h/2e^2$ was predicted to be universally equal to half of the resistance quantum in the case of one conducting channel [20], independently of the transparency of the QPC connecting cavity and lead. This result is in striking contrast with the resistance measured in DC experiments and was originally labeled as a *Violation of Kirchhoff's Laws for a Coherent RC Circuit* [20]. Reference [21] extensively reviews the original theoretical predictions and their experimental confirmation, in a non-interacting and self-consistent setting, which we also review and put in relation with their Hamiltonian formulation in Appendix B.

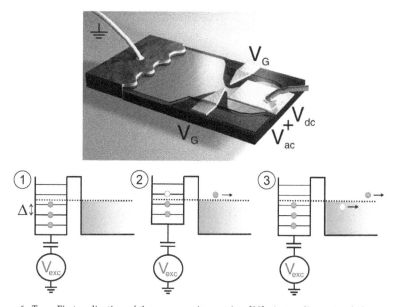

Figure 6. Top—First realization of the mesoscopic capacitor [20]: A two-dimensional electron gas (2DEG) in the Quantum Hall regime is coupled to a quantum cavity via a gate-controlled QPC. Bottom—Working principle of a single electron emission [22–26]. A gate potential moves the quantized levels of the cavity above and below the Fermi surface of the coupled reservoir. Electron/hole emission in steps 2 and 3 follows from moving occupied/empty orbitals above/below the Fermi surface.

Below, we discuss how the LFL approach challenges and extends the above studies. In particular:

1. The total capacitance C is given by the static charge susceptibility $\chi_c = -e^2\partial\langle N\rangle/\partial V_g$ of the cavity and does not generally correspond to a series of a geometric and quantum contribution, proportional to the density of states in the cavity. For instance, in Kondo regimes, the charge susceptibility of the cavity remains small, because of frozen charge fluctuations, while the density of states increases below the Kondo temperature [49]. This effect was directly probed in a recent experiment with a quantum dot device embedded in circuit-QED architecture [145];

2. A LFL low energy behavior implies universality of the charge relaxation resistance in the single channel case. In particular, the universality of R_q stems from a Korringa–Shiba (KS) relation [146]

$$\text{Im}\,[\chi_c(\omega)]|_{\omega\to 0} = \omega\hbar\pi\chi_c^2(\omega = 0),\tag{32}$$

in which $\chi_c(\omega)$ is the Fourier transform of the dynamical charge susceptibility (37);

3. The LFL approach shows various non-trivial dissipative effects triggered by strong correlations. In particular, it predicts a mesoscopic crossover between two universal regimes in which $R_q = h/2e^2 \to h/e^2$ [65] by increasing the dot size, also at charge degeneracy, in which the CBM maps on the Kondo model [101]. It also predicts giant dissipative regimes, described by giant universal peaks in R_q, triggered by the destruction of the Kondo singlet by a magnetic field [67,147];

4. In proper out-of-equilibrium regimes, interactions and inelastic effects become unavoidable and circuit analogies, such as Equation (31), do not capture the dynamic behavior of the mesoscopic capacitor [148]. We show here how previously published data [25] also show a previously overlooked signature of non-trivial many-body dynamics induced by interactions.

4.1. Hamiltonian Description of the Quantum RC Circuit: Differential Capacitance and Korringa–Shiba Relation

Expanding the square in Equation (1) and neglecting constant contributions, $\mathcal{H}_c$ renormalizes the orbital energy ε_d in Equation (12) and adds a quartic term in the annihilation/creation operators d_l, namely

$$\mathcal{H}_c = -eV_g(t)N + E_cN^2. \tag{33}$$

The driving gate voltage V_g couples to the charge occupation of the quantum dot $Q = e\langle N \rangle$. In single-electron emitters, one operates on the time dependent voltage drive $V_g(t)$ to bring occupied discrete levels above the Fermi surface and then trigger the emission of charge, see Figure 6. The current of the device is a derivative in time of the charge leaving the quantum dot, the admittance reads then, in the Fourier frequency representation,

$$\mathcal{A}(\omega) = -i\omega\frac{Q(\omega)}{V_g(\omega)}. \tag{34}$$

We start by considering small oscillations of amplitude ε_ω of the gate voltage:

$$V_g(t) = V_g + \varepsilon_\omega\cos(\omega t). \tag{35}$$

Close to equilibrium, expression (34) is calculated relying on Kubo's linear response theory [149]

$$\mathcal{A}(\omega) = -i\omega e^2\chi_c(\omega), \tag{36}$$

in which $\chi_c(\omega)$ is the Fourier transform of the dynamical charge susceptibility:

$$\chi_c(t - t') = \frac{i}{\hbar}\theta(t - t')\langle[N(t), N(t')]\rangle_0. \tag{37}$$

The notation $\langle\cdot\rangle_0$ refers to quantum averages performed at equilibrium, i.e., without the driving term $V_g(t)$ in Equation (33). The low frequency expansion of $\chi_c(\omega)$ reads:

$$\mathcal{A}(\omega) = -i\omega e^2\{\chi_c + i\text{Im}[\chi_c(\omega)]\} + \mathcal{O}(\omega^2), \tag{38}$$

where we relied on the fact that the even/odd part of the response function (37) coincide with its real/imaginary part, see Appendix C. We also introduce the static charge susceptibility $\chi_c = \chi_c(\omega = 0)$. The expansion (38) matches that of a classical RC circuit (31). Identifying term by term, we find the expression of the charge relaxation resistance and, in particular, that the capacitance C of the mesoscopic capacitor is actually given by a *differential capacitance* C_0:

$$C = C_0 = e^2\chi_c = -e^2\frac{\partial\langle N\rangle}{\partial\varepsilon_d} = \frac{\partial Q}{\partial V_g}, \qquad R_q = \frac{1}{e^2\chi_c^2}\left.\frac{\text{Im}\chi_c(\omega)}{\omega}\right|_{\omega\to 0}. \tag{39}$$

The differential capacitance is proportional to the density of states of *charge* excitations on the dot, which, as mentioned above, generally differs from the local density of states in the presence of strong correlations. Equation (39) provides also the general condition for the universal quantization of the charge-relaxation resistance $R_q = h/2e^2$, namely:

$$\text{Im}\chi_c(\omega)|_{\omega\to 0} = \hbar\pi\omega\chi_c^2. \tag{40}$$

Such kind of relation is known as a Korringa–Shiba (KS) relation [146]. The KS relation establishes that the imaginary part of the dynamic charge susceptibility, describing dissipation in the system, is controlled by the static charge fluctuations on the dot, χ_c.

Additionally, we mention that the relation (40) also affects the phase-shift of reflected or transmitted light through a mesoscopic system in the Kondo LFL regime [150–152]. Such situations have been recently realized with quantum-dot devices embedded in circuit-QED architectures [153–159], in which the driving input signal can be modeled by an AC potential of the form (35).

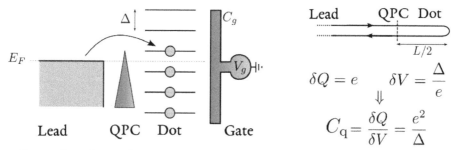

Figure 7. Physical origin of the quantum capacitance C_q. Pauli exclusion forces electrons entering the dot to pay an energy price equal to the local level spacing Δ, resulting in a capacitance $C_q = e^2/\Delta$. On the right, a one-dimensional representation of the mesoscopic capacitor, with a dot of size ℓ.

4.2. The Origin of the Differential Capacitance as a 'Quantum' Capacitance as far as Interactions are Neglected

As far as interactions are neglected, the differential capacitance C_0 is a manifestation of the fermionic statistics of electrons, determined by the Pauli exclusion principle. For this reason it has been originally labeled as a 'quantum' capacitance C_q [51–53]. When an electron is added to the quantum dot, in which energy levels are spaced by Δ, the Pauli exclusion principle does not allow one to fill an occupied energy state, but requires to pay a further energy price Δ, see Figure 7. The capacitance associated to this process is then $C_q = \delta Q/\delta V$. For one electron $\delta Q = e$ and $\delta V = \Delta/e$. Substituting these two expressions, we recover a uniform quantum capacitance:

$$C_q = \frac{e^2}{\Delta}. \tag{41}$$

This expression establishes that the quantum capacitance is proportional to the density of states in the quantum dot at the Fermi energy $C_q = e^2\mathcal{N}(E_F)$, with $\mathcal{N}(E_F) = 1/\Delta$, to be distinguished from ν_0, the density of states of the lead electrons.

Additionally, the quantum capacitance is related to the dwell-time spent by electrons in the cavity. The general relation is derived in Appendix B, but it also results from simple estimates. Considering the representation of the mesoscopic capacitor of Figure 7, in the open-dot limit, the time spent by an electron in the cavity coincides with its time of flight $\tau_f = \ell/v_F$: The ratio between the size of the cavity ℓ and its (Fermi) velocity v_F. The level spacing Δ of an isolated cavity of size ℓ is estimated by linearizing the spectrum close to the Fermi level. The distance in momentum between subsequent levels is h/ℓ, corresponding to $\Delta = hv_F/\ell = h/\tau_f$. Substituting in Equation (41) leads to an equivalent expression for the quantum capacitance:

$$C_q = \frac{e^2}{h}\tau_f. \tag{42}$$

On the experimental side, the level spacing of the quantum dot can be actually estimated and, in the experimental conditions of Ref. [20], it was established to be of the order of $\Delta \sim 15$ GHz, corresponding to a quantum capacitance $C_q \sim 1$ fF. Experimental measurements of C, reported in Figure 8, give an estimate also for C_g, showing that $C_q \ll C_g$. This implies that the level spacing Δ was much larger than the charging energy $E_c = e^2/2C_g$, of the order of fractions of the GHz, apparently justifying the mean-field approach to describe experimental results, with the limitations that we are going to discuss in out-of-equilibrium regimes, see Section 5.

The argument leading to Equation (41) implicitly assumes the perfect transparency of the QPC, namely that the probability amplitude r for a lead electron to be reflected when passing though the QPC to enter the cavity is equal to zero ($r = 0$). In this limit, the density of states on the dot is uniform. Finite reflection $r \neq 0$ is responsible for resonant tunneling processes, leading to oscillatory behavior of the local density (or the dwell-time) of states as a function of the gate potential V_g, in agreement with the experimental findings reported in Figure 8. In Appendix B, we provide a quantitative analysis of this effect by explicitly calculating the differential capacitance C_0, Equation (39), by neglecting the term proportional to $E_c N^2$ in Equation (33).

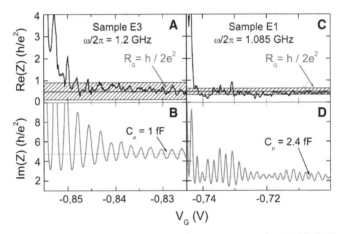

Figure 8. Top—Measurement of the universal charge relaxation resistance from Ref. [20]. The resistance of two different samples (E3 and E1) is given by the real part of their impedance $Z = 1/\mathcal{A}$ as a function of the QPC potential V_G, in Figure 6, which also affected the gate potential V_g. Measurements were carried out for $T = 30mK$ and a magnetic field $B = 1.3T$ polarizing the electrons, resulting in one conducting channel. Uncertainties are indicated by the hatched areas. Bottom—Measurement of the total capacitance $C_\mu = C$, through $\text{Im}(Z) = -1/\omega C$, for the same samples. The oscillatory behavior is related to resonances in the density of states of the dot.

4.3. The Physical Origin of the Universal Charge Relaxation Resistance

The universality of R_q was also verified experimentally in Ref. [20], see Figure 8. In the quantum coherent regime the charge relaxation resistance is *universal*: It does not depend on the microscopic details of the circuit (C_g, r, ...), but only on fundamental constants, namely Planck's constant h and the electron charge e. This is surprising. When applying a DC voltage across a QPC connecting two leads, the QPC behaves as a resistive element of resistance [160,161]:

$$R_{DC} = \frac{h}{e^2 D},\tag{43}$$

where $D = 1 - |r|^2$ is the QPC transparency. This quantity depends on the amplitude r for electrons to be backscattered when arriving at the QPC, which can be tuned by acting on a gate potential. As we considered spinless electrons, the factor 2 in $R_q = h/2e^2$ cannot be related to spin degeneracy and $D = 1$ in Equation (43). It is rather related to the fact that the dot is connected to a single reservoir, in contrast to the source-drain reservoirs present in DC transport experiments [56,162]. In direct transport, each metallic contact is responsible for a quantized contact resistance $R_c = h/2e^2$, the Sharvin–Imry resistance [163,164]. In source-drain experiments, Equation (43) could be recast in the form $R_{QPC} + 2R_c$, with $R_{QPC} = \frac{h}{e^2}\frac{1-D}{D}$ the resistive contribution proper to the QPC. For the case of a mesoscopic capacitor, there is a single reservoir and one would thus expect for the charge relaxation resistance:

$$R_q^{\text{expected}} = R_{\text{QPC}} + R_c = \frac{h}{2e^2} + \frac{h}{e^2}\frac{1-D}{D}. \tag{44}$$

The fact that R_q does not depend on the transparency D, assuming the universal value $h/2e^2$, cannot be attributed to the laws governing DC quantum transport, and this is the reason why one can speak about the *Violation of Kirchhoff's Laws for a Coherent RC Circuit* [20]. The universality of R_q is rather a consequence of the fact that R_q, differently from the contact resistance R_c, is related to energy (Joule) dissipation. As electrons propagate coherently within the cavity, they cannot dissipate energy inside it, but only once they reach the lead. We will illustrate in Section 4.5 how this phenomenon is a direct consequence of the possibility to excite particle-hole pairs by driven electrons. At low frequency, dissipation is thus only possible in the presence of a continuum spectrum, accessible in the metallic reservoirs. The expected resistance (44) is recovered only if electrons lose their phase coherence inside the dot [56,162], see also Refs. [165,166], which take Coulomb blockade effects into account. For instance, in the high temperature limit $k_B T \gg \Delta$, Equation (44) is not recovered. The reason is that, in scattering theory, temperature is fixed by the reservoirs without affecting the coherent/phase-preserving propagation in the mesoscopic capacitor.

4.4. The Open-Dot Limit

We address now the role of the charging energy E_c and come back to our initial example of the open dot limit, considered in Section 2. The possibility to rely on an exact bosonized solution for the model (1)–(2), made possible the derivation of the admittance $\mathcal{A}(\omega)$ in linear-response theory for a fully transparent point contact ($r = 0$) and a finite-sized cavity [65]:

$$\mathcal{A}(\omega) = -i\omega C_g \left(1 - \frac{i\omega\tau_c}{1 - e^{i\omega\tau_f}}\right)^{-1}. \tag{45}$$

This expression is important as it makes possible to study the interplay between two different time-scales, namely the time of flight τ_f of electrons inside the cavity, already present in the previous discussion, and $\tau_c = hC_g/e^2$ the time scale corresponding to the charging energy E_c. We mention that, interestingly, the admittance (45) was also found to describe the coherent transmission of electrons through interacting Mach–Zehnder interferometers [167,168].

What is quite remarkable about the admittance (45) is that, to linear order in ω, the two time scales τ_f and τ_c still combine into the universal charge relaxation resistance $R_q = h/2e^2$ and a series of a geometrical and quantum capacitance $C_q = e^2\tau_f/h$ [51–53] (see also Equation (42)):

$$\frac{1}{C_0} = \left[\frac{1}{C_g} + \frac{h}{e^2\tau_f}\right]. \tag{46}$$

The low-frequency behavior of Equation (45) illustrates how interacting systems behave as if interactions were absent at low energies. What is then also implicit in Equation (45) is that, to observe separate effects on the charge dynamics, induced by free propagation (τ_f) or interactions (τ_c), one has to consider proper out-of-equilibrium/high-frequency regimes. These regimes will be addressed in Section 5.

Nevertheless, interactions still matter even in low-frequency regimes. Consider the infinite-size (metallic) limit for the cavity, $\tau_f \to \infty$. In this limit, also describing the experiment in Figure 2, one implicitly assumes that the driving frequency ω is larger than the internal level spacing of the dot Δ,

$$\hbar\omega \gg \Delta. \tag{47}$$

The discrete spectrum of the dot can thus be treated as a continuum, which allows for energy dissipation also inside the cavity, see Figure 9. In particular, averaging the admittance (45) over a finite bandwidth

$\delta\omega$, such that $\omega \gg \delta\omega \gg \Delta$, one exactly recovers the admittance of a classical RC circuit of capacitance C_g and charge-relaxation resistance $R_q = h/e^2$ [65]:

$$A(\omega) = \frac{-i\omega C_g}{1 - i\omega C_g \frac{h}{e^2}}. \tag{48}$$

The mesoscopic crossover $R_q = h/2e^2 \to h/e^2$ is an exquisite coherent effect triggered by interactions. This phenomenon has fundamentally the same origin of the elastic electron transfer exemplified by the correlation function (3), considered at the very beginning of this review.

Remarkably, the universality of the charge-relaxation resistance holds in the presence of backscattering at the dot entrance, without affecting the mesoscopic crossover $R_q = h/2e^2 \to h/e^2$ [65]. Nevertheless, the possibility to interpret the differential capacitance as a series of two separate geometric and quantum term as in Equation (46), is lost. If we locate the entrance of the dot at $x = 0$, backscattering corrections to the model (1)–(2) read:

$$\mathcal{H}_r = -\hbar r v_F \left[\Psi_R(0)^\dagger \Psi_L(0) + \Psi_L(0)^\dagger \Psi_R(0) \right], \tag{49}$$

and compromise a non-interacting formulation of the problem, even in its bosonized form [65,72,148].

It becomes then important to understand why and to which extent quantities such as the charge-relaxation resistance show universal coherent behavior even in the presence of interactions. The extension of the LFL theory in the quasistatic approximation provides the unified framework to understand the generality of such phenomena.

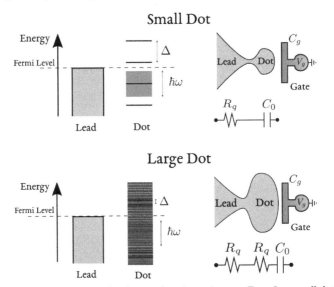

Figure 9. Mesoscopic crossover in the charge relaxation resistance. Top—In a small dot, the level spacing Δ is larger than the driving energy $\hbar\omega$ and energy levels in the dot are not excited. The universal resistance $R_q = h/2e^2$ of the equivalent *RC* circuit is furnished exclusively by the lead electron reservoir. Bottom—Excitation of energy levels inside the dot are permitted in the large dot limit, which acts as a further dissipative reservoir in series to the lead.

4.5. The Tunneling Limit and the Quasi-Static Approximation

In Section 3, we showed that a large class of models of the form (4) are effectively described, in the low-energy limit, by a LFL theory (8), in which the potential scattering coupling constant W

depends on the orbital energy of the dot ε_d. The expansion of the charging energy Hamiltonian (33) made apparent that this energy is renormalized by the gate potential $\varepsilon_d \to \varepsilon_d - eV_g(t)$. For an AC bias voltage, we consider then a periodic function of time oscillating at the frequency ω:

$$\varepsilon_d(t) = \varepsilon_d^0 + \varepsilon_\omega \cos(\omega t). \tag{50}$$

The *quasi-static approximation* consists in substituting Equation (50) directly in Equation (8). This condition assumes that the low energy Hamiltonian (8), derived for the equilibrium problem, follows, without any delay, the orbital oscillations expected from the parent, high-energy, model. The quasi-static approximation is then a statement about a behavior close to adiabaticity.

We consider then the linear response regime and expand the coupling $W(\varepsilon_d)$ in ε_ω. Focusing on the single channel case, the extension to multiple channels being straightforward, Equation (8) becomes:

$$\mathcal{H} = \sum_k \varepsilon_k c_k^\dagger c_k + \left[W(\varepsilon_d^0) + W'(\varepsilon_d^0)\varepsilon_\omega \cos(\omega t) \right] \sum_{kk'} c_k^\dagger c_{k'}. \tag{51}$$

We diagonalize the time independent part of this Hamiltonian [104]:

$$\mathcal{H} = \sum_{kk'} \varepsilon_k a_k^\dagger a_k + \frac{W'(\varepsilon_d^0)}{1 + \left[\pi \nu_0 W(\varepsilon_d^0) \right]^2} \varepsilon_\omega \cos(\omega t) \sum_{kk'} a_k^\dagger a_{k'}, \tag{52}$$

where the operators a and $a^\dagger$ describe the new quasi-particles diagonalizing the time independent part of the Hamiltonian (51). The Friedel sum rule (11) establishes that:

$$\chi_c = \frac{\nu_0 W'(\varepsilon_d^0)}{1 + \left[\pi \nu_0 W(\varepsilon_d^0) \right]^2}, \tag{53}$$

and the Hamiltonian (52) can be cast in the more compact and transparent form:

$$\mathcal{H} = \sum_{kk'} \varepsilon_k a_k^\dagger a_k + \frac{\chi_c}{\nu_0} \varepsilon_\omega \cos(\omega t) \sum_{kk'} a_k^\dagger a_{k'}. \tag{54}$$

This Hamiltonian shows the mechanism responsible for energy dissipation at low energy for the rich variety of strongly interacting systems satisfying the Friedel sum rule and LFL behavior at low energy. The time dependent term pumps energy in the system, which is then dissipated by the creation of particle-hole pairs. Crucially, this term is controlled by the static charge susceptibility χ_c of the quantum dot. The non-interacting Hamiltonian (54) explains why non-interacting results also hold for the universal charge relaxation resistance in the presence of interactions on the quantum dot.

We now illustrate how the Hamiltonian (54) implies the validity of the KS relation and thus universality of the charge-relaxation resistance R_q. The proof was originally devised for spin-fluctuations [169] and we extend it here to the case of charge fluctuations. For drives of the form (50), the power dissipated by the system is proportional to the imaginary part of the dynamic charge susceptibility, see Appendix C,

$$\mathcal{P} = \frac{1}{2} \varepsilon_\omega^2 \omega \mathrm{Im} \chi_c(\omega). \tag{55}$$

A direct calculation of $\mathrm{Im}\chi_c$ is a difficult task and this is where the low-energy model (54) becomes useful. Similarly as for Equation (55), the LFL theory (54) predicts the dissipated power:

$$\mathcal{P} = \frac{1}{2} \varepsilon_\omega^2 \omega \mathrm{Im} \chi_A(\omega), \tag{56}$$

where the linear response function $\chi_A(t - t') = \frac{i}{\hbar}\theta(t - t')\langle[A(t), A(t')]\rangle_0$ is a correlator at different times of the potential scattering operator:

$$A = \frac{\chi_c}{v_0} \sum_{kk'} a_k^\dagger a_{k'},$$ (57)

responsible for the creation of particle-hole pairs. The Fourier transform of the response function reads:

$$\chi_A(\omega) = -\frac{1}{\hbar}\frac{\chi_c^2}{v_0^2} \sum_{pp'} f(\varepsilon_p)[1 - f(\varepsilon_{p'})] \left[\frac{1}{\omega + \frac{\varepsilon_p - \varepsilon_{p'}}{\hbar} + i0^+} - \frac{1}{\omega + \frac{\varepsilon_{p'} - \varepsilon_p}{\hbar} + i0^+} \right],$$ (58)

in which $f(\varepsilon_p) = 1/(e^{\beta\varepsilon_p} + 1)$ is the Fermi distribution. We consider the electron lifetime as infinite, i.e., much longer than the typical time scales τ_c and τ_f. Taking the imaginary part and the continuum limit for the spectrum in the wide-band approximation, one finds, at zero temperature,

$$\mathrm{Im}\chi_A(\omega) = \pi\hbar\omega\chi_c^2.$$ (59)

The two dissipated powers (55) and (59) have to be identical, implying the Korringa–Shiba relation (32), enforcing then a universal value for the charge relaxation resistance $R_q = h/2e^2$.

4.6. The LFL Theory of Large Quantum Dots: The Mesoscopic Crossover $R_q = h/2e^2 \to h/e^2$

The above demonstration has to be slightly adapted to show the mesoscopic crossover $R_q = h/2e^2 \to h/e^2$. This crossover takes place for the CBM (12), in the infinite-size limit of the dot. As implicit in the effective description (21) of the CBM, the dot and the lead constitute two separate Fermi liquids. Sections 2 and 3 illustrated how the energy cost E_c prevents the low-energy transfer of electrons between the dot and lead [77]. The electrons of both these gases are then only backscattered at the lead/dot boundary with opposite amplitudes. In the quasi-static approximation, all the steps carried in the previous discussion apply for the Hamiltonian (21). In this case, the time variation of the orbital energy ε_d also drives particle-hole excitations in the dot. The operator responsible for energy dissipation becomes:

$$A = \frac{\chi_c}{v_0}\left(\sum_{kk'} c_k^\dagger c_{k'} - \sum_{ll'} d_l^\dagger d_{l'} \right),$$ (60)

in which the operators $c_k^\dagger$ and $d_l^\dagger$ create lead and dot electrons of energy $\varepsilon_{k,l}$ respectively. This formulation of the operator A adds a further contribution to Equation (58), analogous to the contribution of particle-hole pairs excited in the lead, namely:

$$-\frac{1}{\hbar}\frac{\chi_c^2}{v_0^2} \sum_{ll'} f(\varepsilon_l)[1 - f(\varepsilon_{l'})] \left[\frac{1}{\omega + \frac{\varepsilon_l - \varepsilon_{l'}}{\hbar} + i0^+} - \frac{1}{\omega + \frac{\varepsilon_{l'} - \varepsilon_l}{\hbar} + i0^+} \right].$$ (61)

The limits $\omega \to 0$ and $\Delta \to 0$ do not commute in the above expression. This fact has a clear physical interpretation: If the frequency is sent to zero before the level spacing, energy cannot be dissipated in the cavity and no additional contribution to $\mathrm{Im}\chi_c(\omega)$ is found. If the opposite limit is taken, the condition (47) is met and the Korringa–Shiba relation is then modified by a factor two:

$$\mathrm{Im}\chi_c(\omega) = 2\pi\hbar\omega\chi_c^2,$$ (62)

which doubles the universal value of the single-channel charge relaxation resistance $R_q = h/e^2$. The relation (62) was originally shown by explicit perturbation theory in the tunneling amplitude, close and away from charge degeneracy points [65]. As summarized in Figure 9, driving at a frequency higher than the dot level spacing induces the creation of particle/hole pairs inside the dot as well, enhancing energy dissipation with respect to the small dot limit $\hbar\omega < \Delta$. As energy can be *coherently* dissipated in two fermionic baths (dot and lead), the dot acts effectively as a further (Joule) resistor in series with the lead, leading to a doubled and still universal charge relaxation resistance.

4.7. The Multi-Channel Case and Universal Effects Triggered by Kondo Correlations

The above discussion also extends to the M channels case, leading to a generalized KS relation:

$$\text{Im}\chi_c(\omega)|_{\omega\to 0} = \hbar\pi\omega \sum_\sigma \chi_\sigma^2, \tag{63}$$

which corresponds to a non-universal expression for the charge relaxation resistance [68,69]

$$R_q = \frac{h}{2e^2} \frac{\sum_\sigma \chi_\sigma^2}{\left(\sum_\sigma \chi_\sigma\right)^2}. \tag{64}$$

This expression is analogous to the one obtained by Nigg and Büttiker [54]. In their derivation leading to Equation (A48), the densities of states, or dwell-times τ_σ, of the σ channel in the dot, replace the susceptibilities χ_σ. The single channel case is remarkable in that the numerator simplifies with the denominator in Equation (64), leading to the universal value $h/(2e^2)$, which is thus physically robust. Otherwise, in the fine-tuned case that all the channel susceptibilities are equal, one finds $R_q = h/2e^2M$.

Spinful systems in the presence of a magnetic field are the simplest ones to study how the charge-relaxation resistance is affected by breaking the symmetry between different conduction channels. Indeed, lifting the orbital level degeneracy by a magnetic field breaks the channel symmetry and the charge relaxation resistance is no longer universal, as it was originally realized in studies of the AIM (13) relying on the Hartree–Fock approximation [54].

Nevertheless, the self-consistent approach misses important and sizable effects triggered by strong Kondo correlations. These were originally observed relying on the Numerical Renormalization Group (NRG) [147]. The numerical results, reported in Figure 10, showed that, for Zeeman splittings of the order of the Kondo temperature T_K, the charge relaxation resistance can reach up to 100 times the universal value of $R_q = h/(4e^2)$, which would be expected in the two-fold spin degenerate case.

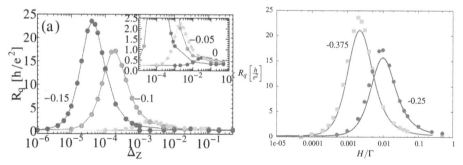

Figure 10. Left—Dependence of R_q on the Zeeman splitting Δ_Z in the Kondo regime from Ref. [147]. These results have been obtained by Numerical Renormalization Group (NRG) calculations with $\Gamma = 0.02$ and $E_c = 0.2$ (both quantities are measured in units of the contact bandwidth D and the definition of the hybridization energy Γ is provided in Appendix A.3). They show that, for Zeeman energies of the order of the Kondo temperature, a giant non-universal peak appears in the charge relaxation resistance. Right—Comparison of R_q as a function of the magnetic field between NRG calculations (dots) (extracted from Ref. [147]) and our Bethe ansatz results (solid lines) for different ε_d/U and $U/\Gamma = 20$, showing excellent agreement [67,68].

The LFL approach allows the analytical quantification and physical interpretation of such giant dissipative phenomenon [67,68]. For two spin channels, the total charge on the dot is the sum of the two spin occupation $\langle N \rangle = \langle N_\uparrow \rangle + \langle N_\downarrow \rangle$. Equation (64) can then be recast in the useful form:

$$R_q = \frac{h}{4e^2} \left(1 + \frac{\chi_m^2}{\chi_c^2} \right) , \tag{65}$$

in which we introduce the usual charge susceptibility: $\chi_c = -\partial \langle N \rangle / \partial \varepsilon_d$ and the *charge-magneto* susceptibility

$$\chi_m = -2 \frac{\partial \langle m \rangle}{\partial \varepsilon_d} . \tag{66}$$

This quantity is twice the derivative of the dot magnetization $\langle m \rangle = (\langle N_\uparrow \rangle - \langle N_\downarrow \rangle)/2$, with respect to the orbital energy ε_d. The charge-magneto susceptibility is an atypical object to study quantum dot systems, where the *magnetic* susceptibility $\chi_H = -\partial \langle m \rangle / \partial H$ is rather considered to study the sensitivity of the local moment of the quantum dot to variations of the magnetic field H. Equation (65) shows that the susceptibility of the *magnetization* of the dot, and not its charge, is responsible for the departure from the universal quantization $h/(4e^2)$ of the charge relaxation resistance. Equation (65) also separates explicitly charge and spin degrees of freedom of the electrons in the quantum dot. They can display very different behaviors in correlated systems, as illustrated in Figure 11 in the Kondo regime, defined for one charge blocked on the dot and Zeeman energies below the Kondo temperature (26).

In particular, Kondo correlations strongly affect the dot magnetization, but not its occupation. The points where χ_m differs from χ_c correspond to non-universal charge relaxation resistances. In Figure 10, the values derived with the LFL approach (65) are compared to those obtained with NRG [147], showing excellent agreement. Additionally, the LFL approach also allows to derive an exact analytical description of this peak, showing a genuinely giant dissipation regime: Simultaneous breaking of the SU(2) ($H \neq 0$) and particle-hole symmetry ($\varepsilon_d \neq -U/2$) trigger a peak in R_q which scales as the 4th(!) power of U/Γ and has its maximum for Zeeman splittings of the order of the Kondo temperature. This effect is caused by the fact that breaking the Kondo singlet by a magnetic field activates spin-flip processes, which dissipate energy through creation of particle-hole pairs [147].

We conclude by discussing the deviations of the differential capacitance C_0 from the local density of states of the cavity, which is clearly apparent in Kondo regimes. The spin/charge separation arising in the AIM allows to observe important physical effects on the differential capacitance of strongly interacting systems. Charge and spin on the dot are carried by different excitations: Holons and spinons. We report in Figure 12 the density of states of these excitations in the particle-hole symmetric case $\varepsilon_d = -U/2$. In the absence of interactions ($U/\Gamma = 0$), they have the same shape, but they start to strongly differ as the interaction parameter U/Γ is increased. They develop well pronounced peaks, but at different energies, signaling the appearance of separated charge and spin states. In the case of holons, the excited charge state appears close to $\varepsilon = U/2$, the energy required to change the dot occupation at particle-hole symmetry. In Ref. [170], it is shown that the density of states of the holons equals the static charge susceptibility χ_c, coinciding then with the differential capacitance C_0. At particle-hole symmetry, this quantity scales to zero as $8\Gamma/\pi U^2$, see Equation (29). Instead, the spinon density of states develops a sharp peak at zero energy, known as the Abrikosov–Suhl resonance [118], signaling the emergence of the strongly correlated Kondo singlet. The differential capacitance C_0 is completely insensitive to this resonance, which dominates the *total* density of states on the dot. Such an effect was distinctly observed in carbon nanotube devices coupled to high-quality-factor microwave cavities [145]. These systems efficiently probe the admittance (30) also in quantum dots with more than two internal degrees of freedom [150,152], such as extensions of the AIM to SU(4) regimes, relevant for quantum dots realized with carbon nanotubes [68,171–177].

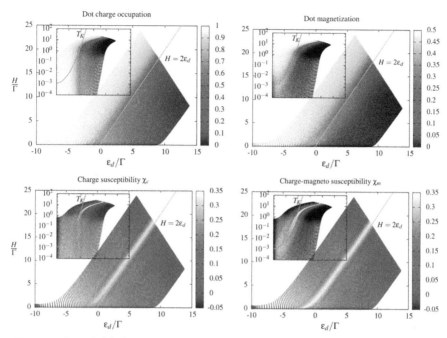

Figure 11. From Ref. [68]. Top—Charge occupation and magnetization of the dot for $U/\Gamma = 20$ as function of the orbital energy ε_d and magnetic field H. The insets show the same quantities on a logarithmic scale. The light green lines in the linear plots correspond to $H = 2\varepsilon_d$, and separate regions with different charge occupations, while the dark green lines in the insets correspond to T_K, Equation (26), and separate regions with different magnetization. The charge is not sensitive to the formation of the Kondo singlet for Zeeman energies below the Kondo temperature (green line), while the magnetization becomes zero. Bottom—Corresponding charge susceptibility and charge-magneto susceptibility. The susceptibilities are in units of $1/\Gamma$. In the insets the same quantities are plotted on a logarithmic scale and the zone of appearance of the giant peak of the charge relaxation resistance can be appreciated. It is the region, following T_K, in which χ_c is close to zero, while χ_m acquires important values because of the formation of the Kondo singlet.

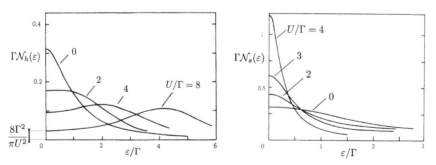

Figure 12. From Ref. [170]. (**Left**)—Density of states of local holons on the dot $\mathcal{N}_h(\varepsilon)$. ε is the excitation energy. A Coulomb peak emerges increasing the interaction parameter U/Γ and vanishes at zero energy as $8\Gamma/\pi U^2$. This quantity coincides with χ_c, plotted in Figure 11. (**Right**)—Density of states of local spinons $\mathcal{N}_s(\varepsilon)$. It behaves as the holonic one for $U/\Gamma = 0$ and develops the Abrikosov–Suhl resonance at zero energy, the signature of the formation of the strongly correlated Kondo singlet state.

The above discussion completes the review of the application of the LFL theory to study the low-energy dynamics of quantum impurity driven systems. Further applications could be envisioned to describe various correlation effects on different aspects of weakly driven interacting quantum-dot systems, as long as they can be described by an effective theory of the form (8). An important case involves the driving of the coupling term $\mathcal{H}_{res-dot} \rightarrow \mathcal{H}_{res-dot}(t)$ in the Hamiltonian (4), which has been implemented experimentally, with important metrologic applications [27–32]. Another interesting perspective concerns the application of the LFL theory to energy transfer [178–180], or coupling quantum-dot systems to mechanical degrees of freedom [181–187], which are described by similar models as quantum-dot devices embedded in circuit-QED devices [150,153–158].

Deviations from universal and coherent behaviors are expected in non-LFL regimes, arising when the reservoirs are Luttinger Liquids [188,189] or in over-screened Kondo impurities, in which the internal degrees of freedom of the bath surpass those of the impurity [85,190].

We have thus illustrated how a coherent and effectively non-interacting LFL theory accounts for strong correlation effects in the dynamics of quantum dot devices. It has to be clarified how interaction are supposed to affect proper out-of-equilibrium regimes. As a direct example, consider again the admittance (45). Its expansion to low-frequencies completely reproduces the self-consistent predictions of Refs. [51–53], but it 'hides' the qualitative difference between the two time scales τ_c and τ_f, associated to interactions and free-coherent propagation respectively. Higher-frequency driving will inevitably unveil this important difference, as we are going to demonstrate by giving a new twist to past experimental data in the next conclusive section.

5. What about Out-Of-Equilibrium Regimes? A New Twist on Experiments

We conclude this review by showing how interaction inevitably dominate proper out-of-equilibrium or fastly-driven regimes. We will focus, also in this case, on the mesoscopic capacitor. In particular, we will show that past experimental measurements, showing fractionalization effects in out-of-equilibrium charge emission from a driven mesoscopic capacitor [25], also manifest previously overlooked signatures of non-trivial many-body dynamics induced by interactions in the cavity.

As a preliminary remark, notice that the circuit analogy (31) does not apply for a non-linear response to a gate voltage change or to fast (high-frequency) drives. An important example is a large step-like change in the gate voltage $V_g(t) = V_g \theta(t)$, $\theta(t)$ being the Heaviside step function, which is relevant to achieve triggered emission of quantized charge [24]. Such a non-linear high-frequency response has been considered extensively for non-interacting cavities [22,24,191–195], where the current response to a gate voltage step at time $t = 0$ was found to be of the form of simple exponential relaxation [22,191,193,195]:

$$I(t) \propto e^{-t/\tau_R} \theta(t). \tag{67}$$

For a cavity in the quantum Hall regime the relaxation time $\tau_R = \tau_f/(1 - |r|^2)$, where τ_f is the time of flight around the edge state of the cavity, see Figures 6 and 7, and r the reflection amplitude of the point contact.

There have been relatively few studies of the out-of-equilibrium behavior of the mesoscopic capacitor in the presence of interactions. The charging energy leads to an additional time scale $\tau_c = 2\pi\hbar C_g/e^2$ for charge relaxation. The limit $1 - |r|^2 \ll 1$ of a cavity weakly coupled to the lead, such that it can effectively be described by a single level, was addressed in Refs. [196–201].

The full characterization of the out-of-equilibrium dynamics behavior of the mesoscopic capacitor, with a close-to-transparent point contact, was carried out in Ref. [148], extending the analysis of Ref. [65] to a non-linear response in the gate voltage V_g. A main result, spectacular in its simplicity, is that for a fully transparent contact ($r = 0$) the linear-response admittance (45) also describes the non-linear response, i.e., the correction terms in Equation (30) vanish for an ideal point contact connecting cavity and lead [136,202–204]. The Fourier transform of the admittance (45) describes the real-time evolution of the charge $Q(t)$ after a step change in the gate voltage.

Figure 13 illustrates that initially, for times up to τ_f, $Q(t)$ relaxes exponentially with time τ_c, whereas at time $t = \tau_f$ the capacitor abruptly enters a regime of exponentially damped oscillations, the period and the exponential decay, controlled by a complex function of τ_f and τ_c, which does not correspond to any time scale extracted from low-frequency circuit analogies. This behavior is not captured by Equation (67), derived in the non-interacting limit. These oscillations correspond to the emission of initially sharp charge density pulses, which are damped and become increasingly wider after every charge oscillation. Such complex dynamics is exquisitely coherent, but totally governed by interactions.

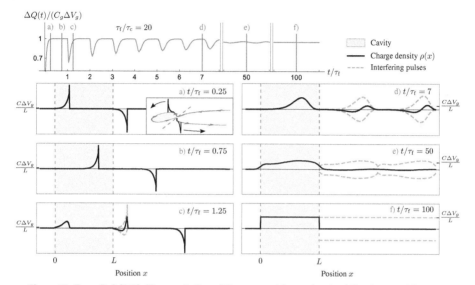

Figure 13. From Ref. [148]. Time-evolution of the current/charge density following a sudden gate voltage shift at time $t = 0$ for large interaction strength, $\tau_f = 20\tau_c$. Top—Charge response $\Delta Q(t)$ as a function of time t. Bottom—Series of snapshots of the current/charge density $j(x,t)$ at different times. In the inset of panel (a), the real-space representation (reproducing the one adopted in Figure 7 with the dot site $\ell = L$) of the mesoscopic capacitor with the profiles of the emitted charge pulses is given. The times at which the snapshots are taken are indicated by vertical dashed lines in the top panel. Notice that the scale changes along the vertical axis in the different panels. At time $t = 0$, two charge pulses of width $\sim v_F \tau_c$ and opposite sign emerge from the point contact (**a,b**), one pulse entering the cavity and one pulse entering the chiral edge of the bulk two-dimensional electron gas. Both pulses have a net charge approaching $C_g \Delta V_g$. The pulse that is emitted into the cavity returns to the point contact at time $t = \tau_f$. As that pulse leaves the cavity, a second pulse-antipulse pair is generated (**c**), partially canceling the original charge pulse that leaves the cavity at $t = \tau_f$. The resulting pulse exiting the cavity is the sum of the dashed profiles. The repetition of this mechanism leads to the widening and lowering of successive pulses (**d,e**) (notice the change of scale between snapshots). Finally, the asymptotic configuration is attained with a charge $C\Delta V_g$ uniformly distributed along the cavity edge (**f**).

Additionally, it is also interesting to consider the effect of a small reflection amplitude r in the point contact. In this case, the charge Q_r acquires nonlinear terms in the gate voltage V_g,

$$Q_r(t) = Q(t) - \frac{e\tilde{r}}{\pi C} \int dt'\, \mathcal{A}(t - t') \sin[2\pi Q(t')/e], \qquad (68)$$

in which $\mathcal{A}(t)$ and $Q(t)$ are the Fourier transform of the admittance and charge for the case of a point contact with perfect transparency, $r = 0$, see Equations (30) and (45). The parameter $\tilde{r}$ involves both the (weak) backscattering amplitude r and temperature T, details can be found in Ref. [148].

Experimental Signatures of the Effects of Interaction in Quantum Cavities Driven out of Equilibrium

The prediction that, in the open dot limit, interactions trigger the emission of a series of subsequent charge density pulses led to the possible explanation of additional effects that relate to a, so far not satisfactorily explained, part of the Hong-Ou-Mandel current noise measurements at the LPA [25,205]. The experimental setup is the solid state realization of the Hong-Ou-Mandel experiment, see the left panel in Figure 14: When two electrons collided at the same time on the QPC from different sources (states 1 and 2), they could not occupy the same state because of Pauli's exclusion principle and their probability to end up in different leads (states 3 and 4) was increased. As a consequence, the current noise was suppressed [132], see the right panels in Figure 14. More generally, $\Delta q(\tau)$ measures the cross-correlation (or overlap) in time of the two incoming currents at the level of the QPC. If the two incoming currents are identical in each input, one should get $\Delta q(\tau = 0) = 0$ and the rest of the curve will reflect on the time trace of the current. However, because of small asymmetries in the two electronic paths and the two electron sources, the noise suppression is not perfect [205]. In Ref. [25], the current noise Δq as a function of the time delay τ with which electrons arrived at the QPC from different sources was measured in more detail for the outer and inner edge of the filling factor $\nu = 2$ (central and right panel in Figure 14). The current in the inner edge channel was induced by inter-edge Coulomb interactions and could be computed with a plasmon-scattering formalism [25,206]. In addition to what this plasmon scattering model predicted, unexpected oscillations as a function of τ were observed. These could be satisfactorily explained by our prediction [148] of further charge emission triggered by interactions in the electron sources [205]. From independent calibration measurements, the total RC time constant of the source could be measured to set the constrain $(\tau_f^{-1} + \tau_c^{-1})^{-1}/2 = \tau_{RC} = 21$ ps. Combining Equation (30) and (45) with the plasmon scattering formalism one could compute the current noise in the ouput of the Hong-Ou-Mandel interferometer $\Delta q(\tau)$ with only one fitting parameter: The ratio τ_f/τ_c. The minimization procedure gave the most-likely result: $\tau_f = 136$ ps. The comparison is shown on the right panels of Figure 14, where the model for $\tau_f = 2\tau_{RC}$ [$\tau_c \to \infty$, i.e., no interaction within the dot] described the fractionalization process due to interedge Coulomb interactions but not the interactions within the dot itself. This provides a reasonable qualitative and quantitative agreement with the experimental data reported in Ref. [25]. On the bottom-left panel in Figure 14 we compared, for $\tau_f = 136$ ps, the charge exiting the dot $Q(t)$ with the applied square pulse sequence on the top-gate $V(t)$ which had a finite rise time of 30 ps. In particular, it could explain the appearance of extra rebounds in Δq for time delays τ between 70 and 450 ps which was not possible with a non-interacting dot ($\tau_f = 0$). This was directly due to the additional effects coming from the interactions within the dot itself and could not be explained by the fractionalization mechanism. Indeed, relying exclusively on the model describing fractionalization, we could not reproduce the pronounced additional rebound for $|\tau| = 200$ ps for $\tau_f < 100$ ps. This highlights the relevance of Coulomb interactions in the open dot dynamics.

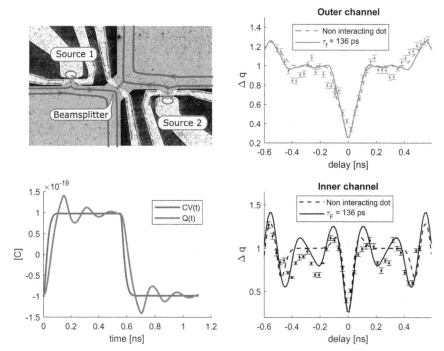

Figure 14. Top-Left—Hong-Ou-Mandel experiment from Ref. [25]: Two single electron sources, as that shown in Figure 6, inject single electron towards the same QPC, which works as a beamsplitter. Bottom-left—Simulation based on Ref. [148] of the charge exiting the open dot when applying a square pulse sequence $V(t)$ with a rise time of 30 ps, one clearly sees the additional pulses coming from interactions. Top-Right—Normalized Hong-Ou-Mandel current noise Δq of the outer edge as a function of the time delay τ with which charge arrives on the QPC from different sources. Noise is suppressed for $\tau = 0$ because of anti-bunching effects, but additional oscillations were observed for $\tau \neq 0$, which the theory in Ref. [148] contributed to explain. The points are the experimental data while the solid and dashed lines are theoretical curves with different fittings for the time of flight τ_f for electrons in the cavity. Bottom-Right—Same as in the central panel but for the inner edge.

6. Conclusions

This review addressed the importance of interactions for the investigation and control of dynamical quantum coherent phenomena in mesoscopic quantum-dot devices.

In Section 2, we discussed how interactions are the essential ingredient allowing long-range quantum state transfer in mesoscopic devices. Notice that the same phenomenon has been suggested to enforce nonlocal phase-coherent electron transfer in wires supporting topologically protected Majorana modes at their edges [207,208]. Such an effect is currently being considered in various Majorana network models for stabilizer measurements in corresponding implementations of topological quantum error correction codes [209–211].

The local Fermi liquid approach, discussed in Section 3, provides the unifying theoretical framework to describe the low-energy dynamics of such various mesoscopic devices. Its application to the various experimental setups mentioned in the Introduction, will be definitively useful to bring further understanding in the complex and rich field of out-of-equilibrium many-body systems. The insight given on universal quantum dissipation phenomena, discussed for the mesoscopic capacitor in Section 4, and, in particular, the novel interaction effects, unveiled in the experiment discussed in Section 5, give two clear examples of the utility of this approach.

Beyond the already mentioned potential for quantum dot devices coupled to microwave cavities [150–159] and to energy transfer [178–180], important extensions of the LFL approach should be envisioned for understanding the properties of mesoscopic devices involving non-Fermi liquids at the place of normal metallic leads. The most important cases would involve superconductors [18,19,212–217] or fractional Quantum Hall edges states, in which quantum noise measurements have been crucial to address and unveil the dynamics of fractionally charged excitations [218–235]. Additionally, the recent realization of noiseless levitons [8–12] paves the way to interesting perspectives to investigate flying anyons [13,220,236] and novel interesting dynamical effects [188,189,237].

Author Contributions: M.F. prepared and finalized the original draft of this review, C.M. gave substantial suggestion for its structure and all the authors equally contributed to revise the manuscript. A.M. and G.F. realized the original data analysis presented in Figure 14. All authors have read and agreed to the published version of the manuscript.

Funding: M.F. acknowledges support from the FNS/SNF Ambizione Grant PZ00P2_174038 and G.F. is funded by ERC consolidator grant "EQuO" (no. 648236).

Acknowledgments: M.F. is indebted to Christopher Bäuerle, Géraldine Haack, Frédéric Pierre, Inès Safi, and Eugene Sukhorukov for important comments and suggestions.

Conflicts of Interest: The authors declare no conflict of interest. The founders had no role in the design of the study; in the collection, analyses, or interpretation of data; in the writing of the manuscript, or in the decision to publish the results.

Abbreviations

The following abbreviations are used in this manuscript:

QPC	quantum point contact
LFL	local Fermi liquid
CBM	Coulomb blockade model
AIM	Anderson impurity model
SW	Schrieffer–Wolff
DC	direct current
2DEG	two-dimensional electron gas
KS	Korringa–Shiba
AC	alternate current
NRG	numerical renormalization group

Appendix A. Scattering Theory, Phase-Shifts and the Friedel Sum Rule

In this Appendix we review some useful results from scattering theory. In Appendix A.1, we provide the definition of the S- and T-matrix in scattering theory. In Appendix A.2, we derive the Friedel sum rule for a non-interacting electron gas with an elastic impurity. In Appendix A.3, we illustrate these concepts on the simple case of a chiral edge state tunnel coupled to a single resonant level. In Appendix A.4, we derive the phase-shift induced on lead electrons by the scattering potential of the LFL, Equation (8).

Appendix A.1. General Definitions

We aim at describing the general situation of Figure A1, which is reproduced by mesoscopic settings, such as a single resonant level coupled to a chiral edge state, which also describes the mesoscopic capacitor in the non-interacting limit, see also Figure 7 in the main text. Consider a wave packet emitted and detected in the distant past and future, namely $t = -\infty$ and $t = +\infty$, and which enters a scattering region at time $t = 0$. Close to detection and emission, it is assumed that the wave packet does not feel the presence of the scatterer, whose interaction range is delimited inside the dashed line in Figure A1. The system is described by a single-particle Hamiltonian of the form:

$$\mathcal{H} = \mathcal{H}_0 + \mathcal{V}, \tag{A1}$$

in which $\mathcal{H}_0$ describes the free propagation of a wave-packet and $\mathcal{V}$ the scatterer. The T-matrix describes the effects of a scatterer on the propagation of a free particle. It is an improper self-energy for the resolvent of the lead electrons, which appears in a modified form of Dyson's equation:

$$G(z) = G_0(z) + G_0(z)T(z)G_0(z), \qquad\qquad G(z) = \frac{1}{z - \mathcal{H}}, \tag{A2}$$

in which z is a complex number. G_0 is the free resolvent describing free electrons:

$$G^0_{kk'}(z) = \langle k| G^0(z) |k'\rangle = \frac{\delta_{kk'}}{z - \varepsilon_k}, \tag{A3}$$

the $|k\rangle$ states being the single particle eigenvectors of the unperturbed Hamiltonian $\mathcal{H}_0$. One can readily show that the T-matrix reads:

$$T(z) = V(\mathbb{I} - G_0(z)\mathcal{V})^{-1}. \tag{A4}$$

The general definition of the phase-shift, a key quantity within scattering theory [238,239], reads:

$$\delta_\varepsilon = \arg\left[T(\varepsilon + i0^+)\right]. \tag{A5}$$

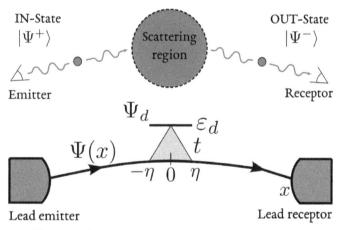

Figure A1. Top—Illustration of the physical situation described by the scattering formalism. Electron wave packets are emitted in the IN-State $|\Psi^+\rangle$ and then measured in the OUT-state $|\Psi^-\rangle$ once they have passed through the scattering region. Bottom—Realization of the scattering setup with a quantum Hall chiral edge state tunnel coupled (in a region of size 2η) with a resonant level of energy ε_d.

We define the IN and OUT states $|\Psi^\pm\rangle$ as the eigenvectors of energy ε of the Hamiltonian of the whole system, including the scattering region, as the states coinciding asymptotically with free plane waves in the past and in the future respectively. The scattering matrix S gives the overlap between these two states:

$$S_{kk'}(\varepsilon) = \langle \Psi^-_k | \Psi^+_{k'} \rangle, \tag{A6}$$

where k/k' are the momenta of the OUT/IN states. The T- and S-matrix are related by the relation [238,239]:

$$S_{kk'} = \delta_{kk'} - 2\pi i \delta(\varepsilon_k - \varepsilon_{k'}) T_{kk'},$$ (A7)

or equivalently, in the energy representation,

$$S(\varepsilon) = \mathbb{I} - 2\pi i v_0 T(\varepsilon).$$ (A8)

The S-matrix is unitary and in the single channel case it is completely defined by a phase $S(\varepsilon) = e^{2i\delta_\varepsilon}$. The phase δ_ε is the phase-shift caused by scattering and in general the condition $S(\varepsilon) = e^{2i\delta_\varepsilon}$ is always verified if we take the definition of the phase-shift directly from the T-matrix:

$$T(\varepsilon) = -\frac{1}{\pi v_0} \sin \delta_\varepsilon e^{i\delta_\varepsilon}.$$ (A9)

There is an interesting connection with the Kondo regime. In Section 3.3.2, we illustrated how particle-hole symmetry enforces that the phase-shift is given by $\delta = \delta_K = \pi/2$. The scattering matrix thus equals the identity. This case is also known as the unitary limit of the Kondo model, in which the transmission probability through a Kondo correlated dot is unity.

Appendix A.2. The Friedel Sum Rule

We show here the Friedel sum rule for non-interacting electrons scattering on an elastic impurity. We first consider the total electron occupation $\langle \mathcal{N} \rangle$ of an electron gas, which reads:

$$\langle \mathcal{N} \rangle = \sum_\alpha \int_{-\infty}^\infty d\omega A_\alpha(\omega) f(\omega),$$ (A10)

the sum on the label α running over all the eigenstates of the Hamiltonian (A1). $A_\alpha(\omega)$ is the spectral function of the state α, defined as:

$$A_\alpha(\omega) = -\frac{1}{\pi} \text{Im} G_{\alpha\alpha}(\omega + i0^+) = -\frac{1}{\pi} \text{Im} \langle \alpha | G(\omega + i0^+) | \alpha \rangle,$$ (A11)

$G_{\alpha\alpha}$ being the *retarded* Green's function associated to the state α defined in Equation (A2). In the absence of the scatterer, $G_{kk}(\omega + i0^+) = G_0(\omega + i0^+) = (\omega + i0^+ - \varepsilon_k)^{-1}$ and Equation (A10) reduces to a sum over the Fermi function $\sum_k f(\varepsilon_k)$ giving the total number of electrons $\langle \mathcal{N} \rangle$ in the system. If the chemical potential is fixed, the introduction of the scatterer modifies the total average number of electrons. The difference with the initial one gives the amount of electrons $\langle N \rangle$ displaced by the scatterer. In the case of a single channel, one obtains:

$$\langle \mathcal{N} \rangle_{\text{with scatterer}} - \langle \mathcal{N} \rangle_{\text{without scatterer}} = \langle N \rangle = -\frac{1}{\pi} \text{Im} \int_{-\infty}^\infty d\omega \text{Tr}[G(\omega) - G_0(\omega)] f(\omega).$$ (A12)

Using Equation (A4) and the fact that $\frac{d}{d\varepsilon} \text{Tr} \log[A(\varepsilon)] = \text{Tr} \left[A^{-1}(\varepsilon) \frac{d}{d\varepsilon} A(\varepsilon) \right]$, one finds:

$$\begin{aligned} \langle N \rangle &= -\frac{i}{2\pi} \int_{-\infty}^\infty d\omega\, f(\omega) \frac{d}{d\omega} \log \det[\mathbb{I} - 2\pi i v_0\, T(\omega + i0^+)] \\ &= -\frac{i}{2\pi} \int_{-\infty}^\infty d\omega\, f(\omega) \frac{d}{d\omega} \log \det S(\omega + i0^+), \end{aligned}$$ (A13)

in which we applied the definition (A8) of the S-matrix. This is the general form of the Friedel sum rule. For $S(\varepsilon) = e^{2i\delta_\varepsilon}$ and zero temperature it gives a direct relation between the charge displaced by the impurity and the phase-shift at the Fermi energy E_F:

$$\langle N \rangle = \frac{\delta_{E_F}}{\pi}. \tag{A14}$$

The extension to M channels requires to add an overall sum over the channel label σ and leads to Equation (10) in the main text.

Appendix A.3. Illustration on the Resonant Level Model

We illustrate now the above concepts on a simple situation, sketched in Figure A1, in which the scatterer is a single resonant level tunnel coupled to chiral electrons propagating on an edge state. Such a situation is an effective representation of the mesoscopic capacitor, see Figures 7 and A2, and it is described by the Coulomb Blockade Model (CBM) (12), that we remind here to the reader:

$$\mathcal{H}_{CBM} = \sum_k \varepsilon_k c_k^\dagger c_k + t \sum_{k,l} \left[c_k^\dagger d_l + d_l^\dagger c_k \right] + \sum_l (\varepsilon_d + \varepsilon_l) d_l^\dagger d_l + E_c \left(N - N_g \right)^2. \tag{A15}$$

In this section, we neglect the last term, corresponding to interactions, and, for simplicity, we retain only a single fermionic level (annihilated by the fermion operator d) for the cavity, with $\varepsilon_l = 0$. One thus obtains the Hamiltonian of a resonant level:

$$\mathcal{H}_{Res} = \sum_k \varepsilon_k c_k^\dagger c_k + t \sum_k \left(c_k^\dagger d + d^\dagger c_k \right) + \varepsilon_d d^\dagger d. \tag{A16}$$

We first calculate the occupation of the cavity by calculating the retarded Green's functions, defined as $G_{dk}(t - t') = -i\theta(t - t') \langle \{d(t), c_k^\dagger(t')\} \rangle$. They are derived by solving the equations of motion in frequency space:

$$(\omega - \varepsilon_d) G_{dd}(\omega) = 1 + t \sum_k G_{kd}(\omega), \qquad (\omega - \varepsilon_k) G_{kk'}(\omega) = \delta_{kk'} + t G_{dk'}(\omega), \tag{A17}$$

$$(\omega - \varepsilon_k) G_{kd}(\omega) = t G_{dd}(\omega), \qquad (\omega - \varepsilon_d) G_{dk} = t \sum_{k'} G_{k'k}(\omega). \tag{A18}$$

Solving the system, the Green's function for the lead electrons reads:

$$G_{kk'}(\omega) = \frac{\delta_{kk'}}{\omega - \varepsilon_k} + \frac{1}{\omega - \varepsilon_k} t^2 G_{dd}(\omega) \frac{1}{\omega - \varepsilon_{k'}}. \tag{A19}$$

Writing Equation (A19) in the form $G = G^0 + G^0 T G^0$, the T-matrix is found to be:

$$T(\omega + i0^+) = t^2 G_{dd}(\omega + i0^+) = \frac{t^2}{\sqrt{(\omega - \varepsilon_d)^2 + \Gamma^2}} e^{i\delta_\omega}, \qquad \delta_\omega = \frac{\pi}{2} - \arctan\left(\frac{\varepsilon_d - \omega}{\Gamma}\right) \tag{A20}$$

in which we introduced the hybridization constant:

$$\Gamma = t^2 \sum_k (\omega + i0^+ - \varepsilon_k)^{-1} \sim \pi v_0 t^2, \tag{A21}$$

which corresponds to the width acquired by the resonant level by coupling to the lead and which depends on the density of states of the lead electrons at the Fermi energy v_0.

We can now determine the number of displaced charges $\langle N \rangle$ as given by Equation (A22) and show the validity of the Friedel sum rule (A14) in this example. In the wide-band limit, the contribution from the second term in Equation (A19) can be neglected. The number of displaced electrons is given solely by the quantum dot Green's function G_{dd}:

$$\langle \mathcal{N} \rangle_{\text{with dot}} - \langle \mathcal{N} \rangle_{\text{without dot}} = \langle N \rangle = -\frac{1}{\pi} \text{Im} \int_{-\infty}^{\infty} d\omega G_{dd}(\omega) f(\omega) = \frac{1}{2} - \frac{1}{\pi} \arctan\left(\frac{\varepsilon_d}{\Gamma}\right), \quad \text{(A22)}$$

which is consistent with the Friedel sum rule (A14) with the phase-shift (A20). Equation (A22) is also meaningful because: i) It shows that, in the wide-band approximation, the number of displaced electrons $\langle N \rangle$ is given by the local Green's function G_{dd}, which can be interpreted as the charge occupation of the quantum dot and ii) the number of displaced electrons depends on the orbital energy ε_d. As a consequence, a time-dependent variation of ε_d drives a current in the system by displacing electrons in the leads.

As an additional illustration, clarifying how the scattering phase-shift appears on single particle wave-functions, we also solve explicitly the same problem in its real-space formulation. We consider chiral fermions $\Psi(x)$ which tunnel on the resonant level of energy ε_d (and wave-function amplitude Ψ_d), from a region of size 2η centered around $x = 0$, useful to properly regularize the calculation and to be sent to zero at the end [240]. For lead electrons at the Fermi energy, that we set to zero ($\omega = 0$ in Equation (A20)), the Schrödinger equation reads (for $x \in [-\eta, \eta]$):

$$0 = -i\hbar v_F \partial_x \Psi(x) + \frac{t}{2\eta} \Psi_d, \qquad 0 = \varepsilon_d \Psi_d + \frac{t}{2\eta} \int_{-\eta}^{\eta} dx' \Psi(x'). \quad \text{(A23)}$$

By integrating the first equation in the interval $[-\eta, x < \eta]$ and inserting the result for $\Psi(x)$ in the second one, one finds:

$$\Psi(x) = \Psi(-\eta) + \frac{t}{2\eta i \hbar v}(x + \eta)\Psi_d, \qquad \Psi_d = \frac{-t}{\varepsilon_d - i\Gamma} \Psi(-\eta). \quad \text{(A24)}$$

Notice that η does not appear on the last equality and can be sent safely to zero. We consider, as boundary condition for $\Psi(x)$, incoming scattering states of the form $\Psi(x < 0) = e^{ikx}/\sqrt{2\pi\hbar v_F} = \sqrt{v_0}e^{ikx}$. One thus finds:

$$|\Psi_d|^2 = \frac{1}{\pi} \frac{\Gamma}{\varepsilon_d^2 + \Gamma^2}, \qquad \Psi(0^+) = \sqrt{v_0}e^{2i\delta_0}, \qquad \delta_0 = \frac{\pi}{2} - \arctan\left(\frac{\varepsilon_d}{\Gamma}\right), \quad \text{(A25)}$$

in agreement with Equation (A20). This short calculation illustrates how, just after scattering with the resonant level, the electron wave-packet at the Fermi energy acquires a phase $e^{2i\delta_0}$ which is fixed by the resonant level occupation via the Friedel sum rule. Notice that, at the resonance condition for the orbital energy ($\varepsilon_d = 0$), $\delta_0 = \pi/2$ and $\Psi(0^+) + \Psi(0^-) = 0$, as for the unitary limit in the Kondo model, in which $\delta_K = \pi/2$, see also the discussion in Section 3.3.2.

Appendix A.4. T-Matrix in the Potential Scattering Hamiltonian

We derive now the phase-shift caused by the potential scattering term on lead electrons in a local Fermi liquid (LFL). It is useful to recall here the LFL Hamiltonian (8):

$$\mathcal{H}_{\text{LFL}} = \sum_{k\sigma} \varepsilon_k c_{k\sigma}^{\dagger} c_{k\sigma} + W(\varepsilon_d, E_c, \dots) \sum_{k \neq k'\sigma} c_{k\sigma}^{\dagger} c_{k'\sigma}. \quad \text{(A26)}$$

We focus on the the single-channel case for simplicity ($M = 1$). The generalization to $\sigma = 1, \dots M$ channels is straightforward. The Hamiltonian (A26) is quadratic and the Green's function of the lead electrons can be readily obtained relying on the path integral formalism [79]. The partition function corresponding to Equation (A26) reads:

$$\mathcal{Z} = \int \mathcal{D}\left[c, c^\dagger\right] e^{-\mathcal{S}_{\text{LFL}}\left[c, c^\dagger\right]}, \tag{A27}$$

where $\mathcal{S}_{\text{LFL}}\left[c, c^\dagger\right]$ is the action of the system, which reads:

$$\mathcal{S}_{\text{LFL}} = \int_0^\beta d\tau \left\{ -\sum_k c_k^\dagger(\tau) G_k^{-1}(\tau) c_k(\tau) + W \sum_{k \neq k'} c_k^\dagger(\tau) c_{k'}(\tau) \right\}, \tag{A28}$$

where we introduced the free propagator $G_k^{-1}(\tau) = -\partial_\tau - \varepsilon_k$ and in which c_k is a Grassmann variable. It is practical to switch to the frequency representation $c_{k\sigma}(\tau) = \frac{1}{\beta}\sum_{i\omega_n} e^{-i\omega_n\tau} c_{k\sigma}(i\omega_n)$, where we defined the fermionic Matsubara frequencies $i\omega_n = (2n+1)\pi/\beta, n \in \mathbb{Z}$. They satisfy the anti-periodicity property $c(\beta) = -c(0)$ and lead to:

$$\mathcal{S}_{\text{LFL}} = \sum_{i\omega_n} \left\{ -\sum_k c_k^\dagger(i\omega_n) G_k^{-1}(i\omega_n) c_k(i\omega_n) + W \sum_{k \neq k'} c_k^\dagger(i\omega_n) c_{k'}(i\omega_n) \right\}, \tag{A29}$$

with $G_k^{-1}(i\omega_n) = i\omega_n - \varepsilon_k$, which recovers the usual retarded/advanced Green's functions by performing the analytical continuation $i\omega_n \rightarrow \omega \pm i0^+$. The full Green's function $G_{kk'}(i\omega_n) = -\langle c_k(i\omega_n) c_k^\dagger(i\omega_n) \rangle$ is derived by expanding the partition function (A27) in the coupling W and by applying Wick's theorem [83]. The pertubation expansion of $G_{kk'}(i\omega_n)$ has the simple form:

$$G_{kk'}(i\omega_n) = \frac{\delta_{kk'}}{i\omega_n - \epsilon_k} + \frac{1}{i\omega_n - \epsilon_k}\frac{1}{i\omega_n - \epsilon_{k'}} W\left[1 + \Sigma(i\omega_n) + \Sigma^2(i\omega_n) + \Sigma^3(i\omega_n) + \dots\right], \tag{A30}$$

in which we introduced the self-energy:

$$\Sigma(i\omega_n) = \sum_p \frac{W(\varepsilon_d)}{i\omega_n - \epsilon_p}. \tag{A31}$$

Using the definition (A2), the T-matrix thus reads:

$$T(z) = \frac{W}{1 - \Sigma(z)}. \tag{A32}$$

Making the analytical continuation $i\omega_n \rightarrow \omega + i0^+$ and considering a constant density of states ν_0 for the lead electrons, we obtain:

$$T(\omega + i0^+) = \frac{W(\varepsilon_d)}{1 + i\pi\nu_0 W} = \frac{W(\varepsilon_d)}{\sqrt{1 + [\pi\nu_0 W(\varepsilon_d)]^2}} e^{i\delta_W}, \tag{A33}$$

in which we introduced the phase-shift:

$$\delta_W = -\arctan\left(\pi\nu_0 W\right). \tag{A34}$$

Applying the definition of the phase-shift given in Equation (A5), one finds Equation (9) in the main text. As a consistency check, substituting Equation (A33) in Equation (A8), we find that the scattering matrix reads $S = e^{2i\delta}$.

Appendix B. Self-Consistent Description- of a 2DEG Quantum *RC* Circuit

In Appendix B.1, we shortly review the self-consistent scattering theory of the mesoscopic capacitor and, in Appendix B.2, we show how equivalent results can be derived with a Hamiltonian formulation.

Appendix B.1. Self-Consistent Theory

In this section, we shortly review the description of the mesoscopic capacitor in the seminal works by Büttiker, Thomas, and Prêtre [21,51–53], based on a self-consistent extension of the Landauer–Büttiker scattering formalism [45–47]. The following discussion is also inspired from Refs. [20,241,242].

Figure A2 illustrates the intuition behind the interpretation of the mesoscopic capacitor as a quantum analog of a classical *RC* circuit. The top metallic gate (a classical metal) and the quantum dot cannot exchange electrons. These two components make up the two plates of a capacitor on which electrons accumulate according to variations of the gate potential V_g. The value of the capacitance depends on the geometry of the contact and, in the experiment of Ref. [20], the *geometrical capacitance* C_g was estimated ~10–100 fF. This capacitance is in series with a quantum point contact. As mentioned in the main text, direct transport measurements [160,161] consider QPCs as resistive elements of resistance $R_{DC} = h/e^2 D$, where D is the QPC transparency. The above considerations suggest the interpretation of the device in Figure A2 as an *RC* circuit. The admittance of a classical *RC* circuit reads:

$$A(\omega) = \frac{-i\omega C}{1 - i\omega RC} = -i\omega C\left(1 + i\omega CR\right) + O\left(\omega^3\right). \tag{A35}$$

The admittance (A35) can be calculated with the scattering formalism [45–47], which requires to be adapted to describe the mesoscopic capacitor in Figure A2. The main problem is the "mixed" nature of the quantum *RC* device: The mesoscopic capacitor is composed of a phase-coherent part (two-dimensional electron gas + quantum dot) in contact to an incoherent top metallic gate. As these constituents do not exchange electrons, preventing a direct current, electron transport is only possible by driving the system. We focus on the case of a gate potential oscillating periodically, as Equation (35) in the main text,

$$V_g(t) = V_g + \varepsilon_\omega \cos(\omega t). \tag{A36}$$

For small oscillation amplitudes ε_ω, the Landauer–Büttiker formalism allows to derive the circuit admittance within linear response theory. In the case of a single conduction mode, the admittance reads [51]:

$$a(\omega) = \frac{e^2}{h} \int d\varepsilon \operatorname{Tr}\left[1 - S^\dagger(\varepsilon)S(\varepsilon + \hbar\omega)\right] \cdot \frac{f(\varepsilon) - f(\varepsilon + \hbar\omega)}{\hbar\omega}, \tag{A37}$$

in which $f(\varepsilon)$ is the Fermi distribution function, in which we fix to zero the value of the Fermi energy. For one channel, the elastic scattering assumption implies that the matrix $S(\varepsilon)$ reduces to a pure phase $S(\varepsilon) = e^{2i\delta_\varepsilon}$, as electrons entering the dot return to the lead with unit probability. The phase-shift δ_ε is related to the dot electron occupation, via the Friedel sum rule (A14). Additionally, this phase is also related to the *dwell-time* that electrons spend in the quantum dot, or Wigner–Smith delay time [89,243]:

$$\frac{\tau(\varepsilon)}{h} = \frac{1}{2\pi i}S^\dagger(\varepsilon)\frac{dS(\varepsilon)}{d\varepsilon} = \frac{1}{\pi}\frac{d\delta_\varepsilon}{d\varepsilon}. \tag{A38}$$

The interpretation of τ as a dwell-time is illustrated in Section 4.2. In the limits $T \to 0$ and $\hbar\omega \to 0$, Equation (A37) becomes:

$$a(\omega) = -i\omega\frac{e^2}{h}\left[\tau + \frac{1}{2}i\omega\tau^2 + O\left(\omega^2\right)\right]. \tag{A39}$$

The dwell-time $\tau = \tau(0)$ is considered at the Fermi energy. Notice that this expression has the same frequency expansion as Equation (A35). Matching term by term, one finds:

$$C_q = \frac{e^2}{h}\tau, \qquad\qquad R_q = \frac{h}{2e^2}. \qquad\qquad (A40)$$

Such relation between time-delays and circuit elements comes from the fact that electrons arriving on the dot at different times are differently phase-shifted, because of the variations in time of the gate potential V_g. This effect causes a local accumulation of charges, which is responsible for the emergence of quantum capacitive effects, corresponding to C_q. The time delay of the electron phase δ with respect to the driving potential $U(t)$ is responsible for energy dissipation, controlled by R_q. The characteristic time that an electron spends in the quantum dot is given by $\tau = 2R_qC_q$, twice the RC time because it includes the charging and relaxation time of the RC circuit. Notice the emergence of a universally quantized relaxation resistance, regardless of any microscopic detail of the quantum RC circuit, in contrast with the resistance $R_{DC} = h/e^2D$, sensitive to the transparency D of the QPC and which would be measured in a DC experiment, see also Equation (43) in the main text.

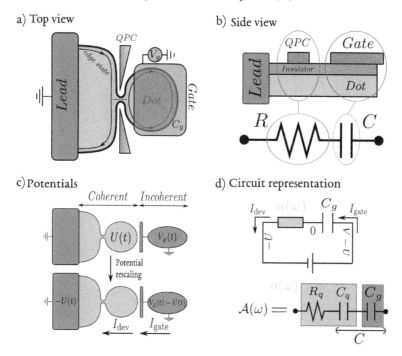

Figure A2. (**a,b**) Schematic representation of the quantum RC circuit. Electrons in the edge states of a two-dimensional electron gas (2DEG) in the integer quantum Hall regime tunnel inside a quantum dot through a QPC. The dot is driven by a top metallic gate. The dot and the gate are separated by an insulator and cannot exchange electrons, thus forming the two plates of a capacitor C. The QPC is a resistive element of resistance R. These two circuit elements are in series and define a quantum coherent RC circuit. (**c,d**) The blue region of the two-dimensional gas is phase coherent. The top metallic gate is an incoherent metal, driven by a time-dependent gate potential $V_g(t)$, which induces an unknown uniform potential $U(t)$ on the dot. The classical circuit analogy in (**d**) is made possible by shifting all energies by $-U(t)$. The whole device behaves as a charge relaxation resistance R_q in series to a total capacitance C: series of a quantum and geometrical capacitance C_q and C_g.

Notice that the geometrical capacitance C_g does not appear in the previous discussion. The admittance Equation (A37) has been derived by applying linear response theory for the driving potential $U(t)$ in the quantum dot, see Figure A2. The potential $U(t)$ does not coincide with the actual driving gate-potential $V_g(t)$. The situation is pictured in Figure A2: The geometric capacitance C_g leads to a potential drop between gate and dot. In a mean-field/Hartree–Fock treatment, the potential $U(t)$ on the dot is assumed to be uniform for each electron. This assumption is equivalent to a Random Phase Approximation (RPA), valid for weak interactions or to leading order in a $1/M$ expansion, with M the number of channels connected to the dot [78]. The potential U can then be determined self-consistently from the constraint of charge/current conservation in the whole device. The current I_{dev} flowing in the coherent part of the device has to equal the current I_{gate} flowing in the incoherent metallic gate:

$$I = I_{dev} = I_{gate} \, . \tag{A41}$$

As all potentials are defined with respect to an arbitrary energy, they can be shifted by $-eU(t)$, setting the potential to zero in the quantum dot. Thus, the currents in the device and in the metallic gate read:

$$I_{dev} = -U(\omega)g(\omega) \, , \qquad I_{gate} = -iC_g\omega\left[U(\omega) - V_g(\omega)\right] . \tag{A42}$$

Applying the current conservation condition (A41), the potential U can be eliminated, and the admittance of the total device is derived:

$$\mathcal{A}(\omega) = -\frac{I}{V_g} = \frac{1}{\frac{1}{a(\omega)} + \frac{1}{-i\omega C_g}} \, . \tag{A43}$$

Recalling the low frequency behavior of $a(\omega)$ in Equation (A39), the above expression shows that the whole device behaves as a RC circuit. Albeit with two capacitances in series, Equation (A43) still gives a universally quantized $R_q = h/2e^2$. The series of C_q and C_g gives the total capacitance C, originally denoted as *electro-chemical capacitance* [21,51–53].

We can consider a simple model for the mesoscopic capacitor to estimate the behavior of the dwell-times τ setting the quantum capacitance (A40). We consider the case of Figure A2, in which electrons propagate in the integer quantum Hall edges inside the quantum dot, see also Figure 7 in the main text. We label ℓ the length of the edge state in the quantum dot and v_F the Fermi velocity of the electron. The dwell-time for electrons of velocity v_F inside the dot is $\tau_f = \ell/v_F$. An electric wave of energy ε acquires a phase $\phi(\varepsilon) = (\varepsilon - eU)\tau_f/\hbar$, when making a tour of the dot. Notice that we had to shift the energy ε of the electron by $-eU$ because of the potential shift schematized in Figure A2. The chiral nature of the edge states allows for a one-dimensional representation of the problem, pictured in the right-top of Figure 7. For a quantum well of size $\ell/2$, close to the Fermi energy, the level spacing $\Delta = hv_F/\ell$ is constant. Thus $\tau_f = h/\Delta$ and, substituting in Equation (A40), leads to the uniform quantum capacitance $C_q = e^2\tau_f/h$, derived heuristically in the main text, see Equation (41). If the reflection amplitude at the entrance of the dot is r and $D = 1 - |r|^2$ the transmission probability, the dot can be viewed as a Fabry–Perot cavity and the phase of the out-coming electron is:

$$S(\varepsilon) = r - De^{i\phi(\varepsilon)} \sum_{q=0}^{\infty} r^q e^{iq\phi(\varepsilon)} = \frac{r - e^{i\phi(\varepsilon)}}{1 - re^{i\phi(\varepsilon)}} = e^{i2\delta_\varepsilon} \, . \tag{A44}$$

Applying Equation (A38) we obtain the local density of states:

$$\mathcal{N}(\varepsilon) = \frac{\tau_f}{h} \frac{1 - r^2}{1 - 2r\cos\left[\frac{2\pi}{h}(\varepsilon - eU)\tau_f\right] + r^2} \, . \tag{A45}$$

This quantity is plotted in Figure A3 and reproduces the oscillatory behavior of the capacitance in Figure 8. In the limit of small transmission ($D \ll 1$ and $r \approx 1$), Equation (A45) reduces to a sum of Lorentzian peaks of width $\hbar\gamma$, $\gamma = D/\tau_f$:

$$\mathcal{N}(\varepsilon) = \frac{2}{\pi\hbar\gamma} \sum_n \frac{1}{1 + \left(\frac{\varepsilon - eU - n\Delta}{\hbar\gamma/2}\right)^2} . \tag{A46}$$

These peaks are the discrete spectrum of the dot energy levels.

The above arguments can be readily generalized to the case with M channels in the lead and in the cavity. In this case, every single channel σ can be considered independently. The admittance (A39) can be then cast in the form:

$$a(\omega) = -i\omega\frac{e^2}{h} \sum_{\sigma=1}^{M} \left[\tau_\sigma + \frac{1}{2}i\omega\tau_\sigma^2 + O(\omega^2)\right] . \tag{A47}$$

The low-frequency expansion of the RC circuit admittance (A35) is then recovered by defining:

$$C_q = \frac{e^2}{h} \sum_{\sigma=1}^{M} \tau_\sigma , \qquad\qquad R_q = \frac{h}{2e^2} \frac{\sum_\sigma \tau_\sigma^2}{\left(\sum_\sigma \tau_\sigma\right)^2} , \tag{A48}$$

in which τ_σ are the dwell-times in the quantum dot of electrons in the σ-th mode.

Appendix B.2. Hamiltonian Description of the Quantum RC Circuit with a Resonant Level Model

In this appendix, we study the resonant level model (A16) as a RC circuit. The generalization of the following calculations to the many-channel case is straightforward and, in particular, we extend to the multi-level case. We thus recover the self-consistent scattering theory analysis discussed in Appendix B.1, corresponding to the limit $C_g \to \infty$ (non-interacting limit). Our aim is the calculation of the dynamical charge susceptibility:

$$\chi_c(t - t') = \frac{i}{\hbar}\theta(t - t') \left\langle \left[N(t), N(t')\right]\right\rangle_0 , \tag{A49}$$

which leads to the admittance of the circuit $A(\omega) = -i\omega e^2 \chi_c(\omega)$, see also Equation (36) in the main text. We make use of the path integral formalism as in Appendix A.4. The partition function corresponding to Equation (A16) reads:

$$\mathcal{Z} = \int \mathcal{D}\left[c, c^\dagger, d, d^\dagger\right] e^{-S_{\text{Res}}\left[c, c^\dagger, d, d^\dagger\right]} , \tag{A50}$$

where $S\left[c, c^\dagger, d, d^\dagger\right]$ is the action of the system, which reads

$$S_{\text{Res}} = \int_0^\beta d\tau \left\{-\sum_k c_k^\dagger(\tau)G_k^{-1}(\tau)c_k(\tau) - d^\dagger(\tau)D^{-1}(\tau)d(\tau) + t\sum_k \left[c_k^\dagger(\tau)d(\tau) + \text{c.c.}\right]\right\} , \tag{A51}$$

with the free propagators:

$$G_k^{-1}(\tau) = -\partial_\tau - \varepsilon_k , \qquad\qquad D^{-1}(\tau) = -\partial_\tau - \varepsilon_d , \tag{A52}$$

in which c_k and d_l are Grassmann variables. It is practical to switch to the Matsubara frequency representation, which leads to:

$$S_{\text{Res}} = \sum_{i\omega_n} \left\{ -\sum_k c_k^\dagger(i\omega_n) G_k^{-1}(i\omega_n) c_k(i\omega_n) - d^\dagger(i\omega_n) D^{-1}(i\omega_n) d(i\omega_n) \right.$$

$$\left. +t \sum_k \left[c_k^\dagger(i\omega_n) d(i\omega_n) + d^\dagger(i\omega_n) c_k(i\omega_n) \right] \right\}, \quad (A53)$$

with $G_k^{-1}(i\omega_n) = i\omega_n - \varepsilon_k$ and $D^{-1}(i\omega_n) = i\omega_n - \varepsilon_d$. Reestablishing dimensions, the Fourier transform of Equation (A49) reads:

$$\chi_c(i\nu_n) = \frac{1}{\hbar} \int_0^{\hbar\beta} d(\tau - \tau') e^{i\nu_n(\tau - \tau')} \langle N(\tau) N(\tau') \rangle_0, \quad (A54)$$

and $\chi_c(\omega)$ is recovered by performing the analytical continuation $i\nu \to \omega + i0^+$. This function is periodic in imaginary time and its Fourier transform is a function of the bosonic Matsubara frequencies $i\nu_n = 2n\pi/\beta$. $N(\tau) = d^\dagger(\tau)d(\tau)$ counts the number of charges on the dot. The cyclic invariance property of the trace implies that $\langle N(\tau)N(\tau')\rangle_0 = f(\tau - \tau')$, allowing us to recast (A54) in the form:

$$\chi_c(i\nu_n) = \frac{1}{\beta} \sum_{i\omega_{1,2}} \left\langle d^\dagger(i\omega_1) d(i\omega_1 + i\nu_n) d^\dagger(i\omega_2) d(i\omega_2 - i\nu_n) \right\rangle. \quad (A55)$$

To calculate this expression, we first perform the Gaussian integral of the lead modes in Equation (A53), leading to the effective action S'_{Res} of the resonant level model:

$$S'_{\text{Res}} = -\sum_{i\omega_n} d^\dagger(i\omega_n) D(i\omega_n) d(i\omega_n), \qquad D^{-1}(i\omega_n) = i\omega_n - \varepsilon_d - t^2 \sum_k G_k(i\omega_n). \quad (A56)$$

In the wide-band approximation the propagator can be written as $D^{-1}(i\omega_n) = i\omega_n - \varepsilon_d + i\Gamma\,\text{sgn}(\omega_n)$, where we introduced the hybridization constant $\Gamma = \pi\nu_0 t^2$. The action (A56) is quadratic and the application of Wick's theorem [83] in Equation (A55) leads to:

$$\chi_c(i\nu_n) = -\frac{1}{\beta} \sum_{i\omega_n} D(i\omega_n) D(i\omega_n + i\nu_n) \to -\frac{1}{\pi\Gamma} \int_{-\infty}^{\infty} dx\, f(\Gamma x + \varepsilon_d) \frac{2x}{(x^2+1)[x^2 - (\omega/\Gamma + i)^2]}, \quad (A57)$$

where the analytical continuation $i\nu \to \omega + i0^+$ has been performed. At zero temperature, the integral can be calculated analytically, leading to:

$$\chi_c(\omega) = \frac{1}{\pi\Gamma} \frac{1}{\frac{\omega}{\Gamma}\left(\frac{\omega}{\Gamma} + 2i\right)} \ln \frac{\varepsilon_d^2 + \Gamma^2}{\varepsilon_d^2 - (\omega + i\Gamma)^2}. \quad (A58)$$

The low frequency expansion of this expression matches the one of a classical RC circuit (A35). Reestablishing correct dimensions $\omega \to \hbar\omega$, the result recovers Equation (A40) obtained within scattering theory:

$$C_0 = \frac{e^2}{h} \nu(\varepsilon_d), \qquad\qquad R_q = \frac{h}{2e^2}, \quad (A59)$$

where $\nu(\varepsilon_d)$ is the density of states associated to the single orbital ε_d

$$\nu(\varepsilon_d) = \frac{1}{\pi} \frac{\Gamma}{\varepsilon_d^2 + \Gamma^2}. \quad (A60)$$

The extension to M channels is straightforward. One can consider a Hamiltonian of the form:

$$H_{\text{Res}-\text{Mch}} = \sum_{k\sigma} \varepsilon_k c_{k\sigma}^\dagger c_{k\sigma} + t \sum_{k\sigma} \left(c_{k\sigma}^\dagger d_\sigma + d_\sigma^\dagger c_{k\sigma} \right) + \sum_\sigma \varepsilon_\sigma d_\sigma^\dagger d_\sigma , \tag{A61}$$

with $\sigma = 1, \ldots, M$ the number of channels. In this model, each channel can be treated independently and one finds a a generalization of Equation (A58):

$$\chi_c(\omega) = \sum_{\sigma=1}^{M} \frac{1}{\pi\Gamma} \frac{1}{\frac{\omega}{\Gamma}\left(\frac{\omega}{\Gamma} + 2i\right)} \ln \frac{\varepsilon_\sigma^2 + \Gamma^2}{\varepsilon_\sigma^2 - (\omega + i\Gamma)^2} . \tag{A62}$$

This expression, when expanded to low frequency, has also the form (A47), where the dwell-times are substituted by the density of states of the channels (A60): $\tau_\sigma \to \nu_\sigma = \nu(\varepsilon_\sigma)$. One thus finds expressions for the differential capacitance and the charge relaxation resistance analog to Equation (A48):

$$C_0 = \frac{e^2}{h} \sum_{\sigma=1}^{M} \nu_\sigma , \qquad\qquad R_q = \frac{h}{2e^2} \frac{\sum_\sigma \nu_\sigma^2}{\left(\sum_\sigma \nu_\sigma\right)^2} . \tag{A63}$$

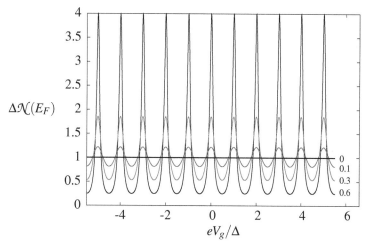

Figure A3. Peaked structure of the local density of states $\mathcal{N}(\varepsilon)$ on the dot as a function of the orbital energy shift controlled by the gate potential eV_g from Equation (A45). $\mathcal{N}(E_F)$ is plotted for different values of the backscattering amplitude r. The progressive opening of the dot drives a transition from a Lorentzian to an oscillatory behavior of C_q, coherent with the experimental measurements illustrated in Figure 8. For a completely transparent dot ($r = 0$) the density of states is uniform, which implies $C_0 = e^2/\Delta$.

Multi-Level Case

In this section we carry out the calculation of the quantum dot density of states in the case of a single channel and an infinite number of equally spaced levels in the quantum dot. The action reads:

$$S = \sum_{i\omega_n} \left\{ -\sum_k c_k^\dagger G_k^{-1}(i\omega_n) c_k - \sum_l d_l^\dagger D_l^{-1}(i\omega_n) d_l + t \sum_{kl} \left[c_k^\dagger d_l + d_l^\dagger c_k \right] \right\} , \tag{A64}$$

with $G_k^{-1}(i\omega_n) = i\omega_n - \varepsilon_k$ and $D_l^{-1}(i\omega_n) = i\omega_n - \varepsilon_l$. The Gaussian integration of the lead electron modes leads to the effective action:

$$S' = \sum_{i\omega_n} \left\{ -\sum_l d_l^\dagger(i\omega_n) D_l^{-1}(i\omega_n) d_l(i\omega_n) + t^2 \sum_k G_k(i\omega_n) \sum_{ll'} d_l^\dagger(i\omega_n) d_{l'}(i\omega_n) \right\}. \tag{A65}$$

Applying Wick's theorem [83] the full propagator of the dot electrons is readily obtained:

$$\mathcal{D}_{ll'}(i\omega_n) = \delta_{ll'} D_l(i\omega_n) + D_l(i\omega_n) D_{l'}(i\omega_n) \frac{\gamma(i\omega_n)}{1 - \gamma(i\omega_n)\Theta(i\omega_n)}, \tag{A66}$$

where we defined:

$$\gamma(i\omega_n) = t^2 \sum_k G_k(i\omega_n), \qquad \Theta(i\omega_n) = \sum_l D_l(i\omega_n). \tag{A67}$$

In the wide band limit $\gamma(i\omega_n) = -i\Gamma\,\mathrm{sgn}(i\omega_n)$. The charge $\langle Q \rangle = e \sum_l \langle d_l^\dagger d_l \rangle$ on the dot is given by:

$$\langle Q \rangle = \frac{e}{\beta} \sum_{l,i\omega_n} e^{i\omega_n 0^+} \mathcal{D}_{ll}(i\omega_n) = \frac{e}{2\pi i} \sum_l \int_{-\infty}^{\infty} d\varepsilon\, f(\varepsilon) \left[\mathcal{D}_{ll}(\varepsilon + i0^+) - \mathcal{D}_{ll}(\varepsilon - i0^+) \right]. \tag{A68}$$

We write the energy spectrum on the dot as $\varepsilon_l = -eV_g + l\Delta$, with $l \in \mathbb{Z}$ and Δ the level spacing. Equation (A66) is then a function of $\varepsilon + eV_g$. Shifting all energies by eV_g, the differential capacitance $C_0 = -\partial \langle Q \rangle / \partial V_g$ is readily obtained at zero temperature:

$$C_0 = \frac{e^2}{2\pi i} \sum_l \left[\mathcal{D}_{ll}(eV_g + i0^+) - \mathcal{D}_{ll}(eV_g - i0^+) \right], \tag{A69}$$

with

$$\begin{aligned}
\mathcal{D}_{ll}(eV_g \pm i0^+) &= \frac{1}{eV_g - l\Delta} \mp \frac{1}{(eV_g - l\Delta)^2} \frac{i\Gamma}{1 \pm i\Gamma \left[\sum_p \frac{1}{eV_g - p\Delta} \right]} \\
&= \frac{1}{\Delta} \left\{ \frac{1}{x+l} \mp \frac{i\Gamma/\Delta}{(x+l)^2} \frac{1}{1 \pm i\pi\frac{\Gamma}{\Delta}\coth(\pi x)} \right\},
\end{aligned} \tag{A70}$$

where $x = eV_g/\Delta$ and we exploited the fact that $\sum_l \frac{1}{x+l} = \Psi_0(1-x) - \Psi_0(x) = \pi\coth(\pi x)$, in which $\Psi_0(x)$ is the digamma function. Substituting this expression in Equation (A69), the sum over levels can be also carried out, leading to:

$$C_0 = e^2 \frac{\pi\Gamma}{2\Delta^2} \frac{1}{\sin^2\left(\pi\frac{eV_g}{\Delta}\right)} \left[\frac{1}{1 + i\pi\frac{\Gamma}{\Delta}\coth\left(\pi\frac{eV_g}{\Delta}\right)} + \frac{1}{1 - i\pi\frac{\Gamma}{\Delta}\coth\left(\pi\frac{eV_g}{\Delta}\right)} \right], \tag{A71}$$

where we relied on the identity: $\sum_l \frac{1}{(l+x)^2} = \Psi_1(1-x) - \Psi_1(x) = \frac{\pi^2}{\sin^2(\pi x)}$, in which $\Psi_n(x)$ is the polygamma function. Some algebra leads to:

$$C_0 = \frac{e^2}{\Delta} \frac{2}{\frac{\Delta}{\pi\Gamma} + \frac{\pi\Gamma}{\Delta} - \left(\frac{\Delta}{\pi\Gamma} - \frac{\pi\Gamma}{\Delta}\right)\cos\left(\frac{2\pi eV_g}{\Delta}\right)}. \tag{A72}$$

This quantity is plotted in Figure A3 as a function of the gate potential V_g and reproduces the oscillations of the capacitance observed in Figure 8, in the main text. As we did not consider any many-body interaction to derive C_0, this quantity corresponds to the quantum capacitance $C_q = e^2 \mathcal{N}(E_F)$ corresponding to the density of states at the Fermi level, see also discussion in

Section 4.2. Such density of states was also derived within scattering theory in Appendix B.1. Indeed, Equations (A45) and (A72) coincide if one makes the identification (It is useful to recall here that $\tau_f = h/\Delta$)

$$\frac{\pi\Gamma}{\Delta} = \frac{(1-r)^2}{1-r^2} = \frac{1-r}{1+r} \qquad \Leftrightarrow \qquad r = \frac{1-\frac{\pi\Gamma}{\Delta}}{1+\frac{\pi\Gamma}{\Delta}}. \tag{A73}$$

Notice that the fully transparent limit coincides with $\pi\Gamma/\Delta = 1$, corresponding to a change of sign of the reflection amplitude r (remind that we assumed r to be a real number). Additionally, if we consider the tunneling limit $\pi\Gamma/\Delta \ll 1$, we can write $r = \sqrt{1-D}$ and, in the low-transparency limit $D \ll 1$ one recovers $\pi\Gamma/\Delta = D/4$, which is consistent with the expectation $D \propto t^2$ in the tunneling limit of the Hamiltonian (A16). Notice also that the relation (A73) implies $r = -1$ in the $\Gamma \to \infty$ limit, which can be explained by the formation of bonding and anti-bonding states at the junction between electrons in the lead and in the dot, suppressing tunneling in the dot [240]. For a single level and one channel, we recover the universal charge relaxation resistance $R_q = h/2e^2$.

Appendix C. Useful Results of Linear Response Theory

In this appendix we remind some useful properties of linear response theory following Ref. [244]. In Appendix C.1, we show that the real/imaginary parts of the dynamical charge susceptibility (A49) are respectively even/odd functions of the frequency, leading to:

$$A(\omega) = -i\omega e^2 \left\{ \chi_c + i\text{Im}\left[\chi_c(\omega)\right] \right\} + \mathcal{O}(\omega^2), \tag{A74}$$

that is Equation (38) in the main text. In Appendix C.2, we demonstrate that the power dissipated by the quantum RC circuit in the linear response regime is given by:

$$\mathcal{P} = \frac{1}{2}\varepsilon_\omega^2 \omega \text{Im}\chi_c(\omega), \tag{A75}$$

that is Equation (55) in the main text.

Appendix C.1. Parity of the Dynamical Charge Susceptibility

The Lehman representation [97] of the dynamical charge susceptibility $\chi_c(\omega)$ (A49) makes explicit its real and imaginary parts. This is obtained from the Fourier transform of Equation (A49):

$$\chi_c(\omega) = \frac{i}{\hbar} \int_{-\infty}^{\infty} d(t-t') e^{i(\omega+i0^+)(t-t')} \theta(t-t') \left\langle [N(t), N(t')] \right\rangle_0, \tag{A76}$$

where the factor $i0^+$ is inserted to regularize retarded functions. Inserting the closure relation with the eigenstates $|n\rangle$ of energy E_n of the time independent Hamiltonian $\mathcal{H}_0$, the average can be written as:

$$\langle N(t)N(t')\rangle_0 = \sum_{n,m} p_n e^{i\omega_{nm}(t-t')} N_{nm}N_{mn}, \tag{A77}$$

where $p_n = e^{-\beta E_n}/Z$ is the Boltzmann weight, $\hbar\omega_{nm} = E_n - E_m$ and $N_{nm} = \langle n|N|m\rangle$ the matrix elements of the dot occupation. In this representation, the Fourier transform (A76) reads:

$$\chi_c(\omega) = -\frac{1}{\hbar}\sum_{nm} p_n N_{nm}N_{mn} \left(\frac{1}{\omega+i0^+ + \omega_{nm}} - \frac{1}{\omega+i0^+ - \omega_{nm}} \right). \tag{A78}$$

Applying the relation $\frac{1}{x\pm i0^+} = P\left[\frac{1}{x}\right] \mp i\pi\delta(x)$, with $P[f(x)]$ the principal value of the function $f(x)$, the real and imaginary part of $\chi_c(\omega)$ are readily obtained:

$$\text{Re}\left[\chi_c(\omega)\right] = -\frac{1}{\hbar}\sum_{nm} p_n N_{nm} N_{mn}\left\{P\left[\frac{1}{\omega + \omega_{nm}}\right] - P\left[\frac{1}{\omega - \omega_{nm}}\right]\right\}, \tag{A79}$$

$$\text{Im}\left[\chi_c(\omega)\right] = \frac{i\pi}{\hbar}\sum_{nm} p_n N_{nm} N_{mn}\left\{\delta(\omega + \omega_{nm}) - \delta(\omega - \omega_{nm})\right\}, \tag{A80}$$

which are respectively an even and odd function of ω. As a consequence, in the low frequency expansion of the dynamical charge susceptibility $\chi_c(\omega) = \chi_c(0) + \omega\partial_\omega\chi_c(\omega)|_{\omega=0} + \mathcal{O}(\omega^2)$, the linear term in ω has to coincide with the imaginary part of $\text{Im}[\chi_c(\omega)]$, leading to Equation (A74).

Appendix C.2. Energy Dissipation in the Linear Response Regime

In the situation addressed in Section 4, the time dependence of orbital energies in the dot is given by $\varepsilon_d(t) = \varepsilon_d^0 + \varepsilon_\omega \cos(\omega t)$. In the time unit, the systems dissipates the energy:

$$\delta W = \delta \langle N \rangle \varepsilon_\omega \cos(\omega t). \tag{A81}$$

In the stationary regime, the average power $\mathcal{P}$ dissipated by the system during the time period T reads:

$$\mathcal{P} = \frac{\varepsilon_\omega}{T}\int_0^T dt \frac{d\langle N(t) \rangle}{dt} \cos(\omega t). \tag{A82}$$

Neglecting constant contributions, $\langle N(t) \rangle$ is given by the dynamical charge susceptibility (A49):

$$\langle N(t) \rangle = \varepsilon_\omega \int_{-\infty}^\infty dt' \chi_c(t - t') \cos(\omega t). \tag{A83}$$

Substituting this expression in Equation (A82), we obtain:

$$\mathcal{P} = -i\omega\frac{\varepsilon_\omega^2}{4}\left[\chi_c(\omega) - \chi_c(-\omega)\right] + i\omega\frac{\varepsilon_\omega^2}{T}\int_0^T dt \frac{\chi_c(-\omega)e^{2i\omega t} - \chi_c(\omega)e^{-2i\omega t}}{4}. \tag{A84}$$

Expressing $\chi_c(\omega) = \text{Re}\left[\chi_c(\omega)\right] + i\text{Im}\left[\chi_c(\omega)\right]$ as the sum of its real and imaginary part and applying the parity properties demonstrated in Appendix C.1, the first term recovers Equation (A75) for the dissipated power:

$$\mathcal{P} = \frac{1}{2}\varepsilon_\omega^2\omega\text{Im}\left[\chi_c(\omega)\right], \tag{A85}$$

while the second term in Equation (A84) reduces to vanishing integrals of $\sin(2\omega t)$ and $\cos(2\omega t)$ over their period. In the case of Section 4.5, describing the energy dissipated by the LFL effective low-energy theory, we can apply the same considerations by replacing $\delta\langle N \rangle$ in Equation (A81) with the average of the operator A, defined in Equation (57). One thus derives Equation (56) in the main text.

References

1. Averin, D.V.; Likharev, K.K. Coulomb blockade of single-electron tunneling, and coherent oscillations in small tunnel junctions. *J. Low Temp. Phys.* **1986**, *62*, 345–373.
2. Grabert, H.; Devoret, M.; Kastner, M. Single Charge Tunneling: Coulomb Blockade Phenomena in Nanostructures. *Phys. Today* **1993**, *46*, 62.
3. Aleiner, I.; Brouwer, P.; Glazman, L. Quantum effects in Coulomb blockade. *Phys. Rep.* **2002**, *358*, 309–440.
4. Loss, D.; DiVincenzo, D.P. Quantum computation with quantum dots. *Phys. Rev. A* **1998**, *57*, 120–126.
5. Vandersypen, L.M.K.; Bluhm, H.; Clarke, J.S.; Dzurak, A.S.; Ishihara, R.; Morello, A.; Reilly, D.J.; Schreiber, L.R.; Veldhorst, M. Interfacing spin qubits in quantum dots and donors—hot, dense, and coherent. *NPJ Quantum Inf.* **2017**, *3*, 1–10.

6. Vinet, M.; Hutin, L.; Bertrand, B.; Barraud, S.; Hartmann, J.M.; Kim, Y.J.; Mazzocchi, V.; Amisse, A.; Bohuslavskyi, H.; Bourdet, L.; et al. Towards scalable silicon quantum computing. In Proceedings of the 2018 IEEE International Electron Devices Meeting (IEDM), San Francisco, CA, USA, 1–5 December 2018; pp. 6.5.1–6.5.4. ISSN: 0163-1918.

7. Xiang, Z.L.; Ashhab, S.; You, J.Q.; Nori, F. Hybrid quantum circuits: Superconducting circuits interacting with other quantum systems. *Rev. Mod. Phys.* **2013**, *85*, 623–653.

8. Levitov, L.S.; Lee, H.; Lesovik, G.B. Electron counting statistics and coherent states of electric current. *J. Math. Phys.* **1996**, *37*, 4845–4866.

9. Ivanov, D.A.; Lee, H.W.; Levitov, L.S. Coherent states of alternating current. *Phys. Rev. B* **1997**, *56*, 6839–6850.

10. Keeling, J.; Klich, I.; Levitov, L.S. Minimal Excitation States of Electrons in One-Dimensional Wires. *Phys. Rev. Lett.* **2006**, *97*, 116403.

11. Dubois, J.; Jullien, T.; Portier, F.; Roche, P.; Cavanna, A.; Jin, Y.; Wegscheider, W.; Roulleau, P.; Glattli, D. Minimal-excitation states for electron quantum optics using levitons. *Nature* **2013**, *502*, 659–663.

12. Jullien, T.; Roulleau, P.; Roche, B.; Cavanna, A.; Jin, Y.; Glattli, D. Quantum tomography of an electron. *Nature* **2014**, *514*, 603–607.

13. Rech, J.; Ferraro, D.; Jonckheere, T.; Vannucci, L.; Sassetti, M.; Martin, T. Minimal Excitations in the Fractional Quantum Hall Regime. *Phys. Rev. Lett.* **2017**, *118*, 076801.

14. Hermelin, S.; Takada, S.; Yamamoto, M.; Tarucha, S.; Wieck, A.D.; Saminadayar, L.; Bäuerle, C.; Meunier, T. Electrons surfing on a sound wave as a platform for quantum optics with flying electrons. *Nature* **2011**, *477*, 435–438.

15. Bertrand, B.; Hermelin, S.; Takada, S.; Yamamoto, M.; Tarucha, S.; Ludwig, A.; Wieck, A.D.; Bäuerle, C.; Meunier, T. Fast spin information transfer between distant quantum dots using individual electrons. *Nat. Nanotechnol.* **2016**, *11*, 672–676.

16. Bertrand, B.; Hermelin, S.; Mortemousque, P.A.; Takada, S.; Yamamoto, M.; Tarucha, S.; Ludwig, A.; Wieck, A.D.; Bäuerle, C.; Meunier, T. Injection of a single electron from static to moving quantum dots. *Nanotechnology* **2016**, *27*, 214001.

17. Takada, S.; Edlbauer, H.; Lepage, H.V.; Wang, J.; Mortemousque, P.A.; Georgiou, G.; Barnes, C.H.W.; Ford, C.J.B.; Yuan, M.; Santos, P.V.; et al. Sound-driven single-electron transfer in a circuit of coupled quantum rails. *Nat. Commun.* **2019**, *10*, 4557.

18. van Zanten, D.M.T.; Basko, D.M.; Khaymovich, I.M.; Pekola, J.P.; Courtois, H.; Winkelmann, C.B. Single Quantum Level Electron Turnstile. *Phys. Rev. Lett.* **2016**, *116*, 166801.

19. Basko, D.M. Landau-Zener-Stueckelberg Physics with a Singular Continuum of States. *Phys. Rev. Lett.* **2017**, *118*, 016805.

20. Gabelli, J.; Fève, G.; Berroir, J.M.; Plaçais, B.; Cavanna, A.; Etienne, B.; Jin, Y.; Glattli, D.C. Violation of Kirchhoff's Laws for a Coherent RC Circuit. *Science* **2006**, *313*, 499–502.

21. Gabelli, J.; Fève, G.; Berroir, J.M.; Plaçais, B. A coherent RC circuit. *Rep. Prog. Phys.* **2012**, *75*, 126504.

22. Fève, G.; Mahé, A.; Berroir, J.M.; Kontos, T.; Plaçais, B.; Glattli, D.C.; Cavanna, A.; Etienne, B.; Jin, Y. An On-Demand Coherent Single-Electron Source. *Science* **2007**, *316*, 1169–1172.

23. Mahé, A.; Parmentier, F.D.; Bocquillon, E.; Berroir, J.M.; Glattli, D.C.; Kontos, T.; Plaçais, B.; Fève, G.; Cavanna, A.; Jin, Y. Current correlations of an on-demand single-electron emitter. *Phys. Rev. B* **2010**, *82*, 201309.

24. Parmentier, F.D.; Bocquillon, E.; Berroir, J.M.; Glattli, D.C.; Plaçais, B.; Fève, G.; Albert, M.; Flindt, C.; Büttiker, M. Current noise spectrum of a single-particle emitter: Theory and experiment. *Phys. Rev. B* **2012**, *85*, 165438.

25. Freulon, V.; Marguerite, A.; Berroir, J.M.; Plaçais, B.; Cavanna, A.; Jin, Y.; Fève, G. Hong-Ou-Mandel experiment for temporal investigation of single-electron fractionalization. *Nat. Commun.* **2015**, *6*, 1–6.

26. Marguerite, A.; Cabart, C.; Wahl, C.; Roussel, B.; Freulon, V.; Ferraro, D.; Grenier, C.; Berroir, J.M.; Plaçais, B.; Jonckheere, T.; et al. Decoherence and relaxation of a single electron in a one-dimensional conductor. *Phys. Rev. B* **2016**, *94*, 115311.

27. Leicht, C.; Mirovsky, P.; Kaestner, B.; Hohls, F.; Kashcheyevs, V.; Kurganova, E.V.; Zeitler, U.; Weimann, T.; Pierz, K.; Schumacher, H.W. Generation of energy selective excitations in quantum Hall edge states. *Semicond. Sci. Technol.* **2011**, *26*, 055010.

28. Battista, F.; Samuelsson, P. Spectral distribution and wave function of electrons emitted from a single-particle source in the quantum Hall regime. *Phys. Rev. B* **2012**, *85*, 075428.

29. Fletcher, J.D.; See, P.; Howe, H.; Pepper, M.; Giblin, S.P.; Griffiths, J.P.; Jones, G.A.C.; Farrer, I.; Ritchie, D.A.; Janssen, T.J.B.M.; et al. Clock-Controlled Emission of Single-Electron Wave Packets in a Solid-State Circuit. *Phys. Rev. Lett.* **2013**, *111*, 216807.

30. Waldie, J.; See, P.; Kashcheyevs, V.; Griffiths, J.P.; Farrer, I.; Jones, G.A.C.; Ritchie, D.A.; Janssen, T.J.B.M.; Kataoka, M. Measurement and control of electron wave packets from a single-electron source. *Phys. Rev. B* **2015**, *92*, 125305.

31. Kataoka, M.; Johnson, N.; Emary, C.; See, P.; Griffiths, J.P.; Jones, G.A.C.; Farrer, I.; Ritchie, D.A.; Pepper, M.; Janssen, T.J.B.M. Time-of-Flight Measurements of Single-Electron Wave Packets in Quantum Hall Edge States. *Phys. Rev. Lett.* **2016**, *116*, 126803.

32. Johnson, N.; Fletcher, J.; Humphreys, D.; See, P.; Griffiths, J.; Jones, G.; Farrer, I.; Ritchie, D.; Pepper, M.; Janssen, T.; et al. Ultrafast voltage sampling using single-electron wavepackets. *Appl. Phys. Lett.* **2017**, *110*, 102105.

33. Roussely, G.; Arrighi, E.; Georgiou, G.; Takada, S.; Schalk, M.; Urdampilleta, M.; Ludwig, A.; Wieck, A.D.; Armagnat, P.; Kloss, T.; et al. Unveiling the bosonic nature of an ultrashort few-electron pulse. *Nat. Commun.* **2018**, *9*, 2811.

34. Schoelkopf, R.; Wahlgren, P.; Kozhevnikov, A.; Delsing, P.; Prober, D. The radio-frequency single-electron transistor (RF-SET): A fast and ultrasensitive electrometer. *Science* **1998**, *280*, 1238–1242.

35. Bäuerle, C.; Glattli, D.C.; Meunier, T.; Portier, F.; Roche, P.; Roulleau, P.; Takada, S.; Waintal, X. Coherent control of single electrons: A review of current progress. *Rep. Prog. Phys.* **2018**, *81*, 056503.

36. Akkermans, E.; Montambaux, G. *Mesoscopic Physics of Electrons and Photons*; Cambridge University Press: Cambridge, UK, 2007.

37. Altshuler, B.L.; Aronov, A.G.; Khmelnitsky, D.E. Effects of electron-electron collisions with small energy transfers on quantum localisation. *J. Phys. C Solid State Phys.* **1982**, *15*, 7367.

38. Pierre, F.; Gougam, A.B.; Anthore, A.; Pothier, H.; Esteve, D.; Birge, N.O. Dephasing of electrons in mesoscopic metal wires. *Phys. Rev. B* **2003**, *68*, 085413.

39. Huard, B.; Anthore, A.; Birge, N.O.; Pothier, H.; Esteve, D. Effect of Magnetic Impurities on Energy Exchange between Electrons. *Phys. Rev. Lett.* **2005**, *95*, 036802.

40. Mallet, F.; Ericsson, J.; Mailly, D.; Ünlübayir, S.; Reuter, D.; Melnikov, A.; Wieck, A.D.; Micklitz, T.; Rosch, A.; Costi, T.A.; et al. Scaling of the Low-Temperature Dephasing Rate in Kondo Systems. *Phys. Rev. Lett.* **2006**, *97*, 226804.

41. Saminadayar, L.; Mohanty, P.; Webb, R.A.; Degiovanni, P.; Bäuerle, C. Electron coherence at low temperatures: The role of magnetic impurities. *Phys. E Low-Dimens. Syst. Nanostruct.* **2007**, *40*, 12–24.

42. Niimi, Y.; Baines, Y.; Capron, T.; Mailly, D.; Lo, F.Y.; Wieck, A.D.; Meunier, T.; Saminadayar, L.; Bäuerle, C. Quantum coherence at low temperatures in mesoscopic systems: Effect of disorder. *Phys. Rev. B* **2010**, *81*, 245306.

43. Pines, D.; Nozières, P. *Theory Of Quantum Liquids | Normal Fermi Liquids*; CRC Press: Boca Raton, FL, USA, 2018.

44. Coleman, P. *Introduction to Many-Body Physics*; Cambridge University Press: Cambridge, UK, 2015.

45. Landauer, R. Electrical resistance of disordered one-dimensional lattices. *Philos. Mag.* **1970**, *21*, 863–867.

46. Landauer, R. Spatial variation of currents and fields due to localized scatterers in metallic conduction. *IBM J. Res. Dev.* **1988**, *32*, 306–316.

47. Büttiker, M.; Imry, Y.; Landauer, R.; Pinhas, S. Generalized many-channel conductance formula with application to small rings. *Phys. Rev. B* **1985**, *31*, 6207–6215.

48. Nozières, P. A "fermi-liquid" description of the Kondo problem at low temperatures. *J. Low Temp. Phys.* **1974**, *17*, 31–42.

49. Hewson, A.C. *The Kondo Problem to Heavy Fermions*; Cambridge University Press: Cambridge, UK, 1993.

50. Duprez, H.; Sivre, E.; Anthore, A.; Aassime, A.; Cavanna, A.; Gennser, U.; Pierre, F. Transmitting the quantum state of electrons across a metallic island with Coulomb interaction. *Science* **2019**, *366*, 1243–1247.

51. Büttiker, M.; Prêtre, A.; Thomas, H. Dynamic conductance and the scattering matrix of small conductors. *Phys. Rev. Lett.* **1993**, *70*, 4114–4117.

52. Büttiker, M.; Thomas, H.; Prêtre, A. Mesoscopic capacitors. *Phys. Lett. A* **1993**, *180*, 364–369.

53. Prêtre, A.; Thomas, H.; Büttiker, M. Dynamic admittance of mesoscopic conductors: Discrete-potential model. *Phys. Rev. B* **1996**, *54*, 8130–8143.
54. Nigg, S.E.; López, R.; Büttiker, M. Mesoscopic Charge Relaxation. *Phys. Rev. Lett.* **2006**, *97*, 206804.
55. Büttiker, M.; Nigg, S.E. Mesoscopic capacitance oscillations. *Nanotechnology* **2006**, *18*, 044029.
56. Büttiker, M.; Nigg, S.E. Role of coherence in resistance quantization. *Eur. Phys. J. Spec. Top.* **2009**, *172*, 247–255.
57. Mora, C.; Vitushinsky, P.; Leyronas, X.; Clerk, A.A.; Le Hur, K. Theory of nonequilibrium transport in the SU(N) Kondo regime. *Phys. Rev. B* **2009**, *80*, 155322.
58. Mora, C. Fermi-liquid theory for SU(N) Kondo model. *Phys. Rev. B* **2009**, *80*, 125304.
59. Mora, C.; Moca, C.P.; von Delft, J.; Zaránd, G. Fermi-liquid theory for the single-impurity Anderson model. *Phys. Rev. B* **2015**, *92*, 075120.
60. Oguri, A.; Hewson, A.C. Higher-Order Fermi-Liquid Corrections for an Anderson Impurity Away from Half Filling. *Phys. Rev. Lett.* **2018**, *120*, 126802.
61. Oguri, A.; Hewson, A.C. Higher-order Fermi-liquid corrections for an Anderson impurity away from half filling : Equilibrium properties. *Phys. Rev. B* **2018**, *97*, 045406.
62. Oguri, A.; Hewson, A.C. Higher-order Fermi-liquid corrections for an Anderson impurity away from half filling: Nonequilibrium transport. *Phys. Rev. B* **2018**, *97*, 035435.
63. Filippone, M.; Moca, C.P.; Weichselbaum, A.; von Delft, J.; Mora, C. At which magnetic field, exactly, does the Kondo resonance begin to split? A Fermi liquid description of the low-energy properties of the Anderson model. *Phys. Rev. B* **2018**, *98*, 075404.
64. Teratani, Y.; Sakano, R.; Oguri, A. Fermi liquid theory for nonlinear transport through a multilevel Anderson impurity. *arXiv* **2020**, arXiv:2001.08348.
65. Mora, C.; Le Hur, K. Universal resistances of the quantum resistance-capacitance circuit. *Nat. Phys.* **2010**, *6*, 697–701.
66. Filippone, M.; Mora, C. Fermi liquid approach to the quantum RC circuit: Renormalization group analysis of the Anderson and Coulomb blockade models. *Phys. Rev. B* **2012**, *86*, 125311.
67. Filippone, M.; Le Hur, K.; Mora, C. Giant Charge Relaxation Resistance in the Anderson Model. *Phys. Rev. Lett.* **2011**, *107*, 176601.
68. Filippone, M.; Le Hur, K.; Mora, C. Admittance of the SU(2) and SU(4) Anderson quantum RC circuits. *Phys. Rev. B* **2013**, *88*, 045302.
69. Dutt, P.; Schmidt, T.L.; Mora, C.; Le Hur, K. Strongly correlated dynamics in multichannel quantum RC circuits. *Phys. Rev. B* **2013**, *87*, 155134.
70. Clerk, A.A.; Brouwer, P.W.; Ambegaokar, V. Interaction-Induced Restoration of Phase Coherence. *Phys. Rev. Lett.* **2001**, *87*, 186801.
71. Idrisov, E.G.; Levkivskyi, I.P.; Sukhorukov, E.V. Dephasing in a Mach-Zehnder Interferometer by an Ohmic Contact. *Phys. Rev. Lett.* **2018**, *121*, 026802.
72. Matveev, K.A. Coulomb blockade at almost perfect transmission. *Phys. Rev. B* **1995**, *51*, 1743–1751.
73. Haldane, F.D.M. 'Luttinger liquid theory' of one-dimensional quantum fluids. I. Properties of the Luttinger model and their extension to the general 1D interacting spinless Fermi gas. *J. Phys. C Solid State Phys.* **1981**, *14*, 2585.
74. Haldane, F.D.M. Effective Harmonic-Fluid Approach to Low-Energy Properties of One-Dimensional Quantum Fluids. *Phys. Rev. Lett.* **1981**, *47*, 1840–1843.
75. Delft, J.v.; Schoeller, H. Bosonization for beginners—Refermionization for experts. *Ann. Der Phys.* **1998**, *7*, 225–305.:4<225::AID-ANDP225>3.0.CO;2-L.
76. Giamarchi, T. *Quantum Physics in One Dimension*; Oxford University Press: New York, NY, USA, 2004; Volume 121.
77. Aleiner, I.L.; Glazman, L.I. Mesoscopic charge quantization. *Phys. Rev. B* **1998**, *57*, 9608–9641.
78. Brouwer, P.W.; Lamacraft, A.; Flensberg, K. Nonequilibrium theory of Coulomb blockade in open quantum dots. *Phys. Rev. B* **2005**, *72*, 075316.
79. Altland, A.; Simons, B. *Condensed Matter Field Theory*; Cambridge University Press: Cambridge, UK, 2006.
80. Clerk, A.A. Aspects of Andreev Scattering and Kondo Physics in Mesoscopic Systems. Ph.D. Thesis, Cornell University, Ithaca, NY, USA, 2011.

81. Schuster, R.; Buks, E.; Heiblum, M.; Mahalu, D.; Umansky, V.; Shtrikman, H. Phase measurement in a quantum dot via a double-slit interference experiment. *Nature* **1997**, *385*, 417–420.
82. Edlbauer, H.; Takada, S.; Roussely, G.; Yamamoto, M.; Tarucha, S.; Ludwig, A.; Wieck, A.D.; Meunier, T.; Bäuerle, C. Non-universal transmission phase behaviour of a large quantum dot. *Nat. Commun.* **2017**, *8*, 1710.
83. Wick, G.C. The Evaluation of the Collision Matrix. *Phys. Rev.* **1950**, *80*, 268–272.
84. Büttiker, M. Quantized transmission of a saddle-point constriction. *Phys. Rev. B* **1990**, *41*, 7906–7909.
85. Mora, C.; Le Hur, K. Probing dynamics of Majorana fermions in quantum impurity systems. *Phys. Rev. B* **2013**, *88*, 241302.
86. Friedel, J. On some electrical and magnetic properties of metallic solid solutions. *Can. J. Phys.* **1956**, *34*, 1190–1211.
87. Langreth, D.C. Friedel Sum Rule for Anderson's Model of Localized Impurity States. *Phys. Rev.* **1966**, *150*, 516–518.
88. Rontani, M. Friedel Sum Rule for an Interacting Multiorbital Quantum Dot. *Phys. Rev. Lett.* **2006**, *97*, 076801.
89. Ringel, Z.; Imry, Y.; Entin-Wohlman, O. Delayed currents and interaction effects in mesoscopic capacitors. *Phys. Rev. B* **2008**, *78*, 165304.
90. Brouwer, P.W. Scattering approach to parametric pumping. *Phys. Rev. B* **1998**, *58*, R10135–R10138.
91. Karki, D.B.; Mora, C.; von Delft, J.; Kiselev, M.N. Two-color Fermi-liquid theory for transport through a multilevel Kondo impurity. *Phys. Rev. B* **2018**, *97*, 195403.
92. Anderson, P.W. Localized Magnetic States in Metals. *Phys. Rev.* **1961**, *124*, 41–53.
93. Pustilnik, M.; Glazman, L. Kondo effect in quantum dots. *J. Phys. Condens. Matter* **2004**, *16*, R513–R537.
94. Schrieffer, J.R.; Wolff, P.A. Relation between the Anderson and Kondo Hamiltonians. *Phys. Rev.* **1966**, *149*, 491–492.
95. Goldhaber-Gordon, D.; Shtrikman, H.; Mahalu, D.; Abusch-Magder, D.; Meirav, U.; Kastner, M.A. Kondo effect in a single-electron transistor. *Nature* **1998**, *391*, 156.
96. Bruhat, L.E.; Viennot, J.J.; Dartiailh, M.C.; Desjardins, M.M.; Cottet, A.; Kontos, T. Scaling laws of the Kondo problem at finite frequency. *Phys. Rev. B* **2018**, *98*, 075121.
97. Bruus, H.; Flensberg, K. *Many-Body Quantum Theory in Condensed Matter Physics. An Introduction*; Oxford University Press: New York, NY, USA, 2004.
98. Coqblin, B.; Schrieffer, J.R. Exchange Interaction in Alloys with Cerium Impurities. *Phys. Rev.* **1969**, *185*, 847–853.
99. Grabert, H. Charge fluctuations in the single-electron box: Perturbation expansion in the tunneling conductance. *Phys. Rev. B* **1994**, *50*, 17364–17377.
100. Grabert, H. Rounding of the Coulomb Staircase by the tunneling conductance. *Phys. B Condens. Matter* **1994**, *194*, 1011–1012.
101. Matveev, K.A. Quantum fluctuations of the charge of a metal particle under the Coulomb blockade conditions. *Zh. Eksp. Teor. Fiz.* **1990**, *99*, 1598.
102. Tsvelick, A.; Wiegmann, P. Exact results in the theory of magnetic alloys. *Adv. Phys.* **1983**, *32*, 453–713.
103. Wiegmann, P.B.; Tsvelick, A.M. Exact solution of the Anderson model: I. *J. Phys. C Solid State Phys.* **1983**, *16*, 2281–2319.
104. Krishna-murthy, H.R.; Wilkins, J.W.; Wilson, K.G. Renormalization-group approach to the Anderson model of dilute magnetic alloys. I. Static properties for the symmetric case. *Phys. Rev. B* **1980**, *21*, 1003–1043.
105. Krishna-murthy, H.R.; Wilkins, J.W.; Wilson, K.G. Renormalization-group approach to the Anderson model of dilute magnetic alloys. II. Static properties for the asymmetric case. *Phys. Rev. B* **1980**, *21*, 1044–1083.
106. Affleck, I. Conformal Field Theory Approach to the Kondo Effect. *arXiv* **1995**, arXiv:Cond-mat/9512099.
107. Cardy, J.L. Boundary conditions, fusion rules and the Verlinde formula. *Nucl. Phys. B* **1989**, *324*, 581–596.
108. Affleck, I.; Ludwig, A.W.W. Critical theory of overscreened Kondo fixed points. *Nucl. Phys. B* **1991**, *360*, 641–696.
109. Affleck, I.; Ludwig, A.W.W. Exact conformal-field-theory results on the multichannel Kondo effect: Single-fermion Green's function, self-energy, and resistivity. *Phys. Rev. B* **1993**, *48*, 7297–7321.
110. Lesage, F.; Saleur, H. Perturbation of infra-red fixed points and duality in quantum impurity problems. *Nucl. Phys. B* **1999**, *546*, 585–620.
111. Lesage, F.; Saleur, H. Strong-Coupling Resistivity in the Kondo Model. *Phys. Rev. Lett.* **1999**, *82*, 4540–4543.

112. Haldane, F.D.M. Theory of the atomic limit of the Anderson model. I. Perturbation expansions re-examined. *J. Phys. C* **1978**, *11*, 5015.

113. Haldane, F.D.M. Scaling Theory of the Asymmetric Anderson Model. *Phys. Rev. Lett.* **1978**, *40*, 416–419.

114. Sørensen, E.S.; Affleck, I. Scaling theory of the Kondo screening cloud. *Phys. Rev. B* **1996**, *53*, 9153–9167.

115. Barzykin, V.; Affleck, I. The Kondo Screening Cloud: What Can We Learn from Perturbation Theory? *Phys. Rev. Lett.* **1996**, *76*, 4959–4962.

116. Affleck, I.; Simon, P. Detecting the Kondo Screening Cloud Around a Quantum Dot. *Phys. Rev. Lett.* **2001**, *86*, 2854–2857.

117. Affleck, I.; Borda, L.; Saleur, H. Friedel oscillations and the Kondo screening cloud. *Phys. Rev. B* **2008**, *77*, 180404.

118. Suhl, H. Dispersion Theory of the Kondo Effect. *Phys. Rev.* **1965**, *138*, A515–A523.

119. Glazman, L.; Raikh, M. Resonant Kondo transparency of a barrier with quasilocal impurity states. *JETP Lett.* **1988**, *47*, 105.

120. Goldhaber-Gordon, D.; Göres, J.; Kastner, M.A.; Shtrikman, H.; Mahalu, D.; Meirav, U. From the Kondo Regime to the Mixed-Valence Regime in a Single-Electron Transistor. *Phys. Rev. Lett.* **1998**, *81*, 5225–5228.

121. Takada, S.; Bäuerle, C.; Yamamoto, M.; Watanabe, K.; Hermelin, S.; Meunier, T.; Alex, A.; Weichselbaum, A.; von Delft, J.; Ludwig, A.; et al. Transmission Phase in the Kondo Regime Revealed in a Two-Path Interferometer. *Phys. Rev. Lett.* **2014**, *113*, 126601.

122. Borzenets, I.V.; Shim, J.; Chen, J.C.H.; Ludwig, A.; Wieck, A.D.; Tarucha, S.; Sim, H.S.; Yamamoto, M. Observation of the Kondo screening cloud. *Nature* **2020**, *579*, 210–213.

123. Cragg, D.; Lloyd, P. Potential Scattering and the Kondo Problem. *J. Phys. C* **1978**, *11*, L597.

124. Lloyd, P.; Cragg, D.M. The Nozieres-Wilson relation in the low-temperature Kondo problem. *J. Phys. C* **1979**, *12*, 3289.

125. Cragg, D.M.; Lloyd, P. Universality and the renormalisability of rotationally invariant Kondo Hamiltonians. *J. Phys. C* **1979**, *12*, 3301.

126. Horvatié, B.; Zlatié, V. Equivalence of the perturbative and Bethe-Ansatz solution of the symmetric Anderson Hamiltonian. *J. De Phys.* **1985**, *46*, 1459–1467.:019850046090145900.

127. Nozières, P.; Blandin, A. Kondo effect in real metals. *J. De Phys.* **1980**, *41*, 193–211.

128. Cox, D.L.; Zawadowski, A. Exotic Kondo effects in metals: Magnetic ions in a crystalline electric field and tunnelling centres. *Adv. Phys.* **1998**, *47*, 599–942.

129. Petta, J.; Johnson, A.; Taylor, J.; Laird, E.; Yacoby, A.; Lukin, M.; Marcus, C.; Hanson, M.; Gossard, A. Coherent manipulation of coupled electron spins in semiconductor quantum dots. *Science* **2005**, *309*, 2180–2184.

130. Koppens, F.; Buizert, C.; Tielrooij, K.; Vink, I.; Nowack, K.; Meunier, T.; Kouwenhoven, L.; Vandersypen, L. Driven coherent oscillations of a single electron spin in a quantum dot. *Nature* **2006**, *442*, 766–771.

131. Bocquillon, E.; Parmentier, F.D.; Grenier, C.; Berroir, J.M.; Degiovanni, P.; Glattli, D.C.; Plaçais, B.; Cavanna, A.; Jin, Y.; Fève, G. Electron Quantum Optics: Partitioning Electrons One by One. *Phys. Rev. Lett.* **2012**, *108*, 196803.

132. Bocquillon, E.; Freulon, V.; Berroir, J.M.; Degiovanni, P.; Plaçais, B.; Cavanna, A.; Jin, Y.; Fève, G. Coherence and Indistinguishability of Single Electrons Emitted by Independent Sources. *Science* **2013**, *339*, 1054–1057.

133. Bocquillon, E.; Freulon, V.; Parmentier, F.D.; Berroir, J.M.; Plaçais, B.; Wahl, C.; Rech, J.; Jonckheere, T.; Martin, T.; Grenier, C.; et al. Electron quantum optics in ballistic chiral conductors. *Ann. Der Phys.* **2014**, *526*, 1–30.

134. Grenier, C.; Hervé, R.; Bocquillon, E.; Parmentier, F.; Plaçais, B.; Berroir, J.; Fève, G.; Degiovanni, P. Single-electron quantum tomography in quantum Hall edge channels. *New J. Phys.* **2011**, *13*, 093007.

135. Bocquillon, E.; Freulon, V.; Berroir, J.M.; Degiovanni, P.; Plaçais, B.; Cavanna, A.; Jin, Y.; Fève, G. Separation of neutral and charge modes in one-dimensional chiral edge channels. *Nat. Commun.* **2013**, *4*, 1–7.

136. Safi, I.; Schulz, H.J. Transport in an inhomogeneous interacting one-dimensional system. *Phys. Rev. B* **1995**, *52*, R17040–R17043.

137. Safi, I.; Schulz, H.J. Transport Through a Single-Band Wire Connected to Measuring Leads. In *Quantum Transport in Semiconductor Submicron Structures*; Kramer, B., Ed.; NATO ASI Series; Springer: Dordrecht, The Netherlands, 1996; pp. 159–172.

138. Safi, I. A dynamic scattering approach for a gated interacting wire. *Eur. Phys. J. B Condens. Matter Complex Syst.* **1999**, *12*, 451–455.
139. Oreg, Y.; Finkel'stein, A.M. Interedge Interaction in the Quantum Hall Effect. *Phys. Rev. Lett.* **1995**, *74*, 3668–3671.
140. Blanter, Y.M.; Hekking, F.W.J.; Büttiker, M. Interaction Constants and Dynamic Conductance of a Gated Wire. *Phys. Rev. Lett.* **1998**, *81*, 1925–1928.
141. Fazio, R.; Hekking, F.W.J.; Khmelnitskii, D.E. Anomalous Thermal Transport in Quantum Wires. *Phys. Rev. Lett.* **1998**, *80*, 5611–5614.
142. Sukhorukov, E.V.; Cheianov, V.V. Resonant Dephasing in the Electronic Mach-Zehnder Interferometer. *Phys. Rev. Lett.* **2007**, *99*, 156801.
143. Grenier, C.; Hervé, R.; Fève, G.; Degiovanni, P. Electron quantum optics in quantum Hall edge channels. *Mod. Phys. Lett. B* **2011**, *25*, 1053–1073.
144. Bertrand, B.; Hermelin, S.; Takada, S.; Yamamoto, M.; Tarucha, S.; Ludwig, A.; Wieck, A.; Bäuerle, C.; Meunier, T. Long-range spin transfer using individual electrons. *arXiv* **2015**, arXiv:1508.04307.
145. Desjardins, M.M.; Viennot, J.J.; Dartiailh, M.C.; Bruhat, L.E.; Delbecq, M.R.; Lee, M.; Choi, M.S.; Cottet, A.; Kontos, T. Observation of the frozen charge of a Kondo resonance. *Nature* **2017**, *545*, 71–74.
146. Shiba, H. Korringa Relation for the Impurity Nuclear Spin-Lattice Relaxation in Dilute Kondo Alloys. *Prog. Theor. Phys.* **1975**, *54*, 967–981.
147. Lee, M.; López, R.; Choi, M.S.; Jonckheere, T.; Martin, T. Effect of many-body correlations on mesoscopic charge relaxation. *Phys. Rev. B* **2011**, *83*, 201304.
148. Litinski, D.; Brouwer, P.W.; Filippone, M. Interacting mesoscopic capacitor out of equilibrium. *Phys. Rev. B* **2017**, *96*, 085429.
149. Kubo, R.; Toda, M.; Hashitsume, N. *Statistical Physics: Nonequilibrium Statistical Mechanics*; Springer: Berlin/Heidelberg, Germany, 1992; Volume 2.
150. Schiró, M.; Le Hur, K. Tunable hybrid quantum electrodynamics from nonlinear electron transport. *Phys. Rev. B* **2014**, *89*, 195127.
151. Le Hur, K.; Henriet, L.; Petrescu, A.; Plekhanov, K.; Roux, G.; Schiró, M. Many-body quantum electrodynamics networks: Non-equilibrium condensed matter physics with light. *C. R. Phys.* **2016**, *17*, 808–835.
152. Le Hur, K.; Henriet, L.; Herviou, L.; Plekhanov, K.; Petrescu, A.; Goren, T.; Schiro, M.; Mora, C.; Orth, P.P. Driven dissipative dynamics and topology of quantum impurity systems. *C. R. Phys.* **2018**, *19*, 451–483.
153. Delbecq, M.R.; Schmitt, V.; Parmentier, F.D.; Roch, N.; Viennot, J.J.; Fève, G.; Huard, B.; Mora, C.; Cottet, A.; Kontos, T. Coupling a Quantum Dot, Fermionic Leads, and a Microwave Cavity on a Chip. *Phys. Rev. Lett.* **2011**, *107*, 256804.
154. Delbecq, M.; Bruhat, L.; Viennot, J.; Datta, S.; Cottet, A.; Kontos, T. Photon-mediated interaction between distant quantum dot circuits. *Nat. Commun.* **2013**, *4*, 1400.
155. Liu, Y.Y.; Petersson, K.D.; Stehlik, J.; Taylor, J.M.; Petta, J.R. Photon Emission from a Cavity-Coupled Double Quantum Dot. *Phys. Rev. Lett.* **2014**, *113*, 036801.
156. Bruhat, L.; Viennot, J.; Dartiailh, M.; Desjardins, M.; Kontos, T.; Cottet, A. Cavity photons as a probe for charge relaxation resistance and photon emission in a quantum dot coupled to normal and superconducting continua. *Phys. Rev. X* **2016**, *6*, 021014.
157. Mi, X.; Cady, J.V.; Zajac, D.M.; Stehlik, J.; Edge, L.F.; Petta, J.R. Circuit quantum electrodynamics architecture for gate-defined quantum dots in silicon. *Appl. Phys. Lett.* **2017**, *110*, 043502.
158. Viennot, J.J.; Delbecq, M.R.; Bruhat, L.E.; Dartiailh, M.C.; Desjardins, M.M.; Baillergeau, M.; Cottet, A.; Kontos, T. Towards hybrid circuit quantum electrodynamics with quantum dots. *C. R. Phys.* **2016**, *17*, 705–717.
159. Deng, G.W.; Henriet, L.; Wei, D.; Li, S.X.; Li, H.O.; Cao, G.; Xiao, M.; Guo, G.C.; Schiro, M.; Hur, K.L.; et al. A Quantum Electrodynamics Kondo Circuit with Orbital and Spin Entanglement. *arXiv* **2015**, arXiv:1509.06141.
160. Van Wees, B.J.; van Houten, H.; Beenakker, C.W.J.; Williamson, J.G.; Kouwenhoven, L.P.; van der Marel, D.; Foxon, C.T. Quantized conductance of point contacts in a two-dimensional electron gas. *Phys. Rev. Lett.* **1988**, *60*, 848–850.

161. Wharam, D.; Pepper, M.; Ahmed, H.; Frost, J.; Hasko, D.; Peacock, D.; Ritchie, D.; Jones, G. Addition of the one-dimensional quantised ballistic resistance. *J.Phys. C* **1988**, *21*, L887.

162. Nigg, S.E.; Büttiker, M. Quantum to classical transition of the charge relaxation resistance of a mesoscopic capacitor. *Phys. Rev. B* **2008**, *77*, 085312.

163. Imry, Y. *Directions in Condensed Matter Physics*; Grinstein, G., and Mazenco, G., Eds.; World Scientific: Singapore, 1986.

164. Landauer, R. Electrical transport in open and closed systems. *Z. Phys. B* **1987**, *68*, 217–228.

165. Rodionov, Y.I.; Burmistrov, I.S.; Ioselevich, A.S. Charge relaxation resistance in the Coulomb blockade problem. *Phys. Rev. B* **2009**, *80*, 035332.

166. Rodionov, Y.I.; Burmistrov, I.S. Out-of-equilibrium admittance of single electron box under strong Coulomb blockade. *JETP Lett.* **2010**, *92*, 696–702.

167. Ngo Dinh, S.; Bagrets, D.A.; Mirlin, A.D. Nonequilibrium functional bosonization of quantum wire networks. *Ann. Phys.* **2012**, *327*, 2794–2852.

168. Ngo Dinh, S.; Bagrets, D.A.; Mirlin, A.D. Analytically solvable model of an electronic Mach-Zehnder interferometer. *Phys. Rev. B* **2013**, *87*, 195433.

169. Garst, M.; Wölfle, P.; Borda, L.; von Delft, J.; Glazman, L. Energy-resolved inelastic electron scattering off a magnetic impurity. *Phys. Rev. B* **2005**, *72*, 205125.

170. Kawakami, N.; Okiji, A. Density of states for elementary excitations in the Kondo problem. *Phys. Rev. B* **1990**, *42*, 2383–2392.

171. Lansbergen, G.P.; Tettamanzi, G.C.; Verduijn, J.; Collaert, N.; Biesemans, S.; Blaauboer, M.; Rogge, S. Tunable Kondo Effect in a Single Donor Atom. *Nano Lett.* **2010**, *10*, 455–460.

172. Tettamanzi, G.C.; Verduijn, J.; Lansbergen, G.P.; Blaauboer, M.; Calderón, M.J.; Aguado, R.; Rogge, S. Magnetic-Field Probing of an SU(4) Kondo Resonance in a Single-Atom Transistor. *Phys. Rev. Lett.* **2012**, *108*, 046803.

173. Borda, L.; Zaránd, G.; Hofstetter, W.; Halperin, B.I.; von Delft, J. SU(4) Fermi Liquid State and Spin Filtering in a Double Quantum Dot System. *Phys. Rev. Lett.* **2003**, *90*, 026602.

174. Le Hur, K.; Simon, P. Smearing of charge fluctuations in a grain by spin-flip assisted tunneling. *Phys. Rev. B* **2003**, *67*, 201308.

175. Zaránd, G.; Brataas, A.; Goldhaber-Gordon, D. Kondo effect and spin filtering in triangular artificial atoms. *Solid State Commun.* **2003**, *126*, 463–466.

176. López, R.; Sánchez, D.; Lee, M.; Choi, M.S.; Simon, P.; Le Hur, K. Probing spin and orbital Kondo effects with a mesoscopic interferometer. *Phys. Rev. B* **2005**, *71*, 115312.

177. Filippone, M.; Moca, C.P.; Zaránd, G.; Mora, C. Kondo temperature of SU(4) symmetric quantum dots. *Phys. Rev. B* **2014**, *90*, 121406.

178. Ludovico, M.F.; Lim, J.S.; Moskalets, M.; Arrachea, L.; Sánchez, D. Dynamical energy transfer in ac-driven quantum systems. *Phys. Rev. B* **2014**, *89*, 161306.

179. Ludovico, M.F.; Battista, F.; von Oppen, F.; Arrachea, L. Adiabatic response and quantum thermoelectrics for ac-driven quantum systems. *Phys. Rev. B* **2016**, *93*, 075136.

180. Romero, J.I.; Roura-Bas, P.; Aligia, A.A.; Arrachea, L. Nonlinear charge and energy dynamics of an adiabatically driven interacting quantum dot. *Phys. Rev. B* **2017**, *95*, 235117.

181. Benyamini, A.; Hamo, A.; Kusminskiy, S.V.; Oppen, F.v.; Ilani, S. Real-space tailoring of the electron–phonon coupling in ultraclean nanotube mechanical resonators. *Nat. Phys.* **2014**, *10*, 151–156.

182. Bode, N.; Kusminskiy, S.V.; Egger, R.; von Oppen, F. Scattering Theory of Current-Induced Forces in Mesoscopic Systems. *Phys. Rev. Lett.* **2011**, *107*, 036804.

183. Micchi, G.; Avriller, R.; Pistolesi, F. Mechanical Signatures of the Current Blockade Instability in Suspended Carbon Nanotubes. *Phys. Rev. Lett.* **2015**, *115*, 206802.

184. Micchi, G.; Avriller, R.; Pistolesi, F. Electromechanical transition in quantum dots. *Phys. Rev. B* **2016**, *94*, 125417.

185. Avriller, R.; Pistolesi, F. Andreev Bound-State Dynamics in Quantum-Dot Josephson Junctions: A Washing Out of the 0-pi Transition. *Phys. Rev. Lett.* **2015**, *114*, 037003.

186. Pistolesi, F. Bistability of a slow mechanical oscillator coupled to a laser-driven two-level system. *Phys. Rev. A* **2018**, *97*, 063833.

187. Schaeverbeke, Q.; Avriller, R.; Frederiksen, T.; Pistolesi, F. Single-Photon Emission Mediated by Single-Electron Tunneling in Plasmonic Nanojunctions. *Phys. Rev. Lett.* **2019**, *123*, 246601.

188. Hamamoto, Y.; Jonckheere, T.; Kato, T.; Martin, T. Dynamic response of a mesoscopic capacitor in the presence of strong electron interactions. *Phys. Rev. B* **2010**, *81*, 153305.

189. Hamamoto, Y.; Jonckheere, T.; Kato, T.; Martin, T. Quantum phase transition of dynamical resistance in a mesoscopic capacitor. *J. Physics Conf. Ser.* **2011**, *334*, 012033.

190. Burmistrov, I.S.; Rodionov, Y.I. Charge relaxation resistance in the cotunneling regime of multichannel Coulomb blockade: Violation of Korringa-Shiba relation. *Phys. Rev. B* **2015**, *92*, 195412.

191. Keeling, J.; Shytov, A.V.; Levitov, L.S. Coherent Particle Transfer in an On-Demand Single-Electron Source. *Phys. Rev. Lett.* **2008**, *101*, 196404.

192. Ol'khovskaya, S.; Splettstoesser, J.; Moskalets, M.; Büttiker, M. Shot Noise of a Mesoscopic Two-Particle Collider. *Phys. Rev. Lett.* **2008**, *101*, 166802.

193. Moskalets, M.; Samuelsson, P.; Büttiker, M. Quantized Dynamics of a Coherent Capacitor. *Phys. Rev. Lett.* **2008**, *100*, 086601.

194. Sasaoka, K.; Yamamoto, T.; Watanabe, S. Single-electron pumping from a quantum dot into an electrode. *Appl. Phys. Lett.* **2010**, *96*.

195. Moskalets, M.; Haack, G.; Büttiker, M. Single-electron source: Adiabatic versus nonadiabatic emission. *Phys. Rev. B* **2013**, *87*, 125429.

196. Splettstoesser, J.; Governale, M.; König, J.; Büttiker, M. Charge and spin dynamics in interacting quantum dots. *Phys. Rev. B* **2010**, *81*, 165318.

197. Contreras-Pulido, L.D.; Splettstoesser, J.; Governale, M.; König, J.; Büttiker, M. Time scales in the dynamics of an interacting quantum dot. *Phys. Rev. B* **2012**, *85*, 075301.

198. Kashuba, O.; Schoeller, H.; Splettstoesser, J. Nonlinear adiabatic response of interacting quantum dots. *EPL (Europhys. Lett.)* **2012**, *98*, 57003.

199. Alomar, M.; Lim, J.S.; Sánchez, D. Time-dependent current of interacting quantum capacitors subjected to large amplitude pulses. *J. Phys. Conf. Ser.* **2015**, *647*, 012049.

200. Alomar, M.I.; Lim, J.S.; Sánchez, D. Coulomb-blockade effect in nonlinear mesoscopic capacitors. *Phys. Rev. B* **2016**, *94*, 165425.

201. Vanherck, J.; Schulenborg, J.; Saptsov, R.B.; Splettstoesser, J.; Wegewijs, M.R. Relaxation of quantum dots in a magnetic field at finite bias—Charge, spin, and heat currents. *Phys. Status Solidi (b)* **2017**, *254*.

202. Maslov, D.L.; Stone, M. Landauer conductance of Luttinger liquids with leads. *Phys. Rev. B* **1995**, *52*, R5539–R5542.

203. Ponomarenko, V.V. Renormalization of the one-dimensional conductance in the Luttinger-liquid model. *Phys. Rev. B* **1995**, *52*, R8666–R8667.

204. Cuniberti, G.; Sassetti, M.; Kramer, B. AC conductance of a quantum wire with electron-electron interactions. *Phys. Rev. B* **1998**, *57*, 1515–1526.

205. Marguerite, A. Two-Particle Interferometry for Quantum Signal Processing. Ph.D. Thesis, Université Pierre et Marie Curie, Paris, France, 2017.

206. Degiovanni, P.; Grenier, C.; Fève, G.; Altimiras, C.; le Sueur, H.; Pierre, F. Plasmon scattering approach to energy exchange and high-frequency noise in $\ensuremath{\nu}=2$ quantum Hall edge channels. *Phys. Rev. B* **2010**, *81*, 121302.

207. Fu, L. Electron Teleportation via Majorana Bound States in a Mesoscopic Superconductor. *Phys. Rev. Lett.* **2010**, *104*, 056402.

208. Shi, Z.; Brouwer, P.W.; Flensberg, K.; Glazman, L.I.; von Oppen, F. Long distance coherence of Majorana wires. *arXiv* **2020**, arXiv:2002.01539.

209. Vijay, S.; Hsieh, T.H.; Fu, L. Majorana Fermion Surface Code for Universal Quantum Computation. *Phys. Rev. X* **2015**, *5*, 041038.

210. Plugge, S.; Landau, L.A.; Sela, E.; Altland, A.; Flensberg, K.; Egger, R. Roadmap to Majorana surface codes. *Phys. Rev. B* **2016**, *94*, 174514.

211. Litinski, D.; Kesselring, M.S.; Eisert, J.; von Oppen, F. Combining Topological Hardware and Topological Software: Color-Code Quantum Computing with Topological Superconductor Networks. *Phys. Rev. X* **2017**, *7*, 031048.

212. Moca, C.P.; Mora, C.; Weymann, I.; Zaránd, G. Noise of a Chargeless Fermi Liquid. *Phys. Rev. Lett.* **2018**, *120*, 016803.

213. Ferrier, M.; Arakawa, T.; Hata, T.; Fujiwara, R.; Delagrange, R.; Weil, R.; Deblock, R.; Sakano, R.; Oguri, A.; Kobayashi, K. Universality of non-equilibrium fluctuations in strongly correlated quantum liquids. *Nat. Phys.* **2016**, *12*, 230–235.

214. Delagrange, R.; Luitz, D.J.; Weil, R.; Kasumov, A.; Meden, V.; Bouchiat, H.; Deblock, R. Manipulating the magnetic state of a carbon nanotube Josephson junction using the superconducting phase. *Phys. Rev. B* **2015**, *91*, 241401.

215. Delagrange, R.; Weil, R.; Kasumov, A.; Ferrier, M.; Bouchiat, H.; Deblock, R. 0_pi quantum transition in a carbon nanotube Josephson junction: Universal phase dependence and orbital degeneracy. *Phys. Rev. B* **2016**, *93*, 195437.

216. Delagrange, R.; Basset, J.; Bouchiat, H.; Deblock, R. Emission noise and high frequency cut-off of the Kondo effect in a quantum dot. *Phys. Rev. B* **2018**, *97*, 041412.

217. Crépieux, A.; Sahoo, S.; Duong, T.Q.; Zamoum, R.; Lavagna, M. Emission Noise in an Interacting Quantum Dot: Role of Inelastic Scattering and Asymmetric Coupling to the Reservoirs. *Phys. Rev. Lett.* **2018**, *120*, 107702.

218. Saminadayar, L.; Glattli, D.C.; Jin, Y.; Etienne, B. Observation of the e/3 Fractionally Charged Laughlin Quasiparticle. *Phys. Rev. Lett.* **1997**, *79*, 2526–2529.

219. Reznikov, M.; Picciotto, R.d.; Griffiths, T.G.; Heiblum, M.; Umansky, V. Observation of quasiparticles with one-fifth of an electron's charge. *Nature* **1999**, *399*, 238–241.

220. Kapfer, M.; Roulleau, P.; Santin, M.; Farrer, I.; Ritchie, D.A.; Glattli, D.C. A Josephson relation for fractionally charged anyons. *Science* **2019**, *363*, 846–849.

221. Bisognin, R.; Bartolomei, H.; Kumar, M.; Safi, I.; Berroir, J.M.; Bocquillon, E.; Plaçais, B.; Cavanna, A.; Gennser, U.; Jin, Y.; et al. Microwave photons emitted by fractionally charged quasiparticles. *Nat. Commun.* **2019**, *10*, 1–7.

222. Chamon, C.D.C.; Freed, D.E.; Wen, X.G. Tunneling and quantum noise in one-dimensional Luttinger liquids. *Phys. Rev. B* **1995**, *51*, 2363–2379.

223. Bena, C.; Safi, I. Emission and absorption noise in the fractional quantum Hall effect. *Phys. Rev. B* **2007**, *76*, 125317.

224. Safi, I.; Joyez, P. Time-dependent theory of nonlinear response and current fluctuations. *Phys. Rev. B* **2011**, *84*, 205129.

225. Safi, I. Driven quantum circuits and conductors: A unifying perturbative approach. *Phys. Rev. B* **2019**, *99*, 045101.

226. Safi, I.; Devillard, P.; Martin, T. Partition Noise and Statistics in the Fractional Quantum Hall Effect. *Phys. Rev. Lett.* **2001**, *86*, 4628–4631.

227. Guyon, R.; Devillard, P.; Martin, T.; Safi, I. Klein factors in multiple fractional quantum Hall edge tunneling. *Phys. Rev. B* **2002**, *65*, 153304.

228. Kim, E.A.; Lawler, M.; Vishveshwara, S.; Fradkin, E. Signatures of Fractional Statistics in Noise Experiments in Quantum Hall Fluids. *Phys. Rev. Lett.* **2005**, *95*, 176402.

229. Bena, C.; Nayak, C. Effects of non-Abelian statistics on two-terminal shot noise in a quantum Hall liquid in the Pfaffian state. *Phys. Rev. B* **2006**, *73*, 155335.

230. Carrega, M.; Ferraro, D.; Braggio, A.; Magnoli, N.; Sassetti, M. Spectral noise for edge states at the filling factor $\upnu = 5/2$. *New J. Phys.* **2012**, *14*, 023017.

231. Ferraro, D.; Carrega, M.; Braggio, A.; Sassetti, M. Multiple quasiparticle Hall spectroscopy investigated with a resonant detector. *New J. Phys.* **2014**, *16*, 043018.

232. Ferraro, D.; Braggio, A.; Merlo, M.; Magnoli, N.; Sassetti, M. Relevance of Multiple Quasiparticle Tunneling between Edge States at $\ensuremath{\nu}=p/(2np+1)$. *Phys. Rev. Lett.* **2008**, *101*, 166805.

233. Crépieux, A.; Devillard, P.; Martin, T. Photoassisted current and shot noise in the fractional quantum Hall effect. *Phys. Rev. B* **2004**, *69*, 205302.

234. Roussel, B.; Degiovanni, P.; Safi, I. Perturbative fluctuation dissipation relation for nonequilibrium finite-frequency noise in quantum circuits. *Phys. Rev. B* **2016**, *93*, 045102.

235. Bartolomei, H.; Kumar, M.; Bisognin, R.; Marguerite, A.; Berroir, J.M.; Bocquillon, E.; Plaçais, B.; Cavanna, A.; Dong, Q.; Gennser, U.; et al. Fractional statistics in anyon collisions. *Science* **2020**, *368*, 173–177.

236. Glattli, D.C.; Nath, J.; Taktak, I.; Roulleau, P.; Bauerle, C.; Waintal, X. Design of a Single-Shot Electron detector with sub-electron sensitivity for electron flying qubit operation. *arXiv* **2020**, arXiv:2002.03947.

237. Wagner, G.; Nguyen, D.X.; Kovrizhin, D.L.; Simon, S.H. Driven quantum dot coupled to a fractional quantum Hall edge. *Phys. Rev. B* **2019**, *100*, 245111.

238. Weinberg, S. *The Quantum Theory of Fields: Foundations*; Cambridge University Press: Cambridge, UK, 1996; Volume 1.

239. Newton, R.G. *Scattering Theory of Waves and Particles*; Dover Publications: Mineola, NY, USA, 2002.

240. Filippone, M.; Brouwer, P.W. Tunneling into quantum wires: Regularization of the tunneling Hamiltonian and consistency between free and bosonized fermions. *Phys. Rev. B* **2016**, *94*, 235426.

241. Gabelli, J. Mise en évidence de la Cohérence Quantique des Conducteurs en Régime Dynamique. Ph.D. Thesis, Université Pierre et Marie Curie-Paris VI, Paris, France, 2006.

242. Fève, G. Quantification du Courant Alternatif: La Boîte Quantique Comme Source D'électrons Uniques Subnanoseconde. Ph.D. Thesis, Université Pierre et Marie Curie-Paris VI, Paris, France, 2006.

243. Wigner, E.P. Lower Limit for the Energy Derivative of the Scattering Phase Shift. *Phys. Rev.* **1955**, *98*, 145–147.

244. Cohen-Tannoudji, C.; Dupont-Roc, J.; Grynberg, G. *Photons and Atoms: Introduction to Quantum Electrodynamics*; Wiley: Hoboken , NJ, USA, 1989.

Article

Quantum Pumping with Adiabatically Modulated Barriers in Three-Band Pseudospin-1 Dirac–Weyl Systems

Xiaomei Chen and Rui Zhu *

Department of Physics, South China University of Technology, Guangzhou 510641, China;
201720127507@mail.scut.edu.cn
* Correspondence: rzhu@scut.edu.cn

Received: 28 November 2018; Accepted: 4 February 2019; Published: 22 February 2019

Abstract: In this work, pumped currents of the adiabatically-driven double-barrier structure based on the pseudospin-1 Dirac–Weyl fermions are studied. As a result of the three-band dispersion and hence the unique properties of pseudospin-1 Dirac–Weyl quasiparticles, sharp current-direction reversal is found at certain parameter settings especially at the Dirac point of the band structure, where apexes of the two cones touch at the flat band. Such a behavior can be interpreted consistently by the Berry phase of the scattering matrix and the classical turnstile mechanism.

Keywords: quantum transport

1. Introduction

After quantized particle transport driven by adiabatic cyclic potential variation was proposed by D. J. Thouless in 1983 [1], such a concept has attracted unceasing interest among researchers concerning its theoretical meaning and potential applications in various fields such as a precision current standard and neural networks [2–4]. Mechanism of the quantum pump can be interpreted consistently by the Berry phase of the scattering matrix in the parameter space within the modulation cycle [2] and the classic turnstile picture [5,6]. Usually, the pumped current is unidirectional when the phase difference between the two driving parameters is fixed. In the turnstile picture, the opening order of the two gates is defined by the driving phase. The first-opened gate let in the particle and the second-opened gate let it out forming a direct current (DC) current after a cycle is completed. However, reversed DC current direction has been discovered in various systems even when the driving phase is fixed such as in monolayer graphene [6] and carbon nanotube-superconductor hybrid systems [7]. This is because that conventionally a "gate" is defined by a potential barrier and higher barriers allow smaller transmission probabilities. However, as a result of the Klein tunneling effect, the potential barrier becomes transparent regardless of its height at certain parameter settings. When higher barrier allows even stronger transmission, the opening and closing of a "gate" in the quantum pump is reversed and so the driven current is reversed with the driving phase difference unchanged. The same phenomenon is also discovered in the superconductive carbon nanotube when Andreev reflection again violates the higher-barrier-lower-transmission convention and reversed the pumped current under the same driving forces. This turnstile interpretation of the reversed pumped current coincides with the Berry phase of the scattering matrix in the parameter space within the modulation cycle. However, a clear comparison between the two mechanisms is lacking, which is one of the motivations of this work.

About the significance of the comparison between the Berry phase picture and the classic turnstile mechanism of the adiabatic quantum pumping, we would like to make some further background remarks.

The classic turnstile mechanism and the Berry-phase-of-scattering-matrix picture of adiabatic quantum pumping are proposed based on different physical origin. The former is from classic mechanics and the latter is from quantum mechanics. General agreement between them is certainly a surprising result because they have at least the following differences on the conceptual level.

(1) In quantum mechanics, the leftward and rightward transmission probabilities of both the symmetric and asymmetric double-barrier structure (The former means the height and width of the two barriers are exactly the same. The latter means the height and width of the two barriers are different.) are exactly the same if a typical two-lead device is considered. The difference between the leftward and rightward transmission is in the phase factor of the transmission amplitudes. Such a phase difference gives rise to a nontrivial Berry phase formed by cyclic modulation of the two barriers with $V_1 = V_{1\omega}\cos(\omega t + \varphi)$ and $V_2 = V_{2\omega}\cos(\omega t)$ when time-reversal symmetry is not conserved such as excluding $\varphi = 0$ or π. In the turnstile picture, the two barriers are treated separately like two gates. The opening and closing of the two gates is determined by the transmission probability of the corresponding barrier potential. When higher barrier generates smaller transmission probability, the opening and closing of the gate is defined conventionally: lifting the barrier means closing the gate and lowering the barrier means opening the gate. Because the charge carrier density in a typical semiconductor can be up to $10^{10} \sim 10^{15}/\text{cm}^3$ at room temperature, only a small change in the barrier height and hence in the transmission probability can justify the definition of the "opening" and "closing" of the gate for charge carriers. While the same gate-modulation is applied, charge carriers are driven unidirectionally to one of the reservoirs like the turnstile in daily life. No phase factor is involved in the picture at all.

(2) In the quantum interpretation of parametric pump, time-reversal symmetry is a vital factor. The most prominent case is when the driving phase difference $\varphi = \pi$. In this case, time-reversal symmetry is conserved as the two parameters vary periodically. Because of this, a DC current is forbidden even when the classic turnstile gives rise to the largest mass flow when the phase lag is the largest. At this point, the classic and quantum models become incomparable, which is out of our present discussion.

Therefore, we feel a confirmation of the agreement between the two mechanisms is a significant step forward to understand the underlying physics of adiabatic quantum pumping. In preparation of this work, we have proved it in various parameter settings in different systems such as two-dimensional electron gas and graphene besides the present pseudospin-1 Dirac–Weyl system by calculating term by term Equation (10). Although we could not provide a general proof, up to now, no numerical evaluation violates such a conclusion.

After the idea of the adiabatic quantum pump (also called Thouless pump and parametric pump) is proposed, such a mechanism has been investigated in various transport devices such as a single spin in diamond [8], quantum-dot structures [9], Rashba nanowires [10], Mach–Zehnder interferometers [11], the magnetic nanowire with double domain walls [12], magnetic-barrier-modulated two dimensional electron gas [13], mesoscopic rings with Aharonov–Casher and Aharonov–Bohm effect [14], magnetic tunnel junctions [15], and monolayer graphene [6,16,17]. Correspondingly, theoretical techniques have been put forward for the treatment of the quantum pumps such as the scattering matrix formalism [18], non-equilibrium Green's function [19–22], and the quantum master equation approach [9]. In this work, we use the scattering matrix approach for alternating current (AC) transport, which defines the Berry phase formed within the looped trajectory of the two varying parameters [2,18,23].

Recently, after realization of the monolayer graphene, which is characterized as a pseudospin-1/2 Dirac–Weyl fermionic material, a family of general pseudospin-s ($s = 1/2, 1, 3/2, \cdots$) Dirac–Weyl fermionic materials has been proposed by sharing similar band structure with one or several pairs of Dirac cones. Pseudospin-1 materials with a band structure of two Dirac cones and a flat band through where the cones intersect have attracted intense interest in the physical society currently. Numerical or experimental studies have proposed various host materials of such band structure such as conventional crystal with special space group symmetries [24,25], in the electronic, photonic, and phononic Lieb lattice [26–33], kagome lattice [31,34], dice or T_3 lattice [31,35–45], and K_4 crystal [46]. Along with these progress in material building, various transport properties of the pseudospin-1 Dirac–Weyl fermions have been investigated such as super Klein tunneling effect [47–49], magneto-optics [50], Hall quantization [51], and Hofstadter butterfly [52] in a magnetic field. While the adiabatic quantum pumping process serves as an important platform to detect various properties of novel quantum states, it is worthwhile to apply the idea on newly-emerged pseudospin-1 Dirac–Weyl materials. To understand how their particular transport properties modify the adiabatically-driven pumped current is the other motivation of this work.

The plan of the present work is as follows. In Section 2, the model is introduced and the key formulas for the scattering matrix, Berry phase, and pumped current are given. In Section 3, we present numerical results of the pumped current and discussions of the underlying mechanisms. In Section 4, a rigorous proof of the consistency between the quantum Berry phase picture and the classic turnstile mechanism for adiabatic quantum pumping is provided. A brief summary is given in Section 5. Detailed derivation of the boundary condition and the scattering matrix are provided in Appendices A and B, respectively.

2. Model and Formalism

We consider a two-dimensional (2D) non-interacting pseudospin-1 Dirac–Weyl system modulated by two time-dependent electric potential barriers illustrated in Figure 1. The pseudospin-1 Dirac–Weyl fermions are charged quasiparticles originating from free electrons moving in the three-band structure consisting of gapless tip-to-tip two cones intersected by a flat band, which is shown in Figure 1c. Their dynamics are governed by the dot product of the spin-1 operator and the momentum. Matrices of the spin-1 operator $\hat{S} = (\hat{S}_x, \hat{S}_y, \hat{S}_z)$ in the $\hat{S}_z$-representation (the representation that $\hat{S}_z$ is diagonalized) can be deduced from spin-lifting/lowering operators $\hat{S}_\pm = \hat{S}_x \pm \hat{S}_y$ by $\hat{S}_\pm |S, S_z\rangle = \sqrt{(S \mp S_z)(S \pm S_z + 1)} |S, S_z \pm 1\rangle$ [53]. Simple algebra leads to the results that

$$\hat{S}_x = \frac{1}{\sqrt{2}}\begin{pmatrix} 0 & 1 & 0 \\ 1 & 0 & 1 \\ 0 & 1 & 0 \end{pmatrix}, \quad \hat{S}_y = \frac{1}{\sqrt{2}}\begin{pmatrix} 0 & -i & 0 \\ i & 0 & -i \\ 0 & i & 0 \end{pmatrix}, \quad \hat{S}_z = \begin{pmatrix} 1 & 0 & 0 \\ 0 & 0 & 0 \\ 0 & 0 & -1 \end{pmatrix}. \tag{1}$$

By applying AC gate voltages, Hamiltonian of the pseudospin-1 Dirac–Weyl fermions has the form

$$\hat{H} = -i\hbar v_g \hat{S} \cdot \nabla + V(x,t), \tag{2}$$

where $\hat{S}$ is the spin-1 operator defined in Equation (1), $v_g \approx 10^6$ m/s is the group velocity associated with the slope of the Dirac cone. As shown in Figure 1a, the potential function has the form

$$V(x,t) = \begin{cases} V_0 + V_1(t), & 0 < x < L_1, \\ V_0 + V_2(t), & L_2 < x < L_3, \\ 0, & \text{others,} \end{cases} \tag{3}$$

with $V_1(t) = V_{1\omega}\cos(\omega t + \varphi)$ and $V_2(t) = V_{2\omega}\cos(\omega t)$. The Fermi energy of the two reservoirs to the two sides of the double-barrier structure are equalized to eliminate the external bias and secure energy-conserved tunneling. While the frequency of the potential modulation ω is small compared to the carrier interaction time (Wigner delay time) with the conductor, the quantum pump can be considered "adiabatic" [1,18,23]. In this case, one can employ an instant scattering matrix approach, which depends only parametrically on the time t. The Wigner–Smith delay time can be evaluated by $\tau = \mathrm{Tr}(-i\hbar s^\dagger \frac{\partial s}{\partial E_F})$, with s the scattering matrix defined in Equation (5). Calculations below show $\tau \approx 10^{-14}$ s for all the parameter values. Thus, the adiabatic condition can be well justified when ω is in the order of MHz [3].

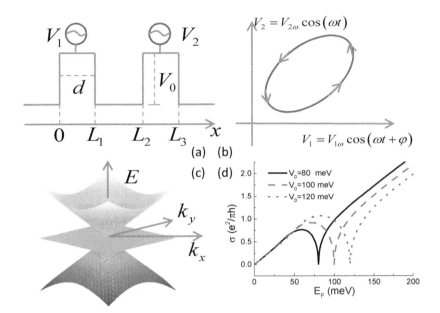

Figure 1. (**a**) schematics of the adiabatic quantum pump. Two time-dependent gate voltages with identical width d and equilibrium strength V_0 are applied to the conductor. Time variation of the two potentials V_1 and and V_2 is shown in panel (**b**). V_1 and V_2 have a phase difference giving rise to a looped trajectory after one driving period; (**c**) two-dimensional band structure of the pseudospin-1 Dirac–Weyl fermions with a flat band intersected two Dirac cones at the apexes; (**d**) conductivity of the pseudospin-1 Dirac–Weyl fermions measured by [54] $\sigma = \frac{e^2 k_F d}{\pi h} \int_{-\pi/2}^{\pi/2} |t(E_F, \theta)|^2 \cos\theta d\theta$ in single-barrier tunneling junction as a function of the Fermi energy for three different values of barrier height V_0. $k_F = E_F / \hbar v_g$ is the Fermi wavevector and t is the transmission amplitude defined in Equation (5). It can be seen that higher barrier allowing larger conductivity occurs at the Dirac point $E_F = V_0$ and around $E_F = V_0/2$ (see the text).

For studying the transport properties, the flux normalized scattering modes in different regions can be expressed in terms of the eigenspinors as

$$
\Psi = \begin{pmatrix} \psi_1 \\ \psi_2 \\ \psi_3 \end{pmatrix} = \begin{cases} a_l \Psi_\rightarrow + b_l \Psi_\leftarrow, & x < 0, \\ a_1 \Psi_{1\rightarrow} + b_1 \Psi_{1\leftarrow}, & 0 < x < L_1, \\ a_2 \Psi_\rightarrow + b_2 \Psi_\leftarrow, & L_1 < x < L_2, \\ a_3 \Psi_{2\rightarrow} + b_3 \Psi_{2\leftarrow}, & L_2 < x < L_3, \\ a_r \Psi_\leftarrow + b_r \Psi_\rightarrow, & x > L_3, \end{cases} \tag{4}
$$

where $k_x = \sqrt{E_F^2/(\hbar v_g)^2 - k_y^2}$ with E_F the quasiparticle energy at the Fermi level of the reservoirs. $\Psi_\rightarrow = \frac{1}{2\sqrt{\cos\theta}} \left(e^{-i\theta}, \sqrt{2}s, e^{i\theta} \right)^\mathrm{T} e^{ik_x x}$ for $E_F \neq 0$ (quasiparticles on the two cone bands) and $\frac{1}{\sqrt{2}} \left(-e^{-i\theta}, 0, e^{i\theta} \right)^\mathrm{T}$ (we also identify it as Ψ_0 for discussions in the next section) for $E_F = 0$ (quasiparticles on the flat band). $\theta = \arctan(k_y/k_x)$, and $s = \mathrm{sgn}(E_F)$. $\Psi_\leftarrow$ can be obtained by replacing k_x with $-k_x$ in $\Psi_\rightarrow$; $\Psi_{i\rightarrow}/\Psi_{i\leftarrow}$ ($i = 1, 2$) can be obtained by replacing k_x with $q_{xi} = \sqrt{(E_F - V_0 - V_i)^2/(\hbar v_g)^2 - k_y^2}$ and s with $s_i' = \mathrm{sgn}(E_F - V_0 - V_i)$ in $\Psi_\rightarrow/\Psi_\leftarrow$. The flux normalization factor $2\sqrt{\cos\theta}$ is obtained [55] by letting $\Psi^\dagger (\partial \hat{H}/\partial k_x) \Psi = 1$. ψ_i ($i = 1, 2, 3$) picks up the i-th row of the spinor wave function in all the five regions. Note that quasiparticles on the flat band contribute no flux in the x-direction. However, it must be taken into account in the pumping mechanisms while the Fermi energy lies close to the Dirac point. We will go to this point again in the next section.

The boundary conditions are that $\psi_1 + \psi_3$ and ψ_2 are continuous at the interfaces, respectively [48]. The derivation of the boundary condition is provided in Appendix A. After some algebra, the instant scattering matrix connecting the incident and outgoing modes can be expressed as

$$\begin{pmatrix} b_l \\ b_r \end{pmatrix} = \begin{pmatrix} r & t' \\ t & r' \end{pmatrix} \begin{pmatrix} a_l \\ a_r \end{pmatrix} = \mathbf{s}(V_1, V_2) \begin{pmatrix} a_l \\ a_r \end{pmatrix}, \tag{5}$$

where $\mathbf{s}$ is parameter-dependent. Detailed derivation of the scattering matrix is provided in Appendix B.

The DC pumped current flowing from the α reservoir at zero temperature could be expressed in terms of the Berry phase of the scattering matrix formed within the looped trajectory of the two varying parameters as [2,18,23]

$$I_{p\alpha} = \frac{\omega e}{2\pi} \int_A \Omega(\alpha) \, dV_1 dV_2, \tag{6}$$

where

$$\Omega(\alpha) = \sum_\beta \text{Im} \frac{\partial s_{\alpha\beta}^*}{\partial V_1} \frac{\partial s_{\alpha\beta}}{\partial V_2}. \tag{7}$$

A is the enclosed area in the V_1-V_2 parameter space. While the driving amplitude is small ($V_{i\omega} \ll V_0$), the Berry curvature can be considered uniform within A and we have

$$I_{p\alpha} = \frac{\omega e \sin \varphi V_{1\omega} V_{2\omega}}{2\pi} \Omega(\alpha). \tag{8}$$

Conservation of current flux secures that the pumped currents flowing from the left and right reservoirs are equal: $I_{pl} = I_{pr}$. The angle-averaged pumped current can be obtained as

$$I_{p\alpha T} = \int_{-\pi/2}^{\pi/2} I_{p\alpha} \cos\theta d\theta. \tag{9}$$

3. Results and Discussion

Previously, we know that transport properties of the pseudospin-1 Dirac–Weyl fermions differs from free electrons in two ways. One is super Klein tunneling, which gives perfect transmission through a potential barrier for all incident angles while the quasiparticle energy equals one half the barrier height [48]. The other is particle-hole symmetry above and below the Dirac point of a potential barrier, which is a shared property with pseudospin-1/2 Dirac–Weyl fermions on monolayer graphene [56]. It gives that the transmission probability closely above and below the Dirac point is mirror symmetric because hole states with identical dispersion to electrons exist within the potential barrier unlike the potential barrier formed by the energy gap in semiconductor heterostructures. These two properties are demonstrated in the conductivity through a single potential barrier shown in Figure 1d. As a result of super Klein tunneling and because the conductivity also depends on the velocity or Fermi wavevector of the charge carriers, the maximum is parabolically-shaped under the present parameter settings and occurs at the Fermi energy larger than half the barrier height. For higher potential barriers, the maximum can be a sharp Λ-shaped peak appearing at the Fermi energy equal to half the barrier height [57]. Because of the existence of the local maximum peak and the V-shape local minimum in the single-barrier transmission probability and hence in the conductivity, it occurs that under certain conditions higher barrier allows larger quasiparticle transmission. The mechanisms of an adiabatic quantum pump in a mesoscopic system can be illustrated consistently by a classic turnstile picture and by the the Berry phase of the scattering matrix in the parameter space [2,5,6]. The turnstile picture can be illustrated within the framework of the single electron approximation and coherent tunneling constrained by the Pauli principle. The two oscillating potential barriers work like two "gates" in a real turnstile. Usually, lower potential allows larger transmissivity and thus defines the opening of one gate. When the two potentials oscillate with a phase difference, the two gates open one by one. Constrained by the Pauli principle, only one electron can occupy the inner single-particle state confined in the quantum well formed by the two potential barriers at one time, electrons flow in a direction determined by the driving phase difference. However, in monolayer graphene and in the pseudospin-1 Dirac–Weyl system, Klein tunneling, super Klein tunneling, and particle-hole symmetry at the Dirac point give rise to a reversal of the transmissivity-barrier height relation. As a result, the direction of the DC pumped current is reversed.

Numerical results of the pumped current are shown in Figure 2. It can be seen from Figure 1d that when the value of E_F is between 70 meV and 100 meV, conductivity through higher potential barriers is larger than that through lower barriers. Angular dependence of the pumped current at Fermi energies selected within this range is shown in Figure 2b. With φ fixed at $\pi/2$, potential barrier V_1 starts lowering first and then it rises and V_2 starts lowering. Usually (like in a semiconductor heterostructure), higher potential barriers give rise to smaller transmission probability. The process can be interpreted as "gate" V_1 "opens" first allowing one particle to enter the middle single-particle state from the left reservoir and then it "closes" and "gate" V_2 "opens" allowing the particle to leave the device and enter the right reservoir. This completes a pump cycle and a DC current is generated. Such is the classical turnstile picture of the pumping mechanism. However, for pseudospin-1 Dirac–Weyl fermions, higher potential barriers give rise to larger transmission probability under certain parameter settings as demonstrated in Figure 1d. In the classical turnstile picture, this means that the definition of "opening" and "closing" of the "gate" is reversed. This is the reason for the negative (direction-reversed) pumped current shown in Figure 2b.

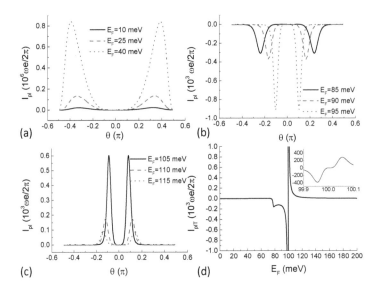

Figure 2. (**a–c**): angular dependence of the pumped for different Fermi energies with the driving phase difference φ fixed; (**d**) angle-averaged pumped current as a function of the Fermi energy. Its inset is the zoom-in close to the Dirac point to show that the large value of the pumped current does not diverge. Other parameters are $V_0 = 100$ meV, $V_{1\omega} = V_{2\omega} = 0.1$ meV, $d = 5$ nm, $L_2 - L_1 = 10$ nm, and $\varphi = \pi/2$.

It can also be seen in Figure 2 that this turnstile picture of quantum pumping works for all parameter settings by comparing with Figure 1d. As a result of particle-hole symmetry above and below the Dirac point of a potential barrier, transmission probability of the pseudospin-1 Dirac–Weyl fermions demonstrate a sharp V-shape local minimum at the Dirac point. It should be noted that, at the Dirac point, eigenspinor wavefunction of the Hamiltonian is Ψ_0 and the transmission probability is exactly zero. We singled out this point in all of our calculations. Below the Dirac point, higher potential barriers allow larger transmission probability. Above the Dirac point, higher potential barriers allow smaller transmission probability. In addition, the difference is very sharp giving rise to a sharp negative pumped current below the Dirac point and a sharp positive pumped current above the Dirac point as shown in Figure 2d. In vast Fermi energy regime, the pumped DC current flows in the same direction for all incident angles as shown in panels (a), (b), and (c) of Figure 2, giving rise to smooth angle-averaged pumped current shown in Figure 2d. It should also be noted that the sharp current peak close to the Dirac point does not diverge and the current has an exact zero value at the Dirac point by taking into account quasiparticles on the flat band, which is a stationary state while the wavevector in the pump–current direction (x-direction in Figure 1a) is imaginary. The finite value of the pump–current peak is shown in the zoom-in inset of Figure 2d.

The previous discussion is based on the classical turnstile mechanism, while the pumped current is evaluated by the Berry phase of the scattering matrix formed from the parameter variation with a looped trajectory (Equation (8)). Such a consistency needs further looking into, which is elucidated in the next section.

4. Consistency between the Turnstile Model and the Berry Phase Treatment

In previous literature, consistency between the turnstile model and the Berry phase treatment is discovered while a clear interpretation is lacking.

Berry curvature $\Omega(\alpha)$ of the scattering matrix $\mathbf{s}$ is defined by Equation (7) with $\mathbf{s}$ defined in Equation (5). t/t' and r/r' are the transmission and reflection amplitudes generated by incidence from the left/right reservoir with $t' = t$ and $r' = -r^*t/t^*$.

Without losing generosity, we consider a conductor modulated by two oscillating potential barriers $X_1 = V_1$ and $X_2 = V_2$ with the same width and equilibrium height. By defining the modulus and argument of t and r as $t = \rho_t e^{i\phi_t}$ and $r = \rho_r e^{i\phi_r}$, we have

$$\Omega(l) = \sum_{i=t,r} \rho_i \frac{d\rho_i}{dV_1}\frac{d\phi_i}{dV_2} - \rho_i \frac{d\phi_i}{dV_1}\frac{d\rho_i}{dV_2}. \tag{10}$$

Analytic dependence of ρ_i and ϕ_i on the parameters V_1 and V_2 cannot be explicitly expressed. We show numerical results of the Berry curvature $\Omega(l)$ and the eight partial derivatives on the right-hand side of Equation (10) in Figure 3. For convenience of discussion, the parameter space in Figure 3a to (i) is divided into four blocks. It can be seen from Figure 3a that $\Omega(l)$ is negative in block II, positive in block III, and nearly zero in blocks I and IV. For the term $\rho_t \frac{d\rho_t}{dV_1}\frac{d\phi_t}{dV_2} - \rho_t \frac{d\phi_t}{dV_1}\frac{d\rho_t}{dV_2}$ in Equation (10), $\rho_t > 0$, $\frac{d\phi_t}{dV_2} \approx \frac{d\phi_t}{dV_1}$ is negative throughout the four blocks and $\frac{d\rho_t}{dV_1} \approx \frac{d\rho_t}{dV_2}$ is positive in block II and negative in block III (see Figure 3b,g). As a result, this term approximates zero in blocks II and III. For the term $\rho_r \frac{d\rho_r}{dV_1}\frac{d\phi_r}{dV_2} - \rho_r \frac{d\phi_r}{dV_1}\frac{d\rho_r}{dV_2}$ in Equation (10), $\rho_r > 0$, $\frac{d\phi_r}{dV_2} - \frac{d\phi_r}{dV_1}$ is positive throughout the four blocks (see Figure 3h,i), $\frac{d\rho_r}{dV_1} \approx \frac{d\rho_r}{dV_2}$ is positive in block III and negative in block II. It can also be seen from Figure 3 that in blocks I and IV the values of the two terms cancel out each other giving rise to nearly zero $\Omega(l)$. Therefore, the combined result of the two terms is that $\Omega(l) > 0$ when $\frac{d\rho_r}{dV_1} > 0$ and $\Omega(l) < 0$ when $\frac{d\rho_r}{dV_1} < 0$. This means that the Berry phase is positive and hence the pumped current is positive when higher potential barrier allows larger reflection probability in block III and that the Berry phase is negative and hence the pump–current direction is reversed when higher potential barrier allows smaller reflection probability in block II. Because $\rho_r^2 + \rho_t^2 = 1$, larger reflection probability means smaller transmission probability, and consistency between the Berry phase picture and the classical turnstile model is numerically proved in the pseudospin-1 Dirac–Weyl system.

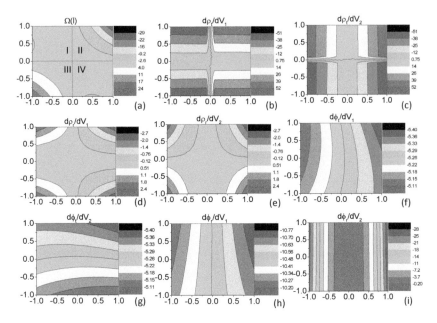

Figure 3. Contours of the Berry curvature $\Omega\,(l)$ and the eight derivatives on the right-hand side of Equation (10) in the V_1-V_2 parameter space. For all the subfigures, the horizonal and vertical axes are V_1 and V_2 in the unit of meV, respectively. The magnitudes of the contours are in the scale of (**a**) 10^{-7}; (**b**) 10^{-4}; (**c**) 10^{-4}; (**d**) 10^{-5}; (**e**) 10^{-5}; (**f**) 10^{-2}; (**g**) 10^{-2}; (**h**) 10^{-2}; (**i**) 10^{-2}; and (**j**) 10^{-5}, respectively. Other parameters are $V_0 = 100$ meV, $d = 5$ nm, $L_2 - L_1 = 10$ nm, $E_F = 100$ meV, and $\theta = 0.5$ in radians. For convenience of discussion, the parameter space in the nine panels is divided into four blocks: I $(-1 < V_1 < 0$ and $0 < V_2 < 1)$, II $(0 < V_1 < 1$ and $0 < V_2 < 1)$, III $(-1 < V_1 < 0$ and $-1 < V_2 < 0)$, and IV $(0 < V_1 < 1$ and $-1 < V_2 < 0)$. The four blocks are illustrated in (**a**).

If we consider normal incidence, consistency between the Berry phase picture and the classic turnstile model becomes straightforward. For normal incidence, derivative of t/r with respect to V_2 is equal to derivative of t'/r' with respect to V_1. Hence, we have

$$\Omega\,(l) = 2\rho_r \frac{d\rho_r}{dV_1}\left(\frac{d\phi_t}{dV_1} - \frac{d\phi_r}{dV_1}\right). \tag{11}$$

From Figure 3, we can see that $\frac{d\phi_t}{dV_1} - \frac{d\phi_r}{dV_1}$ is positive throughout the parameter space. Therefore, the Berry phase has the same sign with $\frac{d\rho_r}{dV_1}$, which demonstrates consistency between the Berry phase picture and the classic turnstile mechanism of the adiabatic quantum pumping.

Besides data shown in Figure 3, which are contours of the Berry curvature $\Omega\,(l)$ and the eight derivatives on the right-hand side of Equation (10) in the parameter window of ± 1 meV for both V_1 and V_2 with $V_0 = 100$ meV, $E_F = 100$ meV, and $\theta = 0.5$ in radians, we have numerically targeted dozens more parameter windows of ± 1 meV for V_1 and V_2 at other values of V_0, E_F, and θ and no obtained results violate consistency of the two mechanisms. Although the main focus of the present manuscript is the pseudospin-1 Dirac–Weyl fermions, we numerically confirmed consistency between the two mechanisms in various parameter settings in different systems such as two-dimensional electron gas, graphene, and the pseudospin-1 Dirac–Weyl system by calculating term by term Equation (10). Although we could not provide a general proof, up to now, no numerical evaluation violates such a conclusion.

We observe in this work and previously that the pump–current direction can be reversed in systems with linear bands such as graphene and pseudospin-1 Dirac–Weyl system. Up to now, similar behavior has not been observed in systems with parabolic band dispersion such as in the semiconductor two-dimensional electron gas. We remark that a quantitative argument of the underlying reason for the dependence of the adiabatic quantum pumping behavior on the band structure as observed numerically is lacking, due to the topological difference of the band structure.

5. Conclusions

In summary, adiabatic quantum pumping in a periodically modulated pseudospin-1 Dirac–Weyl system is studied.   By using two AC electric gate-potentials as the driving parameters, a direction-reversed pumped current is found by the Berry phase of the scattering matrix at certain parameter regimes as a result of super Klein tunneling and particle-hole symmetry close to the Dirac point of the band structure. Such a phenomenon originates from the abnormal transmission behavior of the Dirac–Weyl quasiparticles that sometimes they transmit more through a higher electric potential barrier. As a result, definition of the "opening" and "closing" of a gate is reversed in the classic turnstile picture and hence direction of the pumped DC current is reversed. We also provide rigorous proof of the consistency between the quantum Berry phase picture and the classic turnstile mechanism.

Author Contributions: Conceptualization, R.Z.; Methodology, R.Z.; Validation, R.Z.; Formal Analysis, X.C.; Investigation, X.C.; Resources, X.C.; Data Curation, X.C.; Writing—Original Draft Preparation, R.Z.; Writing—Review & Editing, R.Z.

Funding: The work is supported by the National Natural Science Foundation of China (No. 11004063) and the Fundamental Research Funds for the Central Universities, SCUT (No. 2017ZD099).

Acknowledgments: R.Z. is grateful for enlightening discussions with Pak Ming Hui.

Conflicts of Interest: The authors declare no conflict of interest.

Appendix A. Derivation of the Boundary Condition of the Spinor Wavefunction

In Equation (4), we have defined the spinor wavefunction Ψ of the double-barrier pseudospin-1 Dirac–Weyl system. Within the instant scattering matrix approach, we solve the static Schrödinger equation

$$\hat{H}\Psi = E\Psi, \tag{A1}$$

where $\hat{H}$ is defined in Equation (2) and the time t is taken as a constant. Substituting the spin-1 operator defined in Equation (1), this equation becomes

$$-i\hbar v_g \begin{bmatrix} 0 & \frac{1}{\sqrt{2}}\left(\frac{\partial}{\partial x} - i\frac{\partial}{\partial y}\right) & 0 \\ \frac{1}{\sqrt{2}}\left(\frac{\partial}{\partial x} + i\frac{\partial}{\partial y}\right) & 0 & \frac{1}{\sqrt{2}}\left(\frac{\partial}{\partial x} - i\frac{\partial}{\partial y}\right) \\ 0 & \frac{1}{\sqrt{2}}\left(\frac{\partial}{\partial x} + i\frac{\partial}{\partial y}\right) & 0 \end{bmatrix} \begin{pmatrix} \psi_1 \\ \psi_2 \\ \psi_3 \end{pmatrix}$$

$$+V\left(x,t\right) \begin{pmatrix} \psi_1 \\ \psi_2 \\ \psi_3 \end{pmatrix} = E \begin{pmatrix} \psi_1 \\ \psi_2 \\ \psi_3 \end{pmatrix}, \tag{A2}$$

which is

$$-i\hbar v_g \begin{bmatrix} \frac{1}{\sqrt{2}}\left(\frac{\partial}{\partial x} - i\frac{\partial}{\partial y}\right)\psi_2 \\ \frac{1}{\sqrt{2}}\left(\frac{\partial}{\partial x} + i\frac{\partial}{\partial y}\right)\psi_1 + \frac{1}{\sqrt{2}}\left(\frac{\partial}{\partial x} - i\frac{\partial}{\partial y}\right)\psi_3 \\ \frac{1}{\sqrt{2}}\left(\frac{\partial}{\partial x} + i\frac{\partial}{\partial y}\right)\psi_2 \end{bmatrix} = [E - V\left(x,t\right)] \begin{pmatrix} \psi_1 \\ \psi_2 \\ \psi_3 \end{pmatrix}. \tag{A3}$$

This spinor equation equalizes to three scalar equations

$$-i\hbar v_g \frac{1}{\sqrt{2}} \left(\frac{\partial}{\partial x} - i\frac{\partial}{\partial y} \right) \psi_2 = [E - V(x,t)] \psi_1, \tag{A4}$$

$$-i\hbar v_g \left[\frac{1}{\sqrt{2}} \left(\frac{\partial}{\partial x} + i\frac{\partial}{\partial y} \right) \psi_1 + \frac{1}{\sqrt{2}} \left(\frac{\partial}{\partial x} - i\frac{\partial}{\partial y} \right) \psi_3 \right] = [E - V(x,t)] \psi_2, \tag{A5}$$

and

$$-i\hbar v_g \frac{1}{\sqrt{2}} \left(\frac{\partial}{\partial x} + i\frac{\partial}{\partial y} \right) \psi_2 = [E - V(x,t)] \psi_3. \tag{A6}$$

The boundary condition at x_0 can be obtained from

$$\int_{x_0-\varepsilon}^{x_0+\varepsilon} \hat{H}\Psi dx = \int_{x_0-\varepsilon}^{x_0+\varepsilon} E\Psi dx. \tag{A7}$$

By substituting Equations (A3) and (A4) into Equation (A7), we have

$$\frac{-i\hbar v_g}{\sqrt{2}} \left\{ [\psi_2(x_0+\varepsilon,y) - \psi_2(x_0-\varepsilon,y)] - i\frac{\partial}{\partial y}\psi_2(x_0,y) \, 2\varepsilon \right\} \\ = [E - V(x_0,t)] \psi_1(x_0,y) \, 2\varepsilon. \tag{A8}$$

Because the energy E, the electric potential $V(x,t)$, and the spinor wavefunction $\Psi(x,y)$ is finite, in the limit of $\varepsilon \to 0$, we have the boundary condition

$$[\psi_2(x_0+\varepsilon,y) - \psi_2(x_0-\varepsilon,y)] = 0, \tag{A9}$$

which is that ψ_2 is continuous at $x = x_0$. Similarly, by substituting Equations. (A3) and (A5) into Equation (A7), we have

$$\frac{-i\hbar v_g}{\sqrt{2}} \left\{ [\psi_1(x_0+\varepsilon,y) - \psi_1(x_0-\varepsilon,y)] + i\frac{\partial}{\partial y}\psi_1(x_0,y) \, 2\varepsilon \right. \\ \left. + [\psi_3(x_0+\varepsilon,y) - \psi_3(x_0-\varepsilon,y)] - i\frac{\partial}{\partial y}\psi_3(x_0,y) \, 2\varepsilon \right\} \\ = [E - V(x_0,t)] \psi_2(x_0,y) \, 2\varepsilon. \tag{A10}$$

In the limit of $\varepsilon \to 0$, we have the boundary condition

$$[\psi_1(x_0+\varepsilon,y) + \psi_3(x_0+\varepsilon,y)] - [\psi_1(x_0-\varepsilon,y) + \psi_3(x_0-\varepsilon,y)] = 0, \tag{A11}$$

which is that $\psi_1 + \psi_3$ is continuous at $x = x_0$. Reproducing the procedure in Equations (A3), (A6), and (A7), we can reobtain the boundary condition of ψ_2 in Equation (A9).

Therefore, in the case of pseudospin-1 Dirac–Weyl fermions, the boundary condition is that the second component of the spinor wavefunction ψ_2 is continuous and the first component plus the third component of the spinor wavefunction $\psi_1 + \psi_3$ is continuous. No derivative of the wavefunction is involved in the continuity condition.

Appendix B. Detailed Algebra for Obtaining the Scattering Matrix

We use the transfer-matrix method to obtain the instant scattering matrix **s** defined in Equation (5). Using the obtained spinor wavefunction (4) and the continuity relation (A9) and (A11), we can have the following matrix equations:

$$\mathbf{M}_1 \begin{pmatrix} a_l \\ b_l \end{pmatrix} = \mathbf{M}_2 \begin{pmatrix} a_1 \\ b_1 \end{pmatrix}, \tag{A12}$$

$$\mathbf{M}_3 \begin{pmatrix} a_1 \\ b_1 \end{pmatrix} = \mathbf{M}_4 \begin{pmatrix} a_2 \\ b_2 \end{pmatrix}, \tag{A13}$$

$$\mathbf{M}_5 \begin{pmatrix} a_2 \\ b_2 \end{pmatrix} = \mathbf{M}_6 \begin{pmatrix} a_3 \\ b_3 \end{pmatrix}, \tag{A14}$$

$$\mathbf{M}_7 \begin{pmatrix} a_3 \\ b_3 \end{pmatrix} = \mathbf{M}_8 \begin{pmatrix} a_r \\ b_r \end{pmatrix}, \tag{A15}$$

with

$$\mathbf{M}_1 = \begin{pmatrix} \frac{k_x - ik_y}{k_x + ik_y} + 1 & \frac{-k_x - ik_y}{-k_x + ik_y} + 1 \\ \sqrt{2}\frac{k_x - ik_y}{E_F/\hbar v_g} & \sqrt{2}\frac{-k_x - ik_y}{E_F/\hbar v_g} \end{pmatrix}, \tag{A16}$$

$$\mathbf{M}_2 = \begin{pmatrix} \frac{q_{x1} - ik_y}{q_{x1} + ik_y} + 1 & \frac{-q_{x1} - ik_y}{-q_{x1} + ik_y} + 1 \\ \sqrt{2}\frac{q_{x1} - ik_y}{(E_F - V_0 - V_1)/\hbar v_g} & \sqrt{2}\frac{-q_{x1} - ik_y}{(E_F - V_0 - V_1)/\hbar v_g} \end{pmatrix}, \tag{A17}$$

$$\mathbf{M}_3 = \begin{pmatrix} \left(\frac{q_{x1} - ik_y}{q_{x1} + ik_y} + 1\right) e^{iq_{x1}L_1} & \left(\frac{-q_{x1} - ik_y}{-q_{x1} + ik_y} + 1\right) e^{-iq_{x1}L_1} \\ \sqrt{2}\frac{q_{x1} - ik_y}{(E_F - V_0 - V_1)/\hbar v_g} e^{iq_{x1}L_1} & \sqrt{2}\frac{-q_{x1} - ik_y}{(E_F - V_0 - V_1)/\hbar v_g} e^{-iq_{x1}L_1} \end{pmatrix}, \tag{A18}$$

$$\mathbf{M}_4 = \begin{pmatrix} \left(\frac{k_x - ik_y}{k_x + ik_y} + 1\right) e^{ik_x L_1} & \left(\frac{-k_x - ik_y}{-k_x + ik_y} + 1\right) e^{-ik_x L_1} \\ \sqrt{2}\frac{k_x - ik_y}{E_F/\hbar v_g} e^{ik_x L_1} & \sqrt{2}\frac{-k_x - ik_y}{E_F/\hbar v_g} e^{-ik_x L_1} \end{pmatrix}, \tag{A19}$$

$$\mathbf{M}_5 = \begin{pmatrix} \left(\frac{k_x - ik_y}{k_x + ik_y} + 1\right) e^{ik_x L_2} & \left(\frac{-k_x - ik_y}{-k_x + ik_y} + 1\right) e^{-ik_x L_2} \\ \sqrt{2}\frac{k_x - ik_y}{E_F/\hbar v_g} e^{ik_x L_2} & \sqrt{2}\frac{-k_x - ik_y}{E_F/\hbar v_g} e^{-ik_x L_2} \end{pmatrix}, \tag{A20}$$

$$\mathbf{M}_6 = \begin{pmatrix} \left(\frac{q_{x2} - ik_y}{q_{x2} + ik_y} + 1\right) e^{iq_{x2}L_2} & \left(\frac{-q_{x2} - ik_y}{-q_{x2} + ik_y} + 1\right) e^{-iq_{x2}L_2} \\ \sqrt{2}\frac{q_{x2} - ik_y}{(E_F - V_0 - V_1)/\hbar v_g} e^{iq_{x2}L_2} & \sqrt{2}\frac{-q_{x2} - ik_y}{(E_F - V_0 - V_1)/\hbar v_g} e^{-iq_{x2}L_2} \end{pmatrix}, \tag{A21}$$

$$\mathbf{M}_7 = \begin{pmatrix} \left(\frac{q_{x2} - ik_y}{q_{x2} + ik_y} + 1\right) e^{iq_{x2}L_3} & \left(\frac{-q_{x2} - ik_y}{-q_{x2} + ik_y} + 1\right) e^{-iq_{x2}L_3} \\ \sqrt{2}\frac{q_{x2} - ik_y}{(E_F - V_0 - V_1)/\hbar v_g} e^{iq_{x2}L_3} & \sqrt{2}\frac{-q_{x2} - ik_y}{(E_F - V_0 - V_1)/\hbar v_g} e^{-iq_{x2}L_3} \end{pmatrix}, \tag{A22}$$

$$\mathbf{M}_8 = \begin{pmatrix} \frac{-k_x - ik_y}{-k_x + ik_y} + 1 & \frac{k_x - ik_y}{k_x + ik_y} + 1 \\ \sqrt{2}\frac{-k_x - ik_y}{E_F/\hbar v_g} & \sqrt{2}\frac{k_x - ik_y}{E_F/\hbar v_g} \end{pmatrix}. \tag{A23}$$

It should be noted that it is more convenient to use the k_x/k_y version of the eigenspinors than the $\exp(i\theta)$ version in the numerical treatment because when q_{xi} becomes imaginary, θ becomes ill-defined. In addition, the sign function is avoided in the eigenspinors accordingly. For the wavefunction in the $x > L_3$ region, we used the translated plane wave $\exp[\pm ik_x(x - L_3)]$. From Equations (A12) to (A23), we can have

$$\begin{pmatrix} a_l \\ b_l \end{pmatrix} = \mathbf{M} \begin{pmatrix} a_r \\ b_r \end{pmatrix}, \tag{A24}$$

with

$$\mathbf{M} = \mathbf{M}_1^{-1}\mathbf{M}_2\mathbf{M}_3^{-1}\mathbf{M}_4\mathbf{M}_5^{-1}\mathbf{M}_6\mathbf{M}_7^{-1}\mathbf{M}_8. \tag{A25}$$

Then, the scattering matrix **s** defined in Equation (5) can be obtained by

$$\mathbf{s} = \begin{pmatrix} 0 & -M_{12} \\ 1 & -M_{22} \end{pmatrix}^{-1} \begin{pmatrix} -1 & M_{11} \\ 0 & M_{21} \end{pmatrix}. \tag{A26}$$

Numerical results of the transmission coefficients $T = |t|^2$ are given in Figure A1, which reproduced the results reported in Ref. [57]. From Figure A1, we could find the three characteristic transport properties of the pseudospin-1 Dirac–Weyl fermions: Super Klein tunneling, Klein tunneling, and transmission minimum at the Dirac point. Super Klein tunneling means perfect transmission regardless of the incident angle when the incident energy levels with one half of the electric potential barrier, which is shown in the black solid line of Figure A1a. This is a unique property demonstrated in the pseudospin-1 Dirac–Weyl system. Klein tunneling means perfect transmission at normal incidence regardless of the quasiparticle energy, which is shown in the black solid line of Figure A1b and is also visible in Figure A1a when the horizontal coordinate equates 0. This is a property shared between the pseudospin-1 Dirac–Weyl system and the monolayer graphene, the latter of which also belongs to the pseudospin-1/2 Dirac–Weyl system. Transmission minimum at the Dirac point is shown in the dashed red and dotted blue curves of Figure A1b. This is also a property shared between the pseudospin-1 Dirac–Weyl system and the monolayer graphene. The transmission of normal incidence at the Dirac point is not well-defined, which is overlooked in the numerical treatment.

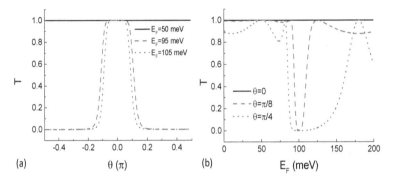

Figure A1. Static transmission probabilities $T = |t|^2$ as a function of the incident angle (**a**) and the Fermi energy (**b**), respectively [57]. Parameters V_0, d, and $L_2 - L_1$ are the same as those in Figure 2.

References

1. Thouless, D.J. Quantization of particle transport. *Phys. Rev. B* **1983**, *27*, 6083 .
2. Xiao, D.; Chang, M.-C.; Niu, Q. Berry phase effects on electronic properties. *Rev. Mod. Phys.* **2010**, *82*, 1959.
3. Switkes, M.; Marcus, C.M.; Campman, K.; Gossard, A.C. An adiabatic quantum electron pump. *Science* **1999**, *283*, 1905.
4. Keller, M.W.; Martinis, J.M.; Kautz, R.L. Rare errors in a well-characterized electron pump: comparison of experiment and theory. *Phys. Rev. Lett.* **1998**, *80*, 4530.
5. Kouwenhoven, L.P.; Johnson, A.T.; van der Vaart, N.C.; Harmans, C.J.P.M.; Foxon, C.T. Quantized current in a quantum-dot turnstile using oscillating tunnel barriers. *Phys. Rev. Lett.* **1991**, *67*, 1626.
6. Zhu, R.; Chen, H. Quantum pumping with adiabatically modulated barriers in graphene. *Appl. Phys. Lett.* **2009**, *95*, 122111.
7. Wei, Y.; Wang, J. Carbon-nanotube-based quantum pump in the presence of a superconducting lead. *Phys. Rev. B* **2002**, *66*, 195419.
8. Ma, W.; Zhou, L.; Zhang, Q.; Li, M.; Cheng, C.; Geng, J.; Rong, X.; Shi, F.; Gong, J.; Du, J. Experimental observation of a generalized Thouless pump with a single spin. *Phys. Rev. Lett.* **2018**, *120*, 120501.
9. Nakajima, S.; Taguchi, M.; Kubo, T.; Tokura, Y. Interaction effect on adiabatic pump of charge and spin in quantum dot. *Phys. Rev. B* **2015**, *92*, 195420.
10. Saha, A.; Rainis, D.; Tiwari, R.P.; Loss, D. Quantum charge pumping through fractional fermions in charge density modulated quantum wires and Rashba nanowires. *Phys. Rev. B* **2014**, *90*, 035422.
11. Alos-Palop, M.; Tiwari, R.P.; Blaauboer, M. Adiabatic quantum pumping of chiral Majorana fermions. *Phys. Rev. B* **2014**, *89*, 045307.
12. Zhu, R.; Berakdar, J. Spin-dependent pump current and noise in an adiabatic quantum pump based on domain walls in a magnetic nanowire. *Phys. Rev. B* **2010**, *81*, 014403.

13. Benjamin, R.; Benjamin, C. Quantum spin pumping with adiabatically modulated magnetic barriers. *Phys. Rev. B* **2004**, *69*, 085318.
14. Citro, R.; Romeo, F. Pumping in a mesoscopic ring with Aharonov-Casher effect. *Phys. Rev. B* **2006**, *73*, 233304.
15. Romeo, F.; Citro, R. Pure spin currents generation in magnetic tunnel junctions by means of adiabatic quantum pumping. *Eur. Phys. J. B* **2006**, *50*, 483.
16. Prada, E.; San-Jose, P.; Schomerus, H. Quantum pumping in graphene. *Phys. Rev. B* **2009**, *80*, 245414.
17. Mohammadkhani, R.; Abdollahipour, B.; Alidoust, M. Strain-controlled spin and charge pumping in graphene devices via spin-orbit coupled barriers. *Europhys. Lett.* **2015**, *111*, 67005.
18. Moskalets, M.; Büttiker, M. Dissipation and noise in adiabatic quantum pumps. *Phys. Rev. B* **2002** *66*, 035306.
19. Wang, B.; Wang, J.; Guo, H. Parametric pumping at finite frequency. *Phys. Rev. B* **2002** *65*, 073306.
20. Wang, B.; Wang, J. Heat current in a parametric quantum pump. *Phys. Rev. B* **2002** *66*, 125310.
21. Wang, B.; Wang, J.; Guo, H. Current plateaus of nonadiabatic charge pump: Multiphoton assisted processes. *Phys. Rev. B* **2003** *68*, 155326.
22. Arrachea, L. Green-function approach to transport phenomena in quantum pumps. *Phys. Rev. B* **2005** *72*, 125349.
23. Brouwer, P.W. Scattering approach to parametric pumping. *Phys. Rev. B* **1998**, *58*, R10135.
24. Bradlyn, B.; Cano, J.; Wang, Z.; Vergniory, M.G.; Felser, C.; Cava, R.J.; Bernevig, B.A. Beyond Dirac and Weyl fermions: Unconventional quasiparticles in conventional crystals. *Science* **2016**, *353*, 558.
25. Orlita, M.; Basko, D.; Zholudev, M.; Teppe, F.; Knap, W.; Gavrilenko, V.; Mikhailov, N.; Dvoretskii, S.; Neugebauer, P.; Faugeras, C.; Barra, A.-L.; Potemski, M. Observation of three-dimensional massless Kane fermions in a zinc-blende crystal. *Nat. Phys.* **2014**, *10*, 233.
26. Slot, M.R.; Gardenier, T.S.; Jacobse, P. H.; van Miert, G.C.P.; Kempkes, S.N.; Zevenhuizen, S.J.M.; Smith, C.M.; Vanmaekelbergh, D.; Swart, I. Experimental realization and characterization of an electronic Lieb lattice. *Nat. Phys.* **2017**, *13*, 672.
27. Guzmán-Silva, D.; Mejía-Cortés, C.; Bandres, M.A.; Rechtsman, M.C.; Weimann, S.; Nolte, S.; Segev, M.; Szameit, A.; Vicencio, R.A. Experimental observation of bulk and edge transport in photonic Lieb lattices. *New J. Phys.* **2014**, *16*, 063061.
28. Mukherjee, S.; Spracklen, A.; Choudhury, D.; Goldman, N.; Öhberg, P.; Andersson, E.; Thomson, R.R. Observation of a localized flat-band state in a photonic Lieb lattice. *Phys. Rev. Lett.* **2015**, *114*, 245504.
29. Vicencio, R.A.; Cantillano, C.; Morales-Inostroza, L.; Real, B.; Mejía-Cortés, C.; Weimann, S.; Szameit, A.; Molina, M.I. Observation of localized states in Lieb photonic lattices. *Phys. Rev. Lett.* **2015**, *114*, 245503.
30. Diebel, F.; Leykam, D.; Kroesen, S.; Denz, C.; Desyatnikov, A.S. Conical diffraction and composite Lieb bosons in photonic lattices. *Phys. Rev. Lett.* **2016**, *116*, 183902.
31. Beugeling, W.; Everts, J.C.; Smith, C.M. Topological phase transitions driven by next-nearest-neighbor hopping in two-dimensional lattices. *Phys. Rev. B* **2012**, *86*, 195129.
32. Niță, M.; Ostahie, B.; Aldea, A. Spectral and transport properties of the two-dimensional Lieb lattice. *Phys. Rev. B* **2013**, *87*, 125428.
33. Goldman, N.; Urban, D.F.; Bercioux, D. Topological phases for fermionic cold atoms on the Lieb lattice. *Phys. Rev. A* **2011**, *83*, 063601.
34. Wang, L.; Yao, D.-X. Coexistence of spin-1 fermion and Dirac fermion on the triangular kagome lattice. *Phys. Rev. B* **2018**, *98*, 161403(R).
35. Horiguchi, T.; Chen, C.C. Lattice Green's function for the diced lattice. *J. Math. Phys.* **1974**, *15*, 659.
36. Vidal, J.; Mosseri, R.; Douçot, B. Aharonov-Bohm cages in two-dimensional structures. *Phys. Rev. Lett.* **1998**, *81*, 5888.
37. Vidal, J.; Butaud, P.; Douçot, B.; Mosseri, R. Disorder and interactions in Aharonov-Bohm cages. *Phys. Rev. B* **2001**, *64*, 155306.
38. Bercioux, D.; Urban, D.F.; Grabert, H.; Häusler, W. Massless Dirac-Weyl fermions in a $\mathcal{T}_3$ optical lattice. *Phys. Rev. A* **2009**, *80*, 063603.
39. Bercioux, D.; Goldman, N.; Urban, D.F. Topology-induced phase transitions in quantum spin Hall lattices. *Phys. Rev. A* **2011**, *83*, 023609.
40. Dey, B.; Ghosh, T.K. Photoinduced valley and electron-hole symmetry breaking in α-T_3 lattice: The role of a variable Berry phase. *Phys. Rev. B* **2018**, *98*, 075422.

41. Biswas, T.; Ghosh, T.K. Magnetotransport properties of the α-T_3 model. *J. Phys. Condens. Matter* **2016**, *28*, 495302.
42. Biswas, T.; Ghosh, T.K. Dynamics of a quasiparticle in the α-T_3 model: role of pseudospin polarization and transverse magnetic field on *zitterbewegung*. *J. Phys.: Condens. Matter* **2018**, *30*, 075301.
43. Dóra, B.; Herbut, I.F.; Moessner, R. Occurrence of nematic, topological, and Berry phases when a flat and a parabolic band touch. *Phys. Rev. B* **2014**, *90*, 045310.
44. Malcolm, J.D.; Nicol, E.J. Frequency-dependent polarizability, plasmons, and screening in the two-dimensional pseudospin-1 dice lattice. *Phys. Rev. B* **2016**, *93*, 165433.
45. Wang, F.; Ran, Y. Nearly flat band with Chern number $C = 2$ on the dice lattice. *Phys. Rev. B* **2011**, *84*, 241103.
46. Tsuchiizu, M. Three-dimensional higher-spin Dirac and Weyl dispersions in the strongly isotropic K_4 crystal. *Phys. Rev. B* **2016**, *94*, 195426.
47. Urban, D.F.; Bercioux, D.; Wimmer, M.; Häusler, W. Barrier transmission of Dirac-like pseudospin-one particles. *Phys. Rev. B* **2011**, *84*, 115136.
48. Xu, H.-Y.; Lai, Y.-C. Revival resonant scattering, perfect caustics, and isotropic transport of pseudospin-1 particles. *Phys. Rev. B* **2016**, *94*, 165405.
49. Fang, A.; Zhang, Z.Q.; Louie, S.G.; Chan, C.T. Klein tunneling and supercollimation of pseudospin-1 electromagnetic waves. *Phys. Rev. B* **2016**, *93*, 035422.
50. Malcolm, J.D.; Nicol, E.J. Magneto-optics of general pseudospin-s two-dimensional Dirac-Weyl fermions. *Phys. Rev. B* **2014**, *90*, 035405.
51. Illes, E.; Carbotte, J.P.; Nicol, E.J. Hall quantization and optical conductivity evolution with variable Berry phase in the α-T_3 model. *Phys. Rev. B* **2015**, *92*, 245410.
52. Illes, E.; Nicol, E.J. Magnetic properties of the α-T_3 model: Magneto-optical conductivity and the Hofstadter butterfly. *Phys. Rev. B* **2016**, *94*, 125435.
53. Griffiths, D.J. *Introduction to Quantum Mechanics*, 2nd ed.; Pearson Education Inc.: London, UK, 2005; Section 4.3, p. 166.
54. Tworzydło, J.; Trauzettel, B.; Titov, M.; Rycerz, A.; Beenakker, C.W.J. Sub-Poissonian shot noise in graphene. *Phys. Rev. Lett.* **2006**, *96*, 246802.
55. Deng, W.Y.; Luo, W.; Geng, H.; Chen, M.N.; Sheng, L.; Xing, D.Y. Non-adiabatic topological spin pumping. *New J. Phys.* **2015**, *17*, 103018.
56. Katsnelson, M.I.; Novoselov, K.S.; Geim, A.K. Chiral tunneling and the Klein paradox in graphene. *Nat. Phys.* **2006**, *2*, 620.
57. Zhu, R.; Hui, P.M. Shot noise and Fano factor in tunneling in three-band pseudospin-1 Dirac–Weyl systems. *Phys. Lett. A* **2017**, *381*, 1971.

Article

Quantum Adiabatic Pumping in Rashba-Dresselhaus-Aharonov-Bohm Interferometer

Yasuhiro Tokura [1,2]

[1] Faculty of Pure and Applied Sciences, University of Tsukuba, 1-1-1, Tennodai, Tsukuba, Ibaraki 305-8571, Japan; tokura.yasuhiro.ft@u.tsukuba.ac.jp

[2] Tsukuba Research Center for Energy Materials Science (TREMS), University of Tsukuba, 1-1-1, Tennodai Tsukuba, Ibaraki 305-8571, Japan

Received: 2 July 2019; Accepted: 8 August 2019; Published: 24 August 2019

Abstract: We investigate the quantum adiabatic pumping effect in an interferometer attached to two one-dimensional leads. The interferometer is subjected to an Aharonov-Bohm flux and Rashba-Dresselhaus spin-orbit interaction. Using Brouwer's formula and rigorous scattering eigenstates, we obtained the general closed formula for the pumping Berry curvatures depending on spin for general interferometers when the external control parameters only modulate the scattering eigenstates and corresponding eigenvalues. In this situation, pumping effect is absent in the combination of the control parameters of Aharonov-Bohm flux and spin-orbit interaction strength. We have shown that finite pumping is possible by modulating both Rashba and Dresselhaus interaction strengths and explicitly demonstrated the spin-pumping effect in a diamond-shaped interferometer made of four sites.

Keywords: spin pump; spin-orbit interaction; quantum adiabatic pump; interferometer; geometric phase

1. Introduction

Coherent transport in mesoscopic systems is of fundamental interest since it allows realization of various phenomena observed in quantum optics in a solid-state system. Furthermore, the electron spin degree of freedom adds an intriguing knob for the manipulation and observation of the transport phenomena. Spin-orbit interaction (SOI) effect [1] is one of the key ingredients in narrow-gap semiconductor devices, whose strength can be controlled by external gates [2], in principle, without changing the electron density. Introducing the effect of SOI to the electron interferometer structure is quite attractive since it enables perfect spin filtering effect [3–5]. Moreover, transient behavior in such an interferometer has been investigated [6].

In addition to passive functional devices such as filters, the active functions, for example, spin-pumping or spin manipulation effect by dynamically modulating the gate voltages [7–9], magnetic field [10–13], or magnetization of the ferromagnets [14–17], has been investigated. In particular, quantum adiabatic pumping (QAP) phenomena [18,19], which stems from geometrical properties of the dynamics, is an active field of research [20–25]. In the non-interacting limit, QAP is related to the scattering matrix of the coherent transport. We have investigated the QAP effect by adiabatically modulating the Aharonov-Bohm (AB) phase [26] of the interferometer as well as the local potential in the interferometer. However, it seems no studies have been made of the adiabatic spin-pumping with purely geometric means such as Aharonov-Casher phase or AB phase. The fundamental question here is whether QAP is possible by only modulating the electron geometric phase.

In this work, we studied spin-QAP in Rashba-Dresselhaus-Aharonov-Bohm interferometer introduced in [3] using Brouwer's formula [19] and derived an explicit formula of the Berry curvature for each spin component. Using the obtained result, we clarified the condition of finite spin-pumping.

In Section 2, we introduce a simple two-terminal setup and the expressions of the scattering amplitudes. Section 3 explains the details of the eigenstates of the scattering problem. Then, with these states, the formula of the QAP is derived in Section 4. It is shown that the modulation of the AB phase cannot induce QAP. Section 5 explains the properties of the diamond-shape interferometer, and is applied to study QAP assuming Rashba SOI and Dresselhaus SOI strengths as control parameters in Section 6. Finally, discussions follow in Section 7 and Appendices are included for the detailed derivations of the formula used in the main text.

We consider a standard setup of scattering problem of spin 1/2 electrons as shown in Figure 1. A coherent scattering region (interferometer) is attached at the site $u = 0$ with the one-dimensional left lead made of sites $u = -1, -2, \ldots$ and is attached at the site $u = 1$ with the one-dimensional right lead made of sites $u = 2, 3, \ldots$. The assumption of one-dimensional leads is not essential as far as the interferometer is coupled to the leads via single mode scattering channels. However, the one-dimensional tight-binding formalism benefits from its simplicity. Although the analysis is standard, the obtained rigorous scattering amplitudes and corresponding scattering eigenstates are essential to clarify the condition and to quantify the quantum adiabatic spin-pumping, as will be shown in the later sections. We introduce the spinor ket vector at site u,

$$|\psi(u)\rangle \equiv \begin{pmatrix} c_{u\uparrow} \\ c_{u\downarrow} \end{pmatrix}, \tag{1}$$

where the two amplitudes $c_{u\sigma}$ for spin $\sigma = \uparrow, \downarrow$ satisfy normalization condition $|c_{u\uparrow}|^2 + |c_{u\downarrow}|^2 = 1$. The total Hamiltonian in the tight-binding approximation is given in general

$$\hat{\mathcal{H}}_{\mathrm{TB}} \equiv \sum_u \epsilon_u |\psi(u)\rangle \langle\psi(u)| + \sum_{uv} \hat{W}_{uv} |\psi(v)\rangle \langle\psi(u)|, \tag{2}$$

where the site index u and v run the entire system. The real parameter ϵ_u is spin-degenerate site energy and $\hat{W}_{uv}$ is a 2×2 hopping matrix satisfying $\hat{W}_{uv}^\dagger = \hat{W}_{vu}$. We assume that the hopping matrix $\hat{W}_{uv}$ is only non-diagonal in the scattering region between $u = 0$ and $u = 1$. We neglect the electron-electron interaction.

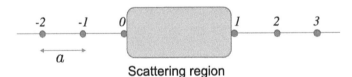

Scattering region

Figure 1. Schematics of the model of a scattering (shaded) region connected with two semi-infinite one-dimensional leads.

2. Model System

In the leads $u \leq -1$, $u \geq 2$, we set $\epsilon_u = 0$ and $\hat{W}_{uv} = -j\mathbb{I}$, where $\mathbb{I}$ is the two-dimensional unit matrix and the real hopping parameter j is only nonzero for nearest-neighbor pair of u, v. With a standard treatment on the tight-binding Hamiltonian, we obtain the eigen-energy $\epsilon_k \equiv -2j\cos(ka)$ and its corresponding eigen-function, $|\psi(u)\rangle \propto e^{ikau} |\chi\rangle$, where k is a real wave-number parameter, a (> 0) is the lattice constant and $|\chi\rangle$ is a certain state vector.

The system of interferometer is represented between $u = 0$ and $u = 1$ sites and we choose $\epsilon_0 = y_0$ and $\epsilon_1 = y_1$ and $\hat{W}_{01} \equiv \hat{W}$ and $\hat{W}_{10} = \hat{W}^\dagger$. The microscopic derivations of y_0, y_1 and $\hat{W}$ for a diamond-shaped interferometer are demonstrated in Section 5. Then the Schrödinger equations at sites $u = 0, 1$ read

$$y_0 \, |\psi(0)\rangle + \hat{W} \, |\psi(1)\rangle - j \, |\psi(-1)\rangle \;=\; \epsilon \, |\psi(0)\rangle \,, \tag{3}$$

$$y_1 \, |\psi(1)\rangle + \hat{W}^\dagger \, |\psi(0)\rangle - j \, |\psi(2)\rangle \;=\; \epsilon \, |\psi(1)\rangle \,. \tag{4}$$

The reflection and transmission amplitude matrices for the electron flux with an energy $\epsilon = \epsilon_k$ injected from the left lead is

$$\hat{r} \;=\; -\mathbb{I} + i\eta_k X_1 \left[\mathbb{I}Y - \hat{W}\hat{W}^\dagger \right]^{-1}, \tag{5}$$

$$\hat{t} \;=\; i\eta_k \hat{W}^\dagger \left[\mathbb{I}Y - \hat{W}\hat{W}^\dagger \right]^{-1}, \tag{6}$$

where $Y \equiv X_0 X_1$ with complex parameters $X_u \equiv \epsilon_k - y_u + j e^{ika}$ ($u = 0, 1$) and we introduced a parameter of energy dimension $\eta_k \equiv 2j \sin(ka)$. The reflection and transmission amplitude matrices for the electron flux injected from the right lead is

$$\hat{r}' \;=\; -\mathbb{I} + i\eta_k X_0 \left[\mathbb{I}Y - \hat{W}^\dagger \hat{W} \right]^{-1}, \tag{7}$$

$$\hat{t}' \;=\; i\eta_k \hat{W} \left[\mathbb{I}Y - \hat{W}^\dagger \hat{W} \right]^{-1}. \tag{8}$$

The details of the derivation of these formulae are given in Appendix A. In the next section, the obtained scattering amplitude matrices are diagonalized and the formulae of the scattering amplitude eigenvalues are given. Then in Section 4, the Berry curvatures for two spin eigenstates, Equations (34) and (35), is given, which allow calculation of QAP spin per cycle.

3. Diagonalization of Hopping Operator $\hat{W}\hat{W}^\dagger$

In this section, we diagonalize the product of hopping operators $\hat{W}$ and $\hat{W}^\dagger$ appearing in the scattering amplitude matrices derived in the previous section. Then we obtain the scattering eigenstates through an interferometer. This is an extension of the discussion in Reference [3]. We consider an interferometer in x-y plane made of two one-dimensional arms, b and c, represented by real coupling parameters γ_b, γ_c and 2×2 unitary matrices, $\hat{U}_b$ and $\hat{U}_c$, showing propagation from the site 0 to 1 via the arms b and c, respectively. We assume following general expressions characterizing the effect of AB phase and Rashba or Dresselhaus SOI:

$$\hat{U}_b \;=\; e^{-i\phi_1} \left(\mathbb{I}\delta + i\boldsymbol{\tau} \cdot \hat{\sigma} \right), \tag{9}$$

$$\hat{U}_c \;=\; e^{i\phi_2} \left(\mathbb{I}\delta' + i\boldsymbol{\tau}' \cdot \hat{\sigma} \right), \tag{10}$$

where $\hat{\sigma}$ is the vector of Pauli spin matrices. $\phi \equiv \phi_1 + \phi_2 = 2\pi(HS)/\Phi_0$ is the AB phase with the magnetic field H in the z direction, the area of the interferometer S, and a magnetic flux quantum Φ_0. Unitarity condition requires the real parameters, δ, δ' and real three-dimensional vectors $\boldsymbol{\tau}, \boldsymbol{\tau}'$ to obey $\delta^2 + |\boldsymbol{\tau}|^2 = \delta'^2 + |\boldsymbol{\tau}'|^2 = 1$. The hopping matrix $\hat{W}$ is given by

$$\hat{W} \;=\; \gamma_b \hat{U}_b + \gamma_c \hat{U}_c. \tag{11}$$

As shown in Appendix B, the matrix factor appearing in the scattering amplitudes for the electron flux injected from the left lead, Equation (5), is

$$\hat{W}\hat{W}^\dagger \;\equiv\; A\mathbb{I} + \boldsymbol{B} \cdot \hat{\sigma}, \tag{12}$$

where

$$A = \gamma_b^2 + \gamma_c^2 + 2\gamma_b\gamma_c \cos\phi \cos\omega, \tag{13}$$

$$B = 2\gamma_b\gamma_c \sin\phi \sin\omega\, \hat{n}. \tag{14}$$

The real parameter ω is determined from $\cos\omega \equiv \delta\delta' + \boldsymbol{\tau} \cdot \boldsymbol{\tau}'$ and the unit vector $\hat{n}$ is defined by

$$\hat{n} = \frac{1}{\sqrt{1-\tau_z^2}}(-\tau_y, \tau_x, \delta). \tag{15}$$

We then introduce two normalized eigenstates of the operator $\hat{n} \cdot \hat{\sigma}$, $|\hat{n}\rangle$ and $|-\hat{n}\rangle$ such that

$$\hat{n} \cdot \hat{\sigma}\,|\hat{n}\rangle = |\hat{n}\rangle, \tag{16}$$

$$\hat{n} \cdot \hat{\sigma}\,|-\hat{n}\rangle = -|-\hat{n}\rangle. \tag{17}$$

Clearly, these are also the eigenstates of the operator $\hat{W}\hat{W}^\dagger$ such that

$$\hat{W}\hat{W}^\dagger\,|\pm\hat{n}\rangle = \lambda_\pm\,|\pm\hat{n}\rangle, \tag{18}$$

with the eigenvalues

$$\lambda_\pm = \gamma_b^2 + \gamma_c^2 + 2\gamma_b\gamma_c \cos(\phi \mp \omega). \tag{19}$$

These eigenvalues are positive since $\lambda_\pm = \langle\pm\hat{n}|\,\hat{W}\hat{W}^\dagger\,|\pm\hat{n}\rangle = \left|\left|\hat{W}^\dagger\,|\pm\hat{n}\rangle\right|\right|^2 \geq 0$.

To study the scattering eigenstates for the electron flux injected from the right, Equation (7), we evaluate $\hat{W}^\dagger\hat{W}$ with similar procedure as above,

$$\hat{W}^\dagger\hat{W} \equiv A\mathbb{I} + \boldsymbol{B}' \cdot \hat{\sigma}, \tag{20}$$

where

$$\boldsymbol{B}' = 2\gamma_b\gamma_c \sin\phi \sin\omega\, \hat{n}', \tag{21}$$

and corresponding unit vector

$$\hat{n}' = \frac{1}{\sqrt{1-\tau_z^2}}(\tau_y, -\tau_x, \delta). \tag{22}$$

Then we introduce two normalized eigenstates of the operator $\hat{n}' \cdot \hat{\sigma}$, $|\pm\hat{n}'\rangle$, which obey

$$\hat{W}^\dagger\hat{W}\,|\pm\hat{n}'\rangle = \lambda_\pm\,|\pm\hat{n}'\rangle, \tag{23}$$

with the same eigenvalues as Equation (18).

The elements of the scattering matrix are now explicitly evaluated with the obtained scattering eigenstates. As detailed in Appendix B, we can show that the transmission amplitude matrices are

$$\hat{t} = t_+\,|\hat{n}'\rangle\langle\hat{n}| + t_-\,|-\hat{n}'\rangle\langle-\hat{n}|, \tag{24}$$

$$\hat{t}' = t_+\,|\hat{n}\rangle\langle\hat{n}'| + t_-\,|-\hat{n}\rangle\langle-\hat{n}'|, \tag{25}$$

where we defined two transmission amplitudes,

$$t_\pm \equiv \frac{i\eta_k\sqrt{\lambda_\pm}}{Y - \lambda_\pm}. \tag{26}$$

Similarly, the reflection amplitude matrices are given by

$$\hat{r} = r_+ |\hat{n}\rangle \langle \hat{n}| + r_- |-\hat{n}\rangle \langle -\hat{n}|, \tag{27}$$

$$\hat{r}' = r'_+ |\hat{n}'\rangle \langle \hat{n}'| + r'_- |-\hat{n}'\rangle \langle -\hat{n}'|, \tag{28}$$

where the reflection amplitudes are

$$r_\pm \equiv -1 + \frac{i\eta_k X_1}{Y - \lambda_\pm}, \tag{29}$$

$$r'_\pm \equiv -1 + \frac{i\eta_k X_0}{Y - \lambda_\pm}. \tag{30}$$

The unitarity condition of the scattering amplitude matrices, $\hat{t}^\dagger \hat{t} + \hat{r}^\dagger \hat{r} = 1$, is confirmed in Appendix C. The unitarity condition $\hat{t}'^\dagger \hat{t}' + \hat{r}'^\dagger \hat{r}' = 1$ can also be checked.

4. Quantum Adiabatic Pump

For a non-interacting system, the response (particle transfer) to the slow modulation of the system's controlling parameters is well described by Brouwer's formula [19], which is expressed by the elements of the scattering matrix. The particles induced in the left lead in one cycle of the adiabatic modulation of two control parameters g_1 and g_2 is

$$n = \sum_\sigma n_\sigma, \tag{31}$$

$$n_\sigma = -\int_S dg_1 dg_2 \Pi_\sigma(g_1, g_2), \tag{32}$$

where S is the area in the two-dimensional control parameter space whose edge corresponds to the trajectory of the cycle. The Berry curvature $\Pi_\sigma(g_1, g_2)$ for spin σ is

$$\Pi_\sigma(g_1, g_2) = \frac{1}{\pi} \Im \langle \sigma | \left\{ \frac{\partial \hat{r}}{\partial g_2} \frac{\partial \hat{r}^\dagger}{\partial g_1} + \frac{\partial \hat{t}'}{\partial g_2} \frac{\partial \hat{t}'^\dagger}{\partial g_1} \right\} | \sigma \rangle, \tag{33}$$

where $\hat{r}$ and $\hat{t}'$ are given in Equations (27) and (25) and $|\sigma\rangle$ is the spinor vector of spin σ.

If we choose the AB phase ϕ and parameters of the interferometers, for example, X_0 or X_1, but not the SOI strengths, we can show that the Berry curvature is finite in general as studied in Reference [26]. In the following, however, we focus on the situation that the control parameters are the AB phase ϕ and Rashba or Dresselhaus SOI strength that modulate the eigenvalue $\lambda_\pm$ as well as the scattering eigenstates $|\pm\hat{n}\rangle, |\pm\hat{n}'\rangle$. To calculate the Berry curvature, we need to evaluate the derivatives of the scattering amplitude matrices, $\hat{r}$ and $\hat{t}'$. Then, as shown in Appendix D, after some manipulations, we have the Berry curvatures for spin components parallel to $\pm\hat{n}$,

$$\Pi_{\hat{n}}(g_1, g_2) = \left(|r_+ - r_-|^2 - |t_+|^2 + |t_-|^2 \right) C_{g_1, g_2}, \tag{34}$$

and

$$\Pi_{-\hat{n}}(g_1, g_2) = \left(-|r_+ - r_-|^2 - |t_+|^2 + |t_-|^2 \right) C_{g_1, g_2}. \tag{35}$$

where the factor at the end is independent of spin and is defined as

$$C_{g_1, g_2} = \frac{1}{4\pi n_z} \frac{1}{1 - \tau_z^2} \left[\frac{\partial \tau_y}{\partial g_1} \frac{\partial \tau_x}{\partial g_2} - \frac{\partial \tau_y}{\partial g_2} \frac{\partial \tau_x}{\partial g_1} \right.$$
$$\left. + \frac{\tau_z}{1 - \tau_z^2} \left\{ \tau_x \left(\frac{\partial \tau_y}{\partial g_1} \frac{\partial \tau_z}{\partial g_2} - \frac{\partial \tau_y}{\partial g_2} \frac{\partial \tau_z}{\partial g_1} \right) + \tau_y \left(\frac{\partial \tau_x}{\partial g_2} \frac{\partial \tau_z}{\partial g_1} - \frac{\partial \tau_x}{\partial g_1} \frac{\partial \tau_z}{\partial g_2} \right) \right\} \right]. \tag{36}$$

This is one of the main results of this work.

The vector $\boldsymbol{\tau}$ is independent of ϕ, but only depends on the SOI strength. Therefore, when one chose the AB phase, $g_1 \equiv \phi$, as one of the control parameters, C_{ϕ,g_2} is identically zero as is evident from Equation (36). Hence we do not expect QAP by modulating the AB phase and SOI strength. It is also obvious that if we chose γ_b or γ_c as one of the control parameters and the other by SOI strength, $C_{\gamma_{b,c},g_2}$ is zero since $\boldsymbol{\tau}$ is independent of γ_b and γ_c and no pumping is expected.

Even for a fixed AB phase, there is still some freedom to choose two control parameters related to the SOI strength since we have two types of SOI interaction mechanisms, Rashba and Dresselhaus SOI. In the next section, we study Rashba-Dresselhaus interferometer in a simple diamond-shape structure made of four sites and choose the strengths of two types of SOI as control parameters.

5. Diamond Interferometer

We consider an electron transport in two-dimensional system on [001] surface with setting x and y axis along the (100) and (010) crystal directions, respectively. The Hamiltonian for the SOI is

$$\hat{\mathcal{H}}_{\mathrm{R}} = \frac{\hbar}{m} k_{\mathrm{R}} (\hat{p}_y \hat{\sigma}_x - \hat{p}_x \hat{\sigma}_y), \tag{37}$$

$$\hat{\mathcal{H}}_{\mathrm{D}} = \frac{\hbar}{m} k_{\mathrm{D}} (\hat{p}_x \hat{\sigma}_x - \hat{p}_y \hat{\sigma}_y), \tag{38}$$

where k_{R} and k_{D} are Rashba and Dresselhaus parameters, respectively. $\hat{p}_\mu$ ($\mu = x, y$) are the momentum and m is the electron effective mass.

The interferometer made of four sites is configured as in Figure 2 which is attached to the leads at site $u = 0$ and $u = 1$ as discussed in Reference [3]. Other two sites constituting the interferometer are $u = b$ and $u = c$, connected with bonds of length L. We also define the opening angle 2β and the relative angle v of the diagonal line to x axis. The Hamiltonian reads

$$\hat{\mathcal{H}}_{\mathrm{IF}} \equiv \sum_u \epsilon_u |\psi(u)\rangle \langle \psi(u)| - \sum_{uv} \tilde{U}_{uv} |\psi(v)\rangle \langle \psi(u)|, \tag{39}$$

for $u, v = 0, c, d, 1$ where ϵ_u is the site energy and $\tilde{U}_{uv} \equiv J_{uv} \hat{U}_{uv}$, J_{uv} is a hopping energy and $\hat{U}_{uv}$ is a 2×2 unitary matrix representing the effect of SOI and AB phase. Total Hamiltonian is $\hat{\mathcal{H}} = \hat{\mathcal{H}}_{\mathrm{IF}} + \hat{\mathcal{H}}_L + \hat{\mathcal{H}}_R$. In the Appendix E, we explain how this problem is reduced to the Schrödinger equations, Equations (3) and (4).

The coordinates of the four sites are $\boldsymbol{r}_0 = (0,0)$, $\boldsymbol{r}_b = (L\cos(v+\beta), L\sin(v+\beta))$, $\boldsymbol{r}_1 = (2L\cos(\beta)\cos(v), 2L\cos(\beta)\sin(v))$, and $\boldsymbol{r}_c = (L\cos(v-\beta), L\sin(v-\beta))$. We define $\alpha_{\mathrm{R}} \equiv k_{\mathrm{R}} L$, $\alpha_{\mathrm{D}} \equiv k_{\mathrm{D}} L$ and $\zeta \equiv \sqrt{\alpha_{\mathrm{R}}^2 + \alpha_{\mathrm{D}}^2}$ and introduce another angle θ, such that $\alpha_{\mathrm{R}} = \zeta \cos\theta$, and $\alpha_{\mathrm{D}} = \zeta \sin\theta$. The unitary matrix for the hopping from site at $(0,0)$ to site at (u_x, u_y) is $\hat{U}_{(0,0),(u_x,u_y)} = \exp[i\boldsymbol{K} \cdot \hat{\boldsymbol{\sigma}}]$ with $\boldsymbol{K} \equiv \alpha_{\mathrm{R}}(u_y, -u_x, 0) + \alpha_{\mathrm{D}}(u_x, -u_y, 0)$ [27]. Therefore, for the hopping from site 0 to b,

$$\boldsymbol{K}_{0b} \cdot \hat{\boldsymbol{\sigma}} = \zeta \sin(\xi_1)\hat{\sigma}_x - \zeta \cos(\xi_2)\hat{\sigma}_y \equiv \zeta \hat{\sigma}_1, \tag{40}$$

with $\xi_1 \equiv \beta + v + \theta$ and $\xi_2 \equiv \beta + v - \theta$. Similarly, for the hopping from site c to 0,

$$\boldsymbol{K}_{c0} \cdot \hat{\boldsymbol{\sigma}} = \zeta \sin(\xi_4)\hat{\sigma}_x + \zeta \cos(\xi_3)\hat{\sigma}_y \equiv \zeta \hat{\sigma}_2, \tag{41}$$

with $\xi_3 \equiv \beta - v + \theta$ and $\xi_4 \equiv \beta - v - \theta$. We introduce factors $F_1 \equiv \sqrt{1 + \sin(2v+2\beta)\sin(2\theta)}$ and $F_2 \equiv \sqrt{1 + \sin(2v - 2\beta)\sin(2\theta)}$ such that $\hat{\sigma}_1^2 = \mathbb{I} F_1^2$ and $\hat{\sigma}_2^2 = \mathbb{I} F_2^2$. Then, for $n = 1, 2$,

$$e^{i\zeta\hat{\sigma}_n} \equiv \mathbb{I} c_n + i s_n \hat{\sigma}_n, \tag{42}$$

where we defined

$$c_n \equiv \cos(F_n\zeta), \qquad s_n \equiv \frac{1}{F_n}\sin(F_n\zeta). \tag{43}$$

Noting that $\hat{U}_{0b} = e^{i\zeta\hat{\sigma}_1}$, $\hat{U}_{b1} = e^{-\frac{i\phi}{2}-i\zeta\hat{\sigma}_2}$, $\hat{U}_{0c} = e^{-i\zeta\hat{\sigma}_2}$ and $\hat{U}_{c1} = e^{\frac{i\phi}{2}+i\zeta\hat{\sigma}_1}$,

$$
\begin{aligned}
\hat{U}_b &\equiv \hat{U}_{0b}\hat{U}_{b1} \\
&= e^{-\frac{i\phi}{2}}\{\mathbb{I}c_1c_2 - ic_1s_2\hat{\sigma}_2 + ic_2s_1\hat{\sigma}_1 + s_1s_2\hat{\sigma}_1\hat{\sigma}_2\} \\
&= e^{-\frac{i\phi}{2}}\left(\mathbb{I}\delta + i\boldsymbol{\tau}\cdot\hat{\boldsymbol{\sigma}}\right),
\end{aligned}
\tag{44}
$$

where

$$
\begin{aligned}
\delta &\equiv c_1c_2 - s_1s_2(\sin(2\nu)\sin(2\theta) + \cos(2\beta)), \\
\tau_x &\equiv -c_1s_2\sin\xi_4 + c_2s_1\sin\xi_1, \\
\tau_y &\equiv -c_1s_2\cos\xi_3 - c_2s_1\cos\xi_2, \\
\tau_z &\equiv s_1s_2\sin(2\beta)\cos(2\theta).
\end{aligned}
\tag{45}
$$

Similarly,

$$
\begin{aligned}
\hat{U}_c &\equiv \hat{U}_{0c}\hat{U}_{c1} \\
&= e^{\frac{i\phi}{2}}\{\mathbb{I}c_2c_1 - is_2c_1\hat{\sigma}_2 + ic_2s_1\hat{\sigma}_1 + s_2s_1\hat{\sigma}_2\hat{\sigma}_1\} \\
&= e^{\frac{i\phi}{2}}\left(\mathbb{I}\delta' + i\boldsymbol{\tau}'\cdot\hat{\boldsymbol{\sigma}}\right),
\end{aligned}
\tag{46}
$$

with $\delta' = \delta$ and $\boldsymbol{\tau}' = (\tau_x, \tau_y, -\tau_z)$. The angle ω is determined by $\cos\omega = \delta\delta' + \boldsymbol{\tau}\cdot\boldsymbol{\tau}' = \delta^2 + \tau_x^2 + \tau_y^2 - \tau_z^2 = 1 - 2\tau_z^2$.

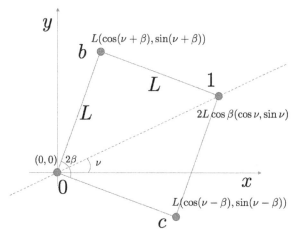

Figure 2. Schematics of the interferometer made of four sites, $0, b, c$, and 1 separated by a length L. The opening angle 2β and relative angle ν from x axis determine the geometric structure.

6. QAP in the Diamond Interferometer

We examine the quantum adiabatic spin-pumping by choosing two SOI strengths $g_1 = \alpha_R$ and $g_2 = \alpha_D$ as control parameters. First we examine the basic property of the function C_{α_R,α_D} defined in

Equation (36) and then evaluate the scattering amplitudes. Using these results, we calculate the Berry curvatures for two spin directions.

6.1. Spin-Independent Function C_{α_R,α_D}

The function C_{α_R,α_D} has symmetries, $C_{\alpha_1,\alpha_2} = C_{\alpha_2,\alpha_1}$, as well as $C_{\alpha_1,\alpha_2} = C_{-\alpha_1,\alpha_2}, C_{\alpha_1,\alpha_2} = C_{\alpha_1,-\alpha_2}$. Moreover, it also obeys the relation $C_{\alpha_R,\alpha_D}|_v = -C_{\alpha_R,\alpha_D}|_{\frac{\pi}{2}-v}$. Therefore, the angle $v = \pi/4$ is rather special. At this angle, C_{α_R,α_D} is identically zero and hence no pumping. One can check this since $F_1 = F_2$ and hence $c_1 = c_2$ and $s_1 = s_2$, then

$$\tau_x = c_1 s_1 (\sin \zeta_1 - \sin \zeta_4) = 2c_1 s_1 \cos \beta \sin(\theta + \frac{\pi}{4}), \tag{47}$$

$$\tau_y = -c_1 s_1 (\cos \zeta_3 + \cos \zeta_2) = -2c_1 s_1 \cos \beta \cos(\theta - \frac{\pi}{4}). \tag{48}$$

Therefore, the relation $\tau_x = -\tau_y$ holds for any β and θ and $C_{\alpha_R,\alpha_D} = 0$.

Because of its symmetric property, we focus on the function C_{α_R,α_D} in the range $0 \leq \alpha_R, \alpha_D \leq \pi$. As an example, we chose $\beta = \pi/5$ and the results for $v = \pi/2$ and $v = 3\pi/8$ are shown in Figure 3. The result for $v = \pi/4$ is uniformly zero as noted above and that for $v = \pi/8$ is similar to that for $v = 3\pi/8$ with reversing the sign of the function. There are areas where the absolute value of C_{α_R,α_D} is enhanced near $(\alpha_R, \alpha_D) = (\frac{\pi}{2}, 0), (0, \frac{\pi}{2})$, which can be understood from Equation (36) since $|\tau_z|$ is very close to one. If we choose $\beta = \pi/4$, the scattering states "flips" at $\pi/2$ when α_R is increased from zero to π with $\alpha_D = 0$ [3]. Then the behavior of $C_{\alpha_R,\alpha_D=0}$ becomes quite singular, which may need further investigation (not being discussed here).

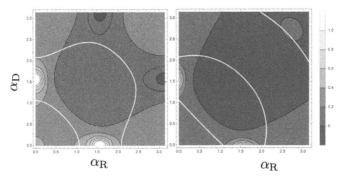

α_D α_R α_R

Figure 3. Contour plot of the function C_{α_R,α_D} depending on the Rashba, α_R, and Dresselhaus, α_D, SOI strength parameters. We chose the geometric angles $\beta = \pi/5$ and $v = \pi/2$ (**left**) and $v = 3\pi/8$ (**right**).

6.2. Spin-Dependent Prefactors

In this section, we examine the scattering amplitudes, $t_\pm$ and $r_\pm$ and the prefactors of the Berry curvatures in Equations (34) and (35). We define these factors as $d_{\hat{n}} = |r_+ - r_-|^2 - |t_+|^2 + |t_-|^2$ and $d_{-\hat{n}} = -|r_+ - r_-|^2 - |t_+|^2 + |t_-|^2$. To be compatible with the analysis in the previous subsection, we focus on the geometry such that $\beta = \pi/5$ and $v = \pi/2$. For simplicity, we chose symmetric setup of the interferometer, where $J_{0b} = J_{b1} = J_{0c} = J_{c1} = j$ and $\epsilon_0 = \epsilon_1 = \epsilon_b = \epsilon_c$. Then, $\gamma_b = \gamma_c = \frac{j^2}{\epsilon_k - \epsilon_0}$. Moreover, in the following calculation we chose $\epsilon_k = -j$. First, we show the result of $d_{-\hat{n}}$ for $\epsilon_0 = 0.9j$ in Figure 4 with choosing the AB phase $\phi = \pi/3$. This function is negatively enhanced near $(\alpha_R, \alpha_D) = (\pi/2, 0)$ and $(0, \pi/2)$. In contrast, the factor $d_{\hat{n}}$ is much smaller as shown in the linear plot for $\alpha_D = 0$. If one chose AB phase $\phi = 5\pi/3$, $d_{-\hat{n}}$ is suppressed and alternatively $d_{\hat{n}}$ is enhanced near $(\alpha_R, \alpha_D) = (\pi/2, 0)$ and $(0, \pi/2)$ (with changing sign of the data in the left Figure 4). The AB phase ϕ and site energy ϵ_0 dependence of $d_{\pm\hat{n}}$ are shown in the left and right of Figure 5, respectively.

Therefore, a large contrast of the QAP in two spin directions can be obtained by choosing $\phi = \pi/3$ and $\epsilon_0 = 0.9j$.

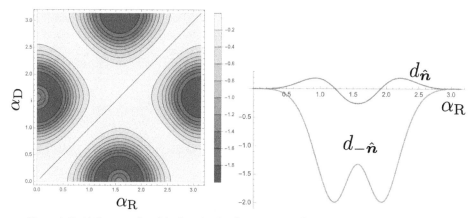

Figure 4. (**Left**) Contour plot of the function $d_{-\hat{n}}$ for $\epsilon_0 = 0.9j$ and $\phi = \pi/3$. (**Right**) Line plot of the functions $d_{\pm\hat{n}}$ as a function of α_R with $\alpha_D = 0$.

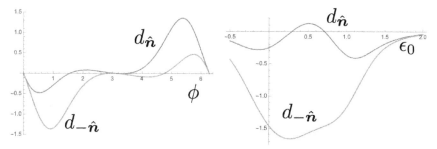

Figure 5. (**Left**) AB phase dependence of the function $d_{\pm\hat{n}}$ for $\epsilon_0 = 0.9j$ and $\alpha_R = \pi/2, \alpha_D = 0$. (**Right**) Site energy dependence of the function $d_{\pm\hat{n}}$ for $f = \pi/3$ and $\alpha_R = \pi/2, \alpha_D = 0$.

6.3. Berry Curvatures

Finally, we calculate the Berry curvature, $\Pi_{-\hat{n}}(\alpha_R, \alpha_D)$ for the spin in the state $|-\hat{n}\rangle$ as shown in the left of Figure 6. Obviously, the Berry curvature becomes large at around $(\alpha_R, \alpha_D) = (\frac{\pi}{2}, 0), (0, \frac{\pi}{2})$. The other spin state is not much pumped as shown in the right of Figure 6.

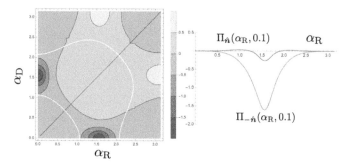

Figure 6. (**Left**) Contour plot of the Berry curvature for $|-\hat{n}\rangle$ with $\beta = \pi/5$ and $\nu = \pi/2$. We set $\phi = \pi/3, \epsilon_k = -j, J_{0b} = J_{b1} = J_{0c} = J_{c1} = j$ and $\epsilon_0 = \epsilon_1 = \epsilon_b = \epsilon_c = 0.9j$. (**Right**) Berry curvatures for two spin directions with $\alpha_D = 0.1$ with the same parameter with the left panel.

7. Discussion

We have derived a general expression of the Berry curvature for an interferometer connected to one-dimensional leads. In this study, we restricted the control parameters in QAP formalism only to modulate the scattering eigenstates and corresponding eigenvalues through the change of the unitary operators for each arm. Then the AB phase, which, despite modifying the scattering eigenvalues, $\lambda_{\pm}$, does not affect the scattering eigenstates and is shown not to function as a control parameter in QAP. In a clear contrast, it has been shown [26] that in combination with the potential modulation, affecting the electron-hopping amplitudes or site energies, QAP by AB phase is possible.

In the current analysis, the control parameters are assumed to purely modulate the phase of the electrons. In real experiments, unintended modulation of hopping amplitudes, J_{uv}, or the site energies, ϵ_b or ϵ_c by the gate voltages may induce additional effects. We demonstrated that by using the two types of the SOI as the two control parameters, spin-QAP is possible. However, in the experiments, independent control of the Rashba SOI and Dresselhaus SOI will be a complicated task. Fortunately, as shown in Figure 6, the area of large Berry curvature is well isolated and the tiny change of Dresselhaus SOI may be sufficient to observe QAP. It would be interesting if other types of SOI interaction [5] could be another control parameter of the QAP.

Funding: This research was funded by JSPS Kakenhi (18K03479).

Acknowledgments: I thank useful discussions with A. Aharony and O. Entin-Wohlman.

Conflicts of Interest: The author declares no conflict of interest.

Abbreviations

The following abbreviations are used in this manuscript:

SOI   Spin-orbit interaction
QAP   Quantum adiabatic pumping
AB   Aharonov-Bohm

Appendix A. Scattering Matrix

In this Appendix, we argue the scattering problem through the interferometer. First, we inject an electron flux with an energy $\epsilon = \epsilon_k$ from the left lead. The wavefunction for $u \leq 0$ is

$$|\psi(u)\rangle = e^{ikua} |\chi_{\text{in}}\rangle + e^{-ikua} |\chi_r\rangle, \tag{A1}$$

and the wavefunction for $u \geq 1$ is

$$|\psi(u)\rangle = e^{ik(u-1)a} |\chi_t\rangle, \tag{A2}$$

where $|\chi_{\text{in}}\rangle$ is the injected wavefunction and $|\chi_r\rangle, |\chi_t\rangle$ are the (un-normalized) wavefunctions of reflection and transmission. In particular, at sites $u = 0, -1$,

$$|\psi(0)\rangle = |\chi_{\text{in}}\rangle + |\chi_r\rangle, \qquad |\psi(-1)\rangle = e^{-ika} |\chi_{\text{in}}\rangle + e^{ika} |\chi_r\rangle, \tag{A3}$$

and at sites $u = 1, 2$,

$$|\psi(1)\rangle = |\chi_t\rangle, \qquad |\psi(2)\rangle = e^{ika} |\chi_t\rangle. \tag{A4}$$

By putting these into Equations (3) and (4), we have

$$(\epsilon_k - y_0) \{|\chi_{\text{in}}\rangle + |\chi_r\rangle\} = \hat{W} |\chi_t\rangle - j \left\{ e^{-ika} |\chi_{\text{in}}\rangle + e^{ika} |\chi_r\rangle \right\}, \tag{A5}$$

$$(\epsilon_k - y_1) |\chi_t\rangle = \hat{W}^\dagger \{|\chi_{\text{in}}\rangle + |\chi_r\rangle\} - je^{ika} |\chi_t\rangle. \tag{A6}$$

Then from Equation (A6),

$$|\chi_t\rangle = \frac{1}{\epsilon_k - y_1 + je^{ika}}\hat{W}^\dagger\{|\chi_{in}\rangle + |\chi_r\rangle\}, \tag{A7}$$

and putting this into Equation (A5), we have

$$(\epsilon_k - y_0 + je^{-ika})|\chi_{in}\rangle + (\epsilon_k - y_0 + je^{ika})|\chi_r\rangle = \hat{W}\frac{1}{\epsilon_k - y_1 + je^{ika}}\hat{W}^\dagger\{|\chi_{in}\rangle + |\chi_r\rangle\}. \tag{A8}$$

Defining complex parameters $X_u \equiv \epsilon_k - y_u + je^{ika}$ ($u = 0, 1$) and noting $\epsilon_k - y_0 + je^{-ika} = X_0 - i\eta_k$, we solve this equation

$$|\chi_r\rangle = -\left[\mathbb{I}X_0 X_1 - \hat{W}\hat{W}^\dagger\right]^{-1}\left[\mathbb{I}X_0 X_1 - \hat{W}\hat{W}^\dagger - \mathbb{I}i\eta_k X_1\right]|\chi_{in}\rangle.$$

Then we have obtained the reflection amplitude matrix

$$\begin{aligned}
\hat{r} &\equiv -\left[\mathbb{I}X_0 X_1 - \hat{W}\hat{W}^\dagger\right]^{-1}\left[\mathbb{I}X_0 X_1 - \hat{W}\hat{W}^\dagger - \mathbb{I}i\eta_k X_1\right]\\
&= -\mathbb{I} + i\eta_k X_1\left[\mathbb{I}Y - \hat{W}\hat{W}^\dagger\right]^{-1},
\end{aligned} \tag{A9}$$

where we have introduced $Y \equiv X_0 X_1$. Using this, transmitted state is calculated with Equation (A7),

$$\begin{aligned}
|\chi_t\rangle &= \frac{1}{X_1}\hat{W}^\dagger\{|\chi_{in}\rangle + |\chi_r\rangle\}\\
&= \frac{1}{X_1}\hat{W}^\dagger\left[\mathbb{I} - \mathbb{I} + i\eta_k X_1\left[\mathbb{I}Y - \hat{W}\hat{W}^\dagger\right]^{-1}\right]|\chi_{in}\rangle\\
&= i\eta_k \hat{W}^\dagger\left[\mathbb{I}Y - \hat{W}\hat{W}^\dagger\right]^{-1}|\chi_{in}\rangle,
\end{aligned}$$

hence the transmission amplitude matrix is

$$\hat{t} = i\eta_k \hat{W}^\dagger\left[\mathbb{I}Y - \hat{W}\hat{W}^\dagger\right]^{-1}. \tag{A10}$$

We alternatively consider the situation that the electron is injected from the right lead. The wavefunction for $u \geq 1$ is

$$|\psi(n)\rangle = e^{-ik(n-1)a}|\chi'_{in}\rangle + e^{ik(n-1)a}|\chi'_r\rangle, \tag{A11}$$

and the wavefunction for $u \leq 0$ is

$$|\psi(n)\rangle = e^{-ikna}|\chi'_t\rangle, \tag{A12}$$

where $|\chi'_{in}\rangle$ is the incoming wavefunction and $|\chi'_r\rangle$, $|\chi'_t\rangle$ are the (un-normalized) wavefunctions of reflection and transmission. At sites $u = 1, 2$,

$$|\psi(1)\rangle = |\chi'_{in}\rangle + |\chi'_r\rangle, \qquad |\psi(2)\rangle = e^{-ika}|\chi'_{in}\rangle + e^{ika}|\chi'_r\rangle, \tag{A13}$$

and at sites $u = 0, -1$,

$$|\psi(0)\rangle = |\chi'_t\rangle, \qquad |\psi(-1)\rangle = e^{ika}|\chi'_t\rangle. \tag{A14}$$

By putting these into Equations (3) and (4), we have

$$(\epsilon_k - y_0)\,|\chi_t'\rangle \;=\; \hat{W}\left\{|\chi_{in}'\rangle + |\chi_r'\rangle\right\} - je^{ika}\,|\chi_t'\rangle, \tag{A15}$$

$$(\epsilon_k - y_1)\left\{|\chi_{in}'\rangle + |\chi_r'\rangle\right\} \;=\; \hat{W}^\dagger\,|\chi_t'\rangle - j\left\{e^{-ika}\,|\chi_{in}'\rangle + e^{ika}\,|\chi_r'\rangle\right\}. \tag{A16}$$

From Equation (A15),

$$|\chi_t'\rangle \;=\; \frac{1}{X_0}\hat{W}\left\{|\chi_{in}'\rangle + |\chi_r'\rangle\right\}, \tag{A17}$$

and putting this into Equation (A16),

$$(X_1 - i\eta_k)\,|\chi_{in}'\rangle + X_1\,|\chi_r'\rangle \;=\; \hat{W}^\dagger\frac{1}{X_0}\hat{W}\left\{|\chi_{in}'\rangle + |\chi_r'\rangle\right\}, \tag{A18}$$

which is solved as

$$
\begin{aligned}
|\chi_r'\rangle &= \left[\mathbb{I}Y - \hat{W}^\dagger\hat{W}\right]^{-1}\left\{-\left(\mathbb{I}Y - \hat{W}^\dagger\hat{W}\right) + i\eta_k X_0\mathbb{I}\right\}|\chi_{in}'\rangle \\
&= \left\{-\mathbb{I} + i\eta_k X_0\left[\mathbb{I}Y - \hat{W}^\dagger\hat{W}\right]^{-1}\right\}|\chi_{in}'\rangle,
\end{aligned}
$$

Therefore, the reflection amplitude matrix is

$$\hat{r}' \;=\; -\mathbb{I} + i\eta_k X_0\left[\mathbb{I}Y - \hat{W}^\dagger\hat{W}\right]^{-1}. \tag{A19}$$

Putting this into Equation (A17),

$$
\begin{aligned}
|\chi_t'\rangle &= \frac{\hat{W}}{X_0}\left\{\mathbb{I} - \mathbb{I} + i\eta_k X_0\left[\mathbb{I}Y - \hat{W}^\dagger\hat{W}\right]^{-1}\right\}|\chi_{in}'\rangle \\
&= i\eta_k\hat{W}\left[\mathbb{I}Y - \hat{W}^\dagger\hat{W}\right]^{-1}|\chi_{in}'\rangle,
\end{aligned}
$$

and hence the transmission amplitude matrix is

$$\hat{t}' \;=\; i\eta_k\hat{W}\left[\mathbb{I}Y - \hat{W}^\dagger\hat{W}\right]^{-1}. \tag{A20}$$

Appendix B. Scattering Eigenstates

In this appendix, we show the details of the calculations of the scattering eigenstates discussed in Section 3. First, we evaluate

$$
\begin{aligned}
\hat{W}\hat{W}^\dagger &= (\gamma_b\hat{U}_b + \gamma_c\hat{U}_c)(\gamma_b\hat{U}_b^\dagger + \gamma_c\hat{U}_c^\dagger) \\
&= (\gamma_b^2 + \gamma_c^2)\mathbb{I} + \gamma_b\gamma_c(\hat{u} + \hat{u}^\dagger),
\end{aligned} \tag{A21}
$$

with $\hat{u} \equiv \hat{U}_b\hat{U}_c^\dagger$ representing the total development around the interferometer in the order $0 \to b \to 1 \to c \to 0$. We study following matrix

$$
\begin{aligned}
\hat{u} &= e^{-i\phi}\left(\mathbb{I}\delta + i\tau\cdot\hat{\sigma}\right)\left(\mathbb{I}\delta' - i\tau'\cdot\hat{\sigma}\right) \\
&= e^{-i\phi}\left\{\mathbb{I}\delta\delta' - i\delta\tau'\cdot\hat{\sigma} + i\delta'\tau\cdot\hat{\sigma} + (\tau\cdot\hat{\sigma})(\tau'\cdot\hat{\sigma})\right\} \\
&= e^{-i\phi}\left\{\mathbb{I}\delta\delta' + i(\delta'\tau - \delta\tau')\cdot\hat{\sigma} + \mathbb{I}\tau\cdot\tau' + i(\tau\times\tau')\cdot\hat{\sigma}\right\}.
\end{aligned}
$$

We introduce a unit vector defined by

$$\hat{n} \equiv \mathcal{N}\left\{\delta'\tau - \delta\tau' + \tau \times \tau'\right\} = (n_x, n_y, n_z), \tag{A22}$$

where $\mathcal{N}$ is a normalization constant. We define z-direction (in spin space) in parallel to $\delta'\tau - \delta\tau'$, namely

$$\mathcal{N}\left\{\delta'\tau - \delta\tau'\right\} \equiv n_z\hat{z}, \tag{A23}$$

with a unit vector $\hat{z}$ in z-direction. Since $\tau \times \tau'$ is orthogonal to $\delta'\tau - \delta\tau'$, we set

$$\mathcal{N}\left\{\tau \times \tau'\right\} \equiv n_x\hat{x} + n_y\hat{y}, \tag{A24}$$

with unit vectors in x, y-directions. Solving Equation (A23) for τ', we have

$$\tau' = \frac{\delta'}{\delta}\tau - \frac{n_z}{\delta\mathcal{N}}\hat{z}, \tag{A25}$$

and hence

$$\tau \times \tau' = -\frac{n_z}{\delta\mathcal{N}}\left(\tau \times \hat{z}\right). \tag{A26}$$

Using this, we obtain for $\tau = (\tau_x, \tau_y, \tau_z)$,

$$n_x = \hat{x} \cdot \mathcal{N}\left\{\tau \times \tau'\right\} = -\frac{n_z}{\delta}\tau_y. \tag{A27}$$

Similarly, we also have $n_y = \frac{n_z}{\delta}\tau_x$. Normalization condition requires

$$1 = \left(-\frac{n_z}{\delta}\tau_y\right)^2 + \left(\frac{n_z}{\delta}\tau_x\right)^2 + n_z^2 = \frac{1 - \tau_z^2}{\delta^2}n_z^2, \tag{A28}$$

hence we determine

$$n_z = \frac{\delta}{\sqrt{1 - \tau_z^2}}, \tag{A29}$$

and

$$\hat{n} = \frac{1}{\sqrt{1 - \tau_z^2}}\left(-\tau_y, \tau_x, \delta\right). \tag{A30}$$

Therefore,

$$\hat{u} = e^{-i\phi}\left\{\mathbb{I}\left(\delta\delta' + \tau \cdot \tau'\right) + i\frac{1}{\mathcal{N}}\hat{n} \cdot \hat{\sigma}\right\}$$
$$\equiv e^{-i\phi}\left\{\mathbb{I}\cos\omega + i\sin\omega\hat{n} \cdot \hat{\sigma}\right\}, \tag{A31}$$

where the real parameter ω is determined from $\cos\omega \equiv \delta\delta' + \tau \cdot \tau'$. The unitarity condition of $\hat{u}$ can be checked by noting

$$\begin{aligned}
\frac{1}{\mathcal{N}^2} &= |\delta'\tau - \delta\tau' + \tau \times \tau'|^2 \\
&= |\delta'\tau - \delta\tau'|^2 + (\tau \times \tau') \cdot (\tau \times \tau') \\
&= \delta^2 + \delta'^2 - 2\delta\delta'\tau \cdot \tau' + (1 - \delta^2)(1 - \delta'^2) - (\tau \cdot \tau')^2 \\
&= 1 - (\delta\delta' + \tau \cdot \tau')^2 = 1 - \cos^2\omega,
\end{aligned} \tag{A32}$$

where we used the relation $(\tau \times \tau') \cdot (\tau \times \tau') = |\tau|^2 |\tau'|^2 - (\tau \cdot \tau')^2$. Now, the operator $\hat{u} + \hat{u}^\dagger$ is calculated as

$$\hat{u} + \hat{u}^\dagger = 2\cos\phi\cos\omega\mathbb{I} + 2\sin\phi\sin\omega\hat{n}\cdot\hat{\sigma}, \tag{A33}$$

hence

$$
\begin{aligned}
\hat{W}\hat{W}^\dagger &= (\gamma_b^2 + \gamma_c^2)\mathbb{I} + 2\gamma_b\gamma_c(\cos\phi\cos\omega\mathbb{I} + \sin\phi\sin\omega\hat{n}\cdot\hat{\sigma}) \\
&\equiv A\mathbb{I} + \mathbf{B}\cdot\hat{\sigma}.
\end{aligned} \tag{A34}
$$

We defined

$$
\begin{aligned}
A &= \gamma_b^2 + \gamma_c^2 + 2\gamma_b\gamma_c\cos\phi\cos\omega, \tag{A35} \\
\mathbf{B} &= 2\gamma_b\gamma_c\sin\phi\sin\omega\hat{n}. \tag{A36}
\end{aligned}
$$

Alternatively, we evaluate

$$
\begin{aligned}
\hat{W}^\dagger\hat{W} &= (\gamma_b\hat{U}_b^\dagger + \gamma_c\hat{U}_c^\dagger)(\gamma_b\hat{U}_b + \gamma_c\hat{U}_c) \\
&= (\gamma_b^2 + \gamma_c^2)\mathbb{I} + \gamma_b\gamma_c(\hat{u}' + \hat{u}'^\dagger),
\end{aligned} \tag{A37}
$$

where $\hat{u}' \equiv \hat{U}_c^\dagger\hat{U}_b$, which represents the total development around the interferometer in the order $1 \to c \to 0 \to b \to 1$. With similar procedure done for $\hat{u}$, we have

$$
\begin{aligned}
\hat{u}' &= e^{-i\phi}\left\{\mathbb{I}\delta\delta' + i(\delta'\tau - \delta\tau')\cdot\hat{\sigma} + \mathbb{I}\tau'\cdot\tau + i(\tau'\times\tau)\cdot\hat{\sigma}\right\} \\
&= e^{-i\phi}\left\{\mathbb{I}\cos\omega + i\sin\omega\hat{n}'\cdot\hat{\sigma}\right\},
\end{aligned} \tag{A38}
$$

where

$$\hat{n}' = \frac{1}{\sqrt{1-\tau_z^2}}(\tau_y, -\tau_x, \delta), \tag{A39}$$

and with the same ω as before. Now, by calculating the factor $\hat{u}' + \hat{u}'^\dagger$, we obtain

$$
\begin{aligned}
\hat{W}^\dagger\hat{W} &= (\gamma_b^2 + \gamma_c^2)\mathbb{I} + 2\gamma_b\gamma_c(\cos\phi\cos\omega\mathbb{I} + \sin\phi\sin\omega\hat{n}'\cdot\hat{\sigma}) \\
&\equiv A\mathbb{I} + \mathbf{B}'\cdot\hat{\sigma},
\end{aligned} \tag{A40}
$$

where

$$\mathbf{B}' = 2\gamma_b\gamma_c\sin\phi\sin\omega\hat{n}'. \tag{A41}$$

As shown in the main text, we introduce two sets of normalized eigenstates of the operators $\hat{n}\cdot\hat{\sigma}$ and $\hat{n}'\cdot\hat{\sigma}$, $|\pm\hat{n}\rangle$ and $|\pm\hat{n}'\rangle$, which obey

$$
\begin{aligned}
\hat{W}\hat{W}^\dagger|\pm\hat{n}\rangle &= \lambda_\pm|\pm\hat{n}\rangle, \\
\hat{W}^\dagger\hat{W}|\pm\hat{n}'\rangle &= \lambda_\pm|\pm\hat{n}'\rangle,
\end{aligned} \tag{A42}
$$

with the same eigenvalues as Equation (19).

It can be shown that these two sets of eigenstates $\{|\pm\hat{n}\rangle, |\pm\hat{n}'\rangle\}$ are related with each other by

$$|\pm\hat{n}'\rangle = \frac{1}{\sqrt{\lambda_\pm}}\hat{W}^\dagger|\pm\hat{n}\rangle, \tag{A43}$$

with double-sign correspondence. This can be checked by the eigen-equation

$$
\begin{aligned}
\hat{W}^{\dagger}\hat{W}\,|\pm\hat{n}'\rangle &= \frac{1}{\sqrt{\lambda_{\pm}}}\hat{W}^{\dagger}(\hat{W}\hat{W}^{\dagger})\,|\pm\hat{n}\rangle \\
&= \frac{1}{\sqrt{\lambda_{\pm}}}\hat{W}^{\dagger}\lambda_{\pm}\,|\pm\hat{n}\rangle \\
&= \lambda_{\pm}\,|\pm\hat{n}'\rangle\,,
\end{aligned}
\tag{A44}
$$

and the normalization condition

$$
\begin{aligned}
\langle\pm\hat{n}'|\pm\hat{n}'\rangle &= \frac{1}{\lambda_{\pm}}\,\langle\pm\hat{n}|\hat{W}\hat{W}^{\dagger}|\pm\hat{n}\rangle \\
&= \langle\pm\hat{n}|\pm\hat{n}\rangle = 1.
\end{aligned}
\tag{A45}
$$

Similarly, we can also prove the relation

$$
|\pm\hat{n}\rangle = \frac{1}{\sqrt{\lambda_{\pm}}}\hat{W}\,|\pm\hat{n}'\rangle\,.
\tag{A46}
$$

We have the spectral decomposition of the matrices $\hat{W}^{\dagger}$ and $\hat{W}$ by

$$
\begin{aligned}
\hat{W}^{\dagger} &= \sqrt{\lambda_{+}}\,|\hat{n}'\rangle\,\langle\hat{n}| + \sqrt{\lambda_{-}}\,|-\hat{n}'\rangle\,\langle-\hat{n}|\,, \tag{A47} \\
\hat{W} &= \sqrt{\lambda_{+}}\,|\hat{n}\rangle\,\langle\hat{n}'| + \sqrt{\lambda_{-}}\,|-\hat{n}\rangle\,\langle-\hat{n}'|\,, \tag{A48}
\end{aligned}
$$

where the second relation is obtained by taking Hermite conjugate of the first relation.

Now, let us turn to discuss the scattering wavefunctions using these eigenstates. For the left incoming states, we choose $|\chi_{\text{in}}\rangle = |\pm\hat{n}\rangle$, then the transmitted states are

$$
\begin{aligned}
|\chi_{\text{t},\pm}\rangle &= \hat{t}\,|\pm\hat{n}\rangle \\
&= i\eta_{k}\hat{W}^{\dagger}\left[\mathbb{I}Y - \hat{W}\hat{W}^{\dagger}\right]^{-1}|\pm\hat{n}\rangle \\
&= i\eta_{k}\hat{W}^{\dagger}\left[\mathbb{I}Y - \mathbb{I}\lambda_{\pm}\right]^{-1}|\pm\hat{n}\rangle \\
&= \frac{i\eta_{k}}{Y - \lambda_{\pm}}\hat{W}^{\dagger}|\pm\hat{n}\rangle \\
&= \frac{i\eta_{k}}{Y - \lambda_{\pm}}\sqrt{\lambda_{\pm}}\,|\pm\hat{n}'\rangle \\
&= t_{\pm}\,|\pm\hat{n}'\rangle\,,
\end{aligned}
\tag{A49}
$$

where we used Equations (5) and (A47) and defined two transmission amplitudes,

$$
t_{\pm} \equiv \frac{i\eta_{k}\sqrt{\lambda_{\pm}}}{Y - \lambda_{\pm}}\,.
\tag{A50}
$$

Using the orthogonality of the eigenstates, the transmission amplitude matrix $\hat{t}$ is expressed as in Equation (24). Similarly, the reflected states are

$$
|\chi_{\text{r},\pm}\rangle = \hat{r}\,|\pm\hat{n}\rangle = r_{\pm}\,|\pm\hat{n}\rangle\,,
\tag{A51}
$$

where the reflection amplitudes are

$$
r_{\pm} \equiv -1 + \frac{i\eta_{k}X_{1}}{Y - \lambda_{\pm}}\,.
\tag{A52}
$$

The reflection amplitude matrix $\hat{r}$ is diagonal and is given in Equation (27). Similarly, for the right incoming states, the transmitted states are

$$|\chi'_{t,\pm}\rangle \;=\; \hat{t}'\,|\pm\hat{n}'\rangle = t_{\pm}\,|\pm\hat{n}\rangle\,, \tag{A53}$$

and hence the transmission amplitude matrix $\hat{t}'$ is given in Equation (25). The reflected states are

$$|\chi'_{r,\pm}\rangle \;=\; \hat{r}'\,|\pm\hat{n}'\rangle = r'_{\pm}\,|\pm\hat{n}'\rangle\,, \tag{A54}$$

where we defined

$$r'_{\pm} \;\equiv\; -1 + \frac{i\eta_k X_0}{Y - \lambda_{\pm}}\,, \tag{A55}$$

and hence the reflection amplitude matrix $\hat{r}'$ is given in Equation (28).

Appendix C. Unitarity of the Scattering Matrix

The scattering matrix needs to satisfy the unitarity condition, such that

$$\hat{t}^{\dagger}\hat{t} + \hat{r}^{\dagger}\hat{r} \;=\; \mathbb{I}. \tag{A56}$$

Using the results of Equations (24) and (27),

$$\hat{t}^{\dagger}\hat{t} + \hat{r}^{\dagger}\hat{r} \;=\; |t_+|^2\,|\hat{n}\rangle\,\langle\hat{n}| + |t_-|^2\,|-\hat{n}\rangle\,\langle -\hat{n}| + |r_+|^2\,|\hat{n}\rangle\,\langle\hat{n}| + |r_-|^2\,|-\hat{n}\rangle\,\langle -\hat{n}|\,.$$

Therefore, if $|t_{\pm}|^2 + |r_{\pm}|^2 = 1$, using the completeness relation of $|\pm\hat{n}\rangle$, the unitarity is confirmed. Let us check this:

$$
\begin{aligned}
|t_{\pm}|^2 + |r_{\pm}|^2 &= \left|\frac{i\eta_k\sqrt{\lambda_{\pm}}}{Y - \lambda_{\pm}}\right|^2 + \left|-1 + \frac{i\eta_k X_1}{Y - \lambda_{\pm}}\right|^2 \\
&= \frac{\eta_k^2\lambda_{\pm}}{|Y - \lambda_{\pm}|^2} + 1 + \frac{i\eta_k X_1^*}{Y^* - \lambda_{\pm}} - \frac{i\eta_k X_1}{Y - \lambda_{\pm}} + \frac{\eta_k^2\,|X_1|^2}{|Y - \lambda_{\pm}|^2} \\
&= 1 + i\eta_k\frac{X_1^*(Y - \lambda_{\pm}) - X_1(Y^* - \lambda_{\pm})}{|Y - \lambda_{\pm}|^2} + \frac{\eta_k^2(\lambda_{\pm} + |X_1|^2)}{|Y - \lambda_{\pm}|^2} \\
&= 1 + \frac{\eta_k}{|Y - \lambda_{\pm}|^2}\left[i\,\{X_1^*(Y - \lambda_{\pm}) - X_1(Y^* - \lambda_{\pm})\} + \eta_k(\lambda_{\pm} + |X_1|^2)\right].
\end{aligned}
$$

The factor in the square bracket is

$$
\begin{aligned}
[\bullet] &= -i(X_1^* - X_1)\lambda_{\pm} + i(X_0\,|X_1|^2 - X_0^*\,|X_1|^2) + \eta_k(\lambda_{\pm} + |X_1|^2) \\
&= [-i(-i\eta_k) + \eta_k]\lambda_{\pm} + [i(i\eta_k) + \eta_k]\,|X_1|^2 = 0,
\end{aligned}
$$

hence $|t_{\pm}|^2 + |r_{\pm}|^2 = 1$ is confirmed.

Appendix D. Derivatives of the Scattering Amplitude Matrices

In this Appendix, we evaluate the Berry curvature, Equation (33), with two control parameters, g_1, g_2, which only modify the scattering eigenvalues $\lambda_{\pm}$ and corresponding eigenvectors $|\pm\hat{n}\rangle, |\pm\hat{n}'\rangle$. We need to calculate the derivatives of the scattering amplitude matrices by a control parameter

$(g = g_1, g_2)$, $\frac{\partial \hat{r}}{\partial g}$ and $\frac{\partial \hat{t}'}{\partial g}$. Since scattering amplitude matrices are expressed with the eigenstates as shown in Equations (24) and (27), we first evaluate the first order derivative of the basis states

$$\frac{\partial}{\partial g}\begin{pmatrix} |\hat{n}\rangle \\ |-\hat{n}\rangle \end{pmatrix} = \begin{pmatrix} a_g & b_g \\ \tilde{b}_g & \tilde{a}_g \end{pmatrix}\begin{pmatrix} |\hat{n}\rangle \\ |-\hat{n}\rangle \end{pmatrix}, \tag{A57}$$

$$\frac{\partial}{\partial g}\begin{pmatrix} |\hat{n}'\rangle \\ |-\hat{n}'\rangle \end{pmatrix} = \begin{pmatrix} a_{g'} & b_{g'} \\ \tilde{b}_{g'} & \tilde{a}_{g'} \end{pmatrix}\begin{pmatrix} |\hat{n}'\rangle \\ |-\hat{n}'\rangle \end{pmatrix}. \tag{A58}$$

Since the basis states are normalized,

$$\frac{\partial}{\partial g}\langle \hat{n}|\hat{n}\rangle = \left(\frac{\partial \langle \hat{n}|}{\partial g}\right)|\hat{n}\rangle + \langle \hat{n}|\frac{\partial |\hat{n}\rangle}{\partial g} = a_g^* + a_g = 0, \tag{A59}$$

and a_g should be pure imaginary. Similarly, $\tilde{a}_g$, $a_{g'}$ and $\tilde{a}_{g'}$ are also pure imaginary. Using the orthogonality condition, we have the relation

$$\frac{\partial}{\partial g}\langle \hat{n}| - \hat{n}\rangle = \left(\frac{\partial \langle \hat{n}|}{\partial g}\right)|-\hat{n}\rangle + \langle \hat{n}|\frac{\partial |-\hat{n}\rangle}{\partial g} = b_g^* + \tilde{b}_g = 0, \tag{A60}$$

and $b_{g'}^* + \tilde{b}_{g'} = 0$.

To have the formula for a_g, b_g and $\tilde{b}_g$, we consider a unit vector $\hat{n} \equiv (n_x, n_y, n_z) = (\sin\Theta\cos\Phi, \sin\Theta\sin\Phi, \cos\Theta)$, with angles Θ and Φ. Since the operator $\hat{n}\cdot\hat{\sigma}$ in the matrix form,

$$\hat{n}\cdot\hat{\sigma} = \begin{pmatrix} \cos\Theta & \sin\Theta e^{-i\Phi} \\ \sin\Theta e^{i\Phi} & -\cos\Theta \end{pmatrix}, \tag{A61}$$

has two eigenvalues $\lambda = \pm 1$, the eigenvector in a form $(c_1, c_2)^t$ satisfies for $\lambda = 1$, $(\cos\Theta - 1)c_1 + (\sin\Theta e^{-i\Phi})c_2 = 0$ and normalization condition, and hence

$$|n\rangle = \left(\frac{n_x - in_y}{\sqrt{2(1-n_z)}}, \frac{\sqrt{1-n_z}}{\sqrt{2}}\right)^t, \tag{A62}$$

and for $\lambda = -1$, $(\cos\Theta + 1)c_1 + (\sin\Theta e^{-i\Phi})c_2 = 0$, hence

$$|-n\rangle = \left(\frac{-n_x + in_y}{\sqrt{2(1+n_z)}}, \frac{\sqrt{1+n_z}}{\sqrt{2}}\right)^t. \tag{A63}$$

Differentiation with g gives

$$\frac{\partial}{\partial g}|n\rangle = \left(\frac{\frac{\partial n_x}{\partial g} - i\frac{\partial n_y}{\partial g}}{\sqrt{2(1-n_z)}} + \frac{(n_x - in_y)\frac{\partial n_z}{\partial g}}{2\sqrt{2}(1-n_z)^{3/2}}, -\frac{\frac{\partial n_z}{\partial g}}{2\sqrt{2}\sqrt{1-n_z}}\right)^t, \tag{A64}$$

and

$$\frac{\partial}{\partial g}|-n\rangle = \left(\frac{-\frac{\partial n_x}{\partial g} + i\frac{\partial n_y}{\partial g}}{\sqrt{2(1-n_z)}} + \frac{(n_x - in_y)\frac{\partial n_z}{\partial g}}{2\sqrt{2}(1+n_z)^{3/2}}, \frac{\frac{\partial n_z}{\partial g}}{2\sqrt{2}\sqrt{1+n_z}}\right)^t. \tag{A65}$$

Therefore,

$$a_g \equiv \langle n | \frac{\partial}{\partial g} | n \rangle = \frac{i}{2(1 - n_z)} \left\{ n_y \frac{\partial n_x}{\partial g} - n_x \frac{\partial n_y}{\partial g} \right\}, \tag{A66}$$

$$\tilde{a}_g \equiv \langle -n | \frac{\partial}{\partial g} | - n \rangle = \frac{i}{2(1 + n_z)} \left\{ n_y \frac{\partial n_x}{\partial g} - n_x \frac{\partial n_y}{\partial g} \right\}, \tag{A67}$$

$$b_g \equiv \langle -n | \frac{\partial}{\partial g} | n \rangle$$

$$= -\frac{1}{2\sqrt{1 - n_z^2}} \left\{ i n_y \frac{\partial n_x}{\partial g} - i n_x \frac{\partial n_y}{\partial g} + \frac{\partial n_z}{\partial g} \right\}, \tag{A68}$$

$$\tilde{b}_g \equiv \langle n | \frac{\partial}{\partial g} | - n \rangle = -b_g^*$$

$$= \frac{1}{2\sqrt{1 - n_z^2}} \left\{ -i n_y \frac{\partial n_x}{\partial g} + i n_x \frac{\partial n_y}{\partial g} + \frac{\partial n_z}{\partial g} \right\}. \tag{A69}$$

We also evaluate $a_{g'}$, $\tilde{a}_{g'}$, $b_{g'}$ and $\tilde{b}_{g'}$ similarly. From Equations (A30) and (A39), we found that following relations hold: $a_{g'} = a_g$, $\tilde{a}_{g'} = \tilde{a}_g$ and $b_{g'} = b_g^*$ and $\tilde{b}_{g'} = -b_g$.

Using these relations, we evaluate the derivatives of the scattering amplitudes:

$$\frac{\partial \hat{r}}{\partial g} = \frac{\partial r_+}{\partial g} |\hat{n}\rangle \langle \hat{n}| + r_+ \frac{\partial |\hat{n}\rangle}{\partial g} \langle \hat{n}| + r_+ |\hat{n}\rangle \frac{\partial \langle \hat{n}|}{\partial g}$$

$$+ \frac{\partial r_-}{\partial g} |-\hat{n}\rangle \langle -\hat{n}| + r_- \frac{\partial |-\hat{n}\rangle}{\partial g} \langle -\hat{n}| + r_- |-\hat{n}\rangle \frac{\partial \langle -\hat{n}|}{\partial g}$$

$$\equiv (|\hat{n}\rangle, |-\hat{n}\rangle) \hat{r}_g \begin{pmatrix} \langle \hat{n}| \\ \langle -\hat{n}| \end{pmatrix}, \tag{A70}$$

with defining a matrix

$$\hat{r}_g = \begin{pmatrix} \frac{\partial r_+}{\partial g} & (r_+ - r_-)b_g^* \\ (r_+ - r_-)b_g & \frac{\partial r_-}{\partial g} \end{pmatrix}. \tag{A71}$$

Similarly,

$$\frac{\partial \hat{t}'}{\partial g} = \frac{\partial t_+}{\partial g} |\hat{n}\rangle \langle \hat{n}'| + t_+ \frac{\partial |\hat{n}\rangle}{\partial g} \langle \hat{n}'| + t_+ |\hat{n}\rangle \frac{\partial \langle \hat{n}'|}{\partial g}$$

$$+ \frac{\partial t_-}{\partial g} |-\hat{n}\rangle \langle -\hat{n}'| + t_- \frac{\partial |-\hat{n}\rangle}{\partial g} \langle -\hat{n}'| + t_- |-\hat{n}\rangle \frac{\partial \langle -\hat{n}'|}{\partial g}$$

$$\equiv (|\hat{n}\rangle, |-\hat{n}\rangle) \hat{t}'_g \begin{pmatrix} \langle \hat{n}'| \\ \langle -\hat{n}'| \end{pmatrix}, \tag{A72}$$

with defining a matrix

$$\hat{t}'_g = \begin{pmatrix} \frac{\partial t_+}{\partial g} & t_+ b_g - t_- b_g^* \\ t_+ b_g - t_- b_g^* & \frac{\partial t_-}{\partial g} \end{pmatrix}. \tag{A73}$$

Obviously, it is convenient to take the scattering eigenstates $|\pm \hat{n}\rangle$ to the spin axis $|\sigma = \pm 1\rangle$ in the Berry curvatures, which is written as

$$\Pi_{\hat{n}/-\hat{n}}(g_1, g_2) = \frac{1}{\pi} \Im \left\{ \left(\hat{r}_{g_2} \hat{r}_{g_1}^\dagger + \hat{t}'_{g_2} \hat{t}'^\dagger_{g_1} \right)_{(1,1)/(2,2)} \right\}. \tag{A74}$$

Then the diagonal components of $\hat{r}_{g_2}\hat{r}_{g_1}^\dagger$ is

$$\left(\hat{r}_{g_2}\hat{r}_{g_1}^\dagger\right)_{(1,1)} = \frac{\partial r_+}{\partial g_2}\frac{\partial r_+^*}{\partial g_1} + |r_+ - r_-|^2 b_{g_2}^* b_{g_1}, \tag{A75}$$

$$\left(\hat{r}_{g_2}\hat{r}_{g_1}^\dagger\right)_{(2,2)} = \frac{\partial r_-}{\partial g_2}\frac{\partial r_-^*}{\partial g_1} + |r_+ - r_-|^2 b_{g_2} b_{g_1}^*, \tag{A76}$$

and the diagonal components of $\hat{t}_{g_2}'\hat{t}_{g_1}'^\dagger$ is

$$\left(\hat{t}_{g_2}'\hat{t}_{g_1}'^\dagger\right)_{(1,1)} = \frac{\partial t_+}{\partial g_2}\frac{\partial t_+^*}{\partial g_1} + (t_+ b_{g_2} - t_- b_{g_2}^*)(t_+^* b_{g_1}^* - t_-^* b_{g_1}), \tag{A77}$$

$$\left(\hat{t}_{g_2}'\hat{t}_{g_1}'^\dagger\right)_{(2,2)} = \frac{\partial t_-}{\partial g_2}\frac{\partial t_-^*}{\partial g_1} + (t_+ b_{g_2} - t_- b_{g_2}^*)(t_+^* b_{g_1}^* - t_-^* b_{g_1}). \tag{A78}$$

As stated before, we restrict the type of control parameters, g, that only change the eigenvalues $\lambda_\pm$ in the transmission amplitudes, $r_\pm$ and $t_\pm$ and corresponding eigenstates, and we take

$$\frac{\partial r_\pm}{\partial g} = \frac{\partial \lambda_\pm}{\partial g}\frac{\partial r_\pm}{\partial \lambda_\pm}, \qquad \frac{\partial t_\pm}{\partial g} = \frac{\partial \lambda_\pm}{\partial g}\frac{\partial t_\pm}{\partial \lambda_\pm}. \tag{A79}$$

Then the factors in the Berry curvature, Equations (A75) and (A77),

$$\frac{\partial r_\pm}{\partial g_2}\frac{\partial r_\pm^*}{\partial g_1} = \frac{\partial \lambda_\pm}{\partial g_2}\frac{\partial \lambda_\pm}{\partial g_1}\left|\frac{\partial r_\pm}{\partial \lambda_\pm}\right|^2, \tag{A80}$$

$$\frac{\partial t_\pm}{\partial g_2}\frac{\partial t_\pm^*}{\partial g_1} = \frac{\partial \lambda_\pm}{\partial g_2}\frac{\partial \lambda_\pm}{\partial g_1}\left|\frac{\partial t_\pm}{\partial \lambda_\pm}\right|^2, \tag{A81}$$

are real and are not contributing to the pumping.

Then, the Berry curvature for the spin $|\hat{n}\rangle$ is

$$\Pi_{\hat{n}}(g_1, g_2) = \frac{1}{\pi}\Im\left\{|r_+ - r_-|^2 b_{g_2}^* b_{g_1} + (t_+ b_{g_2} - t_- b_{g_2}^*)(t_+^* b_{g_1}^* - t_-^* b_{g_1})\right\}$$

$$= \frac{1}{2\pi i}(|r_+ - r_-|^2 - |t_+|^2 + |t_-|^2)\left(b_{g_2}^* b_{g_1} - b_{g_2} b_{g_1}^*\right). \tag{A82}$$

Using Equation (A68) and the relation

$$\frac{\partial n_z}{\partial g} = -\frac{1}{n_z}\left(n_x\frac{\partial n_x}{\partial g} + n_y\frac{\partial n_y}{\partial g}\right), \tag{A83}$$

derived from $n_z = \sqrt{1 - n_x^2 - n_y^2}$, the factor in the last bracket of Equation (A82) is manipulated to

$$b_{g_2}^* b_{g_1} - b_{g_2} b_{g_1}^* = \frac{i}{2n_z}\left(\frac{\partial n_x}{\partial g_2}\frac{\partial n_y}{\partial g_1} - \frac{\partial n_x}{\partial g_1}\frac{\partial n_y}{\partial g_2}\right) \equiv 2\pi i\mathcal{C}_{g_1, g_2}. \tag{A84}$$

From Equation (A30), the derivatives of the elements of $\hat{n}$ by some control parameter g are calculated as

$$\frac{\partial n_x}{\partial g} = -\frac{1}{\sqrt{1 - \tau_z^2}}\left\{\frac{\partial \tau_y}{\partial g} + \frac{\tau_y \tau_z}{1 - \tau_z^2}\frac{\partial \tau_z}{\partial g}\right\}, \tag{A85}$$

$$\frac{\partial n_y}{\partial g} = \frac{1}{\sqrt{1 - \tau_z^2}}\left\{\frac{\partial \tau_x}{\partial g} + \frac{\tau_x \tau_z}{1 - \tau_z^2}\frac{\partial \tau_z}{\partial g}\right\}, \tag{A86}$$

therefore, the factor C_{g_1,g_2} in Equation (A84) is obtained as Equation (36) and the Berry curvatures are given as Equations (34) and (35).

Appendix E. Formulation of Diamond-Shape Interferometer

This section explains the foundation of the Schrödinger Equations (3) and (4). At the four sites in the interferometer, the Schrödinger equation is

$$(\epsilon - \epsilon_u) |\psi(u)\rangle = -\sum_v \tilde{U}_{uv} |\psi(v)\rangle. \tag{A87}$$

Explicitly, at sites $u = 0, 1, b, c$:

$$(\epsilon - \epsilon_0) |\psi(0)\rangle = -\left(\tilde{U}_{0b} |\psi(b)\rangle + \tilde{U}_{0c} |\psi(c)\rangle\right) - j |\psi(-1)\rangle, \tag{A88}$$

$$(\epsilon - \epsilon_1) |\psi(1)\rangle = -\left(\tilde{U}_{b1}^\dagger |\psi(b)\rangle + \tilde{U}_{c1}^\dagger |\psi(c)\rangle\right) - j |\psi(2)\rangle, \tag{A89}$$

$$(\epsilon - \epsilon_b) |\psi(b)\rangle = -\left(\tilde{U}_{0b}^\dagger |\psi(0)\rangle + \tilde{U}_{b1} |\psi(1)\rangle\right), \tag{A90}$$

$$(\epsilon - \epsilon_c) |\psi(c)\rangle = -\left(\tilde{U}_{0c}^\dagger |\psi(0)\rangle + \tilde{U}_{c1} |\psi(1)\rangle\right). \tag{A91}$$

Using Equations (A90) and (A91),

$$|\psi(b)\rangle = -\frac{1}{\epsilon - \epsilon_b} \left[\tilde{U}_{0b}^\dagger |\psi(0)\rangle + \tilde{U}_{b1} |\psi(1)\rangle\right], \tag{A92}$$

$$|\psi(c)\rangle = -\frac{1}{\epsilon - \epsilon_c} \left[\tilde{U}_{0c}^\dagger |\psi(0)\rangle + \tilde{U}_{c1} |\psi(1)\rangle\right]. \tag{A93}$$

By putting these into Equation (A88),

$$\begin{aligned}
(\epsilon - \epsilon_0) |\psi(0)\rangle &= -\tilde{U}_{0b} \left(-\frac{1}{\epsilon - \epsilon_b}\right) \left[\tilde{U}_{0b}^\dagger |\psi(0)\rangle + \tilde{U}_{b1} |\psi(1)\rangle\right] \\
&\quad -\tilde{U}_{0c} \left(-\frac{1}{\epsilon - \epsilon_c}\right) \left[\tilde{U}_{0c}^\dagger |\psi(0)\rangle + \tilde{U}_{c1} |\psi(1)\rangle\right] - j |\psi(-1)\rangle \\
&= \left(\frac{J_{0b}J_{b0}}{\epsilon - \epsilon_b} + \frac{J_{0c}J_{c0}}{\epsilon - \epsilon_c}\right) |\psi(0)\rangle + \left[\frac{\tilde{U}_{0b}\tilde{U}_{b1}}{\epsilon - \epsilon_b} + \frac{\tilde{U}_{0c}\tilde{U}_{c1}}{\epsilon - \epsilon_c}\right] |\psi(1)\rangle - j |\psi(-1)\rangle.
\end{aligned}$$

Then we define real variables

$$\gamma_{uvw} \equiv \frac{J_{uv}J_{uw}}{\epsilon - \epsilon_v}, \tag{A94}$$

and introducing a 2×2 matrix

$$\begin{aligned}
\hat{W} &\equiv \frac{\tilde{U}_{0b}\tilde{U}_{b1}}{\epsilon - \epsilon_b} + \frac{\tilde{U}_{0c}\tilde{U}_{c1}}{\epsilon - \epsilon_c} \\
&= \gamma_{0b1}\hat{U}_{0b}\hat{U}_{b1} + \gamma_{0c1}\hat{U}_{0c}\hat{U}_{c1}, \tag{A95}
\end{aligned}$$

we obtain the relation equivalent to Equation (3)

$$(\epsilon - y_0) |\psi(0)\rangle = \hat{W} |\psi(1)\rangle - j |\psi(-1)\rangle, \tag{A96}$$

where we defined renormalized site energy at $u = 0$, $y_0 \equiv \epsilon_0 + \gamma_{0b0} + \gamma_{0c0}$. By putting Equation (A92) in Equation (A89),

$$
\begin{aligned}
(\epsilon - \epsilon_1) \, |\psi(1)\rangle \;&=\; \tilde{U}_{b1}^\dagger \left(\frac{1}{\epsilon - \epsilon_b} \right) \left[\tilde{U}_{0b}^\dagger \, |\psi(0)\rangle + \tilde{U}_{b1} \, |\psi(1)\rangle \right] \\
&\quad + \tilde{U}_{c1}^\dagger \left(\frac{1}{\epsilon - \epsilon_c} \right) \left[\tilde{U}_{0c}^\dagger \, |\psi(0)\rangle + \tilde{U}_{c1} \, |\psi(1)\rangle \right] - j \, |\psi(2)\rangle \\
&=\; \left(\frac{J_{1b} J_{b0}}{\epsilon - \epsilon_b} \hat{U}_{b1}^\dagger \hat{U}_{0b}^\dagger + \frac{J_{1c} J_{c0}}{\epsilon - \epsilon_c} \hat{U}_{c1}^\dagger \hat{U}_{0c}^\dagger \right) |\psi(0)\rangle + \left(\frac{J_{1b} J_{b1}}{\epsilon - \epsilon_b} + \frac{J_{1c} J_{c1}}{\epsilon - \epsilon_c} \right) |\psi(1)\rangle - j \, |\psi(2)\rangle .
\end{aligned}
$$

Hence, we have the equation equivalent to Equation (4)

$$
(\epsilon - y_1) \, |\psi(1)\rangle \;=\; \hat{W}^\dagger \, |\psi(0)\rangle - j \, |\psi(2)\rangle , \tag{A97}
$$

where we introduced renormalized site energy at $u = 1$, $y_1 \equiv \epsilon_1 + \gamma_{1b1} + \gamma_{1c1}$. We defined $\gamma_b \equiv \gamma_{0b1}$, $\gamma_c \equiv \gamma_{0c1}$ and $\hat{U}_b \equiv \hat{U}_{0b} \hat{U}_{b1}$, $\hat{U}_c \equiv \hat{U}_{0c} \hat{U}_{c1}$.

References

1. Bercioux, D.; Lucignano, P. Quantum transport in Rashba spin-orbit materials: A review. *Rep. Prog. Phys.* **2015**, *78*, 106001.
2. Nitta, J.; Akazaki, T.; Takayanagi, H.; Enoki, T. Gate Control of Spin-Orbit Interaction in an Inverted $In_{0.53}Ga_{0.47}As/In_{0.52}Al_{0.48}As$ Heterostructure. *Phys. Rev. Lett* **1997**, *78*, 1335.
3. Aharony, A.; Tokura, Y.; Cohen, G.Z.; Entin-Wohlman, O.; Katsumoto, S. Filtering and analyzing mobile qubit information via Rashba-Dresselhaus-Aharonov-Bohm interferometers. *Phys. Rev. B* **2011**, *84*, 035323.
4. Matityahu, S.; Aharony, A.; Entin-Wohlman, O.; Tarucha, S. Spin filtering in a Rashba-Dresselhaus-Aharonov-Bohm double-dot interferometer. *New J. Phys.* **2013**, *15*, 125017.
5. Kondo, K. Spin filter effects in an Aharonov-Bohm ring with double quantum dots under general Rashba spin-orbit interactions. *New J. Phys.* **2016**, *18*, 013002.
6. Tu, M.W.-Y.; Aharony, A.; Zhang, W.-M.; Entin-Wohlman, O. Real-time dynamics of spin-dependent transport through a double-quantum-dot Aharonov-Bohm interferometer with spin-orbit interaction. *Phys. Rev. B* **2014**, *90*, 165422.
7. Watson, S.K.; Potok, R.M.; Marcus, C.M.; Umansky, V. Experimental Realization of a Quantum Spin Pump. *Phys. Rev. Lett.* **2003**, *91*, 258301.
8. Sharma, P; Brouwer, P.W. Mesoscopic Effects in Adiabatic Spin Pumping. *Phys. Rev. Lett.* **2003**, *91*, 166801.
9. Averin, D.V.; Möttönen, M.; Pekola, J.P. Maxwell's demon based on a single-electron pump. *Phys. Rev. B* **2011**, *84*, 245448.
10. Zhang, P.; Xue, Q.-K.; Xie, X.C. Spin Current through a Quantum Dot in the Presence of an Oscillating Magnetic Field. *Phys. Rev. Lett.* **2003**, *91*, 196602.
11. Aono, T. Adiabatic spin pumping through a quantum dot with a single orbital level. *Phys. Rev. B* **2003**, *67*, 155303.
12. Nakajima, S.; Taguchi, M.; Kubo T.; Tokura, Y. Interaction effect on adiabatic pump of charge and spin in quantum dot. *Phys. Rev. B* **2015**, *92*, 195420.
13. Hashimoto, K.; Tatara, G.; Uchiyama, C. Nonadiabaticity in spin pumping under relaxation. *Phys. Rev. B* **2017**, *96*, 064439.
14. Tserkovnyak, Y.; Brataas, A.; Bauer, G.E.W. Spin pumping and magnetization dynamics in metallic multilayers. *Phys. Rev. B* **2002**, *66*, 224403.
15. Winkler, N.; Governale, M.; König, J. Theory of spin pumping through an interacting quantum dot tunnel coupled to a ferromagnet with time-dependent magnetization. *Phys. Rev. B* **2013**, *87*, 155428.
16. Chen, K.; Zhang, S. Spin Pumping in the Presence of Spin-Orbit Coupling. *Phys. Rev. Lett.* **2015**, *114*, 126602.
17. Tatara, G.; Mizukami, S. Consistent microscopic analysis of spin pumping effects. *Phys. Rev. B* **2017**, *96*, 064423.

18. Büttiker, M.; Thomas, H.; Prêtre, A. Current partition in multiprobe conductors in the presence of slowly oscillating external potentials. *Z. Phys. B* **1994**, *94*, 133.

19. Brouwer, P.W. Scattering approach to parametric pumping. *Phys. Rev. B* **1998**, *58*, R10135.

20. Levinson, Y.; Entin-Wohlman, O.; Wölfle, P. Pumping at resonant transmission and transferred charge quantization. *Phys. A* **2001**, *302*, 335.

21. Hiltscher, B.; Governale, M.; König, J. Interference and interaction effects in adiabatic pumping through quantum dots. *Phys. Rev. B* **2010**, *81*, 085302.

22. Sau, J.D.; Kitagawa, T.; Halperin, B. Conductance beyond the Landauer limit and charge pumping in quantum wires. *Phys. Rev. B* **2012**, *85*, 155425.

23. Calvo, H.L.; Classen, L.; Splettstoesser, J.; Wegewijs, M.R. Interaction-induced charge and spin pumping through a quantum dot at finite bias. *Phys. Rev. B* **2012**, *86*, 245308.

24. Pluecker, T.; Wegewijs, M.R.; Splettstoesser, J. Gauge freedom in observables and Landsberg's nonadiabatic geometric phase: Pumping spectroscopy of interacting open quantum systems. *Phys. Rev. B* **2017**, *95*, 155431.

25. Nakajima, S.; Tokura, Y. Excess Entropy Production in Quantum System: Quantum Master Equation Approach. *J. Stat. Phys.* **2017**, *169*, 902.

26. Taguchi, M.; Nakajima, S.; Kubo, T.; Tokura, Y. Quantum Adiabatic Pumping by Modulating Tunnel Phase in Quantum Dots. *J. Phys. Soc. Jpn.* **2016**, *85*, 084704.

27. Oreg, Y.; Entin-Wohlman, O. Transmissions through low-dimensional mesoscopic systems subject to spin-orbit scattering. *Phys. Rev. B* **1992**, *46*, 2393.

Article

Nonadiabaticity in Quantum Pumping Phenomena under Relaxation

Kazunari Hashimoto [1,*,†] and **Chikako Uchiyama** [1,2,*,†]

1   Faculty of Engineering, University of Yamanashi, 4-3-11 Takeda, Kofu 400-8511, Japan
2   National Institute of Informatics, 2-1-2 Hitotsubashi, Chiyoda-ku, Tokyo 101-8430, Japan
*   Correspondence: hashimotok@yamanashi.ac.jp (K.H.); hchikako@yamanashi.ac.jp (C.U.)
†   These authors contributed equally to this work.

Received: 8 July 2019; Accepted: 23 August 2019; Published: 28 August 2019

Abstract: The ability to control quanta shown by quantum pumping has been intensively studied, aiming to further develop nano fabrication. In accordance with the fast progress of the experimental techniques, the focus on quantum pumping extends to include the quicker transport. For this purpose, it is necessary to remove the "adiabatic" or "slow" condition, which has been the central concept of quantum pumping since its first proposal for a closed system. In this article, we review the studies which go beyond the conventional adiabatic approximation for open quantum systems to transfer energy quanta and electron spins with using the full counting statistics. We also discuss the recent developments of the nonadiabatic treatments of quantum pumping.

Keywords: nonadiabaticity; quantum heat pumping; spin pumping; relaxation

1. Introduction

According to the rapid development of experimental techniques, the downsizing of devices has been accelerated to extend possibilities to control single-electron current [1], spin-polarized current [2,3] and even thermal transport [4]. The trend is based on the aim to construct electronics with low energy consumption, quantum information processing, as well as quantum metrology [1,5].

Quantum pumping phenomena have attracted intensive attention, since they show the controllability of quantum transfer to extend the possibility of nano fabrications. The first proposal of quantum pumping was given by Thouless to transport electrons between two environments [6,7]. Its essential point is to adiabatically or slowly modulate the potential, which is described with the superposition of two standing waves in an out-of phase way [6–8]. Since the work of Thouless, the "adiabatic" change or "slow" modulation of parameters has played a central role in theoretical treatments of quantum pumping phenomena. However, the fast development of experimental techniques after the first experimental study on electron pumping with a quantum dot [9] requires us to investigate conditions on transferring quanta quicker and more precisely. In the present review, we classify the meaning of "quick" or "slow" in quantum pumping and show a standardized theoretical treatment—called full counting statistics (FCS)—to attain the purpose.

The physical situations referred to by the same term "adiabaticity" are roughly divided into three categories: (1) slow change in potential to allow the application of the adiabatic approximations to wave functions associated with transported particles [6,7]; (2) slow and small change of parameter(s) such as the chemical potential and voltage to allow a linear expansion of the scattering matrix [10–17] or Green functions [18–26] associated with transported electron charge or spin; and (3) slow change of parameters compared with the relaxation time of the relevant system with using FCS [27–32]. Different from the former two treatments, the third succeeds in including explicitly the finite relaxation time in adiabaticity. Because isolating any quantum system from its surroundings is impossible, considering relaxation phenomena is indispensable in implementing quantum pumping systems. With further developments of experimental techniques in mind, removing the adiabatic condition in this third instance is necessary. Before going further, let us provide a quick review of the conventional studies on pumping phenomena from the view point of the above summarized classifications of the adiabaticity.

Sinitsyn and Nemenman treated the classical two-state stochastic system [27] using FCS to represent the pumped quantity with a geometrical phase, first expounded in Reference [33]. The relationship between finite relaxation time and adiabaticity in the context of open quantum systems was first discussed by Ren, Hänggi, and Li [28] by extending the FCS approach in Reference [27]. Considering a two-level system coupled to two environments, they found that the pumping of energy quanta occurs under out-of-phase conditions and sufficiently slow (adiabatic) temperature modulations of the environments even if the bias is averaged out during a period. The power of the FCS approach can be found in the further application to electron charge pumping through single or double quantum dot(s) coupled to two leads [29–31]. They found that the sufficiently slow and out-of-phase modulations of the chemical potential can induce electron charge pumping, which is represented by a geometrical formula. In these instances, the condition for a sufficiently slow (adiabatic) environmental modulation means that the relevant two-level system approaches steady state sufficiently quickly.

In many varieties of quantum pumping phenomena, the generation of spin polarized electron current (spin current) by periodic parameter modulation has attracted a great deal of attention because of its promising applications in spintronics. Referred to as spin pumping, much effort has been made to develop its protocols. Conventionally, the protocols fall roughly into three classes: those using (i) a precession of magnetization in a magnetic material attached to a normal metal [12,13,19–26,34–36]; (ii) a periodic modulation of parameters such as gate voltages and/or tunneling amplitudes in a system consisting of quantum dots subjected to a magnetic field and normal metal leads [14,32,37–40]; and (iii) a periodic modulation of strength of magnetization in addition to parameters in a system consisting of quantum dots attached to a magnetic lead and/or normal metal leads [34,35,41,42]. Among these protocols, those using precession of magnetization—protocol (i)—have attracted intensive studies because it can generate pure spin current in a simple ferromagnet/normal metal heterojunction [3]. So far, the protocol has been mostly studied in situations where the precession of the magnetization is sufficiently slow, which is called adiabatic pumping. It was first proposed by Tserkovnyak et al. [12,13] based on the scattering theory of adiabatic quantum pumping given by Brouwer [11]. Its alternative formalisms based on Green's function [19,20,22–26,36] have also been proposed by several authors. In these studies, the adiabatic contribution to the spin current generation has been obtained as a linear response to the precession, which corresponds to adiabaticity No. (2), with an implicit assumption of an infinite relaxation time. There are a few studies addressing adiabatic spin pumping with a finite relaxation time [34,35], where a slow modulation means smallness of the precession frequency comparing with the tunneling rate.

In the present article, we intend to review our recent studies on the role of nonadiabaticity with a finite relaxation time in quantum pumping of energy quanta and electron spins. For the purpose, we rely on adiabaticity condition No. (3), where the adiabaticity means a slow modulation of parameters compared to the relaxation time of the relevant system—defining the relaxation time of the relevant system as τ_r, we find that the condition for a slow modulation requires τ_r to be much shorter than the period of the temperature modulation, which is written equivalently in terms of the modulation frequency Ω, $\tau_r^{-1} \gg \Omega$. Thus, we consider the nonadiabatic regime up to $\tau_r^{-1} \gtrsim \Omega$ in the following. As a formulation of nonadiabatic pumping, we present our extension of the FCS approach to quantum pumping toward the nonadiabatic regime. By applying the formulation to the pumping phenomena of energy quanta and electron spin, we find the following features: For the former, we demonstrate that nonadiabaticity yields a contribution to the pumped quantity in addition to the terms such as dynamical and geometrical phase terms which were obtained under adiabatic conditions. For the latter, surprisingly, we show that there are no contributions under the adiabatic condition and nonadiabaticity is an essential feature.

In the rest of the paper, we present in Section 2 our formulation describing the pumped quantity based on FCS. We discuss quantum heat pumping in Section 3 and spin pumping in Section 4, followed by a discussion and conclusion in Section 5.

2. General Formalism

We formulate a model of quantum pumping under periodic modulation of a parameter applying the FCS based on two-point projective measurements [43,44]. Let us consider a system consisting of a relevant system (S) and an environment (E) described by the Hamiltonian

$$H = H_0 + H_{SE}, \tag{1}$$

where $H_0 \equiv H_S + H_E$ and H_{SE} is the system–environment interaction. The FCS provides the statistical average of the net amount of a physical quantity, such as energy and particle number, exchanged between the system and the environment during a certain time interval. It is based on a joint probability of outcomes of two successive projective measurements of an observable of the environment Q corresponding to the exchanged quantity. The measurement scheme is—at $t = t_i$, we perform a projective measurement of Q to obtain a measurement outcome q_{t_i}; for $t_i \leq t \leq t_{i+1}$, the system undergoes a unitary time evolution through an interaction between the system and the environment; and at $t = t_{i+1}$, we perform a second projective measurement of Q to obtain another outcome $q_{t_{i+1}}$. The joint probability for the measurement scheme is given by

$$P[q_{t_{i+1}}, q_{t_i}] \equiv \mathrm{Tr}[P_{q_{t_{i+1}}} U(t_{i+1}, t_i) P_{q_{t_i}} W(t_i) P_{q_{t_i}} U^\dagger(t_{i+1}, t_i) P_{q_{t_{i+1}}}], \tag{2}$$

where Tr denotes the trace taken over the total system, $P_{q_t} \equiv |q_t\rangle\langle q_t|$ the projective measurement of Q at t, $U(t, t_i)$ the unitary time evolution operator for the total system, and $W(t_i)$ the initial condition of the total system (see Note [45]). The net amount of exchanged quantity during the time interval $\delta t \equiv t_{i+1} - t_i$ is then given by $\Delta q_i \equiv q_{t_{i+1}} - q_{t_i}$, where its sign is chosen to be positive when the physical quantity is transferred from the system to the environment. The statistics of Δq_i is contained in its probability distribution function

$$P(\Delta q_i) \equiv \sum_{q_{t_{i+1}}, q_{t_i}} \delta(\Delta q_i - (q_{t_{i+1}} - q_{t_i})) P[q_{t_{i+1}}, q_{t_i}], \tag{3}$$

The moments of Δq_i are provided by the moment generating function,

$$Z(\lambda) \equiv \int_{-\infty}^{\infty} P(\Delta q_i) e^{i\lambda \Delta q_i} d\Delta q_i. \tag{4}$$

where λ is the counting field associated with Q, for example, the mean value is computed from

$$\langle \Delta q_i \rangle = \left. \frac{\partial Z(\lambda)}{\partial(i\lambda)} \right|_{\lambda=0}. \tag{5}$$

Our next task is to describe the time evolution of $Z(\lambda)$. Using the Definition (3) and introducing the modified evolution operator $U_\lambda(t, t_i) \equiv e^{i\lambda Q} U(t, t_i) e^{-i\lambda Q}$, the moment generating function $Z(\lambda)$ is expressed as

$$Z(\lambda) = \mathrm{Tr}[W^{(\lambda)}(t_{i+1})], \tag{6}$$

with

$$W^{(\lambda)}(t) \equiv U_{\lambda/2}(t, t_i) \bar{W}(t_i) U^\dagger_{-\lambda/2}(t, t_i), \tag{7}$$

where $\bar{W}(t_i) \equiv \sum_{q_{t_i}} P_{q_{t_i}} W(t_i) P_{q_{t_i}}$ is the diagonal part of $W(t_i)$. For $\lambda = 0$, $W^{(\lambda=0)}(t)$ reduces to the usual reduced density matrix of the total system. By taking the time derivative of $W^{(\lambda)}(t)$, we obtain a modified Liouville–von Neumann equation

$$i\frac{\partial}{\partial t}W^{(\lambda)}(t) = \mathcal{L}^{(\lambda)}W^{(\lambda)}(t), \tag{8}$$

with a modified Liouvillian $\mathcal{L}^{(\lambda)}W^{(\lambda)}(t) \equiv \hbar^{-1}[H, W^{(\lambda)}(t)]_\lambda$, where $[A, B]_\lambda \equiv A^{(\lambda)}B - BA^{(-\lambda)}$, and $A^{(\lambda)} \equiv e^{i\lambda Q/2}Ae^{-i\lambda Q/2}$. In Appendix B, we explain the connection between the formalism of the FCS based on two-point measurements and the formalism by Sinitsyn and Nemenman in Reference [27].

By introducing a projection operator $\mathcal{P} : W^{(\lambda)}(t) \mapsto \mathrm{Tr_E}[W^{(\lambda)}(t)] \otimes \rho_E$, where $\mathrm{Tr_E}$ is the partial trace taken over the environment and ρ_E is a fixed state of the environment, the equation of motion for the reduced operator of the relevant system $\rho^{(\lambda)}(t) \equiv \mathrm{Tr_E}[W^{(\lambda)}(t)]$ can be cast into the form of a time convolutionless (TCL)-type quantum master equation [46–53].

Assuming that the initial state is factorized between system and environment as $W(t_i) = \rho(t_i) \otimes \rho_E$ and the fixed state of the environment ρ_E is the Gibbs state with an inverse temperature β, the TCL master equation including the counting field is expressed as

$$\frac{\partial}{\partial t}\rho^{(\lambda)}(t) = \xi^{(\lambda)}(t)\rho^{(\lambda)}(t). \tag{9}$$

The super-operator $\xi^{(\lambda)}(t)$ is expanded as a sum of "ordered cumulants" of the interaction Hamiltonian H_{SE} up to infinite order. Taking leading terms up to second-order, we have

$$\xi^{(\lambda)}(t)\rho^{(\lambda)}(t) = -\frac{i}{\hbar}[H_S, \rho^{(\lambda)}(t)] - \frac{1}{\hbar^2}\int_0^t d\tau \mathrm{Tr_E}[H_{SE}, [H_{SE}(-\tau), \rho^{(\lambda)}(t) \otimes \rho_E^{eq}]_\lambda]_\lambda. \tag{10}$$

Note that the time dependence of the memory kernel reflects the finiteness of the correlation time of the dot–lead interaction, which allows us to describe the non-Markovian dynamics.

To work with the super-operator, it is convenient to introduce its supermatrix representation, where we represent the density matrix $\rho^{(\lambda)}$ in vector form and the super-operator $\xi^{(\lambda)}$ in matrix form. In this representation, the formal solution of the master equation Equation (9) is expressed as

$$|\rho^{(\lambda)}(t)\rangle\rangle = \mathrm{T}_+ \exp\left[\int_{t_i}^t \Xi^{(\lambda)}(s)ds\right]|\rho^{(\lambda)}(t_i)\rangle\rangle, \tag{11}$$

where $|\rho^{(\lambda)}(t)\rangle\rangle$ represents the vector form of $\rho^{(\lambda)}(t)$, $\mathrm{T}_+ \exp$ the time-ordered exponential, and $\Xi^{(\lambda)}(t)$ the supermatrix form of $\xi^{(\lambda)}(s)$. With the representation, the moment generating function Equation (6) is rewritten as $Z(\lambda) = \mathrm{Tr_S}[\rho^{(\lambda)}(t_{i+1})] = \langle\langle 1|\rho^{(\lambda)}(t_{i+1})\rangle\rangle$, where $\mathrm{Tr_S}$ is the partial trace taken over the relevant system and $\langle\langle 1|$ the vector representation of the partial trace $\mathrm{Tr_S}$. Using the formal solution Equation (11), we recast the expression of the mean value into the form

$$\langle \Delta q_i \rangle = \int_{t_i}^{t_{i+1}} J(s)ds \tag{12}$$

with the inertial flow of the quantity,

$$J(t) \equiv \langle\langle 1|\left[\frac{\partial \Xi^{(\lambda)}(t)}{\partial(i\lambda)}\right]_{\lambda=0}|\rho^{(\lambda=0)}(t)\rangle\rangle. \tag{13}$$

To formulate quantum pumping based on the above framework, we need to accumulate transfers of the physical quantity under a cyclic modulation of system and/or environmental parameters during a period $\mathcal{T}$. For this purpose, we consider a step-like change of the parameters; specifically, dividing the period $\mathcal{T}$ into N intervals, $t_i \le t \le t_{i+1}$ ($i = 1, 2, \cdots, N$) with $t_1 = 0$ and $t_{N+1} = \mathcal{T}$, fixing a value of the parameters during each interval, and changing the value at each t_i discretely. With the total density matrix factorized at each t_i, the mean value as well as the inertial flow of the quantity for each interval are given by Equations (12) and (13), respectively. The time integration of $J(t)$ over one period $\mathcal{T}$ provides the accumulated value of the quantity over one cycle

$$\langle \Delta q \rangle \equiv \int_0^{\mathcal{T}} J(t)dt = \sum_{i=1}^N \langle \Delta q_i \rangle, \tag{14}$$

where $\langle \Delta q_i \rangle$ is the mean value of the net transferred quantity in the ith interval.

3. Pumping of Energy Quanta

3.1. Model

Let us consider the energy transfer via a two-level system between two environmental systems (L and R) consisting of an infinite number of bosons [28,54–56]. With the definition of the lower (higher) level of the two-level system as $|0\rangle$ ($|1\rangle$), respectively, the Hamiltonian of the model is written as

$$\begin{aligned} H_S &= \sum_{m=0,1} \varepsilon_m |m\rangle \langle m|, \quad H_E = \sum_{k,\nu} \hbar \omega_{k,\nu} b^\dagger_{k,\nu} b_{k,\nu}, \\ H_{SE} &= \sum_\nu X_\nu (|0\rangle \langle 1| + |1\rangle \langle 0|), \qquad (\nu = L, R), \end{aligned} \tag{15}$$

where $X_\nu = \sum_k \hbar g_{k,\nu} (b^\dagger_{k,\nu} + b_{k,\nu})$, with $b^\dagger_{k,\nu}$ and $b_{k,\nu}$ the creation and annihilation boson operators of the kth mode of the νth environment. The model scheme is shown in Figure 1A.

We consider the energy transfer from out-of-phase temperature modulations of the two environments, corresponding to the bias averaged out during a period $\mathcal{T}$ [28]. To discuss the nonadiabaticity for this model, we study energy (boson) transfers under cyclic and piecewise modulations of the environmental temperatures T_L and T_R dividing $\mathcal{T}$ into N intervals, $t_i \leq t \leq t_{i+1}$ ($i = 1, \cdots, N$) with $t_1 = 0$ and $t_{N+1} = \mathcal{T}$. We need to discretize the temperature modulation because conventional treatments describing relaxation phenomena require the environmental temperature to remain constant. By changing the number of intervals of the temperature modulation, we compared each time interval with the relaxation time of the relevant two-level system and thus we are able to discuss nonadiabaticity explicitly, for example, from the scale between τ_r^{-1} and Ω. In taking the limit $N \to \infty$, we reveal energy transfer features under a continuous modulation. In Figure 1B, we plot the time dependence of the temperature modulations used in this study calculated for a typical number of discrete time intervals $N = 20$.

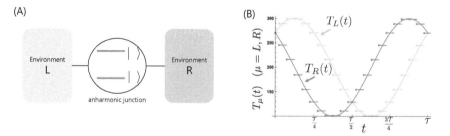

Figure 1. (**A**) Model scheme: a two-level system as an anharmonic junction interacts with two environments (L and R). (**B**) Temperature modulations, $T_L(t) = 200 + 100 \cos(\omega t + \pi/4)$, and $T_R(t) = 200 + 100 \sin(\omega t + \pi/4)$, discretized with $N = 20$.

3.2. FCS Formalism Applied to Pumping of Energy Quanta

We apply the general formalism of the FCS in the former section to this model focusing on weak system–environment coupling and considering long time (Born-Markovian) limits by taking $t \to \infty$ in the upper bound of the integral in Equation (10). In this limit, the super-operator $\xi^{(\lambda)}(t)$ becomes time independent during each interval. We find that the mean value of the transferred quantity between the relevant two-level system and the νth environment in the ith time interval, $\langle \Delta q_i^\nu \rangle$ is written as

$$\langle \Delta q_i^\nu \rangle = \hbar \omega_0 \{ A_i^\nu \int_{t_i}^{t_{i+1}} dt' \rho_{00}(t') - B_i^\nu \delta t \}, \tag{16}$$

with $\omega_0 = (\varepsilon_1 - \varepsilon_0)/\hbar$, $\rho_{00}(t) = \langle 0 | \rho^{(\lambda=0)}(t) | 0 \rangle$, and $\delta t = t_{i+1} - t_i$. A_i^ν and B_i^ν in Equation (16) are coefficients defined as $A_i^\nu = -(k_{d,i}^\nu + k_{u,i}^\nu)$ and $B_i^\nu = -k_{u,i}^\nu$ with rate constants $k_{u,i}^\nu$ and $k_{d,i}^\nu$, which govern the time evolution of $\rho_{00}(t)$ during $t_{i-1} \le t < t_i$. Their explicit expressions are $k_{d,i}^\nu = \Gamma_\nu n_{\nu,i}$ and $k_{u,i}^\nu = \Gamma_\nu (1 + n_{\nu,i})$, where $n_{\nu,i} = 1/(\exp[\hbar \beta_i^\nu \omega_0] - 1)$ is the Bose–Einstein distribution for the inverse temperature β_i^ν of the νth environment during the ith interval and Γ_ν denotes the feature of the system–environment coupling as $\Gamma_\nu = 2\pi h_\nu(\omega_0)$ with the coupling spectral density $h_\nu(\omega) \equiv \sum_k g_{k,\nu}^2 \delta(\omega - \omega_{k,\nu}) = \lambda \omega \exp[-\omega/\omega_c]$, where λ is the coupling strength and ω_c is the cutoff frequency. To obtain $\langle \Delta q_i^\nu \rangle$, we need $\rho_{00}(t)$, the time evolution for which is

$$\dot{\rho}_{00}(t) = -K_{d,i} \rho_{00}(t) + K_{u,i} \rho_{11}(t)$$

with $K_{d,i} = \sum_\nu k_{d,i}^\nu$ and $K_{u,i} = \sum_\nu k_{u,i}^\nu$. The differential equation for $\rho_{00}(t)$ is solved to give

$$\rho_{00}(t) = \rho_{s,i} + e^{\Lambda_i t}(\rho_{00}(t_{i-1}) - \rho_{s,i}), \tag{17}$$

where we denote $\rho_{s,i} = -K_{u,i}/\Lambda_i$ with $\Lambda_i = -(K_{d,i} + K_{u,i})$. Using Equations (12), we find that the total transferred energy during the period is calculated to be

$$\langle \Delta q^\nu \rangle = \sum_{i=1}^{N+1} \langle \Delta q_i^\nu \rangle = \hbar \omega_0 (\mathcal{G}_1^\nu + \mathcal{G}_2^\nu + \mathcal{G}_3^\nu), \tag{18}$$

where

$$\mathcal{G}_1^\nu = \sum_{i=1}^{N+1} (A_i^\nu \rho_{s,i} - B_i^\nu) \delta t, \tag{19}$$

$$\mathcal{G}_2^\nu = \sum_{i=1}^{N} \frac{A_{i+1}^\nu}{\Lambda_{i+1}} (\rho_{s,i+1} - \rho_{s,i}), \tag{20}$$

$$\mathcal{G}_3^\nu = \sum_{i=1}^{N+1} \phi_{0,i}^\nu + \sum_{i=2}^{N} (\rho_{s,i-1} - \rho_{s,i}) \psi_i^\nu + (\rho_{s,n-1} - \rho_{s,n}) \frac{A_n^\nu}{\Lambda_n} e^{\Lambda_n \delta t}, \tag{21}$$

with

$$\phi_{0,i}^\nu = (\rho_{00}(0) - \rho_{s,1}) f^\nu(1,i), \tag{22}$$

$$\psi_i^\nu = \frac{A_i^\nu}{\Lambda_i} e^{\Lambda_i \delta t} + \sum_{m=i}^{N} f^\nu(i, m+1), \tag{23}$$

$$f^\nu(p,q) = \frac{A_q^\nu}{\Lambda_q} e^{\sum_{x=p}^{q-1} \Lambda_x \delta t} (e^{\Lambda_q \delta t} - 1). \tag{24}$$

In the next subsection, we show the physical meanings of these obtained terms.

3.3. Adiabatic and Nonadiabatic Contributions

Taking the Riemann sum on $\mathcal{G}_1^\nu$ and $\mathcal{G}_2^\nu$ by setting $N \to \infty$ and $\delta t \to 0$, we find that they reduce to the dynamical and geometrical phases, respectively. For instance, we obtain the energy transfer with the environment R with setting $\nu = R$ as [56]

$$\mathcal{G}_1^R = \int_0^T dt' \frac{\Gamma_L \Gamma_R (n_L(t') - n_R(t'))}{K}, \tag{25}$$

with $K \equiv \sum_{\nu=L,R} \Gamma_\nu (1 + 2n_\nu(t'))$, and

$$\mathcal{G}_2^R = \int\int dT_L dT_R \left\{ \frac{2\Gamma_L \Gamma_R (\Gamma_L + \Gamma_R)}{K^3} \frac{dn_R}{dT_R} \frac{dn_L}{dT_L} \right\}, \tag{26}$$

which coincide with the ones in Reference [28] and imply that the sum of $\mathcal{G}_1^\nu$ and $\mathcal{G}_2^\nu$ corresponds to the adiabatic contribution.

Considering this point, we find that the nonadiabatic contribution is described with a new extra term in $\langle \Delta q^\nu \rangle$ added to the adiabatic contribution in the form,

$$\langle \Delta q^\nu \rangle = \mathcal{G}_{ad}^\nu + \mathcal{G}_{nad}^\nu, \tag{27}$$

with $\mathcal{G}_{ad}^\nu = \mathcal{G}_1^\nu + \mathcal{G}_2^\nu$ and $\mathcal{G}_{nad}^\nu = \mathcal{G}_3^\nu$. This is consistent with the expression of $\mathcal{G}_3^\nu$, which shows that, when $\rho_{00}(0) = \rho_{s,1}$ and the absolute value of $\Lambda_i \delta t$ is sufficiently large, we can neglect $\mathcal{G}_3^\nu$. The former condition corresponds to the adiabatic approximation in Reference [28], where the population of the relevant system instantaneously approaches the steady state for the temperature setting at an initial time. The term $\mathcal{G}_3^\nu$ shows that the nonadiabatic contribution to the transferred quantity explicitly depends on the initial condition of the relevant system, $\rho_{00}(0)$. Moreover, expanding Equation (13) about δt up to the first order, we find that the nonadiabatic effect described in $\mathcal{G}_3^\nu$ shows a correction to both $\mathcal{G}_1^\nu$ and $\mathcal{G}_2^\nu$. In the following, we present a numerical evaluation of the formulas obtained.

3.4. Numerical Evaluation of the Nonadiabatic Spin Pumping

3.4.1. Population Dynamics

Figure 2 presents the transient time evolution of $\rho_{00}(t)$ during the first period of modulation by changing the time interval δt while keeping the number of divisions N constant at $N = 40$. We set parameters as $\lambda = 0.01$, $\omega_c = 3\omega_0$, and $\hbar\omega_0 = 25$ meV which shows the relaxation time $\bar{\tau}_r \equiv \omega_0 \tau_r \approx 5$. (The value of $\hbar\omega_0$ is chosen to be the same as the typical value for a molecular junction in Reference [28].) Setting the initial condition of the two-level system with the effective inverse temperature as $\bar{\beta}_s = \frac{\hbar\omega_0}{k_B T_s} = \bar{\beta}(0)(\approx 1.07)$ corresponding to the stationary state for the initial temperature setting, we plot the time dependence of the population in the lower state, $\rho_{00}(t)$. The population $\rho_{00}(t)$ under the adiabatic approximation (Figure 2, red line) shows that the relevant system quickly approaches the stationary state corresponding to the temperature setting in each time interval. Setting the interval δt to be much larger than the relaxation time as in Figure 2A corresponding to the lower modulation frequency $\Omega = 0.3$ THz, we find that the relevant system mostly follows the temperature modulation as the stationary state is approached, which shows the feature close to the adiabatic approximation. With decreasing interval δt (Figure 2B,C), we find that the relevant system does not follow the temperature modulation thus exhibiting nonadiabaticity.

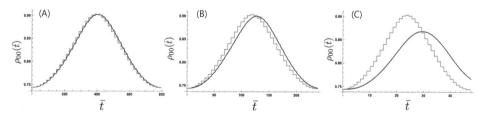

Figure 2. Time dependence of the population in the lower state of the two-level system with changing modulation frequency : (**A**) $\Omega = 0.3$ THz, (**B**) $\Omega = 1$ THz, and (**C**) $\Omega = 5$ THz with $s = 0.01$, $\omega_c = 3\omega_0$, $\hbar\omega_0 = 25$ meV , and $N = 40$. The time variable is scaled with ω_0 as $\bar{t} = \omega_0 t$.

3.4.2. Frequency Dependence

We show in Figure 3 the frequency dependence of the pumped quantity $\hat{I}_{energy} = \frac{1}{T\hbar\omega_0}(\langle \Delta q^R \rangle - \langle \Delta q^L \rangle)$. For comparison, we also exhibit the frequency dependence of the quantity under the adiabatic approximation presented as a geometric phase in Reference [28]. We find that the nonadiabatic term decreases the pumped quantity in the higher frequency region. We also find that the pumped quantity depends on the initial condition of the two-level system. The feature shown in Figure 3 is universal for different settings of these parameters. For example, when we increase τ_r by decreasing the coupling strength with keeping the value of ω_0, we find the similar feature of the frequency dependence ranging up to ~10 GHz which corresponds to the maximum driving frequency of electronic voltage due to the limitation of experimental bandwidth at the present time. The parameter setting of λ in this study is chosen to expect the further acceleration of the recent rapid development of Tera Hz technology in a future.

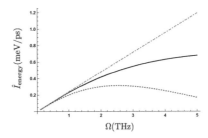

Figure 3. (Color online) Frequency dependence of pumped quantity I_{energy} with $\lambda = 0.01$, $\omega_c = 3\omega_0$, and $\hbar\omega_0 = 25$ meV with changing initial conditions of the two-level system $\bar{\beta}_s$ values: (1) the black line corresponds to $\bar{\beta}_s = \bar{\beta}(0)(\approx 1.07)$ which is the effective inverse temperature of the stationary state for the initial temperature setting, (2) the blue dotted line to $\bar{\beta}_s = 5$; (3) the red dottdashed line represents the frequency dependence of the net geometrical phase [28].

4. Spin Pumping

4.1. A Minimum Model of Spin Pumping

We consider a minimum model of spin pumping involving a quantum dot with dynamic magnetization and an electron lead (Figure 4A). The magnetization of the dot $M(t)$ rotates around the z-axis with a period T. An electron in the quantum dot is spin polarized because of the s–d exchange interaction with magnetization and is represented by the two-component creation and annihilation operators $d^\dagger = (d_\uparrow^\dagger, d_\downarrow^\dagger)$, and d, where $\uparrow$ and $\downarrow$ denote the direction of the electron's spin magnetic moment parallel and antiparallel, respectively, to the z-axis.

The Hamiltonian of the minimum model consists of three terms $H(t) = H_d(t) + H_l + H_t$. $H_d(t)$, describing the dot, is defined by

$$H_d(t) = d^\dagger[\epsilon_d - M(t) \cdot \sigma]d, \tag{28}$$

where ϵ_d is the unpolarized energy of a dot electron, $M(t) \equiv M(\sin\theta \sin\phi(t), \sin\theta \sin\phi(t), \cos\theta)$, and $\sigma = (\sigma_x, \sigma_y, \sigma_z)$ the vector of Pauli matrices. Introducing the eigenstates $|j_\uparrow, j_\downarrow\rangle$ (with $j_{\uparrow(\downarrow)} = 0$ or 1) of the number operator of the dot electron $\sum_\sigma d_\sigma^\dagger d_\sigma$ as a basis, the dot Hamiltonian is represented by the matrix

$$H_d(t) = \begin{array}{cccc} |0,0\rangle & |0,1\rangle & |1,0\rangle & |1,1\rangle \end{array} \\ \begin{pmatrix} 0 & 0 & 0 & 0 \\ 0 & \epsilon_d + M\cos\theta & -Me^{+i\phi(t)}\sin\theta & 0 \\ 0 & -Me^{-i\phi(t)}\sin\theta & \epsilon_d - M\cos\theta & 0 \\ 0 & 0 & 0 & 2\epsilon_d \end{pmatrix}. \tag{29}$$

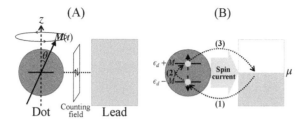

Figure 4. (**A**) The minimum model consists of a ferromagnetic quantum dot attached to an electron lead. The dot has a dynamic magnetization $M(t)$ that rotates around the z-axis with a period $\mathcal{T}$. The number of transferred electrons with spin magnetic moment ↑ (↓) is captured by the counting field. (**B**) Schematic of the spin current generation in the minimum model. The scheme can be summarized as follows: (1) an electron with ↓-spin enters from the lead onto the dot subject to the dot–lead interaction; (2) the spin of the electron is flipped by the precessing magnetization; (3) the electron with ↑-spin moves back from the dot to the lead.

The electron lead is described by the term

$$H_l = \sum_{\sigma=\uparrow,\downarrow} \sum_k \epsilon_k c_{\sigma,k}^\dagger c_{\sigma,k}, \tag{30}$$

where $c_{\sigma,k}$ and $c_{\sigma,k}^\dagger$ with $\sigma=\uparrow$ or $\downarrow$ are annihilation and creation operators of a lead electron with energy ϵ_k and spin-σ. The dot–lead interaction is assumed to be spin conserving with

$$H_t = \sum_\sigma \sum_k \hbar v_k (d_\sigma^\dagger c_{\sigma,k} + c_{\sigma,k}^\dagger d_\sigma), \tag{31}$$

where $\hbar v_k$ is the coupling strength, which we assume to be weak.

Intuitively, the generation of the spin current in the minimum model is summarized by the following scheme (see Figure 4B): (1) an electron with ↓-spin moves from lead to dot under the dot–lead interaction, (2) the spin of the electron is flipped by the precessing magnetization, and (3) an electron with ↑-spin moves back from dot to lead. For spin-current generation, the essential conditions required in setting parameter values are

$$\epsilon_d - M < \mu < \epsilon_d + M \quad \text{and} \quad \beta^{-1} \leq 2M, \tag{32}$$

where β is the inverse temperature of the lead.

4.2. FCS Formalism of the Spin Pumping

In the following, we apply the FCS outlined in Section 2 to evaluate the number of transferred electrons with spin σ from projective measurements of the electron number in the lead represented by $N_\sigma \equiv \sum_k c_{\sigma,k}^\dagger c_{\sigma,k}$. By associating $H_d(t)$, H_l, and H_t with H_S, H_E, and H_{SE}, respectively, and defining an outcome of the projective measurement at time t as $n_{\sigma,t}$, we analyze the electron dynamics under spin pumping.

In order to explicitly examine the influence of the relaxation process on the spin current generation, we discretize the rotation of $M(t)$: divide the period $\mathcal{T}$ into N intervals, $t_i \leq t \leq t_{i+1}$ ($i = 1, \cdots, N$) with $t_1 = 0$ and $t_{N+1} = \mathcal{T}$; fix the direction of $M(t)$ during each interval; and change ϕ at each t_i discretely with substitution $\phi_i = \phi_{i-1} + \delta\phi$ with $\phi_0 = 0$, $\phi_N = 2\pi$ and $\delta\phi \equiv 2\pi/N$ (see Note [57]). The net number of electrons with spin-σ during the *i*th interval can be evaluated from the difference in outcomes $\Delta n_{\sigma,i} = n_{\sigma,t_{i+1}} - n_{\sigma,t_i}$.

By introducing counting fields $\lambda_\uparrow$ and $\lambda_\downarrow$ corresponding to observables $N_\uparrow$ and $N_\downarrow$, respectively, we can evaluate the mean value of transferred electrons,

$$\langle \Delta n_{\uparrow(\downarrow),i} \rangle = \int_{t_i}^{t_{i+1}} J_{\uparrow(\downarrow)}(t),$$ (33)

with an inertial flow of electrons

$$J_{\uparrow(\downarrow)}(t) \equiv \langle\!\langle 1 | \left[\frac{\partial \Xi^{(\lambda_{\uparrow(\downarrow)})}(t)}{\partial(i\lambda_{\uparrow(\downarrow)})} \right]_{\lambda_{\uparrow(\downarrow)}=0} | \rho^{(\lambda_{\uparrow(\downarrow)}=0)}(t) \rangle\!\rangle.$$ (34)

The inertial flow of electrons provides an instantaneous spin current,

$$J_{\text{spin}}(t) \equiv J_\uparrow(t) - J_\downarrow(t),$$ (35)

and its time integration over one period provides a temporal average of the spin current,

$$I_{\text{spin}} \equiv \frac{1}{\mathcal{T}} \int_0^{\mathcal{T}} J_{\text{spin}}(t)dt.$$ (36)

To discuss the role of nonadiabaticity in spin pumping, we focus the Born-Markovian (long-time) limit by taking the limit $t \to \infty$ of the supermatrix $\Xi^{(\lambda)}(t)$ in each interval. In this limit, the matrix elements of $\Xi^{(\lambda)}$ are time-independent during each interval and determined by the direction of M in each interval.

4.3. Absence of Adiabatic Contribution

Let us first show absence of the adiabatic contribution to the spin pumping in the minimum model. In previous studies, the adiabatic regime of the spin pumping in the minimum model has been studied based on the linear expansion of the Green function in the rotation frequency of $M(t)$ [23,24], which we referred to as the adiabaticity No. (2). The purpose of the present subsection is to re-examine the adiabatic contribution of the spin pumping from the view point of the adiabaticity No. (3), where we consider a sufficiently slow rotation of $M(t)$ comparing to the relaxation time, that is, $\Omega \ll \tau_r^{-1}$ following the procedure by Sinitsyn and Nemenman in Reference [27].

Following the procedure, the adiabatic regime is assessed by dividing the cycle of modulation into time intervals $\delta t (\equiv \mathcal{T}/N)$ and assuming a quick approach of the system to its steady state in each interval. In the steady state of the minimum model, we can expect that the quantum dot is occupied by a single electron whose spin is aligned toward the direction of $M(t)$, and, because of the rotational symmetry of the model, the steady state populations of the quantum dot are invariant under the rotation of $M(t)$ around the z-axis. It indicates that no electron transfer occurs in the adiabatic regime. As a result, we can expect absence of the adiabatic contribution to the spin current generation. We provide an analytical proof of the intuitive observation in Appendix C (see also the original argument in Section 4 in Reference [58]).

As a result, we need to include the nonadiabatic effect to obtain a finite spin current. It is in marked contrast to the previous example of the energy pumping, where the adiabatic contribution to the energy pumping $\mathcal{G}_{ad}^v$ is finite.

4.4. Numerical Evaluation of the Nonadiabatic Spin Pumping

We now turn to examine nonadiabaticity in spin pumping. For this purpose, we evaluate numerically the instantaneous spin current $J_{\text{spin}}(t)$ and its temporal average I_{spin}.

To describe the dot–lead coupling, we use the Ohmic spectral density with an exponential cutoff $v(\omega) \equiv \sum_k v_k^2 \delta(\omega - \omega_k) = \lambda\omega \exp[-\omega/\omega_c]$, where λ is the coupling strength and ω_c is the cutoff frequency. For the numerical calculation, we chose $2M$, the energy difference between the spin-$\uparrow$ and

-↓ states in the dot, as an energy unit. We distinguish parameters normalized by their units using an overbar (see Note [59]). Specific values of the normalized parameters are given in the figure captions. As we are focusing on the spin transfer driven by the rotating magnetization, the dot is set in a steady state Equation (A25) at $\bar{t} = 0$ to exclude any transient spin transfer caused by the dot–lead contact.

4.4.1. Electron Transfer Dynamics

Let us first examine electron transfer dynamics under the cyclic rotation of the magnetization to show the generation of the spin current in the nonadiabatic regime. For this purpose, we numerically evaluate the time evolution of the populations $\rho_{jj'}(t) = \langle j, j' | \rho^{(0)}(t) | j, j' \rangle$ (ρ_{00}: empty state, ρ_{10}: half-filled state with spin-↑, ρ_{01}: half-filled state with spin-↓, and ρ_{11}: completely filled state) and corresponding instantaneous electron and spin currents, $J_{\uparrow(\downarrow)}(t)$ and $J_{\text{spin}}(t)$. In Figure 5A,B, we present the time evolution of populations and instantaneous currents for one cycle of the step-like rotation with division number $N = 5$ and time interval $\delta t = 20$. The change in angle at each subsequent t_i is $\delta\phi = 2\pi/5$, that is $\phi_i = \phi_{i-1} + 2\pi/5$ with $\phi_0 = 0$.

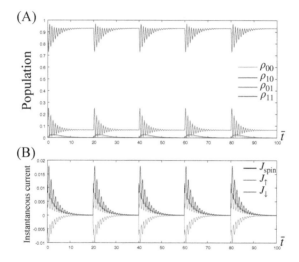

Figure 5. (A) Time evolution of the populations in the dot under the step-like precession of the magnetization with $\delta\bar{t} = 20$ and $N = 5$. The populations deviate from their steady-state values just after a sudden change of the angle ϕ, but then they approach new steady state-values for each ϕ_i. The figure shows the steady-state values of the populations to be invariant. This is because of the rotational symmetry of the system about the z-axis. **(B)** The instantaneous electron and spin currents $J_\uparrow(t)$ (red line), $J_\downarrow(t)$ (blue line) and $J_{\text{spin}}(t)$ (black line) corresponding to the population dynamics in panel (A). The time dependences of the instantaneous currents indicate that electrons starts moving between dot and lead just after the sudden change of ϕ, and $J_\uparrow$ and $J_\downarrow$ have opposing directions. The latter trend show that the instantaneous electron currents are balanced as a result of charge conservation in the lead. In contrast, the instantaneous spin current J_{spin} always takes positive values indicating constant spin current generation. The parameters are set to $\bar{\epsilon}_d = 10$, $\bar{\mu} = 10$, $\bar{\beta} = 100$, $\lambda = 0.01$, $\bar{\omega}_c = 4$, $\theta = 5\pi/6$, and $\delta\phi = 2\pi/5$, which satisfies the condition (32).

In Figure 5A, we find that initially the populations deviate from their steady-state values by changing ϕ at t_i, but then they approach new steady-state values for each ϕ_i with the populations remaining unchanged from their initial values because the steady-state populations are independent of ϕ (see the analytic expression of the steady state, Equation (A25)). In the figure, the time evolution of the components ρ_{01} and ρ_{10} (blue and red lines) exhibit oscillations caused by transitions between states $|0, 1\rangle$ and $|1, 0\rangle$ in consequence of the applied magnetization M (Larmor precession). Its period is given by the inverse of the Larmor frequency $T_L \equiv h/2M = t_u$. The other two components ρ_{00} and ρ_{11} also exhibit transient behavior after changing ϕ but they do not exhibit a Larmor precession because the magnetization contributes transitions including neither $|0,0\rangle$ nor $|1,1\rangle$ (see Equation (29)).

In Figure 5B, the colored lines representing J_σ show that spin-↑ electrons (red line) and spin-↓ electrons (blue line) are moving in opposite directions; the former move from dot to lead, whereas the latter move from lead to dot. These trends show that the instantaneous electron currents $J_\uparrow$ and $J_\downarrow$ are balanced as a result of charge conservation in the lead. In contrast, the instantaneous spin current (black line) always takes positive values, $J_{spin} > 0$, indicating the generation of positive spin current into the lead without an associated charge current, which we call pure spin current.

4.4.2. Frequency Dependence

We next consider the dependence of the spin current on the frequency of precession $\Omega = 2\pi/\mathcal{T}$. Here we change the period of precession $\mathcal{T} = N\delta t$ by varying the time interval δt while the number of divisions remains fixed to $N = 20$. All other parameters and initial conditions are set as before.

In Figure 6, we plot the dependence of the averaged spin current I_{spin} against the normalized frequency $\bar{\Omega} \equiv \Omega/\omega_u$.

The frequency dependence of I_{spin} features two characteristic regimes: a low-frequency regime, where I_{spin} depends linearly on Ω ($\bar{\Omega} \lesssim 0.0025$) and a high-frequency regime, where I_{spin} exhibits oscillations with respect to Ω. These characteristics are explained by comparing the time interval $\delta \bar{t}$ and the relaxation time $\bar{\tau}_r$ of the population of dot electrons ($\bar{\tau}_r \approx 5$ in the present case; see Figure 5A). For lower frequencies, for which $\delta \bar{t} \gg \bar{\tau}_r$, the numerator of the time integral of $J_{spin}(t)$ in Equation (36) becomes constant because the instantaneous spin current has already vanished at a certain $\bar{t} \lesssim \delta \bar{t}$ (see Figure 5A), which results in the linear dependence of I_{spin} on $\bar{\Omega}$. As $\bar{\Omega}$ becomes larger and the time interval satisfies $\delta \bar{t} \lesssim \bar{\tau}_r$, the angle ϕ changes during the relaxation process. In this situation, the electron dynamics exhibits two extreme features; when $\delta \bar{t}$ is an integer multiple of the period of the Larmor precession $h/2M$, we have resonance enhancement of the transition between half-filled states $|0,1\rangle$ and $|1,0\rangle$ by the sudden change of ϕ to exhibit a maximum of I_{spin}, whereas it is anti-resonantly suppressed to exhibit a minimum when $\delta \bar{t}$ is a half-integer multiple of the period [58].

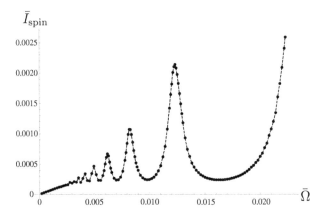

Figure 6. Frequency dependence of the temporal average of spin current I_{spin}. The division number of the step-like precession is now set to $N = 20$. With fixed $\delta \phi = \pi/10$, the frequency is changed by changing δt. The frequency dependence exhibits two characteristic features: the spin current depends linearly on $\bar{\Omega}$ for $\bar{\Omega} \lesssim 0.0025$, whereas it exhibits oscillation with respect to $\bar{\Omega}$ for $\bar{\Omega} \gtrsim 0.0025$. The other parameters are the same as in Figure 5.

Calculating the spin current for different values of θ, we find that the spin polarization of the spin current exhibits a dependence on θ in that for $0 < \theta < \pi/2$ the spin polarization is antiparallel to the z-axis, whereas for $\pi/2 < \theta < \pi$ the spin polarization is parallel to the z-axis. For $\theta = 0$, $\pi/2$, π, the spin current vanishes because the spin flip in the quantum dot does not occur for $\theta = 0$, π or the two half-filled states in the dot $|1,0\rangle$ and $|0,1\rangle$ degenerate for $\theta = \pi/4$ (see Equation (29)).

Finally, we note that the averaged spin current I_{spin} diverges with respect to $\bar{\Omega}$. The divergence is caused by the accumulation of a nonzero impetus of current $J_{\text{spin}}(t)$ just after the sudden change of ϕ (see Figure 5). In Reference [60], we showed that the nonzero impetus of $J_{\text{spin}}(t)$ is an unphysical effect caused by the Born-Markovian approximation, and the divergence is eliminated by taking into account the non-Markovian effect by keeping the upper bound of the time integration in (10) finite.

5. Discussion and Conclusions

In this paper, we reviewed studies which go beyond the conventional adiabatic approximation for open quantum systems to transfer energy quanta and electron spins with using the full counting statistics, which could provide conditions to show quicker transport. We considered a setup consisting of a two-level system representing an anharmonic junction or a quantum dot and its environment(s) representing a canonical or grand canonical ensemble of the energy quanta and the electron to be transferred. We needed to take into account relaxation phenomena in discussing the transfer. In this case, the adiabatic approximation corresponded to the situation where the relevant system such as the two-level system approaches its stationary state faster than the period of modulation, that is, $\tau_r^{-1} \gg \Omega$ with τ_r the relaxation time of the two-level system and Ω the modulation frequency. Because the relaxation time is finite, the condition for which the adiabatic approximation is valid corresponds to the much longer period of the modulation than τ_r. This means that we can analyze systematic features including adiabatic as well as nonadiabatic features by changing the ratio of the modulation period and τ_r. To clarify the relationship between modulation period and τ_r, we discretized the external modulation thereby permitting a systematic analysis of the ratio by changing each interval while retaining the validity of the Born–Markov approximation. For energy quanta pumping, we showed that the nonadiabatic effect contributes a new term to the formula for the pumped quantity under the adiabatic approximation. For spin pumping, we showed that adiabaticity made no contribution but nonadiabaticity is essential. Comparing these features, we showed that the adiabatic contribution can vanish when the stationary state does not depend on the external modulation as for spin pumping. This means that we need to pay attention to the feature of the stationary state in using the adiabatic approximation in describing relaxation phenomena. (We would draw the reader's attention to the differences in the meaning of nonadiabaticity which has been used in the electron charge pumping by modulation of single gate voltage [61,62].)

With the same setup, the role of nonadiabaticity in pumping phenomena involving energy quanta was discussed more extensively under continuous modulation [63] where the relaxation of the two-level system is treated within the Born–Markovian approximation. In recent work of the present authors on the role of the non-Markovian effect on spin pumping phenomena [60], we found that a nonzero impetus of the dynamics of the pumped quantity under the Born–Markovian approximation shows an unphysical effect, especially for higher modulation frequencies or for the short time regimes. Because the instantaneous impetus contributes strongly under continuous modulation, including the non-Markovian effect would also be necessary in pumping phenomena of energy quanta, especially in evaluating the feature under continuous modulation. This situation remains an open problem. In addition, we described in this work the relaxation process with ordered cumulants of up to second order in the system–environment interaction. An extension to higher orders of cumulants is necessary if we are to discuss relaxation phenomena under strong system–environment interactions. The inclusion of cumulants up to infinite order within the Markovian approximation has been discussed for spin pumping phenomena within the linear response regime using the Green functions [19]. To discuss the

Entropy 2019, 21, 842

non-Markovian effect, it would be necessary to include higher-order cumulants, a topic that remains for a future study.

We can find recent extensions of the treatments with full counting statistics into the strong system-environment coupling for heat transfer [64] and electron pumping [65]. The essential idea to go beyond the weak coupling is to use the similarity (unitary) transformations: the polaron transformation (the reaction coordinate mapping) is used in the former (latter) studies, respectively. As mentioned in the general formalism of FCS, it should be noted that we need careful treatments on the joint probability, Equation (2), when we use the similarity (unitary) transformations on the time evolution operator. The transformation of the projective measurement is also necessary to recover the original joint probability (See Reference [45]). It might be necessary to compare the dynamics of transported quantity with and without the transformation of the projective measurement.

Since the treatment of FCS to discuss the nonadiabatic effects on quantum pumping is general, we can apply it to many other cases: One of the most interesting issues is to study the non-adiabatic treatment on the combined effect caused by multiple external parameters such as in References [66–68] where adiabatic transport of charge and/or heat is discussed under time-dependent potential and two reservoirs with biased potentials. We can find other issues to remove the adiabatic approximation in spin pumping via a quantum dot between reservoirs with biased chemical potentials [69] and in the quantum transport and/or quantum pumping under dynamical motion of quantum dot [70] based on the recent developments of experimental techniques on microelectromechanical systems [71]. Further, it would be interesting to discuss the non-adiabatic effect on ac-driven electron systems coupled to multiple reservoirs at finite temperature whose adiabatic treatment is discussed in Reference [72]. We expect that these treatments could provide insights to find new applications, such as the design of nanomachines and understanding of the quantum thermodynamics, as well as quicker transport.

Author Contributions: The authors contributed equally to this work.

Funding: This work was supported by a Grant-in-Aid for Challenging Exploratory Research (Grant No. 16K13853), partially supported by a Grant-in-Aid for Scientific Research on Innovative Areas, Science of Hybrid Quantum Systems (Grant No. 18H04290), and the Open Collaborative Research Program at National Institute of Informatics Japan (FY2018).

Acknowledgments: The authors thank Gen Tatara for valuable discussions.

Conflicts of Interest: The authors declare no conflict of interest.

Appendix A. Derivation of Quantum Master Equation for FCS

When we consider the FCS [44], the density operator $W^{(\lambda)}(t)$ evolves in time in accordance with the modified Liouville–von Neumann equation

$$\dot{W}^{(\lambda)}(t) = -i\mathcal{L}^{(\lambda)}W^{(\lambda)}(t), \tag{A1}$$

where $\mathcal{L}^{(\lambda)}$ is the Liouville operator defined as $\mathcal{L}^{(\lambda)}A = \frac{1}{\hbar}[H_\lambda A - AH_{-\lambda}] \equiv \frac{1}{\hbar}[H, A]_\lambda$ for arbitrary operator A with $H_\lambda = e^{(i/2)\lambda H_E}He^{-(i/2)\lambda H_E} = H_0 + H_{SE,\lambda}$. With these relations, $\mathcal{L}$ is divided into

$$\mathcal{L}^{(\lambda)} = \mathcal{L}_0^{(\lambda)} + \mathcal{L}'^{(\lambda)} \tag{A2}$$

where

$$\mathcal{L}_0^{(\lambda)}A = \frac{1}{\hbar}[H_0 A - AH_0] \ , \ \ \mathcal{L}'^{(\lambda)}A = \frac{1}{\hbar}[H_{SE,\lambda}A - AH_{SE,-\lambda}]. \tag{A3}$$

We eliminate the variables of the environment using a projection operator $\mathcal{P}$, which satisfies the idempotent relation, $\mathcal{P}^2 = \mathcal{P}$. We also introduce a complementary operator $\mathcal{Q} \equiv 1 - \mathcal{P}$. Denoting the relevant and irrelevant parts of the time evolution operator as [52]

$$x(t) \equiv \mathcal{P}e^{-i\mathcal{L}^{(\lambda)}t}, \ \ y(t) \equiv \mathcal{Q}e^{-i\mathcal{L}^{(\lambda)}t}, \tag{A4}$$

with an initial time $t_0 = 0$, we obtain

$$\frac{d}{dt}x(t) = \mathcal{P}(-i\mathcal{L}^{(\lambda)})x(t) + \mathcal{P}(-i\mathcal{L}^{(\lambda)})y(t) , \tag{A5}$$

and

$$\frac{d}{dt}y(t) = \mathcal{Q}(-i\mathcal{L}^{(\lambda)})x(t) + \mathcal{Q}(-i\mathcal{L}^{(\lambda)})y(t) . \tag{A6}$$

The formal solution of Equation (A6) is given by

$$y(t) = \int_0^t e^{-\mathcal{Q}i\mathcal{L}^{(\lambda)}(t-\tau)}\mathcal{Q}(-i\mathcal{L}^{(\lambda)})x(\tau)d\tau + e^{-\mathcal{Q}i\mathcal{L}^{(\lambda)}t}\mathcal{Q}. \tag{A7}$$

Using $x(\tau) = \mathcal{P}e^{i\mathcal{L}^{(\lambda)}(t-\tau)}e^{-i\mathcal{L}^{(\lambda)}t} = \mathcal{P}e^{i\mathcal{L}^{(\lambda)}(t-\tau)}(x(t)+y(t))$ and

$$\theta(t) = 1 - \int_0^t e^{-\mathcal{Q}i\mathcal{L}^{(\lambda)}\tau}\mathcal{Q}(-i\mathcal{L}^{(\lambda)})\mathcal{P}e^{i\mathcal{L}^{(\lambda)}\tau}d\tau \equiv 1 - \sigma(t), \tag{A8}$$

we rewrite the formal solution of $y(t)$ in the form

$$y(t) = \theta(t)^{-1}((1-\theta(t))x(t) + e^{-\mathcal{Q}i\mathcal{L}^{(\lambda)}t}\mathcal{Q}). \tag{A9}$$

By substituting Equation (A9) into Equation (A5), we obtain

$$\frac{d}{dt}x(t) = \mathcal{P}(-i\mathcal{L}^{(\lambda)})\theta(t)^{-1}x(t) + \mathcal{P}(-i\mathcal{L}^{(\lambda)})\theta(t)^{-1}e^{\mathcal{Q}(-i\mathcal{L}^{(\lambda)})t}\mathcal{Q}, \tag{A10}$$

which holds for arbitrary projection operator and initial condition. Using the relation $\theta(t)^{-1} = \sum_{n=0}^{\infty} \sigma(t)^n$, the first term on the right-hand side of Equation (A10) is rewritten as

$$\mathcal{P}(-i\mathcal{L}^{(\lambda)})\theta(t)^{-1}x(t) = \mathcal{P}(-i\mathcal{L}^{(\lambda)})x(t) + \mathcal{P}(-i\mathcal{L}^{(\lambda)})\sigma(t)x(t) + \cdots . \tag{A11}$$

To pick out the lower order of $\mathcal{L}'^{(\lambda)}$, we use the relation

$$e^{-\mathcal{Q}i\mathcal{L}^{(\lambda)}t}\mathcal{Q} = e^{-i\mathcal{L}_0^{(\lambda)}t}\mathcal{Q}T_+\exp\left[\int_0^t dt' e^{i\mathcal{L}_0^{(\lambda)}t'}\mathcal{Q}(-i\mathcal{L}'^{(\lambda)})\mathcal{Q}e^{-i\mathcal{L}_0^{(\lambda)}t'}\right], \tag{A12}$$

and $\mathcal{P}\mathcal{L}_0^{(\lambda)} = \mathcal{L}_0^{(\lambda)}\mathcal{P}$, which gives

$$
\begin{aligned}
\mathcal{P}(-i\mathcal{L}^{(\lambda)})\theta(t)^{-1}x(t) &= \mathcal{P}(-i\mathcal{L}^{(\lambda)})x(t) + \mathcal{P}(-i\mathcal{L}'^{(\lambda)})\int_0^t e^{-i\mathcal{L}_0^{(\lambda)}\tau}\mathcal{Q}(-i\mathcal{L}'^{(\lambda)})\mathcal{P}e^{i\mathcal{L}_0^{(\lambda)}\tau}d\tau x(t) + \cdots \\
&= \mathcal{P}(-i\mathcal{L}^{(\lambda)})x(t) + \mathcal{P}(-i\mathcal{L}'^{(\lambda)})\int_0^t \mathcal{Q}(-i\mathcal{L}'^{(\lambda)}(-\tau))\mathcal{P}d\tau x(t) + \cdots ,
\end{aligned}
\tag{A13}
$$

where we have used the definition

$$\hat{\mathcal{L}}_1^{(\lambda)}(t) = e^{i\mathcal{L}_0^{(\lambda)}t}\mathcal{L}'^{(\lambda)}e^{-i\mathcal{L}_0^{(\lambda)}t}. \tag{A14}$$

When we multiply the initial condition $W^{(\lambda)}(0)$ by the right-hand side of Equation (A8), we obtain the TCL equation for reduced density operator under FCS.

Let us consider a projection operator $\mathcal{P} = \rho_E \mathrm{Tr}_E$ where Tr_E refers to a trace operation over the environment. When we multiply the initial condition of the density operator of the total system, $W(\lambda, 0)$ from the right by $x_S(t)$ in Equation (A2), we obtain

$$x(t)W^{(\lambda)}(0) = \rho_E \mathrm{Tr}_E W^{(\lambda)}(t) \tag{A15}$$

Defining $\mathrm{Tr}_E W^{(\lambda)}(t) \equiv \rho^{(\lambda)}(t)$, we obtain the TCL equation for the reduced density operator under FCS,

$$\frac{d}{dt}\rho^{(\lambda)}(t) = \mathrm{Tr}_E[(-i\mathcal{L})\rho^{(\lambda)}(t)] + \zeta^{(\lambda)}(t)\rho^{(\lambda)}(t) + \psi^{(\lambda)}(t), \tag{A16}$$

with

$$\zeta^{(\lambda)}(t) \equiv \sum_{n=2}^{\infty} \zeta_n^{(\lambda)}(t), \tag{A17}$$

$$\psi^{(\lambda)}(t) \equiv \mathrm{Tr}_E[(-i\mathcal{L})\theta(t)^{-1}e^{\mathcal{Q}(-i\mathcal{L})t}\mathcal{Q}W^{(\lambda)}(0)]. \tag{A18}$$

Equation (9) with (10) is obtained by taking the lower term of $\zeta^{(\lambda)}(t)$ up to second order in $\mathcal{L}'$ and replacing $\zeta_2^{(\lambda)}(t)$ with $\zeta^{(\lambda)}(t)$,

$$\zeta^{(\lambda)}(t)\rho(\lambda,t) = \int_0^t \mathrm{Tr}_E[(-i\mathcal{L}')\mathcal{Q}(-i\mathcal{L}'(-\tau))\rho_E\rho(\lambda,t)]d\tau. \tag{A19}$$

with the factorized initial condition of the total system as $W^{(\lambda)}(0) = \rho_E\rho^\lambda(0)$, which makes the third term on the right-hand side of Equation (A16) vanish. With the assumption $\mathrm{Tr}_E H_{SE,\lambda} = 0$, we have

$$\begin{aligned}
&\zeta^{(\lambda)}(t)\rho^{(\lambda)}(t) \\
&= (\tfrac{i}{\hbar})^2 \mathrm{Tr}_E[H_{SE}[H_{SE}(-\tau), \rho_E\rho^{(\lambda)}(t)]_\lambda]_\lambda, \\
&= (\tfrac{i}{\hbar})^2 \mathrm{Tr}_E[H_{SE,\lambda}H_{SE,\lambda}(-\tau)\rho_E\rho^{(\lambda)}(t) - H_{SE,\lambda}\rho_E\rho^{(\lambda)}(t)H_{SE,-\lambda}(-\tau) \\
&\quad - H_{SE,\lambda}(-\tau)\rho_E\rho^{(\lambda)}(t)H_{SE,-\lambda} + \rho_E\rho^{(\lambda)}(t)H_{SE,-\lambda}(-\tau)H_{SE,-\lambda}],
\end{aligned} \tag{A20}$$

which coincides with Equation (110) in [44].

Appendix B. Connection between Two Formalisms of the FCS

We next explain the relationship between the two formalisms of the FCS provided by Esposito et al. in Reference [44] and by Sinitsyn and Nemenman in Reference [27]. To establish the connection between the two formalisms, we consider the number of quanta N as the observable to be measured to identify the number of quantum transfers from S to E, as stipulated in Reference [27]. For a large class of open quantum systems, the system–environment interaction is described by an interaction Hamiltonian of the form $H_{SE} = V_+ + V_- \equiv A \otimes B^\dagger + A^\dagger \otimes B$, where $B^\dagger$ and B are creation and annihilation operators of a quanta in the environment, respectively, and A is either a Hermitian or non-Hermitian operator acting on the relevant system. $V_+ \equiv A \otimes B^\dagger$ or $V_- = V_+^\dagger$ describes transfer of a quanta from S to E or from E to S. In this instance, the number operator of the quanta is given by $N = B^\dagger B$. The generalized Liouvillian is expressed as $\mathcal{L}^{(\lambda)} = \mathcal{L}_0 + \mathcal{L}_+ e^{i\lambda/2} + \mathcal{L}_- e^{-i\lambda/2}$, where $\mathcal{L}_0 W^{(\lambda)} \equiv \hbar^{-1}(H_0 W^{(\lambda)} - W^{(\lambda)} H_0)$ is the unperturbed Liouvillian, $\mathcal{L}_+ W^{(\lambda)} \equiv \hbar^{-1}(V_+ W^{(\lambda)} - W^{(\lambda)} V_+^\dagger)$ and $\mathcal{L}_- W^{(\lambda)} \equiv \hbar^{-1}(V_- W^{(\lambda)} - W^{(\lambda)} V_-^\dagger)$ are Liouvillians describing the transfer of a quanta from S to E and from E to S, respectively. Using the formal solution of Equation (8) with the given Liouvillian in Equation (6), we obtain a formal expression for the moment generating function

$$Z(\lambda) = \sum_{n=-\infty}^{\infty} P_n e^{in\lambda}, \tag{A21}$$

with P_n the probability of having n net transitions from S to E, e.g.,

$$\begin{aligned}
P_1 = \mathrm{Tr}\Bigg[\Bigg(\int_{t_i}^t dt_1 U_0(t,t_1)(-i\mathcal{L}_+)U_0(t_1,t_i) + \int_{t_i}^t dt_3 \int_{t_i}^{t_3} dt_2 \int_{t_i}^{t_2} dt_1 U_0(t,t_3)(-i\mathcal{L}_+) \\
\times U_0(t_3,t_2)(-i\mathcal{L}_-)U_0(t_2,t_1)(-i\mathcal{L}_+)U_0(t_1,t_i)\cdots\Bigg)\tilde{W}(t_i)\Bigg],
\end{aligned} \tag{A22}$$

where $U_0(t_1,t_2) = \exp[-i\mathcal{L}_0(t_1 - t_2)]$ is the unperturbed time evolution operator. The expression for the moment generating function corresponds to Equation (9) in Reference [27].

Appendix C. An Analytical Proof of the Absence of Adiabatic Contribution to the Spin Pumping

Following the procedure by Sinitsyn and Nemenman in Reference [27], we divide the cycle of precession into intervals δt, which correspond to the step-like changes of M. For the step-like precession, the density matrix of the dot during $t_i \leq t \leq t_{i+1}$ is given by

$$|\rho_i^{(0)}(t)\rangle\rangle = e^{\Xi_i^{(0)}(t-t_i)} \prod_{j=1}^{i-1} e^{\Xi_j^{(0)} \delta t} |\rho_1^{(0)}(0)\rangle\rangle, \tag{A23}$$

where we denote the density matrix and the generator in the ith interval as $|\rho_i^{(0)}(t)\rangle\rangle$ and $\Xi_i^{(0)}$, respectively, and $|\rho_1^{(0)}(0)\rangle\rangle$ is the initial condition in the first interval. Taking the spectral decomposition of $\Xi_i^{(0)}$ and assuming a quick approach of the system to its steady state in each interval in Equation (A23), as in Reference [27], the terms remaining in the decomposition are those that contain the steady state $|u_0^0(t_i)\rangle\rangle$ satisfying $\Xi_i^{(0)}|u_0^0(t_i)\rangle\rangle = 0$. Evaluating the density matrix up to first order in δt, we find the first-order term vanishes, implying the invariance of the steady state populations of the dot under the step-like rotation of ϕ in our model (see Appendix D in Reference [58]). Thus, we find that the density matrix at time t under the adiabatic limit can be approximated by the steady state as $|\rho_0^{(0)}(t)\rangle\rangle \approx |u_0^{(0)}(t_i)\rangle\rangle$. For this steady state $|u_0^{(0)}(t_i)\rangle\rangle$ given by Equation (A25), we also find that there is no electron transfer between dot and lead; specifically, we find that

$$\langle \Delta n_{\sigma,i} \rangle \approx \int_{t_i}^{t_{i+1}} dt' \langle\langle 1| \left[\frac{\partial \Xi_i^{(\lambda_\sigma)}}{\partial(i\lambda_\sigma)} \right]_{\lambda_\sigma=0} |u_0^{(0)}(t_i)\rangle\rangle = 0, \tag{A24}$$

indicating that there is no net electron transfer in the interval and hence the generated spin current represented by Equation (36) is totally absent in the adiabatic limit. We therefore need to include the nonadiabatic effect to generate a finite spin current. The result is in marked contrast to the previous example of the energy pumping, where the adiabatic contribution to the energy pumping $\mathcal{G}_{ad}^v$ is finite.

Appendix D. Steady State of the Minimum Model

The steady state $|u_0^{(\lambda_\sigma=0)}(t_i)\rangle\rangle$, satisfying $\Xi_i^{(\lambda_\sigma=0)}|u_0^{(\lambda_\sigma=0)}(t_i)\rangle\rangle = 0$ is analytically obtained using a graphical method discussed in Reference [73]. In Appendix C of Reference [58], we provide a detailed derivation of the steady state. For use in the present paper, here we simply present the result.

The dynamics of the populations described by the TCL master equation is closed for the six components of the reduced density matrix, $\rho_{00}(t) \equiv \langle 0,0|\rho^{(\lambda_\sigma=0)}(t)|0,0\rangle$, $\rho_{01}(t) \equiv \langle 0,1|\rho^{(\lambda_\sigma=0)}(t)|0,1\rangle$, $\rho_{0110}(t) \equiv \langle 0,1|\rho^{(\lambda_\sigma=0)}(t)|1,0\rangle$, $\rho_{1001}(t) \equiv \langle 1,0|\rho^{(\lambda_\sigma=0)}(t)|0,1\rangle$, $\rho_{10}(t) \equiv \langle 1,0|\rho^{(\lambda_\sigma=0)}(t)|1,0\rangle$, and $\rho_{11}(t) \equiv \langle 1,1|\rho^{(\lambda_\sigma=0)}(t)|1,1\rangle$. By arranging these connected components as $|\rho^{(\lambda_\sigma=0)}(t)\rangle\rangle = [\rho_{00}(t), \rho_{01}(t), \rho_{0110}(t), \rho_{1001}(t), \rho_{10}(t), \rho_{11}(t)]^t$, where $[\cdots]^t$ denotes transposition, an analytic expression of its steady state obtains,

$$|u_0^{(\lambda_\sigma=0)}(t_i)\rangle\rangle = \begin{pmatrix} f^-(\epsilon_\uparrow)f^-(\epsilon_\downarrow) \\ \cos^2\frac{\theta}{2}f^+(\epsilon_\uparrow)f^-(\epsilon_\downarrow) + \sin^2\frac{\theta}{2}f^-(\epsilon_\uparrow)f^+(\epsilon_\downarrow) \\ e^{+i\phi_i}\cos\frac{\theta}{2}\sin\frac{\theta}{2}[f^+(\epsilon_\uparrow)f^-(\epsilon_\downarrow) - f^-(\epsilon_\uparrow)f^+(\epsilon_\downarrow)] \\ e^{-i\phi_i}\cos\frac{\theta}{2}\sin\frac{\theta}{2}[f^+(\epsilon_\uparrow)f^-(\epsilon_\downarrow) - f^-(\epsilon_\uparrow)f^+(\epsilon_\downarrow)] \\ \sin^2\frac{\theta}{2}f^+(\epsilon_\uparrow)f^-(\epsilon_\downarrow) + \cos^2\frac{\theta}{2}f^-(\epsilon_\uparrow)f^+(\epsilon_\downarrow) \\ f^+(\epsilon_\uparrow)f^+(\epsilon_\downarrow) \end{pmatrix}, \tag{A25}$$

where $f^+(\epsilon_k) \equiv \mathrm{Tr}_1[c_{\sigma,k}^\dagger c_{\sigma,k}\rho_1^{eq}]$, $f^-(\epsilon_k) \equiv \mathrm{Tr}_1[c_{\sigma,k}c_{\sigma,k}^\dagger\rho_1^{eq}]$, $\epsilon_\uparrow \equiv \epsilon_d - M$ and $\epsilon_\downarrow \equiv \epsilon_d + M$. From the expression, we find that the steady-state values of the populations $\rho_{00}, \rho_{01}, \rho_{10}$ and ρ_{11} are independent of angle ϕ. Thus, the steady state populations remain unchanged by changing ϕ.

References and Notes

1. Pekola, J.P.; Saira, O.-P.; Maisi, V.F.; Kemppinen, A.; Möttönen, M.; Pashkin, Y.A.; Averin, D.V. Single-electron current sources: Toward a refined definition of the ampere. *Rev. Mod. Phys.* **2013**, *85*, 1421–1472.
2. Žutić, I.; Fabian, J.; Sarma, S.D. Spintronics: Fundamentals and applications. *Rev. Mod. Phys.* **2004**, *76*, 323.
3. Maekawa, S.; Adachi, H.; Uchida, K.-I.; Ieda, J.I.; Saitoh, E. Spin current: Experimental and theoretical aspects. *J. Soc. Phys. Jpn.* **2013**, *82*, 102002.
4. Cui, L.; Miao, R.; Jiang, C.; Meyhofer, E.; Reddy, P. Perspective: Thermal and thermoelectric transport in molecular junctions. *J. Chem. Phys.* **2017**, *146*, 092201.
5. Keller, M.W.; Zimmerman, N.M.; Eichenberge, A.L. Current status of the quantum metrology triangle. *Metrologia* **2007**, *44*, 505.
6. Thouless, D.J. Quantization of particle transport. *Phys. Rev. B* **1983**, *27*, 6083.
7. Niu, Q.; Thouless, D.J. Quantised adiabatic charge transport in the presence of substrate disorder and many-body interaction. *J. Phys. A Math. Gen.* **1984**, *17*, 2453–2462.
8. Altshuler, B.L.; Glazman, L.I. Pumping electrons. *Science* **1999**, *283*, 1864–1865.
9. Switkes, M.; Marcus, C.M.; Campman, K.; Gossard, A.C. An adiabatic quantum electron pump. *Science* **1999**, *283*, 1905–1908.
10. Büttiker, M.; Thomas, H.; Prêtre, A. Current partition in multiprobe conductors in the presence of slowly oscillating external potentials. *Z. Phys. B Condens. Matter* **1994**, *94*, 133–137.
11. Brouwer, P.W. Scattering approach to parametric pumping. *Phys. Rev. B* **1998**, *58*, R10135.
12. Tserkovnyak, Y.; Brataas, A.; Bauer, G.E.W. Enhanced Gilbert Damping in Thin Ferromagnetic Films. *Phys. Rev. Lett.* **2002**, *88*, 117601.
13. Tserkovnyak, Y.; Brataas, A.; Bauer, G.E.W. Spin pumping and magnetization dynamics in metallic multilayers. *Phys. Rev. B* **2002**, *66*, 224403.
14. Mucciolo, E.R.; Chamon, C.; Marcus, C.M. Adiabatic Quantum Pump of Spin-Polarized Current. *Phys. Rev. Lett.* **2002**, *89*, 146802.
15. Avron, J.E.; Elgart, A.; Graf, G.M.; Sadun, L. Geometry, statistics, and asymptotics of quantum pumps. *Phys. Rev. B* **2000**, *62*, R10618.
16. Moskalets, M.; Büttiker, M. Floquet scattering theory of quantum pumps. *Phys. Rev. B* **2010**, *66*, 205320.
17. Moskalets, M. *Scattering Matrix Approach to Non-Stationary Quantum Transport*; Imperial College Press: London, UK, 2011.
18. Jauho, A.P.; Wingreen, N.S.; Meir, Y. Time-dependent transport in interacting and noninteracting resonant-tunneling systems. *Phys. Rev. B* **1994**, *50*, 5528.
19. Wang, B.; Wang, J.; Guo, H. Quantum spin field effect transistor. *Phys. Rev. B* **2003**, *67*, 092408.
20. Zhang, P.; Xue, Q.K.; Xie, X.C. Spin Current through a Quantum Dot in the Presence of an Oscillating Magnetic Field. *Phys. Rev. Lett.* **2003**, *91*, 196602.
21. Hattori, K. Kondo effect on adiabatic spin pumping from a quantum dot driven by a rotating magnetic field. *Phys. Rev. B* **2008**, *78*, 155321.
22. Fransson, J.; Galperin, M. Inelastic scattering and heating in a molecular spin pump. *Phys. Rev. B* **2010**, *81*, 075311.
23. Chen, K.; Zhang, Z. Spin Pumping in the Presence of Spin-Orbit Coupling. *Phys. Rev. Lett.* **2015**, *114*, 126602.
24. Tatara, G. Green's function representation of spin pumping effect. *Phys. Rev. B* **2016**, *94*, 224412.
25. Tatara, G.; Mizukami, S. Consistent microscopic analysis of spin pumping effects. *Phys. Rev. B* **2017**, *96*, 064423.
26. Tatara, G. Effective gauge field theory of spintronics. *Phys. E Low Dimens. Syst. Nanostruct.* **2019**, *106*, 208–238.
27. Sinitsyn, N.A.; Nemenman, I. The Berry phase and the pump flux in stochastic chemical kinetics. *Europhys. Lett.* **2007**, *77*, 58001.
28. Ren, J.; Hänggi, P.; Li, B. Berry-Phase-Induced Heat Pumping and Its Impact on the Fluctuation Theorem. *Phys. Rev. Lett.* **2010**, *104*, 170601.
29. Sagawa, T.; Hayakawa, H. Geometrical expression of excess entropy production. *Phys. Rev. E* **2011**, *84*, 051110.

30. Yuge, T.; Sagawa, T.; Sugita, A.; Hayakawa, H. Geometrical pumping in quantum transport: Quantum master equation approach. *Phys. Rev. B* **2012**, *86*, 235308.

31. Yuge, T.; Sagawa, T.; Sugita, A.; Hayakawa, H. Geometrical Excess Entropy Production in Nonequilibrium Quantum System. *J. Stat. Phys.* **2013**, *153*, 412–441.

32. Nakajima, S.; Taguchi, M.; Kubo, T.; Tokura, Y. Interaction effect on adiabatic pump of charge and spin in quantum dot. *Phys. Rev. B* **2015**, *92*, 195420.

33. Berry, M.V. Quantal phase factors accompanying adiabatic changes. *Proc. R. Soc. Lond. Math. Phys. Sci.* **1984**, *392*, 45–57.

34. Winkler, N.; Governale, M.; König, J. Theory of spin pumping through an interacting quantum dot tunnel coupled to a ferromagnet with time-dependent magnetization. *Phys. Rev. B* **2013**, *87*, 155428.

35. Rojek, S.; Governale, M.; König, J. Spin pumping through quantum dots. *Phys. Status Solidi B* **2013**, *251*, 1912–1923.

36. Jahn, B.O.; Ottosson, H.; Galperin, M.; Fransson, J. Organic Single Molecular Structures for Light Induced Spin-Pump Devices. *ACS Nano* **2013**, *7*, 1064–1071.

37. Cota, E.; Aguado, R.; Platero, G. ac-Driven Double Quantum Dots as Spin Pumps and Spin Filters. *Phys. Rev. Lett.* **2005**, *94*, 107202.

38. Braun, M.; Murkard, G. Nonadiabatic Two-Parameter Charge and Spin Pumping in a Quantum Dot. *Phys. Rev. Lett.* **2008**, *101*, 036802.

39. Cavaliere, F.; Governale, M.; König, J. Nonadiabatic Pumping through Interacting Quantum Dots. *Phys. Rev. Lett.* **2009**, *103*, 136801.

40. Rojek, S.; König, J.; Shnirman, A. Adiabatic pumping through an interacting quantum dot with spin-orbit coupling. *Phys. Rev. B* **2013**, *87*, 075305.

41. Splettstoesser, J.; Governale, M.; König, J. Adiabatic charge and spin pumping through quantum dots with ferromagnetic leads. *Phys. Rev. B* **2008**, *77*, 195320.

42. Riwar, R.; Splettstoesser, J. Charge and spin pumping through a double quantum dot. *J. Phys. Rev. B* **2010**, *82*, 205308.

43. Esposito, M.; Harbola, U.; Mukamel, S. Entropy fluctuation theorems in driven open systems: Application to electron counting statistics. *Phys. Rev. E* **2007**, *76*, 031132.

44. Esposito, M.; Harbola, U.; Mukamel, S. Nonequilibrium fluctuations, fluctuation theorems, and counting statistics in quantum systems. *Rev. Mod. Phys.* **2009**, *81*, 1665.

45. In dealing with a certain open quantum system with a strong system-bath coupling, one may intend to apply a similarity (unitary) transformation, represented by S and S^{-1}, to the total system Hamiltonian. However, in the full counting statistics, it needs a careful treatment. The ultimate goal of the full counting statistics is to calculate a certain statistical quantity provided by the two point measurement described by the joint probability, Equation (2). By application of the similarity transformation, the time evolution operator as well as the projection operators in Equation (2) are transformed as $\mathrm{Tr}[(P_{q_{t_{i+1}}} S)(S^{-1} U(t_{i+1}, t_i) S)(S^{-1} P_{q_{t_i}}) W(t_i)(P_{q_{t_i}} S)(S^{-1} U^\dagger(t_{i+1}, t_i) S)(S^{-1} P_{q_{t_{i+1}}})]$. This means that we need to pay attention that the projection operator should also be transformed to recover the original joint probability. However, we can avoid the difficulty when the similarity transformation commutes with the projection, $[S, P_{q_t}] = 0$.

46. Kubo, R. Stochastic Liouville Equations. *J. Math. Phys.* **1963**, *4*, 174–183.

47. Hänggi, P.; Thomas, H. Time evolution, correlations, and linear response of non-Markov processes. *Z. Phys. B Condens. Matter* **1977**, *26*, 85–92.

48. Hashitsume, N.; Shibata, F.; Shingu, M. Quantal master equation valid for any time scale. *J. Stat. Phys.* **1977**, *17*, 155–169.

49. Shibata, F.; Takahashi, F.; Hashitsume, N. A generalized stochastic liouville equation. Non-Markovian versus memoryless master equations. *J. Stat. Phys.* **1977**, *17*, 171–187.

50. Chaturvedi, S.; Shibata, F. Time-convolutionless projection operator formalism for elimination of fast variables. Applications to Brownian motion. *Z. Phys.* **1979**, *35*, 297–308.

51. Shibata, F.; Arimitsu, T. Expansion Formulas in Nonequilibrium Statistical Mechanics. *J. Phys. Soc. Jpn.* **1980**, *49*, 891–897.

52. Uchiyama, C.; Shibata, F. Unified projection operator formalism in nonequilibrium statistical mechanics. *Phys. Rev. E* **1999**, *60*, 2636.

53. Breuer, H.-P.; Petruccione, F. *The Theory of Open Quantum Systems*; Oxford University Press: New York, NY, USA, 2002.

54. Segal, D.; Nitzan, A. Heat rectification in molecular junctions. *J. Chem. Phys.* **2005**, *122*, 194704.

55. Segal, D. Heat flow in nonlinear molecular junctions: Master equation analysis. *Phys. Rev. B* **2006**, *73*, 205415.

56. Uchiyama, C. Nonadiabatic effect on the quantum heat flux control. *Phys. Rev. E* **2014**, *89*, 052108.

57. The steplike rotation of the magnetization is experimentally feasible by applying a periodic pulse train of magnetic field on the quantum dot along z-axis. If the magnetic field is strong enough and the pulse interval is sufficiently longer than the Larmor period of the magnetization under the magnetic field, dynamic of the magnetization is well described by the steplike motion.

58. Hashimoto, K.; Tatara, G.; Uchiyama, C. Nonadiabaticity in spin pumping under relaxation. *Phys. Rev. B* **2017**, *96*, 064439.

59. We introduce a unit energy $\epsilon_u \equiv 2M$, a unit angular frequency $\omega_u \equiv 2M/\hbar$, and a unit time $t_u \equiv 2\pi/\omega_u$, and normalize energy, angular frequency, inverse temperature and time as $\bar{\omega} \equiv \omega/\omega_u$, $\bar{\omega} \equiv \omega/\omega_u$, $\bar{\epsilon} \equiv \epsilon/\epsilon_u$, $\bar{\beta} \equiv \beta/\beta_u$, and $\bar{t} \equiv t/t_u$, respectively.

60. Hashimoto, K.; Tatara, G.; Uchiyama, C. Nonadiabaticity in spin pumping under relaxation. *Phys. Rev. B* **2019**, *99*, 205304.

61. Kaestner, B.; Kashcheyevs, V. Non-adiabatic quantized charge pumping with tunable-barrier quantum dots: a review of current progress. *Rep. Prog. Phys.* **2015**, *78*, 103901.

62. Potanina, E.; Brandner, K.; Flindt, C. Optimization of quantized charge pumping using full counting statistics. *Phys. Rev. B* **2019**, *99*, 035437.

63. Watanabe, K.L.; Hayakawa, H. Non-adiabatic effect in quantum pumping for a spin-boson system. *Prog. Theor. Exp. Phys.* **2014**, *2014*, 113A01.

64. Wang, C.; Ren, J.; Cao, J. Unifying quantum heat transfer in a nonequilibrium spin-boson model with full counting statistics. *Phys. Rev. A* **2017**, *95*, 023610.

65. Restrepo, S.; Böhling, S.; Cerrillo, J.; Schaller, G. Electron pumping in the strong coupling and non-Markovian regime: A reaction coordinate mapping approach. *Phys. Rev. B* **2019**, *100*, 035109.

66. Entin-Wohlman, O.; Aharony, A. Adiabatic transport in nanostructures. *Phys. Rev. B* **2002**, *65*, 195411.

67. Crépieux, A.; Šimkovic, F.; Cambon, B.; Michelini, F. Enhanced thermopower under a time-dependent gate voltage. *Phys. Rev. B* **2011**, *83*, 153417.

68. Ludovico, M.F.; Battista, F.; von Oppen, F.; Arrachea, L. Adiabatic response and quantum thermoelectrics for ac-driven quantum systems. *Phys. Rev. B* **2016**, *93*, 075136.

69. Hammar, H.; Jaramillo, J.D.V.; Fransson, J. Spin-dependent heat signatures of single-molecule spin dynamics. *Phys. Rev. B* **2019**, *99*, 115416.

70. Bode, N.; Kusminskiy, S.V.; Egger, R.; von Oppen, F. Current-induced forces in mesoscopic systems: A scattering-matrix approach. *Beilstein J. Nanotechnol.* **2012**, *3*, 144–162.

71. Craighead, H.G. Nanoelectromechanical Systems. Dynamics of energy transport and entropy production in ac-driven quantum electron systems. *Science* **2000**, *290*, 1532–1535.

72. Ludovico, M.F.; Moskalets, M.; Sánchez, D.; Arrachea, L. Dynamics of energy transport and entropy production in ac-driven quantum electron systems. *Phys. Rev. B* **2016**, *94*, 035436.

73. Haken, H. *Synergetics: An Introduction: Nonequilibrium Phase Transitions and Self-Organization in Physics, Chemistry, and Biology*; Springer: New York, NY, USA, 1983.

Article

Generalized Master Equation Approach to Time-Dependent Many-Body Transport

Valeriu Moldoveanu [1,*], Andrei Manolescu [2] and Vidar Gudmundsson [3]

[1] National Institute of Materials Physics, Atomistilor 405A, 077125 Magurele, Romania
[2] School of Science and Engineering, Reykjavik University, Menntavegur 1, IS-101 Reykjavik, Iceland
[3] Science Institute, University of Iceland, Dunhaga 3, IS-107 Reykjavik, Iceland
[*] Correspondence: valim@infim.ro

Received: 17 June 2019; Accepted: 23 July 2019; Published: 25 July 2019

Abstract: We recall theoretical studies on transient transport through interacting mesoscopic systems. It is shown that a generalized master equation (GME) written and solved in terms of many-body states provides the suitable formal framework to capture both the effects of the Coulomb interaction and electron–photon coupling due to a surrounding single-mode cavity. We outline the derivation of this equation within the Nakajima–Zwanzig formalism and point out technical problems related to its numerical implementation for more realistic systems which can neither be described by non-interacting two-level models nor by a steady-state Markov–Lindblad equation. We first solve the GME for a lattice model and discuss the dynamics of many-body states in a two-dimensional nanowire, the dynamical onset of the current-current correlations in electrostatically coupled parallel quantum dots and transient thermoelectric properties. Secondly, we rely on a continuous model to get the Rabi oscillations of the photocurrent through a double-dot etched in a nanowire and embedded in a quantum cavity. A many-body Markovian version of the GME for cavity-coupled systems is also presented.

Keywords: time-dependent transport; electron–photon coupling; open quantum systems

1. Introduction

Few-level open systems stand as everyday 'lab rats' in corner stone experiments and future technologies in nanoelectronics [1] and quantum optics [2]. Generically, they are electronic systems with a discrete spectrum (e.g., artificial atoms [3], nanowires or superconducting qubits [4]) connected to particle reservoirs or embedded in bosonic baths. Depending on the nature of the environment (i.e., fermionic or bosonic) to which the open systems are coupled, their theoretical investigation started with two toy-models, namely the single-level Hamiltonian of quantum transport and the Jaynes–Cummings (JC) Hamiltonian of a two-level system (TLS).

Surprisingly or not, studying the sequential tunneling transport regime or the optical properties of quantum emitters eventually boils down to solve formally similar Markovian master equations (MEs) for the so called reduced density operator (RDO). The latter defines the non-unitary evolution of the small system in the presence of the infinite degrees of freedom of the reservoirs. Such MEs are derived by tracing out the reservoir's degrees of freedom and are known from the early days of condensed matter and quantum optics (see the seminal works of Bloch [5], Wangsness [6] and Redfield [7]). The master equation cleverly bypasses the fact that the Liouville–von-Neumann (LvN) equation of the coupled systems (i.e., the open system and the reservoirs) is impossible to solve and takes advantage of the fact that all observables associated to the small and open system can be calculated as statistical averages w.r.t. the RDO.

Indeed, the RDO associated to the Jaynes–Cummings model has been a central object in quantum optics [8,9] (e.g., in the study of lasing and for the calculation of photon correlation functions). In this context the master equation (ME) approach goes as follows: (i) one studies an atomic few-level system whose eigenvalues and eigenfunctions are supposed to be known; (ii) the dissipation in the system (e.g., cavity losses or various non-radiative recombination processes) is included through the so called 'jump' operators; (iii) the occupation of atomic levels changes due to photon emission or absorption, but the particle number is conserved as the system is not coupled to particle reservoirs; (iv) under the Markov approximation the ME acquires a Lindblad form, usually solved in the steady-state regime.

The above scenario changes when one aims to derive a quantum master equation describing transport phenomena. (i) The Coulomb interaction effects on the spectrum and eigenstates of the system cannot be always neglected, especially for confined systems like quantum dots or nanowires; this requires a many-body derivation of the master equation; (ii) The tunneling between source/drain probes prevents the charge conservation in the central system and the main quantity of interest is the electronic current; (iii) Finally, the steady-state regime does not cover the whole physics and cannot even be guaranteed in general; moreover, the validity of the rotating-wave (RWA) and Markov approximations must be established more carefully [10,11]. In fact it turns out that when applied to transport processes the master equation must rather be solved in its non-Markovian version.

Such generalized master equations which take into account the memory effects have been mostly derived and implemented for time-dependent transport in non-interacting [12,13] and interacting [14] quantum dots, nanowires, and rings [15]. It turns out that the generalized master equation (GME) method is a valuable tool for modeling and monitoring the dynamics of specific many-body states as well as for investigating time-dependent propagation along a sample [16] or capturing charge sensing effects [17] and counting statistics in electrostatically parallel QDs [18]. In particular, Harbola et al. [19] showed that a Lindblad form of the quantum master equation is still recovered in the high bias limit and by assuming the RWA.

Since then, a lot of theoretical work has been done to improve and refine the quantum master equation formalism. A formally exact memory-kernel for the Anderson model was derived and calculated using real-time path integral Monte Carlo methods [20]. A hierarchical quantum master equation approach with good convergence at not too low temperature was put forward by Härtle et al. [21]. As for molecular transport calculations one can rely on the GME written in terms of the many-body states of the isolated molecule [22,23]. A recent review on non-markovian effects in open systems is also available [24].

As we shall see below the implementation of GME approach to many-level systems with specific geometries poses considerable technical difficulties. These are related to the many-body structure of the central interacting system, to the accurate description of the contact regions and, more importantly, to the evaluation of the non-Markovian kernels which become complicated objects once we go beyond non-interacting single-level models.

A second useful extension of the ME method emerged in the context of cavity quantum electrodynamics. Here the system under study is a hybrid one, as the electronic system is still coupled to source/drain reservoirs (i.e., leads) but also interacts with a quantum cavity mode, the latter being subjected to dissipation into leaky modes described by a bosonic bath. Such systems are currently used in state-of-the-art measurements in cavity quantum electrodynamics [25–29]. Again, the many-body nature of the problem is essential, as the electron-photon coupling leads to the formation of dressed states whose dynamics in the presence of both particle and dissipative bosonic reservoirs is far from being trivial. The relevant reduced density operator now acts in the many-body electron-photon Fock space and describes the dynamics of dressed-states. This fact brings new technical difficulties in the derivation [30,31] and implementation of ME [32,33]. Let us also mention here recent studies on ground state electroluminescence [34,35] and on cavity enhanced transport of charge [36].

In view of the abovementioned comments, the aim of this work is: (i) to briefly review the development of the generalized master equation approach to time-dependent many-body transport in the presence of both fermionic and bosonic environments and (ii) to illustrate in a unified framework how the method really works, from formal technicalities to numerical implementation. In Section 2 we shall therefore derive a non-Markovian master equation which describes the dynamics and the transport properties of rather general 'hybrid' system consisting in an electronic component S_1 which is connected to particle reservoirs (i.e., leads) and a second subsystem S_2. The latter, although not coupled to particle reservoirs, interacts with system S_1 or with some leaky modes described as bosonic baths. Then we specialize this master equation to several systems of interest. More precisely, in Section 3 we recall GME results on transient charging of excited states and Coulomb-coupled quantum dots. Section 4 deals with thermoelectric transport. Applications to transport in cavity quantum electrodynamics are collected in Sections 5 and 6. We conclude in Section 7.

2. Formalism

2.1. Generalized Master Equation for Hybrid Systems

Non-Markovian master equations for open systems have been derived in many recent textbooks or review papers via projection methods (e.g., Nakajima–Zwanzig formalism or time-convolutionless approach [37]). Nonetheless it is still instructive to outline here some theoretical and computational difficulties one encounters when solving transport master equation for interacting many-level systems.

From the formal point of view the projection technique is quite general and the derivation of a master equation for the RDO does not depend on a specific model (i.e., on the geometry and spectrum of the central system or on the correlation functions of fermionic/bosonic reservoirs). In general, as long as one can write down a system-reservoir coupling Hamiltonian H_{SR} a master equation can be derived.

For the sake of generality we shall consider a hybrid system S made of an electronic structure S_1 which is coupled to n_r particle reservoirs characterized by chemical potentials and temperatures $\{\mu_l, T_l\}$, $l = 1, 2, ..., n_r$, and a second subsystem S_2 (i.e., a localized impurity, or an oscillator, or a single-mode quantum cavity). The subsystem S_2 can only be coupled to thermal or photonic baths which are described as a collection of oscillators with frequencies $\{\omega_k\}$. Let $\mathcal{F}_{S_1}$ and $\mathcal{F}_{S_2}$ be the Fock spaces associated to the two systems. Typically $\mathcal{F}_{S_1}$ is a set of interacting many-body configurations of the electronic system whereas $\mathcal{F}_{S_2}$ is made by harmonic oscillator Fock states.

The dynamics of the open system S_1 and of nearby 'detector' system S_2 are intertwined by a coupling V. Under a voltage bias or a temperature difference the system S_1 carries an electronic or a heat current which need to be calculated in the presence of the second subsystem. Conversely, the averaged observables of S_2 (e.g., mean photon number or the spin of a localized impurity) will also depend on the transport properties of S_1. The Hamiltonian of the hybrid structure is:

$$H_S = H_{S_1} + H_{S_2} + V. \tag{1}$$

In this work H_{S_1} will describe various Coulomb-interacting structures: a single quantum dot, a 2D wire or parallel quantum dots. We shall denote by $|\nu\rangle$ and E_ν the many-body configurations and eigenvalues of H_{S_1}, that is one has $H_{S_1}|\nu\rangle = E_\nu|\nu\rangle$. H_{S_1} can be equally expressed in terms of creation and annihilation operators $\{c_{n\sigma}^\dagger, c_{n\sigma}\}$ associated to a spin-dependent single-particle basis $\{\psi_{n\sigma}\}$ of a single-particle Hamiltonian $h_{S_1}^{(0)}$ (see the next sections for specific models), such that:

$$H_{S_1} = H_{S_1}^{(0)} + W, \tag{2}$$

where $H_{S_1}^{(0)}$ is the 2nd quantized form of $h_{S_1}^{(0)}$ and W is the Coulomb interaction. Similarly, the eigenstates and eigenvalues of the second subsystem S_2 will be denoted by $|j\rangle$ and e_j such that $H_{S_2}|j\rangle = e_j|j\rangle$. As for the coupling V one can mention at least three examples: The exchange interaction between a quantum dot and a localized magnetic impurity with total spin S, the electron–photon coupling in a quantum-dot cavity and the electron–vibron coupling in nanoelectromechanical systems [38,39].

The total Hamiltonian of the system coupled to particle and/or bosonic reservoirs R reads as:

$$H(t) = H_S + H_R + H_{SR}(t) := H_0 + H_{SR}(t), \tag{3}$$

where the system-reservoir coupling H_{SR} collects the coupling to the leads (H_T) and the coupling of a bosonic mode to a thermal or leaky bosonic environments (H_E):

$$H_{SR}(t) = H_T(t) + H_E. \tag{4}$$

Note that the interaction with the bosonic environment H_E does not depend on time. The lead-sample tunneling term H_T carries a time-dependence that will be explained below. The Hamiltonian of the reservoirs,

$$H_R = H_{\text{leads}} + H_{\text{bath}} \tag{5}$$

describes at least two semiinfinite leads (left-L and right-R) but could also contain a bosonic or a thermal bath.

This general scheme allows one to recover several relevant settings. If S_1 describes an optically active structure and S_2 defines a photonic mode then V could become either the Rabi or the Jaynes–Cummings electron–photon coupling. The absence of the particle reservoirs simplifies H_S to well known models in quantum optics, while by adding them one can study photon-assisted transport effects (e.g., Rabi oscillation of the photocurrents or electroluminescence). Also, by removing S_2, V and the bosonic dissipation one finds the usual transport setting for a Coulomb interacting purely electronic structure.

Let $\varepsilon^l(q)$ and $\psi_{q\sigma}^l$ be the single particle energies and wave functions of the l-th lead. For simplicity we assume that the states on the leads are spin-degenerate so their energy levels do not depend on the spin index. Using the creation/annihilation operators $c_{q l \sigma}^{\dagger}/c_{q l \sigma}$ associated to the single particle states, we can write:

$$H_{\text{leads}} = \sum_l H_l = \int dq \sum_{\sigma} \varepsilon^l(q) c_{q l \sigma}^{\dagger} c_{q l \sigma}. \tag{6}$$

As for the bosonic bath, it is described by a collection of harmonic oscillators with frequencies ω_k and by corresponding creation/annihilation operators $b_k^{\dagger}/b_k$:

$$H_{\text{bath}} = \sum_k \hbar \omega_k b_k^{\dagger} b_k. \tag{7}$$

The tunneling Hamiltonian has the usual form:

$$H_T(t) = \sum_l \sum_{n\sigma} \int dq \chi_l(t) (T_{qn}^l c_{q l \sigma}^{\dagger} c_{n\sigma} + h.c), \tag{8}$$

where we considered without loss of generality that the tunneling processes are spin conserving. For the simplicity of writing the spin degree of freedom σ will be henceforth tacitly merged with the single-particle index n and restored if needed.

The time-dependent switching functions $\chi_l(t)$ control the time-dependence of the contacts between the leads and the sample; these functions mimic the presence of a time dependent potential barrier. We emphasize that in most studies based on ME method the coupling to the leads is suddenly switched at some initial instant t_0 such that for each lead $\chi_l(t) = \theta(t - t_0)$ where $\theta(x)$ is the Heaviside step function. This choice is very convenient if one imposes the Markov approximation in view of a time-local Master equation. Here we allow for more general switching functions: (i) a smooth coupling to the leads or (ii) time-dependent signals applied at the contacts to the leads. In particular, if the potential barriers oscillate out of phase the system operates like a turnstile pump under a finite constant bias.

The coupling T_{qn}^l describes the tunneling strength between a state with momentum q of the lead l and the state n of the isolated sample with wavefunctions ψ_n. In the next sections we shall show that these matrix elements have to be calculated for each specific model by taking into account the geometry of the system and of the leads.

The associated density operator $\mathcal{W}$ of the open system obeys the Liouville–von Neumann equation:

$$i\hbar\frac{\partial \mathcal{W}(t)}{\partial t} = \mathcal{L}(t)\mathcal{W}(t), \quad \mathcal{W}(t_0) = \rho_S(t_0) \otimes \rho_R, \tag{9}$$

where:

$$\mathcal{L}(t) = \mathcal{L}_0 + \mathcal{L}_{SR}, \quad \mathcal{L}_0\cdot = [H_0, \cdot]. \tag{10}$$

We also introduce the notations:

$$\mathcal{L}_S\cdot = [H_S, \cdot], \quad \mathcal{L}_{SR}\cdot = [H_{SR}, \cdot]. \tag{11}$$

The Nakajima–Zwanzig projection formalism leads to an equation of motion for the reduced density operator $\rho(t) = \mathrm{Tr}_R\{\mathcal{W}\}$. The initial state $\mathcal{W}_0 := \mathcal{W}(t_0)$ factorizes as:

$$\mathcal{W}_0 = \rho_0 \otimes \rho_{\text{leads}} \otimes \rho_{\text{bath}} := \rho_0 \otimes \rho_R, \tag{12}$$

where the equilibrium density operator of the leads reads:

$$\rho_{\text{leads}} = \prod_l \frac{e^{-\beta_l(H_l - \mu_l N_l)}}{\mathrm{Tr}_l\{e^{-\beta_l(H_l - \mu_l N_l)}\}}, \tag{13}$$

and $\beta_l = 1/k_B T_l$, μ_l and N_l denote the inverse temperature, chemical potential and the occupation number operator of the lead l. Similarly,

$$\rho_{\text{bath}} = \prod_k e^{-\hbar\omega_k b_k^\dagger b_k/k_B T}(1 - e^{-\hbar\omega_k/k_B T}). \tag{14}$$

Finally, ρ_0 is simply a projection on one of the states of the hybrid system, and as such its calculation must take into account the effect of the hybrid coupling V (see the discussion in Section 2.2). We now define two projections:

$$P\cdot = \rho_R \mathrm{Tr}_R\{\cdot\}, \quad Q = 1 - P. \tag{15}$$

It is straightforward to check the following properties:

$$P\mathcal{L}_S = \mathcal{L}_S P, \quad P\mathcal{L}_{SR}(t)P = 0. \tag{16}$$

The Liouville Equation (9) splits then into two equations:

$$i\hbar P\frac{\partial \mathcal{W}(t)}{\partial t} = P\mathcal{L}(t)P\mathcal{W}(t) + P\mathcal{L}(t)Q\mathcal{W}(t) \tag{17}$$

$$i\hbar Q\frac{\partial \mathcal{W}(t)}{\partial t} = Q\mathcal{L}(t)Q\mathcal{W}(t) + Q\mathcal{L}(t)P\mathcal{W}(t), \tag{18}$$

and the second equation can be solved by iterations (T being the time-ordering operator):

$$Q\mathcal{W}(t) = \frac{1}{i\hbar}\int_{t_0}^t ds\, T\exp\left\{-\frac{i}{\hbar}\int_s^t ds'\, Q\mathcal{L}(s')\right\} Q\mathcal{L}(s)P\mathcal{W}(s). \tag{19}$$

Inserting Equation (19) in Equation (17) and using the properties of P we get the Nakajima–Zwanzig equation:

$$
\begin{aligned}
i\hbar P \frac{\partial W(t)}{\partial t} &= P\mathcal{L}_S W(t) \\
&+ \frac{1}{i\hbar} P\mathcal{L}_{SR}(t) Q \int_{t_0}^t ds\, T \exp\left\{-\frac{i}{\hbar}\int_s^t ds' Q\mathcal{L}(s')Q\right\} Q\mathcal{L}_{SR}(s) PW(s).
\end{aligned}
\tag{20}
$$

In order to have an explicit perturbative expansion in powers of $H_{SR}(t)$ one has to factorize the time-ordered exponential as follows:

$$
T\exp\left\{-\frac{i}{\hbar}\int_s^t ds' Q\mathcal{L}(s')Q\right\} = \exp\{Q\mathcal{L}_0 Q\}(1+\mathcal{R}),
\tag{21}
$$

where the remainder $\mathcal{R}$ contains infinitely deep commutators with inconveniently embedded projection operators. Usually one considers a truncated version of the Nakajima–Zwanzig equation up to the second order contribution w.r.t. the system-reservoir H_{SR}:

$$
i\hbar\dot{\rho}(t) = \mathcal{L}_S\rho(t) + \frac{1}{i\hbar}\mathrm{Tr}_R\left\{\mathcal{L}_{SR}\int_{t_0}^t ds\, e^{-i(t-s)\mathcal{L}_0}\mathcal{L}_{SR}(s)\rho_R\rho(s)\right\}.
\tag{22}
$$

Now, by taking into account that for any operator A acting on the Fock space of the hybrid system $e^{-it\mathcal{L}_0}A = e^{-itH_0}Ae^{itH_0}$ and denoting by $U_0(t,s) = e^{-i(t-s)H_0}$ the unitary evolution of the disconnected systems we arrive at the well known form of the GME:

$$
\begin{aligned}
i\hbar\dot{\rho}(t) &= [H_S,\rho(t)] - \frac{i}{\hbar}U_0^\dagger(t,t_0)\mathrm{Tr}_R\left\{\int_{t_0}^t ds\, [\tilde{H}_{SR}(t),[\tilde{H}_{SR}(s),\tilde{\rho}(s)\rho_R]]\right\} U_0(t,t_0) \\
&= [H_S,\rho(t)] - \frac{i}{\hbar}U_S^\dagger(t,t_0)\mathrm{Tr}_R\left\{\int_{t_0}^t ds\, [\tilde{H}_{SR}(t),[\tilde{H}_{SR}(s),\tilde{\rho}(s)\rho_R]]\right\} U_S(t,t_0),
\end{aligned}
$$

where in order to get to the last line we removed the evolution operators of the environment from both sides of the trace. At the next step one observes that when performing the trace over the reservoirs and environment degrees of freedom the mixed terms in the double commutator vanish because each of the coupling terms H_T and H_E carries only one creation or annihilation operator for the corresponding reservoir such that:

$$
\mathrm{Tr}_R\{\tilde{c}_{ql}^\dagger(t)\tilde{b}_k(s)\rho_R\} = \mathrm{Tr}_{\text{leads}}\{\tilde{c}_{ql}^\dagger(t)\rho_{\text{leads}}\}\cdot\mathrm{Tr}_{\text{bath}}\{\tilde{b}_k(s)\rho_{\text{bath}}\} = 0.
\tag{23}
$$

Moreover, the time evolution of each term can be simplified due to the commutation relations $[H_{\text{bath}}, H_T] = [H_{\text{leads}}, H_E] = 0$:

$$
\begin{aligned}
\tilde{H}_T(t) &= e^{\frac{i}{\hbar}tH_S}e^{\frac{i}{\hbar}tH_{\text{leads}}}H_T e^{-\frac{i}{\hbar}tH_S}e^{-\frac{i}{\hbar}tH_{\text{leads}}}, &\tag{24} \\
\tilde{H}_E(t) &= e^{\frac{i}{\hbar}tH_S}e^{\frac{i}{\hbar}tH_{\text{bath}}}H_E e^{-\frac{i}{\hbar}tH_S}e^{-\frac{i}{\hbar}tH_{\text{bath}}}. &\tag{25}
\end{aligned}
$$

The GME then reads as:

$$
\begin{aligned}
\dot{\rho}(t) &= -\frac{i}{\hbar}[H_S,\rho(t)] - \frac{1}{\hbar^2}U_S^\dagger(t,t_0)\mathrm{Tr}_{\text{leads}}\left\{\int_{t_0}^t ds\, [\tilde{H}_T(t),[\tilde{H}_T(s),\tilde{\rho}(s)\rho_{\text{leads}}]]\right\} U_S(t,t_0) \\
&- \frac{1}{\hbar^2}U_S^\dagger(t,t_0)\mathrm{Tr}_{\text{bath}}\left\{\int_{t_0}^t ds\, [\tilde{H}_E(t),[\tilde{H}_E(s),\tilde{\rho}(s)\rho_{\text{bath}}]]\right\} U_S(t,t_0) \tag{26} \\
&:= -\frac{i}{\hbar}[H_S,\rho(t)] - \mathcal{D}_{\text{leads}}[\rho,t] - \mathcal{D}_{\text{bath}}[\rho,t]. \tag{27}
\end{aligned}
$$

Equation (27) is the generalized master equation for our hybrid system. It provides the reduced density operator ρ in the presence of particle and bosonic reservoirs and also takes into account the memory effects and the non-trivial role of time-dependent signals applied at the contact regions through the switching functions χ_l. The third term in Equation (27) is needed only if H_{S_2} describes a quantized optical or mechanical oscillation mode. In our work on open QD-cavity systems we always assume that the coupling between the cavity photons and other leaky modes is much smaller that the electron-photon coupling g_{EM} (see Section 5). On the other hand, for our calculations $g_{EM} \ll \hbar\omega$, ω being the frequency of the cavity mode. Then the RWA holds and $\mathcal{D}_{\text{bath}}[\rho, t]$ can be cast in a Lindblad form. Let us stress that in the ultrastrong coupling regime on typically has $g_{EM}/\hbar\omega > 0.2$ and the derivation of the dissipative term is more complicated and involves the dressed states of the QD-cavity system [30]. In order to describe dissipative effects in the ultrastrong coupling regime beyond the RWA one needs more elaborate techniques [40,41].

For further calculations one has to solve the GME as a system of coupled integro-differential equations for the matrix elements of the RDO with respect to a suitable basis in the Fock space $\mathcal{F}_S = \mathcal{F}_{S_1} \otimes \mathcal{F}_{S_2}$. We discuss this issue in the next subsection.

2.2. 'Hybrid' States and Diagonalization Procedure

The starting point in solving the GME is to write down the matrix elements of the system-environment operators H_T and H_E w.r.t. the 'disjointed' basis formed by the eigenstates of H_{S_1} and H_{S_2}, that is $|v, j\rangle := |v\rangle \otimes |j\rangle$. However this strategy does not help much when evaluating the time evolution (see Equations (24) and (25)) as H_S is not diagonal w.r.t. to $|v, j\rangle$ such that one cannot easily write down the matrix elements of the unitary evolution $U_S(t, t_0)$. In fact we are forced to solve the GME by using the eigenstates $|\varphi_p\rangle$ and eigenvalues $\mathcal{E}_p$ of the Hamiltonian H_S. The former are written as:

$$|\varphi_p\rangle = \sum_{v,j} \mathcal{V}_{vj}^{(p)} |v, j\rangle. \tag{28}$$

Here the round bracket notation $|\varphi_p\rangle$ is meant to underline that the state φ_p describes the interacting system S, in the sense that both Coulomb interactions and the coupling to the bosonic modes were taken into account when diagonalizing H_S. This notation also prevents any confusion if the 'free' states $|v, j\rangle$ were also labelled by a single index p'. In that case the above equation is conveniently rewritten as $|\varphi_p\rangle = \sum_{p'} \mathcal{V}_{p'}^{(p)} |p'\rangle$. Note that p is usually a multiindex carrying information on relevant quantum numbers. In most cases of interest the coupling V between the two systems leads to a strong mixing of the unperturbed basis elements $|v, j\rangle$ and is not necessarily small. Therefore we shall not follow a perturbative approach but rather calculate $\mathcal{E}_p$ and the weights $\mathcal{V}_{vj}^{(p)}$ by numerically diagonalizing H_S on a relevant subspace of 'disjointed' states.

Prior to any model specific calculations or numerical implementations it is useful to comment a bit on the two dissipative contributions in Equation (27). It is clear that the evolution operator U_S describes the joint systems S_1 and S_2 and therefore the hybrid interaction cannot be simply neglected neither in $\mathcal{D}_{\text{leads}}$ nor in $\mathcal{D}_{\text{bath}}$; in fact one can easily check that V does not commute with H_E or H_T. Moreover, as has been clearly pointed out by Beaudoin et al. [30], by disregarding the qubit-resonator interaction when calculating $\mathcal{D}_{\text{bath}}$ one ends up with unphysical results. In what concerns the role of V in the leads' contribution, a recent work emphasized that for QD-cavity systems the corresponding master equation must be derived in the basis of dressed-states [31].

The diagonalization of H_S poses serious technical problems because both spaces $\mathcal{F}_{S_1}$ and $\mathcal{F}_{S_2}$ are in principle infinite dimensional. Besides that, the Coulomb interaction in H_{S_1} prevents one to derive the interacting many-body configurations $\{|v\rangle\}$ analytically. We now propose a step-by-step diagonalization procedure leading to a relevant set of interacting states of the full Hamiltonian. The procedure requires several 'intermediate' diagonalization operations:

(D_1) Analytical or numerical calculation of the single-particle states of the Hamiltonian $\hat{h}_{S_1}^{(0)}$ which describes the non-interacting electronic system S_1. As we shall see in the next sections, this step may not be trivial if the geometry of the sample is taken seriously into account. Let us select a subset of

N_{ses} single-particle states $\{\psi_1, \psi_2, ..., \psi_{N_{\text{ses}}}\}$ (if needed this set of states includes the spin degree of freedom). Typically we choose the lowest energy single-particle states but in some cases [12] it is more appropriate to select the subset of states which effectively contribute to the transport (i.e., states located within the bias window).

(D_2) The construction of a second set of N_{mes} non-interacting many-body configurations (NMBS) $\{|\lambda\rangle\}_{\lambda=1,...,N_{\text{mes}}}$ from the N_{ses} single-particle states introduced above. Note that for computational reasons we have to keep $N_{\text{mes}} < 2^{N_{\text{ses}}}$ for larger N_{ses}. Then, if i_n^λ is the occupation number of the single-particle state ψ_n, a non-interacting many-body configuration $|\lambda\rangle$ reads as:

$$|\lambda\rangle = |i_1^\lambda, i_1^\lambda, ..., i_{N_{\text{ses}}}^\lambda\rangle. \tag{29}$$

(D_3) Diagonalization of the Coulomb-interacting electronic Hamiltonian $H_{S_1} = H_{S_1}^{(0)} + W$ on the subspace of non-interacting many-body states from $\mathcal{F}_{S_1}$. As a result one gets N_{mes} interacting many-body states (IMBS) $|\nu\rangle$ and the associated energy levels E_ν introduced in Section 2.1. We also introduce the 'free' energies of $H_{S_1} + H_{S_2}$, that is $\mathcal{E}_{\nu,j}^{(0)} = E_\nu + e_j$. Note that in view of diagonalization the interaction V between the two subsystems must be also written w.r.t. the 'free' states $\{|\nu, j\rangle\}$. If the second subsystem is not needed then the GME must be solved w.r.t. the set $\{|\nu\rangle\}$ and the diagonalization procedure stops here. It is worth pointing out here that even in the absence of bosonic fields and electron-photon coupling, the master equation for Coulomb interacting systems cannot be written in terms of single particle states. In spite of the fact that the unitary evolution U_S is diagonal w.r.t. the many-body basis $\{|\nu\rangle\}$, the matrix elements of the fermionic operators in the interaction picture $\langle\nu|\tilde{c}_{n\sigma}(t)|\nu'\rangle$ depend on energy differences $E_\nu - E_{\nu'}$ which, due to the Coulomb effects, cannot be reduced to the single-particle energy ε_n.

(D_4) Diagonalization of the fully interacting Hamiltonian H_S on a subspace of $\mathcal{F}$ made by the lowest energy N_{mesT} interacting MBS of H_{S_1} and j_{max} eigenstates of H_{S_2}. Remark that after the 1st truncation w.r.t. NMBSs we perform here a 2nd double truncation both w.r.t. IMBS (as $N_{\text{mesT}} < N_{\text{mes}}$) and w.r.t. the states in $\mathcal{F}_{S_2}$.

Once this procedure is performed, one can express the system-environment couplings H_T and H_B in the fully interacting basis and use the eigenvalues $\mathcal{E}_p$ to replace the unitary evolution U_S by the corresponding diagonal matrix $e^{-it\mathcal{E}_p}\delta_{pp'}$. Finally, the GME is to be solved w.r.t. the fully interacting basis (see Section 2.3).

Now let us enumerate and explain the advantages of this stepwise procedure when compared to a single and direct diagonalization of H_S.

(a) Numerical efficiency and accuracy. Both diagonalization methods (stepwise and direct) require a truncation of both bases and are not free of numerical errors which in principle should diminish as the size of the bases increase. It is clear that in the stepwise procedure the N_{mesT} interacting MBSs are derived from a larger set of non-interacting states $\{|\lambda\rangle\}_{\lambda=1,...,N_{\text{mes}}}$. Then the calculated $N_p := N_{\text{mesT}} \times j_{\text{max}}$ fully interacting states are more accurate than the ones obtained by diagonalizing once a $N_p \times N_p$ matrix. On the other hand, enlarging the full space to $N_{\text{mes}} \times j_{\text{max}}$ elements could be challenging in terms of CPU times. Convergence calculations relevant to circuit quantum electrodynamics have been presented in [42]. In particular it was shown that the inclusion of the (usually neglected) diamagnetic term in the electron-photon coupling improves the convergence of the diagonalization procedure.

The size of various effective bases used in the numerical calculation is decided both by physical considerations and convergence tests. Typically, out of the N_{ses} single-electron states we construct the set of non-interacting MBSs containing up to N_e electrons, the size of this set being, of course, $\binom{N_{\text{ses}}}{N_e}$. The accuracy of numerical diagonalization which leads to the interacting many-body configurations with up to N_e electrons is essentially assessed by comparing the spectra associated obtained for different N_{ses}. In particular, if we discretize our open system in a small number of lattice sites we can use all single-electron states as a basis, and we can calculate all many-body electron states (like in the discrete case presented in [12]. Obviously, this is no longer possible for a more complex geometry, and then we need to evaluate the convergence of the results when the basis is truncated.

For the continuous model an extensive discussion on the convergence of the numerical diagonalization w.r.t. the various truncated bases is given in a previous publication [42]. Let us stresonly extensive tests can be performed to resolve this issue. s here that once the geometry of the system and the spatially-dependent interactions are accounted for there is no simple way to count ahead how many states one needs to get stable transport simulations.

(b) Physical interpretation. It is obvious that the Coulomb interaction W mixes only the non-interacting many-body configurations $|\lambda\rangle$ while the hybrid coupling V mixes both λ and j states. For this reason the weights of a non-interacting state $|\lambda, j\rangle$ in a fully interacting state $|\varphi_p\rangle$ (as provided by a single diagonalization) cannot be easily associated with one of the interacting terms. In view of physical discussion it is more intuitive and natural to analyze the dynamics of the Coulomb-interacting system S_1 in the presence of the second subsystem S_2. One such example is a self-assembled quantum dot embedded in a single-mode quantum cavity [31]. In this system the optical transitions couple electron-hole pairs which are genuine interacting many-body states. A second example is a double quantum dot patterned in a 2D quantum wire which is itself placed in a cavity. There the interdot Coulomb interaction affects the optical transitions as well.

On the other hand, the above procedure will not be appropriate if one is interested in including a time-dependent driving term in H_S. This would be the case for a pumping potential or for a modulating optical signal.

Finally we shall comment on the typical sizes of many-body Fock spaces used to model steady-state or transient transport in various systems. One of the simplest yet promising system for solid-state quantum computation is a double quantum dot accomodating at most two electrons on each dot such that the relevant Fock space already comprises 256 interacting many-body states (counting the spin). In this case transport simulations can be obtained even without truncating the basis, especially if the spectral gaps allows one to disregard the contribution of higher energy configurations. However, when studying transport on edge states due to a strong perpendicular magnetic field in 2D systems (e.g., graphene or phosforene) one is forced to consider the low energy bulk-states as well. In our previous numerical studies we find that one has to take into account at least 10 single-particle states; obviously, performing time-dependent simulations for 2^{10} MBSs is quite inconvenient so a truncation is needed. More importantly, a realistic description of complex systems like rings or double dots etched in a 2DEG cannot be obtained with only few single-particle states. Note that the optical selection rules and matrix elements of the electron–photon interaction depend on the these states as well.

In the calculation of one, rather deep, QD embedded in a short quantum wire we are using 52 single-electron states, asking for 52 one-electron states, 1326 two-electron states and 560 three electron states. Of these we take the lowest in energy 512 and tensor multiply by 17 photon states to obtain a basis of 8704 MBS to calculate the dressed MBS. Then for the transport, we select the lowest in energy 128 dressed states and construct the 16,384 dimensional Liouville space. All this choice is taylored for a rather narrow section of a parameter space, if we consider the wire length, the confinement energy and the shape of the QD and the range of the magnetic field.

Markovian or non-markovian master equation method have been also developed for transport simulations in molecular junctions; here a truncation is required w.r.t. to the basis states describing the molecular vibrations. In particular, Schinabeck et al. [43] proposed a hierarchical polaron master equation which was successfully implemented numerically for two molecular orbitals and several tens of vibrational states.

2.3. Numerical Implementation and Observables

The last step before numerical implementation requires the calculation of the system-environment couplings H_T and H_E w.r.t. the full basis $|\varphi_p\rangle$. Clearly, to this end we shall use the unitary transformations $|\lambda\rangle \leftrightarrow |\nu\rangle$ and $|\nu, j\rangle \leftrightarrow |\varphi_p\rangle$ which are already at hand due to the stepwise diagonalization procedure introduced in the previous section. Then let us introduce some generalized 'jump' operators collecting all transitions, between pairs of fully interacting states, generated by tunneling of an electron with momentum q from the l-th lead to the single-particle levels of the electronic system S_1:

$$\mathcal{T}_l(q) = \sum_{p,p'} \mathcal{T}_{pp'}^l(q)|\varphi_p\rangle\langle\varphi_{p'}|, \quad (\mathcal{T}_l(q))_{pp'} = \sum_n T_{qn}^l \langle\varphi_p|c_n^\dagger|\varphi_{p'}\rangle. \tag{30}$$

Then the dissipation operator associated to the particle reservoirs reads:

$$\mathcal{D}_{\text{leads}}[\rho, t] = -\frac{1}{\hbar^2} \sum_{l=\text{L,R}} \int dq\, \chi_l(t)([\mathcal{T}_l, \Omega_{ql}(t)] + h.c.), \tag{31}$$

with the following notation:

$$\Omega_{ql}(t) = U_S(t, t_0) \int_{t_0}^t ds\, \chi_l(s)\Pi_{ql}(s)e^{i((s-t)/\hbar)\varepsilon_l(q)} U_S^\dagger(t, t_0), \tag{32}$$

$$\Pi_{ql}(s) = U_S^\dagger(s, t_0)\left(\mathcal{T}_l^\dagger\rho(s)(1 - f_l) - \rho(s)\mathcal{T}_l^\dagger f_l\right)U_S(s, t_0), \tag{33}$$

and where for simplicity we omit to write the energy dependence of the Fermi function f_l. Similarly, the bosonic operators have to written down w.r.t. the full basis which then leads to the calculation of $\mathcal{D}_{\text{bath}}$. Under the Markov approximation w.r.t. the correlation function of the bosonic reservoir the latter becomes local in time.

The GME is solved numerically by time discretization using the Crank–Nicholson method which allows us to compute the reduced density operator for discrete time steps $\rho(t_n)$, starting with an initial condition corresponding to a given state of the isolated central system, i.e., before the onset of the coupling with the leads. We take advantage of the fact that, by discretizing the time domain, the operator $\Omega_{ql}(t_{n+1})$ obeys a recursive formula generated by the incremental integration between t_n and t_{n+1}, that is:

$$\Omega_{ql}(t_{n+1}) = U_S(t_{n+1}, t_n)\Omega_{ql}(t_n)U_S^\dagger(t_{n+1}, t_n) + \mathcal{A}_{ql}(t_{n+1}, t_n; \rho(t_{n+1}), \rho(t_n)), \tag{34}$$

where the second term of the right-hand side depends on the yet unknown $\rho(t_{n+1})$. For any pair of time steps $\{t_n, t_{n+1}\}$ we initially approximate $\rho(t_{n+1})$ in $\mathcal{A}_{ql}$ by the already calculated $\rho(t_n)$, and perform iterations to recalculate $\rho(t_{n+1})$ via the GME, each time updating $\rho(t_{n+1})$ in $\mathcal{A}_{ql}$, until a convergence test for $\rho(t_{n+1})$ is fulfilled. At any step of the iteration we also calculate and include $\mathcal{D}_{\text{bath}}[\rho, t_m]$ into the iterative procedure; its calculation is much simpler as the Markov approximation w.r.t. the bath degrees of freedom takes care of the time integral so this dissipative term becomes local in time. Finally, we check numerically the conservation of probability and the positivity of the diagonal elements of $\rho(t_m)$, i.e., the populations of fully interacting states $|\varphi_p\rangle$ at the corresponding time step and for each iteration.

There are several reasons to extend the GME method beyond single-level models. (1) The electronic transport at finite bias collects contributions from all the levels within the bias window. This feature leads to the well known stepwise structure of the current-voltage characteristics; (2) In the presence of Coulomb interaction the GME must be derived in the language of many-body states which allows us to perform exact diagonalization on appropriate Fock subspaces; (3) The minimal model which describes the effect of the field-matter coupling in optical cavities with embedded quantum dots requires at least two single-particle levels.

Both the GME and non-equilibrium Green's function formalism (NEGF) rely on the partitioning approach and allow for many-body interaction in the central system, while the leads are assumed to be non-interacting (this assumption leads in particular to the Fermi distribution of the particle reservoirs). There is however a crucial difference between the two methods. The perturbative expansion of the dissipative kernel forces restricts the master equation approach to weak lead-sample tunnelings while the interaction effects are accounted for exactly. In contrast, the Keldysh formalism is not limited to small system-reservoir couplings but the Coulomb effects have to be calculated from appropriate interaction self-energies. Which method fits better is simply decided by the particular problem at hand.

As stated in the Introduction, the advantage of the RDO stems from the fact that it can be used to calculate statistical averages of various observables $\mathcal{O}$ of the hybrid system:

$$\langle \mathcal{O} \rangle = \text{Tr}_{\mathcal{F}}\{\rho(t)\mathcal{O}\}. \tag{35}$$

Useful examples are averages of the photon number operator $\mathcal{N}^{ph} = a^\dagger a$ and of the charge operator $\mathcal{Q} = \sum_n c_n^\dagger c_n$. Also, the average currents in a two-lead geometry (i.e., $l = L, R$ can be identified from the continuity equation:

$$\langle \dot{\mathcal{Q}} \rangle = \text{Tr}_{\mathcal{F}}\{\mathcal{Q}\dot{\rho}(t)\} = J_L(t) - J_R(t). \tag{36}$$

2.4. Coupling between Leads and Central System

The modeling of the central systems and the reservoirs can be performed either by using continuous confining potentials or a spatial grid. Examples are a short parabolic wire [44,45], ring [46,47], parallel wires with a window coupler [48], and wire with embedded dot [44,49] or dots [50]. The coupling between the leads and the central system with length L_x is described by Equation (8), and in order to reproduce scattering effects seen in a Lippmann–Schwinger formalism [15,51,52] the coupling tensor is defined as

$$T_{qn}^l = \int_{\Omega_S^l \times \Omega_l} d\mathbf{r}d\mathbf{r}' \left(\Psi_q^l(\mathbf{r}')\right)^* \Psi_n^S(\mathbf{r})g_{qn}^l(\mathbf{r}, \mathbf{r}') + h.c., \tag{37}$$

for states with wavefunction Ψ_q^l in lead l, and Ψ_n^S in the central system. The domains for the integration of the wavefunctions in the leads are chosen to be

$$\begin{aligned}
\Omega_L &= \left\{(x,y)| \left[-\tfrac{L_x}{2} - 2a_w, -\tfrac{L_x}{2}\right] \times [-3a_w, +3a_w]\right\}, \\
\Omega_R &= \left\{(x,y)| \left[+\tfrac{L_x}{2}, +\tfrac{L_x}{2} + 2a_w\right] \times [-3a_w, +3a_w]\right\},
\end{aligned} \tag{38}$$

and for the system as

$$\begin{aligned}
\Omega_S^L &= \left\{(x,y)| \left[-\tfrac{L_x}{2}, -\tfrac{L_x}{2} + 2a_w\right] \times [-3a_w, +3a_w]\right\}, \\
\Omega_S^R &= \left\{(x,y)| \left[+\tfrac{L_x}{2} - 2a_w, +\tfrac{L_x}{2}\right] \times [-3a_w, +3a_w]\right\}.
\end{aligned} \tag{39}$$

The function

$$g_{qn}^l(\mathbf{r}, \mathbf{r}') = g_0^l \exp\left[-\delta_1^l(x - x')^2 - \delta_2^l(y - y')^2\right] \exp\left(\frac{-|E_n - \epsilon^l(q)|}{\Delta_E^l}\right). \tag{40}$$

with $\mathbf{r} \in \Omega_S^l$ and $\mathbf{r}' \in \Omega_l$ determines the coupling of any two single-electron states by the "nonlocal overlap" of their wave functions in the contact region of the leads and the system, and their energy affinity. A schematic view of the coupling is seen in Figure 1. The parameters δ_1^l and δ_1^l define the spatial range of the coupling within the domains $\Omega_S^l \times \Omega_l$ [44].

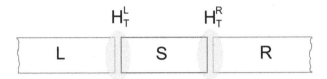

Figure 1. A schematic of the coupling of the system to the leads. The transparent green areas correspond to the contact regions defined by the nonlocal overlap function $g_{qn}^{L,R}$ in $H_T(t)$.

The short quantum wire is considered to have hard wall confinement in the transport direction, the x-direction, at $x = \pm L_x/2$, and parabolic confinement in the y-direction with characteristic energy $\hbar\Omega$. Possibly, the leads and the central system are considered to be placed in a perpendicular homogeneous external magnetic field $\mathbf{B} = B\hat{z}$. Together they lead to a natural length scale, the effective magnetic length a_w with $a_w^2 \Omega_w = \hbar/m$, with $\Omega_w^2 = [(\Omega_0)^2 + (\omega_s)^2]^{1/2}$, and the cyclotron frequency $\omega_c = (eB/m)$. For GaAs with effective mass $m = 0.067 m_e$ relative dielectric constant $\kappa_c = 12.3$ and confinement energy $\hbar\Omega_0 = 2.0$ meV, $a_w = 23.8$ nm. The magnetic field $B = 0.1$ T.

The energy spectrum of the quasi 1D semi-infinite lead l is represented by $\epsilon^l(q)$, with q standing for the momentum quantum number of a standing wave state, and the subband index n^l. The spectrum in the absence of spin orbit interactions can be evaluated exactly analytically [53]. The coupling of the leads and the central systems in the continuous representation conserves parity of the electron states across the tunneling barrier.

The full strength of the continuous approach emerges as it is applied to describe the transport of interacting electrons through 3D photon cavities in the transient time regime or the long time regime ending in a steady state of the system. This will be reported below (see Sections 5 and 6). The numerical calculations can sometimes be simplified by describing the leads and the central system on a discrete spatial lattice, where the geometric details of the central system are usually implemented by hard walls and Dirichlet boundary conditions. The spatial integral of the coupling tensor (37) are then reduced to a set of contact points between the leads and the central system [12,17].

3. Many-Body Effects in the Transient Regime

In this section we review some results on the transient transport in interacting systems described by a lattice model [12,14]. For the sake of generality we extend the GME method by including as well the spin degree or freedom which was previously neglected. The lattice model matches naturally to the partitioning transport setting, facilitates the geometrical description of the central sample (e.g., a parallel quantum dot) and captures the dependence of the tunneling coefficients on the localization of the single-particle wavefunctions at the contact regions. A more realistic description is provided by the continuous model (see the previous section) which requires however a very careful tailoring of the confining potentials.

The results presented in this section are also meant to illustrate the usefulness of the GME approach in describing the transient regime in terms of the dynamical occupations of the interacting many-body configurations. Such a description cannot be recovered within the non-equilibrium Greens' function formalism.

Developing the GME method in the language of interacting many-body states was equally motivated by experimental works. Recording the charging of excited states of QDs in the Coulomb blockade regime constitutes the core of transient current spectroscopy and pump-and-probe techniques [54]. Also, transient currents through split-gate quantum point contacts (QPSs) and Ge quantum dots have been measured some time ago by Nasser et al. [55] and by Lai et al. [56]. Another relevant class of transport phenomena which can be modeled and understood within the GME method is the electron pumping through QDs with tunable-barriers (see e.g., the recent review [57]). In this context we investigated the transient response of a quantum dot submitted to a sequence of rectangular pulses applied at the contact to the input [58] and the turnstile protocol for single-molecule magnets [59].

3.1. Transient Charging of Excited States

We consider a two-dimensional system of length L_x and width L_y described by a lattice with N_x sites on the x axis and N_y sites on the y axis. The total number of sites is denoted by $N_{xy} = N_x N_y$. By setting the two lattice constants a_x and a_y one has $L_x = a_x N_x$ and $L_y = a_y N_y$. Once we know the single-particle eigenstates of the electronic subsytem S_1 we can write down its Hamiltonian $H_{S1} := H_{S_1}^{(0)} + W$ in a second quantized form w.r.t. this basis, that is:

$$
\begin{aligned}
H_{S_1}^{(0)} &= \sum_{n,n'=1}^{N_{xy}} \langle \psi_n | \hat{h}_{S_1}^{(0)} | \psi_{n'} \rangle c_n^\dagger c_{n'} = \sum_n \epsilon_n c_n^\dagger c_n \\
W &= \tfrac{1}{2} \sum_{n,m,n',m'} V_{nmn'm'} c_n^\dagger c_m^\dagger c_{m'} c_{n'}
\end{aligned}
\tag{41}
$$

where the Coulomb matrix elements are given by ($\mathbf{r}, \mathbf{r}'$ are sites of the 2D lattice):

$$
V_{nmn'm'} = \sum_{\mathbf{r},\mathbf{r}'} \psi_n^*(\mathbf{r}) \psi_m^*(\mathbf{r}') V_C(\mathbf{r} - \mathbf{r}') \psi_{n'}(\mathbf{r}) \psi_{m'}(\mathbf{r}').
\tag{42}
$$

The Coulomb potential itself is given by

$$
V_C(\mathbf{r}, \mathbf{r}') = \frac{e^2}{4\pi\epsilon(|\mathbf{r} - \mathbf{r}'| + \eta)},
\tag{43}
$$

where η is a small positive regularization parameter.

Like in the continuous model, the tunneling coefficients T_{qn}^l are associated to a pair of states $\{\psi_q^l, \psi_n\}$ from the lead l and the sample S_1. However the lattice version is much simpler:

$$
T_{qn}^l = V_l \psi_q^{l*}(0_l) \psi_n(i_l),
\tag{44}
$$

where 0_l is the site of the lead l which couples to the contact site i_l in the sample. The wavefunctions of the semi-infinite lead are known analytically:

$$
\psi_q^l(j) = \frac{\sin(q(j+1))}{\sqrt{2\tau \sin q}}, \quad \varepsilon_q = 2\tau \cos q.
\tag{45}
$$

In the above equation τ is the hopping energy of the leads. The integral over q in the tunneling Hamiltonian (see Equation (8) from Section 2) counts the momenta of the incident electrons such that $\varepsilon^l(q)$ scans the continuous spectrum of the semi-infinite leads $\sigma_l \in [-2\tau + \Delta, 2\tau + \Delta]$ where Δ is a shift which is chosen such that σ_l covers the lowest-energy many-body spectrum of the central system. The construction of the coupling coefficients T_{qn}^l shows that a single-particle state which vanishes at the contact sites does not contribute to the currents. This is the case for states which are mostly localized at the center of the sample, while in the presence of a strong magnetic field the currents will be carried by edge states.

In [12] we implemented GME for a non-interacting lattice Hamiltonian, whereas the Coulomb interaction effects were introduced in [14]. In what concerns the geometrical effects we essentially showed that the transient currents depend on the location of the contacts (through the value of the single-particle wavefunctions of the sample at those points) but also on the initial state and on the switching functions $\chi_l(t)$ of the leads. It turns out that the stationary current does not depend on the last two parameters, in agreement with rigorous results [60,61]. We also identified a delay of the output currents which was attributed to the electronic propagation time along the edge states of the Hofstadter spectrum.

The presence of Coulomb interaction brings in specific steady-state features known from previous calculations like the Coulomb blockade and the step-like structure of the current–voltage characteristics. On the other hand the GME method naturally allows a detailed analysis of the time-dependent currents associated to each many-body configuration as well as of the relevant populations.

Since the Hamiltonian H_{S_1} of the interacting system commutes with the total number operator $Q = \sum_n c_n^\dagger c_n$, its eigenstates $|v\rangle$ can still be labelled by the occupation N_v of the non-interacting MBSs from which the state is built. Then the single index v can be replaced by two indices, the particle number N_v and an index $i_v = 0, 1, 2, ...$ for the ground ($i_v = 0$) and excited states ($i_v > 0$, where i_v also counts the spin degeneracy). The notation for the interacting many-body energies is changed accordingly $\mathcal{E}_v \to \mathcal{E}_{N_v}^{(i_v)}$.

We now define some useful quantities for our time-dependent analysis. The charge accumulated on N-electrons states is calculated by collecting the associated populations:

$$q_N(t) = eN \sum_{v, n_v = N} \langle v | \rho(t) | v \rangle, \tag{46}$$

where the sum counts all states whose total occupation $n_v = N$. Similarly one can identify the transient currents $J_{l,N}$ carried by N-particle states. These currents can be traced back form the right hand side of the GME:

$$\langle \dot{Q} \rangle = \sum_N (J_{L,N}(t) - J_{R,N}(t)) = \sum_N \dot{q}_N(t). \tag{47}$$

Throughout this work we shall adopt the following sign convention for the currents associated to each lead: $J_L > 0$ if the electrons flow from the left lead towards the sample and $J_R > 0$ if they flow from the sample towards the right lead.

The sequential tunneling processes change the many-body configurations of the electronic system. The energy required to bring the system to the i-th MBS with N particles is measured w.r.t. the ground state with $N - 1$ electrons ($i = 0, 1, ...$). We introduce two classes of chemical potentials of the sample:

$$\mu_{g,N}^{(i)} = \mathcal{E}_N^{(i)} - \mathcal{E}_{N-1}^{(0)}, \tag{48}$$

$$\mu_{x,N}^{(i)} = \mathcal{E}_N^{(i)} - \mathcal{E}_{N-1}^{(1)}, \tag{49}$$

where $\mu_{g,N}^{(i)}$ characterizes transitions from the ground state $(N - 1)$-particle configuration to various N-particle configurations. In particular $\mu_{g,N}^{(0)}$ describes addition processes involving ground-states with $N - 1$ and N electrons while $\mu_{g,N}^{(i>0)}$ refers to transitions from $(N - 1)$-particle ground state to excited N-particle configurations. The chemical potentials $\mu_{x,N}^{(i)}$ describe transitions from the 1st $(N - 1)$-particle excited states to configurations with N particles. In a transition of this type an electron tunnels on the lowest single particle state to the central system which already contains one electron on the excited single-particle state $|\sigma_2\rangle$. As a result some of the triplet states are being populated. We shall see that these transitions play a role especially in the transient regime.

For numerical calculations we considered a 2D quantum wire of lenght $L_x = 75$ nm and width $L_y = 10$ nm. The lowest two spin-degenerate single-particle levels are $\varepsilon_1 = 0.375$ meV and $\varepsilon_2 = 3.37$ meV. The non-interacting MBSs are described by the spins of the occupied single-particle levels, e.g., $|\sigma_1 \bar{\sigma}_2\rangle$ is a two-particle configuration with a spin σ associated to the lowest single-particle state and a second electron with opposite spin orientation on the energy level ε_2. Besides the usual singlet (S) and triplet (T) states we find that the Coulomb interaction induces the configuration mixing of the antiparallel configurations $| \uparrow_1 \downarrow_1 \rangle$ and $| \uparrow_2 \downarrow_2 \rangle$. More precisely, we get an interacting ground two-particle state mostly made of $| \uparrow_1 \downarrow_1 \rangle$ (whose weight is 0.86) and with a small component (weight 0.14) of state $| \uparrow_2 \downarrow_2 \rangle$. Conversely, $| \uparrow_1 \downarrow_1 \rangle$ is also found in the highest energy two-particle state.

We stress here that the configuration mixing decreases and eventually vanishes if the gap $E_{\uparrow_2\downarrow_2} - E_{\uparrow_1\downarrow_1}$ between two non-interacting energies is much larger that the corresponding matrix element.

In Figure 2a we show the chemical potentials corresponding to interacting MB configurations with up to three electrons. As long as the chemical potential $\mu_{g,N}^{(i)}$ lies within the bias window the corresponding state will contribute both to the transient and steady-state currents. We shall see that if $\mu_{g,N}^{(i)} < \mu_R$ the state $|i, N\rangle$ contributes only to the transient currents. Finally, when $\mu_{g,N}^{(i)} \gg \mu_L$ the state $|i, N\rangle$ is poorly populated and will not contribute to transport. Let us stress here a rather unusual transition from $|\sigma_2\rangle$ to the ground two-particle state which is mostly made of $|\sigma_1\bar{\sigma}_1\rangle$. The corresponding addition energy $\mu_{x,2}^{(0)} = 2$ meV is smaller than the energy required for the usual transition $|\sigma_1\rangle \to |\sigma_1\bar{\sigma}_1\rangle$. This happens because of the Coulomb mixing between $|\uparrow_1\downarrow_1\rangle$ and $|\uparrow_2\downarrow_2\rangle$ which makes possible the transition from the excited single-particle state to the mixed interacting two-particle groundstate. The chemical potential $\mu_{x,2}^{(2)}$ describes the transition from the excited single-particle state $|\sigma_2\rangle$ to the triplet states.

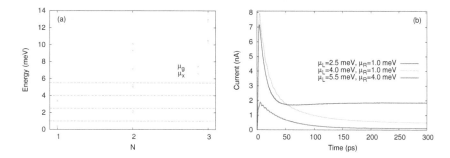

Figure 2. (a) The chemical potentials $\mu_{g,N}^{(i)}$ (red crosses) and $\mu_{x,N}^{(i)}$ (blue crosses) for N-particle configurations, $N = 1, 2, 3$. For a given particle number N the chemical potentials are ordered vertically according to the index $i = 0, 1, ...$ The horizontal lines correspond to specific values of the chemical potentials in the leads (see the discussion in the text); (b) The time-dependent currents in the left lead at different values of the bias window $\mu_L - \mu_R$.

Already by analysing Figure 2 one can anticipate to some extend how the transport takes place in terms of allowed sequential tunneling processes. Suppose that the chemical potentials of the leads are selected such that $\mu_{g,1}^{(0)} < \mu_R < \mu_{g,1}^{(1)} < \mu_L < \mu_{g,2}^{(0)}$ (as an example we set $\mu_R = 1$ meV and $\mu_L = 4$ meV). Then both single-particle levels are available for tunneling but one expects that the double occupancy is excluded because $\mu_L < \mu_{g,2}^{(0)} \sim 5$ meV. According to this scenario, more charge will accumulate on ε_1, the excited states $|\sigma_2\rangle$ will eventually deplete and the steady-state current vanishes in the steady-state. This is the well known Coulomb blockade effect. However, we see in Figure 2b that the steady-state current vanish only when $\mu_L < \mu_{g,1}^{(1)}$ as well, which suggest that the presence of the excited single-particle states within the bias window leads to a partial lifting of the Coulomb blockade. We stress that such an effect cannot be predicted within a single-site model with onsite Coulomb interaction. A third curve shows the current for $\mu_L = 5.5$ meV and $\mu_L = 4$ meV.

Figure 3 presents the evolution of the relevant populations at two values of the bias window. In Figure 3a the population $P_{1g} = P_{\uparrow_1} + P_{\downarrow_1}$ of the ground single-particle states dominates in the steady state. This is expected, as the corresponding chemical potential lies below the bias window so this state will be substantially populated. The other configurations contributing to the steady-state are just the ones which can be populated by tunnelings from the left lead, that is the excited single-particle states and all two-particle states except for the single configuration which cannot be accessed. By looking at Figure 2a one infers that the two-particle states are being populated when one more electron is added from the left lead on the initial excited single-particle state $|\sigma_2\rangle$. In particular, the ground two-particle state is populated only due to the Coulomb-induced configuration mixing.

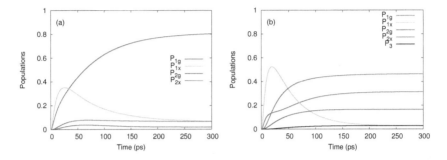

Figure 3. The populations of ground (g) and excited (x) N-particle states ($N = 1, 2, 3$) for two bias windows; (**a**) $\mu_L = 4\,\text{meV}$, $\mu_R = 2\,\text{meV}$; (**b**) $\mu_L = 5.5\,\text{meV}$, $\mu_R = 4\,\text{meV}$. In panel (**a**) P_3 is negligible and was omitted.

A completely different behavior is noticed in Figure 3b. As the bias window is pushed upwards such that $\mu_{g,1}^{(1)} < \mu_R < \mu_{g,2}^{(1)} < \mu_L$ the transitions from the lowest states $|\sigma_1\rangle$ to two-particle states are also activated. Consequently, the population P_{2x} of the excited two-particle configurations exceeds P_{1g} and dominate in the steady-state regime. Note that $P_{2x} > P_{2g}$ because it collects the population of the degenerate triplet states. In the transient regime the excited single particle states are populated much faster than the ground states. This happens because of the different localizations of the single-particle wavefunctions on the contact regions. We find that the wavefunction associated to the 2nd single-particle state has a larger value at the endpoints of the leads. A drop of P_{1x} follows as the ground one-electron states and the other two-particle configurations become active (a similar feature is noticed in Figure 3a). A small populations of the three particle states can be also observed. The steady-state current increases considerably (see Figure 2b) and is due to the two-particle states.

We end this section with a discussion on the partial currents $J_{L,N}$ and $J_{R,N}$ associated to N-particle states. Although they cannot be individually measured, these currents provide further insight into the transport processes, in particular on the way in which the steady-state regime is achieved.

Figure 4a shows that in the steady-state regime the currents carried by the one-particle states $J_{L,1}$ and $J_{R,1}$ achieve a negative value when $\mu_{g,1}^{(0)} < \mu_{g,1}^{(1)} < \mu_R < \mu_{g,2}^{(0)} < \mu_L$, whereas the two-particle currents evolve to a larger positive value such that the total current J_L will be positive as already shown in Figure 2b. When the bias window is shifted down to $\mu_L = 4\,\text{meV}$ and $\mu_R = 2\,\text{meV}$ all transients are mostly positive (see Figure 4b). One observes that the single-particle configurations are responsible for the spikes of the total current J_L and that the two-particle currents display a smooth behavior. These features can be explained by looking at the charge occupations q_N shown in Figure 4c,d. As $\mu_{g,1}^{(0)}$ and $\mu_{g,1}^{(1)}$ are both below $\mu_R = 4\,\text{meV}$ while $\mu_{g,2}^{(0)}$ is well within the bias window, the population of the single-particle states increases rapidly in the transient regime but then also drops in favour of P_2, the total occupation of two-particle states. Such a redistribution of charge among configurations with different particle numbers is less pronounced in Figure 4d, because in this case the smaller contribution of the two-particle states is only due to transitions allowed by $\mu_{x,2}^{(0)}$ and $\mu_{x,2}^{(1)}$ which are now located within the bias window. The slope of q_2 also changes sign in the transient regime and one can check from Figure 4b that on the corresponding time range $J_{R,2}$ slightly exceeds $J_{L,2}$.

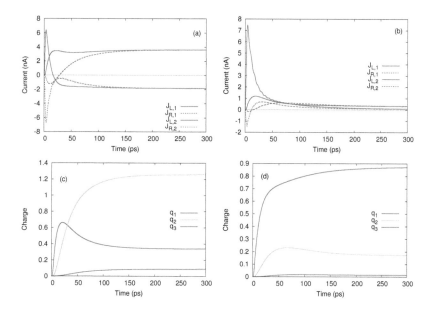

Figure 4. (**a**) The transient currents $J_{L,N}$ and $J_{R,N}$ associated to one and two-particle configurations for $\mu_L = 5.5$ meV, $\mu_R = 4$ meV; (**b**) The same currents for a bias window $\mu_L = 4$ meV, $\mu_R = 2$ meV; (**c**) The charge q_N accumulated on N-particle states at $\mu_L = 5.5$ meV, $\mu_R = 4$ meV; (**d**) q_N for $\mu_L = 4$ meV, $\mu_R = 2$ meV, and q_N are given in units of electron charge e.

The occupation of the three-particle configurations is negligible so q_3 is also small and was included here only for completeness while the associated currents were omitted.

3.2. Coulomb Switching of Transport in Parallel Quantum Dots

After using the GME formalism to describe transient transport via excited states in a single interacting nanowire we now extend its applications to capacitively coupled quantum systems. Besides Coulomb blockade, the electron-electron interaction cause momentum-exchange which leads to the well known Coulomb drag effect in double-layer structures [62] and double quantum dots [63–65] or wires [66]. Also, theoretical calculations on thermal drag between Coulomb-coupled systems were recently presented [67,68].

Here we consider a very simple model for two parallel quantum dots [17] (a sketch of the system is given in Figure 5). Each system is described by a 1D four-sites chain and for simplicity we neglect the spin degree of freedom which will only complicate the discussion of the effects. The diagonalization procedure provides all 256 many-body configurations emerging from the 8 single-particle states. Let us point out that the interdot and intradot interactions are treated on equal footing beyond the single-capcitance model. The hopping energy within the dots is $t_D = 1$ meV and the time unit is expressed in units of $\hbar/t_D$. Then the currents are calculated in units of $et_D/\hbar$. The tunneling rates to the four leads are all equal $V_{La} = V_{Ra} = V_{Lb} = V_{Rb}$.

We shall use the GME method to study the onset of the interdot Coulomb interaction. In order to distinguish the transient features due to mutual capacitive coupling we consider a transport setting in which each dot is connected to the leads at different times. More precisely, one system, say QD_a is open at the initial instant $t_a = 0$ and then reaches a stationary state ($J_{La} = J_{Ra}$) at some later time T_a. The coupling of the nearby system to its leads is switched on at $t_b > T_a$ such that the changes in the current J_a can only be due to mutual Coulomb interaction. Note that the usual Markov–Lindblad version of the master equation simulate the transport when the four leads are coupled suddenly and simultaneously to the double-dot structure.

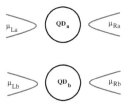

Figure 5. A sketch of the parallel double-dot system. Each QD is coupled to source-drain particle reservoirs described by chemical potentials μ_{Ls} and μ_{Rs}, $s = a, b$. There is no interdot electron tunneling but the systems are correlated via Coulomb interaction.

As before, the interacting many-body configurations can be labeled by to the occupations of each dot according to the correspondence $\mathcal{E}_\nu \rightarrow \mathcal{E}^{(i_\nu)}_{N_{a,\nu}, N_{b,\nu}}$. Here $N_{s,\nu}$ is the number of electrons in the system s associated to a many-body configuration ν. If the two systems are identical the lowest chemical potentials are introduced as:

$$\mu^{(0)}_g(N_a, N_b) = \mathcal{E}^{(0)}_{N_a, N_b} - \mathcal{E}^{(0)}_{N_a-1, N_b} = \mathcal{E}^{(0)}_{N_a, N_b} - \mathcal{E}^{(0)}_{N_a, N_b-1}, \tag{50}$$

because of the degeneracy w.r.t. to the total occupation number $\mathcal{E}^{(0)}_{N_a, N_b} = \mathcal{E}^{(0)}_{N_b, N_a}$. For the parameters chosen here one finds: $\mu^{(0)}_g(1,1) = 3$ meV, $\mu^{(0)}_g(2,0) = 4$ meV and $\mu^{(0)}_g(2,1) = 4.5$ meV. The location of the several chemical potentials w.r.t. the two bias windows already suggests the possible interdot correlation effects. The main point is that the transport channels through one dot also depend on the occupation of the nearby dot. One therefore expects that the currents J_{La} and J_{Ra} also depend on the bias applied on the nearby system.

To discuss this effect we performed transport simulations for two arrangements (A and B) of the bias window $\mu_{Ls} - \mu_{Rs}$. In the A-setup we select the four chemical potentials such that the chemical potentials associated to the many-body configurations relevant for transport obey the inequalities $\mu^{(0)}_g(1,1) < \mu_{Rs} < \mu^{(0)}_g(2,0) < \mu_{Ls} < \mu^{(0)}_g(2,1)$. The scenario is easy to grasp: As QD_a is coupled to the leads and the nearby dot is disconnected and empty, it will accumulate charge and evolve to a steady state where the current is essentially given by tunneling assisted transitions between $\mathcal{E}^{(0)}_{2,0} \leftrightarrow \mathcal{E}^{(0)}_{1,0}$. This behavior is observed in Figure 6a up to $t_b = 150$ ps when QD_b is also coupled to its leads. Note also that the charge occupation of QD_a almost saturates at $Q_a = 1.6$. As expected, for $t > t_b$ a transient current develops in QD_b, but a simultaneous drop the J_{La} and J_{Ra} shows the dynamical onset of the charge sensing effect between the two systems. In the final steady-state the two currents nearly vanish, thus proves their negative correlation due to the mutual Coulomb interaction. The charges $Q_{a,b}$ reach the same value and suggest that in the long time limit the double system contains one electron on each dot. Remark that in the final steady-state the dominant population corresponds to the many-body energy $\mathcal{E}^{(0)}_{1,1}$ which is not favorable for transport through any of the dots as long as $\mu^{(0)}_g(2,1) = \mathcal{E}^{(0)}_{2,1} - \mathcal{E}^{(0)}_{1,1}$ is outside the bias window.

Figure 6b,d present the currents and the charge occupations for the second setup B which is defined by the inequalities $\mu^{(0)}_g(2,0) < \mu_{Rs} < \mu^{(0)}_g(2,1) < \mu_{Ls}$. Following the same reasoning as before one infers that now QD_a will enter the Coulomb blockade regime before $t = t_b$ because there are no transport channels within the bias window. However, the blockade is removed due to the second dot whose charging activates tunneling through $\mu^{(0)}_g(2,1) = \mu^{(0)}_g(1,2)$. This is an example of positive correlations between the two systems. Further discussions can be found in a previous publication [17].

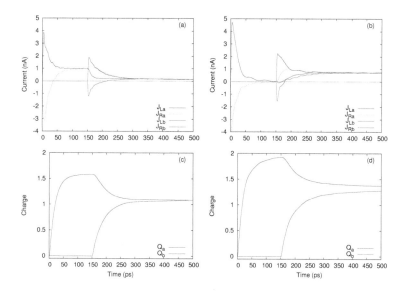

Figure 6. The transient currents in the two systems for different chemical potentials of the leads: (**a**) $\mu_{La} = \mu_{Lb} = 4.25$ meV, $\mu_{Ra} = \mu_{Rb} = 3.75$ meV; (**b**) $\mu_{La} = \mu_{Lb} = 4.75$ meV, $\mu_{Ra} = \mu_{Rb} = 4.35$ meV; (**c,d**) The charge occupations of the two systems associated to the currents in Figure 6a,b. The charges $Q_{a,b}$ are given in units of electron charge e.

4. Thermoelectric Transport

Until now we showed results for the charge transport driven by an electric bias of the leads due to different chemical potentials. The GME formalism allows also, in a straightforward way, the presence of a temperature bias. Instead of different chemical potentials in the left and right leads, $\mu_{L,R}$, one can easily consider different temperatures, $T_{L,R}$, and calculate the resulting currents after switching on the contacts between the leads on the central system. Notice that, like in the case of an electric bias, there is no requirement that the temperature bias is small, such that the nonlinear thermoelectric regime is directly accessible [69]. In addition, since the Coulomb interaction between electrons in the central system is already incorporated via the Fock space, the GME allows the inclusion of Coulomb blocking and other electron correlation effects in the thermoelectric transport [70,71].

The thermoelectric transport at nanoscale is a reach and active topic within the context of the modern quantum thermodynamics, partly motivated by novel ideas on the conversion of wasted heat into electricity, and partly by the characterization of nanoscale system by methods complementary to pure electric transport [72]. For example, an effect specific to nanosystems is the sign change of the thermoelectric current or voltage when the electronic energy spectrum consists of discrete levels. This effect was predicted in the early 90' [73] and detected experimentally for quantum dots [74–76] and molecules [77]. This means that thermoelectric current in a nanoelectronic system may flow from the hotter contact to the colder one, but also from the colder to the hotter, although the second possibility might appear counter-intuitive. A simple explanation of this sign change of the current is that in a nanoscale system with discrete resonances the current can be seen as having two components, one carried by populated states above the Fermi energy, and another one carried by depopulated states below it. By analogy with a semiconductor, the former states correspond to electrons in the conduction band and the later states to holes in the valence band. Whereas an electric bias drives the electric currents due to particles and holes in the same direction, such that they always add up, a thermal bias drives them in opposite directions, such that the net current is their difference, which can be positive, negative, or zero.

We can describe this effect with the GME, first assuming a simple model with unidimensional and discretized leads, and just a single site in between them as central system. By using the Markov approximation one can show analytically that the current in the leads, in the steady state, are obtained as [71]

$$J_{L,R} = \frac{1}{\tau^2} \frac{V_L^2 V_R^2}{V_L^2 + V_R^2} \left[f_L(E) - f_R(E) \right] , \tag{51}$$

where $V_{L,R}$ are the coupling parameters of the leads with the central site, τ is the hopping energy on the leads, and E is the energy of the central site. We see that the sign of the current depends on the difference between the Fermi energies in the leads at the resonance energy,

$$f_l(E) = \frac{1}{e^{(E-\mu_l)/k_B T_l} + 1} , \quad l = L, R . \tag{52}$$

Thus, in the presence of a thermal bias, say $T_L > T_R$, but in the absence of an electric bias, i.e., $\mu_L = \mu_R$, the current is zero and changes sign around $\mu_l = E$. In addition, the current may also vanish if the chemical potential in the leads is sufficiently far from the resonance such that the two Fermi functions are both close to zero or one. Which means that if the central system has more resonant energies the current may also change sign when μ_l is somewhere between two of them.

In Figure 7 we show an example of thermoelectric currents calculated with the GME, using the same model as in Section 3. The lowest single-particle levels having energies $\varepsilon_1 = 0.375$ meV and $\varepsilon_2 = 3.37$ meV are followed by the two-particle singlet state with $E_s = 5.39$ meV and triplet with $E_t = 5.62$ meV, and then by another excited two-body state with zero spin with energy $E_x = 10.5$ meV. We consider temperatures $k_B T_L = 0.5$ meV and $k_B T_R = 0.05$ meV in the left and right lead, respectively (or $T_L = 5.8$ K and $T_R = 0.58$ K), and equal chemical potentials. In Figure 7a one can see the time dependence of the currents in the leads after they are coupled to the central system, for two values of the chemical potentials, 4.8 meV and 5.4 meV, selected on each side of the singlet state. Compared to the results shown in Section 3 here we increased the coupling parameters between the leads and the central system 1.4 times, such that the steady state is reached sooner.

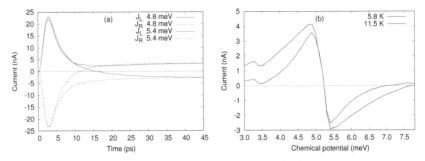

Figure 7. (a) The time evolution of the currents in the left and right leads, $J_{L,R}$, driven by a temperature bias where $T_L = 5.8$ K and $T_R = 0.58$ K. With red color the results for the chemical potential $\mu_L = \mu_R = 48$ meV, and with blue color for $\mu_L = \mu_R = 54$ meV. In the steady state the currents have opposite sign; (b) The current in the steady state for two different temperatures of the left lead, $T_L = 5.8$ K (red) and $T_L = 11.5$ K (blue), for variable chemical potentials $\mu_L = \mu_R$.

As predicted by Equation (51), the currents in the steady state have opposite sign. But in fact, as shown by the red curve of Figure 7b, here we do not resolve the energy interval between the singlet and triplet states with $k_B T_L > E_t - E_s = 0.23$ meV, such that we obtain one single (common) sign change for these two levels (or "resonances"). Next, by increasing the chemical potential within the larger gap between E_s and E_x the current in the steady state approaches zero and changes sign again, for $\mu_l \approx 7.0$ meV, and for $\mu_l \approx 7.8$ meV when the temperature of the hot lead is doubled, $T_L = 11.5$ K.

By varying the chemical potential below the singlet energy E_s we obtain a similar decreasing trend of the current, except that now there is no sign change close to the energy $\varepsilon_2 = 3.37$ meV, but only a succession of minima and maxima. The reason is the level broadening due to the coupling of the central system with the leads [71]. Still, from such data one can observe experimentally the charging energy, as the interval between consecutive maxima, or minima, or mid points between them [77].

In the present review we show only the thermoelectric current, which corresponds to a short-circuit experimental setup, i.e., a circuit without a load. To obtain a voltage with the GME method one has to simulate a load by considering also a chemical potential bias. Thus, one can obtain the open-circuit voltage, which corresponds to that electric bias $\mu_R - \mu_L$ which totally suppresses the thermoelectric current, or the complete I-V characteristic of the "thermoelectric device". Interestingly, the sign change of the thermoelectric current or voltage can also be obtained by increasing the temperature of the hot lead, while keeping the other lead as cold as possible [70,78–80].

A novel example of sign reversal of the thermoelectric current has been recently predicted in tubular nanowires, either with a core-shell structure or made of a topological insulator material, in the presence of a transversal magnetic field [81]. In this case the energy spectra are continuous, but organized in subbands which are nonmonotonic functions of the wavevector along the nanowire, yielding a transmission function nonmonotonic with the energy, and the reversal of the thermoelectric current, even in the presence of moderate perturbations [82,83].

5. Electron Transport through Photon Cavities

5.1. The Electron-Photon Coupling

From the beginning our effort to model electron transport through a nano scale system placed in a photon cavity has been geared towards systems based on a two-dimensional electron gas in GaAs or similar heterostructures. We have emphasized intersubband transitions in the conduction band, active in the teraherz range, in anticipation of experiments in this promising system [84].

Here, subsystem S_1 is a two-dimensional electronic nanostructure placed in a static (classical) external magnetic field. The leads are subjected to the same homogeneous external field. The electronic nanostructure, via split-gate configuration, is parabolically confined in the y-direction with a characteristic frequency Ω_0. The ends of the nanostructure in the x-direction at $x = \pm L_x/2$ are etched, forming a hard-wall confinement of length L_x. The external classical magnetic field is given by $\mathbf{B} = B\hat{\mathbf{z}}$ with a vector potential $\mathbf{A} = (-By, 0, 0)$. The single-particle Hamiltonian reads:

$$
\begin{aligned}
\hat{h}_{S_1}^{(0)} &= \frac{1}{2m}\left(\mathbf{p} + q\mathbf{A}\right)^2 + \frac{1}{2}m\Omega_0^2 y^2 \\
&= \frac{1}{2m}p_x^2 + \frac{1}{2m}p_y^2 + \frac{1}{2}m\Omega_w^2 y^2 + i\omega_c y p_x ,
\end{aligned}
\tag{53}
$$

where m is the effective mass of an electron, $-q$ its charge, $\mathbf{p}$ the canonical momentum operator, $\omega_c = qB/m$ is the cyclotron frequency and $\Omega_w = \sqrt{\omega_c^2 + \Omega_0^2}$ is the modified parabolic confinement. The spin degree of freedom is included with either a Zeeman term added to the Hamiltonian [85], or with Rashba and Dresselhaus spin orbit interactions, additionally [86].

H_{S_2} is simply the free field photon term for one cavity mode and by ignoring the zero point energy can be written as $H_{S_2} = \hbar\omega_p a^\dagger a$ where $\hbar\omega_p$ is the single photon energy and a $(a^\dagger)$ is the bosonic annihilation (creation) operator. The electron-photon interaction term V_{el-ph} can be split into two terms $V_{el-ph} = V_{el-ph}^{(1)} + V_{el-ph}^{(2)}$ where

$$
V_{el-ph}^{(1)} := \sum_{ij}\sum_{i,j}\left\langle \psi_i \left| \frac{q}{2m}\left(\boldsymbol{\pi}\cdot\mathbf{A}_{EM} + \mathbf{A}_{EM}\cdot\boldsymbol{\pi}\right) \right| \psi_j \right\rangle c_i^\dagger c_j
\tag{54}
$$

$$
V_{el-ph}^{(2)} := \sum_{ij}\sum_{i,j}\left\langle \psi_i \left| \frac{q^2}{2m}\mathbf{A}_{EM}^2 \right| \psi_j \right\rangle c_i^\dagger c_j ,
\tag{55}
$$

with $\pi \equiv \mathbf{p} + q\mathbf{A}$ the mechanical momentum. The term in Equation (54) is the paramagnetic interaction, whereas the diamagnetic term is defined by Equation (55). By assuming that the photon wavelength is much larger than characteristic length scales of the system one can approximate the vector potential amplitude to be constant over the electronic system. Let us stress here that, in contrast to the usual dipole approximation, we will not omit the diamagnetic electron-photon interaction term. Then the vector potential is written as:

$$\mathbf{A}_{EM} \simeq \hat{\mathbf{e}} A_{EM} \left(a + a^\dagger \right) = \hat{\mathbf{e}} \frac{\mathcal{E}_c}{q\Omega_w a_w} \left(a + a^\dagger \right) , \tag{56}$$

where $\hat{\mathbf{e}}$ is the unit polarization vector and $\mathcal{E}_c \equiv q A_{EM} \Omega_w a_w$ is the electron-photon coupling strength. For a 3D rectangular Fabry Perot cavity we have $A_{EM} = \sqrt{\hbar / (2\omega_p V \epsilon_0)}$ where V is the cavity volume. Linear polarization in the x-direction is achieved for a $\mathbf{TE}_{011}$ mode, and in the y-direction with a $\mathbf{TE}_{101}$ mode.

Using the approximation in Equation (56), the expressions for the electron-photon interaction in Equations (54) and (55) are greatly simplified by pulling $\mathbf{A}_{EM}$ in front of the integrals. For the paramagnetic term, we get

$$V_{el-ph}^{(1)} \simeq \mathcal{E}_c \left(a + a^\dagger \right) \sum_{ij} g_{ij} c_i^\dagger c_j . \tag{57}$$

where we introduced the dimensionless coupling between the electrons and the cavity mode

$$g_{ij} = \frac{a_w}{2\hbar} \hat{\mathbf{e}} \cdot \int d\mathbf{r} \left[\psi_i^*(\mathbf{r}) \left\{ \pi \psi_j(\mathbf{r}) \right\} + \left\{ \pi \psi_i^*(\mathbf{r}) \right\} \psi_j(\mathbf{r}) \right] . \tag{58}$$

As for the diamagnetic term, we get

$$V_{el-ph}^{(2)} \simeq \frac{\mathcal{E}_c^2}{\hbar \Omega_w} \left[\left(a^\dagger a + \frac{1}{2} \right) + \frac{1}{2} \left(a^\dagger a^\dagger + aa \right) \right] \mathcal{N}^e , \tag{59}$$

where $\mathcal{N}^e$ is the number operator in the electron Fock space. Note that $V_{el-ph}^{(2)}$ does not depend on the photon polarization or geometry of the system in this approximation. We do not use the rotating wave approximation as in our multilevel systems even though a particular electron transition could be in resonance with the photon field we want to include the contribution form others not in resonance.

For the numerical diagonalization of H_S we shall use the lowest $N_{mesT} \ll N_{mes}$ IMBS of H_{S_1} and photon states containing up to N_{EM} photons, resulting in a total of $N_{mesT} \times (N_{EM} + 1)$ states in the 'free' basis $\{|\nu, j\rangle\}$.

5.2. Results

Groups modeling the near resonance interaction of one cavity mode with a two level electronic system have expressed the importance of using a large enough, or the correct type, of a photon basis in the strongly interacting regime [87,88]. In many level systems where wavefunction and geometric effects are accounted for our experience is that convergence in numerical diagonalization is more sensitive to proper truncation of the electronic sector of the Fock many-body space. This reflects the polarizability of the electric charge by a cavity field in the construction of the photon-dressed electronic states. At the same time the inclusion of the diamagnetic interaction curbs the need for states with a very high photon number [42,50,89].

The polarizability of the first photon replica of the two-electron ground state is displayed in Figure 8 as a function of g_{EM}, the photon energy $\hbar\omega$ and its polarization [50].

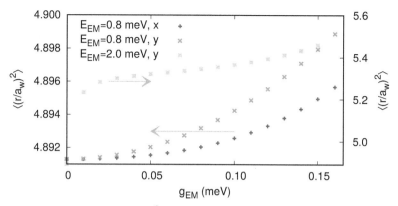

Figure 8. The expectation value $\langle (r/a_w)^2 \rangle$ for the first photon replica of the two-electron ground state in the closed system at $t = 0$ for x- and y-polarization of the photon field. $\hbar\omega = 2.0$ meV, $B = 0.1$ T. Two parallel quantum dots are embedded in the central system.

The polarizability is nonlinear, anisotropic, and largest for the cavity photon close to a resonance with the confinement energy in the y-direction.

A Rabi oscillation of two electrons in the double quantum dot system embedded in the short quantum wire leads to oscillating charge with time in the system. The oscillating probability of charge presence in the contact areas of the short wire thus lead to oscillations in the current leaving the system through the left and right leads [33], see Figure 9.

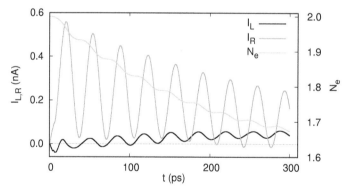

Figure 9. The left (black) and right (gold) currents and the mean electron number (blue) for initially fully entangled Rabi-split singlet two-electron states as the interacting system discharges in the transient regime. $\hbar\omega = 2.0$ meV, $B = 0.1$ T. Two parallel quantum dots are embedded in the central system.

Alternatively, one may view this as the consequence of the Rabi resonance entangling two states with different tunneling probability to the leads.

In the transient or the late transient regime we have used the non-Markovian GME to investigate several results: Thorsten Arnold et al. used a time-convolution-less (TCL) version of the GME to study the effects of magnetic field and photons [46] on the transport of interacting electrons through a quantum ring with spin-orbit interactions in a photon cavity with circular [86] and linear polarization [90]. Aharonov-Bohm oscillations were established in the time-dependent transport through a ring structure with additional vortexes in the contact region of the quantum wire. x-polarized photons with energy 0.3 meV attenuate the Aharonov-Bohm oscillations over a broad range of magnetic field, but y-polarized photons influence the transport in a more complex fashion. The oscillations are generally attenuated, but one oscillation peak is split and the charge current is enhanced at a magnetic field corresponding to a half-integer flux quantum [46]. With the spin-orbit interactions the spin polarization and the spin photo currents of the quantum ring are largest for circularly polarized photon field and a destructive Aharonov–Casher (AC) phase interference. The dip in the charge current caused by the destructive AC phase becomes threefold under the circularly polarized photon field as the interaction of the angular momentum of the electron and the spin angular momentum of the light create a many-body level splitting [86]. The detailed balance between the para- and the diamagnetic electron-photon interactions has been studied for an electron in the quantum ring structure when excited by a short classical dipole pulse [47].

Nzar Rauf Abdullah et al. have used the GME formalism to investigate photon assisted transport [91], photon mediated switching in nanostructures [48,49,92], the balancing of magnetic and forces caused by cavity photons [93], cavity-photon affected thermal transport [94,95], and the influence of cavity photons on thermal spin currents in a system with spin orbit interactions [96,97].

6. Steady-State

The investigation of the time dependent transport of electrons through a photon cavity soon made it clear that for the continuous model the inherent time scales can lead to relaxation times far beyond what is accessible with simple integration of the GME [32,45,46,49,91]. The underlying cause for the diverse relaxation times is on one hand electron tunneling rates affected by the shape or geometry of the system and the condition of weak coupling. Different many-body states can have a high or low probability for electrons to be found in the contact areas of the central system. On the other hand are slow rates of FIR or terahetz active transitions, that are furthermore affected by the geometry of the wavefunctions of the corresponding final and initial states. In addition, the cavity decay, or coupling to the environment, affects relaxation times as we address below [98]. To avoid confusion it is important to remember that we calculate the eigenstates of the closed central system, the interacting electron and photon system, and the opening up of the system to the leads or the external photon reservoir is always a neccessary triggering mechanism for all transitions later in time, photon active or not.

6.1. The Steady-State Limit

In order to investigate the long-time evolution and the steady state of the central system under the influences of the reservoirs we resort to a Markovian version of the GME, whereby we assume memory effects in the kernel of the GME (26) to vanish, relinquishing the reduced density operator local in time enabling the approximation [99]

$$\int_0^\infty ds\, e^{is(E_\beta - E_\alpha - \epsilon)} \approx \pi\delta(E_\beta - E_\alpha - \epsilon), \tag{60}$$

where a small imaginary principle part is ignored. We have furthermore assumed instant lead-system coupling at $t = 0$ with $\chi_l(t) = \theta(t)$, the Heaviside unit step function, in Equation (8) for H_T. In order to transform the resulting Markovian equation into a simpler form we use the vectorization operation [100], that stacks the columns of a matrix into a vector, and its property

$$\text{vec}(A\rho B) = (B^T \otimes A)\text{vec}(\rho) \tag{61}$$

through which the reduced density matrix can always be moved to the right side of the corresponding term, and a Kronecker product has been introduced with the property $B \otimes A = \{B_{\alpha,\beta}A\}$. The Kronecker product of two $N_{\text{mes}} \times N_{\text{mes}}$ matrices results in a $N_{\text{mes}}^2 \times N_{\text{mes}}^2$ matrix, and effectively the vectorization has brought forth that the natural space for the Liouville-von Neumann equation is not the standard Fock space of many-body states, but the larger Liouville space of transitions [101–103].

No further approximations are used to attain the Markovian master equation and due to the complex structure of the non-Markovian GME we have devised a general recipe published elsewhere [99] to facilitate the analytical construction and the numerical implementation. The Markovian master equation has the form

$$\partial_t \text{vec}(\rho) = \mathcal{L}\text{vec}(\rho),$$ (62)

and as the non-Hermitian Liouville operator $\mathcal{L}$ is independent of time the analytical solution of Equation (62) can be written as

$$\text{vec}(\rho(t)) = \{\mathcal{U}[\exp(\mathcal{L}_{\text{diag}}t)]\mathcal{V}\}\text{vec}(\rho(0)),$$ (63)

in terms of the matrices of the column stacked left $\mathcal{U}$, and the right $\mathcal{V}$ eigenvectors of $\mathcal{L}$

$$\mathcal{L}\mathcal{V} = \mathcal{V}\mathcal{L}_{\text{diag}}, \quad \text{and} \quad \mathcal{U}\mathcal{L} = \mathcal{L}_{\text{diag}}\mathcal{U},$$ (64)

obeying

$$\mathcal{U}\mathcal{V} = I, \quad \text{and} \quad \mathcal{V}\mathcal{U} = I.$$ (65)

Special care has to be taken in the numerical implementation of this solution procedure as many software packages use another normalization for $\mathcal{U}$ and $\mathcal{V}$. Calculations in the Liouville space using (63) are memory (RAM) intensive, but bring several benefits: No time integration combined with iterations is needed, thus time points can be selected with other criteria in mind. The solution is thus convenient for long-time evolution, that is not easily accessible with numerical integration. The complicated structure of the left and right eigenvector matrices for a complex system with nontrivial geometry makes Equation (63) the best choice to find the properties of the steady state by monitoring the limit of it as time gets very large. The complex eigenvalue spectrum of the Liouville operator $\mathcal{L}$ reveals information about the mean lifetime and energy of all active transitions in the open system, and the zero eigenvalue defines the steady state.

In the steady state the properties of the system do not change with time, but underneath the "quiet surface" many transitions can be active to maintain it. The best experimental probes to gauge the underlying processes are measurements of noise spectra for a particular physical variable. They are available through the two-time correlation functions of the respective measurable quantities. For a Markovian central system weakly coupled to reservoirs the two-time correlation functions can be calculated applying the Quantum Regression Theorem (QRT) [104,105] stating that the the equation of motion for a two-time correlation function has the same form as the Markovian master equation (62) for the operator [106]

$$\chi(\tau) = \text{Tr}_R \left\{ e^{-iH\tau/\hbar} X\rho_T(0)e^{+iH\tau/\hbar} \right\},$$ (66)

where H is the total Hamiltonian of the system, ρ_T its density operator, and the trace is taken with respect of all reservoirs. For photon correlations $X = a + a^\dagger$ as in [50], or $X = Q\Lambda^l$ for current correlations as in [107], where $Q = \sum_i c_i^\dagger c_i$ is the fermionic charge operator and Λ^l is the Liouville dissipation operator for lead l. The structure of χ (66) indicates that the two-time correlation function is then

$$\langle X(\tau)X(0) \rangle = \text{Tr}_S \{X(0)\chi(\tau)\},$$ (67)

with

$$\text{vec}(\chi(\tau)) = \{\mathcal{U}[\exp(\mathcal{L}_{\text{diag}}t)]\mathcal{V}\}\text{vec}(\chi(0)).$$ (68)

The left side of Equation (67), the two-time correlation function, is written in the Heisenberg picture, in contrast to the Schrödinger picture used elsewhere in the article. The Fourier spectral density for the photon two-time correlation function is denoted by $S(E)$, and for the current-current correlation the corresponding Fourier spectral density denoted by $D_{ll'}(E)$, where l and l' refer to L and R, the Left and Right leads.

6.2. Results

To date we have used the Markovian version of the master equation to investigate properties of the steady state, and how the system with electrons being transported through a photon cavity reaches it. We assume GaAs parameters with effective mass $m = 0.067m_e$, effective relative dielectric constant $\epsilon_r = 12.3$, and effective Landé g-factor $g = -0.44$. The characteristic energy of the parabolic confinement of the semi-infinite leads and the central system in the y-direction is $\hbar\Omega_0 = 2.0$ meV. The length of the short quantum wire is L_x, and the overall coupling coefficient for the leads to the system is $g_{LR}a_w^{3/2} = 0.124$ meV.

We start with a central system made of a finite parabolic quantum wire without any embedded quantum dots. Figure 10 demonstrates that the approach to build and solve the Markovian master Equations (62)–(63) works for an interacting system with 120 many-body states participating in the transport [108].

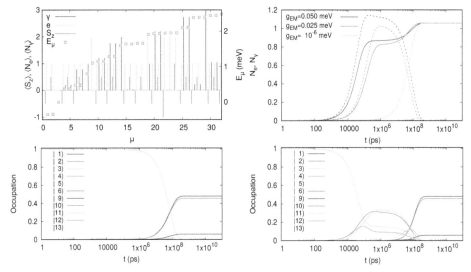

Figure 10. (**upper left**) For the closed system as functions of the number of the eigenstate μ, the many-body energy (squares), the mean photon (γ) and electron content (e), and the mean spin z-component (S_z). The horizontal yellow lines represent the chemical potentials of the left (μ_L) and right leads (μ_R) when the system will be coupled to them. (**upper right**) The mean electron (solid) and photon number (dashed) in the central system as a function of time. The mean occupation of the many-body eigenstates of the system for $g_{EM} = 1 \times 10^{-6}$ meV (**lower left**), and $g_{EM} = 0.05$ meV (**lower right**). $V_g = -1.6$ mV, $\hbar\omega = 0.8$ meV, x-polarization, $\kappa = 1 \times 10^{-5}$ meV, $L_x = 150$ nm, and $B = 0.1$ T. No quantum dots in the short wire.

The upper right panel displays the properties of the lowest 32 many-body states at the plunger gate voltage $V_g = -1.6$ mV. With $\mu_L = 1.4$ meV and $\mu_L = 1.1$ meV there are 8 states below the bias window and five states within it. In the bias window is one spin singlet two-electron state (the two-electron ground state) and two spin components of two one-electron states with a non-integer mean photon content indicating a Rabi splitting. The upper left panel of Figure 10 show the mean electron and photon numbers in the central system when it is initially empty. With a very low coupling, $g_{EM} = 1 \times 10^{-6}$ meV, between the electrons and photons, the charging is very slow with the probability approaching unity around $t \approx 10^8$ s. With increasing g_{EM} the charging becomes faster, and during the phase the mean photon number in the system rises. The lower panels of Figure 10 reveal what is happening. With the low photon coupling (lower left panel) electrons tunnel non-resonantly into the two spin components of the ground state, $|1\rangle$ and $|2\rangle$ as the vacuum state $|3\rangle$ looses occupation, and to a small fraction the two-electron state $|9\rangle$ gets occupied. When the coupling of the electrons and the photons is not vanishingly small (lower right panel) the charging of the system takes a different rout. The finite g_{EM} allows the incoming electron to enter the Rabi-split one-electron states in the bias window as these are a linear combination of electron states with a different photon number. This explains the growing mean number of photons in the system for intermediate times. These states are eigenstates of the central system, but not of the open system, so at a later time they decay into the the the one- and two-electron ground states as before bringing the system into the same Coulomb blocked steady state as before. We thus observe electromagnetically active transitions in the system in an intermediate time regime [108].

The on-set of the steady state regime is difficult to judge only from the shape of the charge being accumulated in the system or the current through or into it as a function of time [85]. For a system of two parallel quantum dots embedded in a short quantum wire ($L_x = 150$) nm the charging and the current as functions of time look the same (see Figures 4 and 5 in ref. [85]), but when the occupation of the eigenstates of the closed system is analyzed, see Figure 11, a clear difference is seen for the approach to the steady state depending on whether the initial state contains only one or no photon [85]. In the case of neither photon nor an electron in the cavity initially an electron tunnels into the system into the two spin components of the one-electron ground state, which happens to be in the bias window for $V_g = -2.0$ mV. Thus, the steady state is a combination of the empty state and these two one-electron states. In the case of one photon and no electron initially in the system an electron tunnels non-resonantly into the 1-electron states $|8\rangle$ and $|9\rangle$ with energy slightly below 2 meV, and thus well above the bias window. The mean photon content of these states is close to unity and at a later time the electron ends up in the two spin components of the one-electron ground state via a radiative transition [85]).

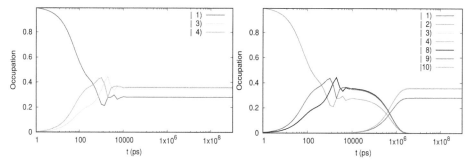

Figure 11. The mean occupation of the many-body eigenstates of the system when the initial state is the ground state $|1\rangle$ (**left**), or the first photon replica of the ground state $|2\rangle$ (**right**). $g_{EM} = 0.05$ meV. $V_g = -2.0$ mV, $\hbar\omega = 0.8$ meV, x-polarization, $L_x = 150$ nm, and $B = 0.1$ T. Two parallel quantum dots embedded in the short wire, but no photon reservoir.

Note that the "irregularly" looking structure around $t \approx 2000$ ps will be addressed below. Please note that the numbering of interacting many-body state depends on the structure of the system, and the plunger gate voltage V_g.

In the steady state all the mean values of the open system have reached a constant value. In order to query about the active underlying processes it is necessary to calculate the spectral densities of the photon or current correlations. We present these for the central system consisting of a short quantum wire ($L_x = 150$) nm with two embedded quantum dots in Figure 12 (see refs. [107,109]).

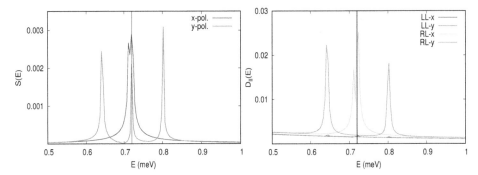

Figure 12. The spectral density $S(E)$ of the emitted cavity radiation for the central system in a steady state (**left**), and the spectral densities for the current–current correlations $D_{ll'}(E)$ (**right**). $g_{EM} = 0.1$ meV, $V_g = -2.0$ mV, $\hbar\omega = 0.72$ meV, $\kappa = 1 \times 10^{-3}$ meV, and $L_x = 150$ nm. Two parallel quantum dots embedded in the short wire.

Importantly we show in Ref. [50] that both the paramagnetic and the diamagnetic electron-photon interactions can lead to a Rabi resonance. The resonance for the diamagnetic interactions is much smaller, but the symmetry of the two parallel quantum dots leads to selection rules where for x-polarized cavity photon field the paramagnetic interaction is blocked, but both are present for the y-polarized field. Here, the active states are the one-electron ground state and the first excited one-electron state, with which the first photon replica of the ground state interacts for $\hbar\omega = 0.72$ meV. The spectral density of the photon-photon two-time correlation function, $S(E)$ seen in the left panel of Figure 12 shows one peak at the energy of the cavity mode $\hbar\omega = 0.72$ meV, and two side peaks for the y-polarization. The central peak is the ground state state electroluminescence and the side peaks are caused by the Rabi-split states [34,109,110]. Here, we observe the ground state electroluminescence even though the electron-photon coupling is not in the ultra strong regime, as we diagonalize the Hamiltonian in a large many-body Fock space instead of applying conventional perturbative calculations.

For the x-polarized cavity field we find a much weaker ground state electroluminescence caused by the diamagnetic electron-photon interaction [109]. In addition, we identify these effects for the fully interacting two-electron ground state, where they are partially masked by many concurrently active transitions. The spectral density for the current–current correlation functions $D_{ll'}(E)$ displayed in the right panel of Figure 12 show only peaks at the Rabi-satellites, as could be expected [107]. An inspection of $D_{ll'}(E)$ over a larger range of energy reveals more transitions active in maintaining the steady state, both radiative transitions and non radiative [107]. Moreover, we notice that when the steady state is not in a Coulomb blocking regime the spectral density of the current-current correlations always shows a background to the peaks with a structure reminiscent of a $1/f$ behavior, that is known in multiscale systems.

An "irregularly" looking structure in the mean occupation, the current current, and the mean number of electrons and photons. This is a general structure seen in all types of central system we have investigated in the continuous model. In Figure 13 we analyze it in a short parabolically confined quantum wire of length $L_x = 180$ nm with two asymmetrically embedded quantum dots [111].

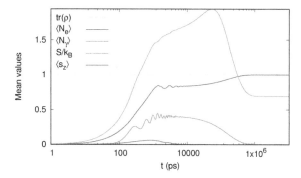

Figure 13. The mean electron (e), photon (γ), z-component of the spin (S_z), trace of the reduced density matrix, and the Réniy-2 entropy (S) as functions of time. $\hbar\omega = 0.373$ meV, x-polarization, $\kappa = 1 \times 10^{-5}$ meV, $g_{EM} = 0.05$ meV, and $L_x = 180$ nm. Two asymmetrically embedded quantum dots in the short wire.

An increased number of time points on the logarithmic scale shows regular oscillations. A careful analysis reveals two independent oscillations: A spatial charge oscillation between the quantum dots with the Rabi frequency in the system, and a still slower nonequilibrium oscillation of the spin populations residing as the system is brought to a steady state [111].

The steady-state Markovian formalism has been used to investigate oscillations in the transport current as the photon energy or the electron–photon coupling strenght are varied with or without flow of photons from the external reservoir [112,113]. Moreover, the formalism has been used to establish the signs of the Purcell effect [114] in the transport current [98].

In light of the experimental interest of using a two-dimensional electron gas in a GaAs heterostructure [84] we have calculated the exact matrix elements for the electron-photon interaction taking into account the spatial variation of the vector field **A** of the electronic system. This is a small correction in most cases but may be important when studying high order transitions or nonperturbational effects caused by the photon field. This has led us to discover a very slow high order transition between the ground states of two slightly dissimilar quantum dots [115].

The fist steps have been taken to investigate thermoelectric effects in the central system coupled to cavity photons, in the steady state. In a short quantum wire with one embedded quantum dot in the resonant regime, an inversion of thermoelectric current is found caused by the Rabi-splitting. The photon field can change both the magnitude and the sign of the thermoelectric current induced by the temperature gradient in the absence of a voltage bias between the leads [116].

7. Summary

It goes without saying that as transport experiments at nanoscale become more involved the formal tools must be suitably extended or adapted. In particular, the unavoidable charging and correlation effects at finite bias pushed the theoretical calculations from the very convenient single-particle (or at most mean-field) Landauer–Büttiker picture to the complicated many-body perturbation theory of the non-equilibrium Keldysh–Green's functions [117].

Here we summarized some results on time-dependent transport in open interacting systems which argue for the similar idea that if one looks for transient effects and dynamics of excited states the simple rate equation approach must be extended to the non-markovian generalized master equation.

The GME we used in all examples is constructed and solved w.r.t the exact many-body states of the central open system and can be therefore implemented numerically without major changes to study both Coulomb-interacting and hybrid systems where the fermion-boson interaction is crucial, like QD-cavity systems or nano-electromechanical systems. A consistent derivation of the GME the full knowledge of the eigenvalues and eigenfunctions of complicated interacting Hamiltonian (e.g., cavity-coupled systems must be described by 'dressed' states). With very few exceptions coming from quantum optics (i.e., the Jaynes-Cummings model for two-level or Λ and V three-level systems) such a task can only be achieved via numerically exact diagonalization of large matrices, especially for elecron-photon systems. To bypass this difficulty we proposed and succesfully used a stepwise diagonalization procedure.

The dynamics of excited states in a quantum wire, the onset of current-current correlations for a pair of electrostatically coupled quantum dots and thermoelectric effects were presented within a simple lattice model which however captures the relevant physics.

When turning to QED-cavity system we developed the GME within a continuous model which accounts for the geometrical details of the sample and of the contact regions. Moreover, the calculations were performed by taking into account both the paramagnetic and diamagnetic contributions to the electron-photon coupling and without relying on the rotating-wave approximation. This is an important step beyond the Jaynes–Cummings model. Also, the number of many-body stated needed in the calculations increased considerably. Thus, the accuracy of the stepwise numerical diagonalization had to be carefully discussed. Finally, for systems with long relaxation time a markovian version of GME was proposed and implemented via a clever vectorization procedure.

We end this review by pointing out possible improvements of the GME method and some of its future applications. At the formal level, perhaps the most challenging upgrade is the inclusion of time-dependent potentials describing laser pulses or microwave driving signals. Provided this is succesfully achieved, one could study transport through driven nano-electromechanical systems (NEMS) or the physics of Floquet states emerging in strongly driven systems [118,119]. Let us mention here that at least for closed systems (i.e., not connected to particle reservoirs) studies based on Floquet master equations for two-level system are already available [120,121]. As for more immediate applications we aim at the theoretical modeling of transport in Tavis–Cummings systems, motivated by the recent observation of state readout in a system of distant coupled quantum dots individually connected to a pair of leads and interacting via cavity photons [29].

Author Contributions: all the authors have a similar contribution to the paper in its concept, research, and manuscript preparation.

Funding: This work was partially supported by the Research Fund of the University of Iceland, the Icelandic Research Fund, grant no. 163082-051, the Icelandic Instruments Fund, and Reykjavik University, grant no. 815051. Some of the computations were performed on resources provided by the Icelandic High Performance Computing Centre at the University of Iceland. V.M. also acknowledge financial support from CNCS-UEFISCDI grant PN-III-P4-ID-PCE-2016-0084 and from the Romanian Core Program PN19-03 (contract no. 21 N/08.02.2019)

Conflicts of Interest: The authors declare no conflict of interest.

References

1. Di Ventra, M. *Electrical Transport in Nanoscale Systems*; Cambridge University Press: Cambridge, UK, 2008.

2. Jahnke, F. *Quantum Optics with Semiconductor Nanostructures*; Woodhead Publishing Series in Electronic and Optical Materials; Elsevier Science: Amsterdam, The Netherlands, 2016.

3. Chow, W.W.; Jahnke, F. On the physics of semiconductor quantum dots for applications in lasers and quantum optics. *Prog. Quantum Electron.* **2013**, *37*, 109–184.

4. Gu, X.; Kockum, A.F.; Miranowicz, A.; Liu, Y.; Nori, F. Microwave photonics with superconducting quantum circuits. *Phys. Rep.* **2017**, *718–719*, 1–102.

5. Bloch, F. Generalized Theory of Relaxation. *Phys. Rev.* **1957**, *105*, 1206–1222.

6. Wangsness, R.K.; Bloch, F. The Dynamical Theory of Nuclear Induction. *Phys. Rev.* **1953**, *89*, 728–739.

7. Redfield, A.G. The Theory of Relaxation Processes. *Adv. Magn. Reson.* **1965**, *1*, 1–32.

8. Scully, M.O.; Zubairy, M.S. *Optics*; Cambridge University Press: Cambridge, UK, 1997.

9. Carmichael, H.J. *Statistical Methods in Quantum Optics 1: Master Equations and Fokker-Planck Equations*; Springer: Berlin, Germany, 2003.

10. Timm, C. Tunneling through molecules and quantum dots: Master-equation approaches. *Phys. Rev. B* **2008**, *77*, 195416.

11. Elenewski, J.E.; Gruss, D.; Zwolak, M. Communication: Master equations for electron transport: The limits of the Markovian limit. *J. Chem. Phys.* **2017**, *147*, 151101.

12. Moldoveanu, V.; Manolescu, A.; Gudmundsson, V. Geometrical effects and signal delay in time-dependent transport at the nanoscale. *New J. Phys.* **2009**, *11*, 073019.

13. Vaz, E.; Kyriakidis, J. Transient dynamics of confined charges in quantum dots in the sequential tunneling regime. *Phys. Rev. B* **2010**, *81*, 085315.

14. Moldoveanu, V.; Manolescu, A.; Tang, C.S.; Gudmundsson, V. Coulomb interaction and transient charging of excited states in open nanosystems. *Phys. Rev. B* **2010**, *81*, 155442.

15. Gudmundsson, V.; Lin, Y.Y.; Tang, C.S.; Moldoveanu, V.; Bardarson, J.H.; Manolescu, A. Transport through a quantum ring, dot, and barrier embedded in a nanowire in magnetic field. *Phys. Rev. B* **2005**, *71*, 235302.

16. Torfason, K.; Manolescu, A.; Molodoveanu, V.; Gudmundsson, V. Excitation of collective modes in a quantum flute. *Phys. Rev. B* **2012**, *85*, 245114.

17. Moldoveanu, V.; Manolescu, A.; Gudmundsson, V. Dynamic correlations induced by Coulomb interactions in coupled quantum dots. *Phys. Rev. B* **2010**, *82*, 085311.

18. Bulnes Cuetara, G.; Esposito, M.; Gaspard, P. Fluctuation theorems for capacitively coupled electronic currents. *Phys. Rev. B* **2011**, *84*, 165114.

19. Harbola, U.; Esposito, M.; Mukamel, S. Quantum master equation for electron transport through quantum dots and single molecules. *Phys. Rev. B* **2006**, *74*, 235309.

20. Cohen, G.; Rabani, E. Memory effects in nonequilibrium quantum impurity models. *Phys. Rev. B* **2011**, *84*, 075150.

21. Härtle, R.; Cohen, G.; Reichman, D.R.; Millis, A.J. Decoherence and lead-induced interdot coupling in nonequilibrium electron transport through interacting quantum dots: A hierarchical quantum master equation approach. *Phys. Rev. B* **2013**, *88*, 235426.

22. Esposito, M.; Galperin, M. Transport in molecular states language: Generalized quantum master equation approach. *Phys. Rev. B* **2009**, *79*, 205303.

23. Galperin, M.; Nitzan, A.; Ratner, M.A. Inelastic transport in the Coulomb blockade regime within a nonequilibrium atomic limit. *Phys. Rev. B* **2008**, *78*, 125320.

24. Breuer, H.P.; Laine, E.M.; Piilo, J.; Vacchini, B. Colloquium: Non-Markovian dynamics in open quantum systems. *Rev. Mod. Phys.* **2016**, *88*, 021002.

25. Kulkarni, M.; Cotlet, O.; Türeci, H.E. Cavity-coupled double-quantum dot at finite bias: Analogy with lasers and beyond. *Phys. Rev. B* **2014**, *90*, 125402.

26. Viennot, J.J.; Delbecq, M.R.; Dartiailh, M.C.; Cottet, A.; Kontos, T. Out-of-equilibrium charge dynamics in a hybrid circuit quantum electrodynamics architecture. *Phys. Rev. B* **2014**, *89*, 165404.

27. Liu, Y.Y.; Stehlik, J.; Eichler, C.; Gullans, M.J.; Taylor, J.M.; Petta, J.R. Semiconductor double quantum dot micromaser. *Science* **2015**, *347*, 285.

28. Viennot, J.J.; Dartiailh, M.C.; Cottet, A.; Kontos, T. Coherent coupling of a single spin to microwave cavity photons. *Science* **2015**, *349*, 408–411.

29. Liu, Y.Y.; Stehlik, J.; Mi, X.; Hartke, T.R.; Gullans, M.J.; Petta, J.R. On-Chip Quantum-Dot Light Source for Quantum-Device Readout. *Phys. Rev. Appl.* **2018**, *9*, 014030.

30. Beaudoin, F.; Gambetta, J.M.; Blais, A. Dissipation and ultrastrong coupling in circuit QED. *Phys. Rev. A* **2011**, *84*, 043832.

31. Dinu, I.V.; Moldoveanu, V.; Gartner, P. Many-body effects in transport through a quantum-dot cavity system. *Phys. Rev. B* **2018**, *97*, 195442.

32. Gudmundsson, V.; Jonasson, O.; Tang, C.S.; Goan, H.S.; Manolescu, A. Time-dependent transport of electrons through a photon cavity. *Phys. Rev. B* **2012**, *85*, 075306.

33. Gudmundsson, V.; Sitek, A.; yi Lin, P.; Abdullah, N.R.; Tang, C.S.; Manolescu, A. Coupled Collective and Rabi Oscillations Triggered by Electron Transport through a Photon Cavity. *ACS Photonics* **2015**, *2*, 930–934.

34. Cirio, M.; De Liberato, S.; Lambert, N.; Nori, F. Ground State Electroluminescence. *Phys. Rev. Lett.* **2016**, *116*, 113601.

35. Cirio, M.; Shammah, N.; Lambert, N.; De Liberato, S.; Nori, F. Multielectron Ground State Electroluminescence. *Phys. Rev. Lett.* **2019**, *122*, 190403.

36. Schachenmayer, J.; Genes, C.; Tignone, E.; Pupillo, G. Cavity-Enhanced Transport of Excitons. *Phys. Rev. Lett.* **2015**, *114*, 196403.

37. Breuer, H.P.; Petruccione, F. *The Theory of Open Quantum Systems*; Oxford University Press: Oxford, UK, 2007.

38. Poot, M.; van der Zant, H.S. Mechanical systems in the quantum regime. *Phys. Rep.* **2012**, *511*, 273–335.

39. Tanatar, B.; Moldoveanu, V.; Dragomir, R.; Stanciu, S. Interaction and Size Effects in Open Nano-Electromechanical Systems. *Phys. Status Solidi (b)* **2019**, *256*, 1800443.

40. Settineri, A.; Macrí, V.; Ridolfo, A.; Di Stefano, O.; Kockum, A.F.; Nori, F.; Savasta, S. Dissipation and thermal noise in hybrid quantum systems in the ultrastrong-coupling regime. *Phys. Rev. A* **2018**, *98*, 053834.

41. Zueco, D.; García-Ripoll, J. Ultrastrongly dissipative quantum Rabi model. *Phys. Rev. A* **2019**, *99*, 013807.

42. Jonasson, O.; Tang, C.S.; Goan, H.S.; Manolescu, A.; Gudmundsson, V. Nonperturbative approach to circuit quantum electrodynamics. *Phys. Rev. E* **2012**, *86*, 046701.

43. Schinabeck, C.; Erpenbeck, A.; Härtle, R.; Thoss, M. Hierarchical quantum master equation approach to electronic-vibrational coupling in nonequilibrium transport through nanosystems. *Phys. Rev. B* **2016**, *94*, 201407.

44. Gudmundsson, V.; Gainar, C.; Tang, C.S.; Moldoveanu, V.; Manolescu, A. Time-dependent transport via the generalized master equation through a finite quantum wire with an embedded subsystem. *New J. Phys.* **2009**, *11*, 113007.

45. Gudmundsson, V.; Jonasson, O.; Arnold, T.; Tang, C.S.; Goan, H.S.; Manolescu, A. Stepwise introduction of model complexity in a generalized master equation approach to time-dependent transport. *Fortschr. Phys.* **2013**, *61*, 305.

46. Arnold, T.; Tang, C.S.; Manolescu, A.; Gudmundsson, V. Magnetic-field-influenced nonequilibrium transport through a quantum ring with correlated electrons in a photon cavity. *Phys. Rev. B* **2013**, *87*, 035314.

47. Arnold, T.; Tang, C.S.; Manolescu, A.; Gudmundsson, V. Excitation spectra of a quantum ring embedded in a photon cavity. *J. Opt.* **2015**, *17*, 015201.

48. Abdullah, N.R.; Tang, C.S.; Gudmundsson, V. Time-dependent magnetotransport in an interacting double quantum wire with window coupling. *Phys. Rev. B* **2010**, *82*, 195325.

49. Abdullah, N.R.; Tang, C.S.; Manolescu, A.; Gudmundsson, V. Delocalization of electrons by cavity photons in transport through a quantum dot molecule. *Phys. E Low-dimens. Syst. Nanostruct.* **2014**, *64*, 254–262.

50. Gudmundsson, V.; Sitek, A.; Abdullah, N.R.; Tang, C.S.; Manolescu, A. Cavity-photon contribution to the effective interaction of electrons in parallel quantum dots. *Ann. Phys.* **2016**.

51. Vargiamidis, V.; Valassiades, O.; Kyriakos, D.S. Lippmann-Schwinger equation approach to scattering in quantum wires. *Phys. Stat. Sol.* **2003**, *236*, 597.

52. Bardarson, J.H.; Magnusdottir, I.; Gudmundsdottir, G.; Tang, C.S.; Manolescu, A.; Gudmundsson, V. Coherent electronic transport in a multimode quantum channel with Gaussian-type scatterers. *Phys. Rev. B* **2004**, *70*, 245308.

53. Arnold, T. The Influence of Cavity Photons on the Transient Transport of Correlated Electrons through a Quantum Ring with Magnetic Field and Spin-Orbit Interaction. Ph.D. Thesis, University of Iceland, Reykjavík, Iceland, 2014.

54. Fujisawa, T.; Austing, D.G.; Tokura, Y.; Hirayama, Y.; Tarucha, S. Electrical pulse measurement, inelastic relaxation, and non-equilibrium transport in a quantum dot. *J. Phys. Condens. Matter* **2003**, *15*, R1395–R1428.

55. Naser, B.; Ferry, D.K.; Heeren, J.; Reno, J.L.; Bird, J.P. Large capacitance in the nanosecond-scale transient response of quantum point contacts. *Appl. Phys. Lett.* **2006**, *89*, 083103,

56. Lai, W.T.; Kuo, D.M.; Li, P.W. Transient current through a single germanium quantum dot. *Phys. E Low-dimens. Syst. Nanostruct.* **2009**, *41*, 886–889.

57. Kaestner, B.; Kashcheyevs, V. Non-adiabatic quantized charge pumping with tunable-barrier quantum dots: A review of current progress. *Rep. Prog. Phys.* **2015**, *78*, 103901.

58. Moldoveanu, V.; Manolescu, A.; Gudmundsson, V. Theoretical investigation of modulated currents in open nanostructures. *Phys. Rev. B* **2009**, *80*, 205325.

59. Moldoveanu, V.; Dinu, I.V.; Tanatar, B.; Moca, C.P. Quantum turnstile operation of single-molecule magnets. *New J. Phys.* **2015**, *17*, 083020. doi:10.1088/1367-2630/17/8/083020.

60. Moldoveanu, V.; Cornean, H.D.; Pillet, C.A. Nonequilibrium steady states for interacting open systems: Exact results. *Phys. Rev. B* **2011**, *84*, 075464.

61. Cornean, H.D.; Neidhardt, H.; Zagrebnov, V.A. The Effect of Time-Dependent Coupling on Non-Equilibrium Steady States. *Ann. Henri Poincaré* **2009**, *10*, 61–93.
62. Narozhny, B.N.; Levchenko, A. Coulomb drag. *Rev. Mod. Phys.* **2016**, *88*, 025003.
63. Sánchez, R.; López, R.; Sánchez, D.; Büttiker, M. Mesoscopic Coulomb Drag, Broken Detailed Balance, and Fluctuation Relations. *Phys. Rev. Lett.* **2010**, *104*, 076801.
64. Kaasbjerg, K.; Jauho, A.P. Correlated Coulomb Drag in Capacitively Coupled Quantum-Dot Structures. *Phys. Rev. Lett.* **2016**, *116*, 196801.
65. Lim, J.S.; Sánchez, D.; López, R. Engineering drag currents in Coulomb coupled quantum dots. *New J. Phys.* **2018**, *20*, 023038.
66. Zhou, C.; Guo, H. Coulomb drag between quantum wires: A nonequilibrium many-body approach. *Phys. Rev. B* **2019**, *99*, 035423.
67. Sánchez, R.; Thierschmann, H.; Molenkamp, L.W. Single-electron thermal devices coupled to a mesoscopic gate. *New J. Phys.* **2017**, *19*, 113040.
68. Bhandari, B.; Chiriacò, G.; Erdman, P.A.; Fazio, R.; Taddei, F. Thermal drag in electronic conductors. *Phys. Rev. B* **2018**, *98*, 035415.
69. Sánchez, D.; López, R. Nonlinear phenomena in quantum thermoelectrics and heat. *C. R. Phys.* **2016**, *17*, 1060–1071.
70. Sierra, M.A.; Sánchez, D. Strongly nonlinear thermovoltage and heat dissipation in interacting quantum dots. *Phys. Rev. B* **2014**, *90*, 115313.
71. Torfason, K.; Manolescu, A.; Erlingsson, S.I.; Gudmundsson, V. Thermoelectric current and Coulomb-blockade plateaus in a quantum dot. *Physica E* **2013**, *53*, 178–185.
72. Sánchez, D.; Linke, H. Focus on Thermoelectric Effects in Nanostructures. *New J. Phys.* **2014**, *16*, 110201.
73. Beenakker, C.W.J.; Staring, A.A.M. Theory of the thermopower of a quantum dot. *Phys. Rev. B* **1992**, *46*, 9667–9676.
74. Staring, A.A.M.; Molenkamp, L.W.; Alphenaar, B.W.; van Houten, H.; Buyk, O.J.A.; Mabesoone, M.A.A.; Beenakker, C.W.J.; Foxon, C.T. Coulomb-Blockade Oscillations in the Thermopower of a Quantum Dot. *Europhys. Lett.* **1993**, *22*, 57.
75. Dzurak, A.; Smith, C.; Pepper, M.; Ritchie, D.; Frost, J.; Jones, G.; Hasko, D. Observation of Coulomb blockade oscillations in the thermopower of a quantum dot. *Solid State Commun.* **1993**, *87*, 1145–1149.
76. Svensson, S.F.; Persson, A.I.; Hoffmann, E.A.; Nakpathomkun, N.; Nilsson, H.A.; Xu, H.Q.; Samuelson, L.; Linke, H. Lineshape of the thermopower of quantum dots. *New J. Phys.* **2012**, *14*, 033041.
77. Reddy, P.; Jang, S.Y.; Segalman, R.A.; Majumdar, A. Thermoelectricity in Molecular Junctions. *Science* **2007**, *315*, 1568–1571.
78. Svensson, S.F.; Hoffmann, E.A.; Nakpathomkun, N.; Wu, P.M.; Xu, H.Q.; Nilsson, H.A.; Sánchez, D.; Kashcheyevs, V.; Linke, H. Nonlinear thermovoltage and thermocurrent in quantum dots. *New J. Phys.* **2013**, *15*, 105011.
79. Zimbovskaya, N.A. The effect of Coulomb interactions on nonlinear thermovoltage and thermocurrent in quantum dots. *J. Chem. Phys.* **2015**, *142*, 244310.
80. Stanciu, A.E.; Nemnes, G.A.; Manolescu, A. Thermoelectric Effects in Nanostructured Quantum Wires in the Non-Linear Temperature Regime. *Rom. J. Phys.* **2015**, *60*, 716.
81. Erlingsson, S.; Manolescu, A.; Nemnes, G.; Bardarson, J.; Sanchez, D. Reversal of Thermoelectric Current in Tubular Nanowires. *Phys. Rev. Lett.* **2017**, *119*, 036804.
82. Thorgilsson, G.; Erlingsson, S.I.; Manolescu, A. Thermoelectric current in tubular nanowires in transverse electric and magnetic fields. *J. Phys. Conf. Ser.* **2017**, *906*, 012021.
83. Erlingsson, S.I.; Bardarson, J.H.; Manolescu, A. Thermoelectric current in topological insulator nanowires with impurities. *Beilstein J. Nanotechnol.* **2018**, *9*, 1156.
84. Zhang, Q.; Lou, M.; Li, X.; Reno, J.L.; Pan, W.; Watson, J.D.; Manfra, M.J.; Kono, J. Collective non-perturbative coupling of 2D electrons with high-quality-factor terahertz cavity photons. *Nat. Phys.* **2016**, *12*, 1005.
85. Gudmundsson, V.; Abdullah, N.R.; Sitek, A.; Goan, H.S.; Tang, C.S.; Manolescu, A. Time-dependent current into and through multilevel parallel quantum dots in a photon cavity. *Phys. Rev. B* **2017**, *95*, 195307.
86. Arnold, T.; Tang, C.S.; Manolescu, A.; Gudmundsson, V. Impact of a circularly polarized cavity photon field on the charge and spin flow through an Aharonov-Casher ring. *Phys. E Low-dimens. Syst. Nanostruct.* **2014**, *60*, 170–182.

87. Feranchuk, I.D.; Komarov, L.I.; Ulyanenkov, A.P. Two-level system in a one-mode quantum field: Numerical solution on the basis of the operator method. *J. Phys. A Math. Gen.* **1996**, *29*, 4035.

88. Li, X.H.; Wang, K.L.; Liu, T. Ground State of Jaynes–Cummings Model: Comparison of Solutions with and without the Rotating-Wave Approximation. *Chin. Phys. Lett.* **2009**, *26*, 044212.

89. Jonasson, O.; Tang, C.S.; Goan, H.S.; Manolescu, A.; Gudmundsson, V. Quantum magneto-electrodynamics of electrons embedded in a photon cavity. *New J. Phys.* **2012**, *14*, 013036.

90. Arnold, T.; Tang, C.S.; Manolescu, A.; Gudmundsson, V. Effects of geometry and linearly polarized cavity photons on charge and spin currents in a quantum ring with spin-orbit interactions. *Eur. Phys. J. B* **2014**, *87*, 113.

91. Abdullah, N.R.; Tang, C.S.; Manolescu, A.; Gudmundsson, V. Electron transport through a quantum dot assisted by cavity photons. *J. Phys. Condens. Matter* **2013**, *25*, 465302.

92. Abdullah, N.R.; Tang, C.S.; Manolescu, A.; Gudmundsson, V. Optical switching of electron transport in a waveguide-QED system. *Phys. E Low-dimens. Syst. Nanostruct.* **2016**, *84*, 280–284.

93. Abdullah, N.R.; Tang, C.S.; Manolescu, A.; Gudmundsson, V. Competition of static magnetic and dynamic photon forces in electronic transport through a quantum dot. *J. Phys. Condens. Matter* **2016**, *28*, 375301.

94. Abdullah, N.R.; Tang, C.S.; Manolescu, A.; Gudmundsson, V. Effects of photon field on heat transport through a quantum wire attached to leads. *Phys. Lett. A* **2018**, *382*, 199–204.

95. Abdullah, N.R.; Tang, C.S.; Manolescu, A.; Gudmundsson, V. Cavity-Photon Controlled Thermoelectric Transport through a Quantum Wire. *ACS Photonics* **2016**, *3*, 249–254,

96. Abdullah, N.R.; Tang, C.S.; Manolescu, A.; Gudmundsson, V. Spin-dependent heat and thermoelectric currents in a Rashba ring coupled to a photon cavity. *Phys. E Low-dimens. Syst. Nanostruct.* **2018**, *95*, 102–107.

97. Abdullah, N.R.; Arnold, T.; Tang, C.S.; Manolescu, A.; Gudmundsson, V. Photon-induced tunability of the thermospin current in a Rashba ring. *J. Phys. Condens. Matter* **2018**, *30*, 145303.

98. Abdulla, N.R.; Tang, C.S.; Manolescu, A.; Gudmundsson, V. Manifestation of the Purcell effect in current transport through a dot-cavity-QED system. *Nanomaterials* **2019**, *9*, 1023.

99. Jonsson, T.H.; Manolescu, A.; Goan, H.S.; Abdullah, N.R.; Sitek, A.; Tang, C.S.; Gudmundsson, V. Efficient determination of the Markovian time-evolution towards a steady-state of a complex open quantum system. *Comput. Phys. Commun.* **2017**, *220*, 81.

100. Petersen, K.B.; Pedersen, M.S. The Matrix Cookbook. Version 20121115, 2012, Available online: http://www2.imm.dtu.dk/pubdb/views/publication_details.php?id=3274 (accessed on 25 July 2019).

101. Weidlich, W. Liouville-space formalism for quantum systems in contact with reservoirs. *Z. Phys.* **1971**, *241*, 325.

102. Nakano, R.; Hatano, N.; Petrosky, T. Nontrivial Eigenvalues of the Liouvillian of an Open Quantum System. *Int. J. Theor. Phys.* **2010**, *50*, 1134–1142.

103. Petrosky, T. Complex Spectral Representation of the Liouvillian and Kinetic Theory in Nonequilibrium Physics. *Prog. Theor. Phys.* **2010**, *123*, 395–420.

104. Swain, S. Master equation derivation of quantum regression theorem. *J. Phys. A Math. Gen.* **1981**, *14*, 2577.

105. Walls, D.; Milburn, G.J. *Quantum Optics*; Springer: Berlin/Heidelberg, Germany, 2008.

106. Goan, H.S.; Chen, P.W.; Jian, C.C. Non-Markovian finite-temperature two-time correlation functions of system operators: Beyond the quantum regression theorem. *J. Chem. Phys.* **2011**, *134*, 124112.

107. Gudmundsson, V.; Abdullah, N.R.; Sitek, A.; Goan, H.S.; Tang, C.S.; Manolescu, A. Current correlations for the transport of interacting electrons through parallel quantum dots in a photon cavity. *Phys. Lett. A* **2018**, *382*, 1672–1678.

108. Gudmundsson, V.; Jonsson, T.H.; Bernodusson, M.L.; Abdullah, N.R.; Sitek, A.; Goan, H.S.; Tang, C.S.; Manolescu, A. Regimes of radiative and nonradiative transitions in transport through an electronic system in a photon cavity reaching a steady state. *Ann. Phys.* **2016**, *529*, 1600177.

109. Gudmundsson, V.; Abdullah, N.R.; Sitek, A.; Goan, H.S.; Tang, C.S.; Manolescu, A. Electroluminescence caused by the transport of interacting electrons through parallel quantum dots in a photon cavity. *Ann. Phys.* **2018**.

110. De Liberato, S.; Gerace, D.; Carusotto, I.; Ciuti, C. Extracavity quantum vacuum radiation from a single qubit. *Phys. Rev. A* **2009**, *80*, 053810.

111. Gudmundsson, V.; Gestsson, H.; Abdullah, N.R.; Tang, C.S.; Manolescu, A.; Moldoveanu, V. Coexisting spin and Rabi-oscillations at intermediate time in electron transport through a photon cavity. *Beilstein J. Nanotechnol.* **2018**, *10*, 606–616.

112. Abdullah, N.R.; Tang, C.S.; Manolescu, A.; Gudmundsson, V. Oscillations in electron transport caused by multiple resonances in a quantum dot-QED system in the steady-state regime. *arXiv* **2019**, arXiv:1903.03655,

113. Abdullah, N.R.; Tang, C.S.; Manolescu, A.; Gudmundsson, V. The photocurrent generated by photon replica states of an off-resonantly coupled dot-cavity system. *arXiv* **2019**, arXiv:1904.04888,

114. Purcell, E.M. Spontaneous Emission Probabilities at Radio Frequencies. *Phys. Rev.* **1946**, *69*, 681.

115. Gudmundsson, V.; Abdullah, N.R.; Tang, C.S.; Manolescu, A.; Moldoveanu, V. Cavity-photon induced high order transitions between ground states of quantum dots. *arXiv* **2019**, arXiv:1905.10883,

116. Abdullah, N.R.; Tang, C.S.; Manolescu, A.; Gudmundsson, V. Thermoelectric Inversion in a Resonant Quantum Dot-Cavity System in the Steady-State Regime. *Nanomaterials* **2019**, *9*.

117. Stefanucci, G.; van Leeuwen, R. *Nonequilibrium Many-Body Theory of Quantum Systems: A Modern Introduction*; Cambridge University Press: Cambridge, UK, 2013.

118. Deng, C.; Orgiazzi, J.L.; Shen, F.; Ashhab, S.; Lupascu, A. Observation of Floquet States in a Strongly Driven Artificial Atom. *Phys. Rev. Lett.* **2015**, *115*, 133601.

119. Koski, J.V.; Landig, A.J.; Pályi, A.; Scarlino, P.; Reichl, C.; Wegscheider, W.; Burkard, G.; Wallraff, A.; Ensslin, K.; Ihn, T. Floquet Spectroscopy of a Strongly Driven Quantum Dot Charge Qubit with a Microwave Resonator. *Phys. Rev. Lett.* **2018**, *121*, 043603.

120. Pagel, D.; Fehske, H. Non-Markovian dynamics of few emitters in a laser-driven cavity. *Phys. Rev. A* **2017**, *96*, 041802.

121. Szczygielski, K.; Alicki, R. Markovian theory of dynamical decoupling by periodic control. *Phys. Rev. A* **2015**, *92*, 022349.

Article

Stochastic Schrödinger Equations and Conditional States: A General Non-Markovian Quantum Electron Transport Simulator for THz Electronics

Devashish Pandey [1], Enrique Colomés [1], Guillermo Albareda [1,2,3,*] and Xavier Oriols [1,*]

[1] Departament d'Enginyeria Electrònica, Universitat Autònoma de Barcelona, 08193 Bellaterra (Barcelona), Spain; devashishpandey21@gmail.com (D.P.); e.colomescapon@gmail.com (E.C.)
[2] Max Planck Institute for the Structure and Dynamics of Matter, 22761 Hamburg, Germany
[3] Institute of Theoretical and Computational Chemistry, Universitat de Barcelona, 08028 Barcelona, Spain
* Correspondence: guillermo.albareda@mpsd.mpg.de (G.A.); xavier.oriols@uab.cat (X.O.)

Received: 25 October 2019; Accepted: 21 November 2019; Published: 25 November 2019

Abstract: A prominent tool to study the dynamics of open quantum systems is the reduced density matrix. Yet, approaching open quantum systems by means of state vectors has well known computational advantages. In this respect, the physical meaning of the so-called conditional states in Markovian and non-Markovian scenarios has been a topic of recent debate in the construction of stochastic Schrödinger equations. We shed light on this discussion by acknowledging the Bohmian conditional wavefunction (linked to the corresponding Bohmian trajectory) as the proper mathematical object to represent, in terms of state vectors, an arbitrary subset of degrees of freedom. As an example of the practical utility of these states, we present a time-dependent quantum Monte Carlo algorithm to describe electron transport in open quantum systems under general (Markovian or non-Markovian) conditions. By making the most of trajectory-based and wavefunction methods, the resulting simulation technique extends to the quantum regime, the computational capabilities that the Monte Carlo solution of the Boltzmann transport equation offers for semi-classical electron devices.

Keywords: conditional states; conditional wavefunction; Markovian and Non-Markovian dynamics; stochastic Schrödinger equation; quantum electron transport

---

1. Introduction

Thanks to its accuracy and versatility, the Monte Carlo solution of the Boltzmann transport equation has been, for decades, the preferred computational tool to predict the DC, AC, transient, and noise performances of semi-classical electron devices [1]. In the past decade, however, due to the miniaturization of electronic devices (with active regions approaching the de Broglie wavelength of the transport electrons), a majority of the device modeling community has migrated from semi-classical to fully quantum simulation tools, marking the onset of a revolution in the community devoted to semiconductor device simulation. Today, a number of quantum electron transport simulators are available to the scientific community [2–4]. The amount of information that these simulators can provide, however, is mainly restricted to the stationary regime and therefore their predicting capabilities are still far from those of the traditional Monte Carlo solution of the semi-classical Boltzmann transport equation [1]. This limitation poses a serious problem in the near future as electron devices are foreseen to operate at the Terahertz (THz) regime. At these frequencies, the discrete nature of electrons in the active region is expected to generate unavoidable fluctuations of the current that could interfere with the correct operation of such devices both for analog and digital applications [5].

A formally correct approach to electron transport beyond the quasi-stationary regime lies on the description of the active region of an electron device as an open quantum system [6,7]. As such, one can then borrow any state-of-the-art mathematical tool developed to study open quantum systems [8,9]. A preferred technique has been the stochastic Schrödinger equation (SSE) approach [10–17]. Instead of directly solving equations of motion for the reduced density matrix, the SSE approach exploits the state vector nature of the so-called conditional states to alleviate some computational burden (and ensure a complete positive map by construction [18]). Even if this technique allows to always reconstruct the full density matrix, a discussion on whether dynamical information can be directly extracted from such conditional states in non-Markovian scenarios has appeared recently in the literature [19,20]. This debate is very relevant to us as we are interested in computing not only one-time expectation values (i.e., DC performance) but also dynamical properties (i.e., AC, transient, and noise), such as multi-time correlation functions, at THz frequencies. At these frequencies the environment correlations are expected to decay on a time-scale comparable to the time-scale relevant for the system evolution [21]. Furthermore, the displacement current becomes important at very high frequencies and a self-consistent solution of the Maxwell equations and the Schrödinger equation is necessary [21,22].

Some light on how to utilize the SSE technique to access dynamical information without the need of reconstructing the reduced density matrix has already been shed by Wiseman and Gambetta by acknowledging the Bohmian conditional wavefunction as the proper mathematical tool to describe general open quantum systems in non-Markovian scenarios [23,24]. In this work we reinforce this idea by showing that the Bohmian conditional wavefunction, together with the corresponding Bohmian trajectory, is an exact decomposition and recasting of the unitary time-evolution of a closed quantum system that yields a set of coupled, non-Hermitian, equations of motion that allows to describe the evolution of arbitrary subsets of the degrees of freedom on a formally exact level. Furthermore, since the measurement process is defined as a routine interaction between subsystems in Bohmian mechanics, conditional states can be used to describe either the measured or unmeasured dynamics of an open quantum system. As an example of the practical utility of the conditional wavefunctions, we present here a Monte Carlo simulation scheme to describe quantum electron transport in open systems that is valid both for Markovian or non-Markovian regimes and that guarantees a dynamical map that preserves complete positivity [25–29].

This paper is structured as follows. In Section 2 we provide a brief account on the SSE approach and on how nanoscale electron devices can be understood as open quantum systems. Section 3 focuses on the physical interpretation of the conditional states (i.e., system states conditioned on a particular value of the environment) in the contexts of the orthodox and Bohmian quantum mechanical theories. Section 4 provides an overall perspective on the points raised in the previous sections and puts into practice the conditional wavefunction concept to build a general purpose electron transport simulator, called BITLLES, beyond the steady state (Markovian) regime. As an example of the use of conditional states, numerical simulations of the THz current in a graphene electron device are presented in Section 5. Final comments and conclusions can be found in Section 6.

2. Electron Devices as Open Quantum Systems

In this section we introduce the SSE approach to open quantum systems and discuss how it can be used to reconstruct the reduced density matrix. We then explain how a nanoscale electron device can be understood as an open quantum system and how the SSE approach can be applied to predict its performance.

2.1. Open Quantum Systems

As usual, we start with a closed quantum system (see Figure 1a). This system is represented by a pure state, $|\Psi(t)\rangle$, which evolves unitarily according to the time-dependent Schrödinger equation:

$$i\hbar\frac{\partial|\Psi(t)\rangle}{\partial t} = \hat{H}|\Psi(t)\rangle. \tag{1}$$

Finding a solution to Equation (1) is inaccessible for most practical scenarios due to the large number of degrees of freedom involved. Therefore, it is a common practice to partition the system into two subsets of degrees of freedom, viz., open system and environment [6]. The open system can be described by a reduced density matrix:

$$\hat{\rho}_{sys}(t) = \text{Tr}_{env}\left[|\Psi(t)\rangle\langle\Psi(t)|\right],\qquad(2)$$

where Tr_{env} denotes the trace over the environment degrees of freedom. The reduced density matrix $\hat{\rho}_{sys}$ can be shown to obey, in most general circumstances, a non-Markovian master equation [30,31]:

$$\frac{\partial\hat{\rho}_{sys}(t)}{\partial t} = -i\left[\hat{H}_{int}(t), \hat{\rho}_{sys}(t)\right] + \int_{t_0}^{t}\hat{\mathcal{K}}(t,t')\hat{\rho}_{sys}(t')dt',\qquad(3)$$

where $\hat{H}_{int}(t)$ is a system Hamiltonian operator in some interaction picture and $\hat{\mathcal{K}}(t,s)$ is the "memory kernel" superoperator, which operates on the reduced state $\hat{\rho}_{sys}(t)$ and represents how the environment affects the system. If the solution to Equation (3) is known then the expectation value of any observable $\hat{A}$ of the system can be evaluated as:

$$\langle\hat{A}(t)\rangle = \text{Tr}_{sys}[\hat{\rho}_{sys}(t)\hat{A}].\qquad(4)$$

Unfortunately, solving Equation (3) is not an easy task. The effect of $\hat{\mathcal{K}}(t,s)$ on $\hat{\rho}_{sys}(t)$ cannot be explicitly evaluated in general circumstances. Moreover, even if the explicit form of $\hat{\mathcal{K}}(t,s)$ is known, the solution to Equation (3) is very demanding as the density matrix $\hat{\rho}_{sys}(t)$ scales very poorly with the number of degrees of freedom of the open system. Finally, if one is aiming at computing multi-time correlations functions, then it is necessary to incorporate the effect (backaction) of the successive measurements on the evolution of the reduced density matrix, which is, in general non-Markovian regimes, a very complicated task both from the practical and conceptual points of view.

2.2. Stochastic Schrodinger Equations

A breakthrough in the computation of the reduced density matrix in Equation (2) came from the advent of the SSE approach [32]. The main advantage behind the SSE approach is that the unknown to be evaluated is in the form of a state vector (of N_{sys} degrees of freedom) rather than a matrix (of size N_{sys}^2) and thus there is an important reduction of the associated computational cost. In addition, it provides equations of motion that, by construction, ensure a complete positive map [18] so that the SSE approach guarantees that the density matrix always yields a positive probability density, a requirement that is not generally satisfied by other approaches that are based on directly solving Equation (3) [33].

The central mathematical object in the SSE approach to open quantum systems is the conditional state of the system:

$$|\psi_q(t)\rangle = \frac{\left(\langle q|\otimes\hat{I}_{sys}\right)|\Psi(t)\rangle}{\sqrt{P(q,t)}},\qquad(5)$$

where $P(q,t) = \langle\psi_q(t)|\psi_q(t)\rangle = \langle\Psi(t)|\hat{I}_{sys}\otimes|q\rangle\langle q|\otimes\hat{I}_{sys}|\Psi(t)\rangle$ and $|q\rangle$ are the eigenstates of the so-called unraveling observable $\hat{Q}$ belonging to the Hilbert space of the environment. To simplify the discussion, and unless indicated, q represents the collection of degrees of freedom of the environment. Using the eigenstates $|q\rangle$ as a basis for the environment degrees of freedom, it is then easy to rewrite the full state $|\Psi(t)\rangle$ as:

$$|\Psi(t)\rangle = \int dq\sqrt{P(q,t)}|q\rangle\otimes|\psi_q(t)\rangle,\qquad(6)$$

which can be simply understood as a Schmidt decomposition of a bipartite (open system plus environment) state. Thus, a complete set of conditional states can be always used to reproduce the reduced density matrix at any time as:

$$\hat{\rho}_{\text{sys}}(t) = \int dq P(q,t) |\psi_q(t)\rangle\langle\psi_q(t)|. \tag{7}$$

Let us note that no specific (Markovian or non-Markovian) assumption was required to write Equation (7). In fact, the above definition of the reduced density matrix simply responds to the global unitary evolution in Equation (1), which (as depicted in Figure 1a) does not include the effect of any measuring apparatus.

2.3. Nanoscale Electron Devices as Open Quantum Systems

At first sight, one could be inclined to say that a nanoscale electron device perfectly fits into the above definition of open quantum system. The open system would then be the device's active region and the environment (including the contacts, the cables, ammeter, etc.) the so called reservoirs or contacts (see Figure 1a). In addition, the observable of interest $\hat{A}$ in Equation (4) would be, most probably, the current operator $\hat{I}$. As long as we are interested only in single-time expectation values, i.e., static or stationary properties, this picture (and the picture in Figure 1a) is perfectly valid. Therefore, the SSE approach introduced in Equations (5)–(7) can be easily adopted to simulate electron devices and hence to predict their static performance.

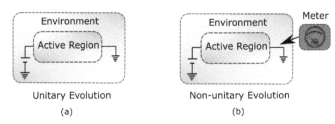

Figure 1. Panel (**a**): Schematic representation of an open quantum system, which can be partitioned into active region and environment. The evolution of the entire device is described by the state $|\Psi(t)\rangle$ that evolves unitarily according to the time-dependent Schrödinger equation. Panel (**b**): Schematic representation of a measured open quantum system, which can be partitioned into meter, active region, and environment. The evolution of the device plus environment wavefunction is no longer unitary due to the (backaction) effect of the measuring apparatus.

However, if one aims at computing dynamical properties such as time-correlation functions, e.g., $\langle I(t + \tau)I(t)\rangle$, then a valid question is whether such an expectation value is expected to be measurable at the laboratory. If so, what would then be the effect of the measurement of I at time t on the measurement of I at a later time $t + \tau$? Figure 1b schematically depicts this question by drawing explicitly the measuring apparatus (or meter). As it is well known, the action of measuring in quantum mechanics is not innocuous. It is quite the opposite: in many relevant situations, extracting information from a system at time t has a non-negligible effect on the subsequent evolution of the system and hence also on what is measured at a later time $t + \tau$. Therefore, as soon as we are concerned about dynamic information (i.e., time-correlation functions), we need to ask ourselves whether an approach to open quantum systems such as the SSE approach can be of any help. In the next section we will answer this question and understand whether the conditional states $|\psi_q(t)\rangle$ defined in Equation (5) do include the backaction of the measuring apparatus depicted in Figure 1b.

3. Interpretation of Conditional States in Open Quantum Systems

The conditional states in Equation (5) were first interpreted as a simple numerical tool [32], that is, exploiting the result in Equation (7) as a numerical recipe to evaluate any expectation value of interest. This interpretation is linked to the assumption that the operator $\hat{A}$ in Equation (4) is the physically relevant operator (associated to a real measuring apparatus), while the operator $\hat{Q}$ associated to the definition of the conditional state in Equation (5) is only a mathematical object with no attached physical reality, i.e., it merely represents a basis. In more recent times, however, it has been generally accepted that the conditional states in Equation (5) can be interpreted as the states of the system conditioned on a type of sequential (sometimes referred to as continuous) measurement [34] of the operator $\hat{Q}$ of the environment (now representing a physical measuring apparatus that substitutes the no longer needed operator $\hat{A}$) [6,12,35]. From a practical point of view, this last interpretation is very attractive as it would allow to link the conditional states, $|\psi_q(t)\rangle$, at different times and thus compute time-correlation functions without the need of introducing the measuring apparatus or of reconstructing the full density matrix. Whether or not this later interpretation is physically sound in general circumstances is the focus of our discussion in the next subsections.

3.1. The Orthodox Interpretation of Conditional States

Let us start by discussing, in the orthodox quantum mechanics theory, what is the physical meaning of the conditional states that appear in Equation (5). When the full closed system follows the unitary evolution of Figure 1a, the conditional state $|\psi_q(t)\rangle$ can be understood as the (renormalized) state that the system is left in after projectively measuring the property Q of the environment (with outcome q). This can be easily seen by noting that the superposition in Equation (6) is, after a projective measurement of Q, reduced (or collapsed) to the product state

$$|\Psi_q(t)\rangle = \sqrt{P(q,t)}|q\rangle \otimes |\psi_q(t)\rangle. \tag{8}$$

It is important to notice that the conditional state $|\psi_{q'}(t')\rangle$ at a later time, $t' > t$, can be equivalently defined as the state of the system when the superposition in Equation (6) is measured at time t' and yields the outcome q'. This interpretation, however, is only valid if no previous measurement (in particular at t) has been performed, as depicted in Figure 2a. Otherwise, the collapse of the wavefunction at time t, yielding the state $\sqrt{P(q,t)}|q\rangle \otimes |\psi_q(t)\rangle$, should be taken into account in the future evolution of the system, which would not be the same as if the measurement had not been performed at the previous time. Therefore, the equation of motion of the conditional states, as defined in Equation (5), cannot be, in general, the result of a sequential measurement protocol such as the one depicted in Figures 1b or 2b. This conclusion seems obvious if one recalls that our starting point was Figure 1a, where there is no measurement.

3.1.1. Orthodox Conditional States in Markovian Scenarios

Even if the conditional states solution of the SSE cannot be generally interpreted as the result of a sequential measurement, such an interpretation has been proven to be very useful in practice for scenarios that fulfill some specific type of Markovian conditions. We are aware that there is still some controversy on how to properly define Markovianity in the quantum regime (see, e.g., Ref. [18]), so it is our goal here only to acknowledge the existence of some regimes (i.e., particular observation time intervals) of interest where the role of the measurement of the environment has no observable effects. In this regime, Figure 1a,b as well as Figure 2a,b can be thought to be equivalent.

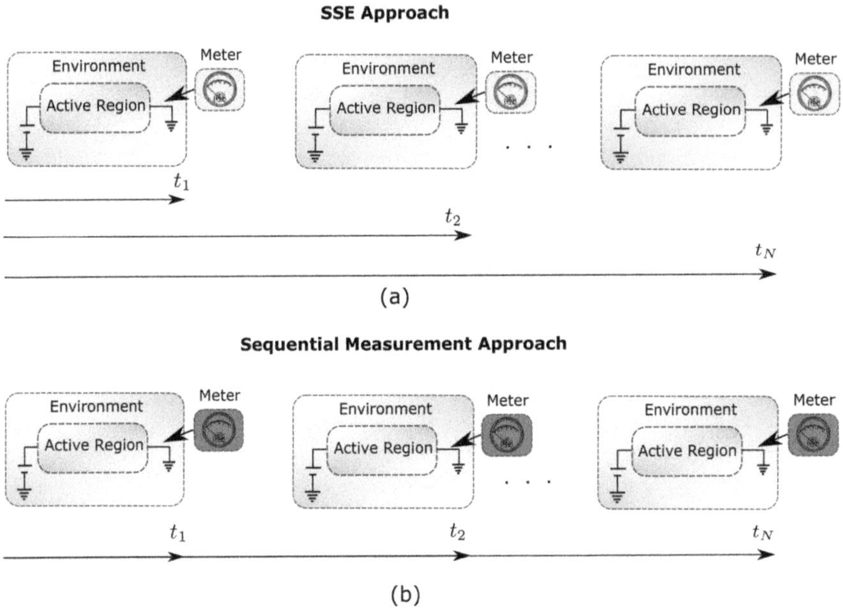

SSE Approach

(a)

Sequential Measurement Approach

(b)

Figure 2. Panel (**a**): Schematic representation of the SSE approach. The states of the system conditioned on a particular value of the environment at time t, $|\psi_q(t)\rangle$, can be given a physical meaning only if no measurement has been performed at a previous time $t' < t$. This approach can be always used to reconstruct the correct reduced density matrix of the system at any time but cannot be used to link in time the conditional states for non-Markovian scenarios. Panel (**b**): Schematic representation of a sequential measurement. The wavefunction of the system plus environment is measured sequentially. In this picture, the link between the states of the full system plus environment at different times is physically motivated.

In our pragmatical definition of Markovianity the entanglement between system and environment decays in a time scale t_D that is much smaller than the observation time interval τ, i.e., $t_D \ll \tau$. In this regime, the environment itself can be thought of as a type of measuring operator (as appears in generalized quantum measurement theory [36]) that keeps the open system in a pure state after the measurement. The open system can be then seen as an SSE in which the stochastic variable q_t (sampled from the distribution $P(q_t, t)$) is directly the output of a sequential measurement of the environment. The stochastic trajectory of this conditioned system state generated by the (Markovian) SSE is often referred to as a quantum trajectory [6,12,35] and can be used, for example, to evaluate time-correlation functions of the environment as:

$$\langle Q(t)Q(t+\tau)\rangle \overset{t_D \ll \tau}{=} \int\int P(q_t,t)P(q_{t+\tau},t+\tau)q_t q_{t+\tau}dq_t dq_{t+\tau} = \langle Q(t)\rangle\langle Q(t+\tau)\rangle. \qquad (9)$$

Let us emphasize that the stochastic variables q_t and $q_{t+\tau}$ in Equation (9) are sampled, separately, from the probability distributions $P(q_t, t) = \langle \psi_q(t)|\psi_q(t)\rangle$ and $P(q_{t+\tau}, t+\tau) = \langle \psi_q(t+\tau)|\psi_q(t+\tau)\rangle$. Therefore, as we have schematically depicted in Figure 3, no matter how the trajectories $\{q_t\}$ are connected in time, one always obtains the correct time-correlation function $\langle Q(t)Q(t+\tau)\rangle$.

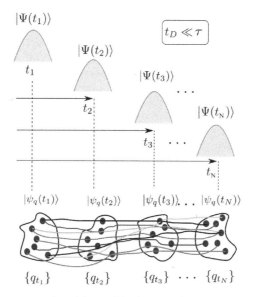

Figure 3. Schematic representation of the combined system plus environment wavefunction (blue Gaussians) measured at different times that result in a state of the system $|\psi_q(t)\rangle$ conditioned to the set of environment values $\{q_t\}$ shown in dark blue circles. In the Markovian regime there exists no specific recipe about how the different q_t's must be connected in time (colored solid lines). No matter how these points are connected in time, one always gets the right expectation value in Equation (9).

It is important to realize that we started our discussion on the physical meaning of the Markovian SSE with an open system whose environment is not being measured (see Figures 1a and 2a). Noticeably, we have ended up discussing an environment that is being measured at every time interval τ (see Figure 2b). How is that possible? Well, the reason is that measuring the environment at time t does not affect the system conditional states at a later time τ when the built-in correlations in the environment due to the measurement at time t decay in a time interval t_D much smaller than the time interval between measurements τ. In other words, Figure 1a,b as well as Figure 2a,b are not distinguishable when $t_D \ll \tau$. In this sense, the Markovian regime has some similarities with a classical system, where it is accepted that information can be extracted without perturbation.

3.1.2. Orthodox Conditional States in Non-Markovian Scenarios

For nanoscale devices operating at very high (THz) frequencies, the relevant dynamics and hence the observation time interval τ are both below the picoseconds time-scale and the previous assumption of Markovianity, i.e., $t_D \ll \tau$, starts to break down. Under the condition $t_D \sim \tau$, non-Markovian SSE have been proposed which allow an alternative procedure for solving the reduced state $|\psi_q(t)\rangle$ [17,33,37–41]. However, non-Markovian SSEs constructed from Equation (5), unlike the Markovian SSEs, suffer from interpretation issues [17]. In the non-Markovian regime, the perturbation of the environment due to the quantum backaction of a measurement at time t would not be washed out in the time lapse $\tau \sim t_D$ and hence the joint probability distribution would not become separable, i.e., $P(q_t, q_{t+\tau}) \neq P(q_t)P(q_{t+\tau})$. As a direct consequence, connecting in time the different solutions q_t and $q_{t+\tau}$ (sampled independently from the probability distributions $P(q_t, t)$ and $P(q_{t+\tau}, t + \tau)$ as in Figure 3 to make a trajectory "would be a fiction" [17,19,20]. Here, the word "fiction" means that the time-correlations computed in Equation (9) are wrong, i.e., the expectation value in Equation (9) would simply be different from the experimental result.

According to D'Espagnat the above discussion can be rephrased in terms of the so-called proper and improper mixtures [42]. Following D'Espagnat arguments, the reduced density matrix in Equation (7) is an improper mixture because it has been constructed by tracing out the degrees of freedom of the environment. On the contrary, a proper mixture is a density matrix constructed to simultaneously define several experiments where a closed system is described, at each experiment, by different pure states. Due to our ignorance, we do not know which pure state corresponds to which experiment, so we only know the probabilities of finding a given pure state. D'Espagnat argues that the ignorance interpretation of the proper density matrix, cannot be applied in the improper density matrix discussed here (See Appendix A). To understand why under a Markovian regime open systems can be described by pure states (using a proper mixture), we remind that Markovianity implies conditions on the observation time. For a given correlation time t_D, a given open system can be in the Markovian or non-Markovian regimes depending on the time of observation τ. That is, for small enough observation times all open systems are non-Markovian and hence must be understood as an improper mixture. On the contrary, for large enough observation times, open systems can behave as closed systems (with a negligible entanglement with the environment) and be effectively represented by pure states.

3.2. The Bohmian Interpretation of Conditional States

So, under non-Markovian (i.e., the most general) conditions, the conditional states $|\psi_q(t)\rangle$ can be used to reconstruct the reduced density matrix as in Equation (7) but cannot be used to provide further information on its own. This interpretation problem is rooted in the fact that orthodox quantum mechanics only provides reality to objects whose properties (such as q) are being directly measured. However, as explained in the previous subsection, it is precisely the fact of introducing the measurement of q (without including the pertinent backaction on the system evolution) which prevents the conditional states $|\psi_q(t)\rangle$ of the non-Markovian SSE from being connected in time for $t_D \sim \tau$. In this context, a valid question regarding the interpretation of $|\psi_q(t)\rangle$ is whether or not we can obtain information of, e.g., the observable Q without perturbing the state of the system. The answer given by orthodox quantum mechanics is crystal clear: this is not possible (except for Markovian conditions) because information requires a measurement, and the measurement induces a perturbation. Notice, however, that the assumption that only measured properties are real is not something forced on us by experimental facts, but it is a deliberate choice of the orthodox quantum theory. Therefore, we here turn to a nonorthodox approach: the Bohmian interpretation of quantum mechanics [43–48].

A fundamental aspect of the Bohmian theory is that reality (of the properties) of quantum objects does not depend on the measurement. That is, the values of some observables, e.g., the value of the positions of the particles of the environment, exist independently of the measurement. If q is the collective degree of freedom of the position of the particles of the environment and x is the collective degree of freedom of the position of particles of the system; then, the Bohmian theory defines an experiment in the laboratory by means of two basic elements: (i) the wavefunction $\langle q, x | \Psi(t) \rangle = \Psi(x, q, t)$ and (ii) an ensemble of trajectories $Q^i(t), X^i(t)$ of the environment and of the system. We use a superindex i to denote that each time an experiment is repeated, with the same preparation for the wavefunction $\Psi(x, q, t)$, the initial positions of the environment and system particles can be different. They are selected according to the probability distribution $|\Psi(X^i, Q^i, 0)|^2$ [44]. The equation of motion for the wavefunction $\Psi(x, q, t)$ is the time-dependent Schrödinger equation in Equation (1), while the equations of motion for the environment and system trajectories $Q^i(t), X^i(t)$ are obtained by time-integrating the velocity fields $v_q(x, q, t) = J_q(x, q, t) / |\Psi(x, q, t)|^2$ and $v_x(x, q, t) = J_x(x, q, t) / |\Psi(x, q, t)|^2$ respectively. Here, $J_q(x, q, t)$ and $J_x(x, q, t)$ are the standard current densities of the environment and the system respectively. We highlight the (nonlocal) dependence of the Bohmian velocities of the particles of the environment on the particles of the system, and vice-versa. This shows just the entanglement between environment and system at the level of the Bohmian trajectories. According to the continuity equation

$$\frac{d|\Psi(x,q,t)|^2}{dt} + \nabla_x(v_x(x,q,t)|\Psi(x,q,t)|^2) + \nabla_q(v_q(x,q,t)|\Psi(x,q,t)|^2) = 0, \tag{10}$$

the ensemble of trajectories $\{Q(t), X(t)\} = \{Q^1(t), X^1(t), Q^2(t), X^2(t)....Q^M(t), X^M(t)\}$ with $M \to \infty$ can be used to reproduce the probability distribution $|\Psi(x,q,t)|^2$ at any time. Thus, by construction, the computation of ensemble values from the orthodox and Bohmian theories are fully equivalent, at the empirical level.

From the full wavefunction $\langle x,q|\Psi(t)\rangle = \Psi(x,q,t)$ (solution of Equation (1)) and the trajectories $Q^i(t), X^i(t)$, one can then easily construct the Bohmian conditional wavefunction of the system and environment as $\tilde{\psi}_{Q^i(t)}(x,t) = \Psi(x,Q^i(t),t)$, and $\tilde{\psi}_{X^i(t)}(q,t) = \Psi(X^i(t),q,t)$ respectively. Notice that this Bohmian definition of conditional states does not require to specify if the system is measured or not because the ontological nature of the trajectories $\{Q(t), X(t)\}$ does not depend on the measurement. Consequently, the conditional wavefunctions $\tilde{\psi}_{Q^i(t)}(x,t)$, with the corresponding Bohmian trajectories, contain all the required information to evaluate dynamical properties of the system no matter whether Markovian or non-Markovian conditions are being considered. This can be seen by noticing that the velocity of the trajectory $X^i(t)$ given by $v_q(X^i(t), Q^i(t))$ can be equivalently computed either from (the x−spatial derivatives of) the global wavefunction $\Psi(x,Q,t)$ evaluated at $X^i(t)$ and $Q^i(t)$ or from (the x-spatial derivative of) the conditional wavefunction $\tilde{\psi}_{Q^i(t)}(x,t)$ evaluated at $X^i(t)$. In other words, the Bohmian velocities computed from $\Psi(x,Q,t)$ or $\tilde{\psi}_{Q^i(t)}(x,t)$ are identical. Thus, in a particular experiment i and for a given time t, the dynamics of the Bohmian trajectory $X^i(t)$ can be computed either from $\tilde{\psi}_{Q^i(t)}(x,t)$ or from $\Psi(x,q,t)$.

The Bohmian conditional wavefunction of the system can now be connected to the orthodox conditional wavefunction in Equation (5) by imposing $Q^i(t) \equiv q_t$. Then one can readily write:

$$|\tilde{\psi}_{q_t}(t)\rangle = P(q_t,t)|\psi_{q_t}(t)\rangle. \tag{11}$$

At first sight, one can think that the difference between the Bohmian and orthodox conditional states is just a simple renormalization constant $P(q_t,t)$ (see Appendix B for a more detailed explanation of the role of this renormalization constant). However, the identity in Equation (11) has to be understood as to be satisfied at any time t, which implies that the following identity should prevail:

$$Q^i(t) \equiv q_t, \qquad \forall t \tag{12}$$

We emphasize the importance of Equation (12) in ensuring the accomplishment of Equation (11). If we consider another experiment $Q^j(t) \equiv q'_t$, we have to define another conditional state $|\tilde{\psi}_{q'_t}(t)\rangle$. It can happen that, at a particular time $t \equiv t_1$, both conditional states become identical i.e., $|\tilde{\psi}_{q_{t_1}}(t_1)\rangle = |\tilde{\psi}_{q'_{t_1}}(t_1)\rangle$. However, this does not imply that both conditional wavefunctions can identically be used in the computation of time-correlations. This is because every Bohmian trajectory has a fundamental role in describing the history of the Bohmian conditional state for one particular experiment. Therefore, the trajectory $Q^i(t)$ uniquely describes the evolution of the conditional wavefunction $|\tilde{\psi}_{q_t}(t)\rangle$ for one experiment (labeled by the index i in the Bohmian language) the same way as the trajectory $Q^j(t)$ and the conditional wave function $|\tilde{\psi}_{q'_t}(t)\rangle$ describes the experiment labeled by j. As we said, $|\tilde{\psi}_{q_{t_1}}(t_1)\rangle = |\tilde{\psi}_{q'_{t_1}}(t_1)\rangle$ are the same orthodox conditional states, but do not necessarily represent the same Bohmian conditional wavefunction. This subtle difference explains why SSEs cannot be connected in time and used to study the time-correlation of non-Markovian open system whereas the same can be done through the Bohmian conditional states, without any ambiguity.

The mathematical definition of the measurement process in Bohmian mechanics and in the orthodox quantum mechanics differs substantially [44]. In the orthodox theory a collapse (or reduction) law, different from the Schrödinger equation, is necessary to describe the measurement process [45]. Contrarily, in Bohmian mechanics the measurement is treated as any other interaction as far as the degrees of freedom of the measuring apparatus are taken into account [44]. Therefore, while in the orthodox theory the conditional states $|\psi_{q_t}(t)\rangle$ cannot be understood without the perturbation of the full wavefunction $\Psi(x, q, t)$, in Bohmian mechanics the states $|\tilde{\psi}_{q_t}(t)\rangle$ do have a physical meaning even when the full wavefunction $\Psi(x, q, t)$ is unaffected by the measurement of the environment [23]. Interestingly, this introduces the possibility of defining what we call "unmeasured (Bohmian) conditional states" when it is assumed that there is no measurement or that the measurement of q_t at time t has a negligible influence on the subsequent evolution of the conditional state.

Importantly, the Bohmian conditional states and the corresponding Bohmian trajectories can be used not only to reconstruct the reduced density matrix in Equation (7) at any time but the environment trajectories $\{Q(t)\}$ allow us to correctly predict any dynamic property of interest including time-correlation functions, e.g.,

$$\langle Q(t)Q(t+\tau)\rangle = \frac{1}{M}\sum_{i=1}^{M} Q^i(t)Q^i(t+\tau) = \int\int P(q_t, q_{t+\tau})q_t q_{t+\tau}dq_t dq_{t+\tau}, \tag{13}$$

where $M \to \infty$ is the number of experiments (Bohmian trajectories) considered in the ensemble and we have defined $P(q_t, q_{t+\tau}) = \frac{1}{M}\sum_{i=1}^{M}\delta(q_t - Q^i(t))\delta(q_{t+\tau} - Q^i(t+\tau))$. As it is shown in Figure 4, the evaluation of Equation (13) and any other dynamic property when $t_D \sim \tau$ can be done only by connecting the (Bohmian) trajectories at different times in accordance with the continuity equation in Equation (10). This is in contrast with the evaluation of the dynamics in the Markovian regime where any position of the environment at time t_1 can be connected to another position of the environment at time t_2 (see Figure 3) and hence we can write $\langle Q(t)Q(t+\tau)\rangle \overset{t_D \ll \tau}{=} \frac{1}{M^2}\sum_{i,j}^{M} Q^i(t)Q^j(t+\tau)$. This very relevant point was first explained by Gambetta and Wiseman [23,24].

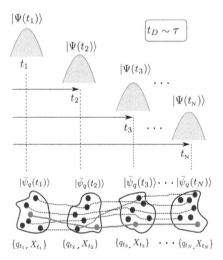

Figure 4. Schematic representation of the combined system+environment wavefunction (blue Gaussians) that is measured at different times and results in a Bohmian conditional state $|\tilde{\psi}_q(t)\rangle$ conditioned to the set of environment values $\{q_t\}$ shown in dark blue circles. In the non-Markovian regime only those values from the set of values satisfying the continuity equation in Equation (10) can be linked in time to form a trajectory (shown as connected red circles). Dashed lines represent connections that do not follow the continuity equation and hence cannot be used to evaluate any dynamic property.

Although the Bohmian theory can also provide measured properties of the system that coincide with the orthodox results in Figure 2b, let us emphasize once more the merit of the unmeasured properties provided by the Bohmian theory, which remains mainly unnoticed in the literature. As it has been already explained, in the orthodox theory, measuring a particular value of the environment property q at time t cannot be conceived without the accompanying perturbation of the wavefunction $\Psi(x,q,t)$. Under non-Markovian conditions, it is precisely this perturbation that prevents the conditional states of the system $|\psi_{q_t}(t)\rangle$ from being connected in time to form a trajectory. Contrarily, in Bohmian mechanics, the existence of the environment trajectories $\{Q(t)\}$, even in the absence of any measurement, allows the possibility of connecting in time the conditional states $|\tilde{\psi}_{q_t}(t)\rangle$ even when $t_D \sim \tau$.

Note that in the Bohmian framework, where the measurement apparatus is simply represented by an additional number of degrees of freedom interacting with the system (i.e., without requiring any additional collapse law), a discussion about measured and unmeasured properties of quantum systems is pertinent [49]. At a practical level, the measurement of many classical systems implies non-negligible perturbations. In particular, electronic devices at high frequencies are paradigmatic examples where such perturbations occur. It is well-known that the experimental setup (for e.g., a coaxial cable) connecting the electronic device to the meter induces dramatic perturbations in high-frequency measurements. An important task for device engineers is to determine what part of the measured signal is due to the intrinsic behaviour of the electron device and what part is due to rest of the experimental setup. When trying to predict the "intrinsic" behaviour of the electronic devices, the coaxial cables are modelled by "parasitic" capacitors or inductors to account for their "spurious" effect. Even the measurement of the whole experimental setup is repeated twice, with and without the "intrinsic" device under test (DUT), to subtract the results and determine experimentally the "intrinsic" properties of the electronic device alone. Such "intrinsic" properties of the electronic devices are what we define in this manuscript as the unmeasured properties of quantum systems.

4. Bohmian Conditional Wavefunction Approach to Quantum Electron Transport

The different notions of reality invoked by the orthodox quantum theory and Bohmian mechanics lead to practical differences in the abilities that these theories can offer to provide information about quantum dynamics. Specifically, we have shown that contrarily to orthodox quantum mechanics, Bohmian mechanics allows to physically interpret (i.e., link in time) the conditional states of the SSE approach in general non-Markovian scenarios. The reason is that whereas in the Bohmian theory the reality of the current is independent of any measurement, the orthodox theory gives reality to the electrical current only when it is being measured (this is the so-called eigenstate–eigenvalue link). From the practical point of view, this has a remarkable consequence. In the Bohmain approach the total current can be defined in terms of the dynamics of the electrons (Bohmian) trajectories without the need to define a measurement operator. As it will be shown in this section, the possibility of computing the total current at high frequencies without specifying the measurement operator is certainly a great advantage of the Bohmian approach in front of the orthodox one [44]. In particular, one can then avoid cumbersome questions like, is the measurement operator of the electrical current strong or weak? If weak, how weak? How often do such operator acts on the system? Every picosecond, every femtosecond? At high frequencies, how we introduce the contribution of the displacement current in the electrical current operator?

In this section we provide a brief summary of the path that the authors of this work followed for developing an electron transport simulator based on the use of Bohmian conditional states. The resulting computational tool is called BITLLES [28,29,50–56]. Let us start by considering an arbitrary quantum system. The whole system, including the open system, the environment, and the measuring apparatus, is described by a Hilbert space $\mathcal{H}$ that can be decomposed as $\mathcal{H} = \mathcal{H}_x \otimes \mathcal{H}_q$ where $\mathcal{H}_x$ is the Hilbert space of the open system and $\mathcal{H}_q$ the Hilbert space of the environment. If needed, the Hamiltonian $\mathcal{H}_q$ can include also the degrees of freedom of the measuring apparatus as explained in Section 3.2. We define $x = \{x_1, x_2...x_n\}$ as the degrees of freedom of n electrons in the open system, while q collectively defines the degrees of freedom of the environment (and possibly the measuring apparatus). The open system plus environment Hamiltonian can then be written as:

$$\hat{H} = \hat{H}_q \otimes \hat{I}_x + \hat{I}_q \otimes \hat{H}_x + \hat{V} \tag{14}$$

where $\hat{H}_x$ is the Hamiltonian of the system, $\hat{H}_q$ is the Hamiltonian of the environment (including the apparatus if required), and $\hat{V}$ is the interaction Hamiltonian between the system and the environment. We note at this point that the number of electrons n in the open system can change in time and so the size of the Hilbert spaces $\mathcal{H}_x$ and $\mathcal{H}_q$ can depend on time too.

The equation of motion for the Bohmian conditional states $\langle x | \tilde{\psi}_{q_t}(t) \rangle = \tilde{\psi}_{q_t}(x, t)$ in the position representation of the system can be derived by projecting the many-body (system-environment) Schrödinger equation into a particular trajectory of the environment $q_t \equiv Q(t)$, i.e., [26,57]:

$$i\hbar \frac{d\tilde{\psi}_{q_t}(x,t)}{dt} = \langle q_t | \otimes \langle x | \hat{H} | \Psi(t) \rangle + i\hbar \nabla_q \langle q | \otimes \langle x | \Psi(t) \rangle |_{q=q_t} \frac{dq_t}{dt}. \tag{15}$$

Equation (15) can be rewritten as:

$$i\hbar \frac{d\tilde{\psi}_{q_t}(x,t)}{dt} = \left[-\frac{\hbar^2}{2m} \nabla_x^2 + U_{q_t}^{eff}(x,t) \right] \tilde{\psi}_{q_t}(x,t), \tag{16}$$

where

$$\tilde{U}_{q_t}^{eff}(x,t) = U(x,t) + V(x,q_t,t) + \mathcal{A}(x,q_t,t) + i\mathcal{B}(x,q_t,t). \tag{17}$$

In Equation (17), $U(x,t)$ is an external potential acting only on the system degrees of freedom, $V(x,q_t,t)$ is the Coulomb potential between particles of the system and the environment evaluated at a given trajectory of the environment, $\mathcal{A}(x,q_t,t) = \frac{-\hbar^2}{2m} \nabla_q^2 \Psi(x,q,t)/\Psi(x,q,t)|_{q=q_t}$ and $\mathcal{B}(x,q_t,t) = \hbar \nabla_q \Psi(x,q,t)/\Psi(x,q,t)|_{q=q_t} \dot{q}_t$ (with $\dot{q}_t = dq_t/dt$) are responsible for mediating the so-called kinetic and advective correlations between system and environment [26,57]. Equation (16) is non-linear and describes a non-unitary evolution.

In summary, Bohmian conditional states can be used to exactly decompose the unitary time-evolution of a closed quantum system in terms of a set of coupled, non-Hermitian, equations of motion [26,57–59]. An approximate solution of Equation (16) can always be achieved by making an educated guess for the terms $\mathcal{A}$ and $\mathcal{B}$ according to the problem at hand. Specifically, in the BITLLES simulator the first and second terms in Equation (17) are evaluated through the solution of the Poisson equation [29]. The third and fourth terms are modeled by a proper injection model [60] as well as proper boundary conditions [56,61] that include the correlations between active region and reservoirs. Electron-phonon decoherence effects can be also effectively included in Equation (16) [25].

In an electron device, the number of electrons contributing to the electrical current are mainly those in the active region of the device. This number fluctuates as there are electrons entering and leaving the active region. This creation and destruction of electrons leads to an abrupt change in the degrees of freedom of the many body wavefunction which cannot be treated with a Schrödinger-like equation for $\tilde{\psi}_{q_t}(x,t)$ with a fixed number of degrees of freedom. In the Bohmian conditional approach, this problem can be circumvented by decomposing the system conditional wavefunction $\tilde{\psi}_{q_t}(x,t)$ into a set of conditional wavefunctions for each electron. More specifically, for each electron x_i, we define a single particle conditional wavefunction $\tilde{\tilde{\psi}}_{q_t}(x_i, \bar{X}_i(t), t)$, where $\bar{X}_i(t) = \{X_1(t), .., x_{i-1}(t), x_{i+1}, .., X_n(t)\}$ are the Bohmian positions of all electrons in the active region except x_i, and the second tilde denotes the single-electron conditional decomposition that we have considered on top of the conditional decomposition of the system-environment wavefunction. The set of equations of motion of the resulting $n(t)$ single-electron conditional wavefunctions inside the active region can be written as:

$$i\hbar \frac{d\tilde{\psi}_{q_t}(x_1, \bar{X}_1(t), t)}{dt} = \left[-\frac{\hbar^2}{2m} \nabla^2_{x_1} + \tilde{U}^{eff}_{q_t}(x_1, \bar{X}_1(t), t) \right] \tilde{\psi}_{q_t}(x_1, \bar{X}_1(t), t) \tag{18}$$

$$\vdots$$

$$i\hbar \frac{d\tilde{\psi}_{q_t}(x_n, \bar{X}_n(t), t)}{dt} = \left[-\frac{\hbar^2}{2m} \nabla^2_{x_n} + \tilde{U}^{eff}_{q_t}(x_n, \bar{X}_n(t), t) \right] \tilde{\psi}_{q_t}(x_n, \bar{X}_n(t), t). \tag{19}$$

That is, the first conditional process is over the environment degrees of freedom and the second conditional process is over the rest of electrons within the (open) system.

We remind here that, as shown in Figure 2b, the active region of an electron device (acting as the open system) is connected to the ammeter (that acts as the measuring apparatus) by a macroscopic cable (that represents the environment). The electrical current provided by the ammeter is then the relevant observable that we are interested in. Thus, the evaluation of the electrical current seems to require keeping track of all the degrees of freedom, i.e., of the system and the environment, which is of course a formidable computational task (see (d) Table 1). At THz frequencies, however, the electrical current is not only the particle current but also the displacement current. It is well-known that the total current defined as the particle current plus the displacement current is a divergence-less vector [21,22]. Consequently, the total current evaluated at the end of the active region is equal to the total current evaluated at the cables. So the variable of the environment associated to the total current, $q_t \equiv I(t)$, can be equivalently computed at the borders of the open system. The reader is referred to Ref. [62] for a discussion on how $I(t)$ can be defined in terms of Bohmian trajectories with the help of a quantum version of the Ramo–Schokley–Pellegrini theorem [63]. In particular, it can be shown that the total (particle plus displacement) current in a two-terminal devices can be written as [63]:

$$I(t) = \frac{e}{L} \sum_{i=1}^{n(t)} v_{x_i}(X_i(t), \bar{X}_i(t), t) = \frac{e}{L} \sum_{i=1}^{n(t)} \text{Im} \left(\frac{\nabla_{x_i} \tilde{\psi}_{q_t}(x_i, \bar{X}_i(t), t)}{\tilde{\psi}_{q_t}(x_i, \bar{X}_i(t), t)} \right) \Bigg|_{x_i = X_i(t)}, \tag{20}$$

where L is the distance between the two (metallic) contacts, e is the electron charge, and $v_{x_i}(X_i(t), \bar{X}_i(t), t)$ is the Bohmian velocity of the i-th electron inside the active region. Let us note that $I(t)$ is the electrical current given by the ammeter (although computed by the electrons inside the open system). Since the cable has macroscopic dimensions, it can be shown that the measured current at the cables is just equal to the unmeasured current (taking into account only the simulation of electrons inside the active region) plus a source of (nearly white) noise which is only relevant at very high frequencies [62]. The basic argument is that the (non-simulated) electrons in the metallic cables have a very short screening time. In other words, the electric field generated by an electron in the cable spatially decreases very rapidly due to the presence of many other mobile charge carriers in the cable that screen it out. Thus, the contribution of this outer electron to the displacement current at the border of the active region is negligible [64].

Summarizing, for the computation of the current at THz frequencies, the degrees of freedom of the environment can be neglected without any appreciable deviation from the correct current value [62]. This introduces an enormous computational simplification as shown (e) in Table 1. This is, for the specific scenarios that we are interested in, the computation cost of the Bohmian conditional wavefunction approach has the same computational cost as the orthodox SSE approach (see Table 1). Yet, in contrast to the orthodox conditional states, which can be used only to evaluate the dynamics of quantum systems in the Markovian regime, the Bohmian conditional states provide direct information on the dynamics of both Markovian or non-Markovian systems.

Table 1. An estimation of the computational cost (in memory) of different approaches mentioned in the text. Here N_{sys} and N_{env} are the number of degrees of freedom of the system and the environment while M denotes the number of elements required.

		Computational Element	N° of Trajectories	N° of Degrees of Freedom	Computational Cost
(a)	Full wave function	$\Psi(x, q, t)$	–	N_{sys} ; N_{env}	$N_{sys} \times N_{env}$
(b)	Density Matrix	$\rho(x, x')$	–	N_{sys}	N_{sys}^2
(c)	Orthodox Conditional state (SSE)	$\psi_{q_t}(x, t)$; q_t	M	N_{sys}	$M(N_{sys} + 1)$
(d)	Bohmian Conditonal state	$\tilde{\psi}_{q_t}(x, t)$; $\tilde{\psi}_{x_t}(q, t)$; q_t ; x_t	M	N_{sys} ; N_{env}	$M(N_{sys} + N_{env} + 2)$
(e)	Bohmian Conditonal state (used in Section 5)	$\tilde{\psi}_{q_t}(x, t)$; x_t	M	N_{sys}	$M(N_{sys} + 1)$

5. Numerical Results

In this section we present numerical results obtained with the BITLLES simulator (see Section 4) that demonstrate the ability of the Bohmian conditional wavefunction approach to provide dynamics information for both Markovian and non-Markovian scenarios. We simulate a two-terminal electron device whose active region is a graphene sheet contacted to the outer by two (ohmic) contacts. Graphene is a $2D$ material that has attracted a lot of attention recently because of its high electron mobility. It is a gapless material with linear energy band, which differs from the parabolic energy bands of traditional semiconductors. In graphene, the conduction and valence bands coincide at an energy point known as the Dirac point. Thus, the dynamics of electrons is no longer governed by an (effective mass) Schrödinger equation but by the Dirac equation, allowing transport from the valence to the conduction band (and vice versa) through Klein tunneling. A Bohmian conditional bispinor (instead of a conditional scalar wavefunction) is used to describe electrons inside the device. The change from a wavefunction to a bispinor does not imply any conceptual difficulty but just a mere increment of the computational cost. More details can be found in Appendix C.

In particular, we want to simulate electron transport in graphene at very high frequencies (THz) taking into account the electromagnetic environment of the electron device. Typically, nanoscale devices are small enough to assume that, even at THz frequencies, the electric field is much more relevant than the magnetic field. Therefore, only the Gauss law (first Maxwell's equations) is enforced to be fulfilled in a self-consistent way (i.e., taking into account the actual charge distribution in the active region). However, the environment of nanoscale devices is commonly a metallic element of macroscopic dimensions. In there, the magnetic and electric fields become both relevant, acting as active (detecting or emitting) THz antennas. For the typical electromagnetic modes propagating in the metals, the magnetic and electric fields are translated into the language of currents and voltages and the whole antenna is modeled as a part of an electric circuit. In this work, the graphene device interacts with an environment that is modeled by a Resistor (R) and a capacitor (C) connected in series through ideal cables (see the schematic plots in Figure 5a–c).

The active region of the graphene device is simulated with the Bohmian conditional wavefunction approach explained in the previous section, while the RC circuit is simulated using a time-dependent finite-difference method. We consider the system plus environment to be in equilibrium. Specifically, the self-consistent procedure to get the current is as follows: an initial (at time $t = 0$) zero voltage is applied at the source ($V_S(0) = 0$) and drain ($V_D(0) = 0$) contacts of the graphene active region. At room temperature this situation yields a non-zero current from Equation (20) (i.e., $I(0) \neq 0$) because of thermal noise. Such current $I(0)$ enters the RC circuit and leads to a new voltage $V_S(dt) \neq 0$ at the next time step dt (where dt represents the time step that defines the interaction between the RC circuit and the quantum device which was set to $dt = 0.5\text{fs}$). The new source $V_S(dt) \neq 0$ and fixed drain $V_D(dt) = 0$ voltages now lead to a new value of the current $I(dt) \neq 0$ in (20) which is different from zero not only because of thermal noise but also because there is now a net bias $(V_D(dt) - V_S(dt) \neq 0)$. This new current $I(dt)$ is used (in the RC circuit) to get a new $V_S(2dt)$ that is introduced back in the device to obtain $I(2dt)$ and so on. Importantly, as the system and environment are in equilibrium, the expectation value of $I(t)$ is zero at any time, i.e., $\langle I(t) \rangle = 0 \ \forall t$.

We consider three different environments (with different values of the capacitance). In Figure 5a we plot the total (particle plus displacement) electrical current at the end of the active region when $R = 0$ and $C = \infty$. The same information is shown in Figure 5b,c for two different values of the capacitance $C = 2.6 \times 10^{-17}$ F and $C = 1.3 \times 10^{-17}$ F. In all cases the value of the resistance is $R = 187 \ \Omega$, and we assumed the current $I(t)$ to be positive when it goes from drain to source.

The effect of the RC circuit is, mainly, to attenuate the current fluctuations, which are originated due to thermal noise. This can be seen by comparing Figure 5a with Figure 5b,c. The smaller the capacitance the smaller the current fluctuations. This can be explained as follows: when the net current is positive, the capacitor in the source starts to be charged and so the voltage at the source increases trying to counteract the initially positive current. Therefore, the smaller the capacitance the faster the RC circuit reacts to a charge imbalance.

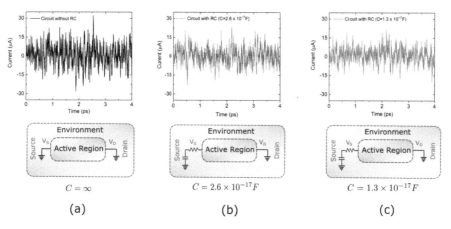

Figure 5. Total (particle plus displacement) electrical current $I(t)$ evaluated at the ammeter as a function of time for a graphene device connected to three different RC circuits with $R = 187 \ \Omega$. The values of the capacitances are: (**a**) $C = \infty$, (**b**) $C = 2.6 \times 10^{-17}$ F and (**c**) $C = 1.3 \times 10^{-17}$ F.

In Figure 6 we plot the total (particle plus displacement) current–current correlations as a function of the observation time τ for the three scenarios in Figure 5. Correlations at very small observation times provide information of the variance of the current, which, as explained above, is reduced as the value of the capacitance is increased. Numerical simulations (not shown here) exhibit that the role of the resistor R is less evident because the active region itself has a much larger (than $R = 187 \ \Omega$) associated resistance. Numerically the distinction between Markovian and non-Markovian dynamics boils down to the comparison of time correlations as defined in Equations (9) and (13). Since there is no net bias applied to the graphene device (i.e., it is in equilibrium), an ensemble average of the current (over an infinite set of trajectories like the one depicted in Figure 5) yields $\langle I(t) \rangle = 0 \ \forall t$. Time correlation functions computed in Equation (9) are thus zero by construction, i.e., $\langle I(t) \rangle \langle I(t + \tau) \rangle = 0 \ \forall t, \tau$. Therefore, the non-Markovian dynamics occurring at very high-frequencies (below the ps time-scale in Figure 6 expressly shows) fixes the correlation time of the environment at $t_D \sim$ ps. Although all three values of the capacitance C in Figure 6 yield the same order of magnitude for $t_D \sim$ ps, it seems also true that the smaller the value of the capacitance, the smaller t_D.

Current–current correlations shown in Figure 6 can be better understood by assessing the transit time of electrons. For a velocity of roughly 10^6 m/s inside an active region of $L = 40$ nm length is roughly $\tau_T = L/v_x = 0.04$ ps. Positive correlations correspond to transmitted electrons traveling from drain to source (as well as electrons traversing the device from source to drain). While $0 < t < \tau_T$ electrons are transiting inside the active region, such electrons provide always a positive (or negative) current as seen in expression (20). In other words, if we have a positive current at time t because electrons are traveling from drain to source, we can expect also a positive current at times t' satisfying $t < t' < t + \tau_T$. The negative correlations belong to electrons that are being reflected. They enter in the active region with a positive (negative) velocity and, after some time τ_R inside the device, they are reflected and have negative (positive) velocities until they leave the device after spending roughly $2\tau_R$ in the active region. Thus, during the time $\tau_R < t < 2\tau_R$ which will be different for each electron depending on the time when they are reflected, we can expect negative correlations. Interestingly, during the 4 ps simulation the number of Bohmian trajectories reflected are double in the black ($C = \infty$) simulation than in the red one ($C = 1.3 \times 10^{17}$ F). This can be explained in a similar way as we explained the reduction of the current fluctuations. The fluctuations of the electrical current imply also fluctuations of the charge inside the active region, which are translated (through the Gauss law) into fluctuations of the potential profile. Thus, the larger the noisy current, the larger the noisy internal potential profile. This implies a larger probability of being reflected by the Klein tunneling phenomenon. Therefore, if one aims at describing the dynamics of nanoscale devices with a time-resolution τ that is comparable to (or goes beyond) the electron transit time τ_T, a non-Markovian approach is necessary. This is so because the total current $I(t)$ (which has contributions from the displacement and the particle currents) shows correlations at times that are smaller than the electron transit time.

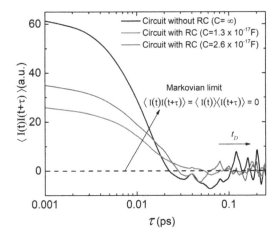

Figure 6. Total current–current correlation as a function of time for the three different experiments in Figure 5. The zero is indicated by a dashed line to show the tendency of the total current, understood as a property of the environment, to vanish at long times τ. Zero autocorrrelation implies an independence between $I(t)$ and $I(t + \tau)$ which is typical for Markovian scenarios. This is not true for the short τ considered here which are the representatives of the non-Markovian dynamics.

6. Conclusions and Final Remarks

Theoretical approaches to open quantum systems that rely on the manipulation of state vectors instead of a reduced density matrix have well known computational advantages. Two major benefits are the substantial reduction of the dimensionality of the involved mathematical objects and the preservation of complete positivity [18]. However, substituting density matrices by state vectors constitutes also an attempt to achieve a more detailed description of the dynamics of open quantum systems [6,19]. It is well recognized, for example, that the continuous measurement of an open quantum system with associated Markovian dynamics can be described by means of a SSE (see Table 2 O4). The conditional state solution to such an equation over some time interval can be linked to a "quantum trajectory" [12,19] of one property of the environment. Thus, the conditional state can be interpreted as the state of the open system evolving while its environment is under continuous monitoring. This is true in general for Markovian systems, no matter whether or not the environment is being actually measured (i.e., it is valid for both Figure 1a,b). This fact is of great importance for designing and experimentally implementing feedback control in open quantum systems [35]. If this interpretation could also be applied to non-Markovian SSEs [33,37], then this would be very significant for quantum technologies, especially in condensed matter environments (e.g., electron devices), which are typically non-Markovian [6].

Table 2. Validity of Bohmian vs. orthodox conditional states to provide dynamic information of open quantum system depending on the relation between the environment decoherence time t_D and the observation period τ. Here (un)measured refers to unmeasured and measured indistinctively.

Validity of Conditional States to Provide Dynamic Information	Non-Markovian -Measured- $t_D > \tau = 0$	Non-Markovian -Unmeasured- $t_D > \tau = 0$	Non-Markovian -(Un)measured- $t_D \sim \tau > 0$	Markovian -(Un)measured- $t_D \ll \tau$
Orthodox	(O1) ✓	(O2) ✗	(O3) ✗	(O4) ✓
Bohmian	(B1) ✓	(B2) ✓	(B3) ✓	(B4) ✓

Unfortunately, for non-Markovian conditions, the above interpretation is only possible for the rather exotic scenario where the environment is being continuously monitored and the system is strongly coupled to it. As no correlation between the system and the environment can build up, the evolved system is kept in a pure state. This is the well-known quantum Zeno regime [65,66], under which conditional states can be trivially used to describe the frozen properties of the system (see Table 2 O1). Without the explicit consideration of the measurement process (as in Figure 1a), however, the postulates of the orthodox theory restrict the amount of dynamical information that can be extracted from state vectors (see Table 2 O2). In most general conditions, for $\tau > 0$ and non-Markovian dynamics, while conditional states can be used to reconstruct the reduced density matrix, they cannot be used to evaluate time-correlations (see Table 2 O3) [20,23]. This is not only true when the environment is being measured (as in Figure 1b), but also when it is not measured (as in Figure 1a).

Therefore, we turned to a nonorthodox approach: the Bohmian interpretation of quantum mechanics. The basic element of the Bohmian theory (as in other quantum theories without observers) is that the intrinsic properties of quantum systems do not depend on whether the system is being measured or not. Such ontological change is, nevertheless, fully compatible with the predictions of orthodox quantum mechanics because a measurement-independent reality of quantum objects is not in contradiction with non-local and contextual quantum phenomena. Yet, the ontological nature of the trajectories in Bohmian mechanics introduces the possibility of evaluating dynamic properties in terms of conditional wavefunctions for Markovian and non-Markovian dynamics, no matter whether the environment is being actually measured or not (see Table 2, B1–B4 and Figure 7a,b).

In summary, the Bohmian conditional states lend themselves as a rigorous theoretical tool to evaluate static and dynamic properties of open quantum systems in terms of state vectors without the need of reconstructing a reduced density matrix. Formally, the price to be paid is that for developing a SSE-like approach based on Bohmian mechanics one needs to evaluate both the trajectories of the environment and of the system see (d) Table 1. Nonetheless, we have seen that this additional computational cost can be substantially reduced in practical situations. For THz electron devices (see Section 5), for example, we showed that invoking current and charge conservation one can easily get rid of the evaluation of the environment trajectories. This reduces substantially the computational cost associated to the Bohmian conditional wave function approach (as shown (e) in Table 1). Let us also notice that here we have always assumed that the positions of the environment are the variables that the states of the system are conditioned to. However, it can be shown that the mathematical equivalence of the SSEs with state vectors conditioned to other "beables" of the environment (different from the positions) is also possible. It requires using a generalized modal interpretation of quantum phenomena, instead of the Bohmian theory. A review on the modal interpretation can be found in [67,68].

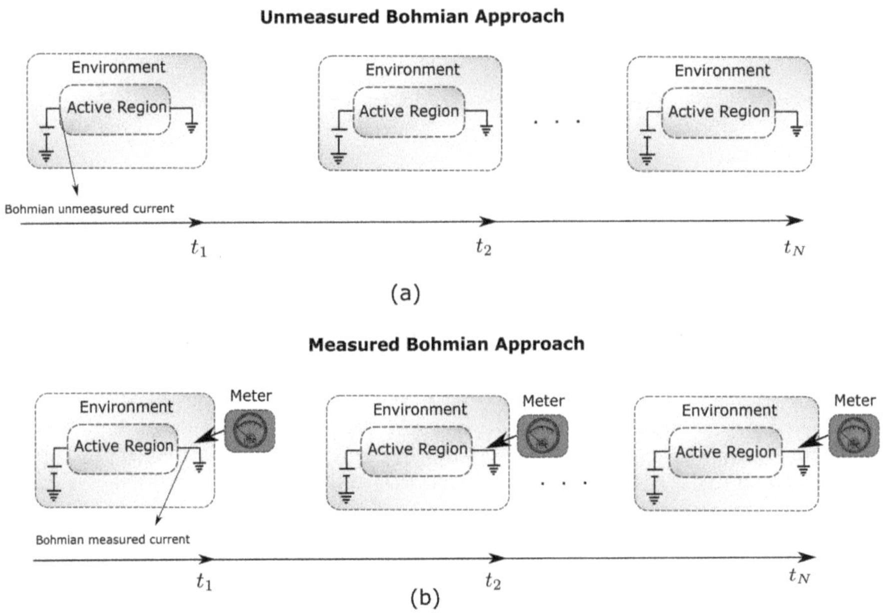

Figure 7. (**a**) Figure depicting the Unmeasured Bohmian approach in which the computation of any property (electric current in an electron device) is independent of the measuring apparatus. (**b**) The continuous measurement of the electric current through an ammeter (measuring apparatus) can be also described in Bohmian mechanics by including the degrees of freedom of the measuring apparatus.

As an example of the practical utility of the Bohmian conditional states, we have introduced a time-dependent quantum Monte Carlo algorithm, called BITLLES, to describe electron transport in open quantum systems. We have simulated a graphene electron device coupled to an RC circuit and computed its current–current correlations up to the THz regime where non-Markovian effects are relevant. The resulting simulation technique allows to describe not only DC and AC device's characteristics but also noise and fluctuations. Therefore, BITLLES extends to the quantum regime the computational capabilities that the Monte Carlo solution of the Boltzmann transport equation has been offering for decades for semi-classical devices.

Entropy **2019**, *21*, 1148

Author Contributions: Conceptualization, D.P., G.A. and X.O. ; methodology, D.P., G.A. and X.O.; software, E.C. and X.O.; validation, D.P., G.A., E.C.and X.O.; formal analysis, D.P., G.A. and X.O.; investigation, D.P., G.A. and X.O.; resources, D.P., G.A. and X.O.; data curation, D.P., E.C., G.A., and X.O.; writing–original draft preparation, D.P., G.A. and X.O.; writing–review and editing, D.P., E.C., G.A. and X.O.; visualization, D.P., E.C., G.A. and X.O.; supervision, G.A. and X.O.; project administration, G.A. and X.O.; funding acquisition, X.O and G.A.

Funding: We acknowledge financial support from Spain's Ministerio de Ciencia, Innovación y Universidades under Grant No. RTI2018-097876-B-C21 (MCIU/AEI/FEDER, UE), the European Union's Horizon 2020 research and innovation programme under grant agreement No Graphene Core2 785219 and under the Marie Skodowska-Curie grant agreement No 765426 (TeraApps). G.A. also acknowledges financial support from the European Unions Horizon 2020 research and innovation programme under the Marie Skodowska-Curie Grant Agreement No. 752822, the Spanish Ministerio de Economa y Competitividad (Project No. CTQ2016-76423-P), and the Generalitat de Catalunya (Project No. 2017 SGR 348).

Acknowledgments: The authors are grateful to Howard Wiseman and Bassano Vacchini for enlightening comments.

Conflicts of Interest: The authors declare no conflict of interest.

Appendix A. D'Espagnat Distinction between "Proper" and "Improper" Mixtures

An alternative explanation on the difficulties of state vectors in describing open quantum system comes from the distinction between "proper" and "improper" mixtures by D'Espagnat.

- The "proper" mixture is simply a mixture of different pure states of a closed system. We define such pure states as $|\psi_q\rangle$ with $q = 1, .., N$. We know that each of these states represent the closed system in one of the repeated experiments, but we ignore which state corresponds to each experiment. We only know the probability $P(q)$ that one experiment is represented by the pure state $|\psi_q\rangle$. Then, if we are interested in computing some ensemble value of the system, over all experiments, von Neumann introduced the mixture $\rho = \int P(q)|\psi_q\rangle\langle\psi_q|dq$. It is important to notice that we are discussing here human ignorance (not quantum uncertainty). The system is always in a well-defined state (for all physical computations), but we (the humans) ignore what the state is in each experiment.
- The "improper" mixture refers to the density matrix that results from a trace reduction of a pure sate (or statistical operator) of a whole system that includes the system and the environment. The reduced density of the system alone is given by tracing out the degrees of freedom of the environment, giving the result in Equation (7), which is mathematically (but not physically) equivalent to the results of the "proper" mixture constructed from our "ignorance" of which state represents the system.

D'Espagnat claims that the ignorance interpretation of the "proper" mixture cannot be given to the "improper" mixture. The D'Espagnat's argument is as follows. Let us assume a pure global system (inclding the open system and the environment) described by Equation (1). Then, if we accept that the physical state of the system is given by $|\psi_q(t)\rangle$, we have to accept that the system-plus-environment is in the physical state $|q\rangle \otimes |\psi_q(t)\rangle$ with probability $P(q)$. The ignorance interpretation will then erroneously conclude that the that the global system is in a mixed state, not in a pure state as assumed in Equation (1). The error is assuming that the system is in a well-defined state that we (the humans) ignore it. This is simply not true. D'Espagnat results shows that a conditional state cannot be a description of an open system with all the static and dynamic information that we can get from the open system.

In addition, let us notice that the conclusion of D'Espagnat applies to any open quantum system without distinguishing between Markovian or non-Markovian scenarios. However, indeed, there is no contradiction between the D'Espagnat conclusion and the attempt of the SSE of using pure states to describe Markovian open quantum systems for static and dynamic properties. Both are right. D'Espagnat discussion is a formal (fundamental) discussion about conditional states, while the discussion about Markovian scenarios is a practical discussion about simplifying approximation when extracting information of the system at large τ.

Finally, let us notice that the D'Espagnat conclusions do not apply to Bohmian mechanics because the Bohmian definition of a quantum system involves a wave function plus trajectories. The conditional state $\tilde{\psi}_{Q^i(t)}(x,t) = \Psi(x, Q^i(t), t)$, together with the environment and system trajectory $Q^i(t)$ and $X^i(t)$, contains all the (static and dynamic) information of the open system in this i-th experiment. An ensemble over all experiments prepared with the same global wavefunction $\Psi(x, q, t)$ requires an ensemble of different environment and system trajectories $Q^i(t)$ and $X^i(t)$ for $i = 1, 2, ..., M$ with $M \to \infty$.

Appendix B. Orthodox and Bohmian Reduced Density Matrices

The orthodox and Bohmian definitions of a quantum state are different. The first uses only a wave function, while the second uses the same wave function plus trajectories. It is well known that both reproduce the same ensemble values by construction. Here, we want to discuss how the orthodox density matrix (without trajectories) can be described by the Bohmian theory with trajectories.

We consider a system plus environment defined by a Hilbert space $\mathcal{H}$ that can be decomposed as $\mathcal{H} = \mathcal{H}_x \otimes \mathcal{H}_q$ where x is the collective positions of particles of the system while q are the collective positions of the particles of the environment. The expectation value of any observable O_x of the system can be computed as $\langle O_x \rangle = \langle \Psi | \hat{O}_x \otimes \hat{I}_q | \Psi \rangle$ with $\hat{I}_q$ the identity operator for the environment. We describe the typical orthodox procedure to define the orthodox reduced density matrix by tracing out all degrees of freedom of the environment:

$$\rho(x, x', t) = \int dq \Psi^*(x', q, t) \Psi(x, q, t) \tag{A1}$$

From Equation (A1) the mean value of the observable O_x can be computed as,

$$\langle O_x \rangle = \int dx \left(O_x \rho(x, x', t)|_{x'=x} \right) \tag{A2}$$

In this appendix, we want to describe Equations (A1) and (A2) in terms of the Bohmian conditional wavefunctions and trajectories described in the text. The conditonal wavefunction associated to the system during the i-th experiment conditioned on a particular value of the environment $Q^i(t)$ is defined as $\psi_{Q^i(t)}(x, t) = \Psi(x, Q^i(t), t)$, being $\Psi(x, q, t) = \langle x, q | \Psi \rangle$ the position representation of the global state. We start from the general expression for the ensemble value in the position representation as,

$$\langle O_x \rangle = \int dx \int dq\, \Psi^*(x, q, t) O_x \Psi(x, q, t) \tag{A3}$$

Multiplying and dividing by $\Psi^*(x, q, t)$ we get,

$$
\begin{aligned}
\langle O_x \rangle &= \int dx \int dq\, \frac{|\Psi(x, q, t)|^2 O_x \Psi(x, q, t)}{\Psi^*(x, q, t)} \\
&= \frac{1}{M} \sum_{i=1}^{M} \int dx \int dq \frac{\delta[(x - X^i(t)]\delta[(q - Q^i(t)] O_x \Psi(x, q, t)}{\Psi^*(x, q, t)} \\
&= \frac{1}{M} \sum_{i=1}^{M} \int dx \frac{O_x \Psi(x, Q^i(t), t)}{\Psi^*(x, Q^i(t), t)} \delta(x - X^i(t))
\end{aligned}
\tag{A4}
$$

where we have used the quantum equilibrium condition $|\Psi(x,q,t)|^2 = \frac{1}{M}\sum_{i=1}^{M}\delta[(x-X^i(t)]\delta[(q-Q^i(t)]$ with $M \to \infty$. Now, we multiply and divide by $\Psi^*(x,Q^i(t),t)$ to get,

$$
\begin{aligned}
\langle O_x \rangle &= \frac{1}{M}\sum_{i=1}^{M}\int dx \frac{O_x \Psi^*(x,Q^i(t),t)\Psi(x,Q^i(t),t)}{|\Psi(x,Q^i(t),t)|^2}\delta(x-X^i(t)) \\
&= \int dx \left[O_x \sum_{i=1}^{M} P_i \tilde{\psi}^{i*}(x',t)\tilde{\psi}^i(x,t) \right]_{x'=x=X^i(t)}
\end{aligned} \tag{A5}
$$

where $P_i = 1/M$ can be interpreted as the probability associated to each $i = 1,2,...,M$ experiment and we have defined:

$$
\tilde{\psi}^i(x,t) \equiv \frac{\Psi(x,Q^i(t),t)}{\Psi(X^i(t),Q^i(t),t)}. \tag{A6}
$$

Now, once we arrive at Equation (A5), one can be tempted to define a type of Bohmian reduced density matrix in terms of the conditional wavefunctions for $i = 1,2,...,M$ experiments as,

$$
\rho_w(x',x,t) = \sum_{i=1}^{M} P_i \tilde{\psi}^{i*}(x',t)\tilde{\psi}^i(x,t) = \sum_{i=1}^{M} P_i \frac{\Psi^*(x',Q^i(t),t)}{\Psi^*(X^i(t),Q^i(t),t)}\frac{\Psi(x,Q^i(t),t)}{\Psi(X^i(t),Q^i(t),t)}, \tag{A7}
$$

where we have arbitrarily eliminated the role of the trajectories. However, strictly speaking Equation (A1) is not equal to Equation (A7). If we include all $i = 1,2,...,M$ experiments in the computation of (A7), there are trajectories $Q^i(t)$ and $Q^j(t)$ that at the particular time t can be represented by the same conditional wavefunction $\tilde{\psi}^i(x,t) = \tilde{\psi}^j(x,t)$ if $Q^i(t) = Q^j(t)$. Such over-summation due to the repetition of the same trajectories is not present in (A1).

To simplify the subsequent discussion, let us assume that q is one degree of freedom in a 1D space. Let us cut such 1D space into small intervals of length Δq. Each interval is defined as $j\,\Delta q < q < (j+1)\,\Delta q$ and it is labelled by the index j. Then, we can define $G^j(t)$ as the number of positions $Q^i(t)$ that are inside the j-interval at time t as:

$$
G^j(t) = \sum_{i=1}^{M} \int_{j\,\Delta q}^{(j+1)\,\Delta q} \delta[q - Q^i(t)]dq \tag{A8}
$$

With this definition, assuming that Δq is so small that all $Q^i(t)$ inside the interval and all the corresponding Bohmian conditional wave functions $\Psi(x,Q^i(t),t)$, system positions $X^i(t)$, and probabilities P_i are almost equivalent, and given by q^j, $\Psi(x,q^j,t)$, x^j and P_j respectively, we can change the sum over $i = 1,...,M$ experiments into a sum over $j = ...,-1,0,1,...$ spatial intervals to rewrite Equation (A7) as:

$$
\rho_w(x',x,t) = \sum_{j=-\infty}^{j=+\infty} G^j(t)P_j \frac{\Psi^*(x',q^j,t)}{\Psi^*(x^j,q^j,t)}\frac{\Psi(x,q^j,t)}{\Psi(x^j,q^j,t)} \approx \int dq^j N^j(t)\Psi^*(x',q^j,t)\Psi(x,q^j,t) \tag{A9}
$$

where $N^j(t) = G^j(t)P_j / \left(\Psi^*(x^j,q^j,t)\Psi(x^j,q^j,t)\right)$. So, finally, a proper normalization of the Bohmian conditional states allows us to arrive to Equation (A1) from Equation (A7). Such normalization is already discussed in quation (11) in the text. The moral of the mathematical developments of this appendix is that open systems are more naturally described in terms of density matrix than in terms of conditional states when using the orthodox theory, while the contrary happens when using the Bohmian theory. Because of the additional variables of the Bohmian theory, the conditional states are a natural Bohmian tool to describe open systems.

Appendix C. Equations of Motion for Single-Electron Conditional States in Graphene

As said in the text, graphene dynamics are given by the Dirac equation and not by the usual Schrödinger one. The presence of the Dirac equation on the description of the dynamics of electrons in graphene is not due to any relativistic correction but to the presence of a linear energy-momentum dispersion (in fact, the graphene Fermi velocity $v_f = 10^6 \ m/s$ is faster than the electron velocity in typical parabolic band materials but still some orders of magnitude slower than the speed of light). Thus, the conditional wavefunction associated to the electron is no longer a scalar but a bispinor. In particular, the initial bispinor is defined (located outside of the active region) as:

$$\begin{pmatrix} \psi_1(x,z,t) \\ \psi_2(x,z,t) \end{pmatrix} = \begin{pmatrix} 1 \\ se^{i\theta_{\vec{k}_c}} \end{pmatrix} \Psi_g(x,z,t), \tag{A10}$$

where $\Psi_g(x,z,t)$ is a Gaussian function with central momentum $\vec{k}_c = (k_{x,c}, k_{z,c})$, $s = 1$ ($s = -1$) if the electron is in the conduction band (valence band) and $\theta_{\vec{k}_c} = atan(k_{z,c}/k_{x,c})$. The wave packet can be considered as a Bohmian conditional wavefunction for the electron, a unique tool of Bohmian mechanics that allows to tackle the many-body and measurement problems in a computationally efficient way [25,62]. The two components are solution of the mentioned Dirac equation:

$$i\hbar \frac{\partial}{\partial t} \begin{pmatrix} \psi_1 \\ \psi_2 \end{pmatrix} = \begin{pmatrix} V(x,z,t) & -i\hbar v_f \frac{\partial}{\partial x} - \hbar v_f \frac{\partial}{\partial z} \\ -i\hbar v_f \frac{\partial}{\partial x} + \hbar v_f \frac{\partial}{\partial z} & V(x,z,t) \end{pmatrix} \begin{pmatrix} \psi_1 \\ \psi_2 \end{pmatrix} = -i\hbar v_f \left(\vec{\sigma} \cdot \vec{\nabla} + V \right) \begin{pmatrix} \psi_1 \\ \psi_2 \end{pmatrix}, \tag{A11}$$

where $v_f = 10^6 \ m/s$ is the mentioned Fermi velocity and $V(x,z,t)$ is the electrostatic potential. $\vec{\sigma}$ are the Pauli matrices:

$$\vec{\sigma} = (\sigma_x, \sigma_z) = \left(\begin{pmatrix} 0 & 1 \\ 1 & 0 \end{pmatrix}, \begin{pmatrix} 0 & -i \\ i & 0 \end{pmatrix} \right). \tag{A12}$$

Usually, in the literature, one finds σ_z as σ_y, however, since we defined the graphene plane as the XZ one, the notation here is different. Then, we can obtain a continuity equation for the Dirac equation and then we can easily identify the Bohmian velocities of electrons as [44]

$$\vec{v}(\vec{r},t) = \frac{v_f \psi(\vec{r},t)^\dagger \vec{\sigma} \psi(\vec{r},t)}{|\psi(\vec{r},t)|^2}. \tag{A13}$$

By time integrating (A13) we can obtain the quantum Bohmian trajectories. The initial positions of the trajectories must be distributed according to the modulus square of the initial wavefunction, i.e., satisfying the quantum equilibrium hypothesis and thus certifying the same empirical results as the orthodox theory [44]. All this formalism was introduced in the BITLLES simulator in order to correctly model graphene and other linear band structure materials.

References

1. Jacoboni, C.; Reggiani, L. The Monte Carlo method for the solution of charge transport in semiconductors with applications to covalent materials. *Rev. Mod. Phys.* **1983**, *55*, 645.
2. Available online: http://www.nextnano.de (accessed on 22 November 2019).
3. Available online: http://www.tibercad.org (accessed on 22 November 2019).
4. Available online: http://vides.nanotcad.com/vides/ (accessed on 22 November 2019).
5. Albareda, G.; Jiménez, D.; Oriols, X. Intrinsic noise in aggressively scaled field-effect transistors. *J. Stat. Mech. Theory Exp.* **2009**, *2009*, P01044.
6. Breuer, H.P.; Petruccione, F. *The Theory of Open Quantum Systems*; Oxford University Press: Oxford, UK, 2002.
7. Smirne, A.; Breuer, H.P.; Piilo, J.; Vacchini, B. Initial correlations in open-systems dynamics: The Jaynes-Cummings model. *Phys. Rev. A* **2010**, *82*, 062114.

8. De Vega, I.; Alonso, D. Dynamics of non-Markovian open quantum systems. *Rev. Mod. Phys.* **2017**, *89*, 015001.

9. Vacchini, B. Non-Markovian dynamics for bipartite systems. *Phys. Rev. A* **2008**, *78*, 022112.

10. Gisin, N. Stochastic quantum dynamics and relativity. *Helv. Phys. Acta* **1989**, *62*, 363–371.

11. Pearle, P. Combining stochastic dynamical state-vector reduction with spontaneous localization. *Phys. Rev. A* **1989**, *39*, 2277.

12. Carmichael, H. *An Open Systems Approach to Quantum Optics: Lectures Presented at the Université Libre de Bruxelles, October 28 to November 4 1991*; Springer Science & Business Media: Berlin/Heidelberg, Germany, 2009; Volume 18.

13. Van Kampen, N.G. *Stochastic Processes in Physics and Chemistry*; Elsevier: Radarweg/Amsterdam, The Netherlands 1992; Volume 1.

14. De Vega, I. Non-Markovian stochastic Schrödinger description of transport in quantum networks. *J. Phys. B At. Mol. Opt. Phys.* **2011**, *44*, 245501.

15. Goetsch, P.; Graham, R. Linear stochastic wave equations for continuously measured quantum systems. *Phys. Rev. A* **1994**, *50*, 5242.

16. Gatarek, D.; Gisin, N. Continuous quantum jumps and infinite-dimensional stochastic equations. *J. Math. Phys.* **1991**, *32*, 2152–2157.

17. Gambetta, J.; Wiseman, H. Non-Markovian stochastic Schrödinger equations: Generalization to real-valued noise using quantum-measurement theory. *Phys. Rev. A* **2002**, *66*, 012108.

18. Rivas, A.; Huelga, S.F.; Plenio, M.B. Quantum non-Markovianity: Characterization, quantification and detection. *Rep. Prog. Phys.* **2014**, *77*, 094001.

19. Diósi, L. Non-Markovian continuous quantum measurement of retarded observables. *Phys. Rev. Lett.* **2008**, *100*, 080401.

20. Wiseman, H.M.; Gambetta, J.M. Pure-state quantum trajectories for general non-Markovian systems do not exist. *Phys. Rev. Lett.* **2008**, *101*, 140401.

21. Eisenberg, B.; Oriols, X.; Ferry, D. Dynamics of current, charge and mass. *Comput. Math. Biophys.* **2017**, *5*, 78–115.

22. Oriols, X.; Ferry, D. Quantum transport beyond DC. *J. Comput. Electron.* **2013**, *12*, 317–330.

23. Gambetta, J.; Wiseman, H. Interpretation of non-Markovian stochastic Schrödinger equations as a hidden-variable theory. *Phys. Rev. A* **2003**, *68*, 062104.

24. Gambetta, J.; Wiseman, H. Modal dynamics for non-orthogonal decompositions. *Found. Phys.* **2003**, *34*, 419–448.

25. Colomés, E.; Zhan, Z.; Marian, D.; Oriols, X. Quantum dissipation with conditional wave functions: Application to the realistic simulation of nanoscale electron devices. *Phys. Rev. B* **2017**, *96*, 075135.

26. Oriols, X. Quantum-Trajectory Approach to Time-Dependent Transport in Mesoscopic Systems with Electron-Electron Interactions. *Phys. Rev. Lett.* **2007**, *98*, 066803.

27. Alarcón, A.; Yaro, S.; Cartoixa, X.; Oriols, X. Computation of many-particle quantum trajectories with exchange interaction: Application to the simulation of nanoelectronic devices. *J. Phys. Condens. Matter* **2013**, *25*, 325601.

28. Albareda, G.; López, H.; Cartoixa, X.; Suné, J.; Oriols, X. Time-dependent boundary conditions with lead-sample Coulomb correlations: Application to classical and quantum nanoscale electron device simulators. *Phys. Rev. B* **2010**, *82*, 085301.

29. Albareda, G.; Suñé, J.; Oriols, X. Many-particle hamiltonian for open systems with full coulomb interaction: Application to classical and quantum time-dependent simulations of nanoscale electron devices. *Phys. Rev. B* **2009**, *79*, 075315.

30. Nakajima, S. On quantum theory of transport phenomena: Steady diffusion. *Prog. Theor. Phys.* **1958**, *20*, 948–959.

31. Zwanzig, R. Ensemble method in the theory of irreversibility. *J. Chem. Phys.* **1960**, *33*, 1338–1341.

32. Dalibard, J.; Castin, Y.; Mølmer, K. Wave-function approach to dissipative processes in quantum optics. *Phys. Rev. Lett.* **1992**, *68*, 580.

33. Strunz, W.T.; Diósi, L.; Gisin, N. Open system dynamics with non-Markovian quantum trajectories. *Phys. Rev. Lett.* **1999**, *82*, 1801.

34. Barchielli, A.; Gregoratti, M. Quantum measurements in continuous time, non-Markovian evolutions and feedback. *Philos. Trans. R. Soc. A Math. Phys. Eng. Sci.* **2012**, *370*, 5364–5385.

35. Wiseman, H.M.; Milburn, G.J. *Quantum Measurement and Control*; Cambridge University Press: Cambridge, UK, 2009.

36. Kraus, K. *States, Effects and Operations: Fundamental Notions of Quantum Theory*; Springer: Berlin/Heidelberg, Germany, 1983.

37. Diósi, L.; Gisin, N.; Strunz, W.T. Non-Markovian quantum state diffusion. *Phys. Rev. A* **1998**, *58*, 1699.

38. Gambetta, J.; Wiseman, H. Perturbative approach to non-Markovian stochastic Schrödinger equations. *Phys. Rev. A* **2002**, *66*, 052105.

39. Budini, A.A. Non-Markovian Gaussian dissipative stochastic wave vector. *Phys. Rev. A* **2000**, *63*, 012106.

40. Bassi, A.; Ghirardi, G. Dynamical reduction models with general Gaussian noises. *Phys. Rev. A* **2002**, *65*, 042114.

41. Bassi, A. Stochastic Schrödinger equations with general complex Gaussian noises. *Phys. Rev. A* **2003**, *67*, 062101.

42. d'Espagnat, B. *Conceptual Foundations of Quantum Mechanics*; CRC Press: Boca Raton, FL, USA, 2018.

43. Benseny, A.; Albareda, G.; Sanz, Á.S.; Mompart, J.; Oriols, X. Applied bohmian mechanics. *Eur. Phys. J. D* **2014**, *68*, 286.

44. Pladevall, X.O.; Mompart, J. *Applied Bohmian Mechanics: From Nanoscale Systems to Cosmology*; CRC Press: Boca Raton, FL, USA, 2019.

45. Holland, P.R. *The Quantum Theory of Motion: An Account of the De Broglie-Bohm Causal Interpretation of Quantum Mechanics*; Cambridge University Press: Cambridge, UK, 1995.

46. Bohm, D. A suggested interpretation of the quantum theory in terms of "hidden" variables. I. *Phys. Rev.* **1952**, *85*, 166.

47. Dürr, D.; Goldstein, S.; Tumulka, R.; Zanghi, N. Bohmian mechanics and quantum field theory. *Phys. Rev. Lett.* **2004**, *93*, 090402.

48. Dürr, D.; Goldstein, S.; Zanghi, N. Quantum equilibrium and the origin of absolute uncertainty. *J. Stat. Phys.* **1992**, *67*, 843–907.

49. Pandey, D.; Sampaio, R.; Ala-Nissila, T.; Albareda, G.; Oriols, X. Unmeasured Bohmian properties and their measurement through local-in-position weak values for assessing non-contextual quantum dynamics. 2019, submitted.

50. Alarcón, A.; Albareda, G.; Traversa, F.L.; Benali, A.; Oriols, X. Nanoelectronics: Quantum electron transport. In *Applied Bohmian Mechanics: From Nanoscale Systems to Cosmology*; Pan Standford Publishing: Singapore, 2012; pp. 365–424.

51. Albareda, G.; Marian, D.; Benali, A.; Yaro, S.; Zanghì, N.; Oriols, X. Time-resolved electron transport with quantum trajectories. *J. Comput. Electron.* **2013**, *12*, 405–419.

52. Albareda, G.; Marian, D.; Benali, A.; Alarcón, A.; Moises, S.; Oriols, X., Electron Devices Simulation with Bohmian Trajectories. In *Simulation of Transport in Nanodevices*; John Wiley and Sons, Ltd.: Hoboken, NJ, USA, 2016; Chapter 7, pp. 261–318.

53. Marian, D.; Colomés, E.; Oriols, X. Time-dependent exchange and tunneling: Detection at the same place of two electrons emitted simultaneously from different sources. *J. Phys. Condens. Matter* **2015**, *27*, 245302.

54. Marian, D.; Colomés, E.; Zhan, Z.; Oriols, X. Quantum noise from a Bohmian perspective: Fundamental understanding and practical computation in electron devices. *J. Comput. Electron.* **2015**, *14*, 114–128.

55. Alarcón, A.; Oriols, X. Computation of quantum electron transport with local current conservation using quantum trajectories. *J. Stat. Mech. Theory Exp.* **2009**, *2009*, P01051.

56. López, H.; Albareda, G.; Cartoixà, X.; Suñé, J.; Oriols, X. Boundary conditions with Pauli exclusion and charge neutrality: application to the Monte Carlo simulation of ballistic nanoscale devices. *J. Comput. Electron.* **2008**, *7*, 213–216.

57. Albareda, G.; Appel, H.; Franco, I.; Abedi, A.; Rubio, A. Correlated electron-nuclear dynamics with conditional wave functions. *Phys. Rev. Lett.* **2014**, *113*, 083003.

58. Norsen, T.; Marian, D.; Oriols, X. Can the wave function in configuration space be replaced by single-particle wave functions in physical space? *Synthese* **2015**, *192*, 3125–3151.

59. Albareda, G.; Kelly, A.; Rubio, A. Nonadiabatic quantum dynamics without potential energy surfaces. *Phys. Rev. Mater.* **2019**, *3*, 023803.

60. Zhan, Z.; Kuang, X.; Colomés, E.; Pandey, D.; Yuan, S.; Oriols, X. Time-dependent quantum Monte Carlo simulation of electron devices with two-dimensional Dirac materials: A genuine terahertz signature for graphene. *Phys. Rev. B* **2019**, *99*, 155412.

61. Albareda, G.; Benali, A.; Oriols, X. Self-consistent time-dependent boundary conditions for static and dynamic simulations of small electron devices. *J. Comput. Electron.* **2013**, *12*, 730–742.

62. Marian, D.; Zanghì, N.; Oriols, X. Weak values from displacement currents in multiterminal electron devices. *Phys. Rev. Lett.* **2016**, *116*, 110404.

63. Albareda, G.; Traversa, F.; Benali, A.; Oriols, X. Computation of quantum electrical currents through the Ramo–Shockley–Pellegrini theorem with trajectories. *Fluct. Noise Lett.* **2012**, *11*, 1242008.

64. Pandey, D.; Bellentani, L.; Villani, M.; Albareda, G.; Bordone, P.; Bertoni, A.; Oriols, X. A Proposal for Evading the Measurement Uncertainty in Classical and Quantum Computing: Application to a Resonant Tunneling Diode and a Mach-Zehnder Interferometer. *Appl. Sci.* **2019**, *9*, 2300.

65. Misra, B.; Sudarshan, E.G. The Zeno's paradox in quantum theory. *J. Math. Phys.* **1977**, *18*, 756–763.

66. Home, D.; Whitaker, M. A conceptual analysis of quantum Zeno; paradox, measurement, and experiment. *Ann. Phys.* **1997**, *258*, 237–285.

67. Colodny, R.G. *Paradigms and Paradoxes: The Philosophical Challenge of The Quantum Domain*; University of Pittsburgh Press: London, UK, 1972; Volume 5.

68. Cabello, A.; Gu, M.; Gühne, O.; Larsson, J.Å.; Wiesner, K. Thermodynamical cost of some interpretations of quantum theory. *Phys. Rev. A* **2016**, *94*, 052127.

Review

Entropy Production in Quantum Is Different

Mohammad H. Ansari [1,*], **Alwin van Steensel** [1] **and Yuli V. Nazarov** [2]

[1] Jülich-Aachen Research Alliance Institute (JARA) and Peter Grünberg Institute (PGI-2), Forschungszentrum Jülich, D-52425 Jülich, Germany

[2] Department of Quantum Nanoscience, Kavli Institute of Nanoscience, TU Delft, Lorentzweg 1, 2628CJ Delft, The Netherlands

* Correspondence: m.ansari@fz-juelich.de

Received: 30 June 2019; Accepted: 28 August 2019; Published: 31 August 2019

Abstract: Currently, 'time' does not play any essential role in quantum information theory. In this sense, quantum information theory is underdeveloped similarly to how quantum physics was underdeveloped before Erwin Schrödinger introduced his famous equation for the evolution of a quantum wave function. In this review article, we cope with the problem of time for one of the central quantities in quantum information theory: entropy. Recently, a replica trick formalism, the so-called 'multiple parallel world' formalism, has been proposed that revolutionizes entropy evaluation for quantum systems. This formalism is one of the first attempts to introduce 'time' in quantum information theory. With the total entropy being conserved in a closed system, entropy can flow internally between subsystems; however, we show that this flow is not limited only to physical correlations as the literature suggest. The nonlinear dependence of entropy on the density matrix introduces new types of correlations with no analogue in physical quantities. Evolving a number of replicas simultaneously makes it possible for them to exchange particles between different replicas. We will summarize some of the recent news about entropy in some example quantum devices. Moreover, we take a quick look at a new correspondence that was recently proposed that provides an interesting link between quantum information theory and quantum physics. The mere existence of such a correspondence allows for exploring new physical phenomena as the result of controlling entanglement in a quantum device.

Keywords: time evolution; quantum information; entropy production; Renyi entropy; quantum thermodynamics

1. Introduction

Entropy is one of the central quantities in thermodynamics and, without its precise evaluation, one cannot predict what new phenomena are to be expected in the thermodynamics of a device. In quantum theory, entropy is defined as a nonlinear function of the density matrix, i.e., $S = -\text{Tr}\hat{\rho}\ln\hat{\rho}$, in the units of the Boltzmann constant k_B. The mere nonlinearity indicates that entropy is not physically observable because, by definition, observables are linear in the density matrix. Let us further describe this statement. Here, we do not assume that the density matrix is a physical quantity. The reason is that evaluating all components of a many-body density matrix requires many repetitions of the same experiment with the same initial state. Not only is this difficult but also the fact that measurement changes quantum states prevents exact evaluation. A physical quantity, such as energy or charge, can be measured in the lab in real time and can be defined in quantum theory to linearly depend on the density matrix. This is not true for entropy and therefore we cannot assume it is a physical quantity directly measurable in the lab.

In fact, the precise time evolution of entropy is still an open problem and has not been properly addressed in the literature [1–3]. A consistent theory of quantum thermodynamics can only be achieved after finding nontrivial relations between the quantum of information and physics. In recent years, exquisite mesoscopic scale control over quantum states has led technology to the quantum realm. This has motivated exploring new phenomena such as exponential speed up in computation as well as power extraction from quantum coherence [4–8]. Recently, there have been attempts to implement quantum versions of heat engines using superconducting qubits [9]. However, recent developments in realizing quantum heat engines, such as in References [10–12], rely on semiclassical stochastic entropy production after discretizing energy. A long-lasting question is how the superposition of states transfers heat and how much entropy is produced as the result of such a transfer.

A quantum heat engine (QHE) is a system with several discrete quantum states and, similar to a common heat engine, is connected to several environments kept at different temperatures. In fact, a number of large heat baths in these engines share some degrees of freedom quantum mechanically. Such a system is supposed to transfer heat according to the laws of quantum mechanics. The motivation for research in QHE originates from differences they may controllably make on the efficiency and output powers. Let us consider the example of two heat baths A and B, both coupled through a quantum system q that contains discrete energies and allows for the superposition of states with long coherence time. Let us clarify that, in this paper, we study the flow of thermodynamic Renyi and von Neumann entropies between the heat baths and quantum system q. Therefore, other entropies are beyond the scope of this paper. This quantum system coupled to the two large heat baths is in fact a physical quantum system that is energetically coupled to the reservoirs and allows for stationary flow of heat as well as a flow of thermodynamic entropy from one reservoir to another. We will see in the next section that, similar to physical quantities such as energy and charge, the total entropy of a closed system is a conserved quantity and does not change in time. However, internally, entropy can flow from one subsystem to another. Therefore, sub-entropies may change in time and this change may indicate a change in the energy transfer. Some important questions one may ask are: *Does a quantum superposition change entropy?* This is one of the questions that we will address in this almost pedagogical review paper and we will furthermore describe how the information content in entropy can be meaningful in physics.

In a typical engine made of reservoirs A, B and an intermediate quantum system q with discrete energy levels, the change of entropy in one of the reservoirs, say B, between the time 0 and t is $S_B(t) - S_B(0) = -\text{Tr}\{\rho(t)\ln\rho_B(t)\} - \text{Tr}_B\{\rho_B^{eq}\ln\rho_B^{eq}\}$, where in the first term we have safely replaced one of the two partial density matrices with the total density matrix, and accordingly replaced the partial trace with total one. The conservation of entropy tells us that the total entropy maintains its initial value at the separable compound state $\rho(0) = \rho_q(0)\rho_A^{eq}\rho_B^{eq}$, i.e., $-\text{Tr}\{\rho(t)\ln\rho(t)\} = -\text{Tr}_q\{\rho_q(0)\ln\rho_q(0)\} - \sum_{i=A,B}\text{Tr}_i\{\rho_i^{eq}\ln\rho_i^{eq}\}$. After a few lines of algebra one can find that the change of entropy at the reservoir is $S_B(t) - S_B(0) = S_B\left(\rho(t)||\rho_A^{eq}\rho_B(t)\rho_q(0)\right) + \sum_{i=q,A}\text{Tr}_i\{(\rho_i(t) - \rho_i(0))\ln\rho_i(0)\}$, with $S(\rho||\rho') \equiv \text{Tr}\{\rho\ln\rho\} - \text{Tr}\{\rho\ln\rho'\}$ being the relative entropy. Since relative entropy is a positive number [13] and equals zero only for identical density matrices $\rho = \rho'$, the first part of the entropy flow is positive and irreversible. This satisfies the classical laws of thermodynamics. We will show that, in contrast to what has been so far presented in the literature [14], the second term in the entropy flow is *not* heat transfer—the average change of energy at the two times $Q_B \equiv \langle H(0)\rangle_B - \langle H(t)\rangle_B$. Instead, it is the difference of incoherent and coherent heat transfers [15], i.e., $(Q_{B,\text{incoh}}(t) - Q_{B,\text{coh}}(t)) - (Q_{B,\text{incoh}}(0) - Q_{B,\text{coh}}(0))$. This is the new result that heavily modifies the flow of entropy in some quantum heat engines and leads to some recent new physics [16–19].

In this review paper, we look at some of the simplest and most important quantum heat engines. Depending on the external drive or internal degeneracy, the exact evaluation of entropy is indeed very different from what has been presented in the literature so far. We will describe how to precisely evaluate entropy and its flow by using a replica trick that properly allows for the mathematically involved nonlinearity. We introduce a new class of correlations that allow information transfers and are different from physical correlations. For equilibrium systems, these informational correlations satisfy a generalized form of Kubo–Martin–Schwinger (KMS) relation [20,21]. This part of the analysis will be presented in a self-contained fashion after reviewing some of the classical and quantum definitions of entropy and introducing our replica trick for evaluating the time evolution of generalized Keldysh contours. We describe a short protocol for evaluating Keldysh diagrams and in some examples perform the evaluation of a number of diagrams. We present results of example quantum devices such as a two-level quantum heat engine, a photocell, as well as a resonator, each one mediating heat transfer between two large heat baths. Finally, we briefly report on the new correspondence that makes entropy flow directly measurable in the lab by monitoring physical quantities, i.e., the statistics of energy transfer.

2. Classical Systems

2.1. Classical Entropy

Many systems in classical physics carry entropy. Some of the most studied systems are: charge transport at a point contact [22,23], energy transport in heat engines [24], and a gravitational hypersurface falling into a black hole [25–28]. Let us for simplicity of the discussion review classical entropy by means of the example of charge transport through a point contact. Consider for this purpose two large conductor plates connected at a point, the so-called 'point contact system'. This classical point contact either transmits a charged particle with probability p or blocks the transmission with probability $1 - p$. Let us consider N attempts take place. For $N \gg 1$ it is most likely that, in pN out of N times, the particles are successfully transferred and, in $(1 - p) N$ out of N times, they are not. For unmarked particles, the order of events does not matter, therefore the number of possibilities with pN transfers out of N attempts is

$$\mathcal{N} = \begin{pmatrix} N \\ pN \end{pmatrix} \approx \frac{N^N}{(pN)^{pN} \left[(1 - p) N \right]^{(1-p)N}} = \frac{1}{p^{pN} (1 - p)^{(1-p)N}}. \tag{1}$$

This number rapidly grows with N. In order to keep the number small, we take its logarithm. This defines the so-called Shannon entropy, i.e., $S_{Shannon} = \log_2 \mathcal{N} = -N \left[p \log_2 p + (1 - p) \log_2 (1 - p) \right]$.

The linear dependence of the Shannon entropy on the number of attempts N indicates its additivity. The definition of entropy can be generalized to account for extended geometries such as a $k + 1$-path terminal that connects any reservoir to k others. In this case, k probabilities contribute to understanding the possibility of transmission from a reservoir to any one of the other k reservoirs, thus entropy is generalized to $S_{Shannon} = -N \sum_{n=1}^{k} p_n \log_2 p_n$. This entropy may vary in time. One possible reason for such variation could be due to time-dependent probabilities $p_n(t)$. Another possibility for time evolution of entropy could be the presence of some bias in controlling the system. For example, consider that, after one successful transfer, the transmission is reduced or closed for a rather long time before it opens again to another transfer attempt. The entropy of such a system depends on whether or not a success transfer has taken place in the past.

In fact, in this paper, what we call entropy production refers to the time variation of partial entropy associated with a part of a closed system. Moreover, as stated in the Introduction, in this paper, we are only interested in the time variation in thermodynamic systems such as heat baths; therefore, our focus is only on thermodynamic entropies and its time evolution, namely 'entropy production'. In this section, although we discuss Shannon entropy $S_{Shannon}$, we have to distinguish between the Shannon entropy, which can be measured as a number of bits, and the rest of the paper in which we study von Neumann thermodynamic entropy measured in the unit Joule per Kelvin. The Shannon entropy and the thermodynamic entropy are related by the Boltzmann constant k_B, i.e., $S_{Thermodynamic} = k_B S_{Shannon}$. Without the loss of generality, we use the convention that $k_B = 1$, although the reader should keep in mind that, in this paper, we are interested in finding changes in thermodynamic entropy flow as the result of energy exchange processes.

2.2. Renyi Entropy

Alfred Renyi introduced the generalization of Shannon entropy that maintains the additive property [29]. For a finite set of k probabilities p_i with $i = 1, \cdots, k$, the Renyi entropy of degree M is defined as

$$\mathcal{S}_M = \frac{1}{1 - M} \log \sum_i (p_i)^M , \tag{2}$$

with positive entropy order $M > 0$. The symbol $\mathcal{S}_M$ indicates that this is the original definition of Renyi entropy to make it distinct from the simplified definition S_M we use in this paper. The constant prefactor $1/(1 - M)$ in Equation (2) has certain advantages. One of the advantages is that it helps to compactify the definition of some other entropies using Equation (2); i.e., the analytical continuation of Renyi entropy in the limit of M approaching 1 (∞) defines Shannon (min) entropy. Another advantage of the prefactor is that it allows for interpretation of the quantity as the number of bits (thanks to one of the referees for pointing out these remarks).

Here, we present a simplified version of the definition. The logic behind such simplification is that the calculation in the limits requires L'Hopital's rule; i.e., $S_{Shannon, \, min} = \lim_{M \to 1, \infty} S_M^R = - \lim_{M \to 1, \infty} d(\log \sum_i (p_i)^M)/dM$. We define a rescaled Renyi entropy, which is different from the original definition by a prefactor $1/(M - 1)$:

$$S_M = - \log \sum_i (p_i)^M . \tag{3}$$

The reason to define the simplified formula is that evaluating entropy itself is beyond the scope of this paper. Instead, we need to find the time derivative of the entropy (i.e., entropy flow). Due to the presence of a logarithm in Equation (3), any contact prefactor in the definition of entropy will be canceled out from the numerator and denominator of entropy flow. The only trouble is that we must keep in mind that the Shannon entropy can be reproduced after taking the dS_M/dM in the limit of $M \to 1$. In fact, given that $dx^M/dM = d \exp(M \ln x)/dM = x^M \ln x$, one can write

$$\lim_{M \to 1} \frac{dS_M}{dM} = - \lim_{M \to 1} \frac{\sum_i (p_i)^M \ln p_i}{\sum_i (p_i)^M} = - \sum_i p_i \ln p_i = S_{Shannon}. \tag{4}$$

In the rest of the paper, we use the simplified definition. However, given that the difference between the two definitions is marginal, only a constant factor, the reader may decide to use either definition, subject to the discussion above.

In a point contact, given that Renyi entropy is additive for independent attempts, the total Renyi entropy after N uncorrelated attempts will be $S_M = -N \log(p^M + (1-p)^M)$. In a classical heat reservoir, the Renyi entropy is more closely related to free energy. Consider a bath at temperature T with a large number of energy states ϵ_i. The corresponding Gibbs probabilities are $p_i = \exp(-\epsilon_i T)/Z(T)$ and $Z(T) \equiv \sum_i p_i$ is the corresponding partition function. The Renyi entropy of the heat bath is $S_M = - \ln(\sum_i \exp(-M\epsilon_i T)) + M \ln Z(T)$. The free energy will be $F(T) = -T \ln Z(T)$, which is related to the Renyi entropy as $S_M = (M/T)(F(T) - F(T/M))$, i.e., the free energy difference at temperatures T and T/M.

3. Quantum

3.1. Von Neumann and Renyi Entropy

Let us now consider that a large system A with many degrees of freedom interacts with a small quantum system q. This can be thought of as the two share some degrees of freedom. The two exchange some energy via those shared degrees of freedom. Quantumness indicates that q carries a discrete energy spectrum and can be found in superposition between energy levels. Let ρ be the density matrix of the compound system. The partial density matrix of A is defined by tracing out the system q from ρ, i.e., $\rho_A = \text{Tr}_q\rho$. The von Neumann entropy for system A in the Boltzmann constant unit is defined as

$$S^{(A)} = -\text{Tr}_A\rho_A \ln \rho_A \tag{5}$$

and the generalization of entropy in quantum theory will naturally give rise to defining the following quantum Renyi entropy for system A:

$$S_M^{(A)} = -\ln \text{Tr}_A (\rho_A)^M . \tag{6}$$

The density matrix of the isolated compound system evolves between the times t' and $t > t'$ using a unitary transformation that depends on the time difference $U(t - t')$. Therefore, one can evaluate $\text{Tr}_A (\rho_A)^M$ using the unitary transformation to trace it back to the time t'; i.e.,

$$\begin{aligned} \text{Tr} \left(\rho(t) \right)^M &= \text{Tr} \left\{ \left(U(t - t') \rho(t') U^\dagger (t - t') \right)^M \right\} = \text{Tr} \left\{ U(t - t') \rho(t')^M U^\dagger (t - t') \right\} \\ &= \text{Tr}\rho(t')^M . \end{aligned}$$

After taking the logarithm from both sides, one finds that the Renyi entropy remains unchanged between the two times t and t'. In other words, in a closed system, similar to energy and charge, Renyi entropy is a conserved quantity:

$$\frac{dS_M}{dt} = 0. \tag{7}$$

Let us consider for now that there is no interaction between A and q. One can expect naturally that partial entropies are conserved as the result of no interaction because each subsystem can evolve with an independent unitary operator:

$$\frac{dS_M^{(A)}}{dt} = \frac{dS_M^{(q)}}{dt} = 0. \tag{8}$$

Interesting physical systems interact. Therefore, let us now consider that A and q interact. Consider that the total Hamiltonian is $H = H_A + H_q + H_{Aq}$. For interacting systems, there is an important difference between conserved physical and information quantities. For physical quantities, the conservation holds in the whole system as well as in each subsystem. As far as Renyi entropies are concerned, there is a conservation law for the total Renyi entropy $\ln S_M^{(A+q)}$; however, this quantity is only approximately equal to the sum $\ln S_M^{(A)} + \ln S_M^{(q)}$, up to the terms proportional to the volume of the system. Therefore, no exact conservation law can be expected for the extensive quantity summation: $\ln S_M^{(A)} + \ln S_M^{(q)}$ [30]. The reason is that, although the evolution of the entire system is governed by a unitary operator, the subsystem evolves non-unitarily. In the limit of weak coupling $|H_{Aq}|/|H_A + H_q| \ll 1$, the entropy of entire system can only be approximated with the sum of two partial entropies, thus the sum of partial entropies can only approximately satisfy a conservation, i.e., $dS_M^{(A)}/dt + dS_M^{(q)}/dt \approx 0$. Outside of the validity of the weak coupling approximation, we must expect that, although the total entropy conserves, the interacting parts have entropy flows different from each other:

$$\frac{dS_M^{(A)}}{dt} \neq -\frac{dS_M^{(q)}}{dt}. \tag{9}$$

This makes the conservation of Renyi entropy different from the conservation of physical quantities. The root for the difference is in fact in the nonlinear dependence on the density matrix, namely 'non-observability' of entropy [31].

3.2. Replica Trick

Calculating the full reduced density matrix for a general system is the subject of active research. Here, we use a different method that is reminiscent of the 'replica trick' in disorder systems. The trick has been introduced in the context of quantum field theory by Wilczek [32] and Cardy [33] and later in the context of quantum transport by Nazarov [31]. The key point is that, if we can evaluate $\text{Tr}\rho^M$ for any $M \geq 1$, we are able to evaluate the von Neumann entropy using the following relation:

$$S^{(A)} = \lim_{M \to 1} \frac{d}{dM} S_M^{(A)} = \lim_{M \to 1} \frac{d}{dM} \text{Tr}_A \left(\rho_A\right)^M. \tag{10}$$

One can see that there is no need to take the logarithm of $\text{Tr}_A \left(\rho_A\right)^M$. This is only a mathematical simplification in the vicinity of $M \to 1$, i.e., when we want to reproduce von Neumann entropy by analytically continuing the derivative of the Renyi entropy. Otherwise, the presence of the logarithm is essential for the definition of the Renyi entropy. It might be useful to further comment that the Renyi entropy without the logarithm has many names such as Tsallis entropy or power entropy, etc. However, the presence of the logarithm is necessary for what we call the Renyi entropy. Otherwise, we would have $\lim_{M \to 1} \text{Tr}\rho^M = 1$, which, in this important limit, cannot be a true measure of information.

However, calculating $\text{Tr}_A \left(\rho_A\right)^M$ for a real or complex number M is a hopeless task. The 'replica trick' does the following: compute $\text{Tr}_A \left(\rho_A\right)^M$ only for integer M and then analytically continue it to a general real or even complex number.

3.3. Time Evolution of Entropy

Let us mention that we limit our analysis here only to weak coupling. In this regime, the dynamics of a quantum system are reversible and can be formulated in terms of the density matrix evolution. This time evolution depends on the the time-dependent Hamiltonian $H(t) = H_A + H_B + H_{AB}$ as follows:

$$\frac{d\rho}{dt} = \frac{i}{\hbar} \left[H(t), \rho(t)\right]. \tag{11}$$

We transform the basis to the interaction frame by using defining a unitary operator with the non-interaction part of the Hamiltonian $U(t) = \exp\left[-i\left(H_A + H_B\right)t\right]$. The density matrix transforms as $R(t) = U(t)\rho(t)U^\dagger(t)$, thereby not changing its entropy, neither in parts nor in total. In the new basis, Equation (11) becomes

$$\frac{dR}{dt} = \frac{i}{\hbar} \left[U^\dagger(t)H_{AB}(t)U(t), R(t).\right] \tag{12}$$

Let us refer to the interaction Hamiltonian H_{AB} in the new basis as H_I, i.e., $H_I \equiv U^\dagger(t) H_{AB}(t) U(t)$. The solution to the time evolution Equation (12) can be written as

$$R(t) = R_0 + R^{(1)}(t) + \mathcal{O}(2) \tag{13}$$

with

$$R_0 \equiv R(0) \qquad \text{(noninteracting)} \tag{14}$$

$$R^{(1)}(t) \equiv \frac{i}{\hbar} \int_0^t ds\, [H_I(s), R_0] \qquad \text{(1st order)} \tag{15}$$

This solution can (repeatedly) be inserted back into the right side of Equation (12), declaring its cycle of internal interaction:

$$\frac{dR(t)}{dt} = \Delta^{(1)} + \Delta^{(2)} + \mathcal{O}(3) \tag{16}$$

with

$$\Delta^{(1)} \equiv \frac{i}{\hbar} [H_I(t), R_0], \qquad \text{(1st order)} \qquad (17)$$

$$\Delta^{(2)} \equiv -\frac{1}{\hbar^2} \int_0^t ds\, [H_I(t), [H_I(s), R_0]]. \qquad \text{(2nd order).} \qquad (18)$$

In order to find the time evolution of the Renyi and von Neumann entropies, we first notice that the unitary transformation $U(t)$, defining the basis change, also transforms any power of the density matrix, i.e.,

$$R(t)^M = U(t) \left(\rho(t)^M\right) U^\dagger(t). \qquad (19)$$

Now, all we need to do is to generalize the evolution of density matrix to the powers of density matrix $(R(t))^M$. We follow the terminology of Nazarov in [31] and name each copy of replica $R(t)$ in the matrix $(R(t))^M$ a 'world', thus $(R(t))^M$ is the generalized density matrix of M worlds:

$$\frac{d}{dt}\left(R(t)^M\right) = \left[\frac{d}{dt}R(t)\right](R(t))^{M-1} + R(t)\left[\frac{d}{dt}R(t)\right](R(t))^{M-2}$$
$$+ \cdots + (R(t))^{M-2}\left[\frac{d}{dt}R(t)\right]R(t) + (R(t))^{M-1}\left[\frac{d}{dt}R(t)\right].$$

By substituting the solutions of Equations (13) to (18), and limiting the result to second order, we find the the following time evolution of the M-world density matrix:

$$\frac{d}{dt}\left(R(t)^M\right) = \Delta^{(2)} R_0^{M-1} + R_0\Delta^{(2)}R_0^{M-2} + \cdots + R_0^{M-1}\Delta^{(2)}$$
$$+\Delta^{(1)}\left\{R^{(1)}R_0^{M-2} + R_0R^{(1)}R_0^{M-3} + \cdots + R_0^{M-2}R^{(1)}\right\}$$
$$+R_0\Delta^{(1)}\left\{R^{(1)}R_0^{M-3} + R_0R^{(1)}R_0^{M-4} + \cdots + R_0^{M-3}R^{(1)}\right\}$$
$$+R_0^2\Delta^{(1)}\left\{R^{(1)}R_0^{M-4} + R_0R^{(1)}R_0^{M-5} + \cdots + R_0^{M-4}R^{(1)}\right\}$$
$$+\cdots$$
$$+\left\{R^{(1)}R_0^{M-2} + R_0R^{(1)}R_0^{M-3} + \cdots + R_0^{M-2}R^{(1)}\right\}\Delta^{(1)}. \qquad (20)$$

This is how the M-world density matrix evolves in time. The first line in Equation (20) denotes the case where the 2nd order perturbation takes place in one world while the $M-1$ remaining worlds are left non-interacting. All these remaining terms have in common that they don't contain a 2nd order term occurring in a single replica. Instead, these terms contain two 1st order interactions, each acting in a single replica, which together combine to give a 2nd order perturbation term. These new terms have recently been found [34].

If you decide to consider higher perturbative orders, say up to k-th order with $k \leq M$, there will be terms like $R_0^{M-1}\Delta^{(k)}$ in the expansions that have k interactions taking place in one replica, leaving $M-1$ replicas noninteracting as well as terms having k first-order configurations combining to give a kth order interaction term, such as $R_0^{M-k}\left(\Delta^{(1)}\right)^k$. In the case $k > M$, some of the lowest-order interactions will obviously become excluded from the summations.

Let us show the time evolution pictorially using the following diagrams, in which the evolution of $(R(t))^M$ is shown by M parallel lines, each one denoting the time evolution of one world, starting in the past at the bottom and arriving at the present time on the top. In the following diagrams, we show five time-slices by horizontal dashed lines. Blue dots denote the interaction $H_I(t)$ and our diagrams are limited to the 2nd order only. Curly photon-like lines connect the two interactions and represent the correlation function.

The first line of Equation (20) contains all terms that have two interactions in a single world. These two interactions within the same world are called 'self-replica interactions'. They can be illustrated pictorially by the following diagrams in Figure 1 from left to right:

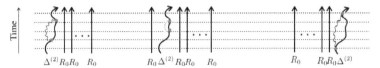

Figure 1. Diagrammatic representation of terms in the first line in Equation (20).

The following diagram in Figure 2 illustrates the typical term $(R_0)^2 \Delta^{(1)} R_0 R^{(1)} (R_0)^{M-4}$ from Equation (20) and pictorially shows the contribution of two first order interactions in two different worlds that together evolve the generalized density matrix of M worlds in the second order.

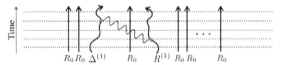

Figure 2. A typical diagram with two first order interactions acting on two different worlds.

A typical higher order digram limited to two-correlation interactions can diagrammatically be shown as below in Figure 3.

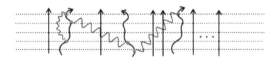

Figure 3. A typical higher order diagram.

3.4. Extended Keldysh Diagrams

In all the above diagrams, quantum states have been represented as labels on the contours. By definition, we know that the density matrix contains both ket and bra states. The second order interactions can, in fact, only take place either between two kets, two bras, or between a ket and a bra. This internal degree of freedom makes it necessary to add more details to our diagrams and represent each replica with the well-known Keldysh contour diagrams [35]. The Keldysh technique permits a natural formulation of the density matrix dynamics in terms of path integrals, which is a generalization of the Feynman–Vernon formalism.

Considering that the time evolution of a quantum system takes place by the Hamiltonian H, kets evolve as $|\psi(t)\rangle = \exp(iHt) |\psi(0)\rangle$ and bras evolve with the opposite phase: $\langle\psi(t)| = \langle\psi(0)| \exp(iHt)$. Based on this simple observation, bras (kets) evolve in the opposite (same) direction of time along the Keldysh contour.

The evolution of the density matrix R from the initial time to the present time can diagrammatically be represented in the following way: one can start at a bra at the present time, move down along the contour to the initial time, pass there through the initial density matrix thereby changing from a bra to a ket, and finally move upwards to end with a ket at the present time. Taking a trace from the density matrix can be shown diagrammatically by closing the contours at the present time: i.e., we connect the present ket to the present bra. It is of course awkward to do this for the total density matrix, as this will simply yield one at any time; however, taking a trace is meaningful for multiple interacting subsystems.

The two subsystems A and B each require a contour, resulting in a double contour. We assume separability of A and B at the initial time: $R(0) = R_A(0) R_B(0)$. Interaction results in energy exchange, which we represent by a cross between the two contours, somewhere between initial and present times, i.e., $0 < t' < t$. In the case we are interested in the evolution of one of the subsystems, say B, the partial trace over A should be taken, which in the diagram can be done by connecting the present bra and ket of system A, see the right diagram in Figure 4. Further details about this Keldysh representation of quantum dynamics can be found in [16].

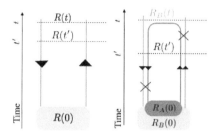

Figure 4. The Keldysh diagram for the time evolution of: (**left**) one world made of one subsystem, (**right**) a world made of two interacting subsystems. Each contour represents a subsystem and the crosses denote interactions.

In order to evaluate the time evolution of the von Neumann and Renyi entropies, we need extended Keldysh contours in multiple parallel worlds (replicas). For this purpose, we consider multiple copies of the Keldysh diagram, one for each world, and add the initial state of the density matrix in each world along the contour at the initial time. The overall trace will get the contours of different worlds connected.

In the second order, one can find:

$$
\begin{aligned}
\frac{d}{dt} S_M^{(B)} &= -\frac{1}{S_M^{(B)}} \mathrm{Tr}_B \left\{ \Delta_B^{(2)} R_B(0)^{M-1} + R_0 \Delta_B^{(2)} R_B(0)^{M-2} + \cdots + R_B(0)^{M-1} \Delta_B^{(2)} \right\} \\
&\quad - \frac{1}{S_M^{(B)}} \mathrm{Tr}_B \left\{ \Delta_B^{(1)} \left[R_B^{(1)} R_B(0)^{M-2} + \cdots + R_B(0)^{M-2} R_B^{(1)} \right] \right. \\
&\quad + R_B(0) \Delta_B^{(1)} \left[R_B^{(1)} R_B(0)^{M-3} + \cdots + R_B(0)^{M-3} R_B^{(1)} \right] + \cdots \\
&\quad \left. + \left[R_B^{(1)} R_B(0)^{M-21} + \cdots + R_B(0)^{M-2} R_B^{(1)} \right] \Delta_B^{(1)} \right\}.
\end{aligned}
\tag{21}
$$

The first line contains terms with second-order interactions taking place in only one world. A typical such diagram for $M = 3$ has been shown in Figure 5.

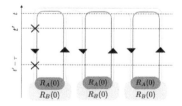

Figure 5. A diagram with two energy exchanges in one replica and no interaction in others.

The rest of the lines other than the first line in Equation (21) denote maximally no more than first-order interaction in a replica. The diagram in Figure 6 shows a typical such term.

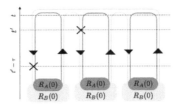

Figure 6. A diagram with two replicas taking over 1st order interactions and the others remain intact.

3.5. Calculating the Diagrams

The main reason why the time evolution of entropy in Equation (21) has been diagrammatically represented is that, due to the multiplicity in time ordering interactions, these extended Keldysh diagrams can help to correctly determine all possible symmetries that may simplify the problem. We need to express all 'single-world' interactions that carry the highest order perturbation as well as all 'cross-world' terms with lower orders of perturbation.

We assume the interaction Hamiltonian does not implicitly depend on time through its parameters; instead, the time dependence is globally assigned in the rotating frame and state evolutions. The explicit formulation of quantum dynamics and keeping track of symmetries between different diagrams have resulted in the following rules for the evaluations of the diagrams:

1. With each system having its own contours in each world, label each separate segment of these contours, according to the state of the associated bra or ket of that segment. The state of the bras and kets change after an interaction, at the initial time and at the final time.
2. Starting from the present time in any of the worlds, say the leftmost world, and encompassing the contours, the following operators or changes must be added along the contour:

    (a) Every interaction on a ket contour will be $(i/\hbar) H_I(t')$ and will be $(-i/\hbar) H_I(t')$ on a bra contour.
    (b) After passing an interaction, the states must change. The new states remain the same until a new interaction is encountered, or if the initial time or the final time is reached.
    (c) A contour arriving at the initial time will capture the initial density matrix in the interaction picture R_0.

3. In general, the result should be integrated over the individual interaction times, i.e., $\int_0^\infty \int_0^\infty dt_1 dt_2$, subject to time order between them. This can be simplified for a small quantum system coupled to a large reservoir kept at a fixed temperature. The reason being that the correlation function of absorption and decay of particles only depends on the time difference between the two interactions [36]. In this case, the double integral over dt_1 and dt_2 can be be simplified to only contain a single integral over the time difference between the two interactions, i.e., $\int_0^\infty d\tau$.

3.6. Quantum Entropy Production

Let us consider that two large heat reservoirs A and B, each one containing many degrees of freedom and kept at a temperature, are coupled to one another via only a few numbers of shared degrees of freedom. The Hamiltonian can be written as $H = H_A + H_B + H_{AB}$ with H_{AB} representing the coupled degrees of freedom.

In order to compute the flow of a quantity between A and B, that quantity should be conserved in the combined system $A + B$. As we discussed in the first section of this paper, Renyi entropy is a conserved quantity in a closed system, therefore $d \ln S_M^{(A+B)}/dt = 0$. However, one should notice that there is a difference between the conservation of physical quantities such as energy and the conservation of entropy. Because physical quantities linearly depend on the density matrix, when it is conserved for a closed system, internally it can flow from a subsystem to another one such that its production in a subsystem is exactly equal to the negative sign of its removal from the other subsystem.

However, entropy is not so. In fact, due to nonlinear dependence of entropy on the density matrix, when it is conserved for a bipartite closed system, it is not equally added and subtracted from the subsystem due to the non-equality in Equation (9).

Below, we will present some example systems with rather general Hamiltonians and, using the diagram rules, we evaluated all entropy production diagrams.

3.6.1. Example 1: Entropy in a Two-Level Quantum Heat Engine

In Ref. [15], we used the extended Keldysh technique and evaluated entropy flow for the simplest quantum heat engine in which a two-level system couples two heat baths kept at different temperatures, see Figure 7. After taking all physical and informational correlations into account, we found that the exact evaluation in the second order is much different from what physical correlations predict. Here, we reproduce the exact result by giving a pedagogical use of the diagram evaluation described above.

Let us consider two heat baths that are kept at different temperatures weakly interact by exchanging the quantum energy ω_0. Such a quantum system can be thought of as a two-level system that couples the two heat baths through shared excitations and de-excitations. The Hilbert space of the two-level system contains the states $|0\rangle$ and $|1\rangle$. The free Hamiltonian contains heat bath energy levels $E_\alpha^{(A)}$s and $E_\beta^{(B)}$s and quantum system energies E_n with $n = 0, 1$, i.e., $H_0 = \sum_\alpha E_\alpha^{(A)} |\alpha\rangle\langle\alpha| + \sum_\beta E_\beta^{(B)} |\beta\rangle\langle\beta| + \sum_{n=0,1} E_n |n\rangle\langle n|$.

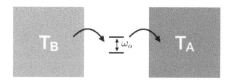

Figure 7. A two-level system quantum heat bath.

We assume the so-called 'transversal' interaction is taken into account between A/B and the two-level system q. This means that they interact via exchanging the quantum of energy ω_0. Of course, we can generalize the discussion to longitudinal interactions in which no energy is exchanged; however, since such interactions are not of immediate interest for heat transfer in quantum heat devices. we ignore them.

This interaction we assume for the heat bath has the following general form: $H_{int} = \sum_{n,m=0,1} |n\rangle\langle m| \left[\hat{X}_{nm}^{(A)}(\omega_0) + \hat{X}_{nm}^{(B)}(\omega_0)\right]$ subject to $E_m \neq E_n$ and $\hat{X}_{nm}$ representing energy absorption/decay in heat baths. The summation in H_{int} can be generalized to an arbitrary number of heat baths interacting at shared degrees of freedom.

Moreover, the entire system including the two-level system is externally driven. The classical heat baths are naturally not influenced effectively by the driving field; however, the driving can pump in and out energy to the two-level system by the following Hamiltonian $H_{dr} = \Omega \cos(\omega_{dr} t) (|0\rangle\langle 1| + |1\rangle\langle 0|)$.

For simplicity, we take the Hamiltonian into the rotating frame that makes excitation/relaxation with the frequency ω_{dr}. In this frame, the excited and ground states are transformed as follows: $|1\rangle_R = \exp(i\omega_{dr} t) |1\rangle$ and $|0\rangle_R = |0\rangle$. This will introduce the unitary transformation $U_R = \exp(i\omega_{dr} t |1\rangle\langle 1|)$ on the Hamiltonian, i.e., $H_R = U_R H U_R^\dagger + i (\partial U_R / \partial t) U_R^\dagger$. A few lines of simplification will result in the following Hamiltonian in the rotating frame:

$$H_R \equiv H_0 + V_{qA} + V_{qB} + V_{AB} + V_{dr},$$

$$H_0 = E_0 |0\rangle\langle 0| + (E_1 - \omega_{dr}) |1\rangle\langle 1| + \sum_\alpha E_\alpha^{(A)} |\alpha\rangle\langle\alpha| + \sum_\alpha E_\alpha^{(B)} |\alpha\rangle\langle\alpha|,$$

$$V_{qA} = |0\rangle\langle 1| \hat{X}_{01}^{(A)}(t) e^{i\omega_{dr}t} + |1\rangle\langle 0| \hat{X}_{10}^{(A)}(t) e^{-i\omega_{dr}t} \equiv \sum_{n,m=0,1(n\neq m)} |n\rangle\langle m| \hat{X}_{nm}^{(A)}(t) e^{i\omega_{dr}\eta_{nm}t}, \quad (22)$$

$$V_{qB} = |0\rangle\langle 1| \hat{X}_{01}^{(B)}(t) e^{i\omega_{dr}t} + |1\rangle\langle 0| \hat{X}_{10}^{(B)}(t) e^{-i\omega_{dr}t} \equiv \sum_{n,m=0,1(n\neq m)} |n\rangle\langle m| \hat{X}_{nm}^{(B)}(t) e^{i\omega_{dr}\eta_{nm}t},$$

$$V_{AB} = 0, \quad V_{dr} = \frac{\Omega}{2} (|0\rangle\langle 1| + |1\rangle\langle 0|),$$

with $\eta_{01} = -\eta_{10} = 1$ and $\eta_{00} = \eta_{11} = 0$. Given the fact that there is no direct exchange of energy between A and B, the density matrix can be represented as $R = R_{qA} \otimes R_B + R_A \otimes R_{qB}$ in an interaction picture, thus determining entropy flow in the heat bath B will depend on the quantum system and the heat bath B, although indirectly the heat bath A will influence the quantum system. In general, $d(R_B)^M/dt = Tr_q \left\{ d(R_{qB})^M/dt \right\}$. Let us recall that this quantity determines the flow of von Neumann entropy and, using Equation (10), it can be simplified to $dS^{(B)}/dt = \lim_{M\to 1} d\left(Tr_B Tr_q \left\{ (dR_{qB}/dt)(R_{qB})^{M-1} + \cdots + (R_{qB})^{M-1} (dR_{qB}/dt) \right\} \right) /dM$. Each term in the sum is evaluated in the interaction picture using $dR/dt = (-i)[V,R]$. One can show that the external driving will cause the density matrix to evolve as $dR_{nm}/dt]_{dr} = (i\Omega/2)(R_{n0}\delta_{m1} + R_{n1}\delta_{m0} - \delta_{n0}R_{1m} - \delta_{n1}R_{0m})$.

The interaction Hamiltonian evolves quantum states and below we evaluate the entropy flow in the $M = 3$ example to the second order perturbation theory. As discussed above, there are in general two types of diagrams in the second order: (1) 'self-interacting' diagrams with second order interaction taking place in one replica, and (2) cross-world-interacting terms in which two different replicas take on each 1st order interaction. The self-interacting diagrams for the two-level system are listed in Figure 8.

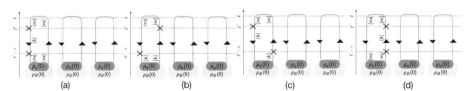

Figure 8. Self-interacting diagrams for interaction between a quantum system and a heat bath.

These diagrams correspond to the following flows, respectively:

$(a):$
$$\frac{(-1)\int_0^\infty d\tau Tr_B \left\{ \sum_{m,k=0,1(m\neq k)} \hat{X}_{mk}^{(B)}(t') \hat{X}_{km}^{(B)}(t'-\tau) \hat{R}_B \hat{R}_{mm} e^{-i\omega_{dr}\eta_{km}\tau} e^{i\omega_{dr}(\eta_{mk}+\eta_{km})t'} \hat{R}_B^2 \right\}}{Tr_B\left(\hat{R}_B^3\right)},$$

$(b):$
$$\frac{(+1)\int_0^\infty d\tau Tr_B \left\{ \sum_{m,k=0,1(m\neq k)} \hat{X}_{mk}^{(B)}(t'-\tau) \hat{R}_B \hat{R}_{kk} \hat{X}_{km}^{(B)}(t') e^{-i\omega_{dr}\eta_{mk}\tau} e^{i\omega_{dr}(\eta_{mk}+\eta_{km})t'} \hat{R}_B^2 \right\}}{Tr_B\left(\hat{R}_B^3\right)},$$

$(c):$
$$\frac{(+1)\int_0^\infty d\tau Tr_B \left\{ \sum_{m,k=0,1(m\neq k)} \hat{X}_{mk}^{(B)}(t') \hat{R}_B \hat{R}_{kk} \hat{X}_{km}^{(B)}(t'-\tau) e^{-i\omega_{dr}\eta_{km}\tau} e^{i\omega_{dr}(\eta_{mk}+\eta_{km})t'} \hat{R}_B^2 \right\}}{Tr_B\left(\hat{R}_B^3\right)},$$

$(d):$
$$\frac{(-1)\int_0^\infty d\tau Tr_B \left\{ \sum_{m,k=0,1(m\neq k)} \hat{R}_B \hat{R}_{mm} \hat{X}_{mk}^{(B)}(t'-\tau) \hat{X}_{km}^{(B)}(t') e^{-i\omega_{dr}\eta_{mk}\tau} e^{i\omega_{dr}(\eta_{mk}+\eta_{km})t'} \hat{R}_B^2 \right\}}{Tr_B\left(\hat{R}_B^3\right)}.$$

In all these terms, there is a time dependent factor $e^{i\omega_{dr}(\eta_{mk}+\eta_{km})t'}$ which is identical to 1 because we always have the following relation valid: $\eta_{mk} = -\eta_{km}$. We assume that heat baths are large and, at equilibrium, therefore the correlation function is the same at all times t' and only depends on the time difference τ between the creation and annihilation of a photon. In the heat bath B, the equilibrium correlation is defined as $S^{(B)}_{mn,pq}(\tau) \equiv \mathrm{Tr}_B \left(\hat{X}^{(B)}_{mn}(0) \hat{X}^{(B)}_{pq}(\tau) R_B \right)$. The Fourier transformation of the correlation defines the following frequency-dependent correlation: $S^{(B)}_{mn,pq}(\omega) = \int_{-\infty}^{\infty} d\tau \mathrm{Tr}_B \left(\hat{X}^{(B)}_{mn}(0) \hat{X}^{(B)}_{pq}(\tau) R_B \right) \exp(i\omega\tau)$. Therefore, in the case of $M = 1$ (i.e., the absence of the last term R_B^2), the diagrams a–d can be rewritten in terms of $S^{(B)}_{mn,pq}(\omega)$. For example, the diagram (a) for the case of $M = 1$ can be simplified to $-\sum_{m,k=0,1(m\neq k)} \hat{R}_{mm} \int_0^{\infty} d\tau \mathrm{Tr}_B \left\{ \hat{X}^{(B)}_{mk}(0) \hat{X}^{(B)}_{km}(\tau) \hat{R}_B e^{-i\omega_{dr}\eta_{km}\tau} \right\}$ in which the integral is half of the domain in Fourier transformation and therefore it can be proved to simplify to $-\sum_{m,k=0,1(m\neq k)} \hat{R}_{mm} \left[(1/2) S^{(B)}_{mk,km}(\omega_{dr}\eta_{mk}) + i\Pi_{mk,km}(\omega_{dr}\eta_{mk}) \right]$ with $\Pi_{mn,pq} \equiv (i/2\pi) \int dv S^{(B)}_{mn,pq}(v) / (\omega - v)$. What is left to be determined is the frequency-dependent correlation function $S^{(B)}_{mn,pq}(\omega)$, which turns out to become completely characterized by the set of reduced frequency-dependent susceptibilities defined as $\tilde{\chi}^{(B)}_{mn,pq}(\omega) \equiv \left(\chi^{(B)}_{mn,pq}(\omega) - \chi^{(B)}_{pq,mn}(-\omega) \right)/i$, with the dynamical susceptibility in the environment being $\chi^{(B)}_{mn,pq}(\omega) \equiv (-i) \int_{-\infty}^0 \mathrm{Tr}_B \left\{ \left[\hat{X}^{(B)}_{mn}(\tau), \hat{X}^{(B)}_{pq}(0) \right] R_B \right\} \exp(-i\omega\tau)$. The fluctuation–dissipation theorem provides a link between the equilibrium correlation and the reduced dynamical susceptibility in the classical thermal bath B at temperature T_B. This relation is usually called the Kubo–Martin–Scwinger (KMS) relation: $S^{(B)}_{mn,pq}(\omega) = n_B(\omega/T_B) \tilde{\chi}^{(B)}_{mn,pq}(\omega)$ with $n_B(\omega/T_B) = 1/(\exp(\omega T_B) - 1)$ being the Bose distribution and k_B the Boltzmann constant.

▶ **Generalized KMS**

In the presence of replicas, similarly, the generalized correlations are defined. For the case in which there are M replicas in total and between creation and annihilations there are N replicas with $0 \leq N \leq M$, the generalized correlation function is defined as

$$S^{N,M(B)}_{mn,pq}(\tau) \equiv \frac{\mathrm{Tr}_B \left(\hat{X}^{(B)}_{mn}(0) \hat{R}_B^N \hat{X}^{(B)}_{pq}(\tau) \hat{R}_B^{M-N} \right)}{\mathrm{Tr}_B \left(\hat{R}_B^M \right)}. \tag{23}$$

Similarly, one can show that

$$\frac{\int_0^{\infty} d\tau \mathrm{Tr}_B \left\{ \hat{X}^{(B)}_{mn}(0) \hat{R}_B^N \hat{X}^{(B)}_{pq}(\tau) \hat{R}_B^{M-N} e^{i\omega\tau} \right\}}{\mathrm{Tr}_B \left(\hat{R}_B^M \right)} = \frac{S^{N,M(B)}_{mn,pq}(\omega)}{2} + i\Pi^{N,M(B)}_{mn,pq}(\omega), \tag{24}$$

with the definition $\Pi^{N,M(B)}_{mn,pq}(\omega) \equiv (i/2\pi) \int dv S^{N,M(B)}_{mn,pq}(v) / (\omega - v)$. One can also check from definitions that, for any heat bath, the following identities: $S^{N,M}_{mn,pq}(-\omega) = S^{M-N,M}_{pq,mn}(\omega)$, $\Pi^{N,M}_{mn,pq}(-\omega) = -\Pi^{M-N,M}_{pq,mn}(\omega)$, and $\tilde{\chi}_{mn,pq}(-\omega) = -\tilde{\chi}_{pq,mn}(\omega)$.

Fourier transformation of this generalized correlation will define the frequency-dependent generalized correlation and, following the same mathematics as above, one can show at equilibrium thermal bath of temperature T_B that all correlation functions can be determined through a generalized KMS relation:

$$S^{N,M(B)}_{mn,pq}(\omega) = n_B \left(\frac{\omega}{T_B} \right) \tilde{\chi}^{(B)}_{mn,pq}(\omega) e^{N\frac{\omega}{k_B T_B}}. \tag{25}$$

Further details can be found in [34]. ◀

Entropy **2019**, *21*, 854

Using these definitions as well as Equation (25), the sum of diagrams (a)–(d) in Figure 8 can be further simplified to

$$
\sum_{m,k=0,1(m\neq k)} \hat{R}_{mm} \left\{ -\left(\frac{1}{2}S_{km,mk}^{3,3\,(B)}\left(\omega_{dr}\eta_{mk}\right) + i\Pi_{km,mk}^{3,3\,(B)}\left(\omega_{dr}\eta_{mk}\right)\right) \right.
$$
$$
\left. -\left(\frac{1}{2}S_{mk,km}^{0,3\,(B)}\left(\omega_{dr}\eta_{km}\right) + i\Pi_{mk,km}^{0,3\,(B)}\left(\omega_{dr}\eta_{km}\right)\right) \right\},
$$
$$
\sum_{m,k=0,1(m\neq k)} \hat{R}_{kk} \left\{ +\left(\frac{1}{2}S_{mk,km}^{1,3\,(B)}\left(\omega_{dr}\eta_{km}\right) + i\Pi_{mk,km}^{1,3\,(B)}\left(\omega_{dr}\eta_{km}\right)\right) \right.
$$
$$
\left. +\left(\frac{1}{2}S_{km,mk}^{2,3\,(B)}\left(\omega_{dr}\eta_{mk}\right) + i\Pi_{km,mk}^{2,3\,(B)}\left(\omega_{dr}\eta_{mk}\right)\right) \right\},
$$
$$
= \sum_{m,k=0,1(m\neq k)} -S_{mk,km}^{0,3\,(B)}\left(\omega_{dr}\eta_{km}\right)\hat{R}_{mm} + S_{mk,km}^{1,3\,(B)}\left(\omega_{dr}\eta_{km}\right)\hat{R}_{kk}. \tag{26}
$$

In total, there are M number of terms similar to the last line in Equation (26) associated with similar diagrams at M worlds. It is important to notice that these self-replica correlated terms are determined in fact only by physical correlations and they make already known results for the flow of von Neumann entropy in the heat bath [37]. To see this more in more detail, one can expand the summation and use the KMS relation and its generalized version in Equation (25). After generalizing the result for M replicas, taking derivative with respect to M and analytically continuing the result to $M \to 1$, the incoherent part of flow in von Neumann entropy is

$$
\left.\frac{dS^{(B)}}{dt}\right|_{\text{incoherent}} = -\frac{1}{T_B}\left(\Gamma_\uparrow^{(B)}p_0 - \Gamma_\downarrow^{(B)}p_1\right), \tag{27}
$$

with $\Gamma_\uparrow^{(B)} \equiv \tilde{\chi}\left(n_B\left(\omega_{dr}/T_B\right)+1\right)$ and $\Gamma_\downarrow^{(B)} \equiv \tilde{\chi}n_B\left(\omega_{dr}/T_B\right)$, $\tilde{\chi} \equiv \tilde{\chi}_{10,01}$, and $p_n \equiv R_{nn}$. These are only self-interacting replicas, which are incomplete as they ignore the following diagrams.

The new diagrams are the cross-world interactions. As discussed previously, cross-world diagrams cannot transfer physical quantities as they rely on the fact that entropy depends nonlinearly on the density matrix and therefore it is not a physical observable quantity. Some of these types of diagrams are shown in Figure 9—for the case that one interaction takes place in the leftmost replica and the second interaction in the middle replica, thus leaving the third replica intact.

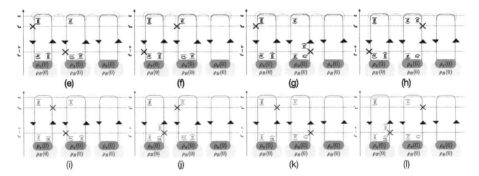

Figure 9. Cross-replica interacting diagrams for a quantum system and a heat bath.

$$(e): \quad -\int_0^\infty d\tau \text{Tr}_B \left\{ \sum_{m,n,k,l} \hat{X}_{mk}^{(B)}(t') \hat{R}_B \hat{R}_{mk} \hat{X}_{nl}^{(B)}(t'-\tau) \hat{R}_B \hat{R}_{nl} e^{-i\omega_{dr}\eta_{nl}\tau} \delta_{E_{nl},E_{km}} \hat{R}_B \right\} / \text{Tr}_B\left(\hat{R}_B^3\right),$$

$$(f): \quad -\int_0^\infty d\tau \text{Tr}_B \left\{ \sum_{m,n,k,l} \hat{X}_{mk}^{(B)}(t'-\tau) \hat{R}_B \hat{R}_{mk} \hat{X}_{nl}^{(B)}(t') \hat{R}_B \hat{R}_{nl} e^{-i\omega_{dr}\eta_{mk}\tau} \delta_{E_{mk},E_{ln}} \hat{R}_B \right\} / \text{Tr}_B\left(\hat{R}_B^3\right),$$

$$(g): \quad \int_0^\infty d\tau \text{Tr}_B \left\{ \sum_{m,n,k,l} \hat{X}_{mk}^{(B)}(t') \hat{R}_B \hat{R}_{mk} \hat{R}_B \hat{R}_{ln} \hat{X}_{ln}^{(B)}(t'-\tau) e^{-i\omega_{dr}\eta_{ln}\tau} \delta_{E_{ln},E_{km}} \hat{R}_B \right\} / \text{Tr}_B\left(\hat{R}_B^3\right),$$

$$(h): \quad \int_0^\infty d\tau \text{Tr}_B \left\{ \sum_{m,n,k,l} \hat{X}_{mk}^{(B)}(t'-\tau) \hat{R}_B \hat{R}_{mk} \hat{R}_B \hat{R}_{ln} \hat{X}_{ln}^{(B)}(t') e^{-i\omega_{dr}\eta_{mk}\tau} \delta_{E_{mk},E_{nl}} \hat{R}_B \right\} / \text{Tr}_B\left(\hat{R}_B^3\right),$$

$$(i): \quad \int_0^\infty d\tau \text{Tr}_B \left\{ \sum_{m,n,k,l} \hat{R}_B \hat{R}_{km} \hat{X}_{km}^{(B)}(t') \hat{X}_{nl}^{(B)}(t'-\tau) \hat{R}_B \hat{R}_{nl} e^{-i\omega_{dr}\eta_{nl}\tau} \delta_{E_{nl},E_{mk}} \hat{R}_B \right\} / \text{Tr}_B\left(\hat{R}_B^3\right),$$

$$(j): \quad \int_0^\infty d\tau \text{Tr}_B \left\{ \sum_{m,n,k,l} \hat{R}_B \hat{R}_{km} \hat{X}_{km}^{(B)}(t'-\tau) \hat{X}_{nl}^{(B)}(t') \hat{R}_B \hat{R}_{nl} e^{-i\omega_{dr}\eta_{km}\tau} \delta_{E_{km},E_{ln}} \hat{R}_B \right\} / \text{Tr}_B\left(\hat{R}_B^3\right),$$

$$(k): \quad -\int_0^\infty d\tau \text{Tr}_B \left\{ \sum_{m,n,k,l} \hat{R}_B \hat{R}_{km} \hat{X}_{km}^{(B)}(t') \hat{R}_B \hat{R}_{ln} \hat{X}_{nl}^{(B)}(t'-\tau) e^{-i\omega_{dr}\eta_{ln}\tau} \delta_{E_{ln},E_{mk}} \hat{R}_B \right\} / \text{Tr}_B\left(\hat{R}_B^3\right),$$

$$(l): \quad -\int_0^\infty d\tau \text{Tr}_B \left\{ \sum_{m,n,k,l} \hat{R}_B \hat{R}_{km} \hat{X}_{km}^{(B)}(t'-\tau) \hat{R}_B \hat{R}_{ln} \hat{X}_{ln}^{(B)}(t') e^{-i\omega_{dr}\eta_{km}\tau} \delta_{E_{km},E_{nl}} \hat{R}_B \right\} / \text{Tr}_B\left(\hat{R}_B^3\right),$$

where we used the following identity $e^{i\omega_{dr}(\eta_{mn}+\eta_{pq})t'} = \delta_{E_{mn},E_{qp}}$.

One can evaluate all diagrams associated with a general number of replicas using the above example. After carefully analyzing all diagrams and proper simplifications—see [34]—the flow of Renyi entropy dS_M/dt in the heat bath B can be found, and consequently the so-called coherent part of entanglement (von Neumann) entropy can be found as follows:

$$\left.\frac{dS^{(B)}}{dt}\right|_{\text{coherent}} = -\frac{\Gamma_\downarrow^{(B)} - \Gamma_\uparrow^{(B)}}{T_B} |R_{01}|^2. \tag{28}$$

This is the new part of the entropy flow that comes from the generalized KMS correlations. We call this part the coherent part because it is nonzero for degenerate states or equivalently a two-level system driven by their detuning frequency.

Therefore, the entanglement entropy flow is naturally separated into two parts and therefore it is equal to the sum between the two parts:

$$\begin{aligned} \frac{dS^{(B)}}{dt} &= \left.\frac{dS^{(B)}}{dt}\right|_{\text{incoherent}} + \left.\frac{dS^{(B)}}{dt}\right|_{\text{coherent}}, \\ &= -\frac{1}{T_B}\left(\Gamma_\uparrow p_0 - \Gamma_\downarrow p_1\right) - \frac{\Gamma_\downarrow - \Gamma_\uparrow}{T_B}|R_{01}|^2, \end{aligned} \tag{29}$$

in which the first term on the second line is what in textbooks has so far been mistakenly taken as total entropy flow.

As we can see, Equation (29) is not directly related to energy flow—which here corresponds to the incoherent part instead of a finite flow that depends on the quantum coherence $(R_{01})^2$.

Consider that the two-level system with energy difference ω_o is driven at the same frequency, i.e., $H = \Omega \cos(\omega_o t)$ and weakly coupled to two heat reservoirs at temperatures T_A and T_B.

From Equation (1) of Ref. [34], one can find the following time evolution equations for the density matrix and setting them to zero determines the stationary solutions:

$$\frac{dR_{11}}{dt} = -\frac{i\Omega}{2}(R_{01} - R_{10}) - \Gamma_{\downarrow}R_{11} + \Gamma_{\uparrow}R_{00} = 0,$$

$$\frac{dR_{01}}{dt} = -\frac{i\Omega}{2}(R_{11} - R_{00}) - \frac{1}{2}(\Gamma_{\downarrow} + \Gamma_{\uparrow})R_{01} = 0, \quad R_{00} + R_{11} = 1,$$

which finds the stationary ground state population $R_{00} = (\Gamma_{\downarrow}(\Gamma_{\downarrow} + \Gamma_{\uparrow}) + \Omega^2)/((\Gamma_{\downarrow} + \Gamma_{\uparrow})^2 + 2\Omega^2)$ and the stationary off-diagonal density matrix element $R_{10} = -i\Omega(1 - 2R_{00})/(\Gamma_{\downarrow} + \Gamma_{\uparrow})$, with $\Gamma_{\downarrow} \equiv \Gamma_{\downarrow}^{(A)} + \Gamma_{\downarrow}^{(B)}$ and $\Gamma_{\uparrow} \equiv \Gamma_{\uparrow}^{(A)} + \Gamma_{\uparrow}^{(B)}$. By considering that B is a probe environment with zero temperature, substituting all solutions in Equation (28), the incoherent and coherent parts of entropy flow in the probe environment have been plotted in Figure 10 for different driving amplitudes and ω_0/T_A.

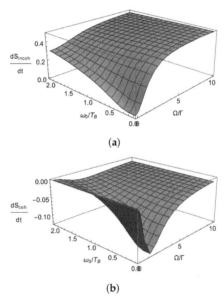

(a)

(b)

Figure 10. Entropy production in a probe bath that is kept at zero temperature and is coupled to a two-level system depicted in Figure 7. The entropy is the sum of two parts: the incoherent and the coherent parts. (**a**) the incoherent part of entropy is nothing new and can be determined by standard correlations. It is positive by the convention that entropy enters from a higher temperature bath (via the two-level system); (**b**) the coherent part of entropy is a previously unknown part as it comes from the informational correlations between different replicas. This part depends quadratically on the off diagonal density. Quite nontrivially, this part of entropy is negative and summing it with the incoherent part will result in a positive flow yet with much smaller magnitude for entropy at small driving amplitudes.

3.6.2. Example 2: Entropy in a Four-Level Quantum Photovoltaic Cell

Scovil and Schulz–DuBois first introduced a model of a quantum heat engine (SSDB heat engine) in which a single three-level atom, consisting of a ground and two excited states, is in contact with two heat baths [38,39]. A large enough difference between the heat bath temperatures can create population inversion between the two excited states and a coherent light output. One hot photon is absorbed and one cold photon is emitted; therefore, a laser photon is produced. The SSDB heat engine model gives a clear demonstration of the quantum thermodynamics. However, we notice that some detailed properties of this lasing heat engine, e.g., the threshold behavior and the statistics of the output light, are still not well studied. There are a number of applications for the model, such as light-harvesting biocells, photovoltaic cells, etc.

Since then, the model has been modified to describe other systems such as light-harvesting biocells, photovoltaic cells, etc.

Recently, in Ref. [40], one of us studied the entropy flow using the replica trick for a 4-level photovoltaic cell with two degenerate ground states and two excited states, see Figure 11. This heat engine was first proposed by Schully in [11] and recently studied in many further details by Schully and others [17,41].

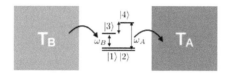

Figure 11. A four-level doubly degenerate photovoltaic cell.

After finding all extended Keldysh diagrams for an arbitrary Renyi degree M, evaluating all self-interacting and cross-interacting diagrams and simplifying the results, the von Neumann entropy flow in heat bath A becomes [40]:

$$\frac{dS}{dt}\bigg|_A = \frac{1}{T_A}\left\{ \gamma p_4 - \omega_A \tilde{\chi}_{42}\bar{n}\left(\frac{\omega_A}{T_A}\right) p_2 - \omega_A \tilde{\chi}_{41}\bar{n}\left(\frac{\omega_A}{T_A}\right) p_1 \right.$$
$$- \tilde{\chi}_{14,42}\left[\omega_A \bar{n}\left(\frac{\omega_A}{T_A}\right) + \omega_A \bar{n}\left(\frac{\omega_A}{T_A}\right) \right] \mathrm{Re} R_{12}$$
$$\left. - \frac{1}{2}\sum_{i=1,2} \omega_A \tilde{\chi}_{14,42}|R_{12}|^2 \right\}. \tag{30}$$

The first two lines can be found using physical correlations. The last line, however, which plays an essential role in the entropy evaluation, can be obtained only through informational correlations. Here, the state probabilities are $p_x \equiv R_{xx}$ with x being $1,2,3,4$ and depending on the characteristics of all heat baths. The dynamical response function is $\tilde{\chi}_{\alpha i} \equiv \tilde{\chi}_{i\alpha,\alpha i}(\omega_{i\alpha})$ with $i = 1,2$ and $\alpha = 3,4$, and $\tilde{\chi}_{1\alpha,\alpha 2} = \sqrt{\tilde{\chi}_{\alpha 1}\tilde{\chi}_{\alpha 2}}$. Moreover, $\gamma \equiv \sum_{i=1,2}[\bar{n}(\omega_A/T_A) + 1]\omega_A \tilde{\chi}_{3i}$.

In order to evaluate the stationary value of the entropy flow in this heat bath, we must solve the quantum master equation for the density matrix time evolution. This can be found in Ref. [40]. The solution is such that the coupling between the environment and the quantum system introduces decoherence in quantum states. Energy exchange between the heat bath and a quantum system introduces a limited coherence time, namely τ_1, for quantum state probabilities. The phase of a quantum state can fluctuate and, depending on environmental noise, the lifetime of quantum state can be limited to τ_2. These two coherence times affect all elements of the density matrix. From solving the quantum Bloch equation, one can see that the only stationary solution in the off-diagonal part is the imaginary part of R_{12} whose real part of exponential decay due to dephasing is: $\mathrm{Im}R_{12} \sim \exp(t/\tau_2)$.

One can substitute the stationary solution of the density matrix in Equation (30) and the flow of entropy in the heat bath changes depending on the dephasing time—see Figure 2a,b in [40]. In fact, increasing the dephasing time will increase the contribution of the coherent part of the entropy flow, i.e., information correlations. This will reduce the total entropy flow in the heat bath, which will equivalently increase the output power in this photovoltaic cell.

3.6.3. Example 3: Entropy in a Quantum Resonator/Cavity Heat Engine

Using a rather different technique—i.e., the correspondence between entropy and statistics of energy transfer that we discuss in the next section—in [15,16], we calculated entropy production for a resonator/cavity coupled two different environments kept at two different temperatures, see Figure 12. One of the two baths is a probe environment at a temperature of zero for which we calculate the flow of entropy.

Knowing how entropy flows as the result of interactions between the resonator, cavity and other parts of the circuit can help to obtain important information about the possibility of leakage or dephasing in the system and ultimately give rise to modifications of quantum circuits [4]. A good understanding of cavities/resonators is beneficial to search for the nature of non-equilibrium quasiparticles in quantum circuits [42,43]. This can help with detecting light particles like muons whose tunnelling in a quantum circuit can signal a sudden jump in the entropy flow [44–46]. Given that entropy flow can be measured by the full counting statistics of energy transfer, see the next section, it is important to keep track of entropy flow in a resonator.

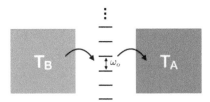

Figure 12. A quantum cavity heat engine.

Again, we use the standard technique that we described above. Let us consider a single harmonic oscillator of frequency ω_0 and Hamiltonian $\hat{H} = \omega_0(\hat{a}^\dagger \hat{a} + 1/2)$, which is coupled to a number of environments at different temperatures with different coupling strengths. We concentrate on a probe environment that is weakly coupled to the oscillator. In addition, the oscillator is driven by an external force at frequency Ω. We calculate the Renyi flow and consequently the von Neumann entropy flow of the probe environment. The coupling Hamiltonian between the harmonic oscillator and the probe reservoir is $\hat{H}(t) = \hat{X}(t)\hat{a}^\dagger(t) + h.c.$, with $\hat{X}$ being the probe reservoir operator. The Fourier transform of the correlator is: $S_{mn}(\omega) = \int \exp(-i\omega t) S_{mn}(t) d\omega/2\pi$. Due to the conservation of energy, the energy exchange occurs either with quantum Ω or with quantum ω_0.

We note that the time dependence of the average of two operators can be written as $\langle \hat{a}^\dagger(t)\hat{a}(t') \rangle = \langle\langle \hat{a}^\dagger \hat{a} \rangle\rangle e^{i\omega_0(t-t')} + \langle \hat{a}(t) \rangle \langle \hat{a}^\dagger(t') \rangle$, where the time dependence of $\langle a(t) \rangle$ is due to the driving force and therefore oscillates at frequency Ω: $\langle a(t) \rangle = \langle a \rangle_+ \exp(i\Omega t) + \langle a \rangle_- \exp(-i\Omega t)$. This corresponds to the fact that the oscillator can oscillate both at its own frequency and at the frequency of external force.

Obtaining the entropy flows from the extended Keldysh correlators is straightforward. The generalized KMS relation in Equation (25) helps to describe the correlators in the thermal bath B in terms of their dynamical susceptibility. The result can be summarized as follows:

$$\frac{dS_M^{(B)}}{dt} = \frac{M\bar{n}(M\omega_0/T_B)\tilde{\chi}}{\bar{n}((M-1)\omega_0/T_B)\bar{n}(\omega_0/T_B)} \left\{ \langle\langle a^\dagger a \rangle\rangle e^{\frac{\omega_0}{T_B}} - \langle\langle aa^\dagger \rangle\rangle \right\},$$

where we defined $T_{\text{resonator}}$ to be the effective temperature of the harmonic oscillator $\langle\langle aa^\dagger \rangle\rangle = \bar{n}(\omega_0/T_{\text{resonator}}) + 1$ and $\langle\langle a^\dagger a \rangle\rangle = \bar{n}(\omega_0/T_{\text{resonator}})$. Taking the derivative with respect to M and analytically continuing the result in the limit of $M \to 1$ will determine the thermodynamic entropy flow:

$$\frac{dS_M^{(B)}}{dt} = \frac{1}{T_B} \left\{ \bar{n}(\omega_0/T_{\text{resonator}}) - \bar{n}(\omega_0/T_B) \right\}. \tag{31}$$

The entropy flow changes sign at the onset temperature $T_{\text{resonator}} = T_B$. Moreover, after the exact evaluation of the incoherent part of the entropy flow, one should notice that it contains some terms proportional to $\langle a \rangle$ and $\langle a\dagger \rangle$. These terms oscillate with the external drive and are nonzero. However, they are all cancelled out by the coherent part of entropy flow such that the overall flow will only depend on the temperatures, and not on the driving force. Therefore, the entropy flow is robust in the sense that it only depends on the temperatures of the probe and harmonic oscillator and is completely insensitive to the external driving force.

The insensitivity of entropy flow to external driving force is interesting and a direct result of including coherent flow of entropy that is absent in semi-classical analysis. The difference can put the coherent entropy flow into an experimental verification.

In the absence of cross-replica correlators, the thermodynamic entropy of a probe environment, coupled to a thermal bath via a resonator, will dramatically depend on the amplitude of the external driving. If no such dependence on the driving amplitude is found, then this is an indication that they are absent; they are in fact eliminated by quantum coherence!

4. Linking Information to Physics: A New Correspondence

As discussed above, the Renyi entropies in quantum physics are considered unphysical, i.e., non-observable quantities, due to their nonlinear dependence on the density matrix. Such quantities cannot be determined from immediate measurements; instead, their quantification seems to be equivalent to determining the density matrix. This requires reinitialization of the density matrix between many successive measurements. Therefore, the Renyi entropy flows between the systems are conserved measures of nonphysical quantities. An interesting and nontrivial question is: Is there any relation between the Renyi entropy flows and the physical flows?

An idea of such a relation was first put forward by Levitov and Klich in [23], where they proposed that entanglement entropy flow in electronic transport can be quantified from the measurement of the full counting statistics (FCS) of charge transfers [22,47–49]. The validity of this relation is restricted to zero temperature and obviously to the systems where interaction occurs by means of charge transfer. Recently, we presented a relation that is similar in spirit [15]. We derived a correspondence for coherent and incoherent second-order diagrams in a general time-dependent situation.

This relation gives an exact correspondence between the informational measure of Renyi entropy flows and physical observables, namely, the full counting statistics of energy transfers [47,50].

We consider reservoir B and quantum system q. We assume that B is infinitely large and is kept in thermal equilibrium at temperature T_B. System q is arbitrary as it may carry several degrees of freedom as well as infinitely many. It does not have to be in thermal equilibrium and is in general subject to time-dependent forces. It is convenient to assume that these forces are periodic with a period of τ; however, the period does not explicitly enter the formulation of our result, which is also valid for aperiodic forces. The only requirement is that the flows of physical quantities have stationary limits. The stationary limits are determined after averaging instant flows over a period and—for aperiodic forces—by averaging over a sufficiently long time interval. In the case of energetic interactions, energy transfer is statistical. The statistics can be described by the generating function of the full counting statistics (FCS), namely 'FCS Keldysh actions'.

Recently, in Ref. [15], we proved that the flow of thermodynamic entropy as well as the flow of Renyi entropy between two heat baths via a quantum system is exactly equivalent to the difference between two FCS Keldysh actions of incoherent and coherent energy transfers. In the limit of long τ and for a typical reservoir B with temperature T_B, the incoherent and coherent FCS Keldysh actions are $f_i(\xi, T_B)$ and $f_c(\xi, T_B)$, with ξ being the counting field of energy transfer. These generating functions can be determined using Keldysh diagrams, see [16]. After their evaluation, one finds the statistical m-th cumulant function C_m by taking the derivative of the generating function in the limit of zero counting function, i.e., $C_m = \lim_{\xi \to 0} \partial^m f / \partial \xi^m$.

In fact, any physical quantity should depend on the cumulants and consequently on a zero counting field. However, informational measures are exceptional. Detailed analysis shows that the flow of Renyi entropy of degree M in the reservoir B at equilibrium temperature T_B is exactly, and unexpectedly, the following: $dS_M(T_B)/dt = M[f_i(\xi^*, T_B/M) - f_c(\xi^*, T/M)]$ with $\xi^* \equiv i(M - 1)/T_B$. Notice that in this correspondence the temperature on the left side is T_B while it is T_B/M

on the right side. In addition, it is important to notice that the entropy is evaluated by using the generating function of full counting statistics at nonzero counting field ζ^*. This relation is valid in the weak-coupling limit where the interaction between the systems can be treated perturbatively.

5. Discussion

Currently, 'time' does not play any essential role in quantum information theory. In this sense, quantum theory is underdeveloped similarly to how quantum physics was underdeveloped before Schrödinger introduced his wave equation. In this review article, we discussed a fascinating extension of the Keldysh formalism that consistently copes with the problem of time for one of the central quantities in quantum information theory: entropy. We characterized the flows of conserved entropies (both Renyi and von Neumann entropies) and illustrated them diagrammatically to introduce new correlators that have been absent so far in the literature.

Given that entropy is not an observable, as it is a nonlinear function of the density matrix, one can use a probe environment to make an indirect measurement of the entropy in light of the new correspondence between entropy and full counting statistics of energy transfer. This can be done equally well for the imaginary and real values of the characteristic parameter. The measurement procedures may be complex, yet they are feasible and physical. The correspondence can have many other advantages. For instance, a complete understanding of entropy flows may help to identify the sources of fidelity loss in quantum communication and may help to develop methods to control or even prevent them.

Author Contributions: Y.V.N. contributed in conceptualization and idea; A.v.S. in editing, and M.H.A. wrote the original draft.

Funding: The authors declare no funding support.

Conflicts of Interest: The authors declare no conflict of interest.

References

1. Polkovnikov, A.; Sengupta, K.; Silva, A.; Vengalatorre, M. Colloquium: Non- equilibrium dynamics of closed interacting quantum systems. *Rev. Mod. Phys.* **2011**, *83*, 863.
2. Santos, L.F.; Polkovnikov, A.; Rigol, M. Entropy of Isolated Quantum Systems after a Quench. *Phys. Rev. Lett.* **2011**, *107*, 040601.
3. Jaeger, G. *Quantum Information*; Springer: Berlin, Germany, 2007.
4. Ansari, M.H. Superconducting qubits beyond the dispersive regime. *Phys. Rev. B* **2019**, *100*, 024509.
5. Lagoudakis, K.G.; McMahon, P.L.; Fischer, K.A.; Puri, S.; Müller, K.; Dalacu, D.; Poole, P.J.; Reimer, M.E.; Zwiller, V.; Yamamoto, Y.; et al. Initialization of a spin qubit in a site-controlled nanowire quantum dot. *New J. Phys.* **2014**, *16*, 023019.
6. Ansari, M.H., Exact quantization of superconducting circuits. *Phys. Rev. B* **2019**, *100*, 024509.
7. Paik, H.; Mezzacapo, A.; Sandberg, M.; McClure, D.T.; Abdo, B.; Córcoles, A.D.; Dial, O.; Bogorin, D.F.; Plourde, B.L.T.; Steffen, M.; et al. Experimental demonstration of a resonator-induced phase gate in a multiqubit circuit-qed system. *Phys. Rev. Lett.* **2016**, *117*, 250502.
8. Ansari, M.H.; Wilhelm, F.K. Noise and microresonance of critical current in Josephson junction induced by Kondo trap states. *Phys. Rev. B* **2011**, *84*, 235102.
9. Pekola, J.P.; Khaymovich, I.M. Thermodynamics in single-electron circuits and superconducting qubits. *Annu. Rev. Condens. Matter Phys.* **2019**, *10*, 193.
10. Uzdin, R.; Levy, A.; Kosloff, R. Equivalence of Quantum Heat Machines, and Quantum-Thermodynamic Signatures. *Phys. Rev. X* **2015**, *5*, 031044.
11. Scully, M.O.; Chapin, K.; Dorfman, K.; Kim, M.; Svidzinsky, A. Quantum heat engine power can be increased by noise-induced coherence. *Proc. Natl. Acad. Sci. USA* **2011**, *108*, 15097–15100.
12. Linden, N.; Popescu, S.; Skrzypczyk, P. How small can thermal machines be? The smallest possible refrigerator. *Phys. Rev. Lett.* **2010**, *105*, 13.
13. Frank, R.L.; Lieb, E.H. Monotonicity of a relative Rényi entropy. *J. Math. Phys.* **2013**, *54*, 122201.

14. Esposito, M.; Esposito, M., Lindenberg, K.; Van den Broeck, C. Entropy production as correlation between system and reservoir. *New J. Phys.* **2010**, *12*, 013013.

15. Ansari, M.H.; Nazarov, Y.V. Exact correspondence between Renyi entropy flows and physical flows. *Phys. Rev. B* **2015**, *91*, 174307.

16. Ansari, M.H.; Nazarov, Y.V. Keldysh formalism for multiple parallel worlds. *J. Exp. Theor. Phys.* **2016**, *122*, 389–401.

17. Li, S.W.; Kim, M.B.; Agarwal, G.S.; Scully, M.O. Quantum statistics of a single-atom Scovil–Schulz-DuBois heat engine. *Phys. Rev. A* **2017**, *96*, 063806.

18. Utsumi, Y. Optimum capacity and full counting statistics of information content and heat quantity in the steady state. *Phys. Rev. B* **2019**, *99*, 115310.

19. Utsumi, Y. Full counting statistics of information content. *Eur. Phys. J. Spec. Top.* **2019**, *227*, 1911.

20. Kubo, R. Statistical-mechanical theory of irreversible process. I. General theory and simple applications to magnetic and conduction problems. *J. Phys. Soc. Jpn.* **1957**, *12*, 570–586.

21. Martin, P.; Schwinger, J. Theory of many-particle systems. I. *Phys. Rev.* **1959**, *115*, 1342.

22. Levitov, L.S.; Lee, H.W.; Lesovik, G.B. Electron counting statistics and coherent states of electric current. *J. Math. Phys.* **1996**, *37*, 4845–4866.

23. Klich, I.; Levitov, L.S. Quantum noise as an entanglement meter. *Phys. Rev. Lett.* **2009**, *102*, 100502.

24. Nazarov, Y.V.; Kindermann, M. Full counting statistics of a general quantum mechanical variable. *Eur. Phys. J. B* **2003**, *35*, 413–420.

25. Ansari, M.H. The Statistical Fingerprints of Quantum Gravity. Ph.D. Thesis, University of Waterloo, Waterloo, ON, Canada, 2008.

26. Ansari, M.H. Spectroscopy of a canonically quan-tized horizon. *Nucl. Phys. B* **2007**, *783*, 179–212.

27. Ansari, M.H. Generic degeneracy and entropy in loop quantum gravity. *Nucl. Phys. B* **2008**, *795*, 635–644.

28. Ansari, M.H. Quantum amplification effect in a horizon fluctuation. *Phys. Rev. D* **2010**, *81*, 104041.

29. Renyi, A. On Measures of Entropy and Information. In Proceedings of the 4th Berkeley Symposium on Mathematics and Statistical Probability, Berkeley, CA, USA, 20 June–30 July 1960; pp. 547–561.

30. Deutsch, D.; Hayden, P. Information flow in entangled quantum systems. *Proc. Soc. Lond. A* **2000**, *456*, 1759–1774.

31. Nazarov, Y.V. Flows of Rényi entropies. *Phys. Rev. B* **2011**, *84*, 205437.

32. Holzhey, C.; Larsen, F.; Wilczek, F. Geometric and renormalized entropy in conformal field theory. *Nucl. Phys. B* **1994**, *424*, 443.

33. Calabrese, P.; Cardy, J. Entanglement entropy and conformal field theory. *J. Phys. A* **2009**, *42*, 504005.

34. Ansari, M.H.; Nazarov, Y.V. Rényi entropy flows from quantum heat engines. *Phys. Rev. B* **2015**, *91*, 104303.

35. Keldysh, L.V. Diagram technique for nonequilibrium processes. *Zh. Eksp. Teor. Fiz.* **1964**, *47*, 1515–1527.

36. Nazarov, Y.V.; Blanter, Y.M. *Quantum Transport: Introduction to Nanoscience*; Cambridge University Press: Cambridge, UK, 2009.

37. Alicki, R.; Kosloff, R. Introduction to Quantum Thermodynamics: History and Prospects. *arXiv* **2018**, arXiv:1801.08314.

38. Scovil, H.E.D.; Schulz–DuBois, E.O. Three-level masers as heat engines. *Phys. Rev. Lett.* **1959**, *2*, 262.

39. Geusic, J.E.; Schulz-DuBios, E.O.; Scovil, H.E.D. Quantum equivalent of the carnot cycle. *Phys. Rev.* **1967**, *156*, 343.

40. Ansari, M.H. Entropy production in a photovoltaic cell. *Phys. Rev. B* **2017**, *95*, 174302.

41. Mitchison, M.T. Quantum thermal absorption machines: Refrigerators, engines and clocks. *arXiv* **2019**, arXiv:1902.02672.

42. Houzet, M.; Serniak, K.; Catelani, G.; Devoret, M.H.; Glazman, L.I. Photon-assisted charge-parity jumps in a superconducting qubit. *arXiv* **2019**, arXiv:1904.06290.

43. Ansari, M.H. Rate of tunneling nonequilibrium quasiparticles in superconducting qubits. *Supercond. Sci. Technol.* **2015**, *28*, 045005.

44. Bal, M.; Ansari, M.H.; Orgiazzi, J.L.; Lutchyn, R.M.; Lupascu, A. Dynamics of parametric fluctuations induced by quasiparticle tunneling in superconducting flux qubits. *Phys. Rev. B* **2015**, *91*, 195434.

45. Ansari, M.H.; Wilhelm, F.K.; Sinha, U.; Sinha, A. The effect of environmental coupling on tunneling of quasiparticles in Josephson junctions. *Supercond. Sci. Technol.* **2013**, *26*, 035209.

46. Jafari-Salim, A.; Eftekharian, A.; Majedi, A.H.; Ansari, M.H. Stimulated quantum phase slips from weak electromagnetic radiations in superconducting nanowires. *AIP Advances* **2016**, *6*, 125013.
47. Kindermann, M.; Pilgram, S. Statistics of heat transfer in mesoscopic circuits. *Phys. Rev. B* **2004**, *69*, 155334.
48. Buttiker, M.; Imry, Y.; Landauer, R.; Pinhas, S. Generalized many-channel conductance formula with application to small rings. *Phys. Rev. B* **1985**, *31*, 6207.
49. Buttiker, M. Four-terminal phase-coherent conductance. *Phys. Rev. Lett.* **1986**, *57*, 1761.
50. Heikkilä, T.T.; Nazarov, Y.V. Statistics of temperature fluctuations in an electron system out of equilibrium. *Phys. Rev. Lett.* **2009**, *102*, 130605.

Article

Power, Efficiency and Fluctuations in a Quantum Point Contact as Steady-State Thermoelectric Heat Engine

Sara Kheradsoud [1], Nastaran Dashti [2], Maciej Misiorny [2], Patrick P. Potts [1], Janine Splettstoesser [2] and Peter Samuelsson [1,*]

[1] Physics Department and NanoLund, Lund University, S-221 00 Lund, Sweden
[2] Department of Microtechnology and Nanoscience (MC2), Chalmers University of Technology, S-412 96 Göteborg, Sweden
* Correspondence: peter.samuelsson@teorfys.lu.se

Received: 11 May 2019; Accepted: 29 July 2019; Published: 8 August 2019

Abstract: The trade-off between large power output, high efficiency and small fluctuations in the operation of heat engines has recently received interest in the context of thermodynamic uncertainty relations (TURs). Here we provide a concrete illustration of this trade-off by theoretically investigating the operation of a quantum point contact (QPC) with an energy-dependent transmission function as a steady-state thermoelectric heat engine. As a starting point, we review and extend previous analysis of the power production and efficiency. Thereafter the power fluctuations and the bound jointly imposed on the power, efficiency, and fluctuations by the TURs are analyzed as additional performance quantifiers. We allow for arbitrary smoothness of the transmission probability of the QPC, which exhibits a close to step-like dependence in energy, and consider both the linear and the non-linear regime of operation. It is found that for a broad range of parameters, the power production reaches nearly its theoretical maximum value, with efficiencies more than half of the Carnot efficiency and at the same time with rather small fluctuations. Moreover, we show that by demanding a non-zero power production, in the linear regime a stronger TUR can be formulated in terms of the thermoelectric figure of merit. Interestingly, this bound holds also in a wide parameter regime beyond linear response for our QPC device.

Keywords: thermoelectricity; heat engines; quantum transport; mesoscopic physics; fluctuations; thermodynamic uncertainty relations

---

1. Introduction

Nanoscale thermodynamics has attracted considerable attention during the last three decades. Key motivations are the prospect of on-chip cooling and power production as well as an enhanced thermoelectric performance arising from unique properties of nanoscale systems, such as quantum size effects and strongly energy-dependent transport properties [1–9]. Among various nanoscale systems, quantum point contacts (QPC) [10] are arguably the simplest devices which show a thermoelectric response [11]. A requirement of such a response is an energy-dependent transmission probability [12,13], which breaks the electron-hole symmetry. Within non-interacting scattering theory, the transmission probability fully determines the thermoelectric response of a two-terminal device. The QPC and similar devices provide a particularly interesting thermoelectric platform as their transmission probability may approximate a step function, maximizing the power generation [14,15]. This feature is in contrast to the case of a quantum dot, where the transmission probability may approximate a Dirac delta distribution, maximizing the efficiency of heat-to-power conversion [16–20].

Most previous studies on the thermoelectric properties of QPCs focused on the linear response regime [11,12,21–24]. In this regime, the optimal performance of thermodynamic devices was extensively investigated, especially the efficiency at maximum power which is limited by the Curzon-Ahlborn efficiency [25–27]. There are however several works considering various aspects of the thermoelectric response in the non-linear regime [14,15,28–37]. This includes a Landauer–Büttiker scattering approach to the weakly non-linear regime [35,37], detailed investigations of the relation between power and efficiency when operating the QPC as a heat engine or refrigerator [14,15,36,37] as well as the full statistics of efficiency fluctuations [28].

Here, we review the thermoelectric effect of a QPC acting as a steady-state thermoelectric heat engine. We focus on the non-linear-response regime and analyze the output power and the efficiency for different parameter regimes, varying the smoothness of the step in the transmission probability of the QPC. In addition to a high efficiency and power production, it is desirable to have as little fluctuations as possible in the output of a heat engine. However, these three quantities, which we will analyze as three independent performance quantifiers, are often restricted by a thermodynamic uncertainty relation (TUR), preventing the design of an efficient and powerful heat engine with little fluctuations [38–44]. In this paper, we use a TUR-related coefficient as an additional combined performance quantifier, accounting for power output, efficiency, and fluctuations together. While TURs have been rigorously proven for time-homogeneous Markov jump processes with local detailed balance [39,41], they are not necessarily fulfilled in systems well described by scattering theory [45]. Nevertheless, we find the TUR to be valid in a temperature- and voltage-biased QPC. We note that recently, it has been shown that a weaker, generalized TUR applies whenever a fluctuation theorem holds [46,47]. Here, we show further constraints on the TUR under the restriction that the thermoelectric element *produces* power, necessary to define a useful performance quantifier. Interestingly, in linear response, this constraint can be related to the figure of merit, ZT.

This paper is structured as follows. In Section 2, we introduce the model of a QPC with smooth energy-dependent transmission, as well as the transport quantities and resulting performance quantifiers of interest. The latter are then analyzed for the QPC with different degrees of smoothness of the transmission function, namely the output power in Section 3, the efficiency in Section 4, the (power) fluctuations in Section 5, and the combined performance quantifier deduced from the TUR in Section 6.

2. Model System and Transport Theory

We consider the two-terminal setup shown in Figure 1, with a single-mode QPC connected to a left (L) and a right (R) electronic reservoir, characterized by electrochemical potentials $\mu_L = \mu_0 - eV_L$ and $\mu_R = \mu_0 - eV_R$, and kept at temperatures $T_L = T_0$ (cold reservoir) and $T_R = T_0 + \Delta T$ (hot reservoir), respectively. Here, V_L and V_R are externally applied voltages, μ_0 denotes the electrochemical potential in the absence of voltage bias, T_0 corresponds to the background temperature and $\Delta T \geq 0$ stands for the temperature difference due to heating of the right reservoir. In the following, we always set μ_0 as the reference energy.

2.1. Quantum Point Contact

We employ the established model for a QPC [12] and describe the energy-dependent transmission probability as

$$D(E) = \frac{1}{1 + \exp\left(\frac{-E+E_0}{\gamma}\right)} . \tag{1}$$

This is a step-like function of the energy E, see Figure 1b, where E_0 and γ denote the position and width in energy of the step, respectively. For a vanishingly small width, $\gamma \to 0$, the transmission probability reduces to a step function, $D(E) \to \theta(E - E_0)$.

In experiments with 2DEGs, the width or smoothness of the QPC barrier γ, typically takes values of the order of 1 meV (corresponding to temperatures of the order of 10 K) [11,32,48,49].

The results presented in this paper are equally valid for different types of conductors, where the transmission function has a (smooth) step-like behavior, such as quantum wires with interfaces or controlled by finger gates. Here, smoothness parameters γ of values down to several μeV are expected (corresponding to temperatures of the order of 10–100 mK) [29].

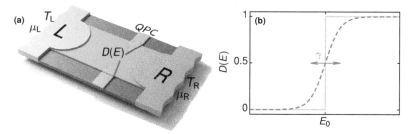

Figure 1. (**a**) Schematic depiction of the system, with a quantum point contact (QPC) connected to two electronic reservoirs, L and R, with electrochemical potentials μ_L and μ_R and temperatures T_L and T_R, respectively. (**b**) Transmission probability $D(E)$ shown as a function of energy, Equation (1), with a step positioned at energy E_0 and energy-smearing width, or smoothness, γ. The solid line shows the limit of vanishing width $\gamma \to 0$.

2.2. Non-Linear-Transport Theory

The transport properties of the system are described by scattering theory [21]. In the non-linear-transport regime, the scattering properties of the QPC become dependent on the applied voltages V_L and V_R of the reservoirs and V_g of the QPC-defining split gate [50] and possibly also on the temperature bias [5,35,51]. Since the details of this dependence will not be of importance for our analysis, we for simplicity consider a basic model with the QPC-potential capacitively coupled with equal strength, $C_L = C_R = C$, to the two terminals L and R. This leads to a modification of the transmission probability as $D(E) \to D(E + e[(V_R + V_L)C + V_g C_g]/[2C + C_g])$, where C_g is the split gate-QPC capacitance. In the following, we absorb the gate potential dependence into the step energy $E_0 + eC_g V_g/(2C + C_g) \to E_0(V_g) \equiv E_0$. This modification of the transmission probability guarantees a gauge-invariant formulation of the problem with observable quantities only dependent on the potential differences $V = V_L - V_R$, $V_L - V_g$ and $V_R - V_g$. We here refrain from including the effect of a large temperature difference in the treatment of the transmission probability $D(E)$, which is not required by fundamental principles such as gauge invariance and which has been little addressed so far, and postpone its study to future work.

For the study of the average currents of interest, namely charge current, I_α, and heat current, J_α, we now consider a symmetric biasing $V_L = -V_R = V/2$. We can then write the average currents that are flowing out of reservoir α as

$$I_\alpha = -\frac{e}{h} \int_{-\infty}^{\infty} dE\, D(E)\left[f_\alpha(E) - f_{\bar{\alpha}}(E)\right], \tag{2}$$

and

$$J_\alpha = \frac{1}{h} \int_{-\infty}^{\infty} dE\, (E + \tau_\alpha eV/2) D(E)\left[f_\alpha(E) - f_{\bar{\alpha}}(E)\right] \tag{3}$$

Here, $\bar{\alpha}$ should be understood as follows: $\bar{L} = R$ and $\bar{R} = L$, whereas $\tau_L = 1$ and $\tau_R = -1$. In Equations (2) and (3), we have introduced the Fermi distribution functions $f_\alpha(E)$,

$$f_\alpha(E) = \left[1 + \exp\left(\frac{E + \tau_\alpha eV/2}{k_B T_\alpha}\right)\right]^{-1} \quad \text{for} \quad \alpha = \text{L,R.} \tag{4}$$

While current conservation ensures $I_L = -I_R \equiv I$, energy conservation results in $J_L = -J_R - IV$.

To analyze the fluctuations in the system we also need the zero-frequency charge-current noise, given by [52]

$$S_I = \frac{e^2}{h} \int_{-\infty}^{\infty} dE \Big\{ D(E) \left[f_L(E)(1 - f_L(E)) + f_R(E)(1 - f_R(E)) \right]$$
$$+ D(E)\left[1 - D(E)\right] \left[f_L(E) - f_R(E) \right]^2 \Big\}. \tag{5}$$

In addition to the study of the noise, it is often convenient to analyze the Fano factor

$$F = \frac{S_I}{|2eI|}, \tag{6}$$

being a measure of how much the noise deviates from the one of Poissonian statistics (for which $F = 1$).

2.3. Thermodynamic Laws and Performance Quantifiers

The laws of thermodynamics set very general constraints on the quantities introduced above and on the performance quantifiers, which we are going to study in this paper. We describe these quantities within scattering theory, known to correctly reproduce the laws of thermodynamics [5]. The first law of thermodynamics guarantees energy conservation and can be written as

$$J_L + J_R = P. \tag{7}$$

Here, we have introduced the electrical power produced,

$$P = -VI, \tag{8}$$

where $-VI > 0$ if the current flows against the applied bias. Please note that throughout the work, we limit our analysis of performance quantifiers to the relevant regime of positive power production. The second law of thermodynamics states that the entropy production σ is non-negative. In our two-terminal geometry, it can be written as

$$\sigma = -\frac{J_L}{T_L} - \frac{J_R}{T_R} \geq 0. \tag{9}$$

This expression determines the direction of energy flows through the system. It equals zero in case that a process is *reversible*.

To determine the performance of the QPC as a heat engine, we now consider three independent quantities and combine them with each other. The first performance quantifier is given by the electrical power, Equation (8), which following the first law, Equation (7), is fully produced from heat.

The second performance quantifier we consider is given by the efficiency

$$\eta = \frac{P}{J_R} = -\frac{VI}{J_R}, \tag{10}$$

where J_R is the heat current that flows out of the hot reservoir. As long as power is positive, the efficiency is bounded by the second law of thermodynamics, Equation (9),

$$0 \leq \eta \leq \eta_C \quad \text{with} \quad \eta_C = 1 - \frac{T_L}{T_R} = \frac{\Delta T}{T_0 + \Delta T}, \tag{11}$$

where η_C denotes the Carnot efficiency. The dissipation arising from an inefficient heat to work conversion is quantified by the entropy, which thereby relates efficiency and produced electrical power to each other

$$\sigma = \frac{P}{T_0} \cdot \frac{\eta_C - \eta}{\eta}. \tag{12}$$

It is desirable to have a thermoelectric heat engine which not only produces large power, at high efficiency, but also minimizes fluctuations. The third independent performance quantifier is therefore provided by the low-frequency power fluctuations

$$S_P = V^2 S_I. \tag{13}$$

Interestingly, a trade-off between these quantities in the form of a TUR usually exists, as discussed in more detail in Section 6. This trade-off is typically written in the form of [38,39]

$$Q_{TUR} \equiv \frac{I^2}{S_I} \cdot \frac{k_B}{\sigma} \le \frac{1}{2}, \tag{14}$$

where we have introduced the coefficient Q_{TUR}. While this inequality is not always fulfilled for systems well described by scattering theory, see e.g., the discussion in [45,53], we find it to be respected in our system for all parameter values. Importantly this coefficient can be cast into the form [42]

$$Q_{TUR} = P \frac{\eta}{\eta_c - \eta} \cdot \frac{k_B T_0}{S_P}, \tag{15}$$

where we used Equations (12) and (13). Thus, under the constraint of positive power production and efficiency, we identify Q_{TUR} as a convenient combined performance quantifier, accounting for power production, efficiency and power fluctuations together, where $1/2$ sets the optimum value.

2.4. Linear-Response Regime

To compare with the much more studied linear-transport regime, we here present the relevant transport properties in this limit. Specifically, with small applied voltage and thermal bias, we can write the heat and charge current in the convenient matrix form [21],

$$\begin{pmatrix} I \\ J \end{pmatrix} = \begin{pmatrix} G & L \\ M & K \end{pmatrix} \begin{pmatrix} V \\ \Delta T \end{pmatrix}, \tag{16}$$

where (only due to linear response!) $J = J_L = -J_R$, and the matrix elements are defined as

$$G = \frac{e^2}{h} \mathcal{I}_0, \quad L = -\frac{M}{T_0} = \frac{e}{h} k_B \mathcal{I}_1, \quad K = -\frac{1}{h} (k_B^2 T_0) \mathcal{I}_2, \tag{17}$$

with

$$\mathcal{I}_n = \int_{-\infty}^{\infty} dE \, D(E) \left(\frac{E}{k_B T_0} \right)^n \left(-\frac{\partial f_0(E)}{\partial E} \right). \tag{18}$$

Here $f_0(E)$ is the Fermi-Dirac distribution in Equation (4) with $V_\alpha = 0$ and $T_\alpha = T_0$. In the same limit, the charge-current noise reduces to the equilibrium noise, given by $S_I = 2k_B T_0 G$, in accordance with the fluctuation-dissipation relation.

Another performance quantifier, which is often used in the linear response, is the figure of merit ZT. It is given by [5]

$$ZT = \frac{L^2}{GK - L^2 T_0} T_0, \tag{19}$$

in terms of the response coefficients given above.

3. Power Production

To characterize the performance of the engine, we first consider the power P produced. The power as a function of applied bias V, for different values of the step energy E_0 and temperature difference ΔT, is shown in Figure 2 for both sharp ($\gamma \to 0$) and smooth ($\gamma = k_B T_0$) transmission step.

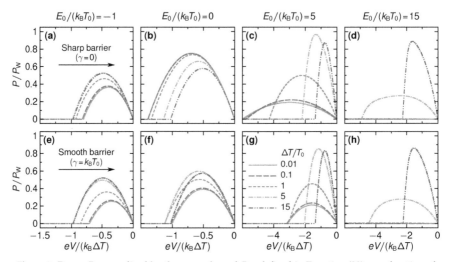

Figure 2. Power P, normalized by the power bound P_W, defined in Equation (23), as a function of applied bias $eV/(k_B \Delta T)$ for a set of step energies $E_0/(k_B T_0)$, shown in different columns, and thermal bias values $\Delta T/T_0$, represented by different styles of lines (the same for all panels, see inset in (**g**)). Panels (**a**–**d**) [(**e**–**h**)] correspond to transmission functions with a step smoothness of $\gamma \to 0$ [$\gamma = k_B T_0$].

As is seen from the figure, a common feature for all P-vs-V curves is that they first increase monotonically from $P = 0$ at $V = 0$ with increasing negative voltage. At some voltage V_{max} the power reaches its maximal value, P_{max}^V, and then decreases monotonically to zero, reached at the stopping voltage V_s. The maximum power with respect to voltage is a function of $E_0/(k_B T_0)$, $\Delta T/T_0$ and $\gamma/(k_B T_0)$, i.e.,

$$P_{max}^V = P_{max}^V \left(\frac{E_0}{k_B T_0}, \frac{\Delta T}{T_0}, \frac{\gamma}{k_B T_0} \right). \tag{20}$$

In addition, we note that in the linear-response regime we have $P_{max}^V = [L^2/(4G)]\Delta T^2$ with $V_{max} = V_s/2 = -(L/[2G])\Delta T$. From Figure 2, it is clear that the power as a function of voltage depends strongly on all parameters $E_0/(k_B T_0)$, $\Delta T/T_0$ and $\gamma/(k_B T_0)$. In particular, going from the linear to the non-linear regime by increasing $\Delta T/T_0$, the maximum power P_{max}^V might increase or decrease depending on the step properties γ and E_0.

3.1. Maximum Power

To further analyze the properties of P_{max}^V, we first recall from the seminal work of Whitney [14,15,54] that the power is bounded from above by quantum mechanical constraints. It was shown that the upper bound is reached for a QPC with a sharp step, $\gamma \to 0$, for which, using Equations (2) and (8), the power becomes

$$P_{sharp} = -\frac{(k_B T_0)^2}{h} \cdot \frac{eV}{k_B T_0} \left\{ \frac{eV}{k_B T_0} - \left(1 + \frac{\Delta T}{T_0}\right) \ln \left[f_R(E_0)\right] + \ln \left[f_L(E_0)\right] \right\}. \tag{21}$$

Maximizing this expression with respect to $eV/(k_B T_0)$ and $E_0/(k_B T_0)$ we find that the maximizing voltage is given by $eV_{max} = -\xi k_B \Delta T$ where $\xi \approx 1.14$ is the solution of $\ln(1 + e^{-\xi}) = -\xi e^{-\xi}/(1 + e^{-\xi})$ [5]. Moreover, the maximizing step energy $E_{0,max}$ and temperature difference ΔT_{max} are related via [14,28]

$$\frac{E_{0,max}}{k_B T_0} = \xi \left(1 + \frac{\Delta T_{max}}{2 T_0} \right). \tag{22}$$

Inserting this expression, together with the relation for the maximizing voltage, into Equation (21) we reach the upper bound for the power established by Whitney [54] and related to the Pendry bound [55],

$$P_W = -\frac{(k_B \Delta T)^2}{h} \xi \ln \left(1 + e^\xi \right) \approx 0.32 \frac{(k_B \Delta T)^2}{h}, \tag{23}$$

which, we emphasize, holds in the linear as well as in the non-linear regime. To relate to this upper bound, in Figure 3a–c we present a set of density plots of P_{max}^V as a function of $E_0/(k_B T_0)$ and $\Delta T/T_0$ for different values of step smoothness parameters γ.

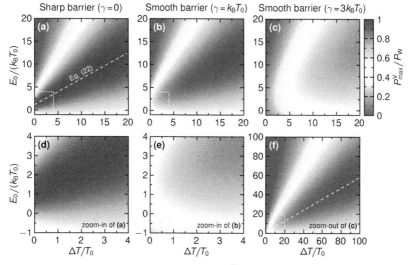

Figure 3. Maximum power with respect to voltage, P_{max}^V, as a function of $E_0/(k_B T_0)$ and $\Delta T/T_0$ presented for three different values of the step smoothness $\gamma/(k_B T_0) = 0, 1, 3$ (**a**–**c**). The white dashed lines in (**a**,**f**) illustrate Equation (22). (**d**,**e**) show close-ups of regions in (**a**,**b**), respectively, indicated with yellow dotted rectangles. On the other hand, (**f**) corresponds to an extended parameter regime of (**c**).

From the figure it is clear that for a sharp step, $\gamma \to 0$, there is a broad range of $E_0/(k_B T_0)$ and $\Delta T/T_0$ around the dashed line in the $(E_0, \Delta T)$-space, given by Equation (22), for which P_{max}^V is close to the theoretical maximum value P_W. For a step smoothness up to $\gamma \sim k_B T_0$, the situation changes only noticeably for small $\Delta T/T_0$. This is illustrated clearly in the close-ups in Figure 3d,e. Increasing the smoothness further, the region with maximum power close to P_W shifts to higher values E_0 and ΔT, although still largely centered around Equation (22), as is shown in Figure 3f.

To provide a more quantitative analysis of this behavior, below we investigate two limiting cases for γ in further detail.

3.1.1. Small Smoothness Parameter $\gamma/(k_B T_0) \ll 1$

In the limit, where the value of the smoothness parameter γ is small, $\gamma/(k_B T_0) \ll 1$, the expression for the transmission probability in Equation (1) can be expanded to leading order in γ as [56]

$$D(E) = \theta(E - E_0) + \gamma^2 \frac{\pi^2}{6} \cdot \frac{d}{dE} \delta(E - E_0). \tag{24}$$

Inserting this into the expression for the charge current, Equation (2), and performing a partial integration for the delta function derivative, we get the power

$$P = P_{\text{sharp}} - \gamma^2 \frac{eV}{h} \cdot \frac{\pi^2}{6} \cdot \frac{d}{dE_0} [f_L(E_0) - f_R(E_0)], \tag{25}$$

with P_{sharp} given in Equation (21). To estimate how the overall maximum power is modified due to finite smoothness we insert into Equation (25) the values for $eV/(k_B T_0)$, $E_0/(k_B T_0)$ and $\Delta T/T_0$ along the line in the $(E_0, \Delta T)$-space, see Figure 3, which gives the bounded power for the sharp barrier. We find

$$P(E_{0,\max}, V_{\max}) = P_W \left\{ 1 - 1.06 \left(\frac{\gamma}{k_B T_0} \right)^2 \frac{1}{1 + \Delta T/T_0} \right\}, \tag{26}$$

noting that $E_{0,\max}$ and ΔT are related via Equation (22). This expression quantifies the effect of the barrier smoothness visible in Figure 3, namely that the maximum power $P_{\max}^V$ in the region along the line in the $(E_0, \Delta T)$ plane defined by Equation (22) is mainly affected for small $\Delta T/T_0$, and approaches P_W in the strongly non-linear regime, $\Delta T/T_0 \gg 1$.

3.1.2. Smoothness $\gamma = k_B T_0$

Also in the case where the barrier gets smoother, such that γ equals the base temperature, $\gamma = k_B T_0$, we can perform an analytical investigation. Focusing on the linear-response regime, $\Delta T/T_0 \ll 1$, where the effect of the smoothness is most pronounced, we can write the power in a compact form as

$$P = -\frac{eV}{h} \left\{ \left[-\mathcal{N}(E_0) - E_0 \frac{d\mathcal{N}(E_0)}{dE_0} \right] eV - \frac{1}{2} E_0^2 \frac{d\mathcal{N}(E_0)}{dE_0} \cdot \frac{\Delta T}{T_0} \right\}. \tag{27}$$

where $\mathcal{N}(E)$ is the Bose-Einstein distribution function, $\mathcal{N}(E) = \{ \exp[E/(k_B T_0)] - 1 \}^{-1}$. As discussed above, in the linear-response regime the maximizing voltage is $V_{\max} = V_s/2$, where the stopping voltage V_s is the voltage that makes the expression in the curly bracket in Equation (27) vanish. Further maximizing over E_0 then gives $E_{0,\max} = 1.6 \, k_B T_0$, which inserted into the power expression gives

$$P_{\max}^{V,E_0} \approx 0.5 \, P_W. \tag{28}$$

From Figure 3 it is clear that both $E_{0,\max}$ and $P_{\max}^{V,E_0}$ are in good agreement with the numerical result.

4. Efficiency

Taking into account the aspect of limited resources, the power output is often not the most significant performance quantifier. A more relevant quantity is then the efficiency of a device. For a heat engine, it is defined as the power output divided by the heat absorbed from the hot bath, Equation (10).

We show the efficiency of the QPC as a steady-state thermoelectric heat engine in Figure 4. Panels (a) to (d) show the efficiency for the sharp barrier as function of voltage $eV/(k_B \Delta T)$ for different temperature differences $\Delta T/T_0$ and step energies $E_0/(k_B T_0)$. For small absolute values of the step energies, see panels (a) and (b) for two examples with $E_0/(k_B T_0) = -1, 0$, the efficiency is rather small with respect to the Carnot efficiency, $\eta/\eta_C \gtrsim 0.25$ and its overall shape only weakly depends on the temperature difference. This is radically different for larger values of E_0: panels (c) and (d) of Figure 4 show a strong increase of the efficiency, which for $E_0/(k_B T_0) = 15$ and large temperature differences can reach about 90% of the Carnot efficiency. Also, the stopping voltage V_s, at which the efficiency is zero and the device stops working as a thermoelectric, is strongly increased, depending on the temperature difference.

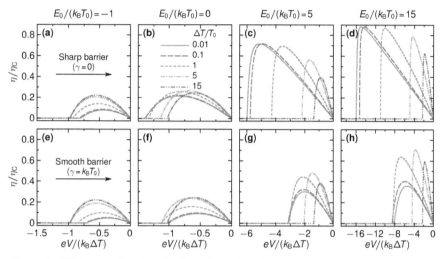

Figure 4. Efficiency as function of voltage for a sharp barrier (**a–d**) and for a smooth barrier, $\gamma = k_B T$ (**e–h**), for selected temperature differences ΔT (see different lines) and step energies E_0 (see different columns).

For large E_0, see panel (d) of Figure 4, and small temperature differences, where large maximum efficiency values are reached, the efficiency-voltage relation takes a close-to-triangular shape. In this regime, we have that $E \pm eV/2 \gg T_0, T_0 + \Delta T$ for all energies above the step energy E_0. Therefore, only the *tails* of the Fermi functions contribute in Equations (2) and (3) and the efficiency in linear response in ΔT can be approximated as

$$\eta = \frac{e|V|}{E_0} \theta\left(eV + E_0 \frac{\Delta T}{T_0}\right).\tag{29}$$

This formula describes well the triangular shape of the curves in panel (d), including the stopping voltage at small ΔT and large E_0, given by $eV_s/k_B\Delta T \approx -E_0/k_B T_0$, from the argument of the Heaviside function θ in Equation (29). We note that for $V \to V_s$ the efficiency $\eta \to \Delta T/T_0 \approx \eta_C$, i.e., the efficiency approaches the Carnot limit, see Equation (11). The mechanism for this is the same as described in Ref. [20]; transport effectively takes place in a very narrow energy interval around E_0, where the distribution functions $f_L(E_0) \approx f_R(E_0)$.

Panels (e) to (h) of Figure 4 show results for the changes in the efficiency for a smooth barrier, $\gamma = k_B T_0$. At temperature differences that are much larger than the smoothness—here the case for $k_B\Delta T/\gamma = 5, 15$—the results for the efficiency are very similar to the case of the sharp barrier. This agrees with the discussion on the power production in the previous section, Section 3. At small temperature differences, however, the efficiency gets strongly reduced by the effect of the smoothness. This is particularly striking for large step energies, see panels (g)–(h) for $E_0/k_B T_0 = 5, 15$, respectively. Here, efficiencies that were close to Carnot efficiency for a sharp barrier get reduced by a factor three due to the barrier smoothness. The reason is that increasing smoothness leads to a broadening of the energy interval where the transport takes place, and hence a breakdown of the mechanism for Carnot efficiency discussed in Ref. [20].

4.1. Maximum Efficiency

We now focus our study on the maximum value of the efficiency that can be reached over the whole range of voltages, η_{max}^{V}, a function of $E_0/(k_B T_0), \Delta T/T_0, \gamma/(k_B T_0)$. The results of this maximization procedure are shown in Figure 5, where panel (a) corresponds to a sharp barrier ($\gamma = 0$) while smooth barriers with $\gamma = k_B T_0$ and $\gamma = 3k_B T_0$ are presented in panels (b) and (c), respectively.

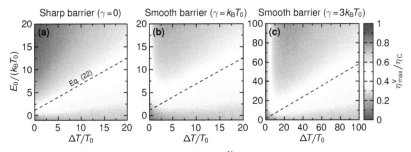

Figure 5. Density plots of the maximum efficiency η_{max}^{V} (that is, maximized over the voltage V) as a function of temperature difference ΔT and step energy E_0, for three different values of barrier smoothness, $\gamma/k_B T_0 = 0, 1, 3$ (a–c). Please note that in panel (c) the axes ranges are strongly enlarged.

Two important results can be immediately seen from these density plots of the efficiency η_{max}^{V} as a function of temperature difference $\Delta T/T_0$ and step energy $E_0/(k_B T_0)$. First, we confirm the observations about the response to small temperature differences made from Figure 4. While for a sharp barrier, efficiencies close to Carnot efficiency are reached in the linear response (close to the stopping voltage, as we know from Figure 4), for even only slightly smoothed barriers this is not the case anymore. For $\gamma/k_B T_0 = 3$, the maximum efficiency in the linear response is even suppressed down towards zero. This clearly shows that whenever the barrier step is not truly sharp, *non-linear response* is required to get a thermoelectric response with large power output and with high efficiency. Second, panels (b) and (c) of Figure 5 show that for temperature differences much larger than the smoothness—or, in other words, with one of the reservoir temperatures being much larger than the smoothness—almost the same (large) efficiency as in the sharp-barrier case is found, as long as the step energy is sufficiently large. Note, however, that these large-efficiency regions are *far* from those regions, which were previously identified as the ones of large power output, and are furthermore limited to regions with very large temperature differences and step energies.

4.2. Power-Efficiency Relations

The relation between power and efficiency for a sharp barrier, $\gamma = 0$, was investigated in detail in Refs. [14,15,28,36]. A convenient way to present the efficiency at a given power output, and vice versa, is in the form of lasso diagrams, as shown in Figure 6.

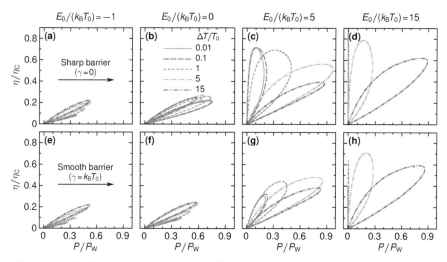

Figure 6. So-called lasso diagrams, showing the efficiency at every power output. The parameter that is changed along the lasso-line is the applied voltage V. We show results for a sharp barrier (**a–d**) and for a barrier with smoothness $\gamma = k_B T_0$ (**e–h**), for selected values of the step energy $E_0/(k_B T_0)$ (see different columns) and temperature differences $\Delta T / T_0$ (see different lines), in analogy to Figures 2 and 4.

At small step energies, $E_0/(k_B T_0) = -1, 0$, the maximum power as well as the maximum efficiency are relatively small. However, maximum efficiency and maximum output power basically happen at the same parameter values. This is advantageous for operation of a thermoelectric device, where one typically must decide whether to optimize the engine operation with respect to efficiency or power output.

This trend continues also for larger step energies, see panels (c) and (d) of Figure 6, as long as the temperature difference is larger or of the order of the step energy, $k_B \Delta T \gtrsim E_0$ (meaning that $T_0 + \Delta T > E_0/k_B > T_0$). In this case, the power output is close to its maximum value $P \approx P_W$, while the efficiency still takes values of up to the order of $\eta \approx 0.6 \, \eta_C$, in agreement with the bounds discussed in Refs. [14,15,28]. These results clearly show the promising opportunities of step-shaped energy-dependent transmissions, as they can possibly be realized in QPCs, for thermoelectric power production.

Please note that the impressively large values for the efficiency at maximum output power do not, however, violate the Curzon–Ahlbohrn [25] bound, η_{CA}, which relates to the Carnot efficiency as

$$\eta_{CA} = \frac{\eta_C}{1 + \left(1 + \Delta T/T_0\right)^{-1/2}}. \tag{30}$$

This predicts a bound on the efficiency at maximum power of $\eta_{CA} = 0.5 \, \eta_C$ in linear response in ΔT. That this bound is respected, can for example be verified by noting that the efficiency at maximum power of the grey solid line for $\Delta T / T_0 = 0.01$ in panel (c) is only slightly above $0.4\eta_C$. Equally, one can check from the green dashed-dotted line in the same panel that the efficiency at maximum power does not exceed the bound for $\Delta T / T_0 = 5$ given by $\eta_{CA} = 0.7 \, \eta_C$.

For step energies that are large with respect to the temperature of *both* reservoirs, $T_0, T_0 + \Delta T < E_0/k_B$, the power output is reduced, the maximum efficiency, however, increases. In the limit of linear response in the temperature difference, efficiencies close to Carnot efficiency are reached at the expense of close-to-zero power output.

5. Power Fluctuations and Inverse Fano Factor

During recent years it has become clear that in addition to the power and the efficiency as performance indicators of a heat engine, the fluctuations of the power output, S_P, should also be considered [40]. A reliable operation of the heat engine, i.e., where fluctuations are limited, is desirable. This is particularly relevant for nanoscale devices, where fluctuations are always a sizable effect. To analyze the effect of power fluctuations, we note that the relevant fluctuations in this QPC steady-state thermoelectric heat engine are the charge-current fluctuations, since $S_P = V^2 S_I$. Therefore, we shift the analysis of power fluctuations to the more straightforward analysis of the Fano factor, see Equation (6).

In Figure 7, we plot the inverse Fano factor $1/F$ as a function of voltage $eV/(k_B \Delta T)$ for different barrier smoothness γ, thermal gradients $\Delta T/T_0$, and step energies $E_0/(k_B T_0)$. Please note that we set the inverse Fano factor to zero outside the parameter range where power is produced, to be able to use it as a performance quantifier. This performance quantifier $1/F$ is desired to be large, meaning that current fluctuations are small with respect to the average. For all parameters, we find that increasing the (negative) voltage decreases the inverse Fano factor $1/F$ (meaning that the Fano factor F increases). This behavior is attributed to the decrease in charge current as the voltage is moved closer to the stopping voltage V_s, while the total noise is less affected. For small voltages, as well as small and negative step energies, increasing the thermal gradient generally increases the inverse Fano factor. These results can be understood from the linear-response expression for the currents and noise, Equations (16)–(18) and below, giving the inverse Fano factor

$$\frac{1}{F} = \left| \frac{eV}{k_B T_0} + \frac{eL}{k_B G} \cdot \frac{\Delta T}{T_0} \right|, \tag{31}$$

where the absolute value can be omitted when focusing on the voltage window in which power is produced. This expression increases with ΔT and decreases as V goes to more negative values. Increasing ΔT thus increases the current without an accompanied increase in fluctuations because $S_I = 2k_B T_0 G$ is independent of the bias in the linear response. For large step energies E_0, the inverse Fano factor no longer increases monotonically in ΔT but a non-monotonic behavior is observed, indicating a more subtle interplay between the fluctuations and the mean value of the current. We note that for almost all parameter values, the inverse Fano factor is substantially smaller than one which can be attributed to the relatively large thermal noise in the present system.

In Figure 8, the inverse Fano factor maximized over the voltage, $(1/F)_{max}^V$, is shown for the same parameters as used in Figures 3 and 5. Please note that the maximization only includes the voltage window where positive electrical power is produced. We find that for all three values of smoothness, the maximum inverse Fano factor increases monotonically with increasing ΔT, saturating at values a bit above unity. The Fano factor is thus slightly below unity, a signature of almost uncorrelated, close-to-Poissonian, charge transfer (for Poissonian statistics, $F = 1$). At small $\Delta T \lesssim T_0$, close to equilibrium, the noise is large even though the average electrical current is small. As noted above, this is purely due to thermal fluctuations, resulting in a small inverse Fano factor.

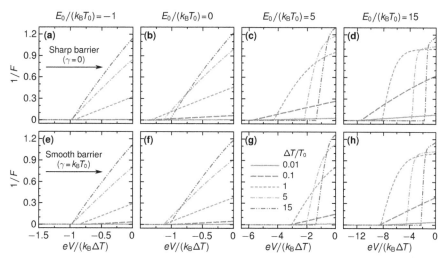

Figure 7. Inverse Fano factor as a function of voltage for sharp barrier (**a–d**) and for smooth barrier (**e–h**), for selected gradients ΔT (see different lines) and step energies E_0 (see different columns). Please note that we set the inverse Fano factor to zero outside the parameter regime where power is produced.

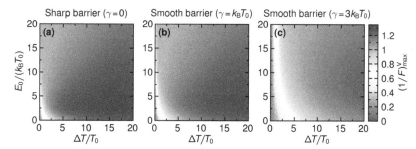

Figure 8. The inverse Fano factor maximized over all those bias values V leading to a non-negative output power, $(1/F)^{\mathrm{V}}_{\mathrm{max}}$, as a function of temperature difference ΔT and step energy E_0, for three different values of barrier smoothness, $\gamma/k_{\mathrm{B}}T_0 = 0, 1, 3$ (**a–c**).

6. Thermodynamic Uncertainty Relation

We now turn to the investigation of the TUR, cf. Equation (14), which provides a combined performance quantifier accounting for power output, efficiency and power fluctuations. We first consider the TUR-coefficient Q_{TUR} in the linear-response regime. Together with Equation (14), we therefore use the relations for power, power fluctuations, entropy and efficiencies, given in Equations (8)–(13). The linear-response expressions for the charge and heat currents occurring in these relations are given in Equation (16) and we furthermore use $S_I = 2k_{\mathrm{B}}T_0 G$. With this, we find

$$Q_{\mathrm{TUR}} = \frac{(GV + L\Delta T)^2}{\Delta T(LT_0 V + K\Delta T) + VT_0(GV + L\Delta T)} \cdot \frac{T_0}{2G}. \tag{32}$$

Maximizing this expression with respect to voltage we find $V_{\mathrm{max}} \to \pm\infty$, resulting in $Q_{\mathrm{TUR}} = 1/2$, and hence, the inequality becoming an equality. However, this voltage is within a voltage regime where power is dissipated ($P < 0$) and not produced; power production ($P \geq 0$) would instead require $V_s \leq V \leq 0$. Thus, this is not of practical relevance for the engine performance. Adding the extra condition that $P \geq 0$ we instead find $V_{\mathrm{max}} \to 0$. The corresponding value of Q_{TUR} on the left-hand side of Equation (32) then becomes $L^2 T_0/(2GK)$. Expressing this in terms of the figure of merit ZT, given in Equation (19), we can write the bound on the operationally meaningful TUR-coefficient in the linear-response regime as

$$Q_{\text{TUR}} \leq \frac{1}{2} \cdot \frac{ZT}{1+ZT}. \tag{33}$$

This shows that in the linear response, the parameters of the steady-state thermoelectric heat engine are actually subjected to a tighter bound than given by Equation (32). Please note that this bound is saturated in the limit $V = 0$, where the power production, the power fluctuations, as well as the efficiency all vanish. Also, only for ideal thermoelectrics, with $ZT \to \infty$, does the bound become $1/2$. As seen in Figure 9d, this maximal bound is actually reached for large step energies E_0.

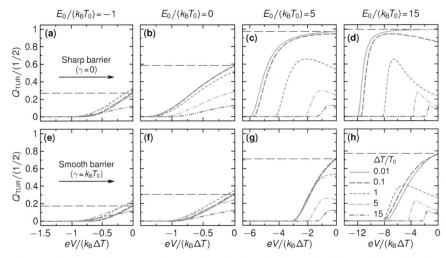

Figure 9. Coefficient Q_{TUR} as a function of voltage for sharp barrier (**a–d**) and for smooth barrier (**e–h**), for selected gradients ΔT (see different lines) and step energies E_0 (see different columns). We note that we set the Q_{TUR} to zero outside the parameter regime where power is produced. The black-dashed line in all panels corresponds to the bound that is given by Equation (33).

The full TUR-coefficient beyond linear response is illustrated in Figures 9 and 10. We find that the inequality $Q_{\text{TUR}} \leq 1/2$ is always respected, even though this is not guaranteed by scattering theory [45]. Interestingly, we find the tighter bound in Equation (33) to be respected for most parameters, even though the inequality is only proven to hold in the linear-response regime and the bound is expressed in terms of linear-response quantities (given by Equation (17)), only. Violations of the bound given in Equation (33) beyond linear response are observed for sufficiently low E_0 and when the temperature difference is of the order of the magnitude of E_0 (cf. Figure 9a for a sharp barrier). The regimes where a violation can occur are extended when the barrier is smooth (cf. Figure 9e,f). These violations agree with the general notion that dissipation increases when moving away from the linear response [39]. Furthermore, from Figure 9, we find that in the linear response, as well as for small and negative E_0, Q_{TUR} decreases monotonically as the (negative) voltage is increased. This reflects the behavior of the inverse Fano factor in Figure 7. Importantly, for sharp step energies E_0, and beyond the linear response, Q_{TUR} is a non-monotonic function of the voltage and takes on its maximum at a point where power production is finite. This non-monotonicity is a consequence of the interplay between the monotonically decreasing inverse Fano factor and the strongly increasing efficiency and power (cf. Figures 2 and 4), as the voltage is changed to more negative values.

Figure 10 shows the TUR-coefficient maximized over voltage, $Q^V_{\text{TUR,max}}$, as a function of the thermal gradient ΔT and the step energy E_0. As for the inverse Fano factor, the maximization only includes the voltage window where power is non-negative. For all values of the barrier smoothness, we find that $Q^V_{\text{TUR,max}}$ generally decreases as a function of ΔT, and a closer inspection reveals small non-monotonic features related to the small violations of Equation (33). This contrasts with the maximized inverse Fano factor, which shows the opposite behavior, cf. Figure 8. The decrease of the fluctuations with ΔT is thus overcompensated by an increase in dissipation which results in the highest values for $Q^V_{\text{TUR,max}}$ being reached in the linear-response regime. This shows that $Q^V_{\text{TUR,max}}$ is maximal in regimes, where the η^V_{max} is large. Note however that the maximal Q_{TUR} is reached at zero voltage, the maximized efficiency η is reached close to the stopping voltage $V_S \neq 0$. Furthermore, no features of the line of optimal power production close to P_W can be identified in the panels of Figure 10.

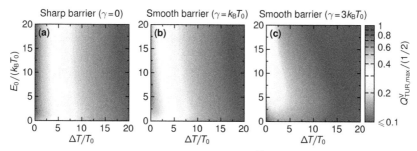

Figure 10. TUR-coefficient maximized over the bias V, $Q^V_{\text{TUR,max}}$, as a function of temperature difference ΔT and step energy E_0, for three different values of barrier smoothness, $\gamma/k_B T_0 = 0, 1, 3$ (a–c).

7. Conclusions and Outlook

In this paper, we have reviewed and extended the analysis of a QPC (or a QPC-like) device, with a transmission probability with a smoothed step-like energy-dependence, as a steady-state thermoelectric heat engine. The interest in a QPC for heat-to-power conversion derives from its optimal performance with respect to the output power, which goes along with rather large efficiencies. We have analyzed the influence of the barrier smoothness on this behavior and found that strong non-linear-response conditions are required to recover a comparable performance.

In addition to the typically studied performance quantifiers—output power and efficiency—we have broadened the analysis by adding the *power fluctuations* as an independent quantification of performance. The bound on the combination of these three quantities set by the recently identified thermodynamic uncertainty relation, suggests investigating this as a combined performance quantifier.

We have shown that the bound of the thermodynamic uncertainty relation is further restricted if one adds the practical constraint of finite (positive!) output power. In the linear response, we quantify this restriction by the figure of merit ZT. Interestingly, we have found that this combined performance quantifier maximized over the voltage has large values in those parameter regions in which the maximized efficiency is large, while regions of maximal output power are not distinguished. However, while efficiencies take their maximal value in regions close to the stopping voltage in which finite power is produced, accounting for fluctuations shifts the optimal performance value to the limit of zero voltage and zero power production.

Whether this result is unique to the QPC as steady-state heat engine or can be generalized for other thermoelectric devices is a topic of further studies. Our analysis also naturally raises the question of how to quantify the performance of the QPC when operated as a refrigerator [14,15]. Given that QPCs are standard components in many mesoscopic experiments and both the currents and noise are experimentally accessible, we anticipate that our results could be tested in experiments in the near future.

Entropy **2019**, *21*, 777

Author Contributions: The project was suggested and conceived by J.S. Numerical maximization of different quantities were mostly performed by S.K. and P.S., analytical estimates were mainly performed by N.D., M.M., and P.S. Thermodynamic interpretation of the TUR aspects was provided by P.P.P. All authors (S.K., N.D., M.M., P.P.P., J.S., P.S.) contributed to the discussion of results and the preparation of the manuscript.

Funding: This research was funded by the Knut and Alice Wallenberg Foundation via an Academy Fellowship (N.D., M.M., J.S.), by the Swedish VR (S.K., J.S., P.S.), and by the European Union's Horizon 2020 research and innovation programme under the Marie Skłodowska-Curie Grant Agreement No. 796700 (P.P.P.).

Acknowledgments: We acknowledge fruitful discussions with Artis Svilans and Heiner Linke and helpful comments on the manuscript by Robert Whitney and David Sánchez.

Conflicts of Interest: The authors declare no interest in conflicts.

References

1. Cahill, D.G.; Ford, W.K.; Goodson, K.E.; Mahan, G.D.; Majumdar, A.; Maris, H.J.; Merlin, R.; Phillpot, S.R. Nanoscale thermal transport. *J. Appl. Phys.* **2003**, *93*, 793.

2. Cahill, D.G.; Braun, P.V.; Chen, G.; Clarke, D.R.; Fan, S.; Goodson, K.E.; Keblinski, P.; King, W.P.; Mahan, G.D.; Majumdar, A.; et al. Nanoscale thermal transport. II. 2003–2012. *Appl. Phys. Rev.* **2014**, *1*, 011305.

3. Giazotto, F.; Heikkilä, T.T.; Luukanen, A.; Savin, A.M.; Pekola, J.P. Opportunities for mesoscopics in thermometry and refrigeration: Physics and applications. *Rev. Mod. Phys.* **2006**, *78*, 217.

4. Sothmann, B.; Sánchez, R.; Jordan, A.N. Thermoelectric energy harvesting with quantum dots. *Nanotechnology* **2015**, *26*, 032001.

5. Benenti, G.; Casati, G.; Saito, K.; Whitney, R.S. Fundamental aspects of steady-state conversion of heat to work at the nanoscale. *Phys. Rep.* **2017**, *694*, 1.

6. Heremans, J.P.; Dresselhaus, M.S.; Bell, L.E.; Morelli, D.T. When thermoelectrics reached the nanoscale. *Nat. Nanotechnol.* **2013**, *8*, 471.

7. Hicks, L.D.; Dresselhaus, M.S. Effect of quantum-well structures on the thermoelectric figure of merit. *Phys. Rev. B* **1993**, *47*, 12727–12731.

8. Hicks, L.D.; Dresselhaus, M.S. Thermoelectric figure of merit of a one-dimensional conductor. *Phys. Rev. B* **1993**, *47*, 16631–16634.

9. Kim, R.; Datta, S.; Lundstrom, M.S. Influence of dimensionality on thermoelectric device performance. *J. Appl. Phys.* **2009**, *105*, 034506.

10. van Wees, B.J.; van Houten, H.; Beenakker, C.W.J.; Williamson, J.G.; Kouwenhoven, L.P.; van der Marel, D.; Foxon, C.T. Quantized conductance of point contacts in a two-dimensional electron gas. *Phys. Rev. Lett.* **1988**, *60*, 848.

11. Van Houten, H.; Molenkamp, L.W.; Beenakker, C.W.J.; Foxon, C.T. Thermo-electric properties of quantum point contacts. *Semicond. Sci. Technol.* **1992**, *7*, B215.

12. Büttiker, M. Quantized transmission of a saddle-point constriction. *Phys. Rev. B* **1990**, *41*, 7906.

13. Su, S.; Guo, J.; Su, G.; Chen, J. Performance optimum analysis and load matching of an energy selective electron heat engine. *Energy* **2012**, *44*, 570.

14. Whitney, R.S. Most Efficient Quantum Thermoelectric at Finite Power Output. *Phys. Rev. Lett.* **2014**, *112*, 130601.

15. Whitney, R.S. Finding the quantum thermoelectric with maximal efficiency and minimal entropy production at given power output. *Phys. Rev. B* **2015**, *91*, 115425.

16. Mahan, G.D.; Sofo, J.O. The best thermoelectric. *Proc. Natl. Acad. Sci. USA* **1996**, *93*, 7436.

17. Bevilacqua, G.; Grosso, G.; Menichetti, G.; Pastori Parravicini, G. Thermoelectric regimes of materials with peaked transmission function. *arXiv* **2018**, arXiv:1809.05449.

18. Josefsson, M.; Svilans, A.; Burke, A.M.; Hoffmann, E.A.; Fahlvik, S.; Thelander, C.; Leijnse, M.; Linke, H. A quantum-dot heat engine operating close to the thermodynamic efficiency limits. *Nat. Nanotechnol.* **2018**, *13*, 920.

19. Wu, P.M.; Gooth, J.; Zianni, X.; Svensson, S.F.; Gluschke, J.G.; Dick, K.A.; Thelander, C.; Nielsch, K.; Linke, H. Large Thermoelectric Power Factor Enhancement Observed in InAs Nanowires. *Nano Lett.* **2013**, *13*, 4080.

20. Humphrey, T.E.; Linke, H. Reversible Thermoelectric Nanomaterials. *Phys. Rev. Lett.* **2005**, *94*, 096601.

21. Butcher, P.N. Thermal and electrical transport formalism for electronic microstructures with many terminals. *J. Phys. Condens. Matter* **1990**, *2*, 4869.

22. Bevilacqua, G.; Grosso, G.; Menichetti, G.; Pastori Parravicini, G. Thermoelectric efficiency of nanoscale devices in the linear regime. *Phys. Rev. B* **2016**, *94*, 245419.

23. Proetto, C.R. Thermopower oscillations of a quantum-point contact. *Phys. Rev. B* **1991**, *44*, 9096.

24. Lindelof, P.E.; Aagesen, M. Measured deviations from the saddle potential description of clean quantum point contacts. *J. Phys. Condens. Matter* **2008**, *20*, 164207.

25. Curzon, F.L.; Ahlborn, B. Efficiency of a Carnot engine at maximum power output. *Am. J. Phys* **1975**, *43*, 22.

26. Van den Broeck, C. Thermodynamic Efficiency at Maximum Power. *Phys. Rev. Lett.* **2005**, *95*, 190602.

27. Esposito, M.; Kawai, R.; Lindenberg, K.; Van den Broeck, C. Efficiency at Maximum Power of Low-Dissipation Carnot Engines. *Phys. Rev. Lett.* **2010**, *105*, 150603.

28. Pilgram, S.; Sánchez, D.; López, R. Quantum point contacts as heat engines. *Physica E* **2016**, *82*, 310.

29. Nakpathomkun, N.; Xu, H.Q.; Linke, H. Thermoelectric efficiency at maximum power in low-dimensional systems. *Phys. Rev. B* **2010**, *82*, 235428.

30. Luo, X.; Li, C.; Liu, N.; Li, R.; He, J.; Qiu, T. The impact of energy spectrum width in the energy selective electron low-temperature thermionic heat engine at maximum power. *Phys. Lett. A* **2013**, *377*, 1566.

31. Çipiloğlu, M.A.; Turgut, S.; Tomak, M. Nonlinear Seebeck and Peltier effects in quantum point contacts. *Phys. Status Solidi B* **2004**, *241*, 2575.

32. Dzurak, A.S.; Smith, C.G.; Martin-Moreno, L.; Pepper, M.; Ritchie, D.A.; Jones, G.A.C.; Hasko, D.G. Thermopower of a one-dimensional ballistic constriction in the non-linear regime. *J. Phys. Condens. Matter* **1993**, *5*, 8055.

33. Bogachek, E.N.; Scherbakov, A.G.; Landman, U. Nonlinear peltier effect in quantum point contacts. *Solid State Commun.* **1998**, *108*, 851.

34. Sánchez, D.; Büttiker, M. Magnetic-Field Asymmetry of Nonlinear Mesoscopic Transport. *Phys. Rev. Lett.* **2004**, *93*, 106802.

35. Sánchez, D.; López, R. Scattering Theory of Nonlinear Thermoelectric Transport. *Phys. Rev. Lett.* **2013**, *110*, 026804.

36. Whitney, R.S. Nonlinear thermoelectricity in point contacts at pinch off: A catastrophe aids cooling. *Phys. Rev. B* **2013**, *88*, 064302.

37. Meair, J.; Jacquod, P. Scattering theory of nonlinear thermoelectricity in quantum coherent conductors. *J. Phys. Condens. Matter* **2013**, *25*, 082201.

38. Barato, A.C.; Seifert, U. Thermodynamic Uncertainty Relation for Biomolecular Processes. *Phys. Rev. Lett.* **2015**, *114*, 158101.

39. Gingrich, T.R.; Horowitz, J.M.; Perunov, N.; England, J.L. Dissipation Bounds All Steady-State Current Fluctuations. *Phys. Rev. Lett.* **2016**, *116*, 120601.

40. Pietzonka, P.; Ritort, F.; Seifert, U. Finite-time generalization of the thermodynamic uncertainty relation. *Phys. Rev. E* **2017**, *96*, 012101.

41. Horowitz, J.M.; Gingrich, T.R. Proof of the finite-time thermodynamic uncertainty relation for steady-state currents. *Phys. Rev. E* **2017**, *96*, 020103.

42. Pietzonka, P.; Seifert, U. Universal Trade-Off between Power, Efficiency, and Constancy in Steady-State Heat Engines. *Phys. Rev. Lett.* **2018**, *120*, 190602.

43. Seifert, U. Stochastic thermodynamics: From principles to the cost of precision. *Physica A* **2018**, *504*, 176.

44. Guarnieri, G.; Landi, G.T.; Clark, S.R.; Goold, J. Thermodynamics of precision in quantum non equilibrium steady states. *arXiv* **2019**, arXiv:1901.10428.

45. Agarwalla, B.K.; Segal, D. Assessing the validity of the thermodynamic uncertainty relation in quantum systems. *Phys. Rev. B* **2018**, *98*, 155438.

46. Hasegawa, Y.; Van Vu, T. Generalized thermodynamic uncertainty relation via fluctuation theorem. *arXiv* **2019**, arXiv:1902.06376.

47. Potts, P.P.; Samuelsson, P. Thermodynamic Uncertainty Relations Including Measurement and Feedback. *arXiv* **2019**, arXiv:1904.04913.

48. Waldie, J.; See, P.; Kashcheyevs, V.; Griffiths, J.P.; Farrer, I.; Jones, G.A.C.; Ritchie, D.A.; Janssen, T.J.B.M.; Kataoka, M. Measurement and control of electron wave packets from a single-electron source. *Phys. Rev. B* **2015**, *92*, 125305.

49. Taboryski, R.; Kristensen, A.; Sørensen, C.B.; Lindelof, P.E. Conductance-quantization broadening mechanisms in quantum point contacts. *Phys. Rev. B* **1995**, *51*, 2282.

50. Christen, T.; Büttiker, M. Gauge-invariant nonlinear electric transport in mesoscopic conductors. *EPL* **1996**, *35*, 523.

51. Sánchez, D.; López, R. Nonlinear phenomena in quantum thermoelectrics and heat. *C. R. Phys.* **2016**, *17*, 1060.

52. Blanter, Y.M.; Büttiker, M. Shot noise in mesoscopic conductors. *Phys. Rep.* **2000**, *336*, 1.

53. Brandner, K.; Hanazato, T.; Saito, K. Thermodynamic Bounds on Precision in Ballistic Multiterminal Transport. *Phys. Rev. Lett.* **2018**, *120*, 090601.

54. Whitney, R.S. Thermodynamic and quantum bounds on nonlinear dc thermoelectric transport. *Phys. Rev. B* **2013**, *87*, 115404.

55. Pendry, J.B. Quantum limits to the flow of information and entropy. *J. Phys. A Math. Gen.* **1983**, *16*, 2161.

56. Lukyanov, V.K. A useful expansion of the Fermi function in nuclear physics. *J. Phys. G Nucl. Part. Phys.* **1995**, *21*, 145.

Article

Thermodynamics and Steady State of Quantum Motors and Pumps Far from Equilibrium

Raúl A. Bustos-Marún [1,2,*] and Hernán L. Calvo [1,3,*]

[1] Instituto de Física Enrique Gaviola (CONICET) and FaMAF, Universidad Nacional de Córdoba, Córdoba 5000, Argentina

[2] Facultad de Ciencias Químicas, Universidad Nacional de Córdoba, Córdoba 5000, Argentina

[3] Departamento de Física, Universidad Nacional de Río Cuarto, Ruta 36, Km 601, Río Cuarto 5800, Argentina

*   Correspondence: rbustos@famaf.unc.edu.ar (R.A.B.-M.); hcalvo@famaf.unc.edu.ar (H.L.C.)

Received: 27 June 2019; Accepted: 16 August 2019; Published: 23 August 2019

Abstract: In this article, we briefly review the dynamical and thermodynamical aspects of different forms of quantum motors and quantum pumps. We then extend previous results to provide new theoretical tools for a systematic study of those phenomena at far-from-equilibrium conditions. We mainly focus on two key topics: (1) The steady-state regime of quantum motors and pumps, paying particular attention to the role of higher order terms in the nonadiabatic expansion of the current-induced forces. (2) The thermodynamical properties of such systems, emphasizing systematic ways of studying the relationship between different energy fluxes (charge and heat currents and mechanical power) passing through the system when beyond-first-order expansions are required. We derive a general order-by-order scheme based on energy conservation to rationalize how every order of the expansion of one form of energy flux is connected with the others. We use this approach to give a physical interpretation of the leading terms of the expansion. Finally, we illustrate the above-discussed topics in a double quantum dot within the Coulomb-blockade regime and capacitively coupled to a mechanical rotor. We find many exciting features of this system for arbitrary nonequilibrium conditions: a definite parity of the expansion coefficients with respect to the voltage or temperature biases; negative friction coefficients; and the fact that, under fixed parameters, the device can exhibit multiple steady states where it may operate as a quantum motor or as a quantum pump, depending on the initial conditions.

Keywords: quantum thermodynamics; steady-state dynamics; nonlinear transport; adiabatic quantum motors; adiabatic quantum pumps; quantum heat engines; quantum refrigerators; transport through quantum dots

---

1. Introduction

In recent years, there has been a sustained growth in the interest in different forms of nanomachines. This was boosted by seminal experiments [1–8], the blooming of new theoretical proposals [9–30], and the latest developments towards the understanding of the fundamental physics underlying such systems [31–45]. Quantum mechanics has proven to be crucial in the description of a broad family of nanomachines, which can be put together under the generic name of "quantum motors" and "quantum pumps" [13,46–53]. They typically consist of an electromechanical device connected to electronic reservoirs and controlled by nonequilibrium sources; see Figure 1. These nonequilibrium sources may include temperature gradients and bias voltages among the reservoirs or even an external driving of the internal parameters of the system. The dimensions of the electronic component of these devices are normally within the characteristic coherence length of the electrons flowing through them, hence the essential role of quantum mechanics in their description.

Various aspects of quantum motors and pumps have been extensively studied in the literature. For example, it has been shown that quantum interferences can be exploited to boost the performance of these devices. Remarkably, some systems operate solely due to quantum interference, e.g., quantum pumps and motors based on chaotic quantum dots [13,39,54], Thouless quantum pumps and motors [13,22,55], or Anderson quantum motors [56], among others. On the other hand, the strong Coulomb repulsion between electrons in quantum-dot-based pumps and motors has shown to enhance the performance (or even induce the activation) of these nanodevices [26,42,57–59]. The effect of decoherence has also been addressed [22,39,54,60], as well as the influence of the friction forces and the system-lead coupling in the dynamics of quantum motors and pumps [22]. Indeed, the thermodynamics of those systems has proven to be a key aspect to study. In the last few years, different individual efforts have coalesced to give rise to a new field dubbed "quantum thermodynamics" [48–53,61], which studies the relations among the different energy fluxes that drive the motion of those machines where quantum mechanics plays a fundamental role.

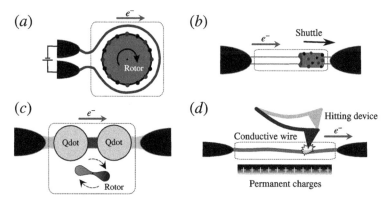

Figure 1. Examples of local systems (enclosed by dashed lines) where the movement of a mechanical piece (in blue) is coupled to the flux of quantum particles traveling from/to infinite reservoirs (black hemiellipses). (**a**) A Thouless' adiabatic quantum motor made of charges periodically arranged on the surface of a rotational piece and interacting with a wire coiled around it [22]. (**b**) An Anderson's adiabatic quantum motor made of a multi-wall nanotube where the outer one, with random impurities, is shorter than the inner one. Another example of it can be made with a rotating piece as in (**a**), but with charges randomly distributed. (**c**) A double quantum dot capacitively coupled to a rotor with positive and negative permanent charges. The dots are assumed to be weakly coupled to the electron reservoirs [42]. (**d**) As a result of an external agent, a tip hits a conductive wire capacitively coupled to permanent charges underneath. This starts the oscillation of the wire, which in turn pumps electrons between the reservoirs [23].

Despite the progress in the theoretical description of quantum motors and pumps, most of the research has focused on parameter conditions that lie close to the thermodynamic equilibrium, i.e., small bias voltages, temperature gradients, or frequencies of the external driving [48–53,61]. This is reasonable since under such conditions, the linear response regime of the nonequilibrium sources gives an accurate description of the problem, greatly simplifying its general treatment. For example, in this regime, it is common to define dimensionless figures of merit made by some combination of linear response coefficients, which give a measurement of the efficiency or the maximum power that quantum devices can achieve. It is also known that such figures of merit fail in nonlinear regime conditions [52]. Although efforts have been made in this direction, currently, there is not a nonlinear version for the figures of merit, and the performance must be calculated from the microscopic details of the system's dynamics. One strategy to deal with such situations is to use phenomenological models where the linear response coefficients are parameterized with respect to the voltage biases, the temperature gradients, or to other relevant parameters of the system; see [52] and the references therein. However, such parameterizations usually hide the physics behind the nonlinearities and require the optimization of a number of variables that grow very fast with the complexity of the model. The weakly nonlinear regime of transport has also been explored within the scattering matrix formalism. Under these conditions, it is enough to expand the response coefficients up to second order of the voltage biases and temperature gradients, which can be done by using standard quantum transport techniques. This approach has been applied to a variety of situations, where it proved to be a valuable strategy; see [48] and the references therein. However, it would be also important to extend this method to more general situations without hindering the description of the physical processes that take part in the nonlinear effects, while keeping the deep connections between the response coefficients.

Regarding the dynamics of quantum motors and pumps, one can notice that most of the works in the literature assume a constant terminal velocity of the driving parameters without a concrete model for them. A typical problem is that when these devices are coupled to nonequilibrium sources, nonconservative current-induced forces (CIFs) appear. These CIFs come, in the first place, by assuming a type of Born–Oppenheimer approximation where the electronic and mechanical degrees of freedom can be treated separately and, secondly, by evaluating the mean value of the force operator [10,11,22,24,26,34,35,37,39,42,44,49–51,62–71]. Because of the delayed response of the electronic degrees of freedom to the mechanical motion, one should include the so-called nonadiabatic corrections with the CIFs. This phenomenon is translated into a possible complex dependency of the CIFs on the velocity of the mechanical degrees of freedom. When the effect of the mechanical velocities on CIFs can be treated in linear response, it is clear whether it is adequate or not to assume a constant terminal velocity [22,39,42]. However, in far-from-equilibrium conditions, this subject has not been fully addressed.

In this article, we discuss two key aspects of far-from-equilibrium quantum motors and pumps: their steady-state dynamics, especially when CIFs present nontrivial dependencies on the terminal velocities; and their nonequilibrium thermodynamical properties, when a linear response description is not enough. We provide a systematic expansion to study the relations between the different energy fluxes that drive the quantum device. These aspects are illustrated in a concrete example where nonlinear effects due to nonequilibrium sources play a major role in the steady-state properties of the system. We show that these nonlinearities may result in, e.g., negative friction coefficients or motor/pump coexistence regimes.

Our work is organized as follows: In Section 2, we present the general model that describes the considered type of systems, and we derive an effective Langevin equation that characterizes the dynamics of the mechanical degrees of freedom, treated classically in the present context. In Section 3, we discuss in general terms the steady-state dynamics of quantum motors and pumps, highlighting some key aspects that differentiate close-to and far-from equilibrium conditions. In Section 4, we derive, on general grounds, the first law of thermodynamics for the kind of systems treated. Then, we expand the different energy fluxes passing through the system in terms of the nonequilibrium sources (temperature gradients, bias voltages, and velocities) for an arbitrary number of reservoirs. In this way, we obtain an order-by-order relation between the different energy fluxes entering and leaving the device. In Section 5, we perform a derivation of the rate of entropy production from first principles. Then, based on the second law of thermodynamics, we discuss the limits of the efficiency for different forms of quantum motors and pumps in general nonequilibrium conditions. In Section 6, we analyze and give physical interpretation to some of the relations obtained in Section 4. Finally, in Section 7, we consider the CIFs for strongly-interacting electrons in a particular example based on a double quantum dot system coupled to a mechanical rotor. We then analyze in detail the effects of higher order terms in the CIFs on the final steady state of the electromechanical system.

2. Current-Induced Forces and Langevin Equation

In this section, we introduce the generic model for the treatment of CIFs and the standard method employed in the description of the dynamics of the mechanical degrees of freedom. As a starting point, we consider as the local system the region where electronic and mechanical degrees of freedom are present and coupled to each other, like the examples shown in Figure 1. Such a local system is generically modeled by the following Hamiltonian:

$$\hat{H}_{\text{local}} = \hat{H}_{\text{s}}(\hat{\boldsymbol{X}}) + \frac{\hat{P}^2}{2m_{\text{eff}}} + U(\hat{\boldsymbol{X}}, t),\tag{1}$$

where $\hat{\boldsymbol{X}} = (\hat{X}_1, ..., \hat{X}_N)$ is the vector of mechanical coordinates and $\hat{\boldsymbol{P}} = (\hat{P}_1, ..., \hat{P}_N)$ collects their associated momenta, m_{eff} is the effective mass related to $\hat{\boldsymbol{X}}$, and $U(\hat{\boldsymbol{X}}, t)$ represents some external potential, of a mechanical nature, that may be acting on the local system. The explicit time dependence on this potential thus emphasizes the fact that an external agent can exert some effective work on the local system. The Hamiltonian $\hat{H}_{\text{s}}$ includes both the electronic degrees of freedom and their coupling to the mechanical ones through:

$$\hat{H}_{\text{s}}(\hat{\boldsymbol{X}}) = \sum_i E_i(\hat{\boldsymbol{X}}) |i\rangle \langle i|,\tag{2}$$

where the sum runs over all possible electronic many-body eigenstates $|i\rangle$. The local system is then coupled to macroscopic reservoirs, and the total Hamiltonian, including the mechanical degrees of freedom, reads:

$$\hat{H}_{\text{total}} = \hat{H}_{\text{local}} + \sum_r \hat{H}_r + \sum_r \hat{H}_{\text{s},r}.\tag{3}$$

Each lead r is described as a reservoir of noninteracting electrons through the Hamiltonian:

$$\hat{H}_r = \sum_{k\sigma} \epsilon_{rk} \hat{c}^\dagger_{rk\sigma} \hat{c}_{rk\sigma},\tag{4}$$

where $\hat{c}^\dagger_{rk\sigma}$ ($\hat{c}_{rk\sigma}$) creates (annihilates) an electron in the r-reservoir with state-index k and spin projection σ. As usual, the reservoirs are assumed to be always in equilibrium, characterized by a temperature T_r and electrochemical potential μ_r. The coupling between the local system and the r-lead is determined by the tunnel Hamiltonian:

$$\hat{H}_{\text{s},r} = \sum_{k\sigma\ell} \left(t_{r\ell} \hat{d}^\dagger_{\ell\sigma} \hat{c}_{rk\sigma} + \text{h.c.} \right),\tag{5}$$

where $t_{r\ell}$ denotes the tunnel amplitude, assumed to be k and σ independent for simplicity, and the fermion operator $\hat{d}^\dagger_{\ell\sigma}$ ($\hat{d}_{\ell\sigma}$) creates (annihilates) one electron with spin σ in the ℓ-orbital of the local system. The tunnel-coupling strengths $\Gamma_{r\ell} = 2\pi\rho_r |t_{r\ell}|^2$ then characterize the rate at which the electrons enter/leave the local system from/to the r-reservoir, where ρ_r is the density of states in the r-lead. Note that $\hat{H}_{\text{s}}$ is defined in the eigenstate basis, while $\hat{H}_{\text{s},r}$ is written in terms of single-particle field operators. The tunnel matrix elements accounting for transitions between different eigenstates can then be obtained from linear superpositions of the above tunnel amplitudes [72].

To obtain an effective description of the dynamics of the mechanical degrees of freedom through a Langevin equation, we start from the Heisenberg equation of motion for the $\hat{\boldsymbol{P}}$ operator, which yields:

$$m_{\text{eff}} \frac{\text{d}\dot{\hat{\boldsymbol{X}}}}{\text{d}t} + \vec{\nabla} U(\hat{\boldsymbol{X}}, t) = -\vec{\nabla} H_{\text{s}}(\hat{\boldsymbol{X}}).\tag{6}$$

The measured value A_{measured} of an observable described by an operator $\hat{A}$ can always be taken as its mean value $A = \langle \hat{A} \rangle$ plus some fluctuation ζ_A around it, i.e., $A_{\text{measured}} = A + \zeta_A$. We will work under the nonequilibrium Born–Oppenheimer approximation [13,22,34,35,37,39,49,73,

74] (or Ehrenfest approximation [33,63,75–77]), where the dynamics of the electronic and mechanical degrees of freedom can be separated and the latter is treated classically. This allows us to neglect the fluctuations of the terms appearing in the left-hand side of Equation (6) and describe the mechanical motion only through the mean value X, which is reasonable for large or massive objects. With this in mind, we obtain the following Langevin equation of motion:

$$m_{\text{eff}}\frac{d\dot{X}}{dt} + F_{\text{ext}} = F + \xi, \tag{7}$$

where $F = -\langle \vec{\nabla} H_s \rangle = i \langle [\hat{H}_s(\hat{X}), \hat{P}] \rangle$ and ξ account for the mean value and the fluctuation of the CIF, respectively (throughout this manuscript, we take $\hbar = 1$ for simplicity). As we shall see later on, the external force applied to the mechanical part of the local system, F_{ext}, plays the role of an eventual "load" force for a quantum motor or a "driving" force for a quantum pump. As this force will be typically opposed to the CIF, we define F_{ext} with a minus sign for better clarity in future discussions. The main task, therefore, relies on the calculation of the CIFs from appropriate formalisms capable of describing the dynamics of the electronic part of the system. Once these forces are calculated, we can use Equation (7) to integrate the classical equations of motion and obtain the effective dynamics of the complete electromechanical system.

In most previous works, F is expanded up to first order in $\dot{X}$, i.e., $F \approx F^{(0)} - \gamma \cdot \dot{X}$. The resulting CIF is then the sum of an adiabatic contribution $F^{(0)}$ and its first nonadiabatic correction $F^{(1)} = -\gamma \dot{X}$, respectively. Under this approximation, Equation (7) turns into:

$$m_{\text{eff}}\frac{d\dot{X}}{dt} + F_{\text{ext}} = F^{(0)} - \gamma \cdot \dot{X} + \xi. \tag{8}$$

Explicit formulas for the calculation of $F^{(0)}$, γ, and ξ in terms of Green functions and scattering matrices were derived in [10,11,34,35,37] and extended in [22,39] to account for decoherent events. Although these expressions were obtained in the context of noninteracting particles, they can be used in effective Hamiltonians derived from first principles calculations [62,64]. In [49–51], the CIFs were obtained from the Floquet–Green's function formalism. The role of Coulomb interactions was addressed through different formalisms and methods like, e.g., many-body perturbation theory based on nonequilibrium Green's functions [44]; modeling the system as a Luttinger liquid [24]; and using a time-dependent slave-boson approximation [26]. In [42], explicit expressions for the CIFs within the Coulomb blockade regime of transport were obtained using a real-time diagrammatic approach [78], which we present in more detail in Section 7 when considering the example of Figure 1c.

3. Mechanical Steady State

In this paper, we will restrict ourselves to systems that perform overall cyclic motions. Immediate examples are shown in Figure 1a,c, where the rotation angle of the rotor can be assigned as the natural mechanical coordinate. On the other hand, the examples shown in Figure 1b,d may also, under certain circumstances, sustain cyclic motion, though the general coordinate could be not so obvious. As a possibility for the quantum shuttle of Figure 1b, the cyclic motion would involve a cyclic reversal of the bias voltage (AC-driven). This AC-driven case, though intriguing, goes beyond the scope of the present manuscript, as we are not considering here time-dependent biases. Another scenario would be that of Figure 1d, where the cyclic motion is in principle attainable by periodically hitting the device. Note that we are not dealing with the steady-state of sets of interacting nanomotors, such as those described in, e.g., [79–81]. Instead, here we are interested in the steady-state of the mechanical part of isolated quantum motors and pumps that interact solely with the electrons of a set of reservoirs and where, at most, Coulomb interactions are only taken into account within the local system.

To discuss the dynamics of cyclic motions in simple terms, we start by projecting Equation (7) on a closed trajectory defined in the space of X. By assuming a circular trajectory, the dynamics can be described by an angle θ, its associated angular velocity $\dot{\theta}$, the moment of inertia $\mathcal{I}$, and the torques $\mathcal{F}$, $\mathcal{F}_{ext}$, and ζ_θ. Using this, we obtain an effective angular Langevin equation equivalent to Equation (7),

$$\frac{d\dot{\theta}}{dt} = \frac{1}{\mathcal{I}}[\mathcal{F} - \mathcal{F}_{ext} + \zeta_\theta]. \tag{9}$$

We assume that after a long waiting time, the system arrives at the steady-state regime where the mechanical motion becomes periodic, and it is then characterized by a time period τ such that $\theta(t+\tau) = \theta(t)$ and $\dot{\theta}(t+\tau) = \dot{\theta}(t)$. Moreover, we will assume that the stochastic force plays a minor role in the above equation, such that it does not affect the mean values of the dynamical variables θ and $\dot{\theta}$, i.e., the mean trajectories with or without the stochastic force approximately coinciding. This occurs, for example, at low temperatures or in mechanical systems with a large moment of inertia [22,39,42]. In the following, we will just ignore ζ_θ for practical purposes (this is the opposite regime of another type of nanomotors, the Brownian motors [1,82]). Under the above assumptions, we integrate both sides of Equation (9) from an initial position θ_i to a final one θ_f and obtain:

$$\frac{\mathcal{I}}{2}\left[\dot{\theta}_f^2 - \dot{\theta}_i^2\right] = \int_{\theta_i}^{\theta_f}[\mathcal{F} - \mathcal{F}_{ext}]\,d\theta. \tag{10}$$

The torques in this equation are, in general, intricate functions of both θ and $\dot{\theta}$.

Therefore, the calculation of the θ-dependent angular velocity usually requires the resolution of a transcendental equation (see, e.g., [42]). Alternatively, one can obtain $\theta(t)$ from the numerical integration of the equation of motion by standard techniques like, e.g., the Runge–Kutta method. All this greatly complicates the study of quantum motors and pumps, to the point where it becomes almost impossible to draw any general conclusion. For this reason, one common simplification consists of taking the terminal velocity as constant during the whole cycle [13,22,24,26,39,49–51,56,74]. Indeed, this description is exact if the external agent compels the constant velocity condition to be fulfilled in a controllable manner, as is often conceived in quantum pumping protocols. However, this is not the case in general, and typically, one expects internal variations for $\dot{\theta}$ in one period. We now address this interesting issue in more detail. First, we take the integral in Equation (10) over the whole period. This gives:

$$\mathcal{W}_{ext} = \mathcal{W}_F, \quad \text{where} \quad \mathcal{W}_F = \int_0^\tau \mathcal{F}\dot{\theta}\,dt, \quad \text{and} \quad \mathcal{W}_{ext} = \int_0^\tau \mathcal{F}_{ext}\dot{\theta}\,dt. \tag{11}$$

The above stationary state condition thus establishes that the work originated from the CIF is always compensated by the external mechanical work in the case that this regime can be reached. Now, let us assume for a moment that the terminal velocity of a nanodevice is constant and positive (we leave the discussion of the effect of the sign of $\dot{\theta}$ for later when treating a concrete example in Section 7.1). If we now expand $\mathcal{F}$ in terms of $\dot{\theta}$, Equation (11) yields:

$$\mathcal{W}_{ext} = \sum_k \left(\int_0^{2\pi} \left.\frac{\partial^k \mathcal{F}}{\partial \dot{\theta}^k}\right|_{\dot{\theta}=0} \frac{d\theta}{k!}\right)\dot{\theta}^k. \tag{12}$$

Two important conclusions can be extracted from the above formal solution. First, there may be conditions where some roots of Equation (12) are complex numbers, meaning that the assumption $\dot{\theta} = $ const. is nonsense, as the periodicity condition required for the steady-state regime would not be fulfilled. Second, for real solutions, it was shown in [22,42] that the moment of inertia $\mathcal{I}$ not only affects the time that it takes the mechanical system to reach the stationary regime, but also the internal range in the angular velocity, i.e., the difference $\Delta\dot{\theta} = \dot{\theta}_{max} - \dot{\theta}_{min}$ in one period. According to Equation (12),

$\dot{\theta}$ is independent of $\mathcal{I}$, while the variation of $\dot{\theta}$ scales with $\mathcal{I}^{-1}$, cf. Equation (10). Then, the ratio $\Delta\dot{\theta}/\dot{\theta}$, which is the relevant quantity in our analysis, should vanish for large $\mathcal{I}$ values, justifying the constant velocity assumption for large or massive mechanical systems.

The value of $\mathcal{F}_{\text{ext}}$ is supposed to be controllable externally, as well as the voltage and temperature biases, which, in turn, affect the current-induced torque $\mathcal{F}$. Therefore, under the above discussed conditions, $\dot{\theta}$ can be thought as a parameter that surely depends on the internal details of the system, but it is also tunable by external "knobs". Let us analyze a concrete example: Consider a local system connected to two leads at the same temperature and with a small bias voltage $eV = \mu_L - \mu_R$. By considering the current-induced torque up to its first nonadiabatic correction, i.e., $\mathcal{F} \approx \mathcal{F}^{(0)} - \gamma\dot{\theta}$, and assuming that $\mathcal{F}_{\text{ext}}$ is independent of $\dot{\theta}$, the following relation must hold, according to the above discussion,

$$\dot{\theta} \approx \frac{Q_{I_R}}{2\pi\bar{\gamma}} \left(V - \frac{\mathcal{W}_{\text{ext}}}{Q_{I_R}} \right) = \frac{Q_{I_R} V_{\text{eff}}}{2\pi\bar{\gamma}}, \tag{13}$$

where $\bar{\gamma}$ is the average electronic friction coefficient along the cycle, Q_{I_R} is the pumped charge to the right lead, and we used Onsager's reciprocal relation between $\mathcal{F}$ and the charge current I_R in the absence of magnetic fields [22,39,42,49]. Alternatively, if we assume that the external torque is of the form $\mathcal{F}_{\text{ext}} = \gamma_{\text{ext}}\dot{\theta}$, one finds:

$$\dot{\theta} \approx \frac{Q_{I_R} V}{2\pi} \frac{1}{(\bar{\gamma} + \bar{\gamma}_{\text{ext}})} = \frac{Q_{I_R} V}{2\pi\gamma_{\text{eff}}}. \tag{14}$$

Note in the above equations that, at least in the present order, the effect of the external forces can be described as a renormalization of the bias voltage V or the electronic friction coefficient γ. Numerical simulations in [22,42] showed that the above equations agree well in general with the steady-state velocities found by integrating the equation of motion. However, at very small voltages, essential differences may appear. There is a critical voltage below which the dissipated energy per cycle cannot be compensated by the work done by the CIF, and thus, $\dot{\theta} = 0$. We dubbed this the "nonoperational" regime of the motor. Moreover, when increasing the bias voltage, there is an intermediate region where a hysteresis cycle appears, and two values of the velocity are possible ($\dot{\theta} = 0$ and those given by the above equations). Although in Section 7.1, we will take $\dot{\theta}$ as constant when discussing a specific example, the reader should keep in mind that this approximation does not always hold, especially at very small voltages or $\mathcal{I}$.

4. Order-by-Order Energy Conservation

In the previous sections, we introduced and discussed the role of the mechanical degrees of freedom, emphasizing certain parameter restrictions, which allowed for the simplification in their dynamics. In this section, we are going to derive, on general grounds, essential relations between the electronic and mechanical degrees of freedom from the point of view of energy conservation. Importantly, we will focus on systematic expansions beyond the standard linear regime of nonequilibrium sources like, e.g., the mechanical velocity, the bias voltage, and the temperature gradient in systems composed by an arbitrary number of reservoirs.

As already pointed out, we are treating the mechanical degrees of freedom classically, such that their effect on the electronic degrees of freedom enters as a parametric dependence in the electronic part of the total Hamiltonian, which now reads:

$$\hat{H} = \sum_r \hat{H}_r + \hat{H}_s + \sum_r \hat{H}_{s,r}. \tag{15}$$

The mechanical part of the local system, when treated classically in Equation (7), introduces an explicit time dependence into $\hat{H}_s$ that, in turn, makes $d\langle\hat{H}\rangle/dt \neq 0$. According to the above Hamiltonian, the total internal energy of the electronic system, $U = \langle\hat{H}\rangle$, can be split into energy contributions from the reservoirs, the local system, and the tunnel couplings. The time variation of the internal energies associated with the different partitions of the system is:

$$\dot{U}_\beta = J_\beta^E + \left\langle \frac{\partial \hat{H}_\beta}{\partial t} \right\rangle, \quad \text{where} \quad J_\beta^E = i \left\langle [\hat{H}, \hat{H}_\beta] \right\rangle \tag{16}$$

which is the mean value of the energy flux entering in the subsystem $\beta = \{r, (s,r), s\}$. The value of $\langle \partial_t \hat{H}_\beta \rangle$ is zero when we evaluate the energy flux in the reservoirs and the tunnel couplings, but equals $-\dot{W}_F$ when β is evaluated in the local system. Although, strictly speaking, W_F is time independent when considering a closed trajectory, we used the symbol $\dot{W}_F$ to denote the power delivered by the CIF, i.e., $\dot{W}_F = -\langle \partial_t \hat{H}_s \rangle = \mathbf{F} \cdot \dot{\mathbf{X}}$. The latter comes from the definition of the CIF given in Equation (7). Note that the energy fluxes fulfill the condition $\sum_r (J_r^E + J_{s,r}^E) + J_s^E = 0$. Therefore, the variation of the total internal energy of the electrons yields the following conservation rule:

$$\dot{U} = \sum_r \dot{U}_r + \sum_r \dot{U}_{s,r} + \dot{U}_s = \sum_r \left(J_r^E + J_{s,r}^E \right) + J_s^E - \dot{W}_F = -\dot{W}_F. \tag{17}$$

Conservation of the total number of particles implies a relation between the particle currents of the reservoirs $\dot{N}_r$ and that of the local system $\dot{N}_s$:

$$\dot{N}_{\text{total}} = \frac{d}{dt} \langle \hat{N}_{\text{total}} \rangle = \sum_r \frac{d}{dt} \langle \hat{N}_r \rangle + \frac{d}{dt} \langle \hat{N}_s \rangle = 0 \quad \Rightarrow \quad \sum_r \dot{N}_r = -\dot{N}_s. \tag{18}$$

This is so since no particle can be assigned to the coupling region. Now, let us assume the system is in the steady-state regime and integrate Equation (17) over a time period τ of the cyclic motion. Under this condition, the above-defined local quantities only depend periodically on time, i.e., $U_s(t + \tau) = U_s(t)$, $U_{s,r}(t + \tau) = U_{s,r}(t)$, and $\dot{N}_s(t + \tau) = \dot{N}_s(t)$. Therefore, the following quantities should evaluate to zero, i.e.,

$$\int_0^\tau \dot{U}_{s,r} \, dt = 0, \quad \int_0^\tau \dot{U}_s \, dt = 0, \quad \text{and} \quad \int_0^\tau \dot{N}_s \, dt = 0, \tag{19}$$

as these are integrals of a total derivative of some periodic function. This means that no energy (or particles) is accumulated/extracted indefinitely within the finite regions defined by the local system or its coupling to the leads. Equation (19) can be used together with Equation (18) to prove the charge current conservation between reservoirs, $\sum_r \int_0^\tau I_r \, dt = 0$, where the charge current of the r reservoir is defined as $I_r = e\dot{N}_r$, with $e > 0$ being minus the electron's charge.

We will take the following definition for the heat current J_r in the reservoir r:

$$J_r = J_r^E - \mu_r \dot{N}_r. \tag{20}$$

In [43,50,83], the authors proposed a different definition for the heat current of the reservoirs, $J_r = J_r^E + (J_{s,r}^E/2) - \mu_r \dot{N}_r$. However, the inclusion or not of half the heat current of the coupling region, $(J_{s,r}^E/2)$, does not make any difference in the present paper, as this quantity integrates to zero over a cycle, Equation (19), and does not contribute to the rate of entropy production, as we will see in the next section, Equation (30). Replacing Equation (20) in Equation (17) and integrating over a period result in:

$$\sum_r (Q_{J_r} + Q_{I_r} \delta V_r) = -W_F, \quad \text{where} \quad Q_{J_r} = \int_0^\tau J_r \, dt, \quad \text{and} \quad Q_{I_r} = \int_0^\tau I_r \, dt, \tag{21}$$

where $\delta V_r = (\mu_r - \mu_0)/e$ and μ_0 is an arbitrary reference's potential. We, in addition, defined the quantities Q_{J_r} and Q_{I_r}, which are, respectively, the total heat and charge pumped to reservoir r in a cycle. Note that in the above equation, we used conservation of the total charge.

Before we continue, we would like to emphasize that the periodic motion imposed by the steady-state regime can be further exploited to reduce the number of mechanical coordinates to an effective description of the CIF. In principle, one can always recognize a generalized coordinate χ, which parameterizes the closed mechanical trajectory. This, for example, can be accounted for by taking $\chi = \Omega \mod(t, \tau)$, such that $\dot{\chi} = \Omega = 2\pi/\tau$. In other words, one can always find a natural scale, Ω in the present case, for all mechanical velocities along the trajectory. We will take this natural scale as the expansion variable for the CIF. The current-induced work then reads $W_F = \int_0^\tau F_\Omega \Omega \, dt$, where $F_\Omega = F \cdot \partial_\chi X$. Of course, with this χ-parameterization, the velocity is constant by definition, but at the expense that now the calculation of F_Ω requires the knowledge of the mechanical trajectory. For circular trajectories and the conditions discussed in Section 3, the coordinate χ would be $|\theta|$. For other cases, it would not be that simple to find χ, and one should first solve the system's equation of motion.

In addition to the mechanical velocity Ω, the CIF and the currents can also be expanded in terms of the remaining nonequilibrium sources, corresponding to voltage and temperature deviations $\delta V = (\delta V_1, \delta V_2, ...)^T$ and $\delta T = (\delta T_1, \delta T_2, ...)^T$, respectively, from their equilibrium values V_{eq} and T_{eq}. For an arbitrary observable R, this general expansion takes the form:

$$R = \sum_{|\alpha| \geq 0} R^\alpha = \sum_{|\alpha| \geq 0} \frac{(\Omega, \delta V, \delta T)^\alpha}{\alpha!} (\partial^\alpha R)_{eq}, \tag{22}$$

where we used the following multi-index notation: $\alpha = (n_\Omega, n_{V_1}, ..., n_{T_1}, ...)$; $\alpha! = n_\Omega! n_{V_1}! ... n_{T_1}!...$; $(\Omega, \delta V, \delta T)^\alpha = \Omega^{n_\Omega} \delta V_1^{n_{V_1}} ... \delta T_1^{n_{T_1}} ...$; and:

$$\partial^\alpha = \frac{\partial^{(n_\Omega + n_{V_1} + ... n_{T_1} + ...)}}{\partial \Omega^{n_\Omega} \partial V_1^{n_{V_1}} ... \partial T_1^{n_{T_1}} ...}. \tag{23}$$

This, in turn, allowed us to recognize in the integral quantities of Equation (21) the following expansion coefficients:

$$W_F^\alpha = \int_0^\tau F_\Omega^\alpha \Omega \, dt, \quad Q_{j_r}^\alpha = \int_0^\tau J_r^\alpha \, dt, \quad \text{and} \quad Q_{I_r}^\alpha = \int_0^\tau I_r^\alpha \, dt, \tag{24}$$

where F_Ω^α, J_r^α, and I_r^α all have the form given by Equation (22). To find a consistent order-by-order conservation relation from Equation (21), we should first note that the involved terms enter in different orders: W_F^α contains an additional Ω from the time-integral as compared with the pumped currents, while the term $Q_{I_r}^\alpha \delta V_r$ is one order higher in δV_r than the other two. Therefore, the following relation must hold for every order α of the expansion:

$$Q_J^\alpha \cdot \mathbf{1} + Q_I^{\alpha(V)} \cdot \delta V = -W_F^{\alpha(\Omega)}, \tag{25}$$

where $\mathbf{1} = (1, 1, ...)^T$, $Q_J^\alpha = (Q_{J_1}^\alpha, Q_{J_2}^\alpha, ...)$, $Q_I^{\alpha(V)} = (Q_{I_1}^{\alpha(V_1)}, Q_{I_2}^{\alpha(V_2)}, ...)$, and we used the shorthand $\alpha(\Omega) = (n_\Omega - 1, n_{V_1}, ..., n_{T_1}, ...)$ and $\alpha(V_r) = (n_\Omega, n_{V_1}, ..., n_{V_r} - 1, ..., n_{T_1}, ...)$. Equation (25) results in being important for a systematic study of far-from-equilibrium systems, as it helps to rationalize how every order of the expansion of an energy flux is connected with the others. The equation is completely general and valid for any value of the relevant quantities Ω, δV, and δT. Importantly, the above conservation rule can be extended to other nonequilibrium sources like, e.g., spin polarization in ferromagnetic leads, such that other types of currents could also be considered in the energy transfer process between the electronic and mechanical parts of the system.

5. Entropy Production And Efficiency

In 1865, Rudolf Clausius proposed a new state function, the thermodynamic entropy S, that turned out to be crucial to study the limits and the efficiency of different physical processes. The thermodynamic entropy is defined as the amount of heat δQ_J that is transferred in a reversible thermodynamic process, $\delta S = \delta Q_{rev}/T$. Here, we are assuming that each reservoir r is in its own equilibrium at a constant temperature T_r and chemical potential μ_r (we are not going to treat time-dependent T_r or μ_r). As the reservoirs are considered to be macroscopic, the aforementioned equilibrium state is not altered by the coupling to the local system. This assumption allows us to associate the reservoirs' heat flux with the variation in their thermodynamic entropy, i.e., $\delta Q_{J_r} = T_r \delta S_r$. Therefore, from Equation (16) and the definition of the heat current given in Equation (20), we can write:

$$\dot{U}_r = T_r \dot{S}_r + \dot{N}_r \mu_r. \tag{26}$$

The information theory entropy, known as the Shannon or von Neumann entropy, times the Boltzmann constant equals the thermodynamic entropy for equilibrium states. There is a debate on whether this equality can be extended to nonequilibrium states; see, e.g., [84–86]. However, for the purpose of this article, only the change of the entropy of the reservoirs is needed, not that of the system. Besides, we will not need to evaluate the entropy from the density matrix. Therefore, the thermodynamic definition of entropy suffices in our case.

Now, let us consider the sum of the internal energy over the set of all reservoirs coupled to the local system,

$$\sum_r \dot{U}_r = \sum_r T_r \dot{S}_r + \sum_r \dot{N}_r \mu_r. \tag{27}$$

If we take $T_r = \delta T_r + T_0$ and $\mu_r = \mu_0 + \delta\mu_r$, add and subtract the change of the internal energy of the local system $\dot{U}_s$ and that of the couplings between the local system and the reservoirs $\sum_r \dot{U}_{s,r}$, one can rearrange the above equation to the following:

$$T_0 \sum_r \dot{S}_r = \left(\dot{U}_s + \sum_r \dot{U}_r + \sum_r \dot{U}_{s,r} \right) - \sum_r \dot{N}_r \delta\mu_r - \sum_r \dot{S}_r \delta T_r - \left(\dot{U}_s + \sum_r \dot{U}_{s,r} \right) - \mu_0 \left(\sum_r \dot{N}_r \right). \tag{28}$$

Note that the values of μ_0 and T_0 are completely arbitrary, and there is no need to identify them with the chemical potential and temperature of the central region, which can be ill-defined far from equilibrium.

Replacing $\dot{S}_r$ by $= J_r/T_r$ in the right-hand side of the above equation, using energy conservation (17) and particle number conservation (18), allows one to rewrite Equation (28) as:

$$T_0 \dot{S}_{res} = -F_\Omega \Omega - \sum_r I_r \delta V_r - \sum_r J_r \left(\frac{\delta T_r}{T_r} \right) - \left(\dot{U}_s + \sum_r \dot{U}_{s,r} \right) + \mu_0 \dot{N}_s, \tag{29}$$

where $\dot{S}_{res} = \sum_r \dot{S}_r$ is the variation of the entropy of the electrons of all reservoirs, and we used $e\dot{N}_r = I_r$. The CIF can be split into "equilibrium" and "nonequilibrium" terms, $F^{(eq)}$ and $F^{(ne)}$, respectively, where one can prove that $F^{(eq)}$ is always conservative [13,22,42,63]. We are interested in the steady-state situation of our local system. As discussed around Equation (19), the change of the internal energy of the electronic part of the local system and that of the coupling region must be zero after a cycle, as energy cannot be accumulated indefinitely within a finite region. The same argument is true for the number of particles accumulated in a cycle, which should be zero. At steady state, we therefore recognize the reversible component of the entropy variation as that given by:

$$\dot{S}_{res}^{(rev)} = -\frac{1}{T_0} \left(F_\Omega^{(eq)} \Omega + \dot{U}_s + \sum_r \dot{U}_{s,r} - \mu_0 \dot{N}_s \right). \tag{30}$$

Obviously, this quantity will not contribute to the total entropy production. Therefore, the rate of entropy production $\dot{S}_{\text{res}}^{(\text{irrev})}$ yields:

$$\dot{S}_{\text{res}}^{(\text{irrev})} = -\frac{1}{T_0}\left(F_{\Omega}^{(\text{ne})}\Omega - I \cdot \delta V - J \cdot \delta\mathcal{T}\right), \tag{31}$$

where $\delta\mathcal{T} = (\delta T_1/T_1, \delta T_2/T_2, ...)^{\text{T}}$, and the currents are defined through $I = (I_1, I_2, ...)$ and $J = (J_1, J_2, ...)$. Integrating Equation (31) over a cycle and taking into account the second law of thermodynamics, one finds:

$$0 \geq \mathcal{W}_F + Q_I \cdot \delta V + Q_J \cdot \delta\mathcal{T}. \tag{32}$$

The above general formula, also valid far from equilibrium, can be used to set efficiency bounds for energy transfer processes between the electronic and mechanical degrees of freedom. For example, if we take $\delta\mathcal{T} = 0$, $Q_I \cdot \delta V < 0$, and $\mathcal{W}_F > 0$, then the system should operate as a nanomotor driven by electric currents, and the following relation holds:

$$1 \geq -\frac{\mathcal{W}_F}{Q_I \cdot \delta V}, \tag{33}$$

while for $Q_I \cdot \delta V > 0$ and $\mathcal{W}_F < 0$, the system operates as a charge pump, and Equation (32) implies:

$$1 \geq -\frac{Q_I \cdot \delta V}{\mathcal{W}_F}. \tag{34}$$

Notice that, because of the steady-state condition, $\mathcal{W}_F$ equals $\mathcal{W}_{\text{ext}}$, where $\mathcal{W}_{\text{ext}}$ can be taken as the output or the input energy, depending on the considered type of process. Therefore, the above formulas describe the efficiency η of the device's process, defined as the ratio between the output and input energies per cycle. It is also interesting to note that the above equations reflect no more than energy conservation in this particular case. A different situation occurs for $\delta V = 0$ and $\delta\mathcal{T} \neq 0$, where Equation (32) yields:

$$1 \geq -\frac{\mathcal{W}_F}{Q_J \cdot \delta\mathcal{T}}, \quad \text{and} \quad 1 \geq -\frac{Q_J \cdot \delta\mathcal{T}}{\mathcal{W}_F}. \tag{35}$$

The first equation thus corresponds to a quantum heat engine and the second one to a quantum heat pump, respectively. Now, because of the factor $\delta\mathcal{T}$, the above formulas differ from what is expected from energy conservation solely. This is clear in a two-lead system, where η is limited by Carnot's efficiency of heat engines and refrigerators, respectively. To illustrate this, let us consider a hot and a cold reservoir and set the temperature of the cold reservoir as the reference. For the heat engine, this gives:

$$Q_J \cdot \delta\mathcal{T} = Q_{J_{\text{hot}}}\left(1 - \frac{T_{\text{cold}}}{T_{\text{hot}}}\right) < 0 \quad \Rightarrow \quad \left(1 - \frac{T_{\text{cold}}}{T_{\text{hot}}}\right) \geq -\frac{\mathcal{W}_F}{Q_{J_{\text{hot}}}}, \tag{36}$$

where the left-hand side of the second equation represents the Carnot limit for heat engines. Other energy transfer processes mixing voltage and temperature biases can also be analyzed in the context of Equation (32) to set the bounds of their associated efficiencies.

6. Pump-Motor Relations

It is clear from Equation (25) that there is an infinite number of relations that can be used to connect the pumped heat or charge and the work done by the CIF. In this section, we give some physical interpretation to the leading orders in the general expansion, which highlights the utility of Equation (25).

We start from the order-by-order relations by taking $|\alpha| = 0$. Importantly, Equation (25) with $|\alpha| = 0$ seems to impose the evaluation of terms with negative coefficients. For example, in the exponent of Q_I, one would be tempted to evaluate $\alpha(V_1) = (0, -1, 0, 0...)$, but this term is not defined in the expansions of Equation (24), and then, it should be taken as zero. Therefore, Equation (25) yields,

$$Q_j^0 \cdot 1 = \sum_r \int_0^\tau (J_r)_{eq} \, dt = 0. \tag{37}$$

This simply reflects the fact that no net heat current occurs at equilibrium. For $|\alpha| = 1$, all relations are summarized in the following three cases:

$$\sum_r \int_0^\tau \left(\frac{\partial J_r}{\partial \Omega} \right)_{eq} \Omega \, dt = 0, \quad \sum_r \int_0^\tau \left(\frac{\partial J_r}{\partial V_i} \right)_{eq} \delta V_i \, dt = 0, \quad \text{and} \quad \sum_r \int_0^\tau \left(\frac{\partial J_r}{\partial T_i} \right)_{eq} \delta T_i \, dt = 0, \tag{38}$$

where we used the fact that equilibrium forces are conservative [13,22,42,50,63], i.e., $\int_0^\tau F_\Omega^{(eq)} \Omega \, dt = 0$, and that there are no net charge currents in equilibrium, thus $\int_0^\tau (I_r)_{eq} dt = 0$. The above relations mean that, at first order, there is a conservation of pumped heat between reservoirs. For $|\alpha| = 2$, there are many relations and cases, but we restrict ourselves to only a few of them. For $n_\Omega = 2$, $n_{V_i} = 0$, and $n_{T_j} = 0$, Equation (25) gives:

$$\sum_r \int_0^\tau \frac{\Omega^2}{2} \left(\frac{\partial^2 J_r}{\partial \Omega^2} \right)_{eq} dt = -\int_0^\tau \left(\frac{\partial F_\Omega}{\partial \Omega} \right)_{eq} \Omega^2 dt. \tag{39}$$

Now, the quantity $(\partial_\Omega F_\Omega)_{eq}$ is minus the electronic friction coefficient at equilibrium. Therefore, this relation shows that the energy dissipated as friction in the motor is delivered as heat to the reservoirs; more precisely, as a second-order pumped heat. For $n_\Omega = 0$, $n_{V_i} = 1$, $n_{V_j} = 1$, and $n_{T_k} = 0$, Equation (25) yields:

$$\sum_r \int_0^\tau \delta V_i \delta V_j \left(\frac{\partial^2 J_r}{\partial V_i \partial V_j} \right)_{eq} dt = -2 \int_0^\tau \delta V_i \delta V_j \left(\frac{\partial I_i}{\partial V_j} \right)_{eq} dt, \tag{40}$$

where we used Onsager's reciprocity relation $(\partial_{V_j} I_i)_{eq} = (\partial_{V_i} I_j)_{eq}$ [42,49]. The quantities $(\partial_{V_i} I_i)_{eq}$ are the linear conductances in the limit of small bias voltages. Therefore, this relation shows that these leakage currents, defined as those currents that cannot be used to perform any useful work, are also dissipated as heat in the reservoirs, a phenomenon known as Joule heating or the Joule law [27,43,50,83]. Finally, for $n_\Omega = 1$, $n_{V_i} = 1$, and $n_{T_k} = 0$, Equation (25) results in:

$$\sum_r \int_0^\tau \Omega \delta V_i \left(\frac{\partial^2 J_r}{\partial \Omega \partial V_i} \right)_{eq} dt + \int_0^\tau \Omega \left(\frac{\partial I_i}{\partial \Omega} \right)_{eq} \delta V_i dt = -\int_0^\tau \delta V_i \left(\frac{\partial F_\Omega}{\partial V_i} \right)_{eq} \Omega dt. \tag{41}$$

Now, using Onsager's reciprocity relation $(-\partial_{V_i} F_\Omega)_{eq} = (\partial_\Omega I_i)_{eq}$ [22,39,42,49], one finds that the first term in the left-hand side of the above equation vanishes, which is an unexpected conservation relation for this second-order pumped heat. We remark that the utility of Equation (25) relies on the fact that it provides a physical interpretation for the connection between different order contributions that participate in the energy conservation rule. One can continue analyzing the other relations for $|\alpha| = 2$ and beyond, but the number of relations and cases grows very fast with $|\alpha|$, and each relation may have its own physical interpretation. The study of higher-order terms in $|\alpha|$ may result in being useful when addressing particular nonlinear effects in the involved energy currents. However, for the purpose of the present article, we believe that the above analysis is enough to illustrate the approach proposed by Equation (25) regarding multi-index expansions.

As the full dynamics of the complete system typically involves a formidable task, many methods in quantum transport treat this problem through a perturbative expansion in Ω. This is the case of the real-time diagrammatic theory we use in Section 7, where the effective dynamics of the electronic part of the system is described through a perturbative expansion in the characteristic frequency of the driving parameters (which in our case is modeled by the mechanical system), while the voltage and temperature biases are treated exactly. In this case, the expansion in Ω comes naturally from the theory itself. In situations like this, it may result in being more useful to simplify the expansions by restricting ourselves to those nonequilibrium sources whose perturbative treatment is inherent to the formalism used, as in [42,87].

Regarding the above discussion, we now use Equation (25) to describe how the different energy contributions are linked in an Ω expansion provided by the theory. The zeroth order terms in Ω reads:

$$Q_J^{(0)} \cdot \mathbf{1} + Q_I^{(0)} \cdot \delta V = 0. \tag{42}$$

This equation shows that the total amount of heat delivered by the leads comes from the bias voltage maintained between them. Its interpretation is similar to that of Equation (40). The heat term can be associated with the leakage energy current, i.e., the energy flowing from source to drain leads without being transferred to the local system. The next order in this expansion yields:

$$Q_J^{(1)} \cdot \mathbf{1} + Q_I^{(1)} \cdot \delta V = -W_F^{(0)}, \tag{43}$$

and can be understood as a generalization of the motor–pump relation, $Q_I^{(1)} \cdot \delta V \approx -W_F^{(0)}$, discussed in [13] for arbitrary bias voltages and temperature gradients. Therefore, according to Equation (43), deviations of the mentioned relation at finite voltages are due to the pumped heat induced by the mechanical motion of the local system. When we evaluate the currents in equilibrium by setting $\delta V = 0$ and $\delta T = 0$, Equation (38) implies $W_F^{(0)} = 0$, meaning that no external work is done in a cycle, and the pumped energy from the leads can again be considered as a leakage current since no net effect on the mechanical system is performed. If, on the other hand, some bias is present (either thermal or electric), the energy transfer from the leads to the local system imprints a mechanical motion, which in turn, produces some useful work. The second-order term in Ω gives:

$$Q_J^{(2)} \cdot \mathbf{1} + Q_I^{(2)} \cdot \delta V = -W_F^{(1)}, \tag{44}$$

and generalizes Equation (39) to finite voltage and temperature biases. The right-hand side of Equation (39) can be interpreted as the dissipated energy of the mechanical system, which is delivered to the electronic reservoirs. However, in the above equation, $W_F^{(1)}$ is not guaranteed to be always negative, and from the point of view of the mechanical system, this can be interpreted as a negative friction coefficient.

Now, we return to the multi-variable expansion of the energy currents to establish which orders should be considered in a consistent calculation of the efficiency of quantum motors and pumps. Assuming that in Equation (25), we take $|\alpha|$ up to some truncation value α_{max}, the order-by-order scheme implies that, for example, the efficiency of the electrically-driven quantum motor should be given by:

$$\eta = -\frac{\sum_{|\alpha|=0}^{\alpha_{max}} W_F^{\alpha(\Omega)}}{\sum_{|\alpha|=0}^{\alpha_{max}} Q_I^{\alpha(V)} \cdot \delta V}. \tag{45}$$

To illustrate this, let us take the case of a local system coupled to left and right reservoirs at voltages V_L and V_R, respectively. We assume the leads are at the same temperature, and we set $\delta V_L = -\delta V_R = \delta V/2$ as the voltage biases. Depending on which kind of expansion we take, one can obtain different expressions for the efficiencies. On the one hand, when expanding in terms of $\delta V = V_L - V_R$ and Ω up to $\alpha_{max} = 2$, the above general expression yields:

$$\eta = -\frac{\mathcal{W}_F^{(0,0)} + \mathcal{W}_F^{(1,0)} + \mathcal{W}_F^{(0,1)}}{\left(Q_I^{(0,0)} + Q_I^{(1,0)} + Q_I^{(0,1)}\right)\delta V} = -\frac{\mathcal{W}_F^{(1,0)} + \mathcal{W}_F^{(0,1)}}{\left(Q_I^{(1,0)} + Q_I^{(0,1)}\right)\delta V}, \tag{46}$$

where the superscripts indicate the order in the expansions in Ω and δV, respectively, and we defined $I = (I_L - I_R)/2$. As we already mentioned, the zeroth order contributions $\mathcal{W}_F^{(0,0)}$ and $Q_I^{(0,0)}$ are simply zero as they correspond to the work done by the conservative part of the CIF and the equilibrium charge current, respectively. On the other hand, when performing an expansion up to $\alpha_{max} = 2$ but only in terms of Ω, Equation (45) turns into:

$$\eta = -\frac{\mathcal{W}_F^{(0)} + \mathcal{W}_F^{(1)}}{\left(Q_I^{(0)} + Q_I^{(1)} + Q_I^{(2)}\right)\delta V}, \tag{47}$$

where now $\mathcal{W}_F^{(0)}$ and $Q_I^{(0)}$ are nonzero in general, since they are not necessarily evaluated at equilibrium. Although we here restricted ourselves to the efficiency of a nanomotor driven by electric currents, its extension to other operational modes of the device can be obtained from Equation (32) in a similar way to that of Equation (45). This procedure then allows us to obtain efficiency expressions for arbitrary expansions in the nonequilibrium sources, which could be useful in the evaluation of the device's performance far from equilibrium.

7. Quantum Motors and Pumps in the Coulomb Blockade Regime

In this section, we consider the CIFs in the so-called Coulomb blockade regime of transport. In this regime, the strong electrostatic repulsion that takes place inside a small quantum dot (usually taken as the local system) highly impacts the device's transport properties, as for small bias voltages, no additional charges can flow through the dot and the current gets completely blocked. The full system dynamics in this strongly interacting regime cannot be described by, e.g., the scattering matrix approach, and one needs to move to some other theoretical framework. A suitable methodology is given by the real-time diagrammatic theory [78], which allows for an effective treatment of the quantum dot dynamics by performing a double expansion in both the tunnel coupling between the dot and the leads and the frequency associated with the external driving parameters. Since then, many extensions and application examples appeared in the context of quantum pumps [12,57–59,87–94] and quantum motors [42].

To lowest order in the tunnel coupling, the dot's reduced density matrix obeys the following master equation:

$$\frac{d}{dt}p = Wp, \tag{48}$$

where the vector $p = \{p_i(t)\}$ describes the dot's occupation probabilities and W is the evolution kernel matrix accounting for the transition rates between the quantum dot states, due to its coupling to the leads. In the context of CIFs, we assume that the time scale of the mechanical motion, characterized by $\dot{X}$, is large as compared to the typical dwell time of the electrons in the local system. This allows for an expansion of the reduced density matrix as $p = \sum_{k \geq 0} p^{(k)}$, with $p^{(k)}$ of order $(\Omega/\Gamma)^k$. Here, Ω and Γ denote the characteristic scales for the velocity of the mechanical degrees of freedom and the tunnel rate of the electronic system, respectively, and we always assume $\Omega < \Gamma$. The above master equation, in turn, takes the following hierarchical structure [12,78]:

$$Wp^{(0)} = 0, \quad \text{and} \quad Wp^{(k)} = \frac{d}{dt}p^{(k-1)}. \tag{49}$$

In the first equation, $p^{(0)}(X)$ represents the steady-state solution the electronic system arrives at when the mechanical system is "frozen" at the point X. As the mechanical coordinate moves in time, this order corresponds to the adiabatic electronic response to the mechanical motion. This needs not to be confused with the steady-state regime of the mechanical system, which obviously takes a much longer time to be reached. The second equation contains higher-order nonadiabatic corrections $p^{(k)}(X, \dot{X})$ due to retardation effects in the electronic response to the mechanical motion. In all these equations, the matrix elements of the kernel are of zeroth order in the mechanical velocities, i.e., $W = W(X)$. The normalization condition on the dot's density matrix implies $e^{\mathsf{T}}p^{(k)} = \delta_{k0}$, with e^{T} the trace over the dot's Hilbert space. This allows for the definition of an invertible pseudo-kernel $\tilde{W}$, whose matrix elements are $\tilde{W}_{ij} = W_{ij} - W_{ii}$, such that we can write:

$$p^{(k)} = \left[\tilde{W}^{-1}\frac{\mathrm{d}}{\mathrm{d}t}\right]^k p^{(0)}. \tag{50}$$

Once we know the different orders of the reduced density matrix, it is possible to compute the expectation value of any observable R after calculation of its corresponding kernel W^R. This implies the same expansion as before, and it reads:

$$R^{(k)} = e^{\mathsf{T}}W^R p^{(k)}. \tag{51}$$

The observables we are going to address here are the charge and heat currents I_r and J_r associated with the r lead and defined in Section 4. In lowest-order in tunneling and under the assumption that coherences are completely decoupled from the occupations, it is possible to obtain simple expressions for the current kernels in terms of the number of particles n_i and energy E_i associated with the dot state $|i\rangle$ [87,95]:

$$W_{ij}^{I_r} = -e(n_i - n_j)[W_r]_{ij}, \qquad W_{ij}^{J_r} = -\left[E_i - E_j - \mu_r(n_i - n_j)\right][W_r]_{ij}. \tag{52}$$

The CIF was derived in Section 2 under the assumption that the mechanical part of the system, characterized by coordinates X and associated momenta P, only interacts with local parameters of the quantum dot through their many-body eigenenergies, cf. Equation (2). This implies that the ν-component of the CIF operator, defined as $\hat{F}_\nu = -\partial\hat{H}_s/\partial X_\nu$, is local in the quantum dot basis. For this local operator, then, we can simply define a diagonal kernel of the form:

$$W_{ij}^{F_\nu} = -\frac{\partial E_i}{\partial X_\nu}\delta_{ij}, \tag{53}$$

such that Equation (51) gives the k order in the Ω expansion for any observable. Importantly, when adding up all contributions from the leads, the above kernel definitions, together with the sum rule $\sum_i W_{ij} = 0$, lead to the following conservation rule for all orders in Ω [87]:

$$\sum_r J_r^{(k)} = -\frac{\mathrm{d}}{\mathrm{d}t}\langle\hat{H}_s\rangle^{(k-1)} - F^{(k-1)}\cdot\dot{X} - \sum_r \frac{\mu_r}{e}I_r^{(k)}. \tag{54}$$

This equation, equivalent to (17), thus expresses the first principle of thermodynamics, which in this case relates the total heat flowing from/into the leads with the variation of the internal energy of the dot and the power contributions due to both mechanical and electrochemical external sources. By considering a system coupled to two reservoirs L and R, with periodic motion characterized by a time period τ, and taking the time integral of the above equation, we recover the frequency expansion of Equation (25):

$$Q_J^{(k)} + Q_I^{(k)}\delta V = -W_F^{(k-1)}, \tag{55}$$

where the bias voltage δV is defined through $\mu_L = -\mu_R = e\delta V/2$, $Q_J = Q_{J_L} + Q_{J_R}$ and $Q_I = (Q_{I_L} - Q_{I_R})/2$. The sign convention employed here implies, for example, that if the left-hand side of Equation (55) is positive, then there is some amount of energy entering into the leads in one cycle, and work is being extracted from the local system.

7.1. Example: Double Quantum Dot Coupled to a Rotor

In this section, we illustrate the discussions of the previous sections in a concrete example based on a double quantum dot (DQD) system locally coupled to a mechanical rotor (see Figure 1c). We assume a capacitive coupling between the dots and the fixed charges in the rotor such that no charge flows between the two subsystems. The DQD system is described as in [42,87,89],

$$\hat{H}_s = \sum_{\ell=L,R} \epsilon_\ell \hat{n}_\ell + U\hat{n}_L\hat{n}_R + \frac{U'}{2}\sum_{\ell=L,R} \hat{n}_\ell(\hat{n}_\ell - 1) - \frac{t_c}{2}\left(\hat{d}^\dagger_{L\sigma}\hat{d}_{R\sigma} + \text{h.c.}\right),\tag{56}$$

where ϵ_ℓ is the on-site energy and $\hat{n}_\ell = \sum_\sigma \hat{d}^\dagger_{\ell\sigma}\hat{d}_{\ell\sigma}$ the number operator in the ℓ-dot, U and U' the inter- and intra-dot charging energies, respectively, and t_c the interdot hopping amplitude. We will work in the strong coupling regime $t_c \gg \Gamma$, such that non-diagonal elements (coherences) in the reduced density matrix are decoupled from the diagonal ones (occupations) and can be disregarded in first order in tunneling. To simplify the analysis (by reducing the number of states in the two-charge block), we work in the limit $U' \to \infty$, such that double occupation in a single dot is energetically forbidden.

Diagonalization of the above DQD Hamiltonian yields the bonding (b) and antibonding (a) basis for the single-electron charge block, and the reduced density matrix reads $p = (p_0, p_{b\uparrow}, p_{b\downarrow}, p_{a\uparrow}, p_{a\downarrow}, p_{\uparrow\uparrow}, p_{\uparrow\downarrow}, p_{\downarrow\uparrow}, p_{\downarrow\downarrow})^T$. These elements thus denote the probability for the DQD to be either empty (p_0), occupied by one electron in the $\ell = b$ (or a) orbital with spin σ ($p_{\ell\sigma}$), or by two electrons ($p_{\sigma\sigma'}$), one of them in the left dot and with spin σ and the other electron in the right dot and with spin σ'. The many-body eigenenergies are therefore $E_0 = 0$ for the empty DQD,

$$E_{b/a} = \frac{\epsilon_L + \epsilon_R}{2} \mp \sqrt{\left(\frac{\epsilon_L - \epsilon_R}{2}\right)^2 + \left(\frac{t_c}{2}\right)^2}\tag{57}$$

for single occupation in the bonding or the antibonding orbital and $E_2 = \epsilon_L + \epsilon_R + U$ for the doubly-occupied DQD, respectively.

To account for the coupling between the electronic and mechanical degrees of freedom, we take as the mechanical coordinate the angle θ describing the orientation of the rotor axis. We assume the following dependence through the on-site energies:

$$\epsilon_L(\theta) = \bar{\epsilon}_L + \lambda x_L = \bar{\epsilon}_L + \delta_\epsilon \cos(\theta), \quad \text{and} \quad \epsilon_R(\theta) = \bar{\epsilon}_R + \lambda x_R = \bar{\epsilon}_R + \delta_\epsilon \sin(\theta),\tag{58}$$

which defines a circular trajectory in energy space of radius $\delta_\epsilon = \lambda r_0$ (r_0 measures the actual radius of the rotor) around the working point $(\bar{\epsilon}_L, \bar{\epsilon}_R)$. For simplicity, in what follows, we will focus on the tangential component of the force only, by assuming that the radial component is always compensated by internal forces in the rotor (e.g., fixed charges along the rotor's axis). By applying this form in Equation (7), we obtain the equation of motion (9) for the angular velocity $\dot{\theta}$, where: $\mathcal{I} = m_{\text{eff}}r_0^2$ is the rotor's moment of inertia, $\mathcal{F} = \sum_i(-\partial_\theta E_i)p_i = r_0 F \cdot \hat{\theta}$ is the current-induced torque, $\mathcal{F}_{\text{ext}}$ is the torque produced by the external force, which is assumed to be constant along the whole trajectory, and ξ_θ accounts for the force fluctuations. The stationary regime is reached once the rotor performs a periodic motion (characterized by a time period τ) with no overall acceleration, i.e., when Equation (11) is fulfilled. We show in Figure 2 some examples of the evolution of the rotor's angular velocity for different initial conditions. This is plotted as a function of the number of cycles performed by the rotor, instead of time, to unify scaling along the horizontal axis. Here, we take a sufficiently large moment of inertia such that the variation $\Delta\dot{\theta}$ over one period is small as compared to the average value of $\dot\theta$ in one cycle. This, in turn, implies a large number of cycles for the rotor to reach the steady-state regime.

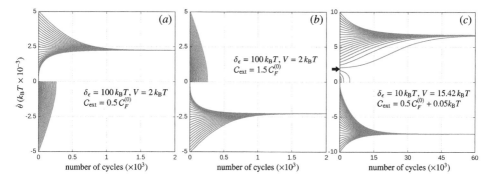

Figure 2. Evolution of the rotor's angular velocity (average over cycle) as a function of the number of cycles for different initial conditions. In (**a**,**b**), we choose $\delta_\epsilon = 100\,k_B T$ and $V = 2\,k_B T$, and we take in (a) $\mathcal{C}_{\mathrm{ext}} = 0.5\,C_F^{(0)}$, so the device operates as a motor, while in (b), we use $\mathcal{C}_{\mathrm{ext}} = 1.5\,C_F^{(0)}$, so the device operates as a pump, cf. Equation (62). (**c**) A situation in which two stable velocities with opposite sign are possible, for $\delta_\epsilon = 10\,k_B T$, $V = 15.42\,k_B T$, and $\mathcal{C}_{\mathrm{ext}} = C_F^{(0)} + 0.05\,k_B T$. The other parameters are: $\delta T = 0$, $U = 20\,k_B T$, $t_c = 10\,k_B T$, $\Gamma_L = \Gamma_R = 0.25\,k_B T$, while for the trajectories, we used as working point $\bar\epsilon_L = \bar\epsilon_R = -6\,k_B T + \delta_\epsilon / \sqrt{2}$.

As discussed in Section 3, when we take $\mathcal{I}$ sufficiently large, $\Delta\dot\theta \to 0$, and it is possible to approximate $\dot\theta$ as constant during the whole cycle. This allows us to take out this quantity from the integrals defining the different orders of the energy currents, as we did in Equation (12). The problem is, however, that the sign in $\dot\theta$ is not a priori known and, with it, the direction of the trajectory for the line integrals defining $\mathcal{W}_F^{(k)}$. To become independent of this issue, we denote by s the sign of $\dot\theta$, and notice that $\mathcal{F}^{(k)} = f_\theta^{(k)} \dot\theta^k$, where $f_\theta^{(k)} = \partial_\theta^k \mathcal{F}|_{\dot\theta=0}/k!$ only depends parametrically on θ, so we have:

$$\mathcal{W}_F^{(k)} = \oint_{\mathcal{C}(s)} F^{(k)} \cdot dX = \int_0^{2\pi s} \mathcal{F}^{(k)} d\theta = s\left[\int_0^{2\pi} f_\theta^{(k)} d\theta\right] \dot\theta^k = s\,C_F^{(k)} \dot\theta^k, \tag{59}$$

where, importantly, the coefficients $C_F^{(k)}$ are independent of s, so the $\dot\theta^k$ and the extra sign s accompanying $C_F^{(k)}$ in the right-hand side of the equation give the correct sign for $\mathcal{W}_F^{(k)}$. As discussed in Section 3, the steady-state condition in Equation (11) can be viewed as the equation for the final velocity that the rotor acquires as a function of the external force, which in the present case turns into:

$$0 = \sum_k C_F^{(k)} \dot\theta^k - \mathcal{C}_{\mathrm{ext}}, \tag{60}$$

where we defined $\mathcal{C}_{\mathrm{ext}} = 2\pi \mathcal{F}_{\mathrm{ext}}$. The stability of the final solution is inherited from Equation (9), and in our case with $\dot\theta$ constant, it is given by:

$$\sum_k C_F^{(k)} k\dot\theta^{k-1} < 0. \tag{61}$$

Notice the specific case where $\mathcal{C}_{\mathrm{ext}} = C_F^{(0)}$, i.e., the external work equals the bias work coming from the first-order currents, cf. Equation (55) for $k = 1$. In this case, there is always a trivial solution, given by $\dot\theta = 0$. Of course, this solution is useless as the system becomes frozen, and there is no energy transfer between the leads and the mechanical system. However, this situation marks one of the transition points between the operation modes of the device. If we now consider an infinitesimal difference between these two coefficients, we can faithfully treat Equation (60) up to linear order in $\dot{\theta}$, such that the solution is:

$$\dot{\theta} = \frac{\mathcal{C}_{ext} - \mathcal{C}_F^{(0)}}{\mathcal{C}_F^{(1)}}, \tag{62}$$

together with the condition $\mathcal{C}_F^{(1)} < 0$, in agreement with Equation (61). This implies that the integral of the friction coefficient over a cycle needs to be positive; otherwise, the rotor cannot reach the stationary regime. As the sign in $\mathcal{C}_F^{(1)}$ is fixed to this order, the sign in $\dot{\theta}$ only depends on the relation between $\mathcal{C}_{ext}$ and $\mathcal{C}_F^{(0)}$. This, together with the stationary condition $\mathcal{W}_{ext} = \mathcal{W}_F$, establishes the operation mode of the device, in the sense that with this information, we can deduce the sign of $\mathcal{W}_{ext} = s\mathcal{C}_{ext}$ and, hence, the direction of the energy flow. If, for example, $\mathcal{W}_{ext} < 0$, the external energy is delivered into the DQD, which in turn goes as energy current to the leads, so the device acts as an energy pump. On the other hand, if $\mathcal{W}_{ext} > 0$, the energy current coming from the leads goes into the DQD, and it is transformed into mechanical work, so the device operates as a motor. These two scenarios are shown in Figure 2, where we take $\mathcal{C}_{ext} > 0$, such that the positive sign in the final velocity implies that the device acts as a motor (see Panel a), while a negative sign implies that the device operates as a pump (see Panel b). Of course, when regarding the device as an energy pump, the above criterion only establishes those regions in $\mathcal{C}_{ext}$ where we can expect some kind of pumping mechanism. Depending then on the specific type of pumping we have in mind, the ranges in $\mathcal{C}_{ext}$ will be subject to additional conditions. For example, if we would like to have this device acting as a quantum charge pump, then we should check those regimes in $\mathcal{C}_{ext}$ where the electrons flow against the bias voltage. This discussion will be reserved to the next section, and for now, we will only define the operation mode from the direction of the energy flow.

Let us now consider second-order effects due to $\mathcal{C}_F^{(2)}$ in Equation (60), where we obtain:

$$\dot{\theta}_{\pm} = -\frac{\mathcal{C}_F^{(1)}}{2\mathcal{C}_F^{(2)}} \pm \sqrt{\left(\frac{\mathcal{C}_F^{(1)}}{2\mathcal{C}_F^{(2)}}\right)^2 + \frac{\mathcal{C}_{ext} - \mathcal{C}_F^{(0)}}{\mathcal{C}_F^{(2)}}}, \tag{63}$$

together with the conditions:

$$\left(\frac{\mathcal{C}_F^{(1)}}{2\mathcal{C}_F^{(2)}}\right)^2 > \frac{\mathcal{C}_F^{(0)} - \mathcal{C}_{ext}}{\mathcal{C}_F^{(2)}}, \qquad \pm \mathcal{C}_F^{(2)} < 0. \tag{64}$$

The first inequality ensures a positive argument in the square root of Equation (63), so $\dot{\theta}$ is a real number, while the second inequality comes from the stability condition given by Equation (61) and tells us which branch one should choose in Equation (63) to get the stable solution: if $\mathcal{C}_F^{(2)} > 0$, then $\dot{\theta}_-$, and if $\mathcal{C}_F^{(2)} < 0$, then $\dot{\theta}_+$. Importantly, far from equilibrium (e.g., $\delta V \gg k_B T$), it is possible to arrive at the odd situation where $\mathcal{C}_F^{(1)} > 0$, and the "dissipated" energy $\mathcal{W}_F^{(1)} = s\mathcal{C}_F^{(1)}\dot{\theta} = \mathcal{C}_F^{(1)}|\dot{\theta}|$ is positive. This, however, does not prevent the rotor from reaching the stationary regime (to this order in $\dot{\theta}$), since the lower-order terms compensate this energy gain. Therefore, one can end up in a situation where the second-order energy current comes from the leads and it is delivered into the mechanical system (thus favoring the motor regime), contrary to the standard situation where the "dissipated" energy flows to the leads. Regarding Equation (63), the sign of $\dot{\theta}$ now depends on the relation between the first term and the square root, together with the sign of $\mathcal{C}_F^{(2)}$, so this analysis is not that simple as in the linear case. However, once we have the $\mathcal{C}_F$-coefficients, we can always determine if the rotor is able to reach the stationary regime and infer whether $\mathcal{W}_{ext}$ is positive or negative and, with it, the operation mode of the device.

For higher orders in $\dot{\theta}$, the above analysis for the operation mode of the device is the same, but more ingredients may come into play due to the (order-dependent) specific solutions for the stationary value of $\dot{\theta}$. Interestingly, by including higher order terms in Equation (60), it could happen that for a fixed choice of the parameters ($\mathcal{F}_{ext}$, δV, δT, etc.), the system presents more than one stable solution, and even with different signs in $\mathcal{W}_{ext}$, so the initial condition on $\dot{\theta}$ decides the operation mode of the device once the rotor reaches the stationary regime. In Figure 2c, we show this nonlinear effect in the dynamics of the rotor's velocity. This was done by taking $\mathcal{F}$ up to third order in the $\dot{\theta}$ expansion, where we find two stable solutions ($\dot{\theta} \sim -7.4 \times 10^{-3} k_B T$ and $6.7 \times 10^{-3} k_B T$) and one unstable solution in between ($\dot{\theta} = 2 \times 10^{-3} k_B T$). Notice that, in this case, if the rotor starts with a positive initial condition below the unstable solution (see gray arrow), then it cannot reach the negative stable solution, as once $\dot{\theta} = 0$, the rotor gets trapped in a local minimum.

In Figure 3, we show how the $\mathcal{C}_F$-coefficients contribute to the current-induced work as a function of the bias voltage for different values of the orbit radius δ_ϵ. According to Equation (59), these coefficients can only be compared upon multiplication with the k-power of some frequency of reference Ω_x. We assess this frequency of reference by performing the following analysis: from Equation (50), we can estimate the maximum allowed value for $\dot{\theta}$ compatible with the frequency expansion. Since $\tilde{W}^{-1} \propto \Gamma^{-1}$, with $\Gamma = \Gamma_L + \Gamma_R$, and:

$$\frac{\mathrm{d}}{\mathrm{d}t} = \dot{\theta}\frac{\partial}{\partial\theta} = \dot{\theta}\sum_i \frac{\partial\epsilon_i}{\partial\theta}\frac{\partial}{\partial\epsilon_i}, \tag{65}$$

we obtain:

$$p^{(k)} = \left[\tilde{W}^{-1}\dot{\theta}\sum_i \frac{\partial\epsilon_i}{\partial\theta}\frac{\partial}{\partial\epsilon_i}\right]^k p^{(0)} \propto \left[\frac{\Omega}{\Gamma}\frac{\delta_\epsilon}{k_B T}\right]^k, \tag{66}$$

where we use the fact that $\partial_\theta\epsilon_i \propto \delta_\epsilon$ and the energy derivative of both $p^{(0)}$ and $\tilde{W}^{-1}$ are proportional to $1/k_B T$. As the consistency of the frequency expansion relies on the convergence of the occupations, this yields the adiabaticity condition [42,87]:

$$\frac{\Omega}{\Gamma} \ll \frac{k_B T}{\delta_\epsilon}, \tag{67}$$

from which we can estimate the maximum allowed frequency Ω_{max} as:

$$\Omega_{max} = \frac{k_B T}{\delta_\epsilon}\Gamma. \tag{68}$$

Obviously, this extreme value sets the point where the expansion could diverge, so to illustrate the different order contributions to the current-induced work, we take, in Figure 3, an intermediate reference frequency $\Omega_x = \Omega_{max}/2$. As we can see, for a long range of the bias voltage ($V \lesssim 10k_B T$), both the zeroth- and first-order terms contribute, while higher order terms are almost negligible. The zeroth-order contributions (black) show a linear dependence whose slope increases with the orbit size, up to some saturation value, related with the quantization of the pumped charge. The first-order terms (red) remain almost constant and negative in this bias regime. This implies that the linear-order treatment given in Equation (62) is enough, as long as $\mathcal{C}_{ext} \sim \mathcal{C}_F^{(0)}$. For $V \gtrsim 10k_B T$, we can observe non-equilibrium effects like "inverse dissipation", i.e., a positive contribution from the first-order term (red) in Figure 3a. Additionally, the higher order terms (blue and green) can become larger than the first two, and they need to be included in the calculation of the final velocity through Equation (60) or in the current-induced force appearing in Equation (9). This, in turn, could lead to several stable solutions whose validity should be determined through a systematic convergence analysis, which is beyond the scope of the present work. For the particular choice of parameters used in Figure 2c, we check if the next-order coefficient ($\mathcal{C}_F^{(4)}$) has a negligible impact on the third-order solutions.

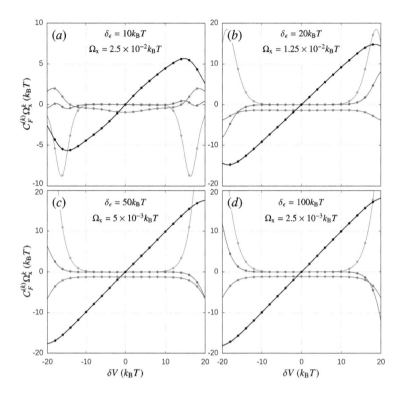

Figure 3. Order-by-order contributions to the current-induced work as a function of the bias voltage for different orbit radius: (a) $\delta_\epsilon = 10$ and $\Omega_x = 2.5 \times 10^{-2}$, (b) $\delta_\epsilon = 20$ and $\Omega_x = 1.25 \times 10^{-2}$, (c) $\delta_\epsilon = 50$ and $\Omega_x = 5 \times 10^{-3}$, and (d) $\delta_\epsilon = 100$ and $\Omega_x = 2.5 \times 10^{-3}$, in units of $k_B T$. We show, in solid lines, the zeroth- (black), first- (red), second- (blue) and third- (green) orders in $\dot\theta$, respectively. The circles accompanying the lines correspond to $(C_J^{(k)} + C_I^{(k)} \delta V)\Omega_x^{(k-1)}$, with $C_I^{(k)}$ and $C_J^{(k)}$ defined in Equation (70), and numerically confirm Equations (55) and (71). The remaining parameters coincide with those of Figure 2.

To clarify the role of C_F-coefficients in the operation regime of the device, let us take for example $\delta_\epsilon = 100\,k_B T$ (see Figure 3d) and $\delta V \sim 2\,k_B T$. As the sign in $C_F^{(1)}$ is negative, the linear-order solution given by Equation (62) is stable. If we start with $C_{\text{ext}} = 0$, then $\dot\theta$ is positive, so for $0 < C_{\text{ext}} < C_F^{(0)}$, the device acts as a motor (see Figure 2a). When $C_{\text{ext}} > C_F^{(0)} > 0$, $\dot\theta$ becomes negative, and hence, the device operates as a pump (see Figure 2b). Additionally, when $C_{\text{ext}} < 0$, the angular velocity is still positive, but as we changed the sign in C_{ext}, we have that $\mathcal{W}_{\text{ext}} < 0$, so the device operates as a pump. This is the other transition point between the two operation modes of the device, i.e., the sign in $\mathcal{W}_{\text{ext}} = s C_{\text{ext}}$ changes with s and the sign of C_{ext}. It is also interesting to notice how the operation regimes change with the sign of the bias voltage. For the force coefficients, we can see in the figure that they have definite parity with respect to the bias voltage:

$$C_F^{(k)}(-\delta V) = (-1)^{k+1} C_F^{(k)}(\delta V),\tag{69}$$

since Ω_x remains constant for a fixed orbit radius. For the chosen parameters $\bar\epsilon_L = \bar\epsilon_R$, $\Gamma_L = \Gamma_R$, and $\delta T = 0$, the transformation $\delta V \to -\delta V$ can be regarded as the inversion operation $(L, R) \to (R, L)$, which changes the sign of $\dot\theta$, i.e., $\dot\theta(-\delta V) = -\dot\theta(\delta V)$. In this sense, if there is some finite temperature gradient, this operation should also involve the sign inversion of δT. To infer how this bias inversion affects the final velocity of the device, we can replace Equation (69) in Equation (60). We can therefore recognize the transformation $\dot{\theta}(-\delta V) = -\dot{\theta}(\delta V)$ if we also change the sign of the external force, such that $\mathcal{C}_{\text{ext}}(-\delta V) = -\mathcal{C}_{\text{ext}}(\delta V)$. The even/odd parity in the k-order coefficient thus implies that the current-induced work is invariant under such a transformation, cf. Equation (59), and the ranges for the operation regimes of the device remain the same if we invert the sign of the external force. Of course, this analysis is no longer valid in more general situations where $\bar{\epsilon}_L \neq \bar{\epsilon}_R$ or $\Gamma_L \neq \Gamma_R$, such that the change in the sign of δV (or δT) cannot be related to the left-right inversion operation.

As an additional test for Equations (55) and (25), we define equivalent coefficients for the amount of transported charge and heat in a cycle:

$$\mathcal{C}_I^{(k)} = \int_0^{2\pi} \frac{\partial^k I_\theta}{\partial \dot{\theta}^k}\bigg|_{\dot{\theta}=0} \frac{d\theta}{k!}, \qquad \mathcal{C}_J^{(k)} = \int_0^{2\pi} \frac{\partial^k J_\theta}{\partial \dot{\theta}^k}\bigg|_{\dot{\theta}=0} \frac{d\theta}{k!}, \tag{70}$$

such that $Q_I^{(k)} = s\mathcal{C}_I^{(k)} \dot{\theta}^{k-1}$, and $Q_J^{(k)} = s\mathcal{C}_J^{(k)} \dot{\theta}^{k-1}$. These contributions can be evaluated independently from the $\mathcal{C}_F$-coefficients, and in Figure 3, these are shown in circles, which gives numerical agreement for the energy conservation principle:

$$\mathcal{C}_J^{(k)} + \mathcal{C}_I^{(k)} \delta V = -\mathcal{C}_F^{(k-1)}, \tag{71}$$

in the considered orders of $\dot{\theta}$. For the considered example in Figure 3, the current coefficients $\mathcal{C}_I$ and $\mathcal{C}_J$ also show (independently) a definite parity with respect to the bias voltage. In fact, as Equation (71) suggests, the heat current coefficients $\mathcal{C}_J^{(k)}$ present the same parity as $\mathcal{C}_F^{(k-1)}$, while the charge current coefficients $\mathcal{C}_I^{(k)}$ need to have the opposite parity since they are multiplied by δV.

Motor-Pump Efficiencies

As we stated above, the sign of the external force, together with the rotor's stationary condition and the first law of thermodynamics, determines the direction of the energy flow and, with it, the operation mode of the device. Obviously, as no other power sources are involved in this example, the efficiency of this energy conversion, defined as $\eta = (\text{output power})/(\text{input power})$, is always equal to one. This, however, only establishes those regions in the parameter space where we can expect the device operating as a quantum motor or a quantum energy pump. In this section, we discuss the particular conditions that appear when the device operates through a specific type of current. In this sense, the motor regime corresponds to the situation in which a transport current (e.g., charge, heat, spin, etc.) flowing through the leads in response to a bias (voltage, temperature, spin polarization, etc.) delivers some amount of energy into the local system, which can be used as mechanical work. The pump regime, on the other hand, corresponds to the inverse operation in which the external work is exploited to produce a current flowing against the imposed bias. This topic was also discussed in [87] for charge and heat currents in a DQD device, where limitations to the efficiency of the considered processes were attributed to the different orders appearing in the frequency expansion of the currents. We here provide a similar analysis in terms of our explicit model for the mechanical system. The inclusion of the external force in the description of the model, as we shall see next, appears as the key ingredient in bridging the motor and pump regimes for a given choice of the bias.

For the device acting as a motor, the output power should be given by $\mathcal{W}_{\text{ext}}/\tau$, under the condition $\mathcal{W}_{\text{ext}} > 0$, but we still need to specify the input power. If we consider that the mechanical rotor is driven by the electric current, i.e., due to some applied bias voltage and no thermal gradient applied, the input power is given by $-Q_I \cdot \delta V/\tau$, and hence, the efficiency of this type of motor is:

$$\eta = -\frac{\mathcal{W}_{\text{ext}}}{Q_I \cdot \delta V} = 1 + \frac{Q_J}{Q_I \delta V}. \tag{72}$$

Equation (33) establishes that the maximum efficiency for this device is $\eta \leq 1$, and the above equation tells us that the heat current produced by the bias voltage reduces the motor's performance. As in this work, we calculate such currents through an expansion in the angular velocity, the efficiency is also limited by this expansion. If in the calculation of $\dot{\theta}$, we consider Equation (60) up to first order in $C_F^{(k)}$, then, as discussed in Section 6, order-by-order energy conservation demands that the currents are to be considered up to second order, and in terms of the current coefficients, this takes the form:

$$\eta = -\frac{C_{\text{ext}}/\delta V}{C_I^{(0)}\dot{\theta}^{-1} + C_I^{(1)} + C_I^{(2)}\dot{\theta}}. \tag{73}$$

In the limits of the motor's operation regime, given by $C_{\text{ext}} = 0$ and $C_{\text{ext}} = C_F^{(0)}$, it is easy to see that the efficiency goes to zero, since for $C_{\text{ext}} = 0$, the numerator in the above expression is zero, and for $C_{\text{ext}} = C_F^{(0)}$ the rotor's velocity goes to zero, so the denominator grows to infinity due to the contribution $C_I^{(0)}/\dot{\theta}$ from the leakage current. The same happens if we consistently include higher order terms in this expression. For example, if we use Equation (63) for the rotor's angular velocity, then we should add $C_I^{(3)}\dot{\theta}^2$ in the denominator.

Away from this region, we enter in the "pumping domain" characterized by a charge current, which opposes the "natural" direction dictated by the bias voltage and $\mathcal{W}_{\text{ext}} < 0$. In this sense, the input power and the output power are inverted with respect to the motor region, and consequently, the efficiency of this "battery charger" device is given by:

$$\eta = -\frac{C_I^{(0)}\dot{\theta}^{-1} + C_I^{(1)} + C_I^{(2)}\dot{\theta}}{C_{\text{ext}}/\delta V}. \tag{74}$$

In this regime there is, however, an additional condition to be fulfilled, which is $Q_I \delta V > 0$. Regarding the different orders in Q_I, it usually happens that close to the transition point $C_{\text{ext}} = C_F^{(0)}$, the charge current still flows in the bias direction, since it is dominated by the leakage current. In this region, we say that the pumping mechanism is "frustrated" as the energy delivered by the rotor is not enough to reverse the direction of the charge current. Going away from this region, the angular velocity acquires some finite value, reducing the zeroth-order contribution $C_I^{(0)}/\dot{\theta}$ to the point where it is equal to the higher order contributions $C_I^{(1)} + C_I^{(2)}\dot{\theta}$, thus marking the activation point of the charge pump. In Figure 4a, we show the efficiency of the device as a function of the external force for different orbit sizes and fixed bias $\delta V = 2k_B T$. In all cases, the device starts from $C_{\text{ext}} = 0$ as a motor, and its efficiency reaches a maximum, which increases with the orbit size. Soon after this point, the motor's efficiency decreases to zero due to the leakage current effect, which becomes dominant at $C_{\text{ext}} = C_F^{(0)}$. From this point, we can observe the gapped region for the frustrated pump, which is more pronounced for small orbits, since the first-order pumped charge $Q_I^{(1)}$ is smaller than its quantized limit, and thereby, it takes a larger value of C_{ext} to compensate the amount of pumped leakage current $Q_I^{(0)}$ in a cycle.

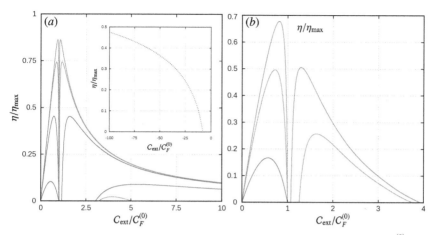

Figure 4. Normalized efficiencies as a function of the normalized external work ($\mathcal{C}_{\text{ext}}/\mathcal{C}_F^{(0)}$) and orbit radius: $\delta_\varepsilon = 10$ (red), 20 (blue), 50 (green), and 100 (orange), in units of k_BT. Panel (**a**) shows the electric motor/pump lowest order efficiencies (solid lines) for the device driven by a bias voltage $\delta V = 2k_BT$ and $\delta T = 0$. The next-order efficiency for the smallest orbit is shown in the dotted red and in the inset for negative $\mathcal{C}_{\text{ext}}$. Panel (**b**) shows the heat engine/refrigerator lowest order efficiencies for the device driven by a temperature gradient $\delta T = 0.5T$ and $\delta V = 0$. The other parameters coincide with those of Figure 2.

Up to this point, we have discussed only the effect of the lowest order terms of the expansion in $\dot{\theta}$, given by Equations (62), (73), and (74). For the smallest orbit, in addition, we show in the dotted red line the next-order efficiency, obtained from Equation (63) and adding the $Q_I^{(3)}$ term to Equations (73) and (74). We can see that for the motor regime, there are no significant changes, but for the pump regime, important differences appear. Firstly, in the $\mathcal{C}_{\text{ext}} > 0$ region, there is a cut-off for the external force in which the pumping mechanism is again frustrated, i.e., the charge current again points in the bias direction. As can be seen in Figure 5a,c, this decreasing in the efficiency is not attributed to the extra heat dissipated to the reservoirs, as one may expect. Note in the figure that the extra contribution to the pumped heat, $Q_J^{(3)}$, is negligible as compared to the lowest order terms. What happens here is that the third-order contribution to the pumped charge $Q_I^{(3)}$ rapidly becomes dominant in the charge pump region, causing a sudden drop in the efficiency and, with it, the appearance of a second frustrated-pump region. Secondly, another higher order effect appears in the $\mathcal{C}_{\text{ext}} < 0$ region. There, the efficiency is nonzero for $\mathcal{C}_{\text{ext}}/\mathcal{C}_F^{(0)} < -8$ (see the inset in Figure 4a), meaning that the pumping mechanism can be activated even when the external force points in the same direction as that of the current-induced force. Given the convention used for $\mathcal{F}_{\text{ext}}$ in Equation (7) and the chosen parameters, in this region, the sign of $\dot{\theta}$ remains the same as that when $\mathcal{F}_{\text{ext}} = 0$. There, the zeroth- and first-order contributions to the charge current flow in the same direction. In the analyzed case, again the third-order term is the one that reverses the direction of the total charge current; see Figure 5b. It is important to mention that the purpose of the present discussion is only to highlight deviations from the linear solution of Equation (62), not to analyze the convergence of the total pumped current. For the larger values of δ_ε used in Figure 4a, we do not show the next-order corrections, as they are negligible in the shown range of $\mathcal{C}_{\text{ext}}$.

An analysis similar to the above one can be carried out for the device driven by a temperature gradient δT, defined through $T_L = T + \delta T/2$ and $T_R = T - \delta T/2$ and no bias voltage applied, such that for $\delta T > 0$, we have $T_{\text{hot}} = T_L$ and $T_{\text{cold}} = T_R$. When $\mathcal{W}_{\text{ext}} > 0$, we have a motor device driven by a heat current in response to a thermal gradient (heat engine), then the input power should be given by $-Q_{J_{\text{hot}}}/\tau$, and Equation (35) implies:

$$\eta = -\frac{\mathcal{W}_{\text{ext}}}{Q_{J_{\text{hot}}}} = 1 + \frac{Q_{J_{\text{cold}}}}{Q_{J_{\text{hot}}}} \le 1 - \frac{T_{\text{cold}}}{T_{\text{hot}}} = \eta_{\text{carnot}}. \tag{75}$$

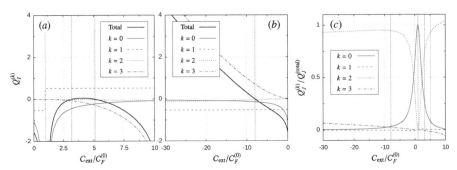

Figure 5. Different order contributions to the pumped charge and heat as a function of the normalized external work for the case $\delta_\varepsilon = 10\,k_BT$ (dotted red in Figure 4a). Panels (**a,b**) show the kth-order pumped charge $Q_J^{(k)} = (Q_{J_L}^{(k)} - Q_{J_R}^{(k)})/2$. The sum of all these contributions is shown in solid black and denoted as $Q_J^{(\text{total})}$. (**c**) Pumped heat contributions to the left and right reservoirs $Q_J^{(k)} = Q_{J_L}^{(k)} + Q_{J_R}^{(k)}$, divided by $Q_J^{(\text{total})}$, i.e., the sum of all contributions from $k = 0$ to $k = 3$. The vertical gray lines mark different transition points: $\mathcal{C}_{\text{ext}}/\mathcal{C}_F^{(0)} = 1$ is the motor/pump (energy pump) transition, while the other lines correspond to transitions between frustrated-pump/charge-pump regimes, i.e., when $Q_J^{(\text{total})}$ changes its sign.

As compared with Equation (73), we can see that the efficiency of the quantum heat engine, now given by:

$$\eta = -\frac{\mathcal{C}_{\text{ext}}}{\mathcal{C}_{J_{\text{hot}}}^{(0)}\dot{\theta}^{-1} + \mathcal{C}_{J_{\text{hot}}}^{(1)} + \mathcal{C}_{J_{\text{hot}}}^{(2)}\dot{\theta}}, \tag{76}$$

is defined in the same range for $\mathcal{C}_{\text{ext}}$ as in the electric motor. Regarding Figure 4b, the engine's normalized efficiency $\eta/\eta_{\text{carnot}}$ looks similar to that of the electric motor. Perhaps the only difference here is that for the smallest orbit $\delta_\varepsilon = 10k_BT$, the efficiency maximum is very low, such that it cannot be appreciated on the employed scale of the plot.

Now, we move to the heat pump region where the device acts as a refrigerator, as we demand that the heat current flows against the direction dictated by δT. Therefore, in addition to the $\mathcal{W}_{\text{ext}} < 0$ condition, the overall amount of pumped heat in the cold reservoir should be negative, i.e., $Q_{J_{\text{cold}}} < 0$. The efficiency of the refrigerator, or coefficient of performance (COP), is then given by:

$$\text{COP} = \frac{Q_{J_{\text{cold}}}}{\mathcal{W}_{\text{ext}}} = \frac{Q_{J_{\text{hot}}}}{Q_J} - 1 \le \frac{T_{\text{cold}}}{T_{\text{hot}} - T_{\text{cold}}} = \text{COP}_{\text{carnot}}, \tag{77}$$

where we can consistently expand $Q_{J_{\text{cold}}}$ in terms of $\dot{\theta}$. In Figure 4b, we show the lowest order contribution from Equation (62), as the next-order calculation does not change significantly the efficiencies in the considered regimes of the parameters. Again, we can observe in Figure 4b a gap region where the device is frustrated since the work delivered by the rotor is not enough to reverse the direction of the heat current. One of the differences with the electric counterpart is that, for the refrigerator, the normalized COP develops a maximum that is always smaller than that of the quantum heat engine, while the obtained efficiency maxima (motor and pump) for a fixed orbit in Figure 4a are very similar. Additionally, for the chosen value $\delta T = 0.5\,T$ and small orbit radius, the device can only work as a heat engine, and the refrigerator cannot be activated even if the external force is large, as happens for $\delta_\varepsilon = 20\,k_BT$ (solid blue line). The reason for this relies on the competition between the different orders in the pumped heat $Q_{J_{cold}}$: the reduction of the leakage pumped current, $Q_{J_{cold}}^{(0)}$, by increasing the magnitude of $\dot{\theta}$, is accompanied by an increase of the second-order contribution, $Q_{J_{cold}}^{(2)}$, such that the first-order term, $Q_{J_{cold}}^{(1)}$, may not be able to compensate these two and the requirement $Q_{J_{cold}} < 0$ cannot be fulfilled; see Figure 6.

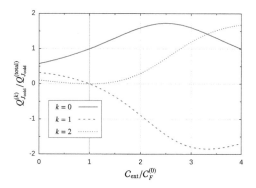

Figure 6. Different order contributions to the pumped heat as a function of the normalized external work for the case $\delta_\varepsilon = 20\,k_B T$ (solid blue line in Figure 4b). Here, $Q_J^{(total)}$ refers to the sum of all contributions.

8. Summary

Throughout this manuscript, we revisited some fundamental aspects related to the physics of quantum motors and pumps. Previous results based on the steady-state properties and energy conservation law were extended to deal with arbitrary nonequilibrium conditions in a systematic way. By considering the dynamics of the mechanical degrees of freedom through a Langevin equation, we were able to treat the motor/pump protocols on the same footing. This allowed us to describe the related energy transfer processes through a single parameter: the external force done on the local system.

In the steady-state regime, we treated in general terms the validity of the constant velocity assumption, in Section 3. For arbitrary orders of the nonadiabatic expansion in the CIFs, this was linked to the separation between the electronic and mechanical dynamic scales through a large moment of inertia. We then performed a general expansion (in terms of nonequilibrium sources) of the energy fluxes that took part in the quantum transport problem. This enabled us to derive an order-by-order scheme for the energy conservation law, Equation (25). This equation may be of help in recognizing the physical processes that enter at each order in the expansion, thereby providing a useful tool for the analysis of nonlinear effects. To illustrate this, we discussed the leading orders of the global expansion and showed how different types of expansions of the energy fluxes change the expressions for the efficiency of quantum motors and pumps.

In Section 7, we introduced a specific example of a quantum motor/pump based on a double quantum dot. There, we discussed in more depth how higher order terms of the CIFs affect the stationary state conditions. We found that multiple solutions for the device's terminal velocity could in principle be available for a fixed choice of parameters (voltage and temperature biases and external force). In such a case, the stability of such solutions imposes an additional constraint on the force coefficients, and the final steady state strongly depends on initial conditions. Interestingly, it is possible to obtain more than one stable solution, each of them belonging to a different operation mode of the device. The treated example is also appealing as it is possible to study the transition between different operational modes by continuously moving the external force. This corresponds to the point at which the steady-state velocity changes its sign and, with it, the direction of the energy flow. When considering a specific type of pumped current (charge or heat), there is an intermediate region where the pumping mechanism is "frustrated". In this situation, the energy delivered by the external force is not enough to reverse the natural direction of the charge or heat currents. We found other interesting features of the studied example such as negative friction coefficients at finite voltages or a definite parity of the expansion coefficients with respect to the bias voltage and the temperature gradient, which is a manifestation of the inversion symmetry in the total energy flux. We also used this example to confirm numerically the order-by-order energy conservation law up to third order in the final velocity. Finally, for heat currents, we found parameter conditions under which the device can never work as a "refrigerator", even for large values of the external force. We explained this behavior in terms of the competition between the different orders that participated in the pumped heat of the cold reservoir, highlighting the importance of the order-by-order conservation laws.

Author Contributions: Both authors contributed equally to this paper.

Funding: This research was funded by Secretaria de Ciencia y Tecnología-Universidad Nacional de Córdoba (Secyt-UNC) and Consejo Nacional de Investigaciones Científicas y Técnicas (CONICET). Both authors are members of CONICET.

Conflicts of Interest: The authors declare no conflict of interest.

Abbreviations

The following abbreviations are used in this manuscript:

CIF   Current-induced force
DQD   Double quantum dot
COP   Coefficient of performance

References

1. Martínez, I.A.; Roldán, E.; Dinis, L.; Petrov, D.; Parrondo, J.M.R.; Rica, R.A. Brownian Carnot engine. *Nat. Phys.* **2016**, *12*, 67, doi:10.1038/nphys3518.
2. Maillet, O.; Erdman, P.A.; Cavina, V.; Bhandari, B.; Mannila, E.T.; Peltonen, J.T.; Mari, A.; Taddei, F.; Jarzynski, C.; Giovannetti, V.; et al. Optimal Probabilistic Work Extraction beyond the Free Energy Difference with a Single-Electron Device. *Phys. Rev. Lett.* **2019**, *122*, 150604, doi:10.1103/PhysRevLett.122.150604.
3. Josefsson, M.; Svilans, A.; Burke, A.M.; Hoffmann, E.A.; Fahlvik, S.; Thelander, C.; Leijnse, M.; Linke, H. A quantum-dot heat engine operating close to the thermodynamic efficiency limits. *Nat. Nanotechnol.* **2018**, *13*, 920–924, doi:10.1038/s41565-018-0200-5.
4. Kim, K.; Xu, X.; Guo, J.; Fan, D.L. Ultrahigh-speed rotating nanomotors assembled from nanoscale building blocks. *Nat. Commun.* **2014**, *5*, 3632.
5. Michl, J.; Sykes, E.C.H. Molecular Rotors and Motors: Recent Advances and Future Challenges. *ACS Nano* **2009**, *3*, 1042.
6. Kudernac, T.; Ruangsupapichat, N.; Parschau, M.; Macia, B.; Katsonis, N.; Harutyunyan, S.R.; Ernst, K.H.; Feringa, B.L. Electrically driven directional motion of a four-wheeled molecule on a metal surface. *Nature* **2011**, *479*, 208.
7. Tierney, H.L.; Murphy, C.J.; Jewell, A.D.; Baber, A.E.; Iski, E.V.; Khodaverdian, H.Y.; McGuire, A.F.; Klebanov, N.; Sykes, E.C.H. Experimental demonstration of a single-molecule electric motor. *Nat. Nanotechnol.* **2011**, *6*, 625–629.
8. Chiaravalloti, F.; Gross, L.; Rieder, K.H.; Stojkovic, S.M.; Gourdon, A.; Joachim, C.; Moresco, F. A rack-and-pinion device at the molecular scale. *Nat. Mater.* **2007**, *6*, 30–33.
9. Cohen, D.; Kottos, T.; Schanz, H. Quantum pumping: The charge transported due to a translation of a scatterer. *Phys. Rev. E* **2005**, *71*, 035202.
10. Mozyrsky, D.; Hastings, M.B.; Martin, I. Intermittent polaron dynamics: Born–Oppenheimer approximation out of equilibrium. *Phys. Rev. B* **2006**, *73*, 035104, doi:10.1103/PhysRevB.73.035104.
11. Pistolesi, F.; Blanter, Y.M.; Martin, I. Self-consistent theory of molecular switching. *Phys. Rev. B* **2008**, *78*, 085127, doi:10.1103/PhysRevB.78.085127.

12. Cavaliere, F.; Governale, M.; König, J. Nonadiabatic Pumping through Interacting Quantum Dots. *Phys. Rev. Lett.* **2009**, *103*, 136801, doi:10.1103/PhysRevLett.103.136801.

13. Bustos-Marún, R.; Refael, G.; von Oppen, F. Adiabatic quantum motors. *Phys. Rev. Lett.* **2013**, *111*, 060802, doi:10.1103/PhysRevLett.111.060802.

14. Mazza, F.; Bosisio, R.; Benenti, G.; Giovannetti, V.; Fazio, R.; Taddei, F. Thermoelectric efficiency of three-terminal quantum thermal machines. *New J. Phys.* **2014**, *16*, 085001.10.1088/1367-2630/16/8/085001.

15. Roßnagel, J.; Abah, O.; Schmidt-Kaler, F.; Singer, K.; Lutz, E. Nanoscale Heat Engine Beyond the Carnot Limit. *Phys. Rev. Lett.* **2014**, *112*, 030602, doi:10.1103/PhysRevLett.112.030602.

16. Perroni, C.A.; Romeo, F.; Nocera, A.; Ramaglia, V.M.; Citro, R.; Cataudella, V. Noise-assisted charge pump in elastically deformable molecular junctions. *J. Phys. Condens. Matter* **2014**, *26*, 365301, doi:10.1088/0953-8984/26/36/365301.

17. Brask, J.B.; Haack, G.; Brunner, N.; Huber, M. Autonomous quantum thermal machine for generating steady-state entanglement. *New J. Phys.* **2015**, *17*, 113029, doi:10.1088/1367-2630/17/11/113029.

18. Campisi, M.; Pekola, J.; Fazio, R. Nonequilibrium fluctuations in quantum heat engines: Theory, example, and possible solid state experiments. *New J. Phys.* **2015**, *17*, 035012, doi:10.1088/1367-2630/17/3/035012.

19. Celestino, A.; Croy, A.; Beims, M.W.; Eisfeld, A. Rotational directionality via symmetry-breaking in an electrostatic motor. *New J. Phys.* **2016**, *18*, 063001, doi:10.1088/1367-2630/18/6/063001.

20. Silvestrov, P.G.; Recher, P.; Brouwer, P.W. Noiseless manipulation of helical edge state transport by a quantum magnet. *Phys. Rev. B* **2016**, *93*, 205130, doi:10.1103/PhysRevB.93.205130.

21. Romero, J.I.; Roura-Bas, P.; Aligia, A.A.; Arrachea, L. Nonlinear charge and energy dynamics of an adiabatically driven interacting quantum dot. *Phys. Rev. B* **2017**, *95*, 235117.10.1103/PhysRevB.95.235117.

22. Fernández-Alcázar, L.J.; Pastawski, H.M.; Bustos-Marún, R.A. Dynamics and decoherence in nonideal Thouless quantum motors. *Phys. Rev. B* **2017**, *95*, 155410, doi:10.1103/PhysRevB.95.155410.

23. Bustos-Marún, R.A. Geometric rectification for nanoscale vibrational energy harvesting. *Phys. Rev. B* **2018**, *97*, 075412, doi:10.1103/PhysRevB.97.075412.

24. Bruch, A.; Kusminskiy, S.V.; Refael, G.; von Oppen, F. Interacting adiabatic quantum motor. *Phys. Rev. B* **2018**, *97*, 195411, doi:10.1103/PhysRevB.97.195411.

25. Roura-Bas, P.; Arrachea, L.; Fradkin, E. Helical spin thermoelectrics controlled by a side-coupled magnetic quantum dot in the quantum spin Hall state. *Phys. Rev. B* **2018**, *98*, 195429, doi:10.1103/PhysRevB.98.195429.

26. Ludovico, M.F.; Capone, M. Enhanced performance of a quantum-dot-based nanomotor due to Coulomb interactions. *Phys. Rev. B* **2018**, *98*, 235409, doi:10.1103/PhysRevB.98.235409.

27. Terrén Alonso, P.; Romero, J.; Arrachea, L. Work exchange, geometric magnetization, and fluctuation-dissipation relations in a quantum dot under adiabatic magnetoelectric driving. *Phys. Rev. B* **2019**, *99*, 115424, doi:10.1103/PhysRevB.99.115424.

28. Lin, H.H.; Croy, A.; Gutierrez, R.; Joachim, C.; Cuniberti, G. Current-induced rotations of molecular gears. *J. Phys. Commun.* **2019**, *3*, 025011, doi:10.1088/2399-6528/ab0731.

29. Manzano, G.; Silva, R.; Parrondo, J.M.R. Autonomous thermal machine for amplification and control of energetic coherence. *Phys. Rev. E* **2019**, *99*, 042135, doi:10.1103/PhysRevE.99.042135.

30. Sánchez, D.; Sánchez, R.; López, R.; Sothmann, B. Nonlinear chiral refrigerators. *Phys. Rev. B* **2019**, *99*, 245304, doi:10.1103/PhysRevB.99.245304.

31. Cohen, D. Quantum pumping and dissipation: From closed to open systems. *Phys. Rev. B* **2003**, *68*, 201303R.

32. Avron, J.E.; Elgart, A.; Graf, G.M.; Sadun, L. Transport and Dissipation in Quantum Pumps. *J. Stat. Phys.* **2004**, *116*, 425, doi:10.1023/B:JOSS.0000037245.45780.e1.

33. Horsfield, A.P.; Bowler, D.R.; Fisher, A.J.; Todorov, T.N.; Sánchez, C.G. Beyond Ehrenfest: Correlated non-adiabatic molecular dynamics. *J. Phys. Condens. Matter* **2004**, *16*, 8251.10.1088/0953-8984/16/46/012.

34. Bennett, S.D.; Maassen, J.; Clerk, A.A. Scattering approach to backaction in coherent nanoelectromechanical systems. *Phys. Rev. Lett.* **2010**, *105*, 217206, doi:10.1103/PhysRevLett.105.217206.

35. Bode, N.; Viola Kusminskiy, S.; Egger, R.; von Oppen, F. Scattering theory of current-induced forces in mesoscopic systems. *Phys. Rev. Lett.* **2011**, *107*, 036804, doi:10.1103/PhysRevLett.107.036804.

36. Esposito, M.; den Broeck, C.V. Second law and Landauer principle far from equilibrium. *EPL (Europhys. Lett.)* **2011**, *95*, 40004, doi:10.1209/0295-5075/95/40004.

37. Thomas, M.; Karzig, T.; Kusminskiy, S.V.; Zaránd, G.; von Oppen, F. Scattering theory of adiabatic reaction forces due to out-of-equilibrium quantum environments. *Phys. Rev. B* **2012**, *86*, 195419, doi:10.1103/PhysRevB.86.195419.

38. Uzdin, R.; Levy, A.; Kosloff, R. Equivalence of Quantum Heat Machines, and Quantum-Thermodynamic Signatures. *Phys. Rev. X* **2015**, *5*, 031044, doi:10.1103/PhysRevX.5.031044.

39. Fernández-Alcázar, L.J.; Bustos-Marún, R.A.; Pastawski, H.M. Decoherence in current induced forces: Application to adiabatic quantum motors. *Phys. Rev. B* **2015**, *92*, 075406, doi:10.1103/PhysRevB.92.075406.

40. Uzdin, R.; Levy, A.; Kosloff, R. Quantum Heat Machines Equivalence, Work Extraction beyond Markovianity, and Strong Coupling via Heat Exchangers. *Entropy* **2016**, *18*, 124, doi:10.3390/e18040124.

41. Pluecker, T.; Wegewijs, M.R.; Splettstoesser, J. Gauge freedom in observables and Landsberg's nonadiabatic geometric phase: Pumping spectroscopy of interacting open quantum systems. *Phys. Rev. B* **2017**, *95*, 155431, doi:10.1103/PhysRevB.95.155431.

42. Calvo, H.L.; Ribetto, F.D.; Bustos-Marún, R.A. Real-time diagrammatic approach to current-induced forces: Application to quantum-dot based nanomotors. *Phys. Rev. B* **2017**, *96*, 165309, doi:10.1103/PhysRevB.96.165309.

43. Ludovico, M.F.; Arrachea, L.; Moskalets, M.; Sánchez, D. Probing the energy reactance with adiabatically driven quantum dots. *Phys. Rev. B* **2018**, *97*, 041416, doi:10.1103/PhysRevB.97.041416.

44. Hopjan, M.; Stefanucci, G.; Perfetto, E.; Verdozzi, C. Molecular junctions and molecular motors: Including Coulomb repulsion in electronic friction using nonequilibrium Green's functions. *Phys. Rev. B* **2018**, *98*, 041405, doi:10.1103/PhysRevB.98.041405.

45. Chen, F.; Miwa, K.; Galperin, M. Current-Induced Forces for Nonadiabatic Molecular Dynamics. *J. Phys. Chem. A* **2019**, *123*, 693–701, doi:10.1021/acs.jpca.8b09251.

46. Büttiker, M.; Thomas, H.; Prêtre, A. Current partition in multiprobe conductors in the presence of slowly oscillating external potentials. *Z. Phys. B* **1994**, *94*, 133–137, doi:10.1007/BF01307664.

47. Brouwer, P.W. Scattering approach to parametric pumping. *Phys. Rev. B* **1998**, *58*, R10135, doi:10.1103/PhysRevB.58.R10135.

48. Sánchez, D.; López, R. Nonlinear phenomena in quantum thermoelectrics and heat. *Comptes Rendus Phys.* **2016**, *17*, 1060–1071, doi:10.1016/j.crhy.2016.08.005.

49. Ludovico, M.F.; Battista, F.; von Oppen, F.; Arrachea, L. Adiabatic response and quantum thermoelectrics for AC-driven quantum systems. *Phys. Rev. B* **2016**, *93*, 075136, doi:10.1103/PhysRevB.93.075136.

50. Ludovico, M.F.; Moskalets, M.; Sánchez, D.; Arrachea, L. Dynamics of energy transport and entropy production in AC-driven quantum electron systems. *Phys. Rev. B* **2016**, *94*, 035436.10.1103/PhysRevB.94.035436.

51. Ludovico, M.F.; Arrachea, L.; Moskalets, M.; Sánchez, D. Periodic Energy Transport and Entropy Production in Quantum Electronics. *Entropy* **2016**, *18*, 419, doi:10.3390/e18110419.

52. Benenti, G.; Casati, G.; Saito, K.; Whitney, R.S. Fundamental aspects of steady-state conversion of heat to work at the nanoscale. *Phys. Rep.* **2017**, *694*, 1–124, doi:10.1016/j.physrep.2017.05.008.

53. Whitney, R.S.; Sánchez, R.; Splettstoesser, J. Quantum Thermodynamics of Nanoscale Thermoelectrics and Electronic Devices. In *Thermodynamics in the Quantum Regime. Fundamental Theories of Physics*; Binder, F., Correa, L., Gogolin, C., Anders, J., Adesso, G., Eds.; Springer: Cham, Switzerland, 2018; Volume 195.

54. Cremers, J.N.H.J.; Brouwer, P.W. Dephasing in a quantum pump. *Phys. Rev. B* **2002**, *65*, 115333, doi:10.1103/PhysRevB.65.115333.

55. Thouless, D.J. Quantization of particle transport. *Phys. Rev. B* **1983**, *27*, 6083.

56. Fernández-Alcázar, L.J.; Pastawski, H.M.; Bustos-Marún, R.A. Nonequilibrium current-induced forces caused by quantum localization: Anderson adiabatic quantum motors. *Phys. Rev. B* **2019**, *99*, 045403, doi:10.1103/PhysRevB.99.045403.

57. Reckermann, F.; Splettstoesser, J.; Wegewijs, M.R. Interaction-Induced Adiabatic Nonlinear Transport. *Phys. Rev. Lett.* **2010**, *104*, 226803, doi:10.1103/PhysRevLett.104.226803.

58. Calvo, H.L.; Classen, L.; Splettstoesser, J.; Wegewijs, M.R. Interaction-induced charge and spin pumping through a quantum dot at finite bias. *Phys. Rev. B* **2012**, *86*, 245308, doi:10.1103/PhysRevB.86.245308.

59. Placke, B.A.; Pluecker, T.; Splettstoesser, J.; Wegewijs, M.R. Attractive and driven interactions in quantum dots: Mechanisms for geometric pumping. *Phys. Rev. B* **2018**, *98*, 085307, doi:10.1103/PhysRevB.98.085307.

60. Moskalets, M.; Büttiker, M. Effect of inelastic scattering on parametric pumping. *Phys. Rev. B* **2001**, *64*, 201305, doi:10.1103/PhysRevB.64.201305.

61. Esposito, M.; Ochoa, M.A.; Galperin, M. Quantum Thermodynamics: A Nonequilibrium Green's Function Approach. *Phys. Rev. Lett.* **2015**, *114*, 080602, doi:10.1103/PhysRevLett.114.080602.
62. Ribeiro, R.M.; Pereira, V.M.; Peres, N.M.R.; Briddon, P.R.; Neto, A.H.C. Strained graphene: Tight-binding and density functional calculations. *New J. Phys.* **2009**, *11*, 115002, doi:10.1088/1367-2630/11/11/115002.
63. Dundas, D.; McEniry, E.J.; Todorov, T.N. Current-driven atomic waterwheels. *Nat. Nanotechnol.* **2009**, *4*, 99, doi:10.1038/NNANO.2008.411.
64. Suárez Morell, E.; Correa, J.D.; Vargas, P.; Pacheco, M.; Barticevic, Z. Flat bands in slightly twisted bilayer graphene: Tight-binding calculations. *Phys. Rev. B* **2010**, *82*, 121407, doi:10.1103/PhysRevB.82.121407.
65. Lü, J.T.; Brandbyge, M.; Hedegård, P. Blowing the Fuse: Berry's Phase and Runaway Vibrations in Molecular Conductors. *Nano Lett.* **2010**, *10*, 1657–1663, doi:10.1021/nl904233u.
66. Cunningham, B.; Todorov, T.N.; Dundas, D. Nonconservative current-driven dynamics: Beyond the nanoscale. *Beilstein J. Nanotechnol.* **2015**, *6*, 2140–2147, doi:10.3762/bjnano.6.219.
67. Lü, J.T.; Christensen, R.B.; Wang, J.S.; Hedegård, P.; Brandbyge, M. Current-Induced Forces and Hot Spots in Biased Nanojunctions. *Phys. Rev. Lett.* **2015**, *114*, 096801, doi:10.1103/PhysRevLett.114.096801.
68. Lü, J.T.; Wang, J.S.; Hedegård, P.; Brandbyge, M. Electron and phonon drag in thermoelectric transport through coherent molecular conductors. *Phys. Rev. B* **2016**, *93*, 205404, doi:10.1103/PhysRevB.93.205404.
69. Gu, L.; Fu, H.H. Current-induced enhancement of DNA bubble creation. *New J. Phys.* **2016**, *18*, 053032, doi:10.1088/1367-2630/18/5/053032.
70. Bai, M.; Cucinotta, C.S.; Jiang, Z.; Wang, H.; Wang, Y.; Rungger, I.; Sanvito, S.; Hou, S. Current-induced phonon renormalization in molecular junctions. *Phys. Rev. B* **2016**, *94*, 035411.10.1103/PhysRevB.94.035411.
71. Christensen, R.B.; Lü, J.T.; Hedegård, P.; Brandbyge, M. Current-induced runaway vibrations in dehydrogenated graphene nanoribbons. *Beilstein J. Nanotechnol.* **2016**, *7*, 68–74, doi:10.3762/bjnano.7.8.
72. Leijnse, M.; Wegewijs, M.R. Kinetic equations for transport through single-molecule transistors. *Phys. Rev. B* **2008**, *78*, 235424, doi:10.1103/PhysRevB.78.235424.
73. Holzbecher, A. Pumping and Motors in the Coulomb Blockade Regime. Master's Thesis, Freie Universität Berlin, Berlin, Germany, 2014.
74. Arrachea, L.; von Oppen, F. Nanomagnet coupled to quantum spin Hall edge: An adiabatic quantum motor. *Phys. E* **2015**, *74*, 96, doi:10.1016/j.physe.2015.08.031.
75. Di Ventra, M.; Pantelides, S.T. Hellmann-Feynman theorem and the definition of forces in quantum time-dependent and transport problems. *Phys. Rev. B* **2000**, *61*, 16207, doi:10.1103/PhysRevB.61.16207.
76. Di Ventra, M.; Chen, Y.C. Are current-induced forces conservative? *Phys. Rev. Lett.* **2004**, *92*, 176803.
77. Todorov, T.N.; Dundas, D. Nonconservative generalized current-induced forces. *Phys. Rev. B* **2010**, *81*, 075416, doi:10.1103/PhysRevB.81.075416.
78. Splettstoesser, J.; Governale, M.; König, J.; Fazio, R. Adiabatic pumping through a quantum dot with Coulomb interactions: A perturbation expansion in the tunnel coupling. *Phys. Rev. B* **2006**, *74*, 085305, doi:10.1103/PhysRevB.74.085305.
79. Antoni, M.; Ruffo, S. Clustering and relaxation in Hamiltonian long-range dynamics. *Phys. Rev. E* **1995**, *52*, 2361–2374, doi:10.1103/PhysRevE.52.2361.
80. Levin, Y.; Pakter, R.; Rizzato, F.B.; Teles, T.N.; Benetti, F.P. Nonequilibrium statistical mechanics of systems with long-range interactions. *Phys. Rep.* **2014**, *535*, 1–60, doi:10.1016/j.physrep.2013.10.001.
81. Atenas, B.; Curilef, S. Dynamics and thermodynamics of systems with long-range dipole-type interactions. *Phys. Rev. E* **2017**, *95*, 022110, doi:10.1103/PhysRevE.95.022110.
82. Spiechowicz, J.; Hänggi, P.; Łuczka, J. Brownian motors in the microscale domain: Enhancement of efficiency by noise. *Phys. Rev. E* **2014**, *90*, 032104, doi:10.1103/PhysRevE.90.032104.
83. Ludovico, M.F.; Lim, J.S.; Moskalets, M.; Arrachea, L.; Sánchez, D. Dynamical energy transfer in AC-driven quantum systems. *Phys. Rev. B* **2014**, *89*, 161306, doi:10.1103/PhysRevB.89.161306.
84. Plenio, M.B.; Vitelli, V. The physics of forgetting: Landauer's erasure principle andinformation theory. *Contemp. Phys.* **2001**, *42*, 25–60, doi:10.1080/00107510010018916.
85. Deutsch, J.M. Thermodynamic entropy of a many-body energy eigenstate. *New J. Phys.* **2010**, *12*, 075021, doi:10.1088/1367-2630/12/7/075021.
86. Gemmer, J.; Steinigeweg, R. Entropy increase in *K*-step Markovian and consistent dynamics of closed quantum systems. *Phys. Rev. E* **2014**, *89*, 042113, doi:10.1103/PhysRevE.89.042113.

87. Juergens, S.; Haupt, F.; Moskalets, M.; Splettstoesser, J. Thermoelectric performance of a driven double quantum dot. *Phys. Rev. B* **2013**, *87*, 245423, doi:10.1103/PhysRevB.87.245423.

88. Splettstoesser, J.; Governale, M.; König, J. Adiabatic charge and spin pumping through quantum dots with ferromagnetic leads. *Phys. Rev. B* **2008**, *77*, 195320, doi:10.1103/PhysRevB.77.195320.

89. Riwar, R.P.; Splettstoesser, J. Charge and spin pumping through a double quantum dot. *Phys. Rev. B* **2010**, *82*, 205308, doi:10.1103/PhysRevB.82.205308.

90. Hiltscher, B.; Governale, M.; Splettstoesser, J.; König, J. Adiabatic pumping in a double-dot Cooper-pair beam splitter. *Phys. Rev. B* **2011**, *84*, 155403, doi:10.1103/PhysRevB.84.155403.

91. Rojek, S.; König, J.; Shnirman, A. Adiabatic pumping through an interacting quantum dot with spin-orbit coupling. *Phys. Rev. B* **2013**, *87*, 075305, doi:10.1103/PhysRevB.87.075305.

92. Riwar, R.P.; Splettstoesser, J.; König, J. Zero-frequency noise in adiabatically driven interacting quantum systems. *Phys. Rev. B* **2013**, *87*, 195407, doi:10.1103/PhysRevB.87.195407.

93. Winkler, N.; Governale, M.; König, J. Theory of spin pumping through an interacting quantum dot tunnel coupled to a ferromagnet with time-dependent magnetization. *Phys. Rev. B* **2013**, *87*, 155428, doi:10.1103/PhysRevB.87.155428.

94. Rojek, S.; Governale, M.; König, J. Spin pumping through quantum dots. *Phys. Status Solidi B* **2014**, *251*, 1912, doi:10.1002/pssb.201350213.

95. Haupt, F.; Leijnse, M.; Calvo, H.L.; Classen, L.; Splettstoesser, J.; Wegewijs, M.R. Heat, molecular vibrations, and adiabatic driving in non-equilibrium transport through interacting quantum dots. *Phys. Status Solidi B* **2013**, *250*, 2315–2329, doi:10.1002/pssb.201349219.

Article

Effective Equilibrium in Out-of-Equilibrium Interacting Coupled Nanoconductors

Lucas Maisel and Rosa López *

Institut de Física Interdisciplinària i de Sistemes Complexos IFISC (CSIC-UIB),
E-07122 Palma de Mallorca, Spain; lucas.maisel@gmail.com
* Correspondence: rosa.lopez-gonzalo@uib.es

Received: 24 October 2019; Accepted: 11 December 2019; Published: 19 December 2019

Abstract: In the present work, we study a mesoscopic system consisting of a double quantum dot in which both quantum dots or artificial atoms are electrostatically coupled. Each dot is additionally tunnel coupled to two electronic reservoirs and driven far from equilibrium by external voltage differences. Our objective is to find configurations of these biases such that the current through one of the dots vanishes. In this situation, the validity of the fluctuation–dissipation theorem and Onsager's reciprocity relations has been established. In our analysis, we employ a master equation formalism for a minimum model of four charge states, and limit ourselves to the sequential tunneling regime. We numerically study those configurations far from equilibrium for which we obtain a stalling current. In this scenario, we explicitly verify the fluctuation–dissipation theorem, as well as Onsager's reciprocity relations, which are originally formulated for systems in which quantum transport takes place in the linear regime.

Keywords: quantum transport; quantum dots; fluctuation–dissipation theorem; Onsager relations

---

1. Introduction

The two paradigms of statistical mechanics for systems that are close to equilibrium are: (i) the Onsager–Casimir reciprocity relations [1]; and (ii) the fluctuation–dissipation theorem (FDT) [2–4]. Both relations are not only inherent to classical systems but are also applicable to the quantum regime. The Onsager–Casimir reciprocity relations state that the Onsager matrix that relates physical fluxes and their conjugate forces is symmetric. For example, considering as forces the electrostatic and thermal gradients, and their associated currents being the electrical and heat fluxes, these relations set an identity between the thermoelectrical conductance (electrical response to a thermal gradient) and the electrothermal conductance (response of the heat current to an electrical bias). On the other hand, the FDT establishes that statistical fluctuations occurring in a system at equilibrium behave similarly to the dissipation that takes place under the action of an external perturbation. Major examples of manifestations of the FDT are found in Einstein's treatment of Brownian motion where the diffusion constant is found to be proportional to the mobility [5] or the Johnson–Nyquist formula for electronic white noise [6]. In the context of quantum transport through electronic nanodevices, the FDT allows us to relate the dissipative response of one current with respect to a variation of its affinity or conjugate force with its spontaneous fluctuations. This property of equilibrium systems is a very important topic when we are interested in controlling dissipation due to currents induced through quantum conductors by external forces.

As mentioned above, the range of validity of the FDT is limited to the linear response regime, i.e., for sufficiently small perturbations. Going beyond this regime requires generalizing this formulation to non-equilibrium conditions. This has been done by introducing additional correlations involving the activity, a magnitude related to the transition rates and the excess of entropy production that is modified antisymmetrically by the external potential that drives the system out of equilibrium [7–10]. In this view, these extensions to the FDT are indeed fluctuation–dissipation relations (FDR) that establish the frequency at which a system produces entropy to the environment between forward and backward processes. The interest of the FDR has been highlighted in the field of quantum transport [1,11–13].

However, here we adopt a different perspective, reported in the work of B. Altaner, M. Polettini, and M. Esposito [14], in which the concept of *stalling currents* is introduced in the context of stochastic thermodynamics. A current that traverses a system can be nullified because of the cancellation of a set of distinct internal processes, and is then called a *stalling current*. Under these conditions, if the perturbative force solely affects the microscopic transitions that contribute to this current, the FDT is restored [14,15]. In addition, we test numerically that Onsager reciprocity relations are additionally satisfied at stalling conditions. We speculate that this property is attained due to the lack of entropy production at stalling conditions forced by the tight coupling between the charge and heat currents (see below). The conclusion is that all contributing elemental transitions being internally equilibrated is equivalent to them being microscopically reversible. One interesting application to the stalling configuration is that, even though correlations are usually difficult to access experimentally, the fact that the FDT is applicable makes it rather easy to obtain such correlations by means of a response function instead.

Our purpose in this work is to implement these conclusions in a nanodevice consisting of two interacting conductors. Such setup was previously investigated by R. Sánchez et al. [16] to analyze the drag effect. The device consists of a parallel double quantum dot system in which the quantum dots interact electrostatically via a mutual capacitance. Besides, each quantum dot is tunnel-connected to two electronic reservoirs. A drag current is encountered in one of the dots, which is unbiased, due to the charge fluctuations provoked by the electrical current driven through the other dot. The detection of this drag current has been demonstrated experimentally [17] showing that high-order tunneling events such as cotunneling have a significant contribution. Besides, a drag current control has been proposed by attaching to the dots different materials with nontrivial energy-dispersion relations [18]. This system has additionally been proposed for the implementation of a Maxwell demon [19], in which one of the dots (the demon) acquires information from the other one, allowing a current to flow opposite to the applied bias voltage in the other dot.

Our goal in this article is to explore the transport properties in an out-of-equilibrium configuration that drives the system into an effective equilibrium in which both the Onsager relations and the FDT are recovered. For this purpose, we compute the electrical and heat currents through each quantum dot. By a numerical search of stalling currents in one of the dots, we check whether or not Onsager relations and the FDT are satisfied. We consider different situations. Firstly, we consider the case where both the electrical and heat flows are cancelled simultaneously under non-equilibrium configurations. This can be achieved only in the so-called strong coupling regime. For this case, we demonstrate that the system indeed behaves as at equilibrium. We also analyze the scenario where only one of the currents vanishes (either the charge or the heat flow), while the other one is kept finite. Finally, we show that the absence of stalling currents prevents the fulfillment of the Onsager relations and the FDT as expected. To conclude, we go beyond the FDT and additionally check the FDRs for the third cumulant in the presence of stalling currents.

2. Theoretical Model

2.1. Description of the System and Underlying Framework

We consider the case of two conductors that are mutually connected via the Coulomb interaction. Each conductor consists of a quantum dot with a single level active for transport. We omit spin indices due to spin degeneracy. Besides, we consider a large on-site Coulomb interaction that prevents the double occupancy in each dot. Each quantum dot is tunnel-coupled to two electronic reservoirs that can be biased with electrostatic and thermal gradients. Each tunneling barrier is modeled by capacitors denoted by C_i with $i = 1, 2, 3, 4$. As mentioned above, the two quantum dots interact electrostatically through a capacitor C. A sketch for this system is depicted in Figure 1b. Under these circumstances, we describe the system using four possible charge states $|0\rangle = |0_u 0_d\rangle$, $|u\rangle = |1_u 0_d\rangle$, $|d\rangle = |0_u 1_d\rangle$, and $|2\rangle = |1_u 1_d\rangle$, where $n_u n_d$ denotes the charge state with n_u electrons in the upper dot and n_d electrons in the lower dot. For simplicity, we consider an isothermal configuration in which all reservoirs are held at a common temperature T. We also keep different bias voltages V_i applied to the four terminals.

We are interested in the charge and heat transport in the sequential tunneling regime, in which the tunneling rate (denoted by Γ) satisfies $\hbar\Gamma \ll k_B T$. In this regime, transport of electrons along each quantum dot occurs in a sequence of one electron transfer event at each time. Electrons can hop into a quantum dot, and then relax before they jump again. This restriction eliminates the transitions $|0(2)\rangle \rightarrow |2(0)\rangle$ and $|u(d)\rangle \rightarrow |d(u)\rangle$. Additionally, we consider that there is no particle transfer from one dot to another by tunneling. The only interaction between the dots is then due to their mutual influence caused by the electrostatic interactions.

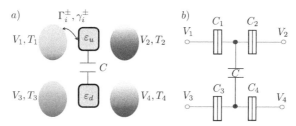

Figure 1. (**a**) Double quantum dot capacitively coupled to four terminals held at potentials V_i and temperatures T_i, for $i = 1, 2, 3, 4$. The transition rates $\Gamma_i^\pm$ and $\gamma_i^\pm$ for each barrier are described in the main text. (**b**) Electrostatic sketch showing the capacitors and voltages involved in the description of the energy levels of the quantum dots.

The theoretical framework employed to describe the quantum transport in our system is called *stochastic thermodynamics* [14,20–22]. Quite generally, we can consider a setup with an arbitrary number of states $n \in \{1, 2, ..., N\}$ and picture each state as a node in a connected network. We draw edges **e** connecting states between which a transition may occur, and require these to be possible in both directions. However, transitions along $\pm$**e** are not required to happen at the same rate or with the same probability. Note that two nodes may be connected with several edges if there are various physical mechanisms through which the system can transition between the associated states. The evolution of the system is modeled as a Markov jump process, i.e., the probability that the system jumps from one state to another is independent of its previous history. This evolution can also be visualized as a random walk on the network. A physical model is defined by prescribing the forward and backward transition rates $w_{\pm\mathbf{e}}$, which evidently may be functions of the physical parameters involved. The fluctuating current along an edge **e**, $j_\mathbf{e}(t) = \sum_k \delta(t - t_k)(\delta_{+\mathbf{e},\mathbf{e}_k} - \delta_{-\mathbf{e},\mathbf{e}_k})$, is a stochastic variable that peaks if the system transitions along the directed edge $\mathbf{e}_k$ at time t_k. Physical currents, i.e., currents associated to the transport of physical quantities such as charge or heat, are weighted currents $J_\alpha = \sum_\mathbf{e} d_\mathbf{e}^\alpha j_\mathbf{e}$, where $d_{+\mathbf{e}}^\alpha = -d_{-\mathbf{e}}^\alpha$ specifies the amount of a physical variable α exchanged with an external reservoir along a transition edge **e**.

When applying the previous theoretical treatment to our particular system, we consider that the tunneling rates depend on the energy of the system. Specifically, we consider the value of Γ_i for the tunneling of electrons between a reservoir i and a quantum dot whenever the other dot is empty, and γ_i when the other dot is occupied. Then, the transition rates (previously called $w_{\pm\mathbf{e}}$) are thus dependent on the dot charge states. The transition rates are defined according to Fermi's golden rule as

$$\Gamma_i^- = \Gamma_i f(\mu_{\ell,0} - qV_i) \tag{1}$$

$$\Gamma_i^+ = \Gamma_i \left(1 - f(\mu_{\ell,0} - qV_i)\right) \tag{2}$$

$$\gamma_i^- = \gamma_i f(\mu_{\ell,1} - qV_i) \tag{3}$$

$$\gamma_i^+ = \gamma_i \left(1 - f(\mu_{\ell,1} - qV_i)\right) \tag{4}$$

where $f(x) = \left(1 + e^{x/k_B T}\right)^{-1}$ is the Fermi–Dirac distribution function, the $-$ superscript stands for the tunneling from the lead to the dot, and $+$ for the reverse process. The transition rates are schematically represented in a network diagram in Figure 2. In our arrangement, we take the *up* dot ($\ell = u$) connected to left and right reservoirs with $i = \{1, 2\}$, and the *down* dot ($\ell = d$) to reservoirs with $i = \{3, 4\}$. Note that the numerical subindex in the previous transition rates thus indicates the reservoir involved in the transition, as shown in Figure 1a. The chemical potential for the dot ℓ, i.e., $\mu_{\ell,0}$ ($\mu_{\ell,1}$), corresponds to the situation in which the *other* dot is empty (occupied).

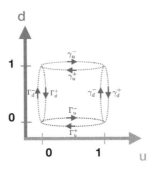

Figure 2. Scheme for the transition rates between the two dot states.

To determine the effective chemical potentials of the dots, we must develop a model that takes into account how their energy levels are influenced by electrostatic interactions. When interactions are properly included as in our description, all currents are gauge invariant, as they depend only on voltage differences. Hereafter, we shorten the notation and define $V_{ij} \equiv V_i - V_j$. Under these considerations, the dot levels become

$$\varepsilon_{u,n} \rightarrow \mu_{u,n} = \varepsilon_u + U(1,0) - U(0,0) + E_C \delta_{1n} \tag{5}$$

$$\varepsilon_{d,n} \rightarrow \mu_{d,n} = \varepsilon_d + U(0,1) - U(0,0) + E_C \delta_{1n} \tag{6}$$

where ε_u and ε_d are the bare energy levels, and $n = 0$ (1) corresponds to the case where *other* dot is empty (occupied). Here, $E_C = 2q^2 C / \left(C_{\Sigma u} C_{\Sigma d} - C^2\right)$ is the charging energy with $C_{\Sigma d(u)} = C_{1(3)} + C_{2(4)} + C$. The chemical potential $\mu_{u(d),n}$ is defined as the change in the electrostatic energy when the charge number $N_{u(d)}$ changes by one when the dot $d(u)$ is either empty ($n = 0$) or occupied by one electron ($n = 1$). The electrostatic energy is computed from $U(N_u, N_d) = \sum_i \int_0^{qN_i} dQ_i' \, \phi_i(Q_i')$ where ϕ_i is the internal potential in each quantum dot obtained by means of elementary electrostatic relations. Then, the arguments of the Fermi functions appearing in the tunneling rates read [16]:

$$\mu_{u,n} - qV_1 = \varepsilon_u + \frac{1}{C_{\Sigma u} C_{\Sigma d} - C^2} \left[\frac{q^2}{2} C_{\Sigma d} + q \left(C_{\Sigma d} C_2 V_{21} + CC_3 V_{31} + CC_4 V_{41}\right)\right] + E_C \delta_{1n} \tag{7}$$

$$\mu_{u,n} - qV_2 = \varepsilon_u + \frac{1}{C_{\Sigma u} C_{\Sigma d} - C^2} \left[\frac{q^2}{2} C_{\Sigma d} + q \left(C_{\Sigma d} C_1 V_{12} + CC_3 V_{32} + CC_4 V_{42}\right)\right] + E_C \delta_{1n} \tag{8}$$

$$\mu_{d,n} - qV_3 = \varepsilon_d + \frac{1}{C_{\Sigma u} C_{\Sigma d} - C^2} \left[\frac{q^2}{2} C_{\Sigma u} + q \left(C_{\Sigma u} C_4 V_{43} + CC_1 V_{13} + CC_2 V_{23}\right)\right] + E_C \delta_{1n} \tag{9}$$

$$\mu_{d,n} - qV_4 = \varepsilon_d + \frac{1}{C_{\Sigma u}C_{\Sigma d} - C^2}\left[\frac{q^2}{2}C_{\Sigma u} + q\left(C_{\Sigma u}C_3 V_{34} + CC_1 V_{14} + CC_2 V_{24}\right)\right] + E_C \delta_{1n} \qquad (10)$$

that now depend only on voltage differences. The four $\mu_{u,n} - V_{1(2)}$, and $\mu_{d,n} - V_{3(4)}$ are the electrochemical potentials. We take V_{12}, V_{13} and V_{34} as the only independent biases, since the rest of voltage differences can be expressed as linear combinations of these values.

As discussed above, we apply the Markov approximation in order to determine the dynamics of the probabilities of finding the system in one of the four states. Specifically, we employ the master equation formalism, where the time evolution of the system is governed by a master equation that gives the probability distribution of the considered stochastic variables in terms of the transition rates between the different states. Defining $\Gamma^{\pm}_{u(d)} = \Gamma^{\pm}_{1(3)} + \Gamma^{\pm}_{2(4)}$, the following relations are found:

$$\begin{pmatrix} \dot{p}_0 \\ \dot{p}_u \\ \dot{p}_d \\ \dot{p}_2 \end{pmatrix} = \begin{pmatrix} -\Gamma^-_u - \Gamma^-_d & \Gamma^+_u & \Gamma^+_d & 0 \\ \Gamma^-_u & -\Gamma^+_u - \gamma^-_d & 0 & \gamma^+_d \\ \Gamma^-_d & 0 & -\gamma^-_u - \Gamma^+_d & \gamma^+_u \\ 0 & \gamma^-_d & \gamma^-_u & -\gamma^+_u - \gamma^+_d \end{pmatrix} \begin{pmatrix} p_0 \\ p_u \\ p_d \\ p_2 \end{pmatrix} \qquad (11)$$

As we are interested in the steady state, we set all $\dot{p}_i = 0$. Considering the normalization condition $\sum_i p_i = 1$, we obtain:

$$p_0 = \frac{1}{\alpha}\left[\Gamma^+_d \gamma^+_u \left(\Gamma^+_u + \gamma^-_d\right) + \Gamma^+_u \gamma^+_d \left(\Gamma^+_d + \gamma^-_u\right)\right] \qquad (12)$$

$$p_u = \frac{1}{\alpha}\left[\Gamma^-_u \Gamma^+_d \left(\gamma^+_u + \gamma^+_d\right) + \gamma^-_u \gamma^+_d \left(\Gamma^-_u + \Gamma^-_d\right)\right] \qquad (13)$$

$$p_d = \frac{1}{\alpha}\left[\Gamma^+_u \Gamma^-_d \left(\gamma^+_u + \gamma^+_d\right) + \gamma^-_u \gamma^+_d \left(\Gamma^-_u + \Gamma^-_d\right)\right] \qquad (14)$$

$$p_2 = \frac{1}{\alpha}\left[\gamma^-_u \gamma^-_d \left(\Gamma^-_u + \Gamma^-_d\right) + \Gamma^-_u \Gamma^+_d \gamma^-_d + \Gamma^+_u \Gamma^-_d \gamma^-_u\right] \qquad (15)$$

with

$$\alpha = \Gamma^-_u \left[\Gamma^+_d \left(\gamma^+_u + \gamma^+_d\right) + \gamma^-_d \left(\gamma_u + \Gamma^+_d\right)\right] + \Gamma^+_u \Gamma^+_d \left(\gamma^+_u + \gamma^+_d\right) + \Gamma_d \gamma^-_d \gamma^-_u + \Gamma_u \gamma^+_d \gamma^-_u + \qquad (16)$$
$$+ \gamma^-_d \left[\Gamma^+_u \left(\gamma_u + \gamma^+_d\right) + \gamma^-_u \gamma_d\right]$$

and $\Gamma_{u(d)} = \Gamma^+_{u(d)} + \Gamma^-_{u(d)}$ (similar for $\gamma_{u(d)}$).

We now compute the electrical current I_1 that flows between the first lead and the upper dot, which, we from now on call *drag current* for historical reasons (note that since generally $V_{12} \neq 0$ it is not a current arising solely from the drag effect). This current is obtained by weighting the transition probabilities with the electron charge q. The result is

$$I_1 = q\left(\Gamma^-_1 p_0 - \Gamma^+_1 p_u + \gamma^-_1 p_d - \gamma^+_1 p_2\right) \qquad (17)$$

Because of electric charge conservation, we immediately know $I_2 = -I_1 = I$ for the current between the second terminal and the *up* dot (we assign a $+$ sign whenever the current flows from a lead into a dot, and a $-$ sign otherwise). We can also compute the heat current by weighting the transitions with the amount of transferred effective energy (the electrochemical potential),

$$J_1 = \tilde{\mu}_{u,0}\left(\Gamma^-_1 p_0 - \Gamma^+_1 p_u\right) + \tilde{\mu}_{u,1}\left(\gamma^-_1 p_d - \gamma^+_1 p_2\right) \qquad (18)$$

where $\tilde{\mu}_{u,n} = \mu_{u,n} - qV_1$. Similar expressions are obtained for the rest of the J_i. Energy conservation leads to $J_1 + J_2 + J_3 + J_4 = I_1 V_{21} + I_3 V_{43}$. These currents were investigated by Sánchez et al. [16] when the *up* dot is at equilibrium with $V_1 = V_2$; a nonzero drag current I_1 then appears when $\Gamma_1 \gamma_2 \neq \gamma_1 \Gamma_2$. This means that the current in the lower terminals (drive system) drives the upper dot towards a non-equilibrium situation by the appearance of a drag current. The drag phenomenon can be clearly understood from the following Joule relation found for this setup

$$J_1 + J_2 + J_c = -I_1 V_{12} \tag{19}$$
$$J_3 + J_4 - J_c = -I_3 V_{34}$$

where

$$J_c = \frac{E_C}{\alpha}(\gamma_u^+ \gamma_d^- \Gamma_u^- \Gamma_d^+ - \gamma_u^- \gamma_d^+ \Gamma_u^+ \Gamma_d^-) \tag{20}$$

is the heat flow between the drag and the drive system. This expression generalizes the relation found in Reference [23] for a three-terminal double quantum dot. In such a system, the drag conductor is connected to two reservoirs, whereas the drive dot is coupled to a single contact. Therefore, the drive subsystem does not support any charge current. Under these considerations, the drive dot carries a heat flow J (with a similar form to Equation (20), which is proportional to the drag charge current when $\Gamma_2 = \gamma_1 = 0$, i.e. in the so-called strong coupling regime. Here, Equation (19) demonstrates the existence of a heat flow J_c between the drag and drive subsystems. This energy flow appears in addition to the heat flows J_1, J_2 through the drag conductor and the heat currents J_3, J_4 in the drive subsystem, even when they are held at common temperature and no particle transfer exists between them.

Returning to our purpose, which is to find a route to an effective equilibrium state, we address the issue of whether the opposite phenomenon to the drag is possible, i.e., if we can achieve a non-equilibrium configuration with $V_1 \neq V_2$ for which the drag effect causes the stalling of the upper currents. Under this novel situation, we check whether our system reaches an "effective linear response regime" by testing the microreversibility property through the Onsager relations and the fulfilment of the FDT. To this end, we focus on the *up* dot and consider three stalling configurations: (i) when both the electrical and the heat flow vanish, i.e., $I_1 = 0$ and $J_1 = 0$, which we call the globally stalled scenario; (ii) when the charge current is nullified, $I_1 = 0$, but there is a finite heat flow $J_1 \neq 0$; and (iii) when there is a finite electrical current $I_1 \neq 0$ but no heat flow, $J_1 = 0$. These situations correspond to the locally charge-stalled and heat-stalled cases, respectively. The simplest manner to achieve the globally stalled case is tuning the system to the strong coupling configuration by setting $\gamma_1 = \Gamma_2 = 0$. Under this situation, electrons can only tunnel in and out of the top-left reservoir if the lower dot is empty, and of the top-right reservoir if the lower dot is occupied.

2.2. Detailed Balance and Behavior at Equilibrium

Before presenting our results, we carefully revise the behavior of systems near thermodynamic equilibrium. In this situation, all existing currents in a system tend to zero on average. This behavior is called *global detailed balance*. According to statistical mechanics, systems subject to these conditions exhibit the property that the correlations of the spontaneous fluctuations and the dissipative response to an external perturbation obey the same rules, which is primarily known as Onsager's regression hypothesis [14]. This important statement is the heart of the fluctuation–dissipation theorem (FDT). If we consider an arbitrary physical current J_α (such as a heat or charge current) and its affinity or conjugate force h_α (which in these cases would correspond to gradients in temperature or electrical potential, respectively), the theorem can be expressed as

$$\partial_{h_\alpha} J_\alpha (x^{eq}) = D_{\alpha,\alpha} (x^{eq}) \tag{21}$$

where $D_{\alpha,\alpha}$ is a generalized diffusion constant proportional to $\langle J_\alpha J_\alpha \rangle$. The vector x contains all the parameters the current may depend on, and satisfies $J_\alpha (x^{eq}) = 0$ for all currents in the system; their conjugate forces are evidently also required to vanish. The previous equation can be generalized in such a way that it expresses the FDT for the combination of two currents and their conjugate forces by changing one index α for a different one and symmetrizing both sides of the expression (see complementary material of Reference [14]).

Another major result in thermodynamics close to equilibrium is found in Onsager's reciprocal relations (RRs), which actually follow from the FDT if the system enjoys the property of being time-reversible [15]. In the following, we restrict ourselves to relations between heat and charge currents, following Onsager's original article [1]. For a system where transport of these quantities exists, the mechanisms are usually not independent, but interfere with each other leading to the well known thermoelectric effects. If we consider a system at equilibrium, small fluctuations or external perturbations may allow for the transport of small quantities of charge and heat while the system is returning to its original state. Onsager established that, in these situations, the responses of a current due to a variation of the other current's conjugate force are equal, i.e. the heat current responds in the same way to a variation of the electrical potential as the charge current to a temperature fluctuation. This result is best visualized by writing the currents in matrix form. For a simple system with a single heat and charge current, we have:

$$
\begin{pmatrix} J_{charge} \\ J_{heat} \end{pmatrix} = \begin{pmatrix} L_{11} & L_{12} \\ L_{21} & L_{22} \end{pmatrix} \begin{pmatrix} \delta\left(\Delta V\right) \\ \delta\left(\Delta T/T\right) \end{pmatrix} \tag{22}
$$

where L_{11} and L_{22} are the electrical and thermal conductances, and $L_{12} = \partial_{\Delta T/T} J_{charge}$ and $L_{21} = \partial_{\Delta V} J_{heat}$ represent the electrothermal and thermoelectrical coefficients that arise from the interference of the two transport mechanisms. Onsager's statement is then equivalent to the requirement that the conductance matrix be symmetric, $L_{12} = L_{12}$. In addition to these relations, the scattering theory formalism ensures that both the thermal and the electrical conductances are semipositive.

Despite these theorems being major cornerstones in our understanding of the behavior of systems obeying global detailed balance, most complex systems live out of equilibrium. Accordingly, similar relations have been sought for systems where detailed balance is explicitly broken, since their finding would allow us to characterize and study out-of-equilibrium systems in a similar manner as when detailed balance is satisfied.

2.3. Local Detailed Balance and Equilibrium-Like Relations

A central assumption in stochastic thermodynamics far from equilibrium, when global detailed balance is not satisfied, is local detailed balance (LDB). It relates the forward and backward transition rates w into and out of a state A by means of a mechanism ν and reads [24]

$$
\frac{w^{\nu}_{A \to B}}{w^{\nu}_{B \to A}} = e^{-\beta_{\nu} \Delta \varepsilon} \tag{23}
$$

where β_{ν} is the inverse temperature of the reservoir involved in the transition and $\Delta\varepsilon$ is the difference between the energies of states A and B. It can be easily checked that the rates in Equations (1)–(4) indeed satisfy the LDB condition.

In Reference [14], it was reported that, if LDB is satisfied in a system driven arbitrarily far from equilibrium, its response to a perturbation or a spontaneous fluctuation may obey a relation similar to the equilibrium FDT if certain additional conditions are fulfilled. More precisely, it has been established that a current J_{α} in such a system obeys Equation (21) with x^{eq} replaced by x^{st}, where x^{st} corresponds to a configuration of the parameters of the current such that $J_{\alpha}\left(x^{st}\right) = 0$, i.e., the considered current stalls. This is valid if the force h_{α} couples exclusively to those transitions that contribute to the conjugate current J_{α}. It is important to notice the difference between this statement and the first FDT valid only near equilibrium, since we now only require a given current to stall internally. This may be a consequence of the appropriate tuning of the rest of the currents in the system, which are no longer required to vanish, and can in fact assume arbitrary magnitudes.

Similarly, Onsager's reciprocal relations have also been extended to non-equilibrium situations, under the condition of a marginal time-reversibility [25]. Again, it is required that the currents stall in order for the RRs to hold far from equilibrium.

3. Results and Discussion

In this section, we present the main results of our work. We verify the RRs and the FDT for a complete understanding of the impact of stalling currents in coupled conductors.

Roots of the Drag Current and Equilibrium-Like Behavior

The aim of our study is to verify the generalized non-equilibrium reciprocity relations and the fluctuation–dissipation relations. As discussed above, they require that the involved currents be at stall in order to hold arbitrarily far from equilibrium. We exclusively focus on situations where the stalling currents are those between the upper dot and the first lead, i.e., the ones in the drag system. Since $I_1 = -I_2$, it is enough for our purposes to seek for roots of I_1. We also only look for roots of J_1, even though $J_1 \neq J_2$. For all the out-of-equilibrium calculations, we consider the isothermal case $T = T_i$, with $i = 1, \ldots, 4$. Since we are only interested in the responses of the currents to small temperature fluctuations in one of the leads (with the rest held constant), we must formally treat the temperatures in each lead as independent of each other for computational means. However, in the end, all derivatives are evaluated at temperature T.

The electric current I_1 [Equation (17)] is a highly nonlinear function of the biases V_{12}, V_{13} and V_{34}. Consequently, the solutions to $I_1 = 0$ must be found by means of numerical analysis in order to verify Onsager's relations and the FDT (further justifications below). To this purpose, we set $\Gamma_i = \gamma_i = \Gamma$ except for $\gamma_1 = 0.1\Gamma$, $k_B T = 5\hbar\Gamma$, $q^2/C_i = 20\hbar\Gamma$, $q^2/C = 50\hbar\Gamma$ and $\varepsilon_u = \varepsilon_d = 0$. Furthermore, we consider natural units where $\hbar = -q = k_B = \Gamma = 1$. Unless otherwise mentioned, these parameters are used in the rest of this work.

We remark that our analysis is purely numerical since the solutions for $I_1 = 0$ require large values of V_{12} at a given set of voltages V_{13} and V_{34}. This fact prevents us from employing a perturbative scheme in terms of the dc voltages. The charge current through the upper dot is composed of the current directly induced by the bias V_{12} and the contribution due to the charge fluctuations caused by the transport in the lower dot. The latter contribution is precisely the drag effect, which is much less significant to the creation of a charge flow through the *up* dot than the effect of a voltage directly applied between the upper terminals. The need for a numerical analysis of this system is hereby justified. To find the roots of the currents for a given set of parameters, we implemented a bisection algorithm (see Appendix A).

Since there is no magnetic field present in our system, its dynamical evolution is time-reversible. Accordingly, microreversibility ensures that the RRs should be satisfied for stalling currents far from equilibrium, as discussed in Reference [15]. In this section, we analyze both the case when the charge and heat currents stall at the same time, as well as the scenario when they do not necessarily vanish simultaneously for the same voltage configuration. The Onsager matrix for our two dot system with four leads should be of dimension 8×8 with elements denoted by $L_{ij,mn}$. In the absence of a magnetic field, Onsager's relations imply $L_{ij,mn} = L_{ji,mn}$. Furthermore, charge conservation laws imply relations such as $I_2 = -I_1$ and therefore more elements of the Onsager matrix are related. At the stalling configuration, we thus check for the fulfilment of the particular relation $L_{12,11} = L_{21,11}$, with

$$L_{12,11} \equiv L_{12} = \frac{\partial I_1}{\partial T_1}, \quad L_{21,11} \equiv L_{21} = \frac{1}{T_1}\frac{\partial J_1}{\partial V_1} \tag{24}$$

As we can see, $I_1 = I_{charge}$ and $J_1 = J_{heat}$ in terms of the example in Equation (22). Here, we consider as conjugate forces the *absolute* potentials and temperatures. This is justified since the thermodynamic variables of the quantum dots do not show up in the currents, and therefore differentiating them with respect to the gradients $\Omega_i - \Omega_{dot}$ yields the same result as differentiating with respect to Ω_i (where Ω represents either a voltage or a temperature). A summary of our first results is presented in Figure 3. We show the coefficients for a given V_{12} as a function of V_{13}. It is understood that the value of V_{34} at each point corresponds to the one where stalling has been numerically found. We consider four cases: (i) the globally stalled configuration depicted in Figure 3a; (ii) the locally charge-stalled case shown in Figure 3b; (iii) the locally heat-stalled scenario in Figure 3c; and (iv) a configuration where none of the currents vanish, as shown in Figure 3d. Firstly, we notice that the RRs are satisfied at the configurations where the current I_1 stalls [cases shown in Figure 3a,b] with J_1 being either zero or not. On the other hand, considering the stalling points of J_1 [see Figure 3c], in general, we do not observe an equality between L_{12} and L_{21}. Even so, there are some exceptions (not shown here) in which the RR are satisfied despite having $I_1 \neq 0$ and $J_1 = 0$. For these cases, however, we checked that they do not follow the FDT.

We now move on to study the validity of the fluctuation–dissipation theorem. In this case, we only consider the FDT for the charge currents. Firstly, we give explicit expressions for the relations between the transport coefficients, i.e., the FDRs. They have been established for the non-equilibrium case. Here, it is instructive to first consider the FDT *near equilibrium*. We consider the following voltage expansion of the currents around the equilibrium point $V_i = 0$:

$$I_\alpha = \sum_\beta G^{eq}_{\alpha,\beta} V_\beta + \sum_{\beta,\gamma} G^{eq}_{\alpha,\beta\gamma} V_\beta V_\gamma + \mathcal{O}\left(V^2\right) \tag{25}$$

where the nth order conductances $G^{eq}_{\mu,v_1...v_n} = \left(\partial^n I_\mu / \partial V_{v_1}...V_{v_n}\right)_{V_i=0}$ are related to nth order FDRs. For instance, at second-order equilibrium FDRs lead to the FDT

$$S^{eq}_{\alpha\beta} = k_B T \left(G^{eq}_{\alpha,\beta} + G^{eq}_{\beta,\alpha} \right) \tag{26}$$

The non-equilibrium FDT is then established to have the same form replacing the equilibrium condition by the stalling condition.

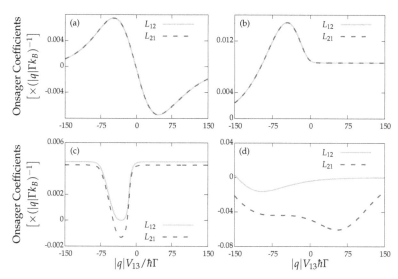

Figure 3. Onsager coefficients L_{12} and L_{21} versus the V_{13} bias voltage at the indicated V_{12} biases: (a) strong coupling configuration ($\Gamma_2 = \gamma_1 = 0$) with $I_1 = 0$ and $J_1 = 0$; (b) $I_1 = 0$ and $J_1 \neq 0$; (c) $I_1 \neq 0$ and $J_1 = 0$; and (d) $I_1 \neq 0$, and $J_1 \neq 0$. $\Gamma_i = \gamma_i = \Gamma$ except for $\gamma_1 = 0.1\Gamma$, $k_B T = 5\hbar\Gamma$, $q^2/C_i = 20\hbar\Gamma$, $q^2/C = 50\hbar\Gamma$ and $\varepsilon_u = \varepsilon_d = 0$. Furthermore, we consider natural units where $\hbar = -q = k_B = \Gamma = 1$.

For our particular device, we investigated the non-equilibrium FDT for two cases. We tested the FDT only for the upper dot charge current, i.e., $I_1 = -I_2$. The results are shown in Figure 4, where we check the FDT for the drag current, i.e.,

$$S_{11} = 2k_B T G_{1,1} \tag{27}$$

as well as the FDT involving the cross-correlations between the drag current (I_1) and the drive current (I_3) contributions, i.e.,

$$S_{13} = k_B T \left(G_{1,3} + G_{3,1} \right) \tag{28}$$

where in both cases the noise $S_{\alpha\beta}$ was computed by applying the Full Counting Statistics (FCS) formalism described in Appendix B. We observe that only the former relation for the drag current is satisfied (Figure 4a,b) since I_1 vanishes but I_3 does not. The FDT involving cross-correlations between I_1 and I_2 also holds since both of the currents stall (not included). These results are independent of whether the heat current vanishes (see Figure 4a for $J_1 = 0$, i.e., the strong coupling regime) or not (see Figure 4b with $J_1 \neq 0$). In the two remaining cases (Figure 4c,d), the fact that the drive current I_3 does not vanish prevents the fulfilment of the FDT for the cross-correlations S_{13}.

To make a complete description of the transport under stalling conditions we now discuss a remarkable result involving the third cumulants of the current. In this case, we talk about fluctuation–dissipation relations instead of the FDT. As mentioned in the Introduction, the FDRs were originally formulated by adding to the transition rates and the excess of entropy production the external potential that drives the system out of equilibrium [8]. In that sense, it is possible to establish relations between the transport coefficients such as nonlinear conductances, non-equilibrium noises, and the third cumulant. In all these cases, the transport coefficients are computed at the non-equilibrium configuration. In particular, López et al. [26] found that the following FDR is satisfied under equilibrium conditions:

$$C_{\alpha\beta\gamma} = (k_B T)^2 \left(G_{\alpha,\beta\gamma} + G_{\beta,\gamma\alpha} + G_{\gamma,\alpha\beta} \right) \tag{29}$$

where $C_{\alpha\beta\gamma} = \langle I_\alpha I_\beta I_\gamma \rangle$ are the third-order cumulants. We ignored the indices referring to the spin degree of freedom appearing in the original paper as in our system we have spin degeneracy due to the absence of a magnetic field. Here, we checked for the fulfilment of the previous relation at stalling conditions far from equilibrium, where all the nonlinear transport coefficients $G_{\alpha,\beta\gamma}$ are computed under non-equilibrium conditions. Note that this can be rewritten as

$$C_{\alpha\beta\gamma} = 3 \left(k_B T \right)^2 G_{(\alpha,\beta\gamma)} \tag{30}$$

where we understand $G_{(\alpha,\beta\gamma)}$ as the symmetrization with respect to the three indices, $G_{(\alpha,\beta\gamma)} = (1/3!) \left(G_{\alpha,\beta\gamma} + G_{\alpha,\gamma\beta} + G_{\beta,\gamma\alpha} + G_{\beta,\alpha\gamma} + G_{\gamma,\alpha\beta} + G_{\gamma,\beta\alpha} \right)$.

We explored the fulfilment of Equation (29) when stalling currents are present in the system, with $G_{\alpha\beta\gamma}$ again computed with help of FCS (see Appendix B). Figure 5 represents the third cumulant fluctuation relations. The case in which $I_1 = 0$ is shown in Figure 5a when $J_1 = 0$ and in Figure 5b when $J_1 \neq 0$. In these two scenarios, the FDRs are fulfilled. However, when the cumulant relation involves currents from both the drive (either I_3 or I_4) and the drag (either I_1 or I_2) subsystems, then the corresponding FDR is no longer satisfied. Finally, for completeness, our last result is shown in Figure 6, where the FDT and third-order cumulant relations are displayed for cases where the system is not in a stalling configuration. As can be seen, none of these relations hold, as expected.

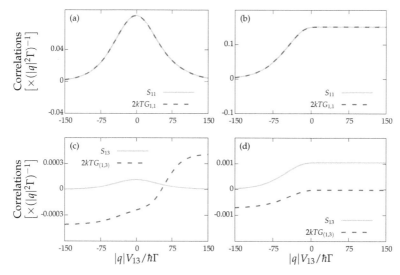

Figure 4. Fluctuation–dissipation theorem $S_{\alpha\beta} = 2k_B T G_{(\alpha,\beta)}$ for: (**a**) the strong coupling regime for the drag current, $I_1 = 0$ and $J_1 = 0$; (**b**) the locally charge-stalled configuration, $I_1 = 0$ and $J_1 \neq 0$; and (**c,d**) the drag and drive currents with $I_1 = 0$ and $J_1 = 0$, and $I_1 = 0$ and $J_1 \neq 0$, respectively. The rest of the parameters are those of Figure 3.

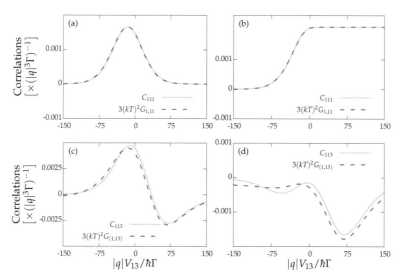

Figure 5. Third-order fluctuation–dissipation relations $C_{\alpha\beta\gamma} = 3\left(k_B T\right)^2 G_{(\alpha,\beta\gamma)}$ for: (**a**) the strong coupling regime for the drag current $I_1 = 0$ and $J_1 = 0$; (**b**) the locally charge-stalled configuration $I_1 = 0$ and $J_1 \neq 0$; and (**c,d**) the drag and drive currents for $I_1 = 0$ and $J_1 = 0$, and $I_1 = 0$ and $J_1 \neq 0$, respectively. The rest of the parameters are those of Figure 3.

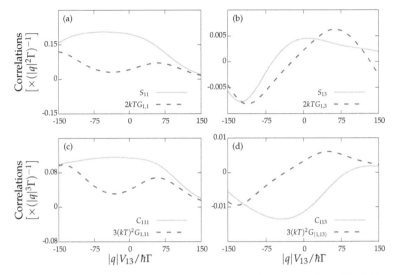

Figure 6. Fluctuation–dissipation theorem in a non-stalled configuration $I_1 \neq 0$ and $J_1 \neq 0$ for: (**a**) the drag current; and (**b**) the drag and drive currents. Fluctuation–dissipation relations in a non-stalled configuration $I_1 \neq 0$ and $J_1 \neq 0$ for: (**c**) the drag current; and (**d**) the drag and drive currents. The rest of the parameters are those of Figure 3.

4. Conclusions

We provide evidence for the validity of the fluctuation–dissipation theorem and Onsager's reciprocal relations far from equilibrium at stalling configurations where $I_1 = 0$. Additionally, we successfully tested the fluctuation relations for the third cumulant in which all the transport coefficients are calculated at stalling but far from equilibrium. The positive results are good news, as they confirm that there are indeed some situations in which a system driven far from equilibrium enjoys near-equilibrium properties, and can therefore be analyzed by means of the well-known theoretical models of equilibrium thermodynamics.

A possible extension to this work is to investigate the behavior of stalling currents and the validity of the non-equilibrium relations with transport coefficients at stalling configurations in cases where the system exhibits purely quantum effects, such as quantum transport under the preservation of phase coherence when higher-order tunneling effects are included.

Author Contributions: Formal analysis, R.L. and L.M.; Investigation, R.L. and L.M.; Methodology, R.L. and L.M.; Project administration, R.L.; Software, L.M.; Supervision, R.L.; Validation, R.L. and L.M.; Writing – original draft, R.L.; Writing – review and editing, R.L. and L.M.

Funding: L.M. acknowledges the SURF@IFISC program. We acknowledge the the the MICINN Grant No. MAT2017-82639 for its financial support.

Acknowledgments: We thank David Sánchez for enriching our work with fruitful discussions and his useful remarks. Finally, we acknowledge the indirect help of Jong Soo Lim for providing some of his previous work, which helped in understanding some of the theoretical framework.

Conflicts of Interest: The authors declare no conflict of interest.

Appendix A. Bisection Method

To find the roots of the drag current at a certain point of parameter space, we make extensive use of the bisection algorithm. Firstly, we fix the value of the biases V_{12} and V_{13}. We are now interested in finding V_{34} such that $I_1 (V_{12}, V_{13}, V_{34}) = 0$. Let $f(x) = I_1 (V_{12}, V_{13}, x)$. We iteratively seek two points a

and b such that $f(a)f(b) < 0$. Since I_1 is continuous, which implies there is a root lying in the interval (a, b). To find this root with a tolerance ϵ, once we have localised such points, where ϵ represents the largest value for the width of the interval centered at the point that we ultimately accept as a root, we proceed as follows:

1. We define $c = (a + b)/2$.
2. If $f(c) = 0$ or $(b - a)/2 < \epsilon$, we accept c as a root and stop.
3. Otherwise if $f(a)f(c) < 0$, we redefine $b = c$ and return to Step 1. If not, we redefine $a = c$ and return to Step 1.

Despite the fact that the width of the interval decreases only linearly with this method, thus making it rather slow, it is always ensured to converge if there is a root lying inside the original interval. Furthermore, for a highly nonlinear function, which may present a behavior that is difficult to picture, such as the drag current, this method is far superior to non-fixed interval algorithms such as the Newton method.

Appendix B. Full Counting Statistics and Computation of Cumulants

Despite this method being extensively discussed in References [27,28], a good understanding of it has been crucial for this work. However, several different approaches are found in the literature, and therefore we consider it adequate to clarify the precise path we have taken. The derivation that follows closely follows the one found in the additional material of Reference [26].

The central quantity we are involved with is $P\left(\{n_1, n_2, n_3, n_4\}; t\right) \equiv P\left(\{n\}; t\right)$, the probability that n_j electrons have been transferred through the terminal j. The associated cumulant generating function (CGF) $\mathcal{F}\left(\{\chi\}; t\right)$ follows from

$$\exp\left[\mathcal{F}(\{\chi\}; t)\right] = \sum_{\{n\}} P\left(\{n\}; t\right) \exp\left(i \sum_j \chi_j n_j\right) \tag{A1}$$

From the CGF, we can obtain the desired cumulants by taking partial derivatives with respect to the counting fields χ_j at $\chi_j = 0$:

$$C_{pqrs} = \partial_{i\chi_1}^p \partial_{i\chi_2}^q \partial_{i\chi_3}^r \partial_{i\chi_4}^s \mathcal{F}(\{\chi\}; t)\Big|_{\chi=0} \tag{A2}$$

Then, the current cumulants in the long-time limit are simply given by

$$\left\langle I_1^p I_2^q I_3^r I_4^s \right\rangle = (-q)^{p+q+r+s} \frac{dC_{pqrs}}{dt}\Big|_{t\to\infty} \tag{A3}$$

However, in general, the expression for the CGF is difficult to obtain and handle, and must be computed recursively. We first introduce the operator $\mathcal{L}(\chi)$ as the Fourier transform of the entries of our matrix $\mathcal{L}(0) \equiv \mathcal{L}$ satisfying $\dot{\rho} = \mathcal{L}\rho$. When only sequential tunneling is considered, it is obtained by adding counting fields to the off-diagonal entries of $\mathcal{L}$, with a plus (minus) sign when the transition corresponds to an electron entering (leaving) the corresponding lead (we skip the details). For our matrix (Equation (11)), it assumes the form

$$\mathcal{L}(\chi) = \begin{pmatrix} -\Gamma_u^- - \Gamma_d^- & \Gamma_1^+ e^{i\chi_1} + \Gamma_2^+ e^{i\chi_2} & \Gamma_3^+ e^{i\chi_3} + \Gamma_4^+ e^{i\chi_4} & 0 \\ \Gamma_1^- e^{-i\chi_1} + \Gamma_2^- e^{-i\chi_2} & -\Gamma_u^+ - \gamma_d^- & 0 & \gamma_3^+ e^{i\chi_3} + \gamma_4^+ e^{i\chi_4} \\ \Gamma_3^- e^{-i\chi_3} + \Gamma_4^- e^{-i\chi_4} & 0 & -\gamma_u^- - \Gamma_d^+ & \gamma_1^+ e^{i\chi_1} + \gamma_2^+ e^{i\chi_2} \\ 0 & \gamma_3^- e^{-i\chi_3} + \gamma_4^- e^{-i\chi_4} & \gamma_1^- e^{-i\chi_1} + \gamma_2^- e^{-i\chi_2} & -\gamma_u^+ - \gamma_d^+ \end{pmatrix} \tag{A4}$$

In the long-time limit, the CGF can be written as

$$\mathcal{F}(\{\chi\};t) = \lambda_0(\chi)t \tag{A5}$$

where $\lambda_0(\chi)$ is the minimum eigenvalue of $\mathcal{L}(\chi)$ [27]. If we are able to compute it, then the current correlations easily follow from Equation (A3) as

$$\left\langle I_1^p I_2^q I_3^r I_4^s \right\rangle = (-q)^{p+q+r+s}\, \partial_{i\chi_1}^p \partial_{i\chi_2}^q \partial_{i\chi_3}^r \partial_{i\chi_4}^s \lambda_0(\chi)\Big|_{\chi=0} \tag{A6}$$

To calculate $\lambda_0(\chi)$, we have employed Flindt's method, which is discussed now. First, we write $\mathcal{L}(\chi)$ as

$$\mathcal{L}(\chi) = \mathcal{L} + \tilde{\mathcal{L}}(\chi) \tag{A7}$$

where $\mathcal{L} = \mathcal{L}(0)$ and $\tilde{\mathcal{L}}(\chi) = \mathcal{L}(\chi) - \mathcal{L}$. Next, we define operators $\mathcal{P} = |0\rangle\langle 0|$ and $\mathcal{Q} = 1 - \mathcal{P}$, where $|0\rangle = (p_0, p_u, p_d, p_2)^T$ and $\langle 0| = (1,1,1,1)$ are the left and right null eigenvectors of $\mathcal{L}$. Clearly, $\mathcal{P}\mathcal{L} = \mathcal{L}\mathcal{P} = 0$ and $\mathcal{Q}\mathcal{L} = \mathcal{L}\mathcal{Q} = \mathcal{L}$. To determine the CGF from Equation (A5), we must solve

$$\mathcal{L}(\chi)|0(\chi)\rangle = \left[\mathcal{L} + \tilde{\mathcal{L}}(\chi)\right]|0(\chi)\rangle = \lambda_0(\chi)|0(\chi)\rangle \tag{A8}$$

By choosing $\langle 0|\, 0(\chi)\rangle = 1$, it follows that

$$\langle 0|\lambda_0(\chi) - \mathcal{L}|0(\chi)\rangle = \lambda_0(\chi) = \langle 0|\tilde{\mathcal{L}}(\chi)|0(\chi)\rangle \tag{A9}$$

and using $\mathcal{Q}$ on $|0(\chi)\rangle$ we also find

$$|0(\chi)\rangle = |0\rangle + \mathcal{Q}|0(\chi)\rangle \tag{A10}$$

From Equation (A8), using that $\mathcal{L}$ and $\mathcal{Q}$ commute and $\mathcal{Q}^2 = \mathcal{Q}$, we obtain

$$\mathcal{Q}|0(\chi)\rangle = \mathcal{Q}\left[\lambda_0(\chi) - \mathcal{L}\right]^{-1}\mathcal{Q}\tilde{\mathcal{L}}(\chi)|0(\chi)\rangle \tag{A11}$$

We now define

$$\mathcal{R}\left[\lambda_0(\chi)\right] = \mathcal{Q}\left[\mathcal{L} - \lambda_0(\chi)\right]^{-1}\mathcal{Q} \tag{A12}$$

and substitute Equation (A11) into Equation (A10) to find

$$|0(\chi)\rangle = |0\rangle - \mathcal{R}\left[\lambda_0(\chi)\right]\tilde{\mathcal{L}}(\chi)|0(\chi)\rangle \tag{A13}$$

so that finally

$$|0(\chi)\rangle = \left\{1 + \mathcal{R}\left[\lambda_0(\chi)\right]\tilde{\mathcal{L}}(\chi)\right\}^{-1}|0\rangle \tag{A14}$$

and therefore using Equation (A9) we arrive at

$$\lambda_0(\chi) = \langle 0|\,\tilde{\mathcal{L}}(\chi)\left\{1 + \mathcal{R}\left[\lambda_0(\chi)\right]\tilde{\mathcal{L}}(\chi)\right\}^{-1}|0\rangle \tag{A15}$$

We now Taylor expand the previous expression. In our case of four counting fields, $\tilde{\mathcal{L}}(\chi)$ is expanded as

$$\begin{aligned}
\tilde{\mathcal{L}}(\chi) = {} & \tilde{\mathcal{L}}^{(1,0,0,0)}(i\chi_1) + \tilde{\mathcal{L}}^{(0,1,0,0)}(i\chi_2) + \tilde{\mathcal{L}}^{(0,0,1,0)}(i\chi_3) + \tilde{\mathcal{L}}^{(0,0,0,1)}(i\chi_4) + \\
& + \frac{1}{2!}\Big[\tilde{\mathcal{L}}^{(2,0,0,0)}(i\chi_1)^2 + \tilde{\mathcal{L}}^{(0,2,0,0)}(i\chi_2)^2 + \tilde{\mathcal{L}}^{(0,0,2,0)}(i\chi_3)^2 + \tilde{\mathcal{L}}^{(0,0,0,2)}(i\chi_4)^2 + \\
& + 2\tilde{\mathcal{L}}^{(1,1,0,0)}(i\chi_1)(i\chi_2) + 2\tilde{\mathcal{L}}^{(1,0,1,0)}(i\chi_1)(i\chi_3) + 2\tilde{\mathcal{L}}^{(1,0,0,1)}(i\chi_1)(i\chi_4) + \\
& + 2\tilde{\mathcal{L}}^{(0,1,1,0)}(i\chi_2)(i\chi_3) + 2\tilde{\mathcal{L}}^{(0,1,0,1)}(i\chi_2)(i\chi_4) + 2\tilde{\mathcal{L}}^{(0,0,1,1)}(i\chi_3)(i\chi_4)\Big] + \mathcal{O}\left(\chi^3\right)
\end{aligned} \tag{A16}$$

where we use $\tilde{\mathcal{L}}(0) = 0$ as follows from the definition. Similarly we obtain for $\mathcal{R}[\lambda_0(\chi)]$ (with $\mathcal{R}(0) \equiv \mathcal{R}$)

$$\mathcal{R}[\lambda_0(\chi)] = \mathcal{R} + \mathcal{R}^{(1,0,0,0)}(i\chi_1) + \mathcal{R}^{(0,1,0,0)}(i\chi_2) + \mathcal{R}^{(0,0,1,0)}(i\chi_3) + \mathcal{R}^{(0,0,0,1)}(i\chi_4) + \mathcal{O}\left(\chi^2\right) \quad (A17)$$

where

$$\tilde{\mathcal{L}}^{(p,q,r,s)} = \partial^p_{i\chi_1}\partial^q_{i\chi_2}\partial^r_{i\chi_3}\partial^s_{i\chi_4}\tilde{\mathcal{L}}(\chi)\Big|_{\chi=0}, \qquad \mathcal{R}^{(p,q,r,s)} = \partial^p_{i\chi_1}\partial^q_{i\chi_2}\partial^r_{i\chi_3}\partial^s_{i\chi_4}\mathcal{R}[\lambda_0(\chi)]\Big|_{\chi=0} \quad (A18)$$

From the definition in Equation (A1), it follows that $\mathcal{F}(\{0\};t) = 0$, so that $\lambda_0(0) = 0$, and therefore $\mathcal{R} = \mathcal{Q}\mathcal{L}^{-1}\mathcal{Q}$. This operator satisfies $\mathcal{L}\mathcal{R} = \mathcal{R}\mathcal{L}$, $\mathcal{R}\mathcal{L}\mathcal{R} = \mathcal{R}$, and $\mathcal{L}\mathcal{R}\mathcal{L} = L$ and is therefore called the *Drazin pseudoinverse*. It can be shown that for a matrix of rank 4 it can be computed as

$$\mathcal{R} = (a_3)^{-2}\mathcal{L}B_2^2 \quad (A19)$$

with the rules

$$\mathcal{L}_0 = 1 \qquad a_0 = 1 \qquad B_0 = I_4 \quad (A20)$$
$$\mathcal{L}_1 = \mathcal{L}B_0 \qquad a_1 = -\mathrm{Tr}(\mathcal{L}_1)/1 \qquad B_1 = \mathcal{L}_1 + a_1 I_4 \quad (A21)$$
$$\mathcal{L}_2 = \mathcal{L}B_1 \qquad a_2 = -\mathrm{Tr}(\mathcal{L}_2)/2 \qquad B_1 = \mathcal{L}_2 + a_2 I_4 \quad (A22)$$
$$\mathcal{L}_3 = \mathcal{L}B_2 \qquad a_3 = -\mathrm{Tr}(\mathcal{L}_3)/3 \quad (A23)$$

Furthermore, it is possible to prove that the first derivatives of $\mathcal{R}$, which are required for the computation of the third-order cumulants, satisfy

$$\mathcal{R}^{(1,0,0,0)} = \mathcal{R}^2\langle 0|\,\tilde{\mathcal{L}}^{(1,0,0,0)}|0\rangle \quad (A24)$$
$$\mathcal{R}^{(0,1,0,0)} = \mathcal{R}^2\langle 0|\,\tilde{\mathcal{L}}^{(0,1,0,0)}|0\rangle \quad (A25)$$
$$\mathcal{R}^{(0,0,1,0)} = \mathcal{R}^2\langle 0|\,\tilde{\mathcal{L}}^{(0,0,1,0)}|0\rangle \quad (A26)$$
$$\mathcal{R}^{(0,0,0,1)} = \mathcal{R}^2\langle 0|\,\tilde{\mathcal{L}}^{(0,0,0,1)}|0\rangle \quad (A27)$$

Finally, we find $\lambda_0(\chi)$ in the form of a power series,

$$\lambda_0(\chi) = \langle 0|\,\tilde{\mathcal{L}}(\chi)\left[1 - \mathcal{R}\tilde{\mathcal{L}}(\chi) + (\mathcal{R}\tilde{\mathcal{L}}(\chi))^2 - ...\right]|0\rangle \quad (A28)$$

or, explicitly,

$$\lambda_0(\chi) = \langle 0|\,\Big\{\tilde{\mathcal{L}}^{(1,0,0,0)}(i\chi_1) + \tilde{\mathcal{L}}^{(0,1,0,0)}(i\chi_2) + \tilde{\mathcal{L}}^{(0,0,1,0)}(i\chi_3) + \tilde{\mathcal{L}}^{(0,0,0,1)}(i\chi_4) +$$
$$+\frac{1}{2!}\Big[\tilde{\mathcal{L}}^{(2,0,0,0)}(i\chi_1)^2 + \tilde{\mathcal{L}}^{(0,2,0,0)}(i\chi_2)^2 + \tilde{\mathcal{L}}^{(0,0,2,0)}(i\chi_3)^2 + \tilde{\mathcal{L}}^{(0,0,0,2)}(i\chi_4)^2 +$$
$$+ 2\tilde{\mathcal{L}}^{(1,1,0,0)}(i\chi_1)(i\chi_2) + 2\tilde{\mathcal{L}}^{(1,0,1,0)}(i\chi_1)(i\chi_3) + 2\tilde{\mathcal{L}}^{(1,0,0,1)}(i\chi_1)(i\chi_4) +$$
$$+ 2\tilde{\mathcal{L}}^{(0,1,1,0)}(i\chi_2)(i\chi_3) + 2\tilde{\mathcal{L}}^{(0,1,0,1)}(i\chi_2)(i\chi_4) + 2\tilde{\mathcal{L}}^{(0,0,1,1)}(i\chi_3)(i\chi_4) -$$
$$- 2\tilde{\mathcal{L}}^{(1,0,0,0)}\mathcal{R}\tilde{\mathcal{L}}^{(1,0,0,0)}(i\chi_1)^2 - 2\tilde{\mathcal{L}}^{(1,0,0,0)}\mathcal{R}\tilde{\mathcal{L}}^{(0,1,0,0)}(i\chi_1)(i\chi_2) -$$
$$- 2\tilde{\mathcal{L}}^{(1,0,0,0)}\mathcal{R}\tilde{\mathcal{L}}^{(0,0,1,0)}(i\chi_1)(i\chi_3) - 2\tilde{\mathcal{L}}^{(1,0,0,0)}\mathcal{R}\tilde{\mathcal{L}}^{(0,0,0,1)}(i\chi_1)(i\chi_4) -$$
$$- 2\tilde{\mathcal{L}}^{(0,1,0,0)}\mathcal{R}\tilde{\mathcal{L}}^{(1,0,0,0)}(i\chi_2)(i\chi_1) - 2\tilde{\mathcal{L}}^{(0,1,0,0)}\mathcal{R}\tilde{\mathcal{L}}^{(0,1,0,0)}(i\chi_2)^2 -$$
$$- 2\tilde{\mathcal{L}}^{(0,1,0,0)}\mathcal{R}\tilde{\mathcal{L}}^{(0,0,1,0)}(i\chi_2)(i\chi_3) - 2\tilde{\mathcal{L}}^{(0,1,0,0)}\mathcal{R}\tilde{\mathcal{L}}^{(0,0,0,1)}(i\chi_2)(i\chi_4) -$$
$$- 2\tilde{\mathcal{L}}^{(0,0,1,0)}\mathcal{R}\tilde{\mathcal{L}}^{(1,0,0,0)}(i\chi_3)(i\chi_1) - 2\tilde{\mathcal{L}}^{(0,0,1,0)}\mathcal{R}\tilde{\mathcal{L}}^{(0,1,0,0)}(i\chi_3)(i\chi_2) -$$
$$- 2\tilde{\mathcal{L}}^{(0,0,1,0)}\mathcal{R}\tilde{\mathcal{L}}^{(0,0,1,0)}(i\chi_3)^2 - 2\tilde{\mathcal{L}}^{(0,0,1,0)}\mathcal{R}\tilde{\mathcal{L}}^{(0,0,0,1)}(i\chi_3)(i\chi_4) -$$

$$-2\tilde{\mathcal{L}}^{(0,0,0,1)}\mathcal{R}\tilde{\mathcal{L}}^{(1,0,0,0)}(i\chi_4)(i\chi_1)-2\tilde{\mathcal{L}}^{(0,0,0,1)}\mathcal{R}\tilde{\mathcal{L}}^{(0,1,0,0)}(i\chi_4)(i\chi_2)-$$

$$-2\tilde{\mathcal{L}}^{(0,0,0,1)}\mathcal{R}\tilde{\mathcal{L}}^{(0,0,1,0)}(i\chi_4)(i\chi_3)-2\tilde{\mathcal{L}}^{(0,0,0,1)}\mathcal{R}\tilde{\mathcal{L}}^{(0,0,0,1)}(i\chi_4)^2\Big]+\mathcal{O}\left(\chi^3\right)\Big\}\,|0\rangle \qquad (A29)$$

We can now apply Equation (A2) to compute all cumulants recursively. To first order, we get (except for a factor of t)

$$C_{1000}=\langle 0|\,\tilde{\mathcal{L}}^{(1,0,0,0)}\,|0\rangle \qquad (A30)$$

$$C_{0100}=\langle 0|\,\tilde{\mathcal{L}}^{(0,1,0,0)}\,|0\rangle \qquad (A31)$$

$$C_{0010}=\langle 0|\,\tilde{\mathcal{L}}^{(0,0,1,0)}\,|0\rangle \qquad (A32)$$

$$C_{0001}=\langle 0|\,\tilde{\mathcal{L}}^{(0,0,0,1)}\,|0\rangle \qquad (A33)$$

To second order, we obtain

$$C_{2000}=\langle 0|\,\tilde{\mathcal{L}}^{(2,0,0,0)}-2\tilde{\mathcal{L}}^{(1,0,0,0)}\mathcal{R}\tilde{\mathcal{L}}^{(1,0,0,0)}\,|0\rangle \qquad (A34)$$

$$C_{0200}=\langle 0|\,\tilde{\mathcal{L}}^{(0,2,0,0)}-2\tilde{\mathcal{L}}^{(0,1,0,0)}\mathcal{R}\tilde{\mathcal{L}}^{(0,1,0,0)}\,|0\rangle \qquad (A35)$$

$$C_{0020}=\langle 0|\,\tilde{\mathcal{L}}^{(0,0,2,0)}-2\tilde{\mathcal{L}}^{(0,0,1,0)}\mathcal{R}\tilde{\mathcal{L}}^{(0,0,1,0)}\,|0\rangle \qquad (A36)$$

$$C_{0002}=\langle 0|\,\tilde{\mathcal{L}}^{(0,0,0,2)}-2\tilde{\mathcal{L}}^{(0,0,0,1)}\mathcal{R}\tilde{\mathcal{L}}^{(0,0,0,1)}\,|0\rangle \qquad (A37)$$

$$C_{1100}=\langle 0|\,\tilde{\mathcal{L}}^{(1,1,0,0)}-\tilde{\mathcal{L}}^{(1,0,0,0)}\mathcal{R}\tilde{\mathcal{L}}^{(0,1,0,0)}-\tilde{\mathcal{L}}^{(0,1,0,0)}\mathcal{R}\tilde{\mathcal{L}}^{(1,0,0,0)}\,|0\rangle \qquad (A38)$$

$$C_{1010}=\langle 0|\,\tilde{\mathcal{L}}^{(1,0,1,0)}-\tilde{\mathcal{L}}^{(1,0,0,0)}\mathcal{R}\tilde{\mathcal{L}}^{(0,0,1,0)}-\tilde{\mathcal{L}}^{(0,0,1,0)}\mathcal{R}\tilde{\mathcal{L}}^{(1,0,0,0)}\,|0\rangle \qquad (A39)$$

$$C_{1001}=\langle 0|\,\tilde{\mathcal{L}}^{(1,0,0,1)}-\tilde{\mathcal{L}}^{(1,0,0,0)}\mathcal{R}\tilde{\mathcal{L}}^{(0,0,0,1)}-\tilde{\mathcal{L}}^{(0,0,0,1)}\mathcal{R}\tilde{\mathcal{L}}^{(1,0,0,0)}\,|0\rangle \qquad (A40)$$

$$C_{0110}=\langle 0|\,\tilde{\mathcal{L}}^{(0,1,1,0)}-\tilde{\mathcal{L}}^{(0,1,0,0)}\mathcal{R}\tilde{\mathcal{L}}^{(0,0,1,0)}-\tilde{\mathcal{L}}^{(0,0,1,0)}\mathcal{R}\tilde{\mathcal{L}}^{(0,1,0,0)}\,|0\rangle \qquad (A41)$$

$$C_{0101}=\langle 0|\,\tilde{\mathcal{L}}^{(0,1,0,1)}-\tilde{\mathcal{L}}^{(0,1,0,0)}\mathcal{R}\tilde{\mathcal{L}}^{(0,0,0,1)}-\tilde{\mathcal{L}}^{(0,0,0,1)}\mathcal{R}\tilde{\mathcal{L}}^{(0,1,0,0)}\,|0\rangle \qquad (A42)$$

$$C_{0011}=\langle 0|\,\tilde{\mathcal{L}}^{(0,0,1,1)}-\tilde{\mathcal{L}}^{(0,0,1,0)}\mathcal{R}\tilde{\mathcal{L}}^{(0,0,0,1)}-\tilde{\mathcal{L}}^{(0,0,0,1)}\mathcal{R}\tilde{\mathcal{L}}^{(0,0,1,0)}\,|0\rangle \qquad (A43)$$

and the same procedure is applied to higher-order cumulants. The expressions become rather cumbersome, but the recipe is clear and easy to use with some algebra.

References

1. Onsager, L. Reciprocal Relations in Irreversible Processes. I. *Phys. Rev.* **1931**, 37, 405–426.
2. Callen, H.B.; Welton, T.A. Irreversibility and Generalized Noise. *Phys. Rev.* **1951**, 83, 34–40.
3. Kubo, R. Statistical-Mechanical Theory of Irreversible Processes. I. General Theory and Simple Applications to Magnetic and Conduction Problems. *J. Phys. Soc. Jpn.* **1957**, 12, 570–586.
4. Weber, J. Fluctuation dissipation theorem. *Phys. Rev.* **1956**, 101, 061620.
5. Einstein, A. Investigations on the Theory of the Brownian Movement. *Ann. der Phys.* **1905**, 17, 549.
6. Johnson, J.B. Thermal Agitation of Electricity in Conductors. *Phys. Rev.* **1928**, 32, 97–109.
7. Blickle, V.; Speck, T.; Lutz, C.; Seifert, U.; Bechinger, C. Einstein Relation Generalized to Nonequilibrium. *Phys. Rev. Lett.* **2007**, 98, 210601.
8. Baiesi, M.; Maes, C.; Wynants, B. Fluctuations and Response of Nonequilibrium States. *Phys. Rev. Lett.* **2009**, 103, 010602.
9. Gomez-Solano, J.R.; Petrosyan, A.; Ciliberto, S.; Chetrite, R.; Gawędzki, K. Experimental Verification of a Modified Fluctuation-Dissipation Relation for a Micron-Sized Particle in a Nonequilibrium Steady State. *Phys. Rev. Lett.* **2009**, 103, 040601.
10. Seifert, U. Generalized Einstein or Green-Kubo Relations for Active Biomolecular Transport. *Phys. Rev. Lett.* **2010**, 104, 138101.
11. Tobiska, J.; Nazarov, Y.V. Inelastic interaction corrections and universal relations for full counting statistics in a quantum contact. *Phys. Rev. B* **2005**, 72, 235328.

12. Saito, K.; Utsumi, Y. Symmetry in full counting statistics, fluctuation theorem, and relations among nonlinear transport coefficients in the presence of a magnetic field. *Phys. Rev. B* **2008**, *78*, 115429.

13. Förster, H.; Büttiker, M. Fluctuation Relations Without Microreversibility in Nonlinear Transport. *Phys. Rev. Lett.* **2008**, *101*, 136805.

14. Altaner, B.; Polettini, M.; Esposito, M. Fluctuation-dissipation relations far from equilibrium. *Phys. Rev. Lett.* **2016**, *117*, 180601.

15. Polettini, M.; Esposito, M. Effective fluctuation and response theory. *J. Stat. Phys.* **2019**, 1–75.

16. Sánchez, R.; López, R.; Sánchez, D.; Büttiker, M. Mesoscopic Coulomb drag, broken detailed balance, and fluctuation relations. *Phys. Rev. Lett.* **2010**, *104*, 076801.

17. Keller, A.J.; Lim, J.S.; Sánchez, D.; López, R.; Amasha, S.; Katine, J.A.; Shtrikman, H.; Goldhaber-Gordon, D. Cotunneling Drag Effect in Coulomb-Coupled Quantum Dots. *Phys. Rev. Lett.* **2016**, *117*, 066602.

18. Lim, J.S.; Sánchez, D.; López, R. Engineering drag currents in Coulomb coupled quantum dots. *New J. Phys.* **2018**, *20*, 023038.

19. Strasberg, P.; Schaller, G.; Brandes, T.; Esposito, M. Thermodynamics of a Physical Model Implementing a Maxwell Demon. *Phys. Rev. Lett.* **2013**, *110*, 040601.

20. Sekimoto, K. Langevin Equation and Thermodynamics. *Prog. Theor. Phys.* **1998**, *130*, 17–27.

21. Jarzynski, C. Equalities and Inequalities: Irreversibility and the Second Law of Thermodynamics at the Nanoscale. *Annu. Rev. Condens. Matter Phys.* **2011**, *2*, 329–351.

22. Esposito, M. Stochastic thermodynamics under coarse graining. *Phys. Rev. E* **2012**, *85*, 041125.

23. Sánchez, R.; Büttiker, M. Optimal energy quanta to current conversion. *Phys. Rev. B* **2011**, *83*, 085428.

24. Esposito, M.; Van den Broeck, C. Three faces of the second law. I. Master equation formulation. *Phys. Rev. E* **2010**, *82*, 011143.

25. Andrieux, D.; Gaspard, P. Fluctuation theorem and Onsager reciprocity relations. *J. Chem. Phys.* **2004**, *121*, 136167.

26. López, R.; Lim, J.S.; Sánchez, D. Fluctuation relations for spintronics. *Phys. Rev. Lett.* **2012**, *108*, 246603.

27. Bagrets, D.; Nazarov, Y.V. Full counting statistics of charge transfer in Coulomb blockade systems. *Phys. Rev. B* **2003**, *67*, 085316.

28. Flindt, C.; Novotný, T.; Braggio, A.; Sassetti, M.; Jauho, A.P. Counting statistics of non-Markovian quantum stochastic processes. *Phys. Rev. Lett.* **2008**, *100*, 150601.

Review

Beyond the State of the Art: Novel Approaches for Thermal and Electrical Transport in Nanoscale Devices

Robert Biele [1,*,†] and Roberto D'Agosta [2,3,*,†]

[1] Institute for Materials Science and Max Bergmann Center of Biomaterials, TU Dresden, 01062 Dresden, Germany

[2] Nano-Bio Spectroscopy Group and European Theoretical Spectroscopy Facility (ETSF), Universidad del Pais Vasco CFM CSIC-UPV/EHU-MPC and DIPC, Av. Tolosa 72, 20018 San Sebastian, Spain

[3] IKERBASQUE, Basque Foundation for Science Maria Diaz de Haro 3, 6 Solairua, 48013 Bilbao, Spain

* Correspondence: robert.biele1@tu-dresden.de (R.B.); roberto.dagosta@ehu.es (R.D.); Tel.: +34-943-015-803 (R.D.)

† Both authors contributed equally to this work.

Received: 5 July 2019; Accepted: 29 July 2019; Published: 2 August 2019

Abstract: Almost any interaction between two physical entities can be described through the transfer of either charge, spin, momentum, or energy. Therefore, any theory able to describe these transport phenomena can shed light on a variety of physical, chemical, and biological effects, enriching our understanding of complex, yet fundamental, natural processes, e.g., catalysis or photosynthesis. In this review, we will discuss the standard workhorses for transport in nanoscale devices, namely Boltzmann's equation and Landauer's approach. We will emphasize their strengths, but also analyze their limits, proposing theories and models useful to go beyond the state of the art in the investigation of transport in nanoscale devices.

Keywords: electronic transport; thermal transport; strongly correlated systems; Landauer-Büttiker formalism; Boltzmann transport equation; time-dependent density functional theory; electron–phonon coupling

---

1. Introduction

Some of the most spectacular advancements in the description of nature have come from the observation that apparently diverse effects are in reality based on the same physical principles. One of these is the realization that whatever interaction between two entities happens through the exchange of some physical quantity, either momentum, energy, spin, or particles. It is, therefore, of paramount importance to be able to describe how this transfer happens at a nanoscopic level, since these principles are usually fundamental to understand more complex systems, such as materials for energy applications, biological systems, and so on [1–4]. For example, it is now clear that our current inability in reproducing solar energy conversion that takes place normally in plants is related to the difficulties in understanding how light is transformed into an electrical current inside a leaf, in particular how the carriers are split and then move away from the light-receptor centers. At the same time, one key point in this energy conversion process is how excess energy is dissipated. Nature has developed for the leaf feedback mechanisms avoiding it being burnt if storing too much energy—these self-regulating mechanisms are yet outside our understanding and only recently with the advancement of both theoretical and experimental methods have some breakthroughs have been made [5,6].

In this review, we will discuss some of the most common approaches of widespread use in the nanoscale transport community. Here, we assume that a "device" of size ranging from 1 to a few hundreds nm is connected to some macroscopic metallic leads that can induce electrical and energy transfer through the device. We will focus on two fundamental models to describe this transport: on the one hand we will discuss the Landauer's formalism for quantum transport. This is valid when the flow crosses the device essentially without being scattered, i.e., in the so-called ballistic regime: an accurate description when the mean free path of the particles, i.e., the distance between two scattering events, is larger than the dimensions of the device. On the other hand, we will consider the Boltzmann's equation for transport. Here, we assume particle are effectively described by a distribution function and they can be scattered by particle–particle interaction, impurities, or by the device. In this case, the mean free path must be small, otherwise the description through an out-of-equilibrium distribution function does not hold. Both theories can be equally applied to the problem of energy and electric current transport, i.e., to the description of the dynamics of electrons and phonons. The ability of treating on essentially equal footing electrons and phonons allows for a comprehensive description of the device maintaining somewhat a consistent level of accuracy. In general, the outcome of the theory are the transport coefficients [1,2,4]. They serve in describing how the device responds to external stimuli such as a thermal gradient or a bias voltage. We will therefore focus on the electrical conductance σ, the Seebeck's coefficient S, and the electron and phonon thermal conductances κ_e and κ_p, respectively.

After having introduced the fundamental models, we will discuss some of their uses, generalizations, and other methods applicable to the intermediate regime where the mean free path is comparable with the dimension of the device. In particular, we will discuss the role of electron–electron and electron–phonon coupling in modifying the transport coefficients. The review is organized as follows: Section 2 introduces Boltzmann's equation and Landauer's quantum transport formalisms. We will discuss their formulation using the most advanced electronic structure methods and discuss critically their advantages and limits. This section also allows us to introduce a consistent notation for the rest of the presentation; In Section 3 we will discuss some recent attempts to go beyond the standard approaches. We will explore, for example, the role of electron–phonon interaction and strong electron correlation in affecting the transport coefficients, and introduce theories and models that allow description of these effects efficiently while maintaining the strength of the general formalisms; Finally, Section 4 contains an outlook of some potential lines of further investigation.

2. Static Approaches: Semiclassical and Quantum Transport Approaches

The calculation of the transport coefficients of a nanoscale device requires an accurate description of both electron and phonon (or, whenever translational invariance is broken, vibrational) properties. This description is used into standard methodologies to evaluate the transport coefficients and afterwards, e.g., the figure-of-merit of thermoelectric energy conversion. In this review, we will describe two of these methods, namely the Boltzmann's transport theory which can be seen as a semiclassical method since it is based on the evaluation of the distribution function through velocities and density of states of the device, and the Landauer's approach to quantum transport, which on the other hand describes a ballistic particle transfer through scattering between states in the leads [1,2,4]. There are many Approaches, however, of high scientific significance due to their accurate description of particle–particle interaction that we will not discuss here. One of these is the rate equations formalism [7], which can be made extremely accurate and describe strongly correlated system, but its wild scaling with the number of states reduces its applicability to either simplified models or small systems.

2.1. Semiclassical Boltzmann Transport

To calculate macroscopic transport coefficients, such as electronic or thermal conductivity, one needs to analyze the microscopic processes happening when electrons are scattered by phonons or impurities in a metal or semiconductor. The idea is to treat the electronic excitations as particles and follow their motion over time. For example, an electron in a metal exposed to an electric field **E** gets accelerated to a velocity **v**, and over time gains the kinetic energy

$$\Delta\epsilon = \mathbf{v}\mathbf{E}t. \tag{1}$$

However, the velocity of the electron cannot grow forever as the electron will be scattered sooner or later with impurities or by phonons, changing its direction and losing part of its kinetic energy. If we do not know the exact microscopic details of how this scattering happens, we might just consider a macroscopic relaxation time τ. This is the average time an electron gets accelerated by the electric field before being scattered, i.e., the time between electron–phonon transitions. After the collision takes place the particle starts again with its acceleration parallel to the electric field until further scattering. The averaged velocity (over time and many different scattering events) due to acceleration and scattering is responsible for the finite electrical conductivity, as the current density is $\mathbf{J} = ne\overline{\mathbf{v}}$, where n is the electron concentration and e its charge.

Let us now consider the conductivity of a metal or semiconductor in more details. An electron in a crystal occupies states according to its distribution function $f(\mathbf{r}, \mathbf{k}, t)$, giving the probability of finding an electron at position $\mathbf{r}$ and time t with the crystal momentum $\mathbf{k}$. At equilibrium, when no fields or temperature gradients are applied, this is given by Fermi function,

$$f_0(\mathbf{k}, \mu, T) = \frac{1}{e^{(\epsilon_n(\mathbf{k}) - \mu)/k_B T} + 1}, \tag{2}$$

where μ is the total chemical potential and $\epsilon_n(\mathbf{k})$ the energy of an electron in the n-th band with momentum $\mathbf{k}$, k_B the Boltzmann's constant and T the temperature. However, if we now apply, for example, a potential bias, electrons will get excited and the actual distribution functions shifts to higher energies. The equal interplay between acceleration on the one hand and collisions and scattering on the other, will possibly set up a steady-state condition. When we switch off the electrical field again, the system relaxes back to its equilibrium state.

In 1872, Boltzmann laid down an equation for f connecting thermodynamics with non-equilibrium kinetics [8]. Although this was long before the birth of modern quantum mechanics, his transport theory consists of a probabilistic description for the one-particle distribution function $f(\mathbf{r}, \mathbf{k}, t)$. This equation for out-of-equilibrium situations is called the semiclassical Boltzmann transport equation (BTE). Here, we will lay down the basic equation and see how to solve it for systems which are close to equilibrium, while in Section 2.2 we review a fully quantum-mechanical treatment for transport properties. As mentioned before, we will treat crystal electrons as semiclassical particles fulfilling the equations

$$\frac{d\mathbf{k}}{dt} = \frac{1}{\hbar}\mathbf{F}_e, \tag{3}$$

$$\mathbf{v} = \frac{1}{\hbar}\nabla_\mathbf{k}\epsilon_n(\mathbf{k}). \tag{4}$$

Here, $\mathbf{F}_e$ is the force acting on the electron and $\mathbf{v}$ its velocity. Considering the total change in time of the distribution function

$$\frac{df(\mathbf{r}, \mathbf{k}, t)}{dt} = \frac{\partial f}{\partial t} + \frac{\partial \mathbf{r}}{\partial t}\cdot\frac{\partial f}{\partial \mathbf{r}} + \frac{\partial \mathbf{k}}{\partial t}\frac{\partial f}{\partial \mathbf{k}} = \frac{\partial f}{\partial t} + \mathbf{v}\cdot\nabla_\mathbf{r}f + \frac{1}{\hbar}\mathbf{F}_e\cdot\nabla_\mathbf{k}f, \tag{5}$$

and setting it equal to the scattering rate, we arrive at the BTE,

$$\frac{\partial f}{\partial t} + \mathbf{v}\cdot\nabla_\mathbf{r}f + \frac{1}{\hbar}\mathbf{F}_e\cdot\nabla_\mathbf{k}f = \left.\frac{\partial f}{\partial t}\right|_{\text{scatt}}. \tag{6}$$

The BTE states that the electron distribution changes due to in- and out-going scattering events which together form $\frac{\partial f}{\partial t}|_{\text{scatt}}$. Although the Boltzmann transport theory is probabilistic, it is still classical as in quantum mechanics one cannot specify the canonical variables **p** and **r** simultaneously, due to the Heisenberg uncertainty principle, $\Delta p \Delta x \geq \hbar$. The assumption here is that the uncertainty in space and momentum is small enough compared to the system size that the electrons can be treated as particles. The scattering term, $\frac{\partial f}{\partial t}|_{\text{scatt}}$, is the most interesting and complicated part of the BTE which accounts for the change of the electron probability due to electron–electron or electron–phonon scattering events, which we can treat quantum mechanically.

Without taking care of the exact form of the underlying scattering mechanisms, the constant relaxation time approximation (CRTA)assumes that the system seeks to return to the equilibrium configuration f_0 in a time τ after being disturbed by external electric or temperature fields. In general, τ, known as relaxation time, should depend on the direction of the scattering events, energy, and on the exact scattering mechanisms. However, the easiest version of the CRTA assumes one constant momentum relaxation time for all modes, direction, and scattering processes and the collision term can be written in the form [9,10]

$$\frac{\partial f(\mathbf{r}, \mathbf{k}, t)}{\partial t}\bigg|_{\text{scatt}} \simeq -\frac{\delta f(\mathbf{r}, \mathbf{k}, t)}{\tau} \tag{7}$$

Here, the rate of change of $f(\mathbf{r}, \mathbf{k}, t)$ due to collisions is assumed to be proportional to the deviation from the equilibrium distribution, $\delta f = f - f_0$. To better understand the meaning of the CTRA, we assume no electric field present (no $\mathbf{F}_e$) and spatial uniformity (gradient with respect to **r** vanishes). Then the BTE, Equation (6), in the CRTA takes the form

$$\frac{\partial f}{\partial t} = -\frac{\delta f}{\tau}, \tag{8}$$

which has the familiar solution for the deviation from equilibrium

$$\delta f(t) = \delta f(0) e^{-t/\tau}. \tag{9}$$

This means that the perturbed system relaxes with a typical timescale τ to its equilibrium distribution. The CRTA is a crude approximation that fails for certain systems as we will see later in this section.

When going beyond the simple CRTA and concentrating on the phonon-limited carrier mobilities in semiconductors, it is common to consider the electron–phonon scattering as the dominant mechanism. Furthermore, we adopt the notation for crystals in k-space, as such, the distribution function depends on the band index n and the momentum **k**, i.e., $f_{n\mathbf{k}}$. Therefore, for time-independent and homogenous fields, the BTE in k-space reads

$$\frac{\partial \mathbf{k}}{\partial t} \frac{\partial f_{n\mathbf{k}}}{\partial \mathbf{k}} = \frac{\partial f_{n\mathbf{k}}}{\partial t}\bigg|_{\text{scatt}}. \tag{10}$$

Neglecting magnetic fields and considering only the presence of electric fields, the BTE for phonon-limited carrier mobilities is given by [11,12]

$$
\begin{aligned}
e\mathbf{E}\frac{\partial f_{n\mathbf{k}}}{\partial \mathbf{k}} = {} & \frac{\Omega}{(2\pi)^2\hbar} \sum_{m,v} \int d\mathbf{q} |g_{mnv}(\mathbf{k}, \mathbf{q})|^2 \\
& \times \Big\{ f_{n\mathbf{k}}[1 - f_{m\mathbf{k}+\mathbf{q}}]\left[(n_{\mathbf{q}v} + 1)\delta(\epsilon_{n\mathbf{k}} - \epsilon_{m\mathbf{k}+\mathbf{q}} - \hbar\omega_{\mathbf{q}v}) + n_{\mathbf{q}v}\delta(\epsilon_{n\mathbf{k}} - \epsilon_{m\mathbf{k}+\mathbf{q}} + \hbar\omega_{\mathbf{q}v})\right] \\
& - f_{m\mathbf{k}+\mathbf{q}}[1 - f_{n\mathbf{k}}]\left[(n_{\mathbf{q}v} + 1)\delta(\epsilon_{n\mathbf{k}} - \epsilon_{m\mathbf{k}+\mathbf{q}} + \hbar\omega_{\mathbf{q}v}) + n_{\mathbf{q}v}\delta(\epsilon_{n\mathbf{k}} - \epsilon_{m\mathbf{k}+\mathbf{q}} - \hbar\omega_{\mathbf{q}v})\right] \Big\}.
\end{aligned} \tag{11}
$$

The first two terms in the second line represent scattering of an electron from energy band n with momentum **k** into the state $m, \mathbf{k} + \mathbf{q}$ by either the emission $(n_{\mathbf{q}v} + 1)$ or the absorption $(n_{\mathbf{q}v})$ of a phonon with frequency ω_{qv}. The factor $f_{nk}[1 - f_{mk+q}]$ make sure that the initial electronic state is occupied while the final state is unoccupied. The last two terms represent the backscattering events from state $m, k + q$ towards the state in the energy band n with momentum k. Here, Ω is the volume of the primitive unit cell, the δ-functions ensure energy conservation during the scattering process and n_{qv} is the Bose-Einstein distribution function, giving the probability that a crystal phonon with moment q is present in branch v. The term $g_{mnv}(k, q)$ is the electron–phonon coupling matrix element, which is normally calculated within density functional perturbation theory [13–15]. Taking the derivative with respect to the i-component of the electric field E_i, and linearizing the equation around the equilibrium distribution f^0, we get a direct equation for $\partial_{E_i} f_{nk}$ [12]

$$\frac{\partial f_{nk}}{\partial E_i} = e \frac{\partial f_{nk}^0}{\partial \epsilon_{nk}} v_{nk,i} \tau_{nk} + \frac{\Omega \tau_{nk}}{(2\pi)^2 \hbar} \sum_{m,v} \int dq \frac{\partial f_{mk+q}}{\partial E_i} |g_{mnv}(k, q)|^2 \tag{12}$$
$$\times \left\{ [1 + n_{qv} - f_{nk}^0] \delta(\epsilon_{nk} - \epsilon_{mk+q} + \hbar\omega_{qk}) + [n_{qv} + f_{nk}^0] \delta(\epsilon_{nk} - \epsilon_{mk+q} - \hbar\omega_{qk}) \right\},$$

where the inverse of the electron energy relaxation time due to the electron–phonon interaction is

$$\tau_{nk}^{-1}(T, \mu) = \frac{\Omega}{(2\pi)^2 \hbar} \sum_{m,v} \int dq |g_{mnv}(k, q)|^2 \left\{ [n_{vq} + f_{mk+q}^0] \delta(\epsilon_{nk} - \epsilon_{mk+q} + \hbar\omega_{qv}) \right.$$
$$\left. + [n_{vq} + 1 - f_{mk+q}^0] \delta(\epsilon_{nk} - \epsilon_{mk+q} - \hbar\omega_{qv}) \right\}. \tag{13}$$

Note here that the Fermi distribution for the electrons is a function of both temperature and chemical potential, while the phonon distribution function depends on the temperature. Equation (12) is the linearized BTE (LBTE) for phonon-limited carrier transport and is valid for most semiconductors where the acceleration of the free carriers is smaller than the thermal energy, $eEv\tau \ll k_B T$. We would like to point out that in contrast to the CRTA, here the relaxation time depends on the energy and momentum of the electron. The LBTE needs to be solved self-consistently for the variation of the electron distribution function with respect to the applied field, therefore also called iterative BTE. By neglecting the integral part of the LBTE, one obtains an equation for the distribution function which can be solved directly,

$$\frac{\partial f_{nk}}{\partial E_i} = e \frac{\partial f_{nk}^0}{\partial \epsilon_{nk}} v_{nk,i} \tau_{nk} \tag{14}$$

and is called the self-energy relaxation time approximation (SERTA). This is due to the analogy that the relaxation time τ in Equation (13) is related to the Fan-Migdal electron self-energy by $\tau_{nk}^{-1} = 2\text{Im}\Sigma_{nk}^{FM}$ [13]. In the SERTA, the electron mobility takes the form

$$\mu_{\alpha\beta}(T, \mu) = -\frac{e\Omega}{n_e (2\pi)^3} \sum_{n\in CB} \int dk \frac{\partial f_{nk}^0(T, \mu)}{\partial \epsilon_{nk}} v_{nk,\alpha} v_{nk,\beta} \tau_{nk}(T, \mu). \tag{15}$$

When calculating the mobility within SERTA or LBTE, the relaxation time in Equation (13) is evaluated on a very fine k-grid by integrating over a dense q-grid for the phonons. This direct Brillouin zone sampling is computationally very demanding and hence cannot be applied for complex material screening. For example, the EPW code [16] calculates the electron–phonon coupling on a coarse grid in the BZ and maps it onto a fine grid by using Wannier's function interpolation. This method is however still very costly and therefore other approaches are needed to tackle the calculations of thermoelectric transport properties. One approach solves the BTE within the CRTA, Equation (7): This leads to good results for electrical conductors where the energy relaxation time depends weakly on the electron energy, ϵ, [10,17] and allows for a single constant relaxation time. However, when performing the CRTA one needs to estimate the relaxation time within simplified models, such as the deformation potential (DP) approximation [18–20] or Allen's formalism [21,22]. For most materials, τ is strongly anisotropic, depending on energy and carrier concentration and the CRTA cannot be applied and we need to revert to first-principles computations for predicting τ_{nk}. The so-called electron–phonon averaged (EPA) approximation [23] turns the demanding integral over momentum in Equation (13) into an integration over energy. This is done by replacing quantities that depend on momentum ($|g_{mnv}(\mathbf{k}, \mathbf{q})|^2$, ω_{vq}) by their energy-dependent averages ($g_v^2(\epsilon_{nk}, \epsilon_{mk+q})$, $\overline{\omega}_v$). This allows a much coarser grid in the electron energies therefore reducing computational cost drastically. Technical details and derivations can be found in [23] and the final relaxation time within the EPA is given by

$$\tau^{-1}(\epsilon, \mu, T) = \frac{2\pi\Omega}{2\hbar} \sum_v \left\{ g_v^2(\epsilon, \epsilon + \overline{\omega}_v) \left[n(\overline{\omega}_v, T) + f^0(\epsilon + \overline{\omega}_v) \right] \rho(\epsilon + \overline{\omega}_v) \right.$$

$$\left. + g_v^2(\epsilon, \epsilon - \overline{\omega}_v) \left[n(\overline{\omega}_v, T) + 1 - f^0(\epsilon - \overline{\omega}_v) \right] \rho(\epsilon - \overline{\omega}_v) \right\}, \tag{16}$$

where $\overline{\omega}_v$ is the average phonon mode energy and $\rho(\epsilon)$ is the electron density of states per unit energy and unit volume. The EPA is implemented in the Quantum Espresso suite [24] and Boltztrap code [25].

Figure 1 compares the relaxation time calculated within Equation (13) by the EPW code and within the EPA, Equation (16) for (a) HfCoSb and (b) HfNiSn. We see that the relaxation times in both approaches are in a good agreement, and furthermore that approximating the relaxation times with a constant, as done in CRTA, might fail for predicting the electron conductivity of real materials.

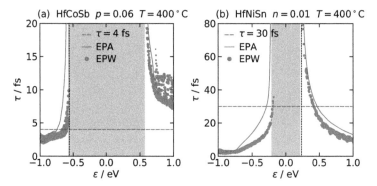

Figure 1. Relaxation time for (**a**) HfCoSb and (**b**) HfNiSn calculated within the EPA (Equation (16)) compared to the complete sampling of the Brillion zone (Equation (13)). Both approaches are in good agreement and the CRTA (dashed blue line) might fail to accurately calculate the conductivity. Reprinted with permission from [23]. Copyright (2018) John Wiley and Sons.

2.2. Landauer–Buttiker and Quantum Transport

The Landauer–Buttiker's formalism (LB) is an elegant and economic way to study transport through molecular devices [26–28]. The idea at its core is deceivingly simple: The overall device is separated into three or more regions of space, one of which the central, usually characterizes the physical properties of the overall system. The other regions serve as sink or reservoir of electrons. The only property that is required from the reservoirs is that they connect smoothly with the central region, thus avoiding backscattering due to the contacts, and that they are either perfect emitters or absorbers, in the sense that one electron entering them cannot leave through the central region again. This requires also that their spectral density is essentially flat over a wide range of energies around the Fermi level of the central system. LB then describes the currents flowing between the leads, through the probability that an electron entering through the lead i is scattered by the central region, and thus leaves through the lead j, $T_{i,j}$. In this respect thus the LB formalism is a scattering theory and the transmission probability is its central ingredient. Finally, one sums over all the allowed electron energies [26–28]. The electrical current is therefore given by

$$I_{i,j} = \frac{2e}{h} \int d\epsilon \left(f_i(\epsilon) - f_j(\epsilon) \right) T_{i,j}(\epsilon), \tag{17}$$

where f_i is the Fermi function describing the electron occupations of the lead i, e is the electron charge, h is the Plank's constant, and the factor 2 takes into account spin degeneracy. Generally, the Fermi functions depend upon the local temperature T_i of the lead i, their chemical potential μ_i, and the bias applied to the lead, V_i. Without loss of generality, in the presence of the bias V_i we can assume that $\mu_i = 0$. It is important to realize the important approximation of the model: we are assuming that the occupation of the electrons entering from the lead i, depends only on the equilibrium physical parameter of the i lead. Finally, we have also assumed that the central system is essentially one dimensional—in this case, the density of states and the electron velocity cancel out to achieve the universal result of Equation (17). A thorough discussion of the derivation and the physical implications of the Landauer transport theory goes way beyond the aim of this review and the topic has been discussed in a large number of publications, including reviews and monographs [1,2,4,26–29].

Given the current I from the different terminals, one can easily calculate the two-terminal conductance, in the linear-response regime $\delta V_{i,j} \rightarrow 0$,

$$\sigma_{i,j} = \lim_{\delta V_{i,j} \to 0} \frac{I_{i,j}}{\delta V_{i,j}} = \frac{2e^2}{h} \int d\epsilon f'(\epsilon) T_{i,j}(\epsilon). \tag{18}$$

where $f'(\epsilon)$ is the first derivative of the Fermi function with respect to the energy, and $\delta V_{i,j} = V_i - V_j$ the bias difference between the leads i and j. The simplest most striking prediction that the LB theory makes is the quantization of the conductance. Indeed, in the low temperature limit, $f' \rightarrow \delta(\epsilon - E_F)$ where E_F is the Fermi energy of the leads, so that [26–28]

$$\sigma_{i,j} = \frac{2e^2}{h} T_{i,j}(E_F). \tag{19}$$

If we assume that $N_{i,j}$ states are fully open to transport current, i.e., their transmission probability goes to 1, we have

$$\sigma_{i,j} = \frac{2e^2}{h} N_{i,j} \tag{20}$$

as it has been carefully verified in many experiments, for example in the quantum Hall effect [30,31] or in the ballistic regime of a quantum point contact [32–34]. Equation (19) is what is normally called the "Landauer's formula" and it is deceivingly simple. However, its physical interpretation has puzzled the community for many years. In particular, one question that arises naturally is from where this conductance originates. Clearly, we are describing an almost ideal system. The electrons flow seamlessly from the leads to the central region, and they are transmitted to the other lead with probability 1, so there is no scattering in these states. In addition, yet, we have a finite conductance thus associated with energy loss and potential drops. The solution of this apparent paradox lies in the different dimensionality of the leads with respect to the central system. The "adjustment" of the wave-function to adapt to the reduced dimensionality of the center causes a charge accumulation at the interfaces between the center and the leads, no matter how smooth these interfaces are. This extra charge that accumulates as soon as we contact the central region, creates a finite bias that opposes the one applied to the reservoir and finite conductance appears.

One of the main advantages of the Landauer's formalism lies in treating on the same footing different physical problems. For example, the study of thermal transport both by electrons and phonons can be easily recast in the form of a scattering problem for the electrons and phonons trough the central region. Therefore, the energy current due to the electrons is evaluated as

$$J_{i,j}^e = \frac{2}{h} \int d\epsilon (\epsilon - \mu) \left(f_i(\epsilon) - f_j(\epsilon) \right) T_{i,j}(\epsilon), \tag{21}$$

while the phonon contribution is given by

$$J_{i,j}^p = \frac{\hbar}{2\pi} \int d\omega \omega \left(n_i(\omega) - n_j(\omega) \right) T_{i,j}(\omega), \tag{22}$$

where n_i is the Bose distribution of the phonons in the lead i kept at temperature T_i. In the following we assume that $T_i = T + \Delta T_i$, where ΔT_i defines the difference between the temperature of the lead i and the reference temperature T. As we did for the electrical conductance, we can calculate the thermal conductance for the phonons and the electrons by simply assuming that a small thermal gradient is present between the leads. Moreover, if one assumes that both the potential and thermal gradient are small, the Landauer formalism reduces to the Onsager linear-response out-of-equilibrium transport theory, with the advantage of providing a reliable way of calculating the Onsager's coefficients [1,35]. Indeed, if one defines the functions (Lorentz's integrals)

$$L_n = \int d\epsilon (\epsilon - \mu)^n f'(\epsilon) T_{i,j}(\epsilon) \tag{23}$$

we have the Onsager's relations (restricting ourselves to only two leads)

$$I = \frac{2e^2}{h} L_0 \Delta V - \frac{2e}{hT} L_1 \Delta T, \tag{24}$$

$$J_e = -\frac{2e}{h} L_1 \Delta V + \frac{2}{hT} L_2 \Delta T. \tag{25}$$

The electron transport coefficients, namely the electrical conductance, the thermal conductance and the Seebeck's coefficient then follow from their physical definitions and are expressed solely in terms of the functions L_n. Indeed, the electrical conductance is $\sigma = I/\Delta V|_{\Delta T \equiv 0} = 2e^2 L_0/h$, the Seebeck's coefficient is $S = -\Delta V/\Delta T|_{I=0} = -L_1/eTL_0$, while the thermal conductance is $\kappa = J/\Delta T|_{I=0} = 2(L_2 - L_1^2/L_0)/hT$. Notice that the physical definition of the transport coefficients is in general not restricted to the linear-response regime, so one could define the Seebeck coefficient as $S = -\Delta V/\Delta T$ also for large ΔT, but the expression in terms of the Onsager's coefficients and Lorentz integrals is valid only in linear response.

Moreover, the formula for the phonon energy current Equation (22) predicts the surprising result that the low energy phonons, usually responsible for large part of the energy transport since they have the longer wavelength, have a quantized thermal conductance,

$$\kappa_p = \frac{k_B^2 \pi^2}{3h} T, \tag{26}$$

per each open channel, as predicted by Rego and Kirczenow [36] and verified by Schwab et al. [37].

Although the Landauer's theory appears quite natural, its physical implications are far reaching as we have seen with the introduction of the quantum of conductance [30–34] and the quantum of thermal conductance that has been observed [37]. However, the maximum strength of the theory is reached when it is coupled with the standard method for electronic structure calculations, namely static Density Functional Theory (DFT), and the non-equilibrium Green's function formalism (NEGF). These two methods made the Landauer's theory and the Boltzmann's transport theory, the base for almost any recent transport calculations. DFT indeed produces a reliable description of the system energy and electron density and of the electronic band structure, both necessary for the evaluation of the transmission probability although its use on transport modeling should be taken with care as we will discuss shortly. Starting from the DFT description the NEGF method can express the transmission coefficients $T_{i,j}$ in terms of the Green's functions of the leads and those of the central system. This combination allows for an almost parameter-free description of quantum transport at the atomistic level and it is the actual reference method for this kind of physical problems.

So far we have not discussed how to calculate the transmission probabilities between the leads i and j, $T_{i,j}$. This is usually a difficult problem since to make an accurate description of the scattering process we need modeling the states of the electrons inside the central region. Fortunately, we can combine the predictive power of DFT, which precisely describes the states of the system with the NEGF formalism which (as we will see) allows calculation of the transmission probability. A detailed derivation of all the results would be outside the scope of this review. There are a significant number of detailed monographs dedicated to the subject and we refer the interested reader to them [29,38]. Given a Hamiltonian H, we formally define its advanced and retarded Green's functions through (we set from now on $\hbar = 1$)

$$\left[-i\frac{d}{dt} - H\right] G^{R(A)}(t,t') = \delta(t-t'),\tag{27}$$

with the conditions $G^R(t,t') = 0$ if $t < t'$ and $G^A(t,t') = 0$ is $t > t'$. An equivalent definition of the retarded and advanced Green's functions is based on their representation in terms of energy, i.e., after a Fourier transform,

$$[(E \pm i\eta) - H] G^{R(A)}(E) = 1\tag{28}$$

where η is an infinitesimal positive quantity that serves to establish the analytical properties of $G^{R(A)}$. When we select a basis set for the Hilbert space, the Green's functions (as well as the Hamiltonian) can be represented as infinite matrices. However, normally only a certain number of states will be relevant for the dynamics (in our case those close to the Fermi energy or electrochemical potential μ) and therefore only a submatrix of the total Green's function will be needed.

For the study of electron transport through a nanoscale device, we need to describe the system (a central region) coupled to at least two external reservoirs. A pictorial representation of a device is given in Figure 2.

Figure 2. A central region (C) is connected to two external energy and particle left and right reservoirs (L and R) by metallic contacts. Current flows between the reservoirs when a temperature or bias gradient is established. The reservoirs are semi-infinite, namely the left proceeds from $-\infty$ to C, the right goes from C to $+\infty$.

If separated, each of these objects can be described by their Hamiltonian and Green's functions. When we couple the central region with the reservoirs, we are effectively introducing an interaction potential that couples states in the reservoirs with states in the leads. We introduce the following notation: H_A with $A = L, R$, or C is the Hamiltonian of the left or right reservoir, or the central region, respectively; V_{AB} with $AB = RC, LC, CR$, or RL represent the coupling between the central region with the left and right reservoirs, respectively. V_{AB} can be for example a tunneling Hamiltonian between the lead and the central system. We assume there is not direct coupling between the reservoirs, so electrons must travel through the central region. In a matrix representation, we can think of the total Hamiltonian as a matrix whose sub-matrices are H_A and V_{AB},

$$H = \begin{pmatrix} H_L & V_{LC} & 0 \\ V_{CL} & H_C & V_{CR} \\ 0 & V_{RC} & H_R \end{pmatrix}. \tag{29}$$

Clearly, one could define a Green's function associated with this Hamiltonian whose dynamical equation can be written as

$$\begin{pmatrix} E - H_L + i\eta & -V_{LC} & 0 \\ -V_{CL} & E - H_C + i\eta & -V_{CR} \\ 0 & -V_{RC} & E - H_R + i\eta \end{pmatrix} \begin{pmatrix} G_L^R & G_{LC}^R & 0 \\ G_{CL}^R & G_C^R & G_{CR}^R \\ 0 & G_{RC}^R & G_R^R \end{pmatrix} = \hat{1}, \tag{30}$$

where we have used the previous notation for the elements of G^R, and $\hat{1}$ is the identity matrix. This system of equations can be solved exactly to first express G_{LC}^R and G_{RC}^R in terms of G_C^R and then, solve for the latter. The exact result is

$$G_C^R = \left[E + i\eta - H_C - \Sigma^R\right]^{-1} = \left[E - H_C - \Sigma^R\right]^{-1} \tag{31}$$

where we have defined the self-energy

$$\Sigma^R = V_{LC}^\dagger (E + i\eta - H_L)^{-1} V_{LC} + V_{RC}^\dagger (E + i\eta - H_R)^{-1} V_{RC}, \tag{32}$$

where with a † we indicate the Hermitian conjugate of V. Notice that in G_C^R the analytic properties are finally determined by those of the self-energy Σ and one can therefore neglect the terms $i\eta$. A similar equation can be derived for G_C^A. It is important to point out that in this theory, the leads enter both in the interaction with the central region through V_{LC} and V_{RC} and their isolated Green's function $G_{L(R)}^R = \left(E + i\eta - H_{L(R)}\right)^{-1}$.

To solve this set of equations, one assumes that there is no particle–particle interaction in the leads. This approximation can be justified by observing that they are thought of as normal metals and thus the screening length is relatively small, thus particles can be treated as weakly interacting if not independent. Within this scheme, $G_{L(R)}$ are uniquely determined and can be used to arrive at the lead self-energies to solve for G_C^R and G_C^A. In these last quantities, however, we cannot neglect particle–particle interaction. Due to the complexity of the problem, the exact many-body Hamiltonian H_C is replaced with the so-called Kohn–Sham (KS) Hamiltonian, where an external single-particle potential (generally unknown) serves to mimic the effect of the interaction (see also the following sections for somewhat deeper discussion) [39–42]. The KS Hamiltonian is tailored to reproduce the exact ground-state energy and density, but it often produces an accurate description of the total Green's function of the isolated central region.

Finally, we connect the Green's function formalism with the Landauer's approach to quantum transport, since the former gives direct access to the transmission probability $T_{i,j}$. This step can be done by, for example, introducing the states for the left and right leads and solve the scattering problem with the central region by using the Green's functions. After some manipulations, one writes the transmission function $T(E)$ as [43]

$$T(E) = \text{Tr}\left(\Gamma_L G_C^A \Gamma_R G_C^R\right), \tag{33}$$

where $\Gamma_{L(R)} = i\left(\Sigma_{L(R)}^R - \Sigma_{L(R)}^A\right)$ is the so-called spectral function of the leads. Other approaches or models to the calculation of the transmission probability are clearly possible, and provide the tools to investigate novel and interesting phenomena, see for example Ref. [44] and the references in that focus issue.

This formalism is suited to describe both thermal and electrical transport since we have never specified the kind of gradient we maintain between the reservoirs. We are therefore entitled to consider both a bias voltage, and a temperature gradient that modifies the particle distribution functions in the reservoirs.

Quite naturally the formalism can be extended to consider the transport of energy through phonons or more generally vibrations. Formally, the only difference lies in replacing the particle Green's function G with the vibration Green's function D which is a solution of

$$\left(\omega^2 + i\eta - H_v \right) D^R(\omega) = 1 \tag{34}$$

where ω is the frequency of the vibration, and H_v the Hamiltonian function describing the vibration dynamics. Following the same steps as before, one can introduce the transmission function of the vibration $T(\omega)$ and derive a formula similar to Equation (33).

It is fair to examine here some of the problems one might face when using DFT + NEGF for the electronic transport calculations:

- The original static DFT is in principle limited to provide the ground-state density and energy and nothing about the excited states of a system. Whatever result one extracts from the theory for example on the band-gap, band structure (and related quantities) should then be checked with independent methods and with experiments. It is well known for example that DFT with its standard approximation tends to underestimate the band-gap of many materials.
- Although the use of the NEGF greatly extends the applicability of the theory, the approximations behind the standard DFT codes makes the results not suitable for strong correlated materials where the effect of the Coulomb interaction is stronger. We will come back on this point later.
- Even if the description of the excited states happened to be accurate, the electron and heat current are dynamical quantities so in principle beyond the static approach of DFT. There are methods to go beyond this limit while remaining in the realm of a time-independent formalism. "Dynamical corrections" need to be included in the theory to take dynamics into account [45,46].

3. Advanced Methods

The Boltzmann's equation and the Landauer-Büttiker formalism can be considered the de-facto standard methods to study transport in nanoscale devices, especially when coupled with accurate electronic structure calculations of the Density Functional Theory. It has become clear however that these methods have severe limitations that reduce the ensemble of systems to which they can be applied. In this section, we will discuss some attempts we made to go beyond and include novel effects outside the limitations of the state of the art. Some of these methods are admittedly in their infancy and a consistent amount of work is necessary before they can be admitted in the mainstream research methods. We hope that this section can motivate the interested readers to enter this enticing field which still bears endless possibilities.

3.1. Time-Dependent Density Functional Theory

Density Functional Theory proved an important point of quantum mechanics, i.e., that some exact properties of a many-body interacting system can be extracted from the study of a Doppelgänger of non-interacting particles [40–42,47]. The simplification brought about cannot be underestimated: it is enough to think that an interacting system requires an exponentially large Hilbert's space for its accurate numerical description, while the same system of non-interacting particles requires only a polynomially large (with the number of states) space. This difference separates being able to treat just a few particles from studying complex molecules or crystal unit cells. There are two prices to pay for this simplification; On the one hand, we are ready to give up the complete information about the many-body system and study only certain quantities. The standard DFT formalism was aimed at the ground-state energy and density. Any other quantity that is then obtained in DFT must then be checked against other models; On the other hand, the particle–particle interaction is *replaced* in the so-called Kohn-Sham (KS) system with a non-linear external potential (known as KS potential) that is assumed to depend only on the single-particle ground-state density [39,47,48]. This potential is clearly unknown and we need to revert to some sort of approximation to make the theory useful. Standard approximations have been developed and admittedly work rather well especially when dealing with atoms and small molecules, but failure is also around the corner. Indeed, some of the standard problems of DFT in dealing with transport calculation derives from a combination of these points: band-gap and transport coefficients are normally not static ground-state properties, and the standard approximations for the KS potential consistently underestimate the electronic band-gap. Finally, there are classes of problems that are outside the realm of standard DFT, such as for example the calculation of spectroscopic quantities. A detailed introduction to DFT would bring us far away from the scope of this review. We can however, recommend excellent introductory material for the interested reader [40–42,47].

There are many ways to go beyond these issues. We can formulate the theory in the time domain, in such a way that we can extract the exact time dynamics of some of the quantities of the many-body interacting system [49–51]. This is the case of Time-Dependent Density Functional Theory. Meanwhile, we can improve the standard approximations of the KS potential, for example introducing corrections that take into account some part of the strong correlation between particles in certain regimes [52].

In the next sections we will introduce the Time-Dependent Density Functional Theory (TDDFT) [40,41] and its further extensions and later the so-called "i-DFT" that can be used to introduce strong correlation effects, such as Coulomb blockade into a TDDFT formulation (see Section 3.2) [53,54].

3.1.1. Time-Dependent Density Functional Theory—Fundamentals

The standard approach of DFT is based on the existence of a one-to-one mapping between the exact ground-state density and the external potential applied to a quantum system [48]. Furthermore, one can extend this mapping between two systems, the many-body interacting "real" system and a many-body non-interacting "Kohn-Sham" system [39] where we introduce an effective external potential (dubbed KS potential). This second mapping allows for calculating quantities, i.e., the ground-state density and energy, belonging to the real system by using the KS system, bringing about a large numerical simplification. DFT is therefore the actual method of choice for calculating electronic structure of materials as well as the energies of complex atoms and molecules [42]. The KS mapping was later extended to the dynamics of the single-particle density by Runge and Gross [55]. The existence of this second mapping lays the foundations for performing, e.g., numerical spectroscopy with so-called ab-initio methods, i.e., without—in principle—any fitting parameter [41]. TDDFT can theoretically be used for studying electrical transport (see for example Ref. [56–60]), but the information it can provide is only partial [61,62].

It appears natural therefore, from the Runge and Gross' theorem (RG), to establish a mapping between the single-particle current density and the external (vector) potential applied to the real system [63]. In principle, this Time-Dependent Current-Density Functional Theory (TDCDFT) should be the workhorse for charge transport calculations, but the lack of suitable approximations for the KS's potential renders the theory of little use at the moment. One the other hand, addressing thermal transport phenomena within DFT would require instead a non-equilibrium theory based on either a local temperature or a local energy density. Recently, a functional theory based on the excess energy density as the basic variable has been presented [64,65] which is suited to study thermoelectrical phenomena in the static and time-dependent case. In the following, we will look only at standard TDDFT.

Entropy **2019**, *21*, 752

The RG theorem proves that for a general time-dependent Hamiltonian H that describes the dynamics of a many-body system,

$$H(t) = T + V_{\text{int}} + V_{\text{ext}}(t) \tag{35}$$

there exists a one-to-one mapping between the single-particle density $n(\mathbf{r}, t)$ and the external potential $V_{\text{ext}}(\mathbf{r}, t)$, given the initial conditions. Here, T is the kinetic energy, V_{int} the particle-particle interaction, and V_{ext} the time-dependent external potential. The proof of the theorem assumes that both density and potential can be expanded in series around the initial time: a detailed proof and the physical assumption on which this is based can be found in [40,41]. It has since been shown that there exists a system of non-interacting particles whose single-particle density is identical at each time to $n(\mathbf{r}, t)$ of the interacting system [66]. The existence of this mapping then allows investigation of the dynamics of a system by looking at its non-interacting Doppelgänger. Clearly, this brought about the same reduction of computational requirements as the standard DFT. However, as was the case with DFT, we are exchanging some of the physical information about the system to reduce the dimensionality of the problem. It should then come as no-surprise that TDDFT cannot for example reproduce the exact single-particle current density $\mathbf{j}(\mathbf{r}, t)$ since some of its contributions are not derivable from the knowledge of the density alone [67]. Indeed, if starting from the continuity equation,

$$\partial_t n(\mathbf{r}, t) = -\nabla \cdot \mathbf{j}(\mathbf{r}, t), \tag{36}$$

we need to conclude that given the density we can determine only the longitudinal part of $\mathbf{j}(\mathbf{r}, t)$ and any term $\mathbf{j}_T$ such that $\mathbf{j}(\mathbf{r}, t) = \mathbf{j}_L(\mathbf{r}, t) + \mathbf{j}_T(\mathbf{r}, t)$ and $\nabla \cdot \mathbf{j}_T \equiv 0$ cannot be obtained from Equation (36). However, some information about the total current can still be obtained from the continuity equation [68].

As with the static DFT, we study the KS Hamiltonian H_{KS},

$$H_{\text{KS}} = T_S + V_{\text{KS}} \tag{37}$$

where $T_S = \sum_i p_i^2 / 2m$ is the kinetic energy of the non-interacting particles of mass m and momenta $\{\mathbf{p}_i\}$, and $V_{\text{KS}} = V_{\text{ext}} + V_{\text{Hxc}}$. The potential V_{Hxc} is the sum of the Hartree and exchange–correlation (xc) potentials that replace the particle–particle interaction. By the RG theorem V_{Hxc} is a functional of the density only, $V_{\text{Hxc}} = V_{\text{Hxc}}[n](\mathbf{r}, t)$, and the dynamics of the single-particle density of the real system evolving with Hamiltonian H is identical to the dynamics of the single-particle density evolving with H_{KS}. Notice that in principle, given the particle–particle interaction of the original many-body problem, V_{Hxc} is a universal functional and does not depend on the external potential V_{ext}. This implies for example that if we are interested in the dynamics of an electronic system, V_{Hxc} is the same either if we are studying an atom, a molecule, a slab, or a bulk, i.e., this potential is transferrable to whatever system we want to study. This high transferability makes finding reliable approximation for V_{Hxc} difficult. Indeed, the universal V_{Hxc} for the electrons contains the physical information needed to go from the standard weakly interacting Landau's Fermi liquid, to the strongly correlated Wigner's crystal or superconducting phases. Notice that for the electrons, a large part of V_{Hxc} is given by the Hartree's (mean-field) interaction, and only a relatively small contribution of the overall energy determines all these interesting phases of matter. Therefore, standard approaches to build some approximation to V_{Hxc} are based on interpolating the numerical solution of the many-body problem with some known high- and low-density limits [69,70]. Many of these approximations are static. To apply them to the dynamics, a common method is the so-called adiabatic local density approximation (ALDA). In this approximation, we take a static $V_{\text{Hxc}}[n]$ and replace the static density with the instantaneous density $n(\mathbf{r}, t)$. Indeed, this neglects the history of the system—while we generally expect that the xc potential to be history dependent. For example, one can show that in this approximation there is not relaxation induced by particle–particle interaction, in contrast with observation [71,72].

3.2. Strongly Correlated System

The Landauer approach coupled with Density Functional Theory through the Non-Equilibrium Green's function formalism provides a reliable and presently standard method to calculate the transport properties of many devices. Countless are the successes of the theory, far beyond its actual range of applicability. Nonetheless there are many reasons to go beyond this state of the art. A simple reason is that we need alternative methods to provide benchmarks for the DFT + LB theory and learn from them. Indeed, the strength of DFT lies also in adapting the KS potential to the cases under investigation. On the other hand, we might need to go beyond some of the fundamental approximations and thus build a different theory.

Meanwhile, a striking example of the limits of the DFT + LB theory is related to one of the most fascinating outcomes of the so-called mesoscopic physics. Imagine that the device we place between two leads is a small molecule or a quantum dot. Both these systems are thought of having just a few states close to the Fermi energy (or the electrochemical potential). Imagine that one electron enters the device and occupies the lowest energy state. The next electron then faces an increased energy barrier, since beside the energy to occupy the lowest available energy states it also musts overcome the Coulomb interaction with the other electron. Normally, for large devices this energy is small since the electron density is "diluted", but when we consider small dots or molecules, the Coulomb interaction might be the dominating energy scale and transport can be "blocked" until either the first electron leaves the device or the second electron has enough energy to overcome the Coulomb interaction. A standard DFT + LB approach to this problem is most likely going to fail. Indeed, DFT describes the electrons through their density and therefore it does not produce the sharp energy transition due to the addition of a single electron. This effect goes normally under the name of "Coulomb blockade", and it is the epitome of a strongly correlated system where essentially the dynamics is dictated by electron interaction and correlation.

However, it is important to point out that these limits are related to our ability of inventing KS potentials able to describe certain physical regime. Per se, DFT can describe the ground-state properties of the system and thus give the exact energy for the single and double occupied electron states. It is our inability of encoding these effects into the KS potential that makes the theory fails. Indeed, progress has been made to include strong correlation into the KS potential into a pure DFT scheme. Here, we will consider the extension of DFT to deal with the Coulomb blockade regime. To do that, we first need to extend the theory to include the transport properties in a more accurate way. As we will see this step corrects the electronic conductance that is calculated from the standard DFT approach.

Our starting point is the observation that generally speaking, the quantities we want to investigate in studying the device of Figure 2 are the local electron density $n(\mathbf{r})$ and the total current I flowing from one reservoir to the other due to a thermal gradient or a bias voltage. For the moment, we will focus on the steady state, i.e., we assume the system has evolved from an initial state and approached, as time goes by, a constant density and current. The external fields that we are applying are the bias voltage V and the gate voltage v which controls the electron density and the total number of particles. We assume that the nuclear potential is not affected by the electron distribution and it is, therefore, constant (we assume uniform temperature). We see V and v as a perturbation and $n(\mathbf{r})$ and I as response. To make a DFT theory for this set of variables, we need to prove that they are uniquely connected. Specifically, that the pair $n(\mathbf{r})$ and I uniquely determine both V and v. Moreover, we are interested in $n(\mathbf{r})$ and I only inside a finite region, $\mathcal{R}$, surrounding the device. The proof of the theorem entails the evaluation of the Jacobian of the mapping $\{n(\mathbf{r}), I\} \rightarrow \{v(\mathbf{r}), V\}$ for any $\mathbf{r} \in \mathcal{R}$. One can prove that around $V = 0$ this mapping is invertible since the Jacobian does not vanish [53]. We can therefore follow the KS construction and find a system of non-interacting particles which can reproduce $n(\mathbf{r})$ and I of the original system by replacing the interaction with the external potentials

$$V_S[n, I] = V[n, I] + V_{xc}[n, I] \tag{38}$$
$$v_S[n, I] = v[n, I] + v_{Hxc}[n, I] \tag{39}$$

where V_{xc} and v_{Hxc} are the xc potentials (in v_{Hxc} we include also the Hartree mean-field potential). $n(\mathbf{r})$ and I in the KS system are determined by

$$n(\mathbf{r}) = 2 \int \frac{d\epsilon}{2\pi} \left[f \left(\epsilon - \frac{V + V_{xc}}{2} \right) A_L(\mathbf{r}, \epsilon) + f \left(\epsilon + \frac{V + V_{xc}}{2} \right) A_R(\mathbf{r}, \epsilon) \right] \tag{40}$$

$$I = 2 \int \frac{d\epsilon}{2\pi} \left[f \left(\epsilon + \frac{V + V_{xc}}{2} \right) - f \left(\epsilon - \frac{V + V_{xc}}{2} \right) \right] T(\epsilon), \tag{41}$$

where $A_{R(L)} = \langle \mathbf{r} | G(\epsilon) \Gamma_{R(L)} G^\dagger(\epsilon) | \mathbf{r} \rangle$ is the right (left) KS partial spectral function, $G(\epsilon)$ the Green's function of the KS system in the energy representation, and $T(\epsilon)$ the transmission function (see Equation (33)). Notice that the right-hand sides are expressed solely in terms of KS quantities, while in the left-hand side we have the many-body quantities. In these expressions, the gate voltage v enters in the KS partial spectral function and the transmission coefficient only. We want now to derive an expression for the electrical conductance σ and the Seebeck's coefficient S solely determined from the KS quantities. We are focusing here to the linear-response regime, i.e., we will take at the end the limit $V \to 0$. In this limit $V_{xc}(n)$ vanishes for any density n, since otherwise we would have a finite current when no external perturbation is present in contradiction with the theorem of uniqueness. We have

$$\sigma = \left. \frac{dI}{dV} \right|_{V=0} = \left(1 + \left. \frac{dV_{xc}}{dV} \right|_{V=0} \right) \int \frac{d\epsilon}{2\pi} f'(\epsilon) T(\epsilon) = \left(1 + \left. \frac{dV_{xc}}{dV} \right|_{V=0} \right) \sigma_S \tag{42}$$

where we have introduced the KS conductance $\sigma_S = \int d\epsilon / 2\pi f'(\epsilon) T(\epsilon)$. To evaluate the term in the round bracket we use the standard chain rules, remembering that our "variables" are $n(\mathbf{r})$ and I,

$$\left. \frac{dV_{xc}}{dV} \right|_{V=0} = \frac{\partial V_{xc}}{\partial I} \frac{\partial I}{\partial V} + \int d\mathbf{r} \frac{\delta V_{xc}}{\delta n(\mathbf{r})} \frac{\partial n(\mathbf{r})}{\partial V} = \frac{\partial V_{xc}}{\partial I} \sigma \tag{43}$$

since the last term vanishes in the linear-response regime, $V_{xc}(n)_{I=0} = 0$. Inserting this expansion into the previous result, we find finally

$$\sigma = \frac{\sigma_S}{1 - \frac{\partial V_{xc}}{\partial I} \sigma_S}. \tag{44}$$

Notice that in general we should expect that $\frac{\partial V_{xc}}{\partial I} \neq 0$, and therefore $\sigma \neq \sigma_S$. Although for the conductance we can find a general correction, for the Seebeck's coefficient we can derive an analytical formula only in the Coulomb blockade regime. In this particular physical condition, we are interested only in the total number of particles, $N = \int_R d\mathbf{r} n(\mathbf{r})$ and we can prove that

$$S = -\frac{\frac{dN}{dT}}{\frac{dN}{d\mu}} \tag{45}$$

if μ is the electrochemical potential. In a similar way as for the conductance, we find for the Seebeck coefficient in the Coulomb blockade regime, the expression

$$S = S_S + \frac{\partial v_{Hxc}}{\partial T} \tag{46}$$

where S_S is the Seebeck's coefficient calculated from the Landauer's approach [73]. We can compare our result with the standard DFT approach and with the exact solution provided by the rate equations [7,74,75]. Other approaches are possible to directly calculate the transmission probability, with and without interaction in the central region [44,76,77]. A direct comparison of the different methods would be desirable, but difficult, since one needs to map each set of parameters appropriately.

Figure 3 report the Seebeck's coefficient as calculated from the exact many-body theory, the rate equations, the correction Equation (46) and the standard DFT approach S_S. We notice that the dynamical correction brings S_S to coincide with the exact results. Notice that N in this case is well reproduced both by the dynamical approach as well as by the standard DFT, therefore in this case is a variable less sensitive to the approximations used. When considering more than one level, we use a constant interaction model. For the case of two levels, we find some discrepancies from the exact theories and the present approach (see Figure 4). This discrepancy originates from using the total number of particles N rather than the single occupation of each states n_1 and n_2 as our basic variables.

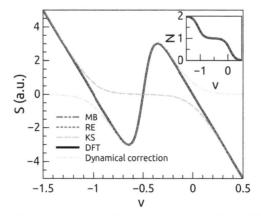

Figure 3. The Seebeck's coefficient for a single quantum dot in the Coulomb blockade regime as a function of the gate voltage v. The exact many-body (MB), the rate equation (RE), and the present theory (DFT) agree quite well in the whole range of the gate voltage v considered. The standard KS gives the correct asymptotic but fails in the central region. In cyan, we plot the effect of the dynamical correction $\frac{\partial v_{Hxc}}{\partial T}$. In the inset, we report the total number of particles N. All the theories agree quite well for this quantity. Reprinted from [73]. Copyright (2015) American Physical Society.

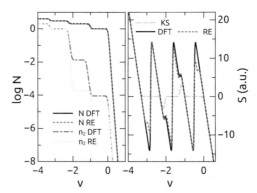

Figure 4. Left pane, the total number of particles N and the occupation of the second (highest) energy level as a function of the gate voltage v. While the first is exactly reproduced by both DFT and rate equations, the second differs. This originates differences also in the Seebeck's coefficient which is sensitive to the single particle occupations (right pane). These differences are however small compared with the correction brought about by the dynamical approach. Reprinted from [73]. Copyright (2015) American Physical Society.

3.3. Open-Quantum Systems

The theory of open-quantum systems (OQS) is a well-established model of the coupling between a system and an external environment [78–80]. Usually, the latter is considered to be a large reservoir of particles, momentum, and energy which it freely exchanges with the system. The coupling between the system and the environment might be quite general, and we can couple more than one environment, with different macroscopic thermodynamic parameters, such as temperature, chemical potential, or pressure, at a time. Usually the theory is used to describe the dynamical relaxation of the system towards some steady state or thermal equilibrium, but it has found also widespread applications in quantum optics, transport modeling, surface hopping in chemical reactions [81], and so forth (see also Section 3.4).

In the following, we are making some standard assumptions about the environment(s):

(1)　They are usually unaffected by the coupling with the systems. This means that the macroscopic parameters that describe the environment are not modified by the coupling with the system. This holds (partially) true if the environment is thought of as made of infinitely many degrees of freedom.

(2)　The external parameters are controllable in time, and due to the previous assumption, there is no-feedback between the environment and the system. For example, this implies that the establishment of an electrical current between two reservoirs at different chemical potentials through a molecular junction does not change the electrochemical potential, the temperature, or charge distribution in the reservoirs.

An efficient way to deal with the environment is therefore by tracing out its degrees of freedom and introduce effective correlation functions, (usually dependent on the thermodynamic parameters) with which we describe the dynamics of the quantum system. The way the tracing is made defines the accuracy of the theory. Normally, one can identify two large families of approximations: in the first (called Markovian) the system dynamics does not have any history. The state is determined solely by the evolution at a certain time t and does not depend on any time $t' < t$. On the other hand, we have non-Markovian dynamics where the state of the system depends on all or some of the previous times, the history of the system, since the initial time t_0. Into this second family, we then distinguish how the kernel generated by the coupling with the environment depends on the actual state of the system. Let us therefore establish some notation and the standard results. In the following, we will work on a formulation based on the density matrix. Alternative formulations based on a vector state in the Hilbert space are possible and will be briefly discussed later [78,80].

We assume that a system S is coupled to an environment B. When there is no coupling, the dynamics of S is described by a set of operators acting on a Hilbert space $\mathcal{H}_S$, while the dynamics of B by operators acting on $\mathcal{H}_B$. We further assume that these Hilbert spaces are disjoint, therefore the total Hilbert space of $S + B$ is given by $\mathcal{H}_S \otimes \mathcal{H}_B$, and thus the state of the total system $S + B$ is represented by a density matrix in this space. It follows that each operator of S commutes with each operator of B. We now assume that there is a coupling between S and B, $H_{SB} = \sum_i^q \lambda_i V_{i,S} \otimes V_{i,B}$ where $V_{i,S(B)}$ is an operator acting on the $\mathcal{H}_{S(B)}$ space, and q the number of operators coupling the system with the bath. Notice that this expansion is always possible due to the commutativity between the operators acting on S and B. For the moment, we focus on the simple case $q = 1$, but the extension to the more general case $q > 1$ is trivial, so in the following we set $\lambda_i = \lambda$. The density matrix of the total system evolves according to the von-Neumann equation,

$$\partial_t \rho(t) = -\frac{i}{\hbar} [H, \rho(t)], \qquad (47)$$

where $H = H_S + H_B + H_{SB}$ is the total Hamiltonian. A basis set for the total Hilbert space is given by $\{|j, k\rangle\}$ where j is a element of the basis for $\mathcal{H}_S$ and k for $\mathcal{H}_B$.

Clearly, if we are interested in the dynamics of the system S only, this equation contains much more information than needed, since it entails the dynamics of the many degrees of freedom of the environment. We wish therefore to obtain an equation of motion for a density matrix, where the coupling with the environment is effectively described by some macroscopic parameters. This is possible when one assumes that the coupling strength λ is sufficiently small and we can use perturbation theory. The exact definition of the "smallness" of λ is actually an open problem, and we will refer the interested reader elsewhere [78–80].

The aim of the theory is to obtain the dynamics of the reduced density matrix, whose element (i, j) is given by

$$(\rho_R)_{i,j} = (\text{tr}_B \rho)_{i,j} = \sum_{k \in \mathcal{H}_B} \langle i, k | \rho | k, j \rangle. \tag{48}$$

The reduced density matrix is therefore a density matrix in the space of the system S, but its dynamics is determined not only by H_S but also by the dynamics of the environment. The derivation of the equation of motion for the reduced density matrix of the system stems from the dynamics of the density matrix of the bath when decoupled from the system, $\rho_B(t) = \exp(-iH_B t)\rho_B(0)\exp(iH_B t)$, the factorization of the initial density matrix $\rho(0) = \rho_S(0) \otimes \rho_B(0)$, and the vanishing of the quantum averages of the bath operators V_B to first order in λ, $\text{tr}_B(\rho_B V_B) \propto \lambda^2$ (this latter assumption can be relaxed and would eventually contribute an effective force that redefines the system Hamiltonian). This standard procedure leads, after some straightforward algebra, to [78,80]

$$\frac{d\rho_S}{dt} = -i[H_S, \rho_S] + \lambda^2 \left[V_S, M^\dagger(t) - M(t) \right], \tag{49}$$

where we have defined

$$M(t) = \int_0^t dt' C(t, t') e^{-iH_S(t-t')} V_S \rho_S(t') e^{iH_S(t-t')} \tag{50}$$

and

$$C(t, t') = \text{tr}_B \left[\rho_B(0) V_B(t) V_B(t') \right] \tag{51}$$

is the bath correlation function, where $V_B(t) = \exp(iH_B t)V_B(0)\exp(-iH_B t)$ is the time evolution of the bath operators. Equation (49) is difficult to solve for two reasons. On the one hand, the density matrix is usually a dense matrix of N^2 elements, if we have N elements in the basis set. This usually requires long computation time and large amounts of memory. On the other hand, it contains the full history of the system and at each time step this history needs to be evaluated to calculate the integrals in Equation (50). To deal with the first problem, one can formally derive an equation of motion for a state vector in the Hilbert space that somewhat resembles a wave-function dynamics [82–84] and scales as N although its physical interpretation is slightly different [85,86]. However, the state vector follows a stochastic dynamic and therefore one needs to average over the realizations of the stochastic noise, balancing the computation gain of the reduced dimensionality. To deal with the second problem, we have two ways: we can neglect completely the history of the system, or retain part of it. In the Markov approximation [78–80] one arrives to the so-called Lindblad equation [87]

$$\frac{d\rho_S}{dt} = -i[H_S, \rho_S] + \frac{\lambda^2}{2}(V_S^\dagger V_S \rho_S + \rho_S V_S^\dagger V_S - 2V_S^\dagger \rho_S V_S). \tag{52}$$

Lindblad proved that this equation is the most general master equation of the Markov's type that preserves trace, positiveness and hermiticity of the density matrix ρ_S at each time steps up to second order in λ. A second approach follows from the observation that up to second order in λ, $\rho_S(t) = \rho_S(0)$. Then, up to the same order of approximation, one replaces $\rho(t')$ with $\rho(t)$, in Equation (50), i.e.,

$$M(t) \approx \int_0^t dt' C(t, t') e^{-iH_S(t-t')} V_S \rho_S(t) e^{iH_S(t-t')}. \tag{53}$$

This operator is now local in time, although one needs to calculate it at each time step. In the absence of any magnetic field, the bath dynamics satisfies time-reversal symmetry, and this is usually sufficient for $C(t, t') = C(t - t')$, especially if $\rho_B(0)$ is the equilibrium statistical density matrix of the bath $\rho_{B,equ} = \exp(-\beta H_B)$ where $\beta = 1/k_B T$ and T the bath temperature, we can separate this integral and arrive at the Redfield master equation [80,88].

To describe thermal transport within OQS one needs to couple two environments, kept at different temperature, locally to the system

$$\frac{d\rho_S}{dt} = -i\left[H_S, \rho_S\right] + \mathcal{L}_L[\rho_S] + \mathcal{L}_R[\rho_S]. \tag{54}$$

Here, $\mathcal{L}_L$ and $\mathcal{L}_R$ describe the left and right environment, respectively. A possible nano-junction attached to two environments is shown in the upper panel of Figure 5. Furthermore, in Figure 5a one sees the voltage drop over the device induced by the applied temperature gradient, ΔT, due to the environmental coupling. One can observe a linear-response regime for small ΔT, a regime of rapid rise in ΔV and for large temperature gradients a region where the voltage drop has reached saturation due to the finite size of the system. The inset of Figure 5 shows the thermopower $S = -d(\Delta V)/d(\Delta T)$ that presents a maximum in response to the thermal gradient at $\Delta T \approx 0.25$ [a.u.].

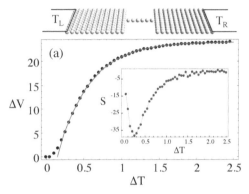

Figure 5. Upper Panel: Nanostructure connected locally to two reservoirs kept at different temperature. Below: Voltage drop, ΔV as a function of the temperature gradient, ΔT. The inset shows the thermopower $S = -\frac{d(\Delta V)}{d(\Delta T)}$. Reprinted (adapted) with permission from [89]. Copyright (2009) American Chemical Society.

Note that all the shown quantities correspond to the steady-state solution of the master equations, where the long-time limit has been reached. In general, the OQS approach is not limited to this regime and can also be used to study time-dependent phenomena in nanoscale devices beyond linear response [78,90,91].

For the purposes of this review, it is relevant to point out that the quantum system is in general a device made of interacting electrons and ions. This, as we have seen (Section 3.1.1), is an incredibly complex problem and the coupling with the external environment does not make it simpler. Fortunately, we can use the theory of open quantum systems with (time-dependent) DFT to extract from a non-interacting open-quantum system information about the dynamics of the interacting one [58,83,92–95]. More surprisingly, it has been shown that the DFT for open-quantum system can be constructed in such a way that the KS system can be made closed. This means that the effects of the external environment can be included in the KS potentials [94,95]. However, it is not clear how one could effectively build the required KS potentials which would be depending on the coupling operators between the system and the environment. These difficulties have so far hindered a full development of the theory and its routine application for the investigation of transport in complex devices and physical conditions.

3.4. Influence of Decoherence onto Thermoelectrical Transport

The observation or measurement process acting on a classical object does not influence its physical properties. However, when entering the nanoscale world and the quantum realm one needs to revisit some classical concepts. In the smallest nanodevices, electrons are usually set up in coherent superpositions, and a global measurement would destroy most of their coherence changing the dynamical response. Decoherence does not only originate from the measurement by a macroscopic observer, but it also occurs when the quantum system interacts locally with another quantum system. This effect of local quantum observation covers a variety of situations, such as, e.g., local electron–phonon coupling and continuous [96] or frequent [97] quantum non-demolition measurements. It has been shown that decoherence influences the efficiency of energy transport in biological [98–100] and molecular devices [100–102] and is responsible for the decoupling of the system from the environment via the Quantum Zeno effect [103,104].

While standard static transport approaches such as the BTE or LB fails in catching up with these dynamical effects, methods such as TDDFT or OQS approaches allow for the time-resolved study of electron transport and energy dissipation. Additionally, methods describing the dynamics within a density matrix formalism are suitable to study the role of coherence in nanoscale transport devices. Indeed, by modeling decoherence as an additional environment within a master equation approach, one can include it into a consistent thermodynamic formalism [105],

$$\mathcal{L}_D[\rho_S] = \lambda_D \left[2|\beta\rangle\langle\beta|\rho_S|\beta\rangle\langle\beta| - |\beta\rangle\langle\beta|\rho_S - \rho_S|\beta\rangle\langle\beta| \right]. \tag{55}$$

Here $|\beta\rangle$ is the ket-vector representing the state of the spatial region where decoherence takes place. This bath can change the coherence of the system and has; in contrast to classical reservoirs that are in a thermal ensemble state, no temperature is associated with it [106]. This formalism has been applied to a ratchet-like device shown in Figure 6a [105].

By changing the on-site energy levels on the parallel horizontal branches of the device (graphically indicated by the size of the spheres), a spatial asymmetry is introduced. In the upper branch the on-site energies increase from left to right in equal proportion, while in the lower branch this is reversed. Therefore, the device acts as a quantum ratchet by the presence of two rectifiers on each branch in opposite direction. This directional transport device driven by thermal and quantum fluctuations has a preferred electronic current direction, in this example clockwise. In the ratchet, a hot (*H*) and cold (*C*) reservoir introduce a thermal gradient in the device while at the same time the decoherence bath (*D*) is acting at site β

$$\frac{d\rho_S}{dt} = -i[H_S, \rho_S] + \mathcal{L}_H[\rho_S] + \mathcal{L}_C[\rho_S] + \mathcal{L}_D[\rho_S]. \tag{56}$$

While in this set-up energy is exchanged via the baths, particle current is then confined inside the device. Figure 6b shows the electronic current at steady state for the upper branch. Please note that by charge conservation, the current in the lower branch is exactly the same but flowing in the opposite direction. Here, red indicates a positive current, from left to right in the upper branch. If no decoherence is applied ($\lambda_D = 0$), the current is flowing in the clockwise direction through the ratchet. Increasing the decoherence, the ring current decreases until its direction gets reversed. It has been found that the local quantum observer cannot only control the particle current but also energy currents in direction and strength inside the device [105]. This demonstrates that in thermoelectric nanodevices the current and heat flows are not only dictated by the temperature and potential gradients, but can also be manipulated by the external control of the coherence of the device. This effect is illustrated in Figure 6c. When looking at the whole picture the water flows uphill, however, when we only observe current flow in the circle (and ignore the rest of the illustration) it seems the water flows down the channel. This apparent paradox mimics the coherent superposition of two quantum states (water flowing up/down). The observation process in specific parts of our system can tune between these two quantum states and hence change the 'physical response of the nanodevice' in a controlled way.

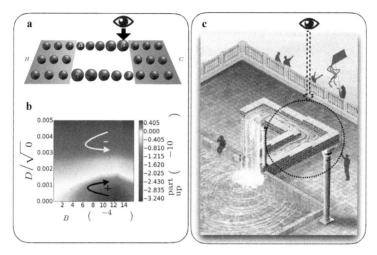

Figure 6. Influence of quantum decoherence on the thermoelectric flows in a quantum ratchet shown in (**a**). (**b**) The electrical current in the upper branch can be seen. The current can change direction as a function of both λ_D and ΔT. (**c**) Artistic illustration (copyright K. Aranburu) of the role of decoherence in a nanodevice: When observing inside the black circle, it appears the water flows down the channel, instead, by looking at the whole painting the water actually flows uphill. This apparent paradox mimics the coherent superposition of two quantum states (water flowing up/down). By observing at specific parts of our system one is able to tune between these two states and hence change the 'physical response of the nanodevice' in a controlled way. Parts (**a**,**b**) are adapted from [105]. Copyright (2017) SpringerNature.

3.5. Time-Dependent Thermal Transport Theory

Modeling thermal transport and thermoelectric effects at the nanoscale is difficult, because besides the intricacies of certain materials, we lack a working definition of temperature at those scales. Indeed, the concepts of equilibrium and thermodynamic limits are difficult to extend to the case of a few electrons or phonons strongly localized in a nanodevice. Clearly, also the idea that the leads/reservoirs have a well-established temperature (distribution function) down to the contact with the device is an abstraction. Many attempts have been tried to remedy this situation and general extensions of some thermodynamics concepts to the nanoscale have been attempted [107,108].

One of such attempts is to completely remove the needs for macroscopic reservoirs and leads by replacing them with an effective radiation of known properties. One might think for example, of replacing the electrostatic bias with the time-dependent vector potential that produces the same electric field. A similar attempt can be made with the temperature gradient. We replace the coupling to the energy reservoirs by two black bodies each radiating according to their own temperature and spectral properties: the electromagnetic field of the blackbody radiation is then included in the system dynamics and establish in this way the needed temperature gradient and transport dynamics. To maintain a steady state and avoid that the energy stored in the device saturates, we assume that the system is weakly coupled to an external environment at a given temperature. This corresponds to an experimental set-up; the quantum systems gets excited out of its equilibrium by thermal radiation and at the same time can release energy to its surrounding. (such as a metal heated up by a laser and starts to glow). Recent progress has been made to include these quantum-electro-dynamical effects into an ab-initio formalism [109]. Being able to study and detect the radiation produced by the transport dynamics could possibly allow the definition of an effective "local temperature" based on the electrodynamics theory.

We assume each blackbody to be located far away from the devices, therefore for all our purposes we will consider the radiation they produce as composed solely of plane waves propagating along the line connecting the device and the blackbody. In Figure 7 the considered nanodevice is shown and the left and right blackbody radiation is indicated by $\mathbf{A}_L$ and $\mathbf{A}_R$ respectively. While the blackbodies radiate according to their respective temperature T_L and T_R, the system is interacting with an environment at temperature T_E. The amplitude of each plane wave of the blackbody field is determined by the temperature of the blackbody itself. We assume therefore the thermal radiation of the form

$$\mathbf{A}_{L(R)}(\mathbf{r},t) = \mathbf{E}_0 \int d\Omega \sqrt{\Omega n\left(\Omega, T_{L(R)}\right)} \sin(kx \pm \Omega t + \varphi(\Omega)), \tag{57}$$

where $n(\Omega, T)$ is the Bose-Einstein distribution, $\mathbf{E}_0 = E_0 \mathbf{e}_y$. Here, E_0 is the strength of the corresponding electric field, where $\mathbf{e}_y$ is the unit vector in the y direction and k is the momentum of the photon with frequency Ω. We have considered a random phase, $\varphi(\Omega) \in [0, 2\pi)$ that ensures the plane waves are incoherently reaching the device. In principle, there are more realistic, but numerically more expensive, ways to model the blackbody thermal radiation [110,111]. A comparison of these methods with the results of Equation (57) is still missing. Due to the coupling with an external environment, we expect that the system reaches a steady state. This coupling is effectively described through a master equation where the environmental power spectrum for $|\omega| < \omega_c$ is

$$C(\omega) = 4|\omega|^3 n(|\omega|, T_E) + \theta(-\omega), \tag{58}$$

where $\theta(\omega)$ is the Heaviside step function, and ω_c is a cutoff frequency determined by the scale of the system. For $|\omega| > \omega_c$, $C(\omega) = 0$. We apply this model to the case of a device described in the tight-binding formalism where the electrons are not interacting. In this case, the Hamiltonian for the system is given by

$$H_s = \sum_{\langle i,j \rangle} T_{i,j} c_i^\dagger c_j, \tag{59}$$

where the operator c_j destroys an electron in the site j and $T_{i,j}$ is the hopping parameters. The blackbody radiation enters here as a Peierls transformation into the Hamiltonian

$$T_{i,j} = t_{i,j} \exp\left[-i \int_{\mathbf{R}_i}^{\mathbf{R}_j} d\mathbf{r} \left(\mathbf{A}_L(\mathbf{r},t) + \mathbf{A}_R(r,t)\right)\right], \tag{60}$$

where $t_{i,j}$ is the hopping parameter in the absence of the radiation. We align the device along the direction joining the two blackbodies. If one considers initially the case where the blackbodies and the environment have the same temperature, we see that the dynamics of the system points towards a steady state with no net energy transport in the long-time limit [109].

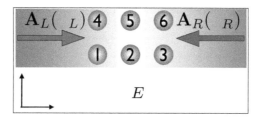

Figure 7. Set-up under consideration. Six tight-binding sites are connected locally to the radiation of two blackbody at temperature, T_L and T_R. Furthermore, the system is connected to an environment at temperature T_E. Reprinted (adapted) from [109]. Copyright (2015) American Physical Society.

In the case of a finite temperature gradient we observe in Figure 8 a turnover of the energy current, j_T, with respect to the flux of energy from the blackbodies. Here, $\langle S \rangle$ is the time-averaged Poynting vector indicating the coupling strength to the blackbodies. Indeed, for small fluxes, the energy current increases when increasing the flux, then reaches a maximum and later vanishes for intense radiations. The reason for this turnover lies in the finite number of states available to carry energy. For small fluxes almost all states are empty and there is the possibility to allocate for a larger number of carriers. When the flux increases, the states fill up, and less and less will be available for conduction, until essentially all the states will by full and no dynamics is possible [109]. We want to point out that this theory can be applied to large temperature gradients: the coupling with the black body radiation is not restricted to linear response, while the system is coupled weakly to the environment [109].

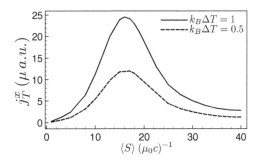

Figure 8. Thermal current along the x direction for different temperature gradients as a function of the strength of the coupling $\langle S \rangle$. For large coupling, the current decreases since the energy level are all filled, and the dynamics is quenched. Reprinted (adapted) [109]. Copyright (2015) American Physical Society.

4. Discussion

The investigation of energy and electrical transport in nanoscale devices has recently attracted a consistent amount of interest for its wide range of potential applications in technology, biomedicine and energy materials to name just a few. A comprehensive look at the field would, therefore, require more than a reasonably short review. We have preferred a bird-eye introduction to what we believe are the actual more promising theoretical methods, the Boltzmann's equation and the Landauer-Büttiker formalism for quantum transport. These methods are somewhat complementary since they apply to different regimes, namely a diffusive regime where scattering events are important, or a ballistic regime where the quantum-mechanical effects are dominant. While these methods might not be the most accurate for any situation, most of the recent progress in our understanding of thermal, energy, and electronic transport at the nanoscale is based upon them or some of their extensions. Both methods benefit from an accurate description of the quantum-mechanical properties of the material under investigation. In this respect, the method of choice is Density Functional Theory, since its practical implementation allow for an almost straightforward interfacing with both the Boltzmann or Landauer theories (the last one through the Non-Equilibrium Green's functions formalism). Therefore, the use of DFT methods for electronic structure calculations (including phonon and vibration properties) combined with Boltzmann's or Landauer's theories is to some extent the state of the art of modern calculations.

We have briefly discussed the need to go beyond this *state of the art* both in terms of the limits of DFT or the limits of the approximations used in the methods themselves. Clearly, this is a very fertile field of investigation and in most cases it requires novel paradigms. We have exposed a few here: the introduction of methods beyond the constant relaxation time approximations ameliorates the description of electron and phonon transport by introducing energy-dependent relaxation times; The static DFT cannot, in principle, be the theory to study transport, an inherent out-of-equilibrium problem. We therefore introduce extension of DFT able to properly take into account the dynamics of the electrons and show how this theory naturally allows for the description of strongly correlated effect in nanoscale molecular transport, in particular how it corrects the calculation of the Seebeck coefficient. We finally introduced the theory of open-quantum system, where the system under investigation is "open" to an external environment with which it could exchange energy, momentum, and particles. The external environment might have a passive role of driving an otherwise out-of-equilibrium system toward a steady or ground state, or an active role in which it drives the system itself allowing for novel quantum-mechanical effects to be exploited. As the OQS approach describes the dynamics of the density operator, this theory can be also used to investigate the influence of coherence in quantum transport devices. This might pave the way for novel strategies to construct quantum devices for thermoelectric energy conversion, photonics, spintronic injection, and sensing, to just name a few.

Clearly, the previous description cannot be complete. For example, we are aware of other DFT methods to include thermoelectric phenomena [64,65] or strong correlations [52,112–114] in the Kohn-Sham system, or an accurate description of vibrational modes and energy transport with classical methods by solving the Newton's equations [115,116], or the inclusion of non-harmonic effects in the non-equilibrium Green's function for the phonons [117]. Each of these topics deserves a complete introduction per se.

We hope this review serves both to the young researcher to have an outlook of the actual state of the art and potential ways beyond it, without striving through the immense amount of literature currently available and in continuous development, and to the experienced scientist as a reference where some of the most fundamental results and future outlooks are collected.

Author Contributions: R.B. and R.D. contributed equally in writing this review.

Funding: This research was funded by the Spanish Ministerio de Economia y Competitividad (MINECO) grant number FIS2016-79464-P (SElecT-DFT) and MINECOG17/A01 (TOWTherm), by the Basque Government (Eusko Jaurlaritza) through the Grupos Consolidados (IT578-13 and IT1249-19). R.B. acknowledges funding from the European Union's Horizon 2020 research and innovation program under the Marie Skłodowska-Curie grant agreement No. 793318. The APC was funded by Dresden University of Technology (TU Dresden).

Conflicts of Interest: The authors declare no conflict of interest. The funders had no role in the design of the study; in the collection, analyses, or interpretation of data; in the writing of the manuscript, or in the decision to publish the results.

Abbreviations

The following abbreviations are used in this manuscript:

ALDA   Adiabatic Local Density Approximation
BTE   Boltzmann Transport Equation
LBTE   Linearized Boltzmann Transport Equation
CRTA   Constant Relaxation Time Approximation
DFT   Density Functional Theory
EPA   Electron–Phonon Averaged approximation
KS   Kohn–Sham
LB   Landauer–Büttiker
NEGF   Non-Equilibrium Green Function
RG   Runge–Gross
TDDFT   Time-Dependent Density Functional Theory
TDCDFT Time-Dependent Current-Density Functional Theory
OQS   Open-Quantum Systems
SERTA   Self-Energy Relaxation Time Approximation
EPW   Electron–Phonon Wannier

References

1. Di Ventra, M. *Electrical Transport in Nanoscale Systems*; Cambridge University Press: New York, NY, USA, 2008.
2. Datta, S. *Electronic Transport in Mesoscopic Systems*; Cambridge University Press: New York, NY, USA, 1997.
3. Datta, S. *Quantum Transport: Atom to Transport*; Cambridge University Press: New York, NY, USA, 2005.
4. Cuniberti, G.; Fagas, G.; Richter, K. (Eds.) *Introducing Molecular Electronics*; Springer: New York, NY, USA, 2005.
5. Holt, N.E.; Zigmantas, D.; Valkunas, L.; Li, X.P.; Niyogi, K.; Fleming, G.R. Carotenoid Cation Formation and the Regulation of Photosynthetic Light Harvesting. *Science* **2005**, *307*, 433–436.
6. Liguori, N.; Xu, P.; van Stokkum, I.H.; van Oort, B.; Lu, Y.; Karcher, D.; Bock, R.; Croce, R. Different carotenoid conformations have distinct functions in light-harvesting regulation in plants. *Nat. Commun.* **2017**, *8*, 1994.
7. Beenakker, C.W.J. Theory of Coulomb-blockade oscillations in the conductance of a quantum dot. *Phys. Rev. B* **1991**, *44*, 1646–1656.
8. Boltzmann, L. Weitere Studien über das Wärmegleichgewicht unter Gasmolekülen. *Weitere Stud. über Das Wärmegleichgewicht Unter Gasmolekülen, Wien. Berichte* **1872**, *66*, 275–370. (In German)
9. Lundstrom, M. *Fundamentals of Carrier Transport*; Cambridge University Press: Cambridge, UK, 2000.
10. Ziman, J.M. *Principles of the Theory of Solids*; Cambridge University Press: Cambridge, UK, 1972.
11. Mahan, G. *Many-Particle Physics*; Springer Science & Business Media: Berlin/Heidelberg, Germany, 1981.
12. Poncé, S.; Margine, E.R.; Giustino, F. Towards predictive many-body calculations of phonon-limited carrier mobilities in semiconductors. *Phys. Rev. B* **2018**, *97*, 121201.
13. Giustino, F. Electron-phonon interactions from first principles. *Rev. Mod. Phys.* **2017**, *89*, 015003.
14. Singh, D.J.; Mazin, I.I. Calculated thermoelectric properties of La-filled skutterudites. *Phys. Rev. B* **1997**, *56*, R1650–R1653.
15. Baroni, S.; de Gironcoli, S.; Dal Corso, A. Phonons and related crystal properties from density-functional perturbation theory. *Rev. Mod. Phys.* **2001**, *73*, 515–562.
16. Noffsinger, J.; Giustino, F.; Malone, B.D.; Park, C.H.; Louie, S.G.; Cohen, M.L. EPW: A program for calculating the electron–phonon coupling using maximally localized Wannier functions. *Comput. Phys. Commun.* **2010**, *181*, 2140–2148.
17. Ahmad, S.; Mahanti, S.D. Energy and temperature dependence of relaxation time and Wiedemann-Franz law on PbTe. *Phys. Rev. B* **2010**, *81*, 165203.
18. Kaasbjerg, K.; Thygesen, K.S.; Jacobsen, K.W. Phonon-limited mobility in n-type single-layer MoS_2 from first principles. *Phys. Rev. B* **2012**, *85*, 115317.
19. Murphy-Armando, F.; Fahy, S. First-principles calculation of carrier-phonon scattering in n-type $Si_{1-x}Ge_x$ alloys. *Phys. Rev. B* **2008**, *78*, 035202.

20. Murphy-Armando, F.; Fagas, G.; Greer, J.C. Deformation Potentials and Electron-Phonon Coupling in Silicon Nanowires. *Nano Lett.* **2010**, *10*, 869–873.

21. Xu, B.; Verstraete, M.J. First principles explanation of the positive seebeck coefficient of lithium. *Phys. Rev. Lett.* **2014**, *112*, 196603.

22. Savrasov, S.Y.; Savrasov, D.Y. Electron-phonon interactions and related physical properties of metals from linear-response theory. *Phys. Rev. B* **1996**, *54*, 16487–16501.

23. Samsonidze, G.; Kozinsky, B. Accelerated Screening of Thermoelectric Materials by First-Principles Computations of Electron-Phonon Scattering. *Adv. Energy Mater.* **2018**, *8*, 1800246.

24. Giannozzi, P.; Baroni, S.; Bonini, N.; Calandra, M.; Car, R.; Cavazzoni, C.; Ceresoli, D.; Chiarotti, G.L.; Cococcioni, M.; Dabo, I.; et al. QUANTUM ESPRESSO: A modular and open-source software project for quantum simulations of materials. *J. Phys. Condens. Matter* **2009**, *21*, 395502.

25. Madsen, G.K.; Singh, D.J. BoltzTraP. A code for calculating band-structure dependent quantities. *Comput. Phys. Commun.* **2006**, *175*, 67–71.

26. Landauer, R. Spatial variation of currents and fields due to localized scatterers in metallic conduction. *IBM J. Res. Dev.* **1957**, *1*, 223–231.

27. Büttiker, M.; Imry, Y.; Landauer, R.; Pinhas, S. Generalized many-channel conductance formula with application to small rings. *Phys. Rev. B* **1985**, *31*, 6207.

28. Büttiker, M. Four-Terminal Phase-Coherent Conductance. *Phys. Rev. Lett.* **1986**, *57*, 1761.

29. Ryndyk, D. *Theory of Quantum Transport at Nanoscale*; Springer Series in Solid-State Sciences; Springer International Publishing: Cham, Switzerland, 2016; Volume 184.

30. Klitzing, K.V.; Dorda, G.; Pepper, M. New method for high-accuracy determination of the fine-structure constant based on quantized hall resistance. *Phys. Rev. Lett.* **1980**, *45*, 494.

31. Tsui, D.C.; Stormer, H.L.; Gossard, A.C. Two-dimensional magnetotransport in the extreme quantum limit. *Phys. Rev. Lett.* **1982**, *48*, 1559.

32. Wharam, D.A.; Thornton, T.J.; Newbury, R.; Pepper, M.; Ahmed, H.; Frost, J.; Hasko, D.; Peacock, D.; Ritchie, D.; Jones, G. One-dimensional transport and the quantisation of the ballistic resistance. *J. Phys. C Solid State Phys.* **1988**, *21*, L209.

33. Van Wees, B.J.; van Houten, H.; Beenakker, C.W.J.; Williamson, J.G.; Kouwenhoven, L.P.; van der Marel, D.; Foxon, C.T. Quantized conductance of point contacts in a two-dimensional electron gas. *Phys. Rev. Lett.* **1988**, *60*, 848–850.

34. Van Wees, B.J.; Kouwenhoven, L.P.; Willems, E.M.M.; Harmans, C.J.P.M.; Mooij, J.E.; van Houten, H.; Beenakker, C.W.J.; Williamson, J.G.; Foxon, C.T. Quantum ballistic and adiabatic electron transport studied with quantum point contacts. *Phys. Rev. B* **1991**, *43*, 12431–12453.

35. Onsager, L. Reciprocal Relations in Irreversible Processes. *Phys. Rev.* **1931**, *37*, 405.

36. Rego, L.G.C.; Kirczenow, G. Quantized Thermal Conductance of Dielectric Quantum Wires. *Phys. Rev. Lett.* **1998**, *81*, 232–235.

37. Schwab, K.; Henriksen, E.A.; Worlock, J.M.; Roukes, M.L. Measurement of the quantum of thermal conductance. *Nature* **2000**, *404*, 974–977.

38. Stefanucci, G.; van Leeuwen, R. *Non-Equilibrium Many-Body Theory of Quantum Systems*; Cambridge University Press: New York, NY, USA, 2013.

39. Kohn, W.; Sham, L.J. Self-Consistent Equations Including Exchange and Correlation Effects. *Phys. Rev.* **1965**, *140*, A1133.

40. Giuliani, G.F.; Vignale, G. *Quantum Theory of the Electron Liquid*; Cambridge University Press: Cambridge, UK, 2005.

41. Ullrich, C. *Time Dependent Density Functional Theory: Concepts and Applications*; Oxford University Press: Oxford, UK, 2012.

42. Capelle, K. A bird's-eye view of density-functional theory. *Braz. J. Phys.* **2006**, *36*, 1318–1343.

43. Meir, Y.; Wingreen, N. Landauer formula for the current through an interacting electron region. *Phys. Rev. Lett.* **1992**, *68*, 2512–2515.

44. Sanchez, D.; Linke, H. Focus on Thermoelectric Effects in Nanostructures. *New J. Phys.* **2014**, *16*, 110201.

45. Sai, N.; Zwolak, M.; Vignale, G.; Di Ventra, M. Dynamical Corrections to the DFT-LDA Electron Conductance in Nanoscale Systems. *Phys. Rev. Lett.* **2005**, *94*, 186810.

46. Vignale, G.; Di Ventra, M. Incompleteness of the Landauer formula for electronic transport. *Phys. Rev. B Condens. Matter Mater. Phys.* **2009**, *79*, 014201.

47. Dreizler, R.M.; Gross, E.K.U. *Density Functional Theory*; Springer: Berlin/Heidelberg, Germany, 1990.

48. Hohenberg, P.; Kohn, W. Inhomogeneous Electron Gas. *Phys. Rev.* **1964**, *136*, B864–B871.

49. D'Agosta, R. Towards a dynamical approach to the calculation of the figure of merit of thermoelectric nanoscale devices. *Phys. Chem. Chem. Phys.* **2013**, *15*, 1758.

50. Kurth, S.; Stefanucci, G.; Almbladh, C.O.; Rubio, A.; Gross, E.K.U. Time-dependent quantum transport: A practical scheme using density functional theory. *Phys. Rev. B* **2005**, *72*, 035308.

51. Stefanucci, G.; Almbladh, C.O. Time-dependent partition-free approach in resonant tunneling systems. *Phys. Rev. B* **2004**, *69*, 195318.

52. Mirtschink, A.; Seidl, M.; Gori-Giorgi, P. Derivative Discontinuity in the Strong-Interaction Limit of Density-Functional Theory. *Phys. Rev. Lett.* **2013**, *111*, 126402.

53. Stefanucci, G.; Kurth, S. Steady-State Density Functional Theory for Finite Bias Conductances. *Nano Lett.* **2015**, *15*, 8020–8025

54. Kurth, S.; Stefanucci, G. Nonequilibrium Anderson model made simple with density functional theory. *Phys. Rev. B* **2016**, *94*, 241103.

55. Runge, E.; Gross, E.K.U. Density-Functional Theory for Time-Dependent Systems. *Phys. Rev. Lett.* **1984**, *52*, 997.

56. Koentopp, M.; Chang, C.; Burke, K.; Car, R. Density functional calculations of nanoscale conductance. *J. Phys. Condens. Matter* **2008**, *20*, 083203.

57. Gebauer, R.; Car, R. Current in Open Quantum Systems. *Phys. Rev. Lett.* **2004**, *93*, 160404.

58. Burke, K.; Car, R.; Gebauer, R. Density Functional Theory of the Electrical Conductivity of Molecular Devices. *Phys. Rev. Lett.* **2005**, *94*, 146803.

59. Gebauer, R.; Car, R. Kinetic theory of quantum transport at the nanoscale. *Phys. Rev. B* **2004**, *70*, 125324.

60. Prodan, E.; Car, R. DC conductance of molecular wires. *Phys. Rev. B* **2007**, *76*, 115102.

61. Di Ventra, M.; Evoy, S.; Heflin, J.R. (Eds.) *Introduction to Nanoscale Science and Technology*; Springer: New York, NY, USA, 2004.

62. D'Agosta, R.; Vignale, G.; Raimondi, R. Temperature dependence of the tunneling amplitude between quantum hall edges. *Phys. Rev. Lett.* **2005**, *94*, 86801.

63. Vignale, G. Mapping from current densities to vector potentials in time-dependent current density functional theory. *Phys. Rev. B* **2004**, *70*, 201102(R).

64. Eich, F.G.; Di Ventra, M.; Vignale, G. Density-functional theory of thermoelectric phenomena. *Phys. Rev. Lett.* **2014**, *112*, 196401.

65. Eich, F.G.; Principi, A.; Di Ventra, M.; Vignale, G. Luttinger-field approach to thermoelectric transport in nanoscale conductors. *Phys. Rev. B* **2014**, *90*, 115116.

66. Ruggenthaler, M.; van Leeuwen, R. Global fixed-point proof of time-dependent density-functional theory. *EPL* **2011**, *95*, 13001.

67. D'Agosta, R.; Vignale, G. Non-V-representability of currents in time-dependent many-particle systems. *Phys. Rev. B* **2005**, *71*, 245103.

68. Ventra, M.D.; Todorov, T.N. Transport in nanoscale systems: The microcanonical versus grand-canonical picture. *J. Phys. Condens. Matter* **2004**, *16*, 8025–8034.

69. Perdew, J.P.; Zunger, A. Self-interaction correction to density-functional approximations for many-electron systems. *Phys. Rev. B* **1981**, *23*, 5048.

70. Perdew, J.P.; Ernzerhof, M.; Burke, K. Generalized Gradient Approximation Made Simple. *Phys. Rev. Lett.* **1996**, *77*, 3865.

71. Wijewardane, H.O.; Ullrich, C.A. Time-dependent Kohn-Sham theory with memory. *Phys. Rev. Lett.* **2005**, *95*, 86401.

72. D'Agosta, R.; Vignale, G. Relaxation in Time-Dependent Current-Density-Functional Theory. *Phys. Rev. Lett.* **2006**, *96*, 016405.

73. Yang, K.; Perfetto, E.; Kurth, S.; Stefanucci, G.; D'Agosta, R. Density functional theory of the Seebeck coefficient in the Coulomb blockade regime. *Phys. Rev. B* **2016**, *94*, 081410.

74. Beenakker, C.W.J.; Staring, A.A.M. Theory of the thermopower of a quantum dot. *Phys. Rev. B* **1992**, *46*, 9667–9676.

75. Zianni, X. Theory of the energy-spectrum dependence of the electronic thermoelectric tunneling coefficients of a quantum dot. *Phys. Rev. B* **2008**, *78*, 165327.

76. Sánchez, R.; Sothmann, B.; Jordan, A.N.; Büttiker, M. Correlations of heat and charge currents in quantum-dot thermoelectric engines. *New J. Phys.* **2013**, *15*, 125001.

77. Sothmann, B.; Sánchez, R.; Jordan, A.N.; Büttiker, M. Powerful energy harvester based on resonant-tunneling quantum wells. *New J. Phys.* **2013**, *15*, 095021.

78. Gardiner, C.W.; Zoller, P. *Quantum Noise*, 2nd ed.; Springer: Berlin/Heidelberg, Germany, 2000.

79. Breuer, H.P.; Petruccione, F. *The Theory of Open Quantum Systems*; Oxford University Press: New York, NY, USA, 2002.

80. Van Kampen, N.G. *Stochastic Processes in Physics and Chemistry*, 3rd ed.; Elsevier: Amsterdam, The Netherlands, 2007.

81. Prezhdo, O.V. Mean field approximation for the stochastic Schrödinger equation. *J. Chem. Phys.* **1999**, *111*, 8366–8377.

82. Gaspard, P.; Nagaoka, M. Non-Markovian stochastic Schrödinger equation. *J. Chem. Phys.* **1999**, *111*, 5676–5690.

83. Biele, R.; D'Agosta, R. A stochastic approach to open quantum systems. *J. Phys. Condens. Matter* **2012**, *24*, 273201.

84. Biele, R.; Timm, C.; D'Agosta, R. Application of a time-convolutionless stochastic Schrödinger equation to energy transport and thermal relaxation. *J. Phys. Condens. Matter* **2014**, *26*, 395303.

85. Strunz, W.T. Stochastic path integrals and open quantum systems. *Phys. Rev. A* **1996**, *54*, 2664–2674.

86. Strunz, W. Path integral, semiclassical and stochastic propagators for Markovian open quantum systems. *J. Phys. A* **1997**, *30*, 4053.

87. Lindblad, G. On the Generators of Quantum Dynamical Semigroups. *Commun. Math. Phys.* **1976**, *48*, 119–130.

88. Redfield, A.G. On the Theory of Relaxation Processes. *IBM J. Res. Dev.* **1957**, *1*, 19–31.

89. Dubi, Y.; Di Ventra, M. Thermoelectric Effects in Nanoscale Junctions. *Nano Lett.* **2009**, *9*, 97–101.

90. Gardiner, C.W.; Collett, M.J. Input and Output in damped quantum systems: Quantum stochastic differential equations and the master equation. *Phys. Rev. A* **1985**, *31*, 3761.

91. Gardiner, C.W. *Handbook of Stochastic Methods*, 3rd ed.; Springer: Berlin/Heidelberg, Germany, 2004; p. 415.

92. Di Ventra, M.; D'Agosta, R. Stochastic Time-Dependent Current-Density-Functional Theory. *Phys. Rev. Lett.* **2007**, *98*, 226403.

93. D'Agosta, R.; Di Ventra, M. Local electron and ionic heating effects on the conductance of nanostructures. *J. Phys. Condens. Matter* **2008**, *20*, 374102.

94. Yuen-Zhou, J.; Tempel, D.G.; Rodríguez-Rosario, C.A.; Aspuru-Guzik, A. Time-Dependent Density Functional Theory for Open Quantum Systems with Unitary Propagation. *Phys. Rev. Lett.* **2010**, *104*, 043001.

95. Yuen-Zhou, J.; Rodríguez-Rosario, C.; Aspuru-Guzik, A. Time-dependent current-density functional theory for generalized open quantum systems. *Phys. Chem. Chem. Phys.* **2009**, *11*, 4509–4522.

96. Mensky, M.B. *Quantum Measurements and Decoherence*; Springer: Dordrecht, The Netherlands, 2000.

97. Erez, N.; Gordon, G.; Nest, M.; Kurizki, G. Thermodynamic control by frequent quantum measurements. *Nature* **2008**, *452*, 724–727.

98. Rebentrost, P.; Mohseni, M.; Kassal, I.; Lloyd, S.; Aspuru-Guzik, A. Environment-assisted quantum transport. *New J. Phys.* **2009**, *11*, 033003.

99. Plenio, M.B.; Huelga, S.F. Dephasing-assisted transport: Quantum networks and biomolecules. *New J. Phys.* **2008**, *10*, 113019.

100. Dubi, Y. Interplay between Dephasing and Geometry and Directed Heat Flow in Exciton Transfer Complexes. *J. Phys. Chem. C* **2015**, *119*, 25252–25259.

101. Segal, D.; Nitzan, A.; Hänggi, P. Thermal conductance through molecular wires. *J. Chem. Phys.* **2003**, *119*, 6840–6855.

102. Dhar, A.; Saito, K.; Hänggi, P. Nonequilibrium density-matrix description of steady-state quantum transport. *Phys. Rev. E* **2012**, *85*, 011126.

103. Misra, B.; Sudarshan, E.C.G. The Zeno's paradox in quantum theory. *J. Math. Phys.* **1977**, *18*, 756–763.

104. Facchi, P.; Nakazato, H.; Pascazio, S. From the Quantum Zeno to the Inverse Quantum Zeno Effect. *Phys. Rev. Lett.* **2001**, *86*, 2699–2703.

105. Biele, R.; Rodríguez-Rosario, C.A.; Frauenheim, T.; Rubio, A. Controlling heat and particle currents in nanodevices by quantum observation. *Npj Quantum Mater.* **2017**, *2*, 38.

106. Rodríguez-Rosario, C.A.; Frauenheim, T.; Aspuru-Guzik, A. Thermodynamics of quantum coherence. *arXiv* **2013**, arXiv:1308.1245.

107. Kosloff, R. Quantum Thermodynamics: A Dynamical Viewpoint. *Entropy* **2013**, *15*, 2100–2128.

108. Yang, C.C.; Mai, Y.W. Thermodynamics at the nanoscale: A new approach to the investigation of unique physicochemical properties of nanomaterials. *Mater. Sci. Eng. R Rep.* **2014**, *79*, 1–40.

109. Biele, R.; D'Agosta, R.; Rubio, A. Time-Dependent Thermal Transport Theory. *Phys. Rev. Lett.* **2015**, *115*, 056801.

110. Donges, A. The coherence length of black-body radiation. *Eur. J. Phys.* **1998**, *19*, 245–249.

111. Bertilone, D.C. On the cross-spectral tensors for black-body emission into space. *J. Mod. Opt.* **1996**, *43*, 207–218.

112. Bergfield, J.P.; Liu, Z.F.; Burke, K.; Stafford, C.A. Bethe Ansatz Approach to the Kondo Effect within Density-Functional Theory. *Phys. Rev. Lett.* **2012**, *108*, 066801.

113. Tröster, P.; Schmitteckert, P.; Evers, F. Transport calculations based on density functional theory, Friedel's sum rule, and the Kondo effect. *Phys. Rev. B* **2012**, *85*, 115409.

114. Stefanucci, G.; Kurth, S. Towards a Description of the Kondo Effect Using Time-Dependent Density-Functional Theory. *Phys. Rev. Lett.* **2011**, *107*, 216401.

115. Frenkel, D.; Smit, B. *Understanding Numerical Simulation*, 2nd ed.; Academic Press San Diego: San Diego, CA, USA, 2002.

116. Donadio, D.; Galli, G. Atomistic Simulations of Heat Transport in Silicon Nanowires. *Phys. Rev. Lett.* **2009**, *102*, 195901.

117. Mingo, N. Anharmonic phonon flow through molecular-sized junctions. *Phys. Rev. B* **2006**, *74*, 125402.

Review

Quantum Phonon Transport in Nanomaterials: Combining Atomistic with Non-Equilibrium Green's Function Techniques

Leonardo Medrano Sandonas [1,2,†], **Rafael Gutierrez** [1,], **Alessandro Pecchia** [3], **Alexander Croy** [1] and **Gianaurelio Cuniberti** [1,2,4,*]

1   Institute for Materials Science and Max Bergmann Center of Biomaterials, TU Dresden, 01062 Dresden, Germany
2   Center for Advancing Electronics Dresden, TU Dresden, 01062 Dresden, Germany
3   Consiglio Nazionale delle Ricerche, ISMN, Via Salaria km 29.6, Monterotondo, 00017 Rome, Italy
4   Dresden Center for Computational Materials Science (DCMS), TU Dresden, 01062 Dresden, Germany
*   Correspondence: gianaurelio.cuniberti@tu-dresden.de
†   Current address: Physics and Materials Science Research Unit, University of Luxembourg, L-1511 Luxembourg, Luxembourg.

Received: 1 July 2019; Accepted: 25 July 2019; Published: 27 July 2019

Abstract: A crucial goal for increasing thermal energy harvesting will be to progress towards atomistic design strategies for smart nanodevices and nanomaterials. This requires the combination of computationally efficient atomistic methodologies with quantum transport based approaches. Here, we review our recent work on this problem, by presenting selected applications of the PHONON tool to the description of phonon transport in nanostructured materials. The PHONON tool is a module developed as part of the Density-Functional Tight-Binding (DFTB) software platform. We discuss the anisotropic phonon band structure of selected puckered two-dimensional materials, helical and horizontal doping effects in the phonon thermal conductivity of boron nitride-carbon heteronanotubes, phonon filtering in molecular junctions, and a novel computational methodology to investigate time-dependent phonon transport at the atomistic level. These examples illustrate the versatility of our implementation of phonon transport in combination with density functional-based methods to address specific nanoscale functionalities, thus potentially allowing for designing novel thermal devices.

Keywords: phonon transport; nanostructured materials; green's functions; density-functional tight binding; Landauer approach, time-dependent transport

1. Introduction

The accelerated pace of technological advances, which has taken place over the last half-century, has driven the continuous search for higher speed and cheaper computing with the concomitant developments of larger integration densities and miniaturization trends based on novel materials and processes [1]. The negative counterpart of this amazing technological developments consists in the increasing problems with the thermal management, which is ultimately leading to limiting the efficiency of many of these technological advances, mostly in the domain of nanoelectronics. As a response to this challenge, a novel field, nanophononics, has emerged, providing a large variety of interesting physical effects and potential applications [2–5]. The major goal of nanophononics is to develop efficient strategies for controlling the *heat flux* in organic and inorganic nanostructured materials, and it was originally aiming at realizing thermal devices such as diodes, logic gates, and thermal transistors [2,6]. However, more recent efforts in the field have triggered radically new applications in nanoelectronics [7,8], renewable energy harvesting [9,10], nano- and optomechanical devices [11], quantum technologies [12,13], and therapies, diagnostics, and medical imaging [14].

An important milestone in the field was the theoretically predicted [15] and subsequently measured *quantization* of the phononic thermal conductance in mesoscopic structures [16] at low temperatures, this result building the counterpart of the well-known quantization of the electrical conductance in quantum point contacts and other nanostructures (with conductance quantum given by e^2/h, e being the electron charge and h Planck constant). Thermal conductance quantization has also been found in smaller nanostructures, such as gold wires, where quantized thermal conductance at room temperature was shown down to single-atom junctions [17,18]. The issue is, however, still a subject of debate (see, e.g., [19] for a recent discussion). In contrast to the electrical conductance quantization, the quantum of thermal conductance κ_0 depends, however, on the absolute temperature T through: $\kappa_0 = \pi^2 k_B^2 T/3h$, with k_B being the Boltzmann constant. This highlights a first important difference between charge and phonon transport. The second one is related to the different energy windows determining the corresponding transport properties: for electrons, the important window lies around the Fermi level, while for phonons, the conductance results from an integral involving the full vibrational spectrum. Working with a broad spectrum of excitations poses major challenges when it comes to designing thermal devices such as cloaks and rectifiers [2,4], or for information processing in phonon-based computing [6].

From the experimental perspective, it is obviously more difficult to tune heat flow than electrical currents. Unlike electrons, phonons are quasi-particles with zero mass and zero charge, thus they cannot be directly controlled through electromagnetic fields in a straightforward way. Moreover, while considerable progress has been achieved in nanoelectronics in the implementation of local electrodes and gates over very short length scales, establishing temperature gradients over nanoscopic length scales remains a considerable challenge. In this respect, for characterizing thermal devices, novel sophisticated experimental techniques have been developed, such as the 3ω method [20] and the frequency domain thermoeflectance [21], pioneered by Cahill et al. [22]. This has led, in turn, to the modification of atomic force microscopes for thermometry [17,23] and to the development of Scanning Thermal Microscopy [18,24].

Turning nanophononics into a practical field with specific applications in energy management requires nanoscale engineering of the thermal transport properties. This approach has been successfully implemented in nanostructured thermoelectric materials. Even though the fundamental tool to understand nanophononics—non-equilibrium thermodynamics—is well established at the macroscale, many open issues remain at the nanoscale, having deep consequences for the development of relevant strategies to control heat transport in low dimensions. Low-dimensional materials have finite cross sections along one or more spatial dimensions and a large surface-to-volume ratio; as a result, their vibrational spectrum and, consequently, the heat transport mechanisms can be dramatically modified (see the recent reviews in [2–4]). Different simulation tools have been used to study heat transport in nanomaterials [4,25]. These approaches can be grouped into three categories, although overlaps between them are clearly possible. The first group includes methodologies based on molecular dynamics (MD) simulations, the equilibrium versions (EMD) based on the Green–Kubo formula [26], and the non-equilibrium versions (NEMD) exploiting Fourier's law [27,28]. The second category includes approaches based on the Boltzmann transport equation (BTE) [29,30] and lattice dynamics (LD) [31]. Finally, the last category covers methodologies relying on the Landauer approach, or more generally on non-equilibrium Green's functions (NEGF) [32–35]. All these methodologies have found extensive application in the prediction of the thermal transport properties of various low-dimensional materials, yielding correct trends and results in good agreement with experimental studies [36–38]. As it turns out, the thermal response is sensitively determined by different parameters such as surface boundaries, overall device geometry, spatial confinement, doping, and structural defects [37,39–43].

Moreover, in non-stationary situations with time-dependent external parameters able to affect the transport characteristics, phonon dynamics becomes crucial. For instance, time-varying temperature fields [44,45] or local heating mediated by laser fields [46,47] can be exploited to exert additional control over thermal transport. Thus, novel non-equilibrium effects such as heat pumping [48,49], cooling [50], and rectification [51,52] have been theoretically proposed. The description of such phenomena requires in many instances to work in the time domain, which is very challenging from a numerical point of view. Although noticeable progress has been achieved in dealing with time-dependent spin [53,54] and electron [55–60] transport, much less attention has been paid to vibrational degrees of freedom [61–63].

Despite the previously delineated methodological advances to model and understand nanoscale thermal transport, there are many basic questions about thermal management of thermoelectric materials, phononic devices, and integrated circuits that must be addressed. In the current paper, we review our recently implemented atomistic models based on the NEGF technique, allowing to address transient and steady quantum phonon transport in low-dimensional systems. We have successfully used our methodology to propose different routes for improved thermal management, eventually leading to realizing novel nanoscale applications.

The paper is organized as follows. In Section 2, the basics of the NEGF approach to compute quantum ballistic transport are introduced. We proceed then to review few selected applications by using the NEGF in combination with a Density-Functional based Tight-Binding approach (DFTB), which allows addressing nanostructures at the atomistic level with considerable accuracy and large computational efficiency. The reviewed applications include 2D materials, BNC heteronanotubes, and molecular junctions. In Section 3, the NEGF formalism previously introduced is expanded to deal with time-dependent thermal transport by exploiting an auxiliary mode approach. This methodology is illustrated for a one-dimensional chain and simple nanoscale junctions based on polyethylene and polyacetylene dimers.

2. DFTB-Based Quantum Transport

2.1. Ballistic Phonon Transport

One of the most powerful methodologies to study quantum (thermal) transport is the non-equilibrium Green's functions (NEGF) formalism [35,64,65]. The NEGF method has its origin in quantum field theory [66], and has been developed to study many-particle quantum systems under both equilibrium and nonequilibrium conditions. Different formulations were derived during the early 1960s [67–69]. Thus, Keldysh developed a diagrammatic approach, Kadanoff and Baym formulated their approach based on equations of motion. Both methods are suitable for studying a dynamic system in nonequilibrium. For instance, by using the Keldysh formalism, one can obtain formal expressions for the current and electron density [70]. The method has also been successfully used to study electron transport properties in open quantum systems [71,72]. Moreover, NEGF has been recently used on thermal transport investigations not only in the ballistic regime [73–75], but also including phonon–phonon scattering [76–79]. In this section, we describe the NEGF formalism to address ballistic phonon transport in nanostructures. Phonon–phonon interactions require the inclusion of extra self-energy terms, which depend on products of single-phonon Green's functions, so that the whole problem must be solved self-consistently; however, this goes beyond the scope of this review (see [35,64] for more details concerning this point).

The main difference between the NEGF formalism and ordinary equilibrium theory is that all time-dependent functions are defined on the so-called Schwinger-Keldysh contour (see Figure 1). However, a simplification occurs when $t_0 \to -\infty$ (Keldysh contour). If the interactions are switch on adiabatically, the contribution from the $[t_0, t_0 + \beta]$ piece vanishes. The information lost by this procedure is related to initial correlations. In many physical situations, such as in steady state transport, it is a plausible approximation to assume that initial correlations have been washed out by the interactions when one reaches the steady state. On the contrary, for the transient response, the role of initial correlations may be important (see Section 3). Here, we consider the Keldysh contour consisting of two branches running from $-\infty$ to ∞ and from ∞ to $-\infty$. Therefore, one can introduce a contour-ordered Green's function as [80]:

$$G(\tau, \tau') = -i\left\langle T_C A(\tau) B^T(\tau') \right\rangle, \tag{1}$$

with T_C as the contour-order operator. Based on it, six real-time Green's functions can be defined [35]:

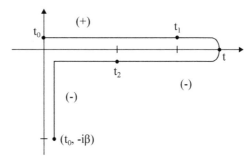

Figure 1. Schwinger/Keldysh contour C in the imaginary time plane, $C = \{t \in C, \Re\ t \in [t_0, \infty] \Im\ t \in [t_0, -\beta]\}$. For more clarity, the different contour branches are displayed slightly off the axes. Time-ordering: time t_2 is later on the contour than time t, and t is larger than t_1 [80].

- The lesser GF, $G^<(t, t') = -i\left\langle A(t')B^T(t) \right\rangle$.
- The greater GF, $G^>(t, t') = -i\left\langle A(t)B^T(t') \right\rangle$.
- The retarded GF, $G^r(t, t') = -i\Theta(t - t')\left\langle [A(t), B^T(t')] \right\rangle$.
- The advanced GF, $G^a(t, t') = i\Theta(t' - t)\left\langle [A(t), B^T(t')] \right\rangle$.
- The time-ordered GF, $G^t(t, t') = \Theta(t - t')G^>(t, t') + \Theta(t' - t)G^<(t, t')$.
- The anti-time-ordered GF, $G^{\bar{t}}(t, t') = \Theta(t' - t)G^>(t, t') + \Theta(t - t')G^<(t, t')$.

$A(t)$ and $B(t)$ are operators in the Heisenberg picture and $\Theta(t)$ is the Heaviside step function. The angular brackets denote trace with the canonical density matrix, i.e., $\langle \cdots \rangle = \mathrm{Tr}(\rho \cdots)$, with $\rho = e^{-\beta H}/\mathrm{Tr}(e^{-\beta H})$ and $\beta = 1/(k_B T)$, and H is the Hamiltonian of the system. The notation $\left\langle [A, B^T] \right\rangle$ represents a matrix and should be understood as $\left\langle AB^T \right\rangle - \left\langle BA^T \right\rangle^T$.

In equilibrium or non-equilibrium steady state, the Green's functions only depend on the time difference, $t - t'$. The Fourier transform of $G^r(t - t') = G^r(t, t')$ is defined as $G^r[\omega] = \int_{-\infty}^{+\infty} G^r(t)e^{i\omega t}dt$. Using the basic definitions, the following linear relations hold in both frequency and time domains [35]:

$$G^r - G^a = G^> + G^<,$$
$$G^t + G^{\bar{t}} = G^> + G^<, \tag{2}$$
$$G^t - G^{\bar{t}} = G^r + G^a.$$

Only three of the six Green's functions are linearly independent. In systems with time translational invariance, the functions G^r and G^a are related by $G^a[\omega] = (G^r[\omega])^\dagger$. Hence, under general non-equilibrium steady-state conditions, only two are independent, with a typical choice of working with G^r and $G^<$, although other combinations are possible. Extra relations are defined in the frequency domain for bosons [81]:

$$G^<[\omega]^\dagger = -G^<[\omega],$$
$$G^r[-\omega] = G^r[\omega]^*, \tag{3}$$
$$G^<[-\omega] = G^>[\omega]^T = -G^t[\omega]^* + G^r[\omega]^T - G^r[\omega]^*.$$

Therefore, based on the last two equations, only the positive frequency part of the functions is needed. Equations (2) and (3) are generally valid for non-equilibrium steady states. However, for systems in thermal equilibrium satisfying the fluctuation–dissipation theorem [82], there is an additional equation relating G^r and $G^<$:

$$G^<[\omega] = f(\omega)\,(G^r[\omega] - G^a[\omega]),\qquad(4)$$

where $f(\omega) = \left(e^{\frac{\hbar\omega}{k_BT}} - 1\right)^{-1}$ is the Bose–Einstein distribution function at temperature T. k_B is the Boltzmann constant. Indeed, the correlation function $G^<$ contains information of fluctuations, while $G^r - G^a$ describes dissipation of the system. $G^>[\omega] = e^{\beta\hbar\omega}G^<[\omega]$ also applies for equilibrium systems and, consequently, there is only one independent Green's function under equilibrium conditions.

In phonon transport calculations, a partitioning scheme is applied, consisting in splitting the whole system in three regions: one central region (also denoted as device region), connected to two thermal baths on the left (L) and right (R) (see Figure 2). At the simplest level, the thermal baths can be considered as a collection of non-interacting harmonic oscillators. All elastic and/or inelastic scattering processes are therefore assumed to be confined to the central (or device) region. Since we focus on thermal transport mediated by the vibrational system, the phonon Hamiltonian of the whole system is given by [80]:

$$H = \sum_{\alpha=L,C,R} H_\alpha + (u^L)^T V^{LC} u^C + (u^C)^T V^{CR} u^R + V_n,\qquad(5)$$

where $H_\alpha = \frac{1}{2}(\dot{u}^\alpha)^T\dot{u}^\alpha + \frac{1}{2}(u^\alpha)^T K^\alpha u^\alpha$ represents the Hamiltonian of the region α; $\alpha = L, C, R$, for the left, center, and right regions, respectively. u^α is a column vector consisting of all the displacement variables in region α, and $\dot{u}^\alpha$ is the corresponding conjugate momentum. The following transformation of coordinates has been considered, $u_j = \sqrt{m_j}x_j$, where x_j is the relative displacement of jth degree of freedom. K^α is the mass-reduced force constant matrix. This matrix is the mass-weighted second derivative of energy with respect to displacement at the equilibrium positions:

$$\left[K_{\gamma\beta}\right]_{ij} = K_{ij}^{\gamma\beta} = \frac{1}{\sqrt{m_i m_j}}\frac{\partial^2 E}{\partial u_i^\gamma u_j^\beta}.\qquad(6)$$

$V^{LC} = (V^{CL})^T$ is the coupling matrix between the left lead to the central region; and similarly for V^{CR}. The last term V_n represents possible many-body interactions, such as phonon–phonon interaction [35]. It contains higher order (higher than 2) derivatives of the energy with respect to the displacements, evaluated at the equilibrium positions.

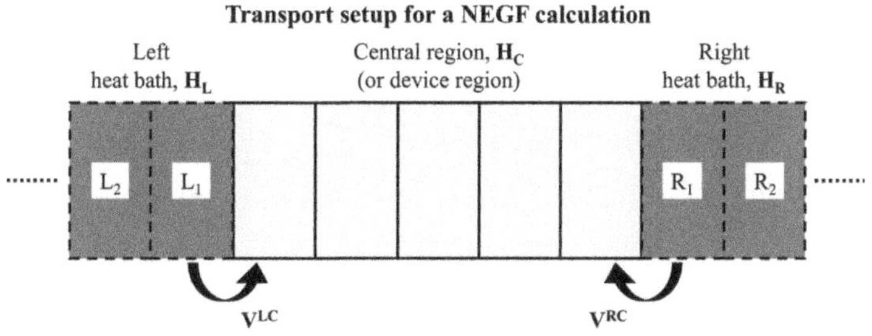

Figure 2. Schematic representation of the common partitioning scheme for phonon transport calculation using Green's function technique. The entire system is split into three regions: central region and, left and right heat baths. Each of this region are characterized by their own Hamiltonian H_α with $\alpha = L, C, R$. The coupling matrices between heat baths and central region are V^{LC} and V^{RC}.

The most important quantity to calculate is the heat flux J, which is defined as the energy transferred from the heat source to the junction in a unit time, and is equal to the energy transferred from the junction to the heat sink in a unit time. Here, it is assumed that no energy is accumulated in the junction. According to this definition, the heat flux out of the left lead is:

$$J_L = -\langle \dot{H}_L(t) \rangle = i\langle [H_L(t), H] \rangle = i\left\langle \left[H_L(t), V^{LC}(t) \right] \right\rangle. \tag{7}$$

In the steady state, energy conservation means that $J_L + J_R = 0$. For simplicity, we set $\hbar = 1$. Using Heisenberg's equation of motion, J_L can be written as:

$$
\begin{aligned}
J_L &= \left\langle (\dot{u}^L)^T(t) V^{LC} u^C(t) \right\rangle \\
&= \lim_{t' \to t} \sum_{j,k} V_{jk}^{LC} \left\langle (\dot{u}_j^L)^T(t') u_k^C(t) \right\rangle.
\end{aligned}
\tag{8}
$$

Thus, the heat flux depends on the expectation value of $(\dot{u}_j^L)^T(t') u_k^C(t)$, which can be written in terms of the lesser Green's function $G_{\dot{C}L}^<(t,t') = -i\langle u^L(t') u^C(t)^T \rangle^T$. Since operators u and $\dot{u}$ are related in Fourier space (frequency domain) as $\dot{u}[\omega] = -i\omega u[\omega]$, the derivative is eliminated and one obtains:

$$J_L = -\frac{1}{2\pi} \int_{-\infty}^{\infty} \mathrm{Tr}\left(V^{LC} G_{CL}^<[\omega] \right) \omega d\omega. \tag{9}$$

The Green's functions of interacting systems can be efficiently obtained by solving their equations of motion (EOM) [64]. In this section, EOMs will only be used to obtain expressions for retarded and lesser GFs of the central region. This topic will be expanded with more details in Section 3. First, we have that the contour-ordered GF $G(\tau, \tau') = -i\langle T_\tau u(\tau) u(\tau')^T \rangle$ satisfies the following equation:

$$-\frac{\partial^2 G(\tau, \tau')}{\partial \tau^2} - KG(\tau, \tau') = I\delta(\tau, \tau') \tag{10}$$

The equation per each region is obtained by partitioning the matrices G and K into submatrices $G^{\alpha,\alpha'}$ and $K^{\alpha,\alpha'}$, $\alpha, \alpha' = L, C, R$. The free Green's function for the decoupled system g is easily obtained by solving:

$$-\frac{\partial^2 g^\alpha(\tau, \tau')}{\partial \tau^2} - K^\alpha g^\alpha(\tau, \tau') = I\delta(\tau, \tau'). \tag{11}$$

The corresponding free GFs in frequency domain are written as:

$$g_\alpha^r[\omega] = \left[(\omega + i\eta)^2 - K^\alpha \right]^{-1}, \tag{12}$$

where η is an infinitesimal positive quantity to single out the correct path around the poles when performing an inverse Fourier transform, such that $g^r = 0$ for $t < 0$. Other Green's functions can be obtained using the general relations among them (see Equations (2) and (3)). Hence, the contour-ordered non-equilibrium GF can be written as:

$$G^{CL}(\tau, \tau') = \int d\tau'' G^{CC}(\tau, \tau'') V^{CL} g^L(\tau'', \tau'), \tag{13}$$

$$G^{CC}(\tau, \tau') = g^C(\tau, \tau') + \int d\tau_1 \int d\tau_2 g^C(\tau, \tau_1) \Sigma(\tau_1, \tau_2) G^{CC}(\tau_2, \tau'), \tag{14}$$

with $\Sigma(\tau_1, \tau_2)$ being the total self-energy including the coupling to the baths and given by:

$$\Sigma(\tau_1, \tau_2) = \Sigma_L(\tau_1, \tau_2) + \Sigma_R(\tau_1, \tau_2) = V^{CL} g^L(\tau_1, \tau_2) V^{LC} + V^{CR} g^R(\tau_1, \tau_2) V^{RC}. \tag{15}$$

g^L and g^R are the GF of the isolated semi-infinite leads. Since the bath-device interaction terms are short-ranged, it is usually only necessary to compute the projection of the bath GF on the layer directly in contact with the device. The resulting GF can be calculated by an iteration method [83] or by decimation techniques [84]. The Dyson equation (see Equation (14)) can be written in the frequency domain as:

$$G_{CC}^r[\omega] = \left((\omega + i\eta)^2 I - K^C - \Sigma^r[\omega]\right)^{-1},$$
$$G_{CC}^<[\omega] = G_C^r[\omega]\Sigma^<[\omega]G_C^a[\omega]. \tag{16}$$

Several physical quantities can be calculated using these relations, e.g., the local density of states (LDOS) is expressed as:

$$\eta_i(\omega) = -\frac{2\omega}{\pi}(\mathrm{Im}G^r[\omega])_{ii}, \tag{17}$$

and the total phonon DOS as $\eta(\omega) = \sum_{i=1}^N \eta_i(\omega)$. The DOS and LDOS give the distribution of phonons in frequency as well as in real space [81]. This is very useful to analyze quantum transport processes, as shown below. Although Equation (17) is obtained for the special case of no phonon–phonon interactions, the same formula is valid in the presence of phonon–phonon interactions described by an interaction self-energy such as in Equation (16). This typical approach assumes that the non-crossing approximation applies, allowing to treat the effect of contacts and interactions as two independent additive contributions. Clearly, this is valid in the limit of small interactions acting only within the central region.

Next, it is useful to introduce the Γ function describing the phonon scattering rate into the thermal baths:

$$\Gamma[\omega] = i(\Sigma^r[\omega] - \Sigma^a[\omega]) = \Gamma_L[\omega] + \Gamma_R[\omega], \tag{18}$$

This function has an important relation with the spectral function, $A[\omega] = G^r[\omega]\Gamma[\omega]G^a[\omega]$. By applying the Langreth theorem [64] to Equation (13), the lesser GF $G_{CL}^<$ turns into:

$$G_{CL}^<[\omega] = G_{CC}^r[\omega]V^{CL}g_L^<[\omega] + G_{CC}^<[\omega]V^{CL}g_L^a[\omega]. \tag{19}$$

Consequently, the heat flux coming from the left lead (see Equation (9)) can be written as:

$$J_L = -\frac{1}{2\pi}\int_{-\infty}^{+\infty} d\omega\omega\,\mathrm{Tr}\left(G^r[\omega]\Sigma_L^<[\omega] + G^<[\omega]\Sigma_L^a[\omega]\right). \tag{20}$$

For simplicity, the subscripts C related to the central region have been dropped. The upperletters are used to identify Green's functions on the central region and lowercase letters for the leads. After symmetrizing with respect to the left and right leads, the heat flux becomes:

$$J = \frac{1}{4}\left(J_L + J_L^* - J_R - J_R^*\right). \tag{21}$$

The final expression reads:

$$J = \int_0^\infty \frac{d\omega}{2\pi}\hbar\omega\tau_{ph}[\omega](f_L - f_R). \tag{22}$$

This result is formally similar to the Landauer equation obtained for electron transport. Here, however, $f_{L,R}$ are the Bose–Einstein distributions for the left and right leads and $\tau_{ph}[\omega]$ is the phonon transmission function, given by:

$$\tau_{ph}[\omega] = \mathrm{Tr}\left(G^r[\omega]\Gamma_L[\omega]G^a[\omega]\Gamma_R[\omega]\right). \tag{23}$$

The retarded GF of the central region connected to the thermal baths is given by:

$$G^r[\omega] = \left[(\omega + i\eta)^2 I - K^C - \Sigma_L^r[\omega] - \Sigma_R^r[\omega] \right]^{-1} \tag{24}$$

Then, the thermal conductance can then be computed according to $\kappa_{ph} = \lim_{\Delta T \to 0} \frac{J}{\Delta T}$, ΔT as the temperature difference between the thermal baths, with $T_L = T + \Delta T/2$ and $T_R = T - \Delta T/2$, respectively. A linear expansion of the Bose–Einstein distribution in ΔT yields [80]:

$$\kappa_{ph} = \frac{1}{2\pi} \int_0^\infty d\omega \, \omega T[\omega] \frac{\partial f(\omega)}{\partial T}. \tag{25}$$

Notice that the thermal conductance can only be obtained by an integration over the whole frequency range of the phonon transmission. In practice, the derivative of the Bose–Einstein distribution will reduce (depending on the temperature) the real integration range. This is in contrast to the Landauer conductance for electrons, where, strictly speaking, only states near the Fermi energy are playing a role.

2.2. Density Functional Tight-Binding

The main quantities to obtain the quantum phonon transport properties by using the NEGF formalism are the mass-reduced force constant matrix in each region K^α ($\alpha = L, C, R$) and the coupling matrices of the left and right bath to the central region, V^{LC} and V^{CR}, respectively [80]. The accuracy of the results depends on the reliability of these quantities to catch the atomistic features of the system. Density-functional theory (DFT) is nowadays the main computational approach used in chemistry and physics to perform quantitative studies on molecules and materials due to its favorable accuracy-to-computational-time ratio [85]. The strong increase in accuracy coming from the development of gradient corrected and hybrid functionals such as PBE [86] and B3LYP [87], which compensate deficiencies of older approximations, has largely contributed to a further increase in popularity. However, hybrid functionals are computationally demanding, limiting DFT to a maximum of a few hundreds of atoms, depending on the chemical species. Classical force fields appear as a reasonable solution to this problem but, in many cases, they suffer of limited transferability and do not yield any information on the electronic structure.

Semiempirical methods appear as another option to DFT, conceptually lying between empirical force fields and first principle approaches, allowing for the treatment of thousands of atoms [88]. These methods can be understood as approximations to more accurate methods (full DFT or Hartree–Fock), but including empirical parameters that are fitted to reproduce reference data. One example of a semiempirical method, which is used in the present work, is the density functional tight-binding (DFTB) approach [89–91]. Here, the basic electronic parameters (Slater–Koster parameters) are consistently obtained from full DFT-based calculations for atom pairs, while the repulsive part of the electronic energy is fitted by means of splines. Based on it, the Hamiltonian and Overlap matrices of a specific system can be decomposed into pair interactions (not only between nearest-neighbors) yielding a generalized tight-binding Hamiltonian. Many studies have been carried out by using the DFTB method, including transport properties of 2D materials [92–94], stability and mechanical properties [95], vibrational signatures [96], computation of molecular absorption spectra [97], and of charge transfer excitation energies [98] (see recent review papers [99–101] for additional topics).

Three different DFTB models have been proposed up to now, which are derived by expanding the DFT total energy functional around a reference density ρ_0 to first, second, and third order, respectively [101]. The choice of the DFTB model depends on the system under study. The non self-consistent DFTB method (or DFTB0) is more appropriate for systems with negligible charge transfer between atoms (typically homonuclear systems or those involving atoms of similar electronegativity as in hydrocarbons [102]). Ionic systems with large inter-atomic charge transfer can also be treated with this method [103]. On the other hand, in systems where a delicate charge balance is crucial such as biological and organic molecules [104,105], a self-consistent charge treatment is required (DFTB2 and DFTB3) [101,106]. Based on the advantages of DFTB for accurately describing large systems involving few thousand of atoms, the force constant matrices of the studied systems are numerically obtained by applying a finite difference method to get the second derivatives of the total energy with respect to the atomic displacements (implemented in the DFTB+ software) [80]. These matrices can also be obtained by density functional perturbation theory, which in the case of DFTB reduces to analytic expressions involving derivatives of only two-center matrix elements [107].

2.3. Application of the DFTB-Based PHONON Tool

From electron transport studies, it is well-known that transport properties of nanoscale systems can be tailored by varying different control parameters. This can include covalent or non-covalent chemistry [108,109], atomic doping [35,110], topological defects [111,112], quantum confinement [113], and mechanical strains [114,115], among others. Similarly, a major focus of research on phonon transport is to identify the major variables allowing for effectively tuning the heat transport properties of nanoscale materials. In this section, we review few of our previous research in this direction using the NEGF-DFTB method [116–121], which is already implemented as a tool in the DFTB+ code (for details of the PHONON tool, see [80]). We focus on 2D orthorhombic materials, BNC heteronanotubes, and phonon filter effects in molecular junctions.

2.3.1. 2D Orthorhombic Materials

The effect of anisotropic atomic structure on the phonon transport of two-dimensional puckered materials was studied by Medrano Sandonas et al. [117]. From this new family of 2D materials [122–125], three representative members, phosphorene, arsenene, and SnS monolayers, which display the main features of this family, were studied. The unit cell of these materials is composed by four atoms, as depicted in Figure 3. Each atom is pyramidally bonded to three neighboring atoms of the same type (phosphorene and arsenene, for homoatomic) or of different type (tin sulfide (SnS), for heteroatomic) forming a puckered-like honeycomb lattice. As shown in Table 1, the lattice constants computed with the DFTB approach quantitatively agree (error $\leq 5\%$) with those obtained at the full DFT level by other authors for all three materials.

We used a standard approach to compute the phonon band structure [80]. This consists in diagonalizing the dynamical matrix at selected k-points, after obtaining them through a Fourier transformation of the real-space force constants (this method is also part of the PHONON tool). Due to the absence of imaginary frequencies, all studied systems can be considered as mechanically stable (see the three lower panels of Figure 3). The acoustic branches display the typical dispersion of 2D materials: longitudinal (LA) and transversal (TA) acoustic branches show linear dispersion as q (wave vector) approaches the Γ point, while out-of-plane ZA branches show on the other hand a quadratic dispersion as a result of the rapid decay of transversal forces. The behavior of the dispersion relation for homoatomic puckered materials is almost identical, except for the maximum frequency of the optical modes, which is a consequence of the difference in mass between As ($\sim$75 u) and P ($\sim$31 u). We also remark that the phonon dispersion for P and As computed with DFTB agrees quite well with DFT results[117]. Only for SnS monolayer the high frequency optical modes are shifted upwards.

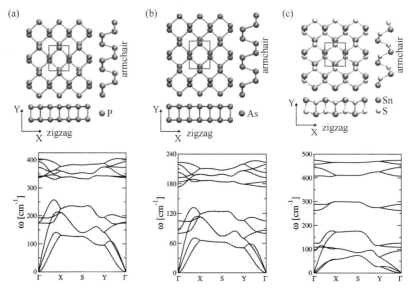

Figure 3. Phonon dispersion for homo-and heteroatomic two-dimensional puckered materials: (a) phosphorene; (b) arsenene; and (c) tin sulfide (SnS) monolayer. We also show the atomistic view of the two-dimensional materials, highlighting the zigzag (ZZ) and armchair (AC) transport directions. The figure is reproduced with permission from Ref. [117]. Copyright 2016 American Chemical Society.

Table 1. Calculated lattice constants of two-dimensional puckered materials along zigzag (ZZ) and armchair (AC) directions. For comparison, the lattice constants from other published theoretical studies are also given. In general, the DFTB lattice parameters agree quite well with those calculated by using DFT method, error $\leq$ 5%. The table is reproduced with permission from Ref. [117]. Copyright 2016 American Chemical Society.

Systems	Transport Direction ZZ [Å]	AC [Å]	Other Works (ZZ, AC) [Å]
Phosphorene	3.49	4.34	(3.28, 4.43) [126] (3.32, 4.58) [127]
Arsenene	3.81	4.75	(3.68, 4.77) [128] (3.69, 4.77) [129]
SnS monolayer	3.93	4.51	(4.03, 4.26) [124] (4.01, 4.35) [130]

Furthermore, based on the group velocities values obtained for ZZ ($\Gamma \rightarrow X$) and AC ($\Gamma \rightarrow Y$) transport directions, we may expect that these materials will display strong anisotropy in their thermal transport. Indeed, the group velocities for the longitudinal acoustic (LA) branch in phosphorene were found to be 8.35 km/s and 4.74 km/s along the Γ-X (ZZ) and Γ-Y (AC) directions, respectively, comparable to DFT results [127,131,132]. The values for arsenene, 5.01 km/s for ZZ and 2.71 km/s for AC, are also in good agreement with those in Ref. [129]. The SnS monolayer displayed group velocities of 6.48 km/s (ZZ) and 2.14 km/s (AC). We note that thermal anisotropy has only been reported for phosphorene [126,132] and arsenene [128,129], but not for SnS monolayers. Accordingly, the largest anisotropy in the thermal conductance was found in SnS monolayers due to the dominant contribution of acoustic modes to thermal transport [117].

2.3.2. Doping Influence on BNC Heteronanotubes

Ternary boron carbonitride nanotubes have recently been in the focus of theoretical and experimental activities because of their excellent mechanical, electrical, and non-linear optical properties which could be controlled by varying their chemical composition [133–135]. Hence, BNC heteronanotubes may play an important role as new generation of thermoelectric materials, and are also of great interest in environmentally relevant issues such as waste heat recovery and solid-state cooling [9,136]. In Ref. [120], we studied the influence of doping on the thermal transport properties of (6,6)-BNC heteronatubes, by considering three different BN doping distribution patterns of a carbon nanotube: helical, horizontal, and random. For this, a (6,6)-CNT of length 43.3 Å was the reference structure (supercell composed by 432 C atoms). Helical BN strips, BN chains (parallel to the transport direction, which corresponds to the z-axis), and BN rings (one ring containing 3B and 3N atoms) were introduced in an otherwise perfect (6,6)-CNT to represent helical, horizontal, and random impurity distributions (see Figure 4a). For a helical distribution, the BN concentration was varied from $c = 11\%$ to $c = 89\%$, while for other cases concentrations ranging from $c = 16\%$ to $c = 84\%$ were studied. The limits of 0% and 100% correspond to pure carbon and hexagonal boron–nitride nanotubes, respectively.

The geometry of the BNC heteronanotubes (BNC-HNT) was optimized with the DFTB method [137,138] with periodic boundary conditions along the z-axis. C-C and B-N bond lengths amount to 1.43 Å and 1.48 Å, respectively. The optimized helical BNC-HNT presented a wave-like profile along the axial direction resulting from the difference between bond lengths at the interfaces (see, e.g., [80,139,140]). Since the doping distribution can be introduced in different ways, the phonon transmission for random and horizontal distributions were averaged over five and three different atomic configurations, respectively. For the transport calculations, the baths are composed of twice the optimized supercell, and the central region includes only one supercell. To have a better understanding of the influence of doping on the transport properties, we introduce the quantity $R_{DOS} = \eta_X(\omega)/\eta_{Total}(\omega)$, where $\eta_{Total}(\omega)$ is the total DOS given by Equation (17), and $\eta_X(\omega)$ can be either the LDOS of C or BN domains.

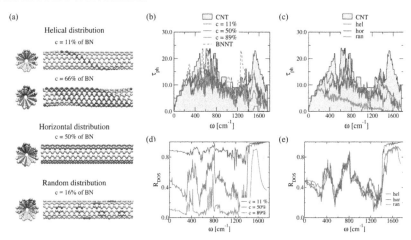

Figure 4. (**a**) Atomistic view of BNC heteronanotubes with helical, horizontal, and random distribution of BN domains. Carbon atoms (cyan), boron atoms (pink), and nitrogen atoms (blue) are shown. (**b**) Variation of the phonon transmission function, τ_{ph}, of helical BNC heteronanotubes after increasing the BN concentration, c. (**c**) Comparison of τ_{ph} for different doping distribution patterns with $c = 50\%$. Variation of R_{DOS} for carbon domains as a function of the vibrational frequency (**d**) for helical BNC heteronanotubes at three different doping concentrations and (**e**) for helical, horizontal, and random BNC heteronanotubes at $c = 50\%$. Reproduced from Ref. [120] with permission from the PCCP Owner Societies.

In Figure 4b, the influence of helical BN stripes on the phonon transmission of a (6,6)-CNT is shown. The high frequency modes ($\omega > 1400$ cm^{-1}) are strongly affected by increasing the BN concentration. These modes correspond to local vibrations related to carbon atoms; this is seen in

R_{DOS} after increasing the doping concentration (see Figure 4d). On the contrary, the transmission of low-frequency vibrations below 200 cm^{-1} is not changed much when varying the disorder concentration. Figure 4c shows the phonon transmission of BNC-HNT with fixed concentration $c = 50\%$ and different BN spatial arrangements. As expected, a random distribution of B and N atoms blocks the transmission over almost the whole frequency spectrum; only low-frequency modes experience less scattering at the localized impurities, so that their transmission is much less affected. Helical and horizontal disorder in BNC-HNT leads to a stronger blocking of the transmission at high frequencies ($\omega > 1400$ cm^{-1}) due to the absence of B-N-C local vibrations in that range (see Figure 4e).

Figure 5 shows the concentration dependence of the phonon thermal conductances, κ_{ph}, for each doping distribution pattern at $T = 300$ K. Horizontal BNC-HNT shows the highest thermal conductance, while the lowest κ_{ph} is obtained for (6,6)-CNT with BN domains randomly distributed. The thermal conductance of helical BNC-HNT remains nearly constant ($\sim$2.5 nW/K) for concentrations between 30% and 80%, and then increases until it reaches the value corresponding to a pristine BNNT, $\sim$3.0 nW/K. An additional case was studied with a helical BNC-HNT connected to CNT leads and, as a consequence of the new contact-device interface, the thermal conductance is continuously suppressed with increasing concentration. Notice that the dominant contribution to the thermal conductance at 300 K mostly derives from long wavelength modes with frequencies ≤ 200 cm^{-1} [120].

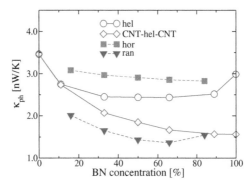

Figure 5. Phonon thermal conductance as a function of the BN concentration for helical, horizontal, and random pattern distributions. Results for helical BNC heteronanotubes connected to two CNT leads are also shown. Reproduced from Ref. [120] with permission from the PCCP Owner Societies.

2.3.3. Selective Molecular-Scale Phonon Filtering

We have recently proposed nano-junctions consisting of two colinear (6,6)-nanotubes (NT) joined by a central molecular structure as a potential molecular-scale phonon filter (see Figure 6a) [121]. The nanotubes act as heat baths kept at the same temperature. The left NT is considered as a reference bath with a broad phonon frequency spectrum, playing the role of a "source" of phonon modes. The molecular system in the central part is as a mode selector, and selected modes are then propagated to the right contact [80]. Besides carbon NTs, boron-nitride (BN) and silicon carbide (SiC) nanotubes as right baths were also considered. The filtering capability of the device thus depends on a mode-specific propagation resulting from the combined effect of molecular vibrations selection rules and the overlap of the contact spectral densities with the molecular region.

The phonon transport problem was treated using the previously described Green's function technique as implemented in the PHONON tool. Figure 6b,c shows the influence of interconnecting chains on the filtering effect for the case where both thermal baths are (6,6)-CNTs with $\omega_D \sim 1800$ cm^{-1}. As bridging molecular systems four parallel chains of ethylene, benzene, and azobenzene were chosen [121]. In Figure 6b, spectral gaps emerge in the transmission induced by the presence of the molecular chains. The overall transmission of the junctions is reduced by roughly a factor of four when compared with the infinite CNT due additional phonon scattering effects at the interfaces.

The influence of azobenzene chains is stronger comparing to the other monomers. Thus, chains with only two monomers already induce phonon gaps and filter out roughly half of the spectral range (see the lower panel of Figure 6b). Contrary to the benzene case, the transmission at low frequencies was strongly reduced, a result probably related to the lower number of modes and additional scattering at lower frequencies induced by larger structural distortions [141,142]. As a result, azobenzene-based junctions display the lowest thermal conductance, κ_{ph} (see Figure 7b for only CNT-based leads). In brief, it becomes clear that channel selection and phonon filtering can be strongly controlled by the chemical composition of the bridge [121].

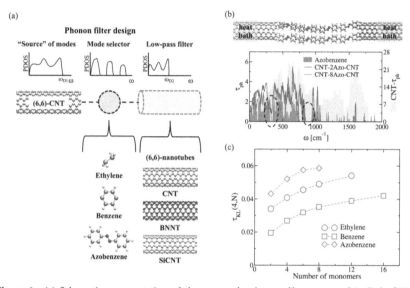

Figure 6. (a) Schematic representation of the nanoscale phonon filter proposed in Ref. [121]. A two-terminal junction is considered, where the role of the thermal baths is played by two semi-infinite (6,6) nanotubes (CNT, BNT, SiCNT) which are bridged by molecular chains consisting of ethylene, benzene, and azobenzene monomers. ω_D represents the Debye frequency in each nanotube. (b) Phonon transmission functions τ_{ph} for benzene-based junctions. We also added the plot corresponding to the phonon transmission function of and infinite CNT (grey) and a single infinite molecular chain of benzene monomers (brown). Highlighted with dashed-line circles are the regions where phonon gaps clearly develop by increasing the chain length. (c) Variation of $\tau_{KL}(j = 4, N)$ as a function of the number of monomers (N) in the different studied molecular junctions by considering both thermal baths made of (6,6)-CNTs. Each junction consists of four molecular chains in parallel, $j = 4$. The figure is reproduced with permission from Ref. [121]. Copyright 2019 American Chemical Society.

To quantify the deviation from the "source of modes" distribution (or degree of filtering), quantified in our case by the transmission spectrum of the CNT, a key design magnitude $\tau_{KL}(j, N)$ was defined as [121]:

$$\tau_{KL}(j, N) = \frac{1}{\omega_D^{CNT}} \int_0^{\omega_D^{CNT}} d\omega \, \tau_{CNT}(\omega) \ln \frac{\tau_{CNT}(\omega)}{\tau_{MJ,j}(\omega)}, \tag{26}$$

with the index j referring to the number of molecular chains in parallel interconnecting the two thermal baths and N the number of monomers in one chain. $\tau_{CNT}(\omega)$ and $\tau_{MJ,j}(\omega)$ are the corresponding transmission functions for an infinite CNT and for the CNT-molecule junction containing j molecular chains in parallel. A perfect filter (zero transmission) yields $\tau_{KL}(j, N) \to \infty$, while no filtering at all yields $\tau_{KL}(j, N) = 0$. As shown in Figure 6c, the azobenzene junctions display the highest τ_{KL} (i.e., highest filter efficiency) due to the efficient blocking of high-frequency modes above 1000 cm^{-1}.

The ethylene-based chains are less efficient, but still display a larger effect than the benzene chains; this is mostly related to two issues: the complete filtering of frequencies larger than 1500 cm^{-1} and the presence of a relatively large phonon gap between 500 cm^{-1} and 750 cm^{-1}. τ_{KL} for benzene-based junctions also increases with the length of the molecular chain and shows a tendency to saturate for $N > 12$ monomers.

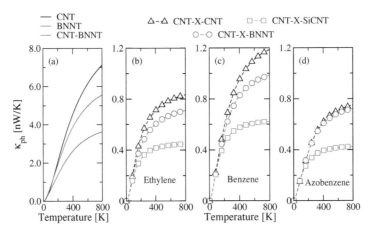

Figure 7. Phonon thermal conductance κ_{ph} as a function of the temperature. (**a**) κ_{ph} values of the infinite CNT and BNNT as well as of the CNT-BNNT junctions. The remaining panels show the thermal conductance in the different junction types with a chain length corresponding to four monomers for: (**b**) ethylene; (**c**) benzene; and (**d**) azobenzene. The figure is reproduced with permission from Ref. [121]. Copyright 2019 American Chemical Society.

Figure 7 highlights the influence of changing the material of the right bath on the phonon filtering. In these calculations, all chains consist of four monomers. The thermal conductance of CNT-BNNT junctions is reduced when compared to the perfect tubes due to interface scattering (see Figure 7a [143–146]). By inserting the molecular system, a reduction of the thermal conductance by roughly a factor of 3–4 is produced, the effect being more pronounced for the azobenzene junction [121]. For azobenzene-based molecular junctions, the thermal conductance barely changes when going from CNT-CNT to CNT-BNNT (see Figure 7d). This is a consequence of the strong suppression of high frequency transport channels in the transmission. The saturation of the thermal conductance is determined by the nanotube with the smaller Debye cutoff (going from CNT ($\omega_D \sim 1800$ cm^{-1}) to BNNT ($\omega_D \sim 1400$ cm^{-1}) to SiCNT ($\omega_D \sim 880$ cm^{-1})), the value of the saturation point is, however, influenced by the specific composition of the molecular chains.

3. Atomistic Framework for Time-Dependent Thermal Transport

A novel atomistic approach able to treat transient phonon transport was recently developed by Medrano Sandonas et al. [147]. The approach is based on the solution of the equation of motion for the phonon density matrix $\sigma(t)$, calculated within the NEGF formalism, by using an auxiliary-mode approach [80,147]. The latter has been previously used for time-resolved electron transport [55–57]. Unlike recent related approaches with limited application range (in terms of an atomistic treatment of the underlying system) [61–63], our method can be efficiently combined with an atomistic description as implemented in standard first-principle or parameterized approaches.

The basic structure of this approach is shown in Figure 8. Two thermal baths made of non-interacting harmonic oscillators in thermal equilibrium are contacted to a scattering region, whose vibrational features are assumed to be well represented by a quadratic Hamilton operator. The total system is described by the Hamiltonian:

$$H = H_C + \sum_{\alpha k} \left(\frac{1}{2} p_{\alpha k}^2 + \frac{1}{2} \omega_{\alpha k}^2 u_{\alpha k}^2 \right) + \sum_{\alpha,k} \frac{1}{2} \left(\mathbf{u}^T \cdot \mathbf{V}_{\alpha k} u_{\alpha k} + u_{\alpha k} \mathbf{V}_{\alpha k}^T \cdot \mathbf{u} \right) . \tag{27}$$

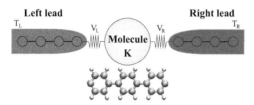

Figure 8. Schematic representation of the target molecular junctions by using the TD-NEGF approach. A molecular system is connected to two harmonic thermal baths, which are the source for the heat flow in the molecule. The figure is reproduced with permission from Ref. [147]. Copyright 2018 American Chemical Society.

The first term $H_C = (1/2)\mathbf{p}^T \cdot \mathbf{p} + (1/2)\mathbf{u}^T \cdot \mathbf{K}_{\text{eff}} \cdot \mathbf{u}$ is the Hamiltonian of the central domain, $\mathbf{u}$ is a column vector consisting of all the displacement variables in the region, and $\mathbf{p}$ contains the corresponding momenta. Both vectors have length N, with N being the number of degrees of freedom in the central region. We chose renormalized displacements $u_i = \sqrt{m_i} x_i$, where m_i is the mass associated to the ith vibrational degree of freedom, and x_i is the actual displacement having the dimension of length [80,147]. The effective force-constant matrix $\mathbf{K}_{\text{eff}} = \mathbf{K} + \mathbf{K}_{\text{ct}}$ has dimension $N \times N$, and includes the force constant matrix of the central region, $\mathbf{K}$, and a counter-term $\mathbf{K}_{\text{ct}}$ [80,147]. The index $\alpha \in \{L, R\}$ labels the left (L) and right (R) heat baths and k denotes their vibrational modes with frequency $\omega_{\alpha k}$. The second term of Equation (27) is the Hamiltonian of the heat bath. The last term represents the interaction between the central region and the baths, given by coupling vectors $\mathbf{V}_{\alpha k}$ which are assumed to vanish before time $t_0 \to -\infty$. Written in this form, the coupling leads to a renormalization of the bare force-constant matrix, which can be canceled by the above introduced counterterm $\mathbf{K}_{\text{ct}} = \sum_{\alpha,k} (\mathbf{V}_{\alpha k} \cdot \mathbf{V}_{\alpha k}^T) / \omega_{\alpha k}^2$ [80,147]. Hence, the coupling to the thermal baths will introduce dissipation but no shift in the vibrational spectrum [148]. The equations of motion (EOM) for the central region ($\mathbf{u}$ and $\mathbf{p}$) and normal modes of one lead (u_k) read [80,147]:

$$\frac{\partial}{\partial t} \begin{pmatrix} \mathbf{u} \\ \mathbf{p} \end{pmatrix} = \mathcal{K}_{\text{eff}} \cdot \begin{pmatrix} \mathbf{u} \\ \mathbf{p} \end{pmatrix} - \sum_k u_k \begin{pmatrix} 0 \\ \mathbf{V}_k \end{pmatrix} . \tag{28}$$

The $2N \times 2N$ dimensional auxiliary matrices (denoted by caligraphic symbols) are given by the expressions [80,147]:

$$\mathcal{I} \equiv \begin{pmatrix} \mathbf{I} & 0 \\ 0 & \mathbf{I} \end{pmatrix}, \quad \mathcal{Q} \equiv \begin{pmatrix} 0 & \mathbf{I} \\ -\mathbf{I} & 0 \end{pmatrix}, \quad \mathcal{K}_{\text{eff}} \equiv \begin{pmatrix} 0 & \mathbf{I} \\ -\mathbf{K}_{\text{eff}} & 0 \end{pmatrix} \equiv \begin{pmatrix} 0 & \mathbf{I} \\ -\mathbf{K} & 0 \end{pmatrix} + \begin{pmatrix} 0 & 0 \\ -\mathbf{K}_{\text{ct}} & 0 \end{pmatrix} .$$

The energy of the central region can be written in terms of the phonon density matrix $\sigma(t) = i\mathcal{G}^<(t,t)$, with $\mathcal{G}^<(t,t)$ being a lesser Green's function [80,147]:

$$E_C(t) = \frac{1}{2} \text{Tr} \left\{ \mathcal{K}_{\text{eff}}^T \cdot \mathcal{Q} \cdot \sigma(t) \right\}, \tag{29}$$

and the total energy is $E_{Tot}(t) = E_C(t) + E_{bath}(t)$. In the absence of external forces, the total energy is conserved, and the heat current coming from the heat baths can be defined as $J(t) = -\frac{\partial}{\partial t}E_C(t)$. Hereafter, $\hbar = 1$ is taken. The time evolution of the heat flux is related to the lesser GF as [62,63]:

$$\mathcal{G}^<(t,t') = -i \begin{pmatrix} \langle \mathbf{u}(t')\mathbf{u}^T(t) \rangle & \langle \mathbf{u}(t')\mathbf{p}^T(t) \rangle \\ \langle \mathbf{p}(t')\mathbf{u}^T(t) \rangle & \langle \mathbf{p}(t')\mathbf{p}^T(t) \rangle \end{pmatrix}. \tag{30}$$

To obtain the time dependence of the lesser GF, Dyson's equation was derived (see Ref. [80] for details). Using Langreth's rules, the lesser GF can be written in terms of retarded and advanced GFs:

$$\mathcal{G}^<(t,t') = \int d\tau_2 \int d\tau_3\, \mathcal{G}^R(t,\tau_2) \cdot \mathcal{S}^<(\tau_2,\tau_3) \cdot \mathcal{G}^A(\tau_3,t'). \tag{31}$$

For the latter, EOMs are given by [80,147]:

$$\frac{\partial}{\partial t}\mathcal{G}^R(t,t') = \delta(t,t')\mathcal{Q} + \mathcal{K}_{eff}\cdot\mathcal{G}^R(t,t') + \mathcal{Q}\cdot\int dt_2\,\mathcal{S}^R(t,t_2)\cdot\mathcal{G}^R(t_2,t'). \tag{32}$$

$$\frac{\partial}{\partial t'}\mathcal{G}^A(t,t') = \delta(t,t')\mathcal{Q}^T + \mathcal{G}^A(t,t')\cdot\mathcal{K}_{eff}^T + \int dt_2\,\mathcal{G}^A(t,t_2)\cdot\mathcal{S}^A(t_2,t')\cdot\mathcal{Q}^T. \tag{33}$$

$\mathcal{S}^{R,A,<}$ denotes the respective self-energy. Setting $t' \to t$ in Equation (31), the EOM for lesser GF becomes:

$$\begin{aligned}
\frac{\partial}{\partial t}\mathcal{G}^<(t,t) = {}& \mathcal{K}_{eff}\cdot\mathcal{G}^<(t,t) + \mathcal{G}^<(t,t)\cdot\mathcal{K}_{eff}^T \\
& + \mathcal{Q}\cdot\left[\int_{\tau_2}^t d\tau_2\,\left(\mathcal{S}^>(t,\tau_2)\mathcal{G}^<(\tau_2,t) - \mathcal{S}^<(t,\tau_2)\mathcal{G}^>(\tau_2,t)\right)\right] \\
& + \left[\int_{\tau_2}^t d\tau_2\,\left(\mathcal{G}^>(t,\tau_2)\mathcal{S}^<(\tau_2,t) - \mathcal{G}^<(t,\tau_2)\mathcal{S}^>(\tau_2,t)\right)\right]\cdot\mathcal{Q}^T,
\end{aligned} \tag{34}$$

where the thermal current matrices $\Pi_\alpha(t)$ are defined as:

$$\Pi_\alpha(t) = \int_{\tau_2}^t d\tau_2\,\left(\mathcal{G}^>(t,\tau_2)\mathcal{S}_\alpha^<(\tau_2,t) - \mathcal{G}^<(t,\tau_2)\mathcal{S}_\alpha^>(\tau_2,t)\right). \tag{35}$$

For harmonic baths, $\mathcal{S}_\alpha^{<,>}(t,t')$ can be obtained as:

$$\mathcal{S}_\alpha^{\lessgtr}(t,t') = -i\int_0^\infty \frac{d\omega}{\pi}\left[\coth\frac{\omega}{2k_B T_\alpha}\cos\omega(t-t') \pm i\sin\omega(t-t')\right]\mathcal{L}_\alpha(\omega). \tag{36}$$

T_α is the bath temperature and $\mathcal{L}_\alpha(\omega)$ is the spectral density of reservoir α [148,149]. The EOM for the phonon density matrix reads [80,147]:

$$\frac{\partial}{\partial t}\sigma(t) = \mathcal{K}_{eff}\cdot\sigma(t) + \sigma(t)\cdot\mathcal{K}_{eff}^T + i\sum_{\alpha\in\{L,R\}}\left(\Pi_\alpha(t)\cdot\mathcal{Q}^T - h.c.\right). \tag{37}$$

3.1. Auxiliary-Mode Approach

An auxiliary-mode approach was used to expand the self-energies $\mathcal{S}_\alpha^{<,>}(t,t)$ in exponential functions to achieve a numerically efficient implementation to calculate the time evolution of $\Pi_\alpha(t)$. The approach has been previously implemented for electrons [55–57] and for vibrations [150,151]. Since the cos and sin functions in Equation (36) can be easily expressed in terms of exponential functions, the only term which has to be considered is the hyperbolic cotangent. Different schemes have been proposed to obtain a suitable pole decomposition of this term [150]. To bypass the slow convergence of the so-called Matsubara decomposition, a more advanced pole decomposition was suggested by

Croy and Saalmann [55], based on a partial fraction decomposition method that displays a faster convergence. However, one needs in this method high-precision arithmetic to compute the poles correctly, so that it is of advantage to have a pole decomposition of the coth function with purely imaginary poles [80]. A Pade decomposition method was recently proposed by Hu et al. [152], which shows very rapid convergence: the hyperbolic cotangent can be written in terms of simple poles as [150]:

$$\coth(x) \approx \frac{1}{x} + \sum_{p=1}^{N_P} \eta_p \left(\frac{1}{x - \zeta_p} + \frac{1}{x - \zeta_p^*} \right), \tag{38}$$

where η_p are residues and ζ_p (with $\mathrm{Im}\,\zeta_p > 0$) are the poles. Here, N_P denotes the number of poles.

Following this auxiliary-mode approach, a spectral density with a Drude regularization (i.e., adding a cut-off frequency ω_c) of the form $\mathcal{L}_\alpha(\omega) = (\omega_c^2 \omega)/(\omega^2 + \omega_c^2)\mathcal{L}_\alpha^{(0)}$ was considered [148]. $\mathcal{L}_\alpha^{(0)} \equiv \mathrm{diag}(\Lambda_\alpha^{(0)}, 0)$ contains the coupling matrix $\Lambda_\alpha^{(0)}$ between the central domain and the leads. This quantity contains the wide-band limit as a special case for $\omega_c \to \infty$. Using a linear combination of Lorentzians, any spectral density can in principle be approximated. The Drude spectral density is inserted into the expression for the lesser/greater self-energies (see Equation (36)), and for $\tau = t - t' > 0$, it turns into [80,147]:

$$\mathcal{S}_\alpha^{\lessgtr}(t,t') = \mathcal{L}_\alpha^{(0)} \left[C_\alpha(\tau) \pm \mathrm{sgn}(\tau)\frac{\omega_c^2}{2}e^{-\omega_c|\tau|} \right]. \tag{39}$$

with:

$$C_\alpha(\tau) = -i \int_{-\infty}^{\infty} \frac{d\omega}{2\pi} \left(\frac{\omega_c^2 \omega}{\omega^2 + \omega_c^2} \right) \coth\frac{\omega}{2k_B T_\alpha} e^{i\omega\tau}, \tag{40}$$

As mentioned above, the goal of the auxiliary-mode approach is to expand the self-energies and $C_\alpha(\tau)$ in terms of exponentials. Using the Pade decomposition of the hyperbolic cotangent in Equation (40), one obtains [80,147]:

$$C_\alpha(\tau) = -ik_B T_\alpha \omega_c e^{-\omega_c \tau} - i\sum_{p=1}^{N_P} R_{\alpha,p} \left(\omega_c e^{-\omega_c \tau} - \chi_{\alpha,p} e^{-\chi_{\alpha,p}\tau} \right),$$

with $R_{\alpha,p} = \dfrac{2k_B T_\alpha \omega_c^2}{\omega_c^2 - \chi_{\alpha,p}^2}\eta_p$ and $\chi_{\alpha,p} = -i2k_B T_\alpha \zeta_p$. Hence:

$$\mathcal{S}_\alpha^{<,>}(t,t') = \frac{\partial}{\partial t}\sum_{p=0}^{N_P} a_{\alpha,p}^{<,>} e^{-b_{\alpha,p}(t-t')}\mathcal{L}_\alpha^{(0)} = \frac{\partial}{\partial t}\mathcal{N}^{<,>}(t,t') \tag{41}$$

where the coefficients $a_{\alpha,p}^{<,>}$ and $b_{\alpha,p}$ are related to the auxiliary-mode decomposition. For $\tau < 0$, the coefficients are $a_{\alpha,p}^{*,<,>}$ and $b_{\alpha,p}^*$ [147].

The phonon current matrices $\Pi_\alpha(t)$ (see Equation (35)) can be written as [80,147]:

$$\Pi_\alpha(t) = \sum_{p=0}^{N_P} \left[\Phi_\alpha^p(t) + a_{\alpha,p}^{*,<}\mathcal{Q}\cdot\mathcal{L}_\alpha^{(0)} \right], \tag{42}$$

with:

$$\Phi_\alpha^p(t) = \int_{t_0}^{t} dt' \left(\mathcal{G}^<(t,t')\cdot\mathcal{K}_{\mathrm{eff}}^T\cdot\mathcal{N}_{\alpha,p}^>(t',t) - \mathcal{G}^>(t,t')\cdot\mathcal{K}_{\mathrm{eff}}^T\cdot\mathcal{N}_{\alpha,p}^<(t',t) \right),$$

The EOM for $\Phi_\alpha^p(t)$ is given by [80,147]:

$$\frac{\partial}{\partial t}\Phi_\alpha^p(t) = \mathcal{A}_\alpha^p \cdot \Phi_\alpha^p(t,t) - \mathcal{B}_\alpha^p \cdot \mathcal{K}_{\text{eff}}^T \cdot \mathcal{L}_\alpha^{(0)} + \mathcal{Q} \cdot \Omega_\alpha^p(t),\tag{43}$$

with $\mathcal{A}_\alpha^p = \mathcal{K} - b_{\alpha,p}^*\mathcal{I}$, $\mathcal{B}_\alpha^p = i\omega_c\delta_{p,0}\sigma(t) + a_{\alpha,p}^{*,<}\mathcal{Q}$. The functions $\Omega_\alpha^p(t) = \sum_{\alpha'}\sum_{p'=0}^{N_p}\Omega_{\alpha'\alpha}^{p'p}(t)$ correlate the main features of both heat baths and their values are obtained by performing the time derivative of $\Omega_{\alpha'\alpha}^{p'p}(t)$, which is expressed as [147]:

$$\frac{\partial}{\partial t}\Omega_{\alpha'\alpha}^{p'p}(t) = C_{\alpha'\alpha}^{p'p}\mathcal{L}_{\alpha'}^{(0)} \cdot \mathcal{Q} \cdot \mathcal{K}_{\text{eff}}^T \cdot \mathcal{L}_\alpha^{(0)} - D_{\alpha'\alpha}^{p'p}\Omega_{\alpha'\alpha}^{p'p}(t)$$

$$- \omega_c\left(\delta_{p',0}\mathcal{L}_{\alpha'}^{(0)} \cdot \mathcal{K}_{\text{eff}} \cdot \Phi_\alpha^p(t) - \delta_{p,0}\Phi_{\alpha'}^{p',\dagger}(t) \cdot \mathcal{K}_{\text{eff}}^T \cdot \mathcal{L}_\alpha^{(0)}\right),\tag{44}$$

where $C_{\alpha'\alpha}^{p'p} = a_{\alpha',p'}^{<}a_{\alpha,p}^{*,>} - a_{\alpha',p'}^{>}a_{\alpha,p}^{*,<}$ and $D_{\alpha'\alpha}^{p'p} = b_{\alpha',p'} + b_{\alpha,p}^*$ are real numbers. Thus, the initial integro-differential equation for the reduced density matrix has been mapped onto a closed set of ordinary differential equations (Equations (37), (43), and (44)), which can be solved using, e.g., a fourth-order Runge–Kutta method [80,147].

The thermal current is calculated as [80,147]:

$$J(t) = -\frac{i}{2}\text{Tr}\left\{\mathcal{K}_{\text{eff}}^T \cdot \mathcal{Q} \cdot \sum_\alpha\left[\Pi_\alpha(t) \cdot \mathcal{Q}^T - h.c.\right]\right\}.\tag{45}$$

3.2. Applications of TD-NEGF Approach

3.2.1. Proof-of-Principle: One-Dimensional Atomic Chain

To illustrate this approach, a one-dimensional (1D) chain with N atoms interacting via force constants with strength $\lambda = 1.0$ eV/μÅ was considered (see Figure 9a) for $N = 4$ [80,147]. The atomic masses of the atoms was set to 1.0 μ, and only nearest-neighbor interactions were considered.

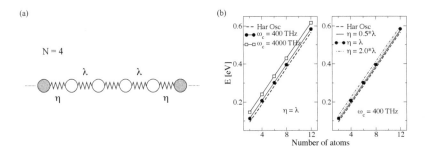

Figure 9. (a) Scheme of the one-dimensional atomic chain studied in this work for the case of $N = 4$ atoms, with the filled atoms representing the beginning of the heat baths. λ and η are the spring force constants between the atoms and the coupling of the central region to the baths. Time-dependent NEGF approach: (b) Variation of the total energy of a dimer at $T_0 = 300$ K after increasing the number of atoms in the one-dimensional atomic chain for different cut-off frequency (left) and η parameter (right). For comparison, we also plotted the energy values corresponding to the ideal harmonic oscillator case (dashed lines). Panel (b) is reproduced with permission from Ref. [147]. Copyright 2018 American Chemical Society.

First, a benchmark of the influence of spectral density parameters on the steady state properties was carried out. In Figure 9b, the dependence of the system energy at 300 K on N for different cut-off frequency ω_c and η parameter is displayed [80,147]. The energy is compared to that obtained for an ideal harmonic oscillator in equilibrium, $E_C = \sum_{i=1}^{N} \hbar\omega_i \left(n(\omega_i) + 1/2\right)$, with $n(\omega)$ being the Bose–Einstein distribution function and ω_i are the frequencies of the isolated central system. For fixed $\eta = \lambda$, increasing ω_c leads to an increase of the system energy, since the spectral density includes more vibrational states. Contrarily, the energy gets closer to the harmonic oscillator values after reducing η for fixed $\omega_c = 400$ THz. This phenomena is expected because, for $\eta \to 0$, the chain can be considered as an isolated harmonic system with no heat injection from the bath and $E_{1Dchain} = E_C$.

A similar effect is also found by analyzing the phonon transmission $\tau_{ph}(\omega)$ calculated along the lines of Section 2.1. Using the spectral density $\Lambda(\omega) = (\omega_c^2\omega)/(\omega^2 + \omega_c^2)\Lambda^{(0)}$, one gets [147]:

$$\Sigma_\alpha^{\lessgtr}(t, t') = -i \int_{-\infty}^{\infty} \frac{d\omega}{2\pi} \left(\frac{\omega_c^2\omega}{\omega^2 + \omega_c^2} \right) \Lambda_\alpha^{(0)} \left[\coth \frac{\omega}{2k_B T_\alpha} \pm 1 \right] e^{i\omega(t-t')}. \tag{46}$$

Hence, the retarded self-energy in time domain is written as:

$$\Sigma_\alpha^r(t, t') = \Theta(t - t') 2i\Lambda_\alpha^{(0)} \int_{-\infty}^{\infty} \frac{d\omega}{2\pi} \left(\frac{\omega_c^2\omega}{\omega^2 + \omega_c^2} \right) e^{i\omega(t-t')}. \tag{47}$$

By performing a Fourier transform of the previous equation and considering the counter-term, the self-energies are given by:

$$\Sigma_\alpha^r(\omega) = \left[i\frac{\omega_c^2\omega}{\omega^2 + \omega_c^2} \right] \Lambda_\alpha^{(0)} = [\Sigma_\alpha^a(\omega)]^\dagger. \tag{48}$$

The retarded Green-function finally reads:

$$G^r(\omega) = \left[\omega^2\mathbf{I} - \mathbf{K} - \Sigma_L^r(\omega) - \Sigma_R^r(\omega) \right]^{-1}, \tag{49}$$

where $\mathbf{K}$ is the force constant matrix of the one-dimensional chain. The transmission function $\tau_{ph}(\omega)$ is computed with Equation (23) and the steady heat flux is calculated by using the Landauer approach (see Equation (22)).

Figure 10a shows $\tau_{ph}(\omega)$ for different N with a cut-off at 400 THz. New transmission peaks appear for larger N due to emergence of new vibrational modes [80]. In addition, it was found that the maximum frequency ω_{max} with non-zero transmission depends on N. Thus, for $N > 8$, ω_{max} remains constant ($\sim$195 THz). For $N \to \infty$, τ_{ph} is constant and is zero for $\omega > \omega_{max}$, i.e., all modes have the same transmission probability $\tau_{ph} = 1.0$ (see also [33]). Figure 10b shows the influence of η for a dimer with $\omega_c = 400$ THz. η has a considerable influence on the phonon transmission. For $\eta \geq \lambda$, τ_{ph} is similar to a Gaussian and the dimer cannot be understood as a weakly coupled system anymore. For $\eta \leq 0.8\lambda$, on the contrary, two main transmission peaks corresponding to the two dimer modes can be resolved. For even weaker coupling, τ_{ph} will yield two delta functions at these frequencies, correctly describing the vibrations of the system. This result confirms the analysis carried out for the system energy and shown in Figure 9b. The influence of the cut-off frequency on τ_{ph} is weak compared to η: increasing ω_c ten times only leads to a slight reduction of the phonon transmission at high frequencies and the frequency spectrum becomes wider (see Figure 10c) [80,147].

Based on the Landauer formalism, the temperature of the heat baths only appears in the Bose–Einstein distribution (see Equation (22)). However, in the TD-NEGF approach, the temperatures appear in the auxiliary-mode expansion of the self-energy. Once the system is in thermal equilibrium, a symmetric temperature bias $\Delta T = T_L - T_R = 2\xi T_0$ ($\xi > 0$) is applied, with T_0 being the mean temperature at which the system was previously equilibrated [80,147]. The left and right baths temperatures are expressed as $T_L = (1 + \xi)T_0$ and $T_R = (1 - \xi)T_0$. Figure 11a shows the steady heat flux as a function of N for different η ($\omega_c = 400$ THz). The heat flux values were obtained for a mean temperature of $T_0 = 300$ K and $\xi = 0.1$. The values in both methods become closer after reducing

η in agreement with the above discussed results in Figures 9b and 10b. Additionally, the heat flux converges for a given N for all ηs [80,147].

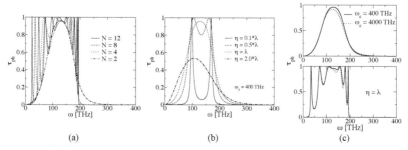

Figure 10. Landauer approach: (**a**) Variation of the phonon transmission function τ_{ph} of an one-dimensional atomic chain as a function of the number of atoms. (**b**) Influence of the coupling parameter on the phonon transmission function τ_{ph} of an atomic dimer. (**c**) Variation of τ_{ph} with respect to the cut-off frequency for $N = 2$ (**top**) and $N = 4$ (**bottom**). The figure is reproduced with permission from Ref. [147]. Copyright 2018 American Chemical Society.

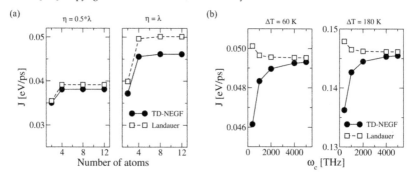

Figure 11. Landauer approach: (**a**) Steady heat flux as a function of the number of atoms in the one-dimensional atomic chain for different η values. (**b**) Cut-off frequency w_c dependence of the steady heat flux for the atomic dimer at various temperatures bias ΔT. For comparison, we also plotted the values obtained using Landauer approach. The figure is reproduced with permission from Ref. [147]. Copyright 2018 American Chemical Society.

The influence of w_c on the steady heat flux was studied for $\eta = \lambda$ (see Figure 11b). One sees that each approach displays a different behavior, i.e., the heat flux increases and decreases after increasing w_c for the TD-NEFG method and the Landauer approach, respectively [80,147]. This effect can be tuned by the value of ΔT. However, independently of ΔT, the heat flux for both approaches become closer with increasing w_c.

3.2.2. Atomistic System: Carbon-Based Molecular Junctions

Medrano Sandonas et al. [147] showed that the TD-NEGF method reviewed in this section can be combined with atomistic methodologies for addressing the phonon dynamics in real systems. Due to the larger number of degrees of freedom, the matrix dimensions considerably increase and, hence, the computational cost. As typical examples, the work in Ref. [147] focused on poly-acetylene (PA, 4 atoms) and poly-ethylene (PE, 6 atoms) dimers connected to thermal baths (see Figure 12a for the case of PA dimer). The main difference between the two systems was the presence of double C bonds in PA compared with single bonds in PE. Both structural optimization and force constant calculations were performed with the Gaussian09 code [153]. $\Lambda_\alpha^{(0)}$ will thus take values corresponding to realistic bonds between the central region and the reservoirs (for more details, see [80]). For both junctions, a cut-off $\omega_c = 100$ THz was used (roughly two times the maximum frequency of the vibrational spectrum). The number of poles in the auxiliary-mode expansion at 100 K, 300 K, and 500 K was 10, 8, and 4, correspondingly [80,147].

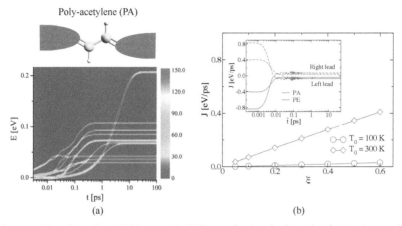

Figure 12. Time-dependent NEGF approach: (**a**) Energy density plot for molecular junctions made of poly-acetylene (PA) dimer. (**b**) Variation of the steady-state heat flux as a function of ξ for the PA dimer at different T_0. Inset: Dynamics of the heat flux for both leads for PA and PE dimers after applying a temperature bias of $\Delta T = 60$ K at $T_0 = 300$ K. The figure is reproduced with permission from Ref. [147]. Copyright 2018 American Chemical Society.

To gain a deeper understanding of the thermal properties in the transient state, the energy density $D(E, t)$ was defined as:

$$D(E,t) = \sum_{i=1}^{N} (1/\gamma\sqrt{2\pi}) \exp\left[-\left((E - E_i)/\gamma\sqrt{2}\right)^2\right] \tag{50}$$

with $\gamma = 0.001$ eV and $\{E_i\}$ the set of eigenvalues of $\mathcal{Z}(t)$. In Figure 12a, the results for PA dimer during thermal equilibration at $T_0 = 300$ K are displayed. As shown in the figure, all modes of the $\mathcal{Z}(t)$ matrix display very low energy at the beginning of the transient. The lowest lying modes gain then energy and reach a maximum at equilibrium. However, the eigenvalues in PE need a longer time to converge as compared to PA [147]. This difference arises from the different coupling strengths to the leads (related to the matrices $\Lambda_\alpha^{(0)}$). Consequently, the magnitude of the oscillations in the heat flux during the transient after applying a temperature bias is different, being larger for PA, as shown in the inset of Figure 12b. This difference in covalently bonded configurations also leads to a larger heat flux for the PA dimer [147]. Moreover, in agreement with linear response, the heat flux behaves nearly linear in ξ for different mean temperatures T_0 (see Figure 12b) [80,147].

4. Summary and Outlook

In the current review, we address selected applications of our recent implementations of quantum transport methodologies in low-dimensional materials. Hereby, we highlight the possibility to perform systematic investigations with atomic resolution, thus addressing material-specific problems for designing potential (nano)phononic devices.

We combined the NEGF formalism with the DFTB methodology to address quantum ballistic transport in various low-dimensional materials with atomistic resolution. This computational approach is implemented as a tool in the DFTB+ software. Although these systems may also be tractable using classical molecular dynamics, extensive parameterizations may be required to study different material combinations (here, machine learning approaches may be of interest). It is therefore more suitable to use the NEGF-DFTB approach, where the chemistry of the problem is naturally included in the first-principle calculation of the Hessian matrix. We showed that 2D puckered materials display strong thermal anisotropy due to their atomic structure, thus transporting heat preferably along the zigzag direction (higher phonon group velocity). As a next application, the influence of BN concentration and defect distribution on the thermal transport of BNC heteronanotubes was considered. Independently of the specific spatial BN distribution, the phonon transmission of pristine (6,6)-CNT is reduced at high frequencies after increasing the BN concentration. As a last application, we demonstrated that the vibrational features of molecular junctions can be exploited in conjunction with an appropriate choice of nanoscale thermal baths to implement a molecule-based phononic filter. This model offers the possibility of engineering different phonon filters based on the rich molecular chemical space. These three reviewed studies clearly demonstrate the potential of the PHONON tool to investigate nanoscale ballistic phonon transport.

In the last section, we present an atomistic method combining time-dependent NEGF with a first-principle based modeling to address phonon dynamics in nanoscale systems. The method is based on solving the equation of motion of the phonon density matrix with an efficient auxiliary-mode approach. The approach was applied to study thermal transport in the transient regime of a 1D chain, providing results in agreement with the Landauer formalism. By using density-functional theory to obtain the force constants and coupling matrices, the phonon dynamics of small molecular junctions was considered. Although the presented study is based on a Drude regularization of the spectral density, realistic scenarios can be easily addressed. This computational approach builds one of the first attempts to deal with time-dependent quantum phonon transport and it will allow studying various topical questions such as heat pumping, on a fully atomistic basis.

We are, however, not yet able to address physical effects such as thermal rectification from a fully quantum picture. Although rectification can be induced by structural asymmetries, phonon–phonon interactions play a dominant role, too. The latter are also crucial when dealing with phonon transport at high temperatures. An implementation combining NEGF with first-principles requires, besides computing the dynamical matrix as the basic input, third and fourth order anharmonic coefficients as well [76,78]. They contribute additional self-energies in the Green's functions of the scattering region, and involve convolutions in frequency space of two-and three phonon Green's functions. As a result, the problem needs to be solved self-consistently, thus considerably increasing the computational effort.

Another issue is the inclusion of electron–phonon coupling in the description of heat transport. Although it has already been implemented within the NEGF approach to address electronic transport [154–158], there are not many atomistic-based studies related to their impact on phonon transport. Since the interaction with the electronic system will provide an additional energy exchange channel, it will be of interest to elucidate how some of the effects discussed in this review as well as in other investigations, such as thermal rectification and phonon filtering, will be modified by the inclusion of electron–phonon interactions.

Author Contributions: L.M.S. wrote the original draft. R.G. wrote and reviewed the final version. A.P., A.C., and G.C. critically revised the manuscript by providing inspiring comments.

Funding: This research was funded by the Deutscher Akademischer Austauschdienst (DAAD) within its doctoral programme scholarship. This work was also partly supported by the German Research Foundation (DFG) within the Cluster of Excellence "Center for Advancing Electronics Dresden".

Acknowledgments: The authors acknowledge very fruitful discussions with Arezoo Dianat, Vladimiro Mujica, and Alvaro Rodriguez Mendez. We acknowledge the Center for Information Services and High Performance Computing (ZIH) at TU Dresden for computational resources.

Conflicts of Interest: The authors declare no conflict of interest.

Abbreviations

The following abbreviations are used in this manuscript:

CNT   Carbon nanotube
DFT   Density functional theory
DFTB   Density functional tight-binding
DOS   Density of states
EOM   Equation of motion
GF   Green's functions
LDOS   Local density of states
NEGF   Non-equilibrium Green's functions
NEMD   Non-equilibrium molecular dynamics
MD   Molecular dynamics
PA   Poly-acetylene
PE   Poly-ethylene
TD   Time-dependent

References

1. Moore, A.L.; Shi, L. Emerging challenges and materials for thermal management of electronics. *Mater. Today* **2014**, *17*, 163–174.
2. Li, N.; Ren, J.; Wang, L.; Zhang, G.; Hänggi, P.; Li, B. Colloquium: Phononics: Manipulating heat flow with electronic analogs and beyond. *Rev. Mod. Phys.* **2012**, *84*, 1045–1066.
3. Dubi, Y.; Di Ventra, M. Colloquium: Heat flow and thermoelectricity in atomic and molecular junctions. *Rev. Mod. Phys.* **2011**, *83*, 131–155.
4. Volz, S.; Ordonez-Miranda, J.; Shchepetov, A.; Prunnila, M.; Ahopelto, J.; Pezeril, T.; Vaudel, G.; Gusev, V.; Ruello, P.; Weig, E.M.; et al. Nanophononics: State of the art and perspectives. *Eur. Phys. J. B* **2016**, *89*, 15.
5. Balandin, A.A.; Nika, D.L. Phononics in low-dimensional materials. *Mater. Today* **2012**, *15*, 266–275.
6. Sklan, S.R. Splash, pop, sizzle: Information processing with phononic computing. *AIP Adv.* **2015**, *5*, 053302.
7. Pop, E. Energy dissipation and transport in nanoscale devices. *Nano Res.* **2010**, *3*, 147–169.
8. Yan, Z.; Liu, G.; Khan, J.M.; Balandin, A.A. Graphene quilts for thermal management of high-power GaN transistors. *Nat. Commun.* **2012**, *3*, 827.
9. Snyder, G.J.; Toberer, E.S. Complex thermoelectric materials. *Nat. Mater.* **2008**, *7*, 105.
10. Biswas, K.; He, J.; Blum, I.D.; Wu, C.I.; Hogan, T.P.; Seidman, D.N.; Dravid, V.P.; Kanatzidis, M.G. High-performance bulk thermoelectrics with all-scale hierarchical architectures. *Nature* **2012**, *489*, 414.
11. Aspelmeyer, M.; Kippenberg, T.J.; Marquardt, F. Cavity optomechanics. *Rev. Mod. Phys.* **2014**, *86*, 1391–1452.
12. McNeil, R.P.G.; Kataoka, M.; Ford, C.J.B.; Barnes, C.H.W.; Anderson, D.; Jones, G.A.C.; Farrer, I.; Ritchie, D.A. On-demand single-electron transfer between distant quantum dots. *Nature* **2011**, *477*, 439.
13. Hermelin, S.; Takada, S.; Yamamoto, M.; Tarucha, S.; Wieck, A.D.; Saminadayar, L.; Bäuerle, C.; Meunier, T. Electrons surfing on a sound wave as a platform for quantum optics with flying electrons. *Nature* **2011**, *477*, 435.
14. Rossignol, C.; Chigarev, N.; Ducousso, M.; Audoin, B.; Forget, G.; Guillemot, F.; Durrieu, M.C. In-vitro picosecond ultrasonics in a single cell. *Appl. Phys. Lett.* **2008**, *93*, 123901.
15. Rego, L.G.C.; Kirczenow, G. Quantized thermal conductance of dielectric quantum wires. *Phys. Rev. Lett.* **1998**, *81*, 232–235.
16. Schwab, K.; Henriksen, E.A.; Worlock, J.M.; Roukes, M.L. Measurement of the quantum of thermal conductance. *Nature* **2000**, *404*, 974–977.
17. Cui, L.; Jeong, W.; Hur, S.; Matt, M.; Klöckner, J.C.; Pauly, F.; Nielaba, P.; Cuevas, J.C.; Meyhofer, E.; Reddy, P. Quantized thermal transport in single-atom junctions. *Science* **2017**, *355*, 1192–1195.
18. Mosso, N.; Drechsler, U.; Menges, F.; Nirmalraj, P.; Karg, S.; Riel, H.; Gotsmann, B. Heat transport through atomic contacts. *Nat. Nanotechnol.* **2017**, *12*, 430.

19. Tavakoli, A.; Lulla, K.; Crozes, T.; Mingo, N.; Collin, E.; Bourgeois, O. Heat conduction measurements in ballistic 1D phonon waveguides indicate breakdown of the thermal conductance quantization. *Nat. Commun.* **2018**, *9*, 4287.

20. Sikora, A.; Ftouni, H.; Richard, J.; Hébert, C.; Eon, D.; Omnés, F.; Bourgeois, O. Highly sensitive thermal conductivity measurements of suspended membranes (SiN and diamond) using a 3ω-Völklein method. *Rev. Sci. Instrum.* **2012**, *83*, 054902; Erratum in **2013**, *84*, 029901

21. Regner, K.T.; Sellan, D.P.; Su, Z.; Amon, C.H.; McGaughey, A.J.H.; Malen, J.A. Broadband phonon mean free path contributions to thermal conductivity measured using frequency domain thermoreflectance. *Nat. Commun.* **2013**, *4*, 1640.

22. Cahill, D.G.; Ford, W.K.; Goodson, K.E.; Mahan, G.D.; Majumdar, A.; Maris, H.J.; Merlin, R.; Phillpot, S.R. Nanoscale thermal transport. *J. Appl. Phys.* **2003**, *93*, 793–818.

23. Kim, K.; Jeong, W.; Lee, W.; Reddy, P. Ultra-high vacuum scanning thermal microscopy for nanometer resolution quantitative thermometry. *ACS Nano* **2012**, *6*, 4248–4257.

24. Kim, K.; Chung, J.; Won, J.; Kwon, O.; Lee, J.S.; Park, S.H.; Choi, Y.K. Quantitative scanning thermal microscopy using double scan technique. *Appl. Phys. Lett.* **2008**, *93*, 203115. doi:10.1063/1.3033545.

25. Zhou, Y.; Fan, Z.; Qin, G.; Yang, J.Y.; Ouyang, T.; Hu, M. Methodology perspective of computing thermal transport in low-dimensional materials and nanostructures: The old and the new. *ACS Omega* **2018**, *3*, 3278–3284.

26. Kubo, R. Statistical-mechanical theory of irreversible processes. I. General theory and simple applications to magnetic and conduction problems. *J. Phys. Soc. Jpn.* **1957**, *12*, 570–586.

27. Müller-Plathe, F. A simple nonequilibrium molecular dynamics method for calculating the thermal conductivity. *J. Chem. Phys.* **1997**, *106*, 6082–6085.

28. Jund, P.; Jullien, R. Molecular-dynamics calculation of the thermal conductivity of vitreous silica. *Phys. Rev. B* **1999**, *59*, 13707–13711.

29. Li, W.; Carrete, J.; Katcho, N.A.; Mingo, N. ShengBTE: A solver of the Boltzmann transport equation for phonons. *Comp. Phys. Commun.* **2014**, *185*, 1747–1758.

30. Fugallo, G.; Lazzeri, M.; Paulatto, L.; Mauri, F. Ab initio variational approach for evaluating lattice thermal conductivity. *Phys. Rev. B* **2013**, *88*, 045430.

31. Dove, M.; Dove, M.; Hochella, M.; Liebermann, R.; Putnis, A. *Introduction to Lattice Dynamics*; Cambridge Topics in Mineral Physics and Chemistry; Cambridge University Press: Cambridge, UK, 1993.

32. Mingo, N.; Yang, L. Phonon transport in nanowires coated with an amorphous material: An atomistic Green's function approach. *Phys. Rev. B* **2003**, *68*, 245406.

33. Zhang, W.; Fisher, T.S.; Mingo, N. The atomistic Green's function method: An efficient simulation approach for nanoscale phonon transport. *Numer. Heat Transf. Part B Fundam.* **2007**, *51*, 333–349.

34. Zhang, W.; Mingo, N.; Fisher, T.S. Simulation of phonon transport across a non-polar nanowire junction using an atomistic Green's function method. *Phys. Rev. B* **2007**, *76*, 195429.

35. Wang, J.S.; Agarwalla, B.K.; Li, H.; Thingna, J. Nonequilibrium Green's function method for quantum thermal transport. *Front. Phys.* **2014**, *9*, 673–697.

36. Wingert, M.C.; Kwon, S.; Hu, M.; Poulikakos, D.; Xiang, J.; Chen, R. Sub-amorphous thermal conductivity in ultrathin crystalline silicon nanotubes. *Nano Lett.* **2015**, *15*, 2605–2611.

37. Wang, Y.; Vallabhaneni, A.; Hu, J.; Qiu, B.; Chen, Y.P.; Ruan, X. Phonon lateral confinement enables thermal rectification in asymmetric single-material nanostructures. *Nano Lett.* **2014**, *14*, 592–596.

38. Zhou, Y.; Gong, X.; Xu, B.; Hu, M. Decouple electronic and phononic transport in nanotwinned structures: A new strategy for enhancing the figure-of-merit of thermoelectrics. *Nanoscale* **2017**, *9*, 9987–9996.

39. Wang, J.; Wang, J.S. Dimensional crossover of thermal conductance in nanowires. *Appl. Phys. Lett.* **2007**, *90*, 241908.

40. Hu, J.; Ruan, X.; Chen, Y.P. Thermal conductivity and thermal rectification in graphene nanoribbons: A molecular dynamics study. *Nano Lett.* **2009**, *9*, 2730–2735.

41. Xu, Y.; Chen, X.; Wang, J.S.; Gu, B.L.; Duan, W. Thermal transport in graphene junctions and quantum dots. *Phys. Rev. B* **2010**, *81*, 195425.

42. Balandin, A.; Wang, K.L. Effect of phonon confinement on the thermoelectric figure of merit of quantum wells. *J. Appl. Phys.* **1998**, *84*, 6149–6153.

43. Kazan, M.; Guisbiers, G.; Pereira, S.; Correia, M.R.; Masri, P.; Bruyant, A.; Volz, S.; Royer, P. Thermal conductivity of silicon bulk and nanowires: Effects of isotopic composition, phonon confinement, and surface roughness. *J. Appl. Phys.* **2010**, *107*, 083503.

44. Park, W.; Romano, G.; Ahn, E.C.; Kodama, T.; Park, J.; Barako, M.T.; Sohn, J.; Kim, S.J.; Cho, J.; Marconnet, A.M.; et al. Phonon conduction in silicon nanobeam labyrinths. *Sci. Rep.* **2017**, *7*, 6233.

45. Seol, J.H.; Jo, I.; Moore, A.L.; Lindsay, L.; Aitken, Z.H.; Pettes, M.T.; Li, X.; Yao, Z.; Huang, R.; Broido, D.; et al. Two-dimensional phonon transport in supported graphene. *Science* **2010**, *328*, 213–216.

46. Wagner, M.R.; Graczykowski, B.; Reparaz, J.S.; El Sachat, A.; Sledzinska, M.; Alzina, F.; Sotomayor Torres, C.M. Two-dimensional phononic crystals: Disorder matters. *Nano Lett.* **2016**, *16*, 5661–5668.

47. Anufriev, R.; Ramiere, A.; Maire, J.; Nomura, M. Heat guiding and focusing using ballistic phonon transport in phononic nanostructures. *Nat. Commun.* **2017**, *8*, 15505.

48. Segal, D.; Nitzan, A. Molecular heat pump. *Phys. Rev. E* **2006**, *73*, 026109.

49. Ren, J.; Hänggi, P.; Li, B. Berry-phase-induced heat pumping and its impact on the fluctuation theorem. *Phys. Rev. Lett.* **2010**, *104*, 170601.

50. Galperin, M.; Saito, K.; Balatsky, A.V.; Nitzan, A. Cooling mechanisms in molecular conduction junctions. *Phys. Rev. B* **2009**, *80*, 115427.

51. Segal, D.; Nitzan, A. Spin-boson thermal rectifier. *Phys. Rev. Lett.* **2005**, *94*, 034301.

52. Segal, D. Heat flow in nonlinear molecular junctions: Master equation analysis. *Phys. Rev. B* **2006**, *73*, 205415. doi:10.1103/PhysRevB.73.205415.

53. Wei, D.; Obstbaum, M.; Ribow, M.; Back, C.H.; Woltersdorf, G. Spin Hall voltages from AC and DC spin currents. *Nat. Commun.* **2014**, *5*, 3768.

54. Bocklage, L. Coherent THz transient spin currents by spin pumping. *Phys. Rev. Lett.* **2017**, *118*, 257202.

55. Croy, A.; Saalmann, U. Propagation scheme for nonequilibrium dynamics of electron transport in nanoscale devices. *Phys. Rev. B* **2009**, *80*, 245311.

56. Popescu, B.; Woiczikowski, P.B.; Elstner, M.; Kleinekathöfer, U. Time-dependent view of sequential transport through molecules with rapidly fluctuating bridges. *Phys. Rev. Lett.* **2012**, *109*, 176802.

57. Popescu, B.S.; Croy, A. Efficient auxiliary-mode approach for time-dependent nanoelectronics. *New J. Phys.* **2016**, *18*, 093044.

58. Kurth, S.; Stefanucci, G.; Almbladh, C.O.; Rubio, A.; Gross, E.K.U. Time-dependent quantum transport: A practical scheme using density functional theory. *Phys. Rev. B* **2005**, *72*, 035308.

59. Zheng, X.; Wang, F.; Yam, C.Y.; Mo, Y.; Chen, G. Time-dependent density-functional theory for open systems. *Phys. Rev. B* **2007**, *75*, 195127.

60. Oppenländer, C.; Korff, B.; Niehaus, T.A. Higher harmonics and ac transport from time dependent density functional theory. *J. Comput. Electron.* **2013**, *12*, 420–427.

61. Biele, R.; D'Agosta, R.; Rubio, A. Time-dependent thermal transport theory. *Phys. Rev. Lett.* **2015**, *115*, 056801.

62. Sena-Junior, M.I.; Lima, L.R.F.; Lewenkopf, C.H. Phononic heat transport in nanomechanical structures: Steady-state and pumping. *J. Phys. A Math. Theor.* **2017**, *50*, 435202.

63. Tuovinen, R.; Säkkinen, N.; Karlsson, D.; Stefanucci, G.; van Leeuwen, R. Phononic heat transport in the transient regime: An analytic solution. *Phys. Rev. B* **2016**, *93*, 214301.

64. Wang, J.S.; Wang, J.; Lü, J.T. Quantum thermal transport in nanostructures. *Eur. Phys. J. B* **2008**, *62*, 381–404.

65. Sevinçli, H.; Roche, S.; Cuniberti, G.; Brandbyge, M.; Gutierrez, R.; Medrano Sandonas, L. Green function, quasi-classical Langevin and Kubo–Greenwood methods in quantum thermal transport. *J. Phys. Condens. Matter* **2019**, *31*, 273003.

66. Stefanucci, G.; van Leeuwen, R. *Nonequilibrium Many-Body Theory of Quantum Systems: A Modern Introduction*; Cambridge University Press: Cambridge, UK, 2013.

67. Martin, P.C.; Schwinger, J. Theory of many-particle systems. I. *Phys. Rev.* **1959**, *115*, 1342–1373.

68. Kadanoff, L.; Baym, G. *Quantum Statistical Mechanics: Green's Function Methods in Equilibrium and Nonequilibrium Problems*; Frontiers in Physics; Benjamin, W.A., Ed.; Reading Mass: New York, NY, USA, 1962.

69. Keldysh, L.V. Diagram technique for nonequilibrium processes. *Zh. Eksp. Teor. Fiz.* **1964**, *47*, 1515–1527.

70. Wagner, M. Expansions of nonequilibrium Green's functions. *Phys. Rev. B* **1991**, *44*, 6104–6117.

71. Jauho, A.P.; Wingreen, N.S.; Meir, Y. Time-dependent transport in interacting and noninteracting resonant-tunneling systems. *Phys. Rev. B* **1994**, *50*, 5528–5544.

72. Datta, S. *Electronic Transport in Mesoscopic Systems*; Cambridge Studies in Semiconductor Physi; Cambridge University Press: Cambridge, UK, 1997.

73. Ozpineci, A.; Ciraci, S. Quantum effects of thermal conductance through atomic chains. *Phys. Rev. B* **2001**, *63*, 125415.

74. Yamamoto, T.; Watanabe, K. Nonequilibrium Green's function approach to phonon transport in defective carbon nanotubes. *Phys. Rev. Lett.* **2006**, *96*, 255503.

75. Dhar, A. Heat transport in low-dimensional systems. *Adv. Phys.* **2008**, *57*, 457–537.

76. Wang, J.S.; Wang, J.; Zeng, N. Nonequilibrium Green's function approach to mesoscopic thermal transport. *Phys. Rev. B* **2006**, *74*, 033408.

77. Wang, J.S.; Zeng, N.; Wang, J.; Gan, C.K. Nonequilibrium Green's function method for thermal transport in junctions. *Phys. Rev. E* **2007**, *75*, 061128.

78. Mingo, N. Anharmonic phonon flow through molecular-sized junctions. *Phys. Rev. B* **2006**, *74*, 125402.

79. Galperin, M.; Nitzan, A.; Ratner, M.A. Heat conduction in molecular transport junctions. *Phys. Rev. B* **2007**, *75*, 155312.

80. Medrano Sandonas, L. Computational Modeling of Thermal Transport in Low-Dimensional Materials. Ph.D. Thesis, Technische Universität Dresden, Dresden, Germany, July 2018.

81. Gang, Z. *Nanoscale Energy Transport and Harvesting: A Computational Study*; Jenny Stanford Publishing: Singapore, 2015.

82. Kubo, R. The fluctuation–dissipation theorem. *Rep. Prog. Phys.* **1966**, *29*, 255.

83. Velev, J.; Butler, W. On the equivalence of different techniques for evaluating the Green function for a semi-infinite system using a localized basis. *J. Phys. Condens. Matter* **2004**, *16*, R637.

84. Sancho, M.P.L.; Sancho, J.M.L.; Sancho, J.M.L.; Rubio, J. Highly convergent schemes for the calculation of bulk and surface Green functions. *J. Phys. F Met. Phys.* **1985**, *15*, 851.

85. Parr, R.; Weitao, Y. *Density-Functional Theory of Atoms and Molecules*; International Series of Monographs on Chemistry; Oxford University Press: Oxford, UK, 1994.

86. Perdew, J.P.; Burke, K.; Ernzerhof, M. Generalized Gradient Approximation Made Simple. *Phys. Rev. Lett.* **1996**, *77*, 3865–3868. doi:10.1103/PhysRevLett.77.3865.

87. Kim, K.; Jordan, K.D. Comparison of Density Functional and MP2 Calculations on the Water Monomer and Dimer. *J. Phys. Chem.* **1994**, *98*, 10089–10094. doi:10.1021/j100091a024.

88. Walter, T. Semiempirical quantum-chemical methods. *Wiley Interdiscip. Rev. Comput. Mol. Sci.* **2013**, *4*, 145–157.

89. Seifert, G.; Eschrig, H.; Bierger, W. An approximation variant of LCAO-X-ALPHA methods. *Z. Phys. Chem.* **1986**, *267*, 529.

90. Foulkes, W.M.C.; Haydock, R. Tight-binding models and density-functional theory. *Phys. Rev. B* **1989**, *39*, 12520–12536.

91. Elstner, M.; Porezag, D.; Jungnickel, G.; Elsner, J.; Haugk, M.; Frauenheim, T.; Suhai, S.; Seifert, G. Self-consistent-charge density-functional tight-binding method for simulations of complex materials properties. *Phys. Rev. B* **1998**, *58*, 7260–7268.

92. Erdogan, E.; Popov, I.H.; Enyashin, A.N.; Seifert, G. Transport properties of MoS_2 nanoribbons: Edge priority. *Eur. Phys. J. B* **2012**, *85*, 33.

93. Ghorbani-Asl, M.; Borini, S.; Kuc, A.; Heine, T. Strain-dependent modulation of conductivity in single-layer transition-metal dichalcogenides. *Phys. Rev. B* **2013**, *87*, 235434.

94. Sevincli, H.; Sevik, C.; Cagin, T.; Cuniberti, G. A bottom-up route to enhance thermoelectric figures of merit in graphene nanoribbons. *Sci. Rep.* **2013**, *3*, 1228.

95. Erdogan, E.; Popov, I.; Rocha, C.G.; Cuniberti, G.; Roche, S.; Seifert, G. Engineering carbon chains from mechanically stretched graphene-based materials. *Phys. Rev. B* **2011**, *83*, 041401.

96. Witek, H.A.; Morokuma, K. Systematic study of vibrational frequencies calculated with the self-consistent charge density functional tight-binding method. *J. Comput. Chem.* **2004**, *25*, 1858–1864.

97. Oviedo, M.B.; Negre, C.F.A.; Sanchez, C.G. Dynamical simulation of the optical response of photosynthetic pigments. *Phys. Chem. Chem. Phys.* **2010**, *12*, 6706–6711.

98. Scholz, R.; Luschtinetz, R.; Seifert, G.; Jägeler-Hoheisel, T.; Körner, C.; Leo, K.; Rapacioli, M. Quantifying charge transfer energies at donor-acceptor interfaces in small-molecule solar cells with constrained DFTB and spectroscopic methods. *J. Phys. Condens. Matter* **2013**, *25*, 473201.

99. Seifert, G.; Joswig, J.O. Density-functional tight binding—An approximate density-functional theory method. *Wiley Interdiscip. Rev. Comput. Mol. Sci.* **2012**, *2*, 456–465.

100. Elstner, M.; Seifert, G. Density functional tight binding. *Philos. Trans. R. Soc. A* **2014**, *372*.

101. Gaus, M.; Cui, Q.; Elstner, M. Density functional tight binding: Application to organic and biological molecules. *Wiley Interdiscip. Rev. Comput. Mol. Sci.* **2014**, *4*, 49–61.

102. Porezag, D.; Frauenheim, T.; Köhler, T.; Seifert, G.; Kaschner, R. Construction of tight-binding-like potentials on the basis of density-functional theory: Application to carbon. *Phys. Rev. B* **1995**, *51*, 12947–12957.

103. Hazebroucq, S.; Picard, G.S.; Adamo, C.; Heine, T.; Gemming, S.; Seifert, G. Density-functional-based molecular-dynamics simulations of molten salts. *J. Chem. Phys.* **2005**, *123*, 134510.

104. Elstner, M. SCC-DFTB: What is the proper degree of self-consistency? *J. Phys. Chem. A* **2007**, *111*, 5614–5621.

105. Liang, R.; Swanson, J.M.J.; Voth, G.A. Benchmark study of the SCC-DFTB approach for a biomolecular proton channel. *J. Chem. Theory Comput.* **2014**, *10*, 451–462.

106. Gaus, M.; Cui, Q.; Elstner, M. DFTB3: Extension of the self-consistent-charge density-functional tight-binding method (SCC-DFTB). *J. Chem. Theory Comput.* **2011**, *7*, 931–948.

107. Witek, H.A.; Irle, S.; Morokuma, K. Analytical second-order geometrical derivatives of energy for the self-consistent-charge density-functional tight-binding method. *J. Chem. Phys.* **2004**, *121*, 5163–5170.

108. Huang, P.; Zhu, H.; Jing, L.; Zhao, Y.; Gao, X. Graphene covalently binding aryl groups: Conductivity increases rather than decreases. *ACS Nano* **2011**, *5*, 7945–7949.

109. Erni, R.; Rossell, M.D.; Nguyen, M.T.; Blankenburg, S.; Passerone, D.; Hartel, P.; Alem, N.; Erickson, K.; Gannett, W.; Zettl, A. Stability and dynamics of small molecules trapped on graphene. *Phys. Rev. B* **2010**, *82*, 165443.

110. Du, A.; Chen, Y.; Zhu, Z.; Lu, G.; Smith, S.C. C-BN single-walled nanotubes from hybrid connection of BN/C nanoribbons: Prediction by ab initio density functional calculations. *J. Am. Chem. Soc.* **2009**, *131*, 1682–1683.

111. Yu, Q.; Jauregui, L.A.; Wu, W.; Colby, R.; Tian, J.; Su, Z.; Cao, H.; Liu, Z.; Pandey, D.; Wei, D.; et al. Control and characterization of individual grains and grain boundaries in graphene grown by chemical vapour deposition. *Nat. Mater.* **2011**, *10*, 443–449.

112. Yazyev, O.V.; Louie, S.G. Electronic transport in polycrystalline graphene. *Nat. Mater.* **2010**, *9*, 806–809.

113. Guo, H.; Lu, N.; Dai, J.; Wu, X.; Zeng, X.C. Phosphorene nanoribbons, phosphorus nanotubes, and van der Waals multilayers. *J. Phys. Chem. C* **2014**, *118*, 14051–14059.

114. Bissett, M.A.; Tsuji, M.; Ago, H. Strain engineering the properties of graphene and other two-dimensional crystals. *Phys. Chem. Chem. Phys.* **2014**, *16*, 11124–11138.

115. Castellanos-Gomez, A.; Singh, V.; van der Zant, H.S.J.; Steele, G.A. Mechanics of freely-suspended ultrathin layered materials. *Ann. Phys.* **2015**, *527*, 27–44.

116. Medrano Sandonas, L.; Gutierrez, R.; Pecchia, A.; Dianat, A.; Cuniberti, G. Thermoelectric properties of functionalized graphene grain boundaries. *J. Self-Assem. Mol. Electron.* **2015**, *3*, 1–20.

117. Medrano Sandonas, L.; Teich, D.; Gutierrez, R.; Lorenz, T.; Pecchia, A.; Seifert, G.; Cuniberti, G. Anisotropic thermoelectric response in two-dimensional puckered structures. *J. Phys. Chem. C* **2016**, *120*, 18841–18849.

118. Medrano Sandonas, L.; Gutierrez, R.; Pecchia, A.; Seifert, G.; Cuniberti, G. Tuning quantum electron and phonon transport in two-dimensional materials by strain engineering: A Green's function based study. *Phys. Chem. Chem. Phys.* **2017**, *19*, 1487–1495.

119. Medrano Sandonas, L.; Sevincli, H.; Gutierrez, R.; Cuniberti, G. First-principle-based phonon transport properties of nanoscale graphene grain boundaries. *Adv. Sci.* **2018**, *5*, 1700365.

120. Medrano Sandonas, L.; Cuba-Supanta, G.; Gutierrez, R.; Landauro, C.V.; Rojas-Tapia, J.; Cuniberti, G. Doping engineering of thermoelectric transport in BNC heteronanotubes. *Phys. Chem. Chem. Phys.* **2019**, *21*, 1904–1911.

121. Medrano Sandonas, L.; Rodriguez Mendez, A.; Gutierrez, R.; Ugalde, J.M.; Mujica, V.; Cuniberti, G. Selective transmission of phonons in molecular junctions with nanoscopic thermal baths. *J. Phys. Chem. C* **2019**, *123*, 9680–9687.

122. Liu, H.; Neal, A.T.; Zhu, Z.; Luo, Z.; Xu, X.; Tomanek, D.; Ye, P.D. Phosphorene: An unexplored 2D semiconductor with a high hole mobility. *ACS Nano* **2014**, *8*, 4033–4041.

123. Zhu, Z.; Guan, J.; Liu, D.; Tománek, D. Designing isoelectronic counterparts to layered group V semiconductors. *ACS Nano* **2015**, *9*, 8284–8290.

124. Fei, R.; Li, W.; Li, J.; Yang, L. Giant piezoelectricity of monolayer group IV monochalcogenides: SnSe, SnS, GeSe, and GeS. *Appl. Phys. Lett.* **2015**, *107*, 173104.

125. Shengli, Z.; Zhong, Y.; Yafei, L.; Zhongfang, C.; Haibo, Z. Atomically thin arsenene and antimonene: Semimetal-semiconductor and indirect-direct band-gap transitions. *Angew. Chem. Inter. Ed.* **2015**, *54*, 3112–3115.

126. Jain, A.; McGaughey, A.J.H. Strongly anisotropic in-plane thermal transport in single-layer black phosphorene. *Sci. Rep.* **2015**, *5*, 8501.

127. Qin, G.; Yan, Q.B.; Qin, Z.; Yue, S.Y.; Hu, M.; Su, G. Anisotropic intrinsic lattice thermal conductivity of phosphorene from first principles. *Phys. Chem. Chem. Phys.* **2015**, *17*, 4854–4858.

128. Kamal, C.; Ezawa, M. Arsenene: Two-dimensional buckled and puckered honeycomb arsenic systems. *Phys. Rev. B* **2015**, *91*, 085423.

129. Zeraati, M.; Vaez Allaei, S.M.; Abdolhosseini Sarsari, I.; Pourfath, M.; Donadio, D. Highly anisotropic thermal conductivity of arsenene: An *ab-initio* study. *Phys. Rev. B* **2016**, *93*, 085424.

130. Kamal, C.; Chakrabarti, A.; Ezawa, M. Direct band gaps in group IV-VI monolayer materials: Binary counterparts of phosphorene. *Phys. Rev. B* **2016**, *93*, 125428.

131. Liu, T.H.; Chang, C.C. Anisotropic thermal transport in phosphorene: Effects of crystal orientation. *Nanoscale* **2015**, *7*, 10648–10654.

132. Fei, R.; Faghaninia, A.; Soklaski, R.; Yan, J.A.; Lo, C.; Yang, L. Enhanced thermoelectric efficiency via orthogonal electrical and thermal conductances in phosphorene. *Nano Lett.* **2014**, *14*, 6393–6399.

133. Stephan, O.; Ajayan, P.M.; Colliex, C.; Redlich, P.; Lambert, J.M.; Bernier, P.; Lefin, P. Doping Graphitic and Carbon Nanotube Structures with Boron and Nitrogen. *Science* **1994**, *266*, 1683–1685.

134. Zhang, J.; Wang, C. Mechanical properties of hybrid boron nitride–carbon nanotubes. *J. Phys. D Appl. Phys.* **2016**, *49*, 155305.

135. Zhong, R.L.; Xu, H.L.; Su, Z.M. Connecting effect on the first hyperpolarizability of armchair carbon-boron-nitride heteronanotubes: Pattern versus proportion. *Phys. Chem. Chem. Phys.* **2016**, *18*, 13954–13959.

136. Alam, H.; Ramakrishna, S. A review on the enhancement of figure of merit from bulk to nano-thermoelectric materials. *Nano Energy* **2013**, *2*, 190–212.

137. Frauenheim, T.; Seifert, G.; Elsterner, M.; Hajnal, Z.; Jungnickel, G.; Porezag, D.; Suhai, S.; Scholz, R. A Self-consistent charge density-functional based tight-binding method for predictive materials simulations in physics, chemistry and biology. *Phys. Status Solidi B* **2000**, *217*, 41–62.

138. Seifert, G. Tight-binding density functional theory: An approximate Kohn-Sham DFT scheme. *J. Phys. Chem. A* **2007**, *111*, 5609–5613.

139. Machado, M.; Kar, T.; Piquini, P. The influence of the stacking orientation of C and BN stripes in the structure, energetics, and electronic properties of BC_2N nanotubes. *Nanotechnology* **2011**, *22*, 205706.

140. Carvalho, A.C.M.; Bezerra, C.G.; Lawlor, J.A.; Ferreira, M.S. Density of states of helically symmetric boron carbon nitride nanotubes. *J. Phys. Condens. Matter* **2014**, *26*, 015303.

141. Li, Q.; Duchemin, I.; Xiong, S.; Solomon, G.C.; Donadio, D. Mechanical tuning of thermal transport in a molecular junction. *J. Phys. Chem. C* **2015**, *119*, 24636–24642.

142. Sasikumar, K.; Keblinski, P. Effect of chain conformation in the phonon transport across a Si-polyethylene single-molecule covalent junction. *J. Appl. Phys.* **2011**, *109*, 114307.

143. Lyeo, H.K.; Cahill, D.G. Thermal conductance of interfaces between highly dissimilar materials. *Phys. Rev. B* **2006**, *73*, 144301.

144. Nika, D.L.; Pokatilov, E.P.; Balandin, A.A.; Fomin, V.M.; Rastelli, A.; Schmidt, O.G. Reduction of lattice thermal conductivity in one-dimensional quantum-dot superlattices due to phonon filtering. *Phys. Rev. B* **2011**, *84*, 165415.

145. Li, X.; Yang, R. Effect of lattice mismatch on phonon transmission and interface thermal conductance across dissimilar material interfaces. *Phys. Rev. B* **2012**, *86*, 054305.

146. Luckyanova, M.N.; Garg, J.; Esfarjani, K.; Jandl, A.; Bulsara, M.T.; Schmidt, A.J.; Minnich, A.J.; Chen, S.; Dresselhaus, M.S.; Ren, Z.; et al. Coherent phonon heat conduction in superlattices. *Science* **2012**, *338*, 936–939.

147. Medrano Sandonas, L.; Croy, A.; Gutierrez, R.; Cuniberti, G. Atomistic framework for time-dependent thermal transport. *J. Phys. Chem. C* **2018**, *122*, 21062–21068.

148. Ulrich, W. *Quantum Dissipative Systems*, 2nd ed.; Series in Modern Condensed Matter Physics; World Scientific Publishing Company: Singapore, 1999.

149. de Vega, I.; Alonso, D. Dynamics of non-Markovian open quantum systems. *Rev. Mod. Phys.* **2017**, *89*, 015001.

150. Ritschel, G.; Eisfeld, A. Analytic representations of bath correlation functions for ohmic and superohmic spectral densities using simple poles. *J. Chem. Phys.* **2014**, *141*, 094101.

151. Meier, C.; Tannor, D.J. Non-Markovian evolution of the density operator in the presence of strong laser fields. *J. Chem. Phys.* **1999**, *111*, 3365.

152. Hu, J.; Xu, R.X.; Yan, Y. Communication: Padé spectrum decomposition of Fermi function and Bose function. *J. Chem. Phys.* **2010**, *133*, 101106.

153. Frisch, M.J.; Trucks, G.W.; Schlegel, H.B.; Scuseria, G.E.; Robb, M.A.; Cheeseman, J.R.; Scalmani, G.; Barone, V.; Mennucci, B.; Petersson, G.A.; et al. *Gaussian−09*; Revision E.01; Gaussian. Inc.: Wallingford, CT, USA, 2009.

154. Pecchia, A.; Di Carlo, A.; Gagliardi, A.; Sanna, S.; Frauenheim, T.; Gutierrez, R. Incoherent electron-phonon scattering in octanethiols. *Nano Lett.* **2004**, *4*, 2109–2114. doi:10.1021/nl048841h.

155. Penazzi, G.; Pecchia, A.; Gupta, V.; Frauenheim, T. A self energy model of dephasing in molecular junctions. *J. Phys. Chem. C* **2016**, *120*, 16383–16392. doi:10.1021/acs.jpcc.6b04185.

156. Markussen, T.; Palsgaard, M.; Stradi, D.; Gunst, T.; Brandbyge, M.; Stokbro, K. Electron-phonon scattering from Green's function transport combined with molecular dynamics: Applications to mobility predictions. *Phys. Rev. B* **2017**, *95*, 245210.

157. Frederiksen, T.; Paulsson, M.; Brandbyge, M.; Jauho, A.P. Inelastic transport theory from first principles: Methodology and application to nanoscale devices. *Phys. Rev. B* **2007**, *75*, 205413. doi:10.1103/PhysRevB.75.205413.

158. Viljas, J.K.; Cuevas, J.C.; Pauly, F.; Häfner, M. Electron-vibration interaction in transport through atomic gold wires. *Phys. Rev. B* **2005**, *72*, 245415. doi:10.1103/PhysRevB.72.245415.

Article

On the Role of Local Many-Body Interactions on the Thermoelectric Properties of Fullerene Junctions

Carmine Antonio Perroni * and Vittorio Cataudella

CNR-SPIN and Physics Department "E. Pancini", Università Federico II, Via Cinthia, I-80134 Napoli, Italy
* Correspondence: perroni@fisica.unina.it

Received: 30 June 2019; Accepted: 31 July 2019; Published: 1 August 2019

Abstract: The role of local electron–vibration and electron–electron interactions on the thermoelectric properties of molecular junctions is theoretically analyzed focusing on devices based on fullerene molecules. A self-consistent adiabatic approach is used in order to obtain a non-perturbative treatment of the electron coupling to low frequency vibrational modes, such as those of the molecule center of mass between metallic leads. The approach also incorporates the effects of strong electron–electron interactions between molecular degrees of freedom within the Coulomb blockade regime. The analysis is based on a one-level model which takes into account the relevant transport level of fullerene and its alignment to the chemical potential of the leads. We demonstrate that only the combined effect of local electron–vibration and electron–electron interactions is able to predict the correct behavior of both the charge conductance and the Seebeck coefficient in very good agreement with available experimental data.

Keywords: molecular junctions; thermoelectric properties; electron–vibration interactions; electron–electron interactions

1. Introduction

In recent years, the field of molecular thermoelectrics has attracted a lot of attention [1–12]. One of the aims is to improve the thermoelectric efficiency of nanoscale devices by controlling the electronic and vibrational degrees of freedom of the molecules. Moreover, useful information on charge and energy transport mechanisms can be extracted by studying the thermoelectric properties of molecular junctions [1,3,4,13,14]. In addition to the charge conductance G, the Seebeck coefficient S is typically measured in these devices. Measurements in junctions with fullerene (C_{60}) have found a high value of thermopower (of the order or even smaller than $-30\,\mu V/K$) [4]. Understanding the thermopower is also important for helping advances in thermoelectric performance of large-area molecular junctions [15,16]. Moreover, recently, the application of an Al gate voltage at $Au–C_{60}–Au$ junction has allowed to achieve the electrostatic control of charge conductance and thermopower with unprecedented control [17]. However, the precise transport mechanisms affecting both G and S remain elusive in these kinds of measurements. Finally, due to experimental challenges [2,18–20], only recently the thermal conductance of single-molecule junctions has been fully characterized [21].

In molecular junctions, relevant contributions to the thermoelectric properties typically result from intramolecular electron–electron and electron–vibration interactions [1,22]. An additional source of coupling between electronic and vibrational degrees of freedom is also provided by the center of mass oscillation of the molecule between the metallic leads [23]. Different theoretical techniques [1,22] have been used to study the effects of local many-body interactions which affect the thermoelectric transport properties [7–9,24–28] in a significant way.

In devices with large molecules such as fullerenes or carbon nanotube quantum dots, a non-perturbative treatment of electron–vibration coupling can be obtained within an adiabatic approach which is based on the slowness of the relevant vibrational modes in comparison with the fast electron dynamics [29–39]. The adiabatic approach can also include a strong Coulomb repulsion allowing the self-consistent calculation of thermoelectric properties of massive molecules, such as fullerenes, within the Coulomb blockade regime [40].

In this paper, the thermoelectric properties of a molecular junction are analyzed focusing on the role of electron–electron and electron–vibration interactions. An adiabatic approach developed in the literature takes into account the interplay between the low frequency center of mass oscillation of the molecule and the electronic degrees of freedom within the Coulomb blockade regime [40]. Parameters appropriate for junctions with C_{60} molecules are considered in this paper. In particular, a one-level model is taken into account since it describes the relevant transport level of fullerene and its alignment to the chemical potential of the metallic leads.

The aim of this paper is to thoroughly investigate both the charge conductance and the Seebeck coefficient since accurate experimental data are available for Au–C_{60}–Au junction in [17] as a function of the voltage gate. We show that an accurate description of the transport properties is obtained in the intermediate regime for the electron–vibration coupling and in the strong coupling regime for the electron–electron interaction. Moreover, we point out that only the combined effect of electron–vibration and electron–electron interactions is able to predict the correct behavior of both the charge conductance and the Seebeck coefficient finding a very good agreement with available experimental data.

The paper is organized as follows. In Section 2, a very general model for many electronic levels and multiple vibrational degrees is considered and the adiabatic approach is exposed. In Section 3, the one-level model is presented. In Section 4, the theoretical results are presented together with the precise comparison with experimental data. Finally, in Section 5, conclusions and final discussions are given.

2. Model and Method

In this section, we introduce a general Hamiltonian for a multilevel molecule including many-body interactions between molecular degrees of freedom: the local electron–electron interaction and the local electron coupling to molecular vibrational modes. The model simulates also the coupling of the molecule to two leads in the presence of a finite bias voltage and temperature gradient. The total Hamiltonian of the system is

$$\hat{\mathcal{H}} = \hat{H}_{mol} + \hat{H}_{leads} + \hat{H}_{leads-mol}, \tag{1}$$

where $\hat{H}_{mol}$ is the Hamiltonian describing the molecular degrees of freedom, $\hat{H}_{leads}$ the leads' degrees of freedom and $\hat{H}_{leads-mol}$ the coupling between molecule and leads.

In this paper, we assume, as usual in the field of molecular junctions, that the electronic and vibrational degrees of freedom in the metallic leads are not interacting [1,41]; therefore, the electron–electron and electron–vibration interactions are effective only on the molecule. In Equation (1), the molecule Hamiltonian $\hat{H}_{mol}$ is

$$\hat{H}_{mol} = \sum_{m,l,\sigma} \hat{c}^{\dagger}_{m,\sigma} \varepsilon^{m,l}_{\sigma} \hat{c}_{l,\sigma} + U \sum_{m,l} \hat{n}^{\dagger}_{m,\uparrow} \hat{n}_{l,\downarrow} + \hat{H}_{osc} + \hat{H}_{int}, \tag{2}$$

where $c_{m,\sigma}$ ($c^{\dagger}_{m,\sigma}$) is the standard electron annihilation (creation) operator for electrons on the molecule levels with spin $\sigma = \uparrow, \downarrow$, where indices m, l can assume positive integer values with a maximum M indicating the total number of electronic levels in the molecule. The matrix $\varepsilon^{m,l}_{\sigma}$ is assumed diagonal in spin space, $\hat{n}_{l,\sigma} = c^{\dagger}_{l,\sigma} c_{l,\sigma}$ is the electronic occupation operator relative to level l and spin σ, and U represents the Coulomb–Hubbard repulsion between electrons. We assume that only the diagonal part of the matrix $\varepsilon^{m,l}_{\sigma}$ is nonzero and independent of the spin: $\varepsilon^{m,m}_{\sigma} = \varepsilon_m$, where ε_m are the energies of the molecule levels.

In Equation (2), the molecular vibrational degrees of freedom are described by the Hamiltonian

$$\hat{H}_{osc} = \sum_s \frac{\hat{p}_s^2}{2m_s} + V(\boldsymbol{X}), \tag{3}$$

where $s = (1, \ldots, N)$, with N being the total number of vibrational modes; m_s is the effective mass associated with the sth vibrational mode; and $\hat{p}_s$ is its momentum operator. Moreover, $V(\boldsymbol{X}) = \frac{1}{2}\sum_s k_s \hat{x}_s^2$ is the harmonic potential (with k_s the spring constants, and the oscillator frequencies $\omega_0^s = \sqrt{k_s/m_s}$), $\hat{x}_s$ is the displacement operator of the vibrational mode s, and $\boldsymbol{X} = (\hat{x}_1, \ldots, \hat{x}_N)$ indicates all the displacement operators.

In Equation (2), the electron–vibration coupling $\hat{H}_{int}$ is assumed linear in the vibrational displacements and proportional to the electron level occupations

$$\hat{H}_{int} = \sum_{s,l} \lambda_{s,l} \hat{x}_s \hat{n}_l, \tag{4}$$

where $s = (1, \ldots, N)$ indicates the vibrational modes of the molecule, $l = (1, \ldots, M)$ denotes its electronic levels, $\hat{n}_l = \sum_\sigma n_{l,\sigma}$ is the electronic occupation operator of the level l, and $\lambda_{s,l}$ is a matrix representing the electron–vibrational coupling.

In Equation (1), the Hamiltonian of the electron leads is given by

$$\hat{H}_{leads} = \sum_{k,\alpha,\sigma} \varepsilon_{k,\alpha} \hat{c}_{k,\alpha,\sigma}^\dagger \hat{c}_{k,\alpha,\sigma}, \tag{5}$$

where the operators $\hat{c}_{k,\alpha,\sigma}^\dagger$ ($\hat{c}_{k,\alpha,\sigma}$) create (annihilate) electrons with momentum k, spin σ, and energy $\varepsilon_{k,\alpha} = E_{k,\alpha} - \mu_\alpha$ in the left ($\alpha = L$) or right ($\alpha = R$) leads. The left and right electron leads are considered as thermostats in equilibrium at the temperatures $T_L = T + \Delta T/2$ and $T_R = T - \Delta T/2$, respectively, with T the average temperature and ΔT temperature difference. Therefore, the left and right electron leads are characterized by the free Fermi distribution functions $f_L(E)$ and $f_R(E)$, respectively, with E the energy. The difference of the electronic chemical potentials in the leads provides the bias voltage V_{bias} applied to the junction: $\mu_L = \mu + eV_{bias}/2$, $\mu_R = \mu - eV_{bias}/2$, with μ the average chemical potential and e the electron charge. In this paper, we focus on the regime of linear response that involves very small values of bias voltage V_{bias} and temperature ΔT.

Finally, in Equation (1), the coupling between the molecule and the leads is described by

$$\hat{H}_{mol-leads} = \sum_{k,\alpha,m,\sigma} (V_{k,\alpha}^m \hat{c}_{k,\alpha,\sigma}^\dagger \hat{c}_{m,\sigma} + h.c.), \tag{6}$$

where the tunneling amplitude between the molecule and a state k in the lead α has the amplitude $V_{k,\alpha}^m$. For the sake of simplicity, we suppose that the density of states $\rho_{k,\alpha}$ for the leads is flat within the wide-band approximation: $\rho_{k,\alpha} \mapsto \rho_\alpha$, $V_{k,\alpha}^m \mapsto V_\alpha^m$. Therefore, the full hybridization width matrix of the molecular orbitals is $\Gamma^{m,n} = \sum_\alpha \Gamma_\alpha^{m,n} = \sum_\alpha \Gamma_\alpha^{m,n}$, with the tunneling rate $\Gamma_\alpha^{m,n} = 2\pi\rho_\alpha V_\alpha^{m*} V_\alpha^n$. In this paper, we consider the symmetric configuration $\boldsymbol{\Gamma}_L = \boldsymbol{\Gamma}_R = \boldsymbol{\Gamma}/2$, where, in the following, bold letters indicate matrices.

In this paper, we consider the electronic system coupled to slow vibrational modes: $\omega_0^s \ll \Gamma^{m,n}$, for each s and all pairs of (m, n). In this limit, we can treat the mechanical degrees of freedom as classical, acting as slow classical fields on the fast electronic dynamics. Therefore, the electronic dynamics is equivalent to a multi-level problem with energy matrix $\varepsilon^m \to \varepsilon^m + \lambda_m x_m$, where x_m are now classical displacements [32,39]. This is called in the literature adiabatic approximation for vibrational degrees of freedom.

Within the adiabatic approximation, one gets Langevin self-consistent equations for the vibrational modes of the molecule [33,39]

$$m_s \ddot{x}_s + k_s x_s = F_s(t) + \tilde{\xi}_s(t), \tag{7}$$

where the generalized force F_s is due to the effect of all electronic degrees of freedom through the electron–vibration coupling [32,39]:

$$F_s^{el}(t) = Tr[i\lambda_s G^<(t,t)], \tag{8}$$

with the trace "Tr", taken over the molecule levels, defined in terms of the lesser molecular matrix Green's function $G^<(t,t')$ with matrix elements $G_{m,l}^<(t,t') = i\langle c_{m,\sigma}^\dagger(t)c_{l,\sigma}(t')\rangle$. Quantum electronic density fluctuations on the oscillator motion are responsible for the fluctuating force $\xi_s(t)$ in Equation (7), which is derived below together with generalized force.

In deriving equations within the adiabatic approximation [39], next, for the sake of simplicity, we do not include explicitly the effect of the Coulomb repulsion on the molecule Hamiltonian. In the next section, we show that, in the case of a single level molecule with large repulsion U, the adiabatic approach works exactly as in the non-interacting case provided that each Green's function pole is treated as a non interacting level [40].

In our notation, G denotes full Green's functions, while $\mathcal{G}$ denotes the strictly adiabatic (or frozen) Green's functions, which are calculated at a fixed value of X. Starting from the Dyson equation [32,39,41], the adiabatic expansion for the retarded Green's function G^R is given by

$$G^R \simeq \mathcal{G}^R + \frac{i}{2}\Big(\partial_E\mathcal{G}^R(\sum_s \lambda_s\dot{x}_s)\mathcal{G}^R - \mathcal{G}^R(\sum_s \lambda_s\dot{x}_s)\partial_E\mathcal{G}^R\Big), \tag{9}$$

where $\mathcal{G}^R(E,X)$ is the strictly adiabatic (frozen) retarded Green's function including the coupling with the leads

$$\mathcal{G}^R(E,X) = [E - \varepsilon(X) - \Sigma^{R,leads}]^{-1}, \tag{10}$$

$\varepsilon(X)$ represents the matrix $\varepsilon_\sigma^{m,l} + \sum_s \lambda_s x_s \delta_{l,m}$ and $\Sigma^{R,leads} = \sum_\alpha \Sigma_\alpha^{R,leads}$ is the total self-energy due to the coupling between the molecule and the leads. For the lesser Green's function $G^<$, the adiabatic approximation involves

$$
\begin{aligned}
G^< \simeq \mathcal{G}^< + \frac{i}{2}\Big[&\partial_E\mathcal{G}^<(\sum_s \lambda_s\dot{x}_s)\mathcal{G}^A - \mathcal{G}^R(\sum_s \lambda_s\dot{x}_s)\partial_E\mathcal{G}^< \\
&+\partial_E\mathcal{G}^R(\sum_s \lambda_s\dot{x}_s)\mathcal{G}^< - \mathcal{G}^<(\sum_s \lambda_s\dot{x}_s)\partial_E\mathcal{G}^A\Big],
\end{aligned} \tag{11}
$$

with $\mathcal{G}^< = \mathcal{G}^R\Sigma^<\mathcal{G}^A$.

The electron–vibration induced forces at the zero order of the adiabatic limit ($G^< \simeq \mathcal{G}^<$) are given by

$$F_s^{el(0)}(X) = -k_s x_s - \int \frac{dE}{2\pi i} tr[\lambda_s \mathcal{G}^<]. \tag{12}$$

The leading order correction to the lesser Green's function $G^<$ provides a term proportional to the vibrational velocity

$$F_s^{el(1)}(X) = -\sum_{s'}\theta_{s,s'}(X)\dot{x}_{s'}, \tag{13}$$

where the tensor θ can be split into symmetric and anti-symmetric contributions [32]: $\theta = \theta^{sym} + \theta^a$, where we have introduced the notation $\{C_{s,s'}\}_{sym,a} = \frac{1}{2}\{C_{s,s'} \pm C_{s',s}\}_{sym,a}$ for symmetric and anti-symmetric parts of an arbitrary matrix C. Indeed, there is a dissipative term θ^{sym} and an orbital, effective magnetic field θ^a in the space of the vibrational modes.

We can now discuss the stochastic forces $\xi_s(t)$ in Equation (7) within the adiabatic approximation. In the absence of electron–electron interactions, the Wick theorem allows writing the noise correlator as

$$\langle \xi_s^{el}(t)\xi_{s'}^{el}(t')\rangle = tr\{\lambda_s G^>(t,t')\lambda_{s'} G^<(t',t)\}, \tag{14}$$

where $G^>(t,t')$ is the greater Green's function with matrix elements $G^>_{m,l}(t,t') = -i\langle c_{m,\sigma}(t)c^\dagger_{l,\sigma}(t')\rangle$. In the adiabatic approximation, one first substitutes the full Green's function G by the adiabatic zero-order Green's function $\mathcal{G}$ and then observes that the electronic fluctuations act on short time scales only. Therefore, the total forces $\xi_s(t)$ are locally correlated in time:

$$\langle \xi^{el}_s(t)\xi^{el}_{s'}(t')\rangle \simeq tr\{\lambda_s\mathcal{G}^>(X,t)\lambda_{s'}\mathcal{G}^<(X,t)\} = D(X)\delta(t-t'),\tag{15}$$

where

$$D_{s,s'}(X) = \int \frac{dE}{2\pi} tr\{\lambda_s\mathcal{G}^<\lambda_{s'}\mathcal{G}^>\}_{sym}.\tag{16}$$

Once the forces and the noise terms are calculated, Equation (7) represents a set of nonlinear Langevin equations in the unknown x_s. Even for the simple case where only one vibrational degree of freedom is present, the stochastic differential equation should be solved numerically in the general non-equilibrium case [33,37,38]. Actually, one can calculate the oscillator distribution functions $P(X,V)$ (where $V = \dot{X} = (v_1,\ldots,v_N)$), and, therefore, all the properties of the vibrational modes. Using this function, one can determine the average O of an electronic or vibrational observable $O(X,V)$:

$$O = \int\int dXdV P(X,V)O(X,V).\tag{17}$$

The electronic observables, such as charge and heat currents, can be evaluated exploiting the slowness of the vibrational degrees of freedom. In a previous paper [39], we discussed the validity of the adiabatic approximation, stressing that it is based on the separation between the slow vibrational and fast electronic timescales. Actually, physical quantities calculated within the adiabatic approach are very reliable in a large regime of electronic parameters since this self-consistent approach is not perturbative in the electron–vibration coupling. Therefore, the approach is able to overcome the limitations of the perturbative theory typically used in the literature [42,43].

3. One-Level Model

In the remaining part of the paper, we consider the simple case where the molecule is modeled as a single electronic level ($M = 1$ in the previous section) locally interacting with a single vibrational mode ($N = 1$ in the previous section). Therefore, the focus is on a molecular level which is sufficiently separated in energy from other orbitals. In particular, we analyze the C_{60} molecule where the lowest unoccupied molecular orbital (LUMO) energy differs from the highest occupied molecular orbital (HOMO) energy for energies of the order of 1 eV [23,44]. Even when the degeneracy of the LUMO is removed by the contact with metal leads, the splitting gives rise to levels which are separated by an energy of the order of a few tenths of eV [44]. Furthermore, the energy of the molecular orbital can be tuned by varying the gate voltage V_G.

One-level transport model has been adopted to interpret experimental data of C_{60} molecular junctions [17] neglecting altogether the effect of electron–electron and electron–vibrations interactions. This model is clearly valid for energies close to the resonance, therefore it is particularly useful in the case of the experiments in [17] where the molecular energy is tuned around the Fermi energy of the leads. Moreover, the one-level model has to be used in the regime of low temperatures, therefore temperatures up to room temperature can be considered for the interpretation of experimental data. Within this model, the energy-dependent transmission function $T(E)$ is assumed to be well approximated by a Lorentzian function:

$$T(E) = \frac{4\Gamma^2}{(E-\epsilon)^2 + 4\Gamma^2},\tag{18}$$

where the molecular level energy ϵ is taken as

$$\epsilon = E_0 - \alpha V_G,\tag{19}$$

with E_0 the energetic separation of the dominant transport level with respect to the chemical potential μ, and α the effectiveness of gate coupling. The expression of ϵ takes clearly into account the tuning of the molecular level by the gate voltage. By using Equation (18), in the limit of low temperature of the Landauer–Büttiker approach valid in the coherent regime [1,41], the gate voltage-dependent electrical conductance G becomes

$$G = \frac{\partial I}{\partial V_{bias}}(V_{bias} = 0, V_G) = G_0\, T(E = \mu), \tag{20}$$

where $G_0 = 2e^2/h$ is the quantum of conductance, with h Planck constant. Moreover, in the same limit, the Seebeck coefficient S is

$$S = -\frac{\pi^2}{3}\frac{k_B}{|e|}k_B T\frac{\partial \ln T(E = \mu)}{\partial E} = \frac{\pi^2}{3}\frac{k_B}{|e|}k_B T\frac{2[\mu - \epsilon]}{[(\mu - \epsilon)^2 + 4\Gamma^2]}, \tag{21}$$

where k_B is the Boltzmann constant. We remark that $k_B/|e| \simeq 86.17\ \mu V/K$ sets the order of magnitude (and, typically, the maximum value in modulus) of the thermopower in molecular junctions.

In the right panel of Figure 1, we report the experimental data of Seebeck coefficient S as a function of the gate voltage V_G taken from [17] for C_{60} junctions. The values of S taken at the temperature $T = 100$ K are quite large in modulus for negative gate. Moreover, the data show a marked change as a function of the gate voltage suggesting that the chemical potential is able to cross a level of the molecule. Since the values of S are negative for small values of V_G and are still negative for zero V_G, the charge transport is dominated by the LUMO level of C_{60}.

Actually, to fit the experimental data shown in the right panel of Figure 1, Equation (21) has been used, getting the positive value $E_0 - \mu = 0.057$ eV [17]. For the optimization of the fit, in the same paper [17], $\Gamma = 0.032$ eV and the gate voltage effectiveness $\alpha = 0.006$ eV/V are also extracted. These three numerical values put in Equation (21) provide the fit curve shown in the right panel of Figure 1. The fit is good, but not excellent.

In the left panel of Figure 1, we report the experimental data of the charge conductance G as a function of the gate voltage V_G taken again from experimental data of [17] for C_{60} junctions. Even if the temperature is not high ($T = 100$ K), the values of G are quite smaller than the conductance quantum G_0. Moreover, if one uses the parameters ($E_0 - \mu = 0.057$ eV, $\Gamma = 0.032$ eV, and $\alpha = 0.006$ eV/V) extracted from the Seebeck data in [17] and reproduced in the right panel of Figure 1, one finds a peak of the conductance for $E_0 - \mu = \alpha V_G$, hence for $V_G \simeq 9$ V. This is in contrast with the peak of G which occurs at $V_G \simeq 5$ V in the experimental data. If we try to describe the experimental data shown in the left panel of Figure 1 by using Equation (20) and the parameters extracted by fitting the Seebeck data, we get the red line reported in the left panel of Figure 1. It is evident that the agreement between theory and data is poor, and, in particular, the maximum observed for V_G around 5 V is not recovered. We remark that $k_B T \simeq 0.0086$ eV represents the smallest energy scale apart from values of V_G very close to the LUMO level. Therefore, the quality of the comparison cannot depend on the low temperature expansion used in Equation (20).

To improve the interpretation of the experimental data, in this paper, we analyze the role of many-body interactions between molecular degrees of freedom. For example, experimental measurements have highlighted that the effects of the electron–vibration interactions are not negligible in junctions with C_{60} molecules and gold electrodes [10,23]. In particular, experimental results for C_{60} molecules [23] provide compelling evidence for a sizable coupling between the electrons and the center of mass vibrational mode. Indeed, previous studies have shown that a C_{60} molecule is held tightly on gold by van der Waals interactions, which can be expressed by the Lennard–Jones form. The C_{60}-gold binding near the equilibrium position can be approximated very well by a harmonic potential with angular frequency ω_0. For C_{60} molecules, the center of mass energy $\hbar\omega_0$ has been estimated to be of the order of 5 meV.

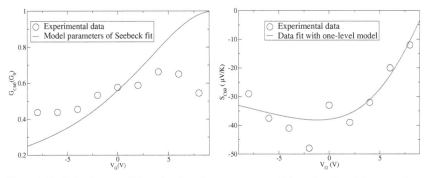

Figure 1. (**Left**) Conductance G (in units of conductance quantum G_0) as a function of the gate voltage V_G (in units of V) at $T = 100$ K from experimental data (black circles, see [17] for fullerene C_{60} junctions) and from a curve (red solid line) obtained by using the parameters of the fit to the Seebeck coefficient. (**Right**) Seebeck coefficient S (in units of μK/V) as a function of the gate voltage V_G (in units of V) at $T = 100$ K from experimental data (black circles), and from a fit (red solid line). For both, see [17] relative to C_{60} molecular junctions.

In this paper, we focus on the center of mass mode as the relevant low frequency vibrational mode for the molecule. The center of mass mode is expected to have the lowest angular frequency ω_0 for large molecules. For fullerene, the energy $\hbar\omega_0$ is still smaller than the thermal energy $k_B T$ corresponding to the temperature $T = 100$ K fixed for the measurements made in [17]. For $k_B T \geq \hbar\omega_0$, the self-consistent adiabatic approach introduced in the previous section can be used for a non-perturbative treatment of the electron–vibration coupling. Equation (7) reduces in this case to a single Langevin equation [33,36]. We hereby report the expression for the displacement dependent electronic spectral function $A(E, x)$

$$A(E,x) = \frac{4\Gamma}{(E - \epsilon - \lambda x)^2 + 4\Gamma^2}. \tag{22}$$

within these assumptions, in Equation (2), the interaction Hamiltonian $\hat{H}_{int}$ reduces to the same interaction term of the single impurity Anderson–Holstein model [1] and the electron–oscillator coupling sets the characteristic polaron energy E_P

$$E_P = \frac{\lambda^2}{2m\omega_0^2}, \tag{23}$$

with m mass of the molecule. Actually, an additional electron injected from the leads compresses the C_{60}-surface bond shortening the C_{60}-surface distance, but not significantly changing the vibrational frequency. Previous studies [10,23] have estimated that the number of vibrational quanta typically excited by the tunnelling electron in fullerene junctions is not large. Therefore, intermediate values of electron–vibration energy E_P corresponding to values comparable with Γ are considered relevant for fullerene molecular junctions. Taking the parameters extracted from the experimental data discussed above, $E_P \simeq 0.030$ eV sets the order of magnitude.

To improve the analysis of the fullerene molecular junction, in this paper, we study also the role of electron–electron interactions acting onto the molecule. Indeed, the conductance gap observed in the data of C_{60} molecules can be interpreted using ideas borrowed from the Coulomb blockade effect [1,23]. Therefore, these features are understood in term of the finite energy required to add (remove) an electron to (from) the molecule. Within the single-level model introduced in the previous section, this charging energy is simulated by fixing the value of the local Hubbard term U in Equation (2). The maximum conductance gap observed in the experimental data [23] indicates that the charging energy of the C_{60} molecule can be around 0.27 eV, therefore experiments set the order of magnitude $U \simeq 0.3$ eV.

To include the Coulomb blockade effect within the adiabatic approach discussed previously, we generalize it to the case in which the electronic level can be double occupied and a strong Coulomb repulsion U is added together with the electron–vibration interaction. The starting point is the observation that, in the absence of electron–oscillator interaction, and in the limit where the coupling of the dot to the leads is small $\Gamma << U$ [41], the single particle electronic spectral function is characterized by two spectral peaks separated by an energy interval equal to U. In the adiabatic regime, one can independently perturb each spectral peak of the molecule [40], obtaining at the zero order of the adiabatic approach

$$A(E, x) = [1 - \rho(x)] \frac{4\Gamma}{(E - \epsilon - \lambda x)^2 + 4\Gamma^2} + \rho(x) \frac{4\Gamma}{(E - \epsilon - \lambda x - U)^2 + 4\Gamma^2}, \tag{24}$$

where $\rho(x)$ is the electronic level density per spin. In our computational scheme, $\rho(x)$ has to be self-consistently calculated for a fixed displacement x of the oscillator through the following integral $\rho(x) = \int_{-\infty}^{+\infty} \frac{dE}{2\pi i} G^<(E, x)$, with the lesser Green function $G^<(E, x) = \frac{i}{2}[f_L(E) + f_R(E)]A(E, x)$. The above approximation is valid if the electron–oscillator interaction is not too large, such that $\Gamma \simeq E_P << U$ and the two peaks of the spectral function can be still resolved [40]. We remark that, in comparison with our previous work [40], parameters appropriate for junctions with C_{60} molecules are considered in this paper focusing on the temperature $T = 100$ K fixed for the measurements made in [17], smaller than the room temperature, where the adiabatic approach can be still adopted. Therefore, the approach is valid in the following parameter regime: $\hbar\omega_0 \leq k_B T < \Gamma \ll U$ [39,40].

Within the adiabatic approach, the actual electronic spectral function $A(E)$ results from the average over the dynamical fluctuations of the oscillator motion, therefore, as a general observable, it is calculated by using Equation (17):

$$A(E) = \int_{-\infty}^{+\infty} dx P(x) A(E, x), \tag{25}$$

where $P(x)$ is the reduced position distribution function of the oscillator. Notice that, in the absence of electron–electron ($U = 0$) and electron–vibration ($E_P = 0$) interactions, the spectral function is proportional to the transmission $T(E)$ given in Equation (18) through the hybridization width Γ: $T(E) = \Gamma A(E)$.

In the linear response regime (bias voltage $V_{bias} \to 0^+$ and temperature difference $\Delta T \to 0^+$), all the electronic transport coefficients can be expressed as integrals of $A(E)$. To this aim, we report the conductance G

$$G = G_0 \Gamma \int_{-\infty}^{+\infty} dE A(E) \left[-\frac{\partial f(E)}{\partial E} \right], \tag{26}$$

where $A(E)$ is the spectral function defined in Equation (25), with $f(E) = 1/(\exp[\beta(E - \mu)] + 1)$ the free Fermi distribution corresponding to the chemical potential μ and the temperature T, and $\beta = 1/k_B T$. The Seebeck coefficient is given by $S = -G_S/G$, where the charge conductance G has been defined in Equation (26), and

$$G_S = G_0 \left(\frac{k_B}{e}\right) \Gamma \int_{-\infty}^{+\infty} dE \frac{(E - \mu)}{k_B T} A(E) \left[-\frac{\partial f(E)}{\partial E} \right]. \tag{27}$$

Then, we calculate the electron thermal conductance $G_K^{el} = G_Q - TGS^2$, with

$$G_Q = G_0 \left(\frac{k_B}{e}\right)^2 \Gamma T \int_{-\infty}^{+\infty} dE \left[\frac{E - \mu}{k_B T}\right]^2 A(E) \left[-\frac{\partial f(E)}{\partial E} \right]. \tag{28}$$

Therefore, in the linear response regime, one can easily evaluate the electronic thermoelectric figure of merit ZT^{el}

$$ZT^{el} = \frac{GS^2T}{G_K^{el}}, \tag{29}$$

which characterizes the electronic thermoelectric conversion. We recall that, in this paper, we do not consider the addition contribution coming from phonon thermal conductance G_K^{ph}.

4. Results

In this section, we discuss the thermoelectric properties within the single-level model analyzing the role of the electron–electron and electron–vibration interactions between the molecular degrees of freedom. We point out that only the combined effect of these interactions is able to provide a good agreement between experimental data and theoretical calculations.

The level density ρ is shown in the upper left panel of Figure 2, the charge conductance G in the upper right panel, the Seebeck coefficient S in the lower left panel, and the electronic thermoelectric figure of merit ZT^{el} in the lower right panel. All quantities are plotted as a function of level energy ϵ at the temperature $T = 100$ K. For all the quantities, we first analyze the coherent regime (black solid lines in the four panels of Figure 2), which means absence of electron–electron and electron–vibration interactions. Then, we study the effect of a finite electron–vibration coupling E_P (red dash lines in the four panels of Figure 2) focusing on the intermediate coupling regime. Finally, we consider the combined effect of electron–vibration and electron–electron interactions for all the quantities (blue dash-dot lines in the four panels of Figure 2) analyzing the experimentally relevant regime of a large Coulomb repulsion U.

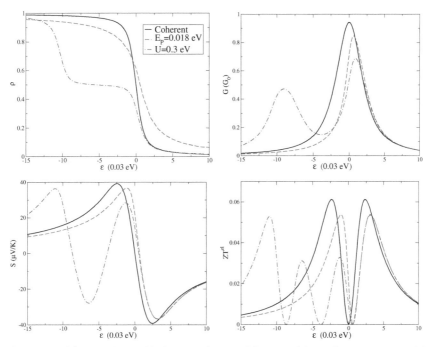

Figure 2. Level density ρ (**Top Left**); charge conductance G (in units of the conductance quantum G_0) (**Top Right**); Seebeck coefficient S (in units of $\mu V/K$) (**Bottom Left**); and electronic thermoelectric figure of merit ZT^{el} (**Bottom Right**) as a function of level energy ϵ (in units of 0.030 eV) at the temperature $T = 100$ K.

The level density ρ per spin reported in the upper left panel of Figure 2 shows the expected decreasing behavior with increasing the level energy ϵ. The electron–vibration interaction induces a shift of the curve of about E_P. In the presence of electron–electron interactions, the behavior is more complex. Actually, in molecular junctions, the strong Coulomb repulsion usually reduces the electronic charge fluctuations and suppresses the double occupation of the electronic levels [1]. For values of ϵ smaller than $-U$, the density is closer to unity, while, for ϵ larger than zero, the density vanishes. For ϵ between $-U$ and 0, there is a plateau with a value of the density close to 0.5. Indeed, these phenomena are characteristic of Coulomb blockade effects.

The conductance G is shown in upper right panel of Figure 2. At low temperatures, this quantity as a function of the level position ϵ provides essentially the spectral function of the molecular level. Indeed, in the coherent low temperature regime, G can be directly related to the transmission with a Lorentzian profile. One of the main effects of an adiabatic oscillator is to shift the conductance peak towards positive energies proportional to the electron–vibration coupling energy E_P. Apparently, another expected effect is the reduction of the peak amplitude. In fact, electron–vibration couplings on the molecule tend to reduce the charge conduction. As a consequence, electron–vibration couplings induces somewhat longer tails far from the resonance. These features, such as the peak narrowing, are common to other theoretical approaches treating electron–vibration interactions, among which that related to the Franck–Condon blockade [45]. Actually, in a previous paper [39], we successfully compared the results of the adiabatic approximation with those of the Franck–Condon blockade formalism in the low density limit where this latter approach becomes essentially exact [1].

We note that a finite electron–electron interaction not only suppresses the electronic conduction for small values of ϵ, but it is also responsible for a second peak centered at $\epsilon \simeq -U$. In fact, there is a transfer of spectral weight from the main peak to the interaction-induced secondary peak. We stress that these features are compatible with experimental data since conductance gap ascribed to Coulomb blockade effects have been measured in fullerene junctions [1,23].

We investigate the properties of the Seebeck coefficient S of the junction in the lower left panel of Figure 2. In analogy with the behavior of the conductance, the main effect of the electron–vibration interaction is to reduce the amplitude of the Seebeck coefficient. Moreover, the shift of the zeroes of S is governed by the coupling E_P as that of the peaks of G. Therefore, with varying the level energy ϵ, if G reduces its amplitude, S increases its amplitude in absolute value, and vice versa. This behavior and the values of S are in agreement with experimental data [4,17]. In the Coulomb blockade regime, S shows a peculiar oscillatory behavior as a function of the energy ϵ, with several positive peaks and negative dips. The energy distance between the peaks (or the dips) is governed by the Hubbard term U. Even in this regime, the Seebeck coefficient S is negligible for the level energies where the electronic conductance presented the main peaks, that is at $\epsilon \simeq 0$ and $\epsilon \simeq -U$. This property turns out to be a result of the strong electron–electron interaction U [26]. In any case, close to the resonance (zero values of the level energy ϵ), the conductance looks more sensitive to many-body interactions, while the Seebeck coefficient appears to be more robust.

The electronic conductance G, Seebeck coefficient S, and electron thermal conductance G_K^{el} combine in giving an electronic figure of merit ZT^{el}. This latter quantity is shown in lower right panel of Figure 2 at the temperature $T = 100$ K. We stress that, due to the low value of the temperature, the quantity ZT^{el} does not show values comparable with unity. However, it is interesting to analyze the effects of many-body interactions on this quantity. A finite value of the electron–vibration coupling E_P leads to a reduction of the height of the figure of merit peaks. It is worth noting that the position of the peaks in ZT^{el} roughly coincides with the position of the peaks and dips of the Seebeck coefficient S. Finally, the electron–electron interactions tend to reduce the amplitude and to further shift the peaks of the figure of merit.

After the analysis of the effects of many-body interactions on the charge conductance and Seebeck coefficient, we can make a comparison with the experimental data available in [17] and shown in Figure 1. These data are plotted again in Figure 3 together with the fit discussed in Figure 1. We recall that for fullerene junctions the one-level model discussed in the previous section is characterized by the following parameters: $E_0 - \mu = 0.057$ eV, $\Gamma = 0.032$ eV, and $\alpha = 0.006$ eV/V. We remark that the level energy ϵ used in the previous discussion is related to the energy E_0 and the gate voltage V_G through Equation (19). Therefore, once the value of E_0 is fixed, one can switch from the energy ϵ to the gate voltage V_G. Before introducing many-body effects, we consider a slight shift of the level position considering the case $E_0 - \mu = 0.065$ eV reported in Figure 3. This energy shift is introduced to counteract the shifts of the peaks (conductance) or zeroes (Seebeck) introduced by many-body interactions which, in addition, reduce the amplitudes of response functions. The aim of this paper is to provide an optimal description for both charge conductance G and Seebeck coefficient S.

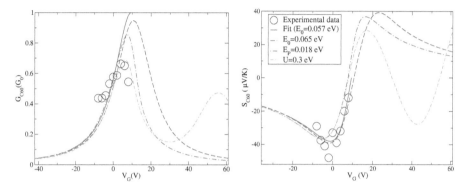

Figure 3. Charge conductance G (in units of the conductance quantum G_0) (**Left**); and Seebeck coefficient S (in units of μV/K) (**Right**) as a function of gate voltage V_G (in units V) at the temperature $T = 100$ K: experimental data (black circles), data fit (red solid line) corresponding to one-level model with energy $E_0 - \mu = 0.057$ eV, coherent results (blue dash line) corresponding to one-level model with energy $E_0 - \mu = 0.065$ eV, effect of the only electron–vibration coupling $E_P = 0.018$ eV (magenta dash-dot line), and effect of additional electron–electron interaction $U = 0.3$ eV (orange double dash-dot line).

Starting from the level energy $E_0 - \mu = 0.065$ eV, in Figure 3, we analyze the effect of the electron–vibration coupling in the intermediate regime $E_P = 0.018$ eV. The shift induced in the zero of the Seebeck coefficient is still compatible with experimental data. Moreover, the electron–vibration interaction shifts and reduces the peak of the charge conductance in an important way. However, this is still not sufficient to get an accurate description of the charge conductance. One could increase the value of the coupling energy E_P, but, this way, the shift of the conductance peak becomes too large with a not marked reduction of the spectral weight.

Another ingredient is necessary to improve the description of both conductance G and Seebeck coefficient S. In our model, the additional Coulomb repulsion plays a concomitant role. Its effects poorly shift the zero of the Seebeck coefficient and slightly modifies the curve far from the zero. Therefore, the description of the Seebeck coefficient remains quite accurate as a function of the gate voltage. On the other hand, it provides a sensible reduction of the conductance amplitude with a not large shift of the peak. Hence, the effects of Hubbard term are able to improve the theoretical interpretation of the experimental data for the conductance G and the Seebeck coefficient S close to the resonance. Far from the resonance, in a wider window of gate voltages, theory predicts the existence of a secondary peak of the conductance and a complex behavior of the Seebeck coefficient due to Coulomb blockade effects. The features are not negligible as a function of the gate voltage.

As far as we know, experimental measurements of the electronic thermal conductance G_K^{el} have become only very recently available [21]. Indeed, it is important to characterize this quantity since it allows determining the thermoelectric figure of merit. Therefore, in Figure 4, we provide the theoretical prediction of the electronic thermal conductance G_K^{el} as a function of V_G starting from the optimized values of the one-level parameters used to describe both charge conductance and Seebeck coefficient in an accurate way. We stress that the plotted thermal conductance is expressed in terms of the thermal conductance quantum $g_0(T) = \pi^2 k_B^2 T / (3h)$ [46]. The main point is that, in the units chosen in Figure 4, the thermal conductance G_K^{el} shows a strong resemblance with the behavior of the charge conductance G in units of the conductance quantum G_0 as a function of the gate voltage V_G. We remark that, at $T = 100$ K, $g_0(T) \simeq 9.456 \times 10^{-11}$ (W/K) $\simeq 100$ pW/K. The values of the thermal conductance G_K^{el} shown in Figure 4 are fractions of $g_0(T)$, therefore they are fully compatible with those estimated experimentally in hydrocarbon molecules [19] (50 pW/K).

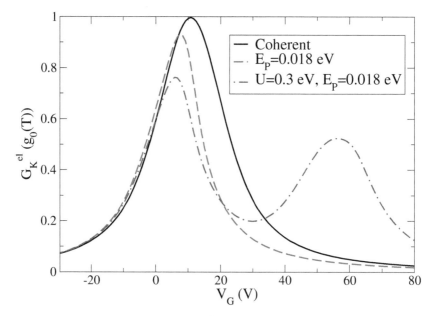

Figure 4. Electronic thermal conductance G_K^{el} in units of thermal conductance quantum $g_0(T)$ $(g_0(T) = \pi^2 k_B^2 T / (3h))$ as a function the voltage gate V_G in units of Volt at the temperature $T = 100$ K: coherent results (black solid line) corresponding to one-level model with energy $E_0 - \mu = 0.065$ eV, effect of the only electron–vibration coupling $E_P = 0.018$ eV (red dash line), and effect of additional electron–electron interaction $U = 0.3$ eV (blue dash-dot line).

5. Conclusions

In this paper, we have theoretically analyzed the role of electron–vibration and electron–electron interactions on the thermoelectric properties of molecular junctions focusing on devices based on fullerene. We have used a self-consistent adiabatic approach which allows a non-perturbative treatment of the electron coupling to low frequency vibrational modes, such as those of the molecular center of mass between metallic electrodes. This approach incorporates Coulomb blockade effects due to strong electron–electron interaction between molecular degrees of freedom. We have analyzed a one-level model which takes into account the LUMO level of fullerene and its alignment to the chemical potential. We have stressed that an accurate description of the transport properties is obtained in the intermediate regime for the electron–vibration coupling and in the strong coupling regime for the electron–electron interaction. Moreover, we have demonstrated that only the combined effect of electron–vibration and electron–electron interactions is able to predict the correct behavior of both the charge conductance and the Seebeck coefficient. The theoretical calculations presented in this paper show a very good agreement with available experimental data of both charge conductance and Seebeck coefficient.

In this paper, we have used a one-level transport model as a starting point to address the role of many-body interactions between molecular degrees of freedom. This model is frequently used in all cases where the energy levels can be tuned around the chemical potential and additional spectral features are absent [4,17]. This is the case of the experiments in [17] for the fullerene junctions analyzed in this paper. The one-level model is expected to be valid for energies close to the Fermi level and low temperatures. Actually, a more realistic description of the molecule and its coupling with metallic leads is needed if more complex transport phenomena take place, in particular interference effects [47,48] recently investigated in molecular junctions. Moreover, inclusion of quantum corrections to oscillator dynamics can be important to explore the effects of additional vibrational modes and further electron–vibration regimes [49] (from adiabatic to anti-adiabatic ones) and their relation with strong electron–electron interactions [50,51].

Author Contributions: Writing—original draft preparation, C.A.P., Supervision, V. C.

Funding: C.A.P. was funded by project Fondo per il finanziamento delle attivita base di ricerca anno 2018.

Conflicts of Interest: The authors declare no conflict of interest.

References

1. Scheer, E.; Cuevas, J.C. *Molecular Electronics: An Introduction to Theory and Experiment*; World Scientific: Singapore, 2010.
2. Dubi, Y.; Di Ventra, M. Heat flow and thermoelectricity in atomic and molecular junctions. *Rev. Mod. Phys.* **2011**, *83*, 131.
3. Reddy, P.; Jang S.-Y.; Segalman, R.A.; Majumdar, A. Thermoelectricity in molecular junctions. *Science* **2007**, *315*, 1568–1571.
4. Yee, S.K.; Malen, J.A.; Majumdar, A.; Segalman, R.A. Thermoelectricity in Fullerene-Metal Heterojunctions. *Nano Lett.* **2011**, *11*, 4089–4094.
5. Finch, C.M.; Garcia-Suarez, V.M.; Lambert, C.J. Giant thermopower and figure of merit in single-molecule devices. *Phys. Rev. B* **2009**, *79*, 033405.
6. Murphy, P.; Mukerjee, S.; Moore, J. Optimal thermoelectric figure of merit of a molecular junction. *Phys. Rev. B* **2008**, *78*, 161406(R).
7. Galperin, M.; Nitzan, A.; Ratner, M.A. Inelastic effects in molecular junction transport: Scattering and self-consistent calculations for the Seebeck coefficient. *Mol. Phys.* **2008**, *106*, 397–404.
8. Koch, J.; von Oppen, F.; Oreg, Y.; Sela, E. Thermopower of single-molecule devices. *Phys. Rev. B* **2004**, *70*, 195107.
9. Leijnse, M.; Wegewijs, M.R.; Flensberg, K. Nonlinear thermoelectric properties of molecular junctions with vibrational coupling. *Phys. Rev. B* **2010**, *82*, 045412.
10. Perroni, C.A.; Ninno, D.; Cataudella, V. Thermoelectric efficiency of molecular junctions. *J. Phys. Condens. Matter* **2016**, *28*, 373001.
11. Cui, L.; Miao, R.; Jiang, C.; Meyhofer, E.; Reddy, P. Perspective: Thermal and thermoelectric transport in molecular junctions. *J. Chem. Phys.* **2017**, *146*, 092201.
12. Park, S.; Kang, H.; Yoon, H.J. Structure-thermopower relationships in molecular thermoelectrics. *J. Mater. Chem. A* **2019**, *7*, 14419–14446.
13. Paulsson, M.; Datta, S. Thermoelectric effect in molecular electronics. *Phys. Rev. B* **2003**, *67*, 241403(R).
14. Datta, S. *Lessons from Nanoelectronics: A New Perspective on Transport*; World Scientific: Singapore, 2012.
15. Park, S.; Yoon, H.J. New Approach for Large-Area Thermoelectric Junctions with a Liquid Eutectic Gallium-Indium Electrode. *Nano Lett.* **2018**, *18*, 7715–7718.
16. Kang, S.; Park, S.; Kang, H.; Cho S.J.; Song, H.; Yoon, H.J. Tunneling and thermoelectric characteristics of N-heterocyclic carbene-based large-area molecular junctions. *Chem. Commun.* **2019**, *55*, 8780–8783.
17. Kim, Y.; Jeong, W.; Kim, K.; Lee, W.; Reddy, P. Electrostatic control of thermoelectricity in molecular junctions. *Nat. Nanotechnol.* **2014**, *9*, 881–885.
18. Wang, R.Y.; Segalman, R.A.; Majumdar, A. Room temperature thermal conductance of alkanedithiol self-assembled monolayers. *Appl. Phys. Lett.* **2006**, *89*, 173113.

19. Wang, Z.; Carter, J.A.; Lagutchev, A.; Koh, Y.K.; Seong, N.-H.; Cahill, D.G.; Dlott, D.D. Ultrafast flash thermal conductance of molecular chains. *Science* **2007**, *317*, 787–790.

20. Meier, T.; Menges, F.; Nirmalraj, P.; Hölscher, H.; Riel, H.; Gotsmann, B. Length-Dependent Thermal Transport along Molecular Chains. *Phys. Rev. Lett.* **2014**, *113*, 060801.

21. Cui, L.; Hur, S.; Akbar, Z.A.; Klöckner, J.C.; Jeong, W.; Pauly, F.; Jang, S.-Y.; Reddy, P.; Meyhofer, E. Thermal conductance of single-molecule junctions. *Nature* **2019**.

22. Galperin, M.; Nitzan, A.; Ratner, M.A. Molecular transport junctions: Vibrational effects. *J. Phys. Condens. Matter* **2007**, *19*, 103201.

23. Park, H.; Park, J.; Lim, A.K.L.; Anderson, E.H.; Alivisatos, A.P.; McEuen, P.L. Nanomechanical oscillations in a single-C_{60} transistor. *Nature* **2000**, *407*, 57–60.

24. Zianni, X. Effect of electron-phonon coupling on the thermoelectric efficiency of single-quantum-dot devices. *Phys. Rev. B* **2010**, *82*, 165302.

25. Entin-Wohlman, O.; Imry, Y.; Aharony, A. Three-terminal thermoelectric transport through a molecular junction. *Phys. Rev. B* **2010**, *82*, 115314.

26. Liu, J.; Sun, Q.-F.; Xie, X.C. Enhancement of the thermoelectric figure of merit in a quantum dot due to the Coulomb blockade effect. *Phys. Rev. B* **2010**, *81*, 245323.

27. Ren, J.; Zhu J.-X.; Gubernatis, J.E.; Wang, C.; Li, B. Thermoelectric transport with electron-phonon coupling and electron-electron interaction in molecular junctions. *Phys. Rev. B* **2012**, *85*, 155443.

28. Arrachea, L.; Bode, N.; von Oppen, F. Vibrational cooling and thermoelectric response of nanoelectromechanical systems. *Phys. Rev. B* **2014**, *90*, 125450.

29. Mozyrsky, D.; Hastings, M.B.; Martin, I. Intermittent polaron dynamics: Born-Oppenheimer approximation out of equilibrium. *Phys. Rev. B* **2006**, *73*, 035104.

30. Pistolesi, F.; Blanter Ya, M.; Martin, I. Self-consistent theory of molecular switching. *Phys. Rev. B* **2008**, *78*, 085127.

31. Hussein, R.; Metelmann, A.; Zedler, P.; Brandes, T. Semiclassical dynamics of nanoelectromechanical systems. *Phys. Rev. B* **2010**, *82*, 165406.

32. Bode, N.; Kusminskiy, S.V.; Egger R.; von Oppen, F. Current-induced forces in mesoscopic systems: A scattering-matrix approach. *Beilstein J. Nanotechnol.* **2012**, *3*, 144–162.

33. Nocera, A.; Perroni, C.A.; Marigliano Ramaglia, V.; Cataudella, V. Stochastic dynamics for a single vibrational mode in molecular junctions. *Phys. Rev. B* **2011**, *83*, 115420.

34. Nocera, A.; Perroni, C.A.; Marigliano Ramaglia, V.; Cataudella, V. Probing nonlinear mechanical effects through electronic currents: The case of a nanomechanical resonator acting as an electronic transistor. *Phys. Rev. B* **2012**, *86*, 035420.

35. Nocera, A.; Perroni, C.A.; Marigliano Ramaglia, V.; Cantele, G.; Cataudella, V. Magnetic effects on nonlinear mechanical properties of a suspended carbon nanotube. *Phys. Rev. B* **2013**, *87*, 155435.

36. Perroni, C.A.; Ninno, D.; Cataudella, V. Electron-vibration effects on the thermoelectric efficiency of molecular junctions. *Phys. Rev. B* **2014**, *90*, 125421.

37. Perroni, C.A.; Nocera, A.; Cataudella, V. Single-parameter charge pumping in carbon nanotube resonators at low frequency. *Europhys. Lett.* **2013**, *103*, 58001.

38. Perroni, C.A.; Romeo, F.; Nocera, A.; Marigliano Ramaglia, V.; Citro, R.; Cataudella, V. Noise-assisted charge pump in elastically deformable molecular junctions. *J. Phys. Condens. Matter* **2014**, *26*, 365301.

39. Nocera, A.; Perroni, C.A.; Marigliano Ramaglia, V.; Cataudella, V. Charge and heat transport in soft nanosystems in the presence of time-dependent perturbations. *Beilstein J. Nanotechnol.* **2016**, *7*, 439–464.

40. Perroni, C.A.; Ninno, D.; Cataudella, V. Interplay between electron-electron and electron-vibration interactions on the thermoelectric properties of molecular junctions. *New J. Phys.* **2015**, *17*, 083050.

41. Haug, H.; Jauho, A.-P. *Quantum Kinetics in Transport and Optics of Semiconductors*; Springer: Berlin, Germany, 2008.

42. Hsu, B.C.; Chiang, C.-W.; Chen, Y.-C. Effect of electron-vibration interactions on the thermoelectric efficiency of molecular junctions. *Nanotechnology* **2012**, *23*, 275401.

43. Kruchinin, S.; Pruschke, T. Thermopower for a molecule with vibrational degrees of freedom. *Phys. Lett. A* **2014**, *378*, 1157–1161.

44. Lu, X.; Grobis, M.; Khoo, K.H.; Louie, S.G.; Crommie, M.F. Spatially Mapping the Spectral Density of a Single C_{60} Molecule. *Phys. Rev. Lett.* **2003**, *90*, 096802.

45. Koch, J.; von Oppen, F.; Andreev, A.V. Theory of the Franck-Condon blockade regime *Phys. Rev. B* **2006**, *74*, 205438.

46. Jezouin, S.; Parmentier, F.D.; Anthore, A.; Gennser, U.; Cavanna, A.; Jin, Y.; Pierre, F. Quantum Limit of Heat Flow Across a Single Electronic Channel. *Science* **2013**, *342*, 601–604.

47. Bai, J.; Daaoub, A.; Sangtarash, S.; Li, X.; Tang, Y.; Zou, Q.; Sadeghi, H.; Liu, S.; Huang, X.; Tan, Z.; et al. Anti-resonance features of destructive quantum interference in single-molecule thiophene junctions achieved by electrochemical gating. *Nat. Mater.* **2019**, *18*, 364–369.

48. Li, Y.; Buerkle, M.; Li, G.; Rostamian, A.; Wang, H.; Wang, Z.; Bowler, D.R.; Miyazaki, T.; Xiang, L.; Asai, Y.; et al. Gate controlling of quantum interference and direct observation of anti-resonances in single molecule charge transport. *Nat. Mater.* **2019**, *18*, 357–363.

49. Perroni, C.A.; Ramaglia, V.M.; Cataudella, V. Effects of electron coupling to intramolecular and intermolecular vibrational modes on the transport properties of single-crystal organic semiconductors. *Phys. Rev. B* **2011**, *84*, 014303.

50. De Filippis, G.; Cataudella, V.; Mishchenko, A.S.; Perroni, C.A.; Nagaosa, N. Optical conductivity of a doped Mott insulator: The interplay between correlation and electron-phonon interaction. *Phys. Rev. B* **2009**, *80*, 195104.

51. Perroni, C.A.; Cataudella, V.; De Filippis, G.; Marigliano, Ramaglia, V. Effects of electron-phonon coupling near and within the insulating Mott phase *Phys. Rev. B* **2005**, *71*, 113107.

Review

Unusual Quantum Transport Mechanisms in Silicon Nano-Devices

Giuseppe Carlo Tettamanzi

Institute of Photonics and Advanced Sensing and School of Physical Sciences, The University of Adelaide, Adelaide SA 5005, Australia; giuseppe.tettamanzi@adelaide.edu.au; Tel.: +61-(0)-883130248

Received: 23 May 2019; Accepted: 9 July 2019; Published: 11 July 2019

Abstract: Silicon-based materials have been the leading platforms for the development of classical information science and are now one of the major contenders for future developments in the field of quantum information science. In this short review paper, while discussing only some examples, I will describe how silicon Complementary-Metal-Oxide-Semiconductor (CMOS) compatible materials have been able to provide platforms for the observation of some of the most unusual transport phenomena in condensed matter physics.

Keywords: mesoscale and nanoscale physics; Complementary Metal Oxide Semiconductor (CMOS) technology; quantum transport

1. Introduction

The analogy between the hydrogen atom and donor and acceptor dopants in semiconductors has been known for a very long time—see References [1,2] and references therein. However, mainly during the latter half of the past century, in experiments related to conventional semiconductor physics, dopant-atoms have been used as passive elements for several kinds of logic building blocks, for example, P-N junctions [1]. Under this original implementation method, many dopants were implanted in semiconductor materials to obtain a primitive control of their band gap and to create a barrier that could, in turn, be used to control the transition between the OFF state and the ON state (or vice versa) of a transistor made with these materials [1]. However, towards the end of the past century, limitations to this primitive approach to band gap engineering became apparent. Successively, many researchers around the world started to fabricate individual devices, which allowed much better control of the band gap via heterojunctions and other sorts of nanostructures [1,3]. A heterojunction is formed when two different semiconducting materials, typically with a small mismatch on the size of their lattice constants, are deposited one on top of the other. By doing so, it is possible to obtain a better control on the band gap of the materials, and consequently, to achieve fundamental advancements in the properties of optoelectronics devices [1,3,4]. While from their optical points of view, heterostructures fabricated via heterojunctions, such as quantum wells and superlattices, have already demonstrated the ability to achieve many of their electronic potentialities [3], zero-dimensional heterostructures (i.e., quantum dots), have also been introduced [3]. The latter potentially give rise to equivalent individual atoms in a semiconductor environment, and can be used as so-called artificial atoms [1,3–5]. In this context, there is a long list of materials, for example see References [6,7], that have been used for the observation of these exciting features of mesoscopic and "many-body" effects. In this short review paper, however, I will limit my discussion on some of the effects recently observed in devices fabricated with CMOS compatible materials [4,5].

2. Introduction to New Results

This review paper discusses how silicon CMOS compatible devices have recently revolutionized not only the fields of quantum information and quantum communications, e.g., see Ref. [4,5], but also those related to mesoscopic physics because of their demonstrated ability to compete with exotic, but not as yet commercially available materials as required for larger-scale applications, such as graphene [6] or carbon nanotubes [7]. Bearing in mind that silicon CMOS compatible devices remain the core platform for most of the traditional information processing infrastructures, this new development represents an impressive achievement [4,5].

To provide further details, this review paper aims at discussing how silicon CMOS compatible ultra-scaled devices can be used for the study of very elusive problems, for example the orbital Kondo effect [8,9], the Kondo-Fano effects [10,11] as well as the mechanism of errors in single electron pump (SEP) devices [12–15].

One of the interesting aspects is that these silicon-based devices possess some peculiar and unique properties, if compared to the ones present in other semiconductor materials, due to the fact that, for silicon-based materials, the electron band gap exists in an indirect form [4,5]. These properties can be summarized by using the schematic picture of Figure 1 and by mentioning that electrons in silicon possess an extra degree of freedom, known as the "valley-orbital pseudo spin degree of freedom" [4,8–11,16,17], which occurrence is related to the fact that, for bulk silicon materials, the energies of electron forms in the conduction band (CB) are not minimized when the crystal momentum **k** is equal to 0 and are 6-fold degenerate at the minimum point [4,8–11,16,17]. The use of the term "pseudo-spin" in conjunction with the terms "valley" and "orbital" is somewhat controversial. Nevertheless, it is often used in many recent publications, see for example References [4,8–11,16,17] However, in this document, I have repetitively made use of this term in the context of silicon valley-orbital degrees of freedom.

Consequently, the pseudo-spin can act as a good quantum number [18,19], and as Figure 1 schematically shows, the level degeneracy observed at the minimum of the CB can be partially or completely lifted by means of confinement effects see References [1–4] and references therein; hence, this degree of freedom can be utilized as a platform for novel quantum logic operations, see References [4,5] and references therein.

As an example, qubits based on the valley-orbital degree of freedom can be better isolated against the deleterious effects of charge noise as opposed to the ones based on more conventional degrees of freedom such as spin or charge [20]. Even though it is true that devices utilizing valley-orbit states are limited by the fact that deterministic control of the valley splitting is far from being routinely achieved [4,20]. In this review paper, I will show that, although the indirect bandgap properties of silicon were for a long time considered as a negative attribute of silicon materials, especially from the optoelectronic point of view, more recently these properties have demonstrated to be silicon material strengths, especially in view of the possible uses of silicon devices for quantum applications [4,5]. This is particularly true because of the numerous significant advancements made in the technologies used in the fabrication of silicon CMOS compatible devices [5]. In view of this, it is now possible to obtain extremely precise control of the quantum properties of electrons confined in silicon CMOS compatible structures even during fabrication [4,5] and it is also possible to engineer devices precise to a single atom level [21–24]. These new enhanced fabrication capabilities have translated into an improved ability for the control of all the quantum degrees of freedom (for example spin, charge, pseudo-spin) of electrons [8–17,20–24] and holes [25] in silicon nanostructures [4,5]. Finally, although there are still a few open questions in this area, for example see Reference [4], the achievements described above have translated into improved control of the electronic signature that can be observed in these devices. This, in turn, explains why these systems have also been used recently as platforms for the observation of some of the most remarkable effects in physics [4,8–15], in the same manner in which they were previously observed in other materials such as carbon nanotubes and graphene [6,7].

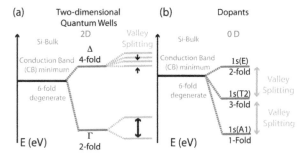

Figure 1. Schematics of the bottom of the conduction band (CB) for a few typical Silicon systems. The effects of confinement, of electrical effects and of structural atomic effects to the CB structure are included in (**a**) for two-dimensional Quantum Well and on (**b**) for an isolated dopant-atom impurity such as arsenic (As) or phosphorous (P), see also Reference [4]. Two-fold spin degeneracies are not included in this illustration [4].

2.1. Special Properties of the Electrons in Silicon CMOS Compatible Devices

As already mentioned above and to be more specific regarding the characteristics of these silicon-based systems, it is important to note the fact that the energetic structure at the minimum of the conduction band of silicon devices fabricated on basic structures of quantum dots and single atom nanostructures can be quite different when compared to the energetic structures observed in bulk materials [1–4], see the schematic picture in Figure 1.

This observation justifies the proven capability of silicon to host nano-devices, e.g., see schematic in Figure 2, which allow a good control of the spin degree of freedom of electrons, due to its weak spin-orbit coupling and to the existence of isotopes of silicon with zero nuclear spin [4]. Hence, by using silicon devices, it is possible to engineer a well-controlled quantum environment that can act as the ground for the observation of several exciting mesoscopic physics effects. These phenomena will be the focus of this review paper.

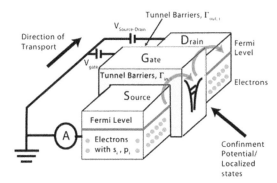

Figure 2. Schematic of a three-terminal device with one or more than one gate controlling the tunneling barrier from the source to the state, $\Gamma_{in,i}$, and the tunneling barrier from state to the drain, $\Gamma_{out,i}$. The leads (source/drain) represents an infinite reservoir of electrons with the all the possible spin (s_i) and valley-orbital (p_i) polarizations. The transport in the device can be controlled by applying a voltage to the V_{gate} terminal that controls the position of the quantum states in the confinement potential respectively to the Fermi level in the source and drain leads. In this configuration, the V_{gate} terminal can also control the transparency of the tunneling barriers $\Gamma_{in,i}$ and $\Gamma_{out,i}$. The $V_{Source-Drain}$ voltage at the drain terminal can control the polarity and the intensity of the current of electrons, while the source terminal is connected to a pico-ammeter. As the figure shows, all the elements of this circuit are connected to the same reference grounding.

Several recent review papers and books, i.e., see References [4,5,26,27] and references therein, have described in detail many of the different approaches that can be used to fabricate these silicon devices. Consequently, my current review will simply refer to the relevant publication and, in instances where it is necessary to elaborate on a particular point, I will provide additional information covering fabrication techniques, without repeating what has already been reported in the relevant publications such as References [4,5,26,27].

In order to understand the peculiar behavior observable with silicon nanostructures [4,5,26,27], it is important to initially describe in detail the structure of the minima of the conduction band introduced above, see Ref. [1–4] and Figure 1.

As silicon has an indirect band gap [1,2] and because of its lattice symmetry, its conduction band minima has six degenerate minima (or valleys) at the point k = 0.85 k_0 of energy vs. wave vector, **k**, diagram in the reciprocal space, as also schematically shown with the black lines in Figure 1, with k_0 being the wave vector that defines the size of the unit cell in the reciprocal lattice space of the material [1,2]. This peculiar band structure, unlike other materials that have a direct band gap, e.g., gallium arsenide (GaAs), produces a situation in that electrons in silicon have an extra degree of freedom, which can be used for their quantum dynamical control; i.e., the valley-orbital pseudo-spin degree of freedom introduced above. As such electrons in silicon are said to be affected by multi-valley physics [4].

Furthermore, because, in these structures, most of the low temperature transport of electrons is located around the minimum of the conduction band, the multi-valley physical properties described in Figure 1 play a critical role for most of the mesoscopic low-temperature effects, as described in this review paper. Moreover, I have no desire to cover in the extensive amount of materials represented by the more than fifty years of research in the fields of silicon microelectronics [1,3] and silicon nano-electronics [4]. However, I would like to discuss some interesting and more recent progress in the field of silicon nano-electronics [4]. Moreover, another interesting aspect is that, as shown in Figure 1, both for donor impurities in silicon, such as arsenic or phosphorous, for two-dimensional (quantum wells) and for zero-dimensional hetero-structures (i.e., quantum dots), due to the severity of effects such as confinement, electric fields with strong gradients, lattice imperfections, and atomic-scale details at the interface between different sections of a device, most of the energy levels degeneracy at the minimum of the conduction bands can be lifted. For a Quantum Well (QW), the 6-fold valley degeneracy is broken by the large in-plane tensile strain to a 2-fold degenerate Γ levels that are below 4-fold degenerate Δ levels [4]. These degeneracies can in turn be broken by all other atomic effects that go under the name of valley-splitting see Figure 1a). For isolated dopant-atoms in the silicon lattice, i.e., single atom transistors [8–13,17,21–24], in the most ideal situation, the final configuration gives a 1-fold degenerate 1s state (A1), below a 3-fold degenerate 1s and another 2-fold 1s state, as shown in Figure 1b), while other intermediate situations are possible [8,9]. As an example, the lower energetic valley-orbital states are fundamental for the observation of the Kondo and the Kondo-Fano effects described in the section below. These properties are also a fundamental ingredient for the successful operation at high frequency (≥GHz) of silicon based quantum electrons pumps [12–15].

The lifting of the valley-degeneracy of the electrons in quantum dot or quantum well devices in silicon can have multiple consequences; first it can allow the observation of novel quantum effects [8–15]. Furthermore, the study of this novel degree of freedom has demonstrated that this valley (pseudo-spin) degree of freedom can also act as a good quantum number [18,28] for electrons, and as such, it could be used for their coherent control in the implementation of quantum gates [4,20,28].

It is also important to clarify that, when electrons are confined in isolated dopant atoms or QD nanostructures an orbital order for the energy levels of the states also appears [4]. This orbital order in the hierarchy of the states can co-exist with the hierarchy of the distribution imposed by the physics of valleys [4]. Sometimes valley and orbital effects can lead to very distinguishable consequences in the structure of the energy levels [4], however, because these effects can typically be influenced in many ways by the microscopic structures of these devices, the most common way for valley/orbital effects to manifest themselves in silicon nano-structures is in a mixed configuration [4]. This explains why, for most silicon nanostructures, the term "valley-orbit degree of freedom" is often in use as opposed to the terms "valley degree of freedom" or "orbital degree of freedom" [20,28]. It is also interesting to briefly mention that the signature of these complex valley-orbit effects imposes selection rules that lead to dramatic alterations of the observed electronic effects [28].

2.2. Kondo Effects in Silicon Nanostructures

The three-terminal nano-device [4,5,8–12,16,17], illustrated in Figure 2 is a good example of the most simplified geometry that can be used to observe the quantum effects in silicon, when operating at sufficiently low temperatures.

As shown in Figure 2, in this type of device, two terminals are used for the source and the drain leads, which therefore dictate the direction of transport according to the polarity of the bias voltage applied on the $V_{Source-Drain}$. The third terminal is typically used to apply a voltage to the V_{Gate} and in turn, to control the transition of the device between the ON state and the OFF state at room temperature [1,29,30] and the position of the quantum level that is present in this system at low temperatures when Coulomb effects are observable [4,29,30]. For geometries, like the one shown below in Figure 2, the V_{Gate} can in principle also control the transparency of the tunneling barriers $\Gamma_{in,i}$ and $\Gamma_{out,i}$. While only more complicated structures, for example see the double gate structure, as discussed later on in this review paper, may offer the possibility for an independent control of the position of the quantum states and of the transparency of $\Gamma_{in,i}$ and of $\Gamma_{out,i}$, even a simple structure like the one described in Figure 2, it is possible to observe many-body effects at sufficient low-temperatures [8–12]. In this review paper, I will often use this geometry as a platform for the description of the interesting effects that can be observed in silicon nano-devices, since the said platform has already demonstrated the capability to offer access to the quantum systems, as illustrated in Figure 1.

The most typical first order regimes of transport, i.e., non-involving virtual states [4,8,9], which can be observed in three terminal nano-devices at low temperatures are the sequential and the coherent regimes of transport [4]. A typical Coulomb Blockade differential conductance ($G = dI_{SD}/dV_{SD}$) signature is shown inside the areas delimited by the pink dashed lines in Figure 3. In this typical Coulomb Blockade situation, only first order transport is expected to arise, and only within those regions limited by the pink lines, while everywhere else no transport should be allowed [4].

However, from Figure 3, it is also clear that, for these systems, it can be affected by many higher order body effects. As such transport can arise in regions represented by the stability diagram in Figure, where it is not expected to, if only first order results are taken into consideration [4]. In fact, in this section, we are mainly interested in discussing in more detail some of these unexpected effects. These involve coherent and incoherent second order charge transitions, such as the ones related to the apparent violation of Heisenberg uncertainty principle and indeed can arise outside the conventional Coulomb Blockade picture. The Kondo, see References [8,9] and references therein, and the co-tunneling, see References [4,31], regimes of transport become observable when the temperature of the system is below a certain critical temperature (e.g., the Kondo Temperature) and when quantum fluctuations in the spin or in the pseudo-spin degrees of freedom, as opposed to temperature ones, become the leading fluctuation effects [4,8–11,32–34].

More about These Kondo Effects

In this section, by observing the schematic of Figure 4, I will describe the main features that can be associated with the different versions of these Kondo effects.

In its most conventional manifestation, the Kondo regime of transport is linked to the availability of the electronic spin degree of freedom only [32]. In this situation, this effect is suppressed as soon as a sufficiently high magnetic field is applied to the system. This is schematically illustrated in the diagram of Figure 4a, which shows that when a dual spin degenerate level is available and if there is no external magnetic field applied, then quantum fluctuation (schematically described in Figure 4 with the dashed loops when observable in the current signature) can be effective in allowing transport in the Coulomb Blockade forbidden regions. Figure 4b also illustrates that, when a sufficient external magnetic field is in place and if the two spin levels are sufficiently separated energetically due to Zeeman splitting, fluctuation effects cannot be sufficiently efficient in generating an observable current [8–11,32,33], i.e., the Kondo effect is indeed suppressed. This effect is also observable in the experiments [32].

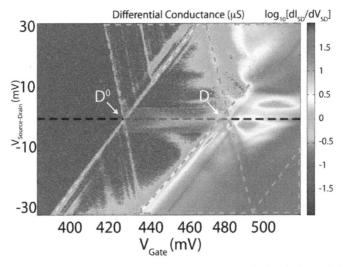

Figure 3. Example of Coulomb Blockade data similar to the ones in [7,8] and taken at 290 mK in a three-terminal device as the one schematically described in Figure 2. As an example, according to the orthodox Coulomb blockade theory, transport should arise only in the region delimited by the triangles contained within the pink dashed lines. Any transport signature outside these regions is linked to higher order effects [4,8,9]. An example of the latter is the signal present in the low bias, i.e., $|V_{Source-Drain}| < 5$ mV, region between the one-electron (i.e., D^0) and the two-electron (D^-) charge states [4]. This signal, also outlined with the rectangular transparent area, is linked to the observation of spin and orbital Kondo fluctuations. The low bias region after the two-electron D^- is most likely linked to the occurrence of the Kondo based on fluctuation of integer degree of freedom [34].

The simple picture introduced in the section above and in Figure 4a,b can become more complicated for systems for which the valley orbit pseudo-spin degree of freedom is also available for Kondo fluctuations, in the same manner as silicon systems, as illustrated in Figure 1 [8–11].

For the latter, a very unusual variety of Kondo, i.e., pure orbital Kondo, can also be observed [8–10,19,33], because valley-orbital pseudo spin is only lightly affected by magnetic field splitting. The occurrence of pure orbital pseudo-spin [19], i.e., not involving fluctuations of the spin of the electrons, Kondo effect in silicon nanostructures, is discussed here below:

(a) For conventional semiconductors, the Kondo effect has only been observed in relation to interactions between the spin of electrons confined within the localized state and the ones of the surrounding free electrons at sufficiently low temperatures (T's), i.e., for Temperature < Kondo temperature (T_{Kondo}), see Reference [32]. This situation is illustrated in Figure 4a, and as shown in Figure 4b, in this situation, the spin-Kondo effect is suppressed when a sufficiently high magnetic field is applied to the system because the Zeeman splitting between the spin up and the spin down of the electrons makes energetically impossible for spin fluctuations to generate virtual states that would open the Kondo transport channel [32]. This situation goes also under the name of conventional SU(2) Spin Kondo effect.

(b) In materials were the pseudo-spin degree of freedom is also available, however, as in the area in Figure 3 outlined by the rectangular shaded shape and as illustrated in Figure 4c, the Kondo effect is somehow different to the one shown in Figure 4a and as described in the above sections. A different situation from the one above has recently been observed and is evident both from the experimental and from the theoretical points of view in silicon CMOS three terminal devices [8,9]. The extension of the Kondo effect to valley-orbital degree of freedom is clearly illustrated in Figure 4c by introducing different colors (black and red) for the two-different valley-orbital levels involved in the effect, i.e., the two lowest states, as shown as degenerate in Figure 4c,d. Consequently, the Kondo effect observed in silicon nanostructures is a more sophisticated phenomenon that goes under the name of SU(4) Kondo effect [4,8,9].

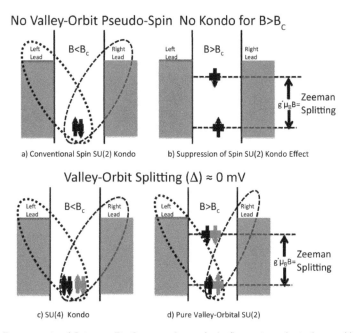

Figure 4. For conventional Quantum Dot Systems where only the fluctuations of spin degree of freedom are available [32], the Kondo effect is suppressed for sufficiently high magnetic fields (B_C) because of the Zeeman splitting makes energetically impossible for Kondo fluctuations to arise. Opposite to this, when observed in an opportunely tuned silicon system, the Kondo Effect can arise as the combined action of the fluctuations of the spin and of the pseudo-spin (Valley-Orbit) degrees of freedom. As the pseudo-spin is typically only lightly affected by the magnetic field, a pure Orbital version [7,8,24,32] of the Kondo effect survive even for $B > B_C$.

What makes these latter results even more interesting is the fact that a previously unobserved Kondo ground state symmetry crossover can be studied in these systems [8], because, as from one side it is possible to saturate the spin degree of freedom by mean of Zeeman splitting, as it is shown in Figure 4b in conventional Kondo systems, and from another side it is possible to be in a situation where the pseudo-spin fluctuations are still in place even if under the effect of a sufficiently high magnetic field, $B > B_c$, with B_c being a certain critical field [8]. As shown in Figure 4d, in this situation, it is possible to saturate only the spin degree of freedom, but also to observe a Kondo effect originating only from the valley-orbital quantum fluctuations, the so-called "pure" SU(2) orbital Kondo effect [8,9,19]. These results imply pure quantum screening of the orbital degree of freedom [8].

Due to the difficulty of accessing these regimes, the observation of these novel effects represents not only an important milestone for quantum many-body physics, but has also opened-up new pathways for silicon quantum electronics such as, for example, valley-orbital quantum bits [4,20]. Of course, these many-body effects cannot be directly used for the implementation of novel quantum information schemes, but their observation is very important, because it has permitted the study and the characterization, in a coherent fashion, of quantum states based on these pseudo-spin degrees of freedom that could, in the future, become the base for novel quantum schemes [20]. What makes the transition between the different versions of the Kondo effects, as illustrated in Figure 4c,d, ever more interesting is that this transition represents a universal phase transition equivalent to the ones observed in other systems having the same symmetry in the quantum degrees of freedom, e.g., some nuclear systems [35]. As shown under Figure 5, in this latter situation, a survival of a $T_K \neq 0$ and a constant evolution of this $T_K \neq 0$ are observed even for $B > B_c$. While for conventional SU(2) Kondo for $B > B_c$ T_K is $= 0$.

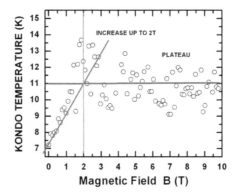

Figure 5. Universal law of the smooth transition between two different versions of the Kondo Effect (because observable in systems very different one from each other [8,35]) under the effect of an external magnetic field (i.e., from an SU(4) Kondo to an SU(2) Kondo effect). Even if this data does present some scattering, nevertheless, an initial linear behavior of the order parameter (Kondo Temperature, T_C) can be observed between 0 Tesla and 2 Tesla. This is most likely followed by a constant value of T_C for any $B > 2$ Tesla.

In conclusion, as also shown in Figure 5, by looking at the evolution of the order parameter of systems that have both spin and pseudo spin degrees of freedom (i.e., T_K in the case of our systems), it is possible to observe a unique universal law for the smooth transition between two different versions of the Kondo effect under the influence of an external magnetic field. These T_K's can be extracted by fitting the curves that describe how the current signature evolves at different temperatures [8]. Even if the data in Figure 5 does present some scattering, nevertheless, the experiment opens up an unique window to the characterization of universal physical behaviors that are expected to be observable in different systems [8,9,35].

2.3. Kondo-Fano Effects in Silicon Nanostructures

Another interesting aspect of the Kondo effect is that it provides the opportunity, when the quantum channel of transport is coherent, to observe phenomena that have for a long time been confined to only the optical side of the physical experiments. In this context, silicon nano-systems like the ones discussed in the previous sections, have recently been able to demonstrate [10,11] that they can manifest the Fano effect, which consists in the interference of a discrete coherent channel with a continuum and which gives rise to characteristically asymmetric peaks in the response [10,11,36]. Another way to describe this effect is by mentioning that specially shaped asymmetrical peaks, in this case current peaks, will appear every time with a path of rapid phase variations with almost constant phase interferences [10,11,36]. In a geometry like the one illustrated in Figure 2, the interference pathway can arise when more than one channel of conduction becomes available between the source and the drain terminals. This multi-channel configuration, in conjunction with the Aharonov-Bohm effect [37], allows the observation of the version of the effect that goes under the name of Fano-Kondo effect [10,11]. In the example discussed in References [10,11], the constant continuum is provided by a sequentially tunneling channel, while the coherent channel is provided by the Kondo effect due to correlated effect in the Coulomb Blockade regime as described in the previous section of this paper. Each of these two channels is linked to the transport via an atomic confinement potential located somewhere in the channel of the three-terminal device. A slave-boson mean-field approximation within the scattering matrix formalism was used to ensure that the data observed in Reference [10] was correctly interpreted within the framework of the Fano-Kondo effect [11].

2.4. Charge Pumping Effects in Silicon Nanostructures: Single Electron Pumps

In 1983, D.J. Thouless published a seminal paper [38] predicting that a periodic perturbation of the confinement potential for electrons in a nanostructure would, in principle, allow the generation of topologically protected band levels. These levels could lead to energetic regions where the transport of electrons from the source to the drain could be precisely controlled, even when the voltage between the source and the drain is zero, which statement is an apparent violation of one of the fundamental principle of electronics, Ohm's law. By taking advantage of Coulomb Blockade phenomena's and of quantum tunneling effects [4,5], this concept was expected to enable the generation of ultra-accurate currents [12–15]. See Figure 6a for an illustration of a typical setup that can be used for the kind of measurements described in this section [12–15].

The quantification of the currents generated under the above-mentioned technique, which goes under the name of quantized pumping, is based on the formula $I_{Source-Drain} = n^*f^*e$, see Figure 6b,c, where e is the elementary charge ($1.6021766208 \times 10^{-19}$C), n is a positive integer indicating the number of transported electrons for each period or cycle, and f is the frequency of perturbation of the confinement potential, with $f = 1/\tau$ and τ being the period. This technique allows the implementation of high performant clocked single-electron sources, i.e., generating sufficiently high currents with sub-parts per million (sub-ppm) uncertainties [12–15]; thanks to shot noise being naturally suppressed in these experiments [39], while adverse temperature and flicker noise effects can be substantially suppressed [12–15]. The basic idea of these experiments aims at the generation of a current flowing through a quantum dot/single atom impurity without applying a voltage between the leads, but by specifically varying potential at one or more gates. Furthermore, as Figure 6a shows, the schematic of the most common devices that are used for charge pumping experiments are slightly more elaborated, if compared to the one discussed in Figure 2, see also the many device geometries described in Reference [5].

In Figure 6a, it is shown that at least two terminals are needed for an independent control of the two gates and, therefore, of an independent control of the in-to state and the out-from state tunnel barriers (i.e., Γ_{in} and Γ_{out}). This is different from the geometry schematically described in Figure 2 where only one terminal was used for the control of both these tunnel barriers. One of the most important aspects of this research is the fact that by operating the voltage gates in an opportune way, it is possible to obtain the appropriate sequential time-evolution of the voltages applied to the two gates, and therefore it is possible to generate an opportune time-evolution of the transparency of the tunnel barriers (i.e., $\Gamma_{in,i}$ and $\Gamma_{out,i}$) that allows the transport of exactly n electrons between the source and the drain each cycle [40]. The number of cycles per seconds ($=f$) will then determine the intensity of the current according to the law $I_{Source-Drain} = n^*f^*e$. This description of the quantum pumping current clarifies why these experiments where the current is made by controlling electrons one-by-one are expected to generate precise/accurate currents. A sequence schematically describing the different sections of an ideal pumping cycle for a single atom pump [12,13] is set out under Figure 7, describing (a) the capture section of the cycle, (b) the isolation section, and (c) the emission section [13].

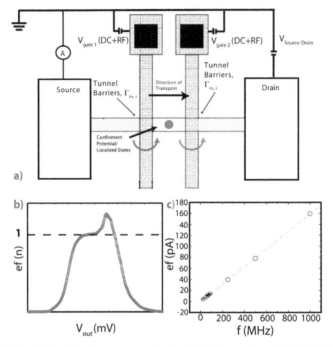

Figure 6. (a) Schematic of a typical measurement setup for a Single-Electron Pump [11–14]. (b) Characteristic *nef* response for *n* = 1 for a Single-Electron Pump [12]. (c) Characteristic *ef* response at different f (from a few MHz to 1000 MHz) for a Single-electron Pump [12].

Finally, if ideal quantum pumping transport is achieved, it is possible to obtain an ultra-fast and accurate circulation of an electrical charge (i.e., electrons) from the source to the drain [12–15,40]. The technical details on the manner in which these experiments can be performed is beyond the scope of my present review, see for example, the experiment descriptions in References [12,14,15]

I would like to conclude this review paper by describing in more detail the physics of the errors/non-ideal behaviors that can be observed in quantum pumps and I would like to discuss how these errors are, in particular, linked to silicon valley-orbital effects. Such errors can be seen as detrimental effects as they can cause the degradation of the currents measured in these charge pumping experiments. Interestingly, those experiments that were originally aimed at obtaining an accurate quantum definition of the ampere [12–15], more recently have opened up new directions in the fields of quantum information science [41–43] and classical signal processing [44].

2.5. Errors during the Operations of Single Electron Pumps

To understand the mechanisms of operations of an ideal single electron pump, it is important to remind ourselves that the main result, which is required when these devices are in operation is to obtain the most accurate possible control of electrons during each one of the fast cycles that are used for the circulation of electrons from the source to the drain via the localized state. Hence, the mainstay of these experiments relies on the ability to obtain a synchronized control of all the tunneling rates; the ones that control the movement of electrons from the source to the ones of the levels of the localized state (i.e., $\Gamma_{in,i}$) and the ones related to the movement of electrons from the localized state to the drain (i.e., $\Gamma_{out,i}$) [12,13]. Of course, the different relaxation rates internal to the localized state will also play a fundamental role in these experiments [12,13,15,25,45–52]. Here, it is important to mention that these experiments are mostly performed with source-drain bias at 0 V or in a region for which changes in the values of the source-drain bias are non-influent [12–15], i.e., the situation has given rise to the assumption that these pumping experiments represent a violation of an ideal definition of Ohm's law [38].

The central idea of these experiments is to obtain optimal dynamical control of the different tunnel barriers that control the movement of electrons from source to drain. In turn, this is expected to lead to an optimization in the transport properties even when high frequencies, f, of operations are in use [12–15]. The conventional model describing the ideal operations of single-electron pumps, based on quantum dot (QD) confinement site, is often described as the Decay-Cascade model [46]. In this model, the pumping cycles can be schematically described with these three cyclical steps: (a) Capture from the source which for a QD is an easy occurrence because the dot is typically semi-open in the first section of the cycle [15,46]; (b) isolation into the localized state of n electrons; and (c) emission to the drain of n electrons, see also Figure 7 for a description of a similar cycle valid for a single atom pump [12,13]. If the ideal picture described above is followed during experiments, then the system will generate a current that is exactly equal to $n*f*e$.

However, I would mention that something could go wrong during each of the steps described above and the quantum pump could lose control of a certain number of electrons for each cycle. In turn, this can be the cause of degradation in the precision/accuracy of the current produced by the single-electron pump, and therefore this can lead to the generation of errors. As set out below, I have endeavored, based on the simplified schematic of Figure 7, to provide a simplified explanation to the most important mechanisms of errors that can be observed in these systems:

(a) The first mechanism of error is the one that becomes important when the temperature of operation (T) of the single electron pump is energetically comparable to the charging energy (E_C) of the confined states, i.e., $E_C \approx K_B T$, with the charging energy being the energy that the system must pay to increase by one electron the number of electrons in the localized state between the two gates, i.e., from N to N + 1. These thermal effects could lead to losing control of many electrons in each step of the cycle described in Figure 7. The probability of the occurrence of this type of errors is linked to the formula $e^{-\frac{E_C}{K_B T}}$ Hence $E_C \gg K_B T$ means that this probability is almost zero. As $K_B T \sim 24$ meV for T = 300 K and ~0.24 meV for T = 3 K, the typical $E_C \sim 30$–50 meV observed in single impurity/atom systems makes that these are immune to temperature errors even when they are operated above the liquid Helium temperatures (≥ 4.2 K). This explains why single-atom based single electron pumps (SAP's) are extremely advantageous, as they do not need to be kept at ultra-low temperatures (<1 K) to operate in an environment completely immune from detrimental temperature effects/errors [12,13]. QD electron pumps are often limited in this sense as E_C for these systems is often limited to less than a few meV and as such require some complicated sub-kelvin temperature of operations to demonstrate their best performances [14]. However, as soon as the E_C of a QD increase to a value like the ones observed naturally in isolated single atom systems, for example by electrostatic confinements [14,47], these errors can also be suppressed in QDs electron pumps operating at temperatures ~4.2 K [14,15].

(b) Another mechanism of errors that can affect single electron pumps, when they operate slightly above the 100 MHz frequencies of excitation, is the one related to non-adiabatic effects [45,48]. These errors are related to the poor efficiency in the achievement of the second step of the pumping cycle, as described in Figure 7b, i.e., the isolation step. This poor efficiency can be observed when the rates that control the back-tunneling of the electrons from the localized state back to the source, Γ_{back}'s, allow the escape of the electrons to the source before the full isolation or before the emission to the drain [46–48]. This kind of error can particularly affect QD electron pumps as, for these systems, electrons are strongly affected when fast perturbations are exciting the system. At high frequencies of operations, these excitations, i.e., non-adiabatic excitations [48], can lead to the delocalization of electrons between the ground-state and the excited states and as for QDs the rates that govern the tunneling between the excites state and the source/drain leads are fast, if compared to the ones between the ground state and the source/drain leads, hence, when electrons are delocalized, their probability of non-completion of the isolation step is much higher than normal [40,48]. Ultimately, this could lead to errors since it means that electrons will not be emitted to the drain and will not complete their cycle [48].

The detrimental mechanisms described above are much less likely to happen in SAPs as for these systems the shape of confinement potential is not as affected by thermal or non-adiabatic effects as much as it is in the case of QDs pumps.

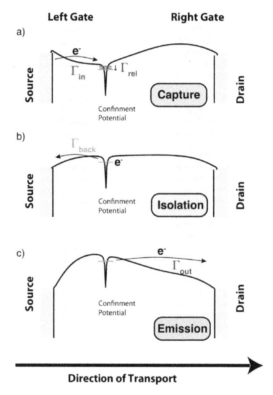

Figure 7. Ideal pumping cycle for a single atom pump [12,13] showing a schematic description for the different steps; (**a**) the capture section of the cycle, (**b**) the isolation, and (**c**) the emission.

Consequently, even when they are operating at 4.2 K and at GHz frequencies of excitation, non-adiabatic effects are not as efficient in causing errors in SAPs, as opposed to QDs pumps [12,13]. Furthermore, for SAPs the relaxation rates ($\Gamma_{relaxation}$, see also schematic in Figure 7a) controlling the relaxation of the electrons from excited states to the ground states are considerably faster [24] when compared to the equivalent ones observed in QD pumps [48]. Here, it is very important to emphasize that the fast relaxation rate effects observed naturally in SAPs are directly linked to the energy spectrum that the multi-valley physics imposes on these silicon systems [12,13,24]. Thus, for SAPs, the isolation step of the pumping cycle can always be reached efficiently. This also leads to a different way to operate these quantum pumps, based on the initial capture via excited state and sub-sequent fast relaxation to a well-isolated ground state [12,13]. As such, it is important to outline that the high performances observed in SAP's are linked to the indirect band gap properties characteristic of silicon materials. It is also important to associate the non-adiabatic effects observed for $f = 1/\tau$ between 100 MHz [48] and a few GHz [45] to a recent set of results hinting to the ability of studying the ultra-fast coherent dynamics of electrons in highly reproducible silicon CMOS compatible devices [45].

(c)   The alternative way to operate a SAP described above can be relatively error-free, unless these systems are excited to frequencies considerably higher than the GHz ones [12,13]. Consequently, the discussion above opens the way to the description of another kind of errors that could arise in QD or in single atom pumps [13] when electrons reach the confinement potential via an excited state and not the ground state. If the $f = 1/\tau$ approaches the values of $\Gamma_{relaxation}$ described in Figure 7, see also Reference [13]. In this situation, the electrons do not have sufficient time to relax to the ground state and the completion of the isolation step is compromised. The picture above can also be used to understand the causes of errors and of the degradation of the precision/accuracy of the measured currents in SAP's [13].

Other complicated versions of the kind of errors described under sections (b) and (c) above have been observed in other systems [50–52] and are often linked to the fact that more than one confinement site is playing a role in the control of electrons [50–52]. It is important to remind that multi-valley physics effects play an important role in the dynamical evolution of all these errors, see for example Reference [13].

(d)   Lastly, I would like to briefly discuss another possible mechanism of error that can cause the degradation of the current and which has recently been observed in a silicon QD system [15]. For a system where a QD pump is operating at ultra-fast frequencies of excitations (up to 3.55 GHz), it has been shown that the ideal behavior of the pump can sometimes be affected by errors that appear when an impurity-trap state can compete with the main QD in the capture and in the emission of the electrons [15]. Note that the eventual presence of impurity-trap states in the gate stack of silicon devices is a well-known fact [1,4,45]. This novel frequency dependent mechanism [15], has not been completely explained, and it is a reminder that for silicon CMOS compatible technology, although extremely controlled and reliable [4,5], it is still possible to observe some unexpected behaviors. It is however comforting to note that the hybrid dot-impurity systems, such as the one discussed in this Refs. [15,45], have recently been able to provide record high performances in term of frequency and accuracy of operations [15,50–53], but have also opened up the way to the use of the quantum pumping technology for novel quantum information schemes [45].

3. Conclusions

In this short review paper, I have discussed some of the most unusual quantum transport effects that have been recently observed in silicon CMOS compatible devices. If from one side valley-orbital effects can be linked to the novel observation of effects such as for example the SU(4) Kondo [8] and the Fano-Kondo ones [10], from another side single-atom transistors and other silicon devices [4,12–15,46,48], because of their immunity to thermal and non-adiabatic detrimental effects, have demonstrated the ability to provide a unique environment for the ultra-fast control of electrons [12–15,46,48]. It is also very interesting to observe that even in the quantum pumping regime, the special valley-orbit energetic structure of silicon nano-materials plays a fundamental role in opening the way to high-performant devices. This is a special outcome as the origin of these powerful effects, i.e., the indirect band gap properties observed for silicon, was in the past considered one of main limitation of these materials [1–4].

Conflicts of Interest: The funders had no role in the design of the review; in the collection, in the analyses, nor in the interpretation of data; in the writing of the manuscript, nor in the decision to publish the results.

References

1. Sze, S.M.; Kwok, K.N. *Physics of Semiconductor Devices*, 3rd ed.; John Wiley & Sons, Inc.: Hoboken, NJ, USA, 2006; pp. 7–75. ISBN 9780471143239.
2. Ramdas, A.K.; Rodriguez, S. Spectroscopy of the solid-state analogues of the hydrogen atom: Donors and acceptors in semiconductors. *Rep. Prog. Phys.* **1981**, *44*, 1297–1387.
3. Cardona, M.; Yu, P.Y. *Fundamentals of Semiconductors*, 4th ed.; Springer: Berlin/Heidelberg, Germany, 2010; ISBN 978-3-642-00710-1.
4. Zwanenburg, F.A.; Dzurak, A.S.; Morello, A.; Simmons, M.Y.; Hollenberg, L.C.L.; Klimeck, G.; Rogge, S.; Coppersmith, S.N.; Eriksson, M.A. Silicon quantum electronics. *Rev. Mod. Phys.* **2013**, *85*, 961–1019.
5. Jehl, X.; Niquet, Y.-M.; Sanquer, M. Single donor electronics and quantum functionalities with advanced CMOS technology. *J. Phys. Condens. Matter* **2016**, *28*, 103001–103018.
6. Neto, A.H.C.; Guinea, F.; Peres, N.M.R.; Novoselov, K.S.; Geim, A.K.; Novoselov, K. The electronic properties of graphene. *Rev. Mod. Phys.* **2009**, *81*, 109–162.
7. Laird, E.A.; Kuemmeth, F.; Steele, G.A.; Grove-Rasmussen, K.; Nygård, J.; Flensberg, K.; Kouwenhoven, L.P. Quantum transport in carbon nanotubes. *Rev. Mod. Phys.* **2015**, *87*, 703–764.
8. Tettamanzi, G.C.; Verduijn, J.; Lansbergen, G.P.; Blaauboer, M.; Calderon, M.J.; Aguado, R.; Rogge, S. Magnetic-Field Probing of an SU(4) Kondo Resonance in a Single-Atom Transistor. *Phys. Rev. Lett.* **2012**, *108*, 046803–046807.
9. Lansbergen, G.P.; Tettamanzi, G.C.; Verduijn, J.; Collaert, N.; Biesemans, S.; Blaauboer, M.; Rogge, S. Tunable Kondo Effect in a Single Donor Atom. *Nano Lett.* **2010**, *10*, 455–460.
10. Verduijn, J.; Tettamanzi, G.C.; Lansbergen, G.P.; Collaert, N.; Biesemans, S.; Rogge, S. Coherent transport through a double donor system in silicon. *Appl. Phys. Lett.* **2010**, *96*, 072110–072112.
11. Verduijn, J.; Agundez, R.R.; Blaauboer, M.; Rogge, S. Non-local coupling of two donor-bound electrons. *New J. Phys.* **2013**, *15*, 033020–033030.
12. Tettamanzi, G.C.; Wacquez, R.; Rogge, S. Charge pumping through a single donor atom. *New J. Phys.* **2013**, *16*, 063036–063052.
13. Van Der Heijden, J.; Tettamanzi, G.C.; Rogge, S. Dynamics of a single-atom electron pump. *Sci. Rep.* **2017**, *7*, 44371.
14. Rossi, A.; Tanttu, T.; Tan, K.Y.; Iisakka, I.; Zhao, R.; Chan, K.W.; Tettamanzi, G.C.; Rogge, S.; Dzurak, A.S.; Möttönen, M. An Accurate Single-Electron Pump Based on a Highly Tunable Silicon Quantum Dot. *Nano Lett.* **2014**, *14*, 3405–3411.
15. Rossi, A.; Klochan, J.; Timoshenko, J.; Hudson, F.E.; Möttönen, M.; Rogge, S.; Dzurak, A.S.; Kashcheyevs, V.; Tettamanzi, G.C. Gigahertz Single-Electron Pumping Mediated by Parasitic States. *Nano Lett.* **2018**, *18*, 4141–4147.
16. Sellier, H.; Lansbergen, G.P.; Caro, J.; Rogge, S.; Collaert, N.; Ferain, I.; Jurczak, M.; Biesemans, S. Transport Spectroscopy of a Single Dopant in a Gated Silicon Nanowire. *Phys. Rev. Lett.* **2006**, *97*, 206805–206808.
17. Lansbergen, G.P.; Rahman, R.; Wellard, C.J.; Woo, I.; Caro, J.; Collaert, N.; Biesemans, S.; Klimeck, G.; Hollenberg, L.C.L.; Rogge, S. Gate-induced quantum-confinement transition of a single dopant atom in a silicon FinFET. *Nat. Phys.* **2008**, *4*, 656–661.
18. Griffiths, D.J. *Introduction to Quantum Mechanics*, 2nd ed.; Pearson Prentice Hall: Upper Saddle River, NJ, USA, 2005; ISBN 0131118927.
19. Trocha, G.C. Orbital Kondo effect in double quantum dots. *Phys. Rev. B* **2010**, *82*, 125323–125332.
20. Mi, X.; Kohler, S.; Petta, J.R. Landau-Zener interferometry of valley-orbit states in Si/SiGe double quantum dots. *Phys. Rev. B* **2018**, *98*, 161404.
21. Pierre, M.; Wacquez, R.; Jehl, X.; Sanquer, M.; Vinet, M.; Cueto, O. Single donor ionization energies in a nanoscale CMOS channel. *Nat. Nanotechnol.* **2010**, *5*, 133–137.
22. Fuechsle, M.; Miwa, J.A.; Mahapatra, S.; Ryu, H.; Lee, S.; Warschkow, O.; Hollenberg, L.C.L.; Klimeck, G.; Simmons, M.Y. A single-atom transistor. *Nat. Nanotechnol.* **2012**, *7*, 242–246.
23. Tan, K.Y.; Chan, K.W.; Möttönen, M.; Morello, A.; Yang, C.; Van Donkelaar, J.; Alves, A.; Pirkkalainen, J.-M.; Jamieson, D.N.; Clark, R.G.; et al. Transport Spectroscopy of Single Phosphorus Donors in a Silicon Nanoscale Transistor. *Nano Lett.* **2010**, *10*, 11–15.

24. Tettamanzi, G.C.; Hile, S.J.; House, M.G.; Fuechsle, M.; Rogge, S.; Simmons, M.Y. Correction to Probing the Quantum States of a Single Atom Transistor at Microwave Frequencies. *ACS Nano* **2017**, *11*, 2444–2451.

25. Van der Heijden, J.; Salfi, J.; Mol, J.A.; Verduijn, J.; Tettamanzi, G.C.; Hamilton, A.R.; Collaert, N.; Rogge, S. Probing the Spin States of a Single Acceptor Atom. *Nano Lett.* **2014**, *14*, 1492–1496.

26. Prati, E.; Shinada, T. *Single-Atom Nano-electronics*; Pan Stanford: Singapore, 2013; ISBN 9789814316316.

27. Collaert, N. *CMOS Nano-Electronics: Innovative Devices, Architectures, and Applications*; Pan Stanford: Singapore, 2012; ISBN 9789814364027.

28. Lansbergen, G.P.; Rahman, R.; Verduijn, J.; Tettamanzi, G.C.; Collaert, N.; Biesemans, S.; Klimeck, G.; Hollenberg, L.C.L.; Rogge, S. Lifetime-Enhanced Transport in Silicon due to Spin and Valley Blockade. *Phys. Rev. Lett.* **2011**, *107*, 136602.

29. Tettamanzi, G.C.; Paul, A.; Lansbergen, G.; Verduijn, J.; Lee, S.; Collaert, N.; Biesemans, S.; Klimeck, G.; Rogge, S. Thermionic Emission as a Tool to Study Transport in Undoped nFinFETs. *IEEE Electron Device Lett.* **2010**, *31*, 150–152.

30. Wacquez, R.; Vinet, M.; Pierre, M.; Roche, B.; Jehl, X.; Cueto, O.; Verduijn, J.; Tettamanzi, G.C.; Rogge, S.; Deshpande, V.; et al. Single dopant impact on electrical characteristics of SOI NMOSFETs with effective length down to 10nm. In Proceedings of the 2010 Symposium on VLSI Technology, Honolulu, HI, USA, 15–17 June 2010.

31. Seo, M.; Roulleau, P.; Roche, P.; Glattli, D.C.; Sanquer, M.; Jehl, X.; Hutin, L.; Barraud, S.; Parmentier, F.D. Strongly Correlated Charge Transport in Silicon Metal-Oxide-Semiconductor Field-Effect Transistor Quantum Dots. *Phys. Rev. Lett.* **2018**, *121*, 027701–027705.

32. Cronenwett, S.M. A Tunable Kondo Effect in Quantum Dots. *Science* **1998**, *281*, 540–544.

33. Rokhinson, L.P.; Guo, L.J.; Chou, S.Y.; Tsui, D.C. Kondo-like zero-bias anomaly in electronic transport through an ultrasmall Si quantum dot. *Phys. Rev. B* **1999**, *60*, R16319–R16321.

34. Sasaki, S.; De Franceschi, S.; Elzerman, J.M.; Van der Wiel, W.G.; Eto, M.; Tarucha, S.; Kouwenhoven, L. Kondo effect in an integer-spin quantum dot. *Nature* **2000**, *405*, 764–767.

35. Gavai, R.V. On the deconfinement transition in SU(4) lattice gauge theory. *Nucl. Phys. B* **2002**, *633*, 127–138.

36. Fano, U. Effects of Configuration Interaction on Intensities and Phase Shifts. *Phys. Rev.* **1961**, *124*, 1866–1878.

37. Aharonov, Y.; Bohm, D. Significance of Electromagnetic Potentials in the Quantum Theory. *Phys. Rev.* **1959**, *115*, 485–491.

38. Thouless, D.J. Quantization of particle transport. *Phys. Rev. B* **1983**, *27*, 6083–6087.

39. Maire, N.; Hohls, F.; Kaestner, B.; Pierz, K.; Schumacher, H.W.; Haug, R.J. Noise measurement of a quantized charge pump. *Appl. Phys. Lett.* **2008**, *92*, 082112–082115.

40. Kaestner, B.; Kashcheyevs, V. Non-adiabatic quantized charge pumping with tunable-barrier quantum dots: A review of current progress. *Rep. Prog. Phys.* **2015**, *78*, 103901.

41. Ubbelohde, N.; Hohls, F.; Kashcheyevs, V.; Wagner, T.; Fricke, L.; Kästner, B.; Pierz, K.; Schumacher, H.W.; Haug, R.J. Partitioning of on-demand electron pairs. *Nat. Nanotechnol.* **2015**, *10*, 46–49.

42. Bauerle, C.; Glattli, D.C.; Meunier, T.; Portier, F.; Roche, P.; Roulleau, P.; Takada, S.; Waintal, X.; Glattli, C. Coherent control of single electrons: A review of current progress. *Rep. Prog. Phys.* **2018**, *81*, 056503.

43. Bocquillon, E.; Parmentier, F.; Grenier, C.; Berroir, J.-M.; DeGiovanni, P.; Glattli, D.C.; Placais, B.; Cavanna, A.; Jin, Y.; Feve, G. Electron Quantum Optics: Partitioning Electrons One by One. *Phys. Rev. Lett.* **2012**, *108*, 196803.

44. Okazaki, Y.; Nakamura, S.; Onomitsu, K.; Kaneko, N.-H. Digital processing with single electrons for arbitrary waveform generation of current. *Appl. Phys. Express* **2018**, *11*, 036701–036704.

45. Yamahata, G.; Ryu, S.; Johnson, N.; Sim, H.-S.; Fujiwara, A.; Kataoka, M. Picosecond coherent electron motion in a silicon single-electron source. *arXiv* **2019**, arXiv:1903.07802.

46. Kashcheyevs, V.; Kaestner, B. Universal Decay Cascade Model for Dynamic Quantum Dot Initialization. *Phys. Rev. Lett.* **2010**, *104*, 186805.

47. Seo, M.; Ahn, Y.-H.; Oh, Y.; Chung, Y.; Ryu, S.; Sim, H.-S.; Lee, I.-H.; Bae, M.-H.; Kim, N. Improvement of electron pump accuracy by a potential-shape-tunable quantum dot pump. *Phys. Rev. B* **2014**, *90*, 085307.

48. Kataoka, M.; Fletcher, J.; See, P.; Giblin, S.P.; Janssen, T.J.B.M.; Griffiths, J.P.; Jones, G.A.C.; Farrer, I.; Ritchie, D.A. Tunable Nonadiabatic Excitation in a Single-Electron Quantum Dot. *Phys. Rev. Lett.* **2011**, *106*, 126801.

49. Roche, B.; Riwar, R.-P.; Voisin, B.; Dupont-Ferrier, E.; Wacquez, R.; Vinet, M.; Sanquer, M.; Splettstoesser, J.; Jehl, X. A two-atom electron pump. *Nat. Commun.* **2014**, *4*, 1581.

50. Yamahata, G.; Nishiguchi, K.; Fujiwara, A. Gigahertz single-trap electron pumps in silicon. *Nat. Commun.* **2014**, *5*, 5038.

51. Wenz, T.; Hohls, F.; Jehl, X.; Sanquer, M.; Barraud, S.; Klochan, J.; Barinovs, G.; Kashcheyevs, V. Dopant-controlled single-electron pumping through a metallic island. *Appl. Phys. Lett.* **2016**, *108*, 213107.

52. Yamahata, G.; Giblin, S.P.; Kataoka, M.; Karasawa, T.; Fujiwara, A. High-accuracy current generation in the nanoampere regime from a silicon single-trap electron pump. *Sci. Rep.* **2017**, *7*, 45137.

53. Clapera, P.; Klochan, J.; Lavieville, R.; Barraud, S.; Hutin, L.; Sanquer, M.; Vinet, M.; Cinins, A.; Barinovs, G.; Kashcheyevs, V.; et al. Design and operation of CMOS-compatible electron pumps fabricated with optical lithography. *IEEE Electron Device Lett.* **2017**, *38*, 414–417.

Article

Enhanced Negative Nonlocal Conductance in an Interacting Quantum Dot Connected to Two Ferromagnetic Leads and One Superconducting Lead

Cong Lee, Bing Dong * and Xiao-Lin Lei

Key Laboratory of Artificial Structures and Quantum Control (Ministry of Education), Department of Physics and Astronomy, Shanghai Jiaotong University, 800 Dongchuan Road, Shanghai 200240, China; kitozumika@sina.cn (C.L.); xllei@sjtu.edu.cn (X.-L.L.)
* Correspondence: bdong@sjtu.edu.cn

Received: 24 August 2019; Accepted: 11 October 2019; Published: 14 October 2019

Abstract: In this paper, we investigate the electronic transport properties of a quantum dot (QD) connected to two ferromagnetic leads and one superconducting lead in the Kondo regime by means of the finite-U slave boson mean field approach and the nonequilibrium Green function technique. In this three-terminal hybrid nanodevice, we focus our attention on the joint effects of the Kondo correlation, superconducting proximity pairing, and spin polarization of leads. It is found that the superconducting proximity effect will suppress the linear local conductance (LLC) stemming from the weakened Kondo peak, and when its coupling Γ_s is bigger than the tunnel-coupling Γ of two normal leads, the linear cross conductance (LCC) becomes negative in the Kondo region. Regarding the antiparallel configuration, increasing spin polarization further suppresses LLC but enhances LCC, i.e., causing larger negative values of LCC, since it is beneficial for the emergence of cross Andreev reflection. On the contrary, for the parallel configuration, with increasing spin polarization, the LLC decreases and greatly widens with the appearance of shoulders, and eventually splits into four peaks, while the LCC decreases relatively rapidly to the normal conductance.

Keywords: superconducting proximity effect; Kondo effect; spin polarization; Anreev reflection

1. Introduction

Recently, electron transport through a hybrid nanodevice, for instance, a quantum dot (QD), connected to normal and superconducting electrodes, has attracted much attention in many experimental [1–20] and theoretical studies [21–29] due to the associated physical challenges and potential applications in spintronics and quantum information. When a QD is connected to a superconductor, superconducting order can leak into it to give rise to pairing correlations and an induced superconducting gap, known as the superconducting proximity effect; this privileges the tunneling of Cooper pairs of electrons with opposite spin, and thereby favors QD states with even numbers of electrons and a zero total spin. At the same time, the local Coulomb repulsion enforces a one-by-one filling of the QD, and thereby induces the Coulomb blockade and even the Kondo effect at very low temperatures, which exhibits the zero-bias anomaly in the differential conductance with odd numbers of electrons residing in the QD. In this case, the superconducting proximity effect competes with the on-site Coulomb correlation [1,6,10,21,24,25,28,29].

It is even more intriguing when the QD additionally connects to a ferromagnetic lead [30,31]. It is known that the effective exchange field induced by the ferromagnetic correlation can cause a spin imbalance inside the QD, and as a result, suppress and/or even split the Kondo peak in the differential conductance [32–37]. Furthermore, spin polarization of the QD, on the one hand, is disadvantageous to the formation of on-dot superconducting pairing. However, the spin polarization in the antiparallel configuration, on the other hand, is favorable to the Andreev reflection (AR) and Cooper pair splitting [30,38]. It is, therefore, very interesting to study how the interplay of the Kondo, superconducting pairing, and ferromagnetic correlations affects the electron tunneling through a QD [39]. In a recent paper, Futterer et al. present a theoretical analysis of the subgap transport of such a three-terminal hybrid system, which consists of n interacting QD attached to two ferromagnetic leads and one superconducting lead [40,41]. They focused on the first-order sequential tunneling by using a master equation and found that the strong on-dot electron–electron interaction, rather than the nonlocal AR, leads to negative values of the nonlocal current response at an appropriately large bias voltage. Moreover, the bias-dependent supercurrent in the superconducting electrode was proposed as a sensitive detector to probe the exchange field of the QD induced by ferromagnetic leads [42]. Thereafter, the tunneling magnetoresistance was calculated for the same system to display a nontrivial dependence on the bias voltage and the level detuning caused by the AR [43]. Very recently, it has been reported, in contrast to [40], that the cross AR is indeed the dominant nonlocal transport channel at a low bias voltage and leads to a negative value of the cross conductance in the three-terminal hybrid nanodevice with two normal electrodes instead [44,45].

In the present work, we extend the finite-U slave boson mean field (SBMF) approach of Kotliar and Ruckenstein [46] with the help of the nonequilibrium Green function (NGF) method to investigate the subgap transport for the same three-terminal hybrid QD as in [40]. This kind of SBMF approach is generally believed to be reliable in describing not only spin fluctuations rigorously but also charge fluctuations to a certain degree in the Kondo regime at zero temperature [46–49]. This nonperturbative approach has been successfully utilized to calculate the linear and nonlinear conductance within a relatively wide dot-level range from the mixed valence to the empty orbital regimes, in which the major characteristics induced by the external magnetic field and the magnetization in Kondo transport arise [49–52]. Furthermore, this approach has been applied to analyze the π-phase transition in a double-QDs Josephson junction caused by competition between Kondo and interdot antiferromagnetic coupling [53]. The main purpose of this paper is to analyze in detail the interplay of the Kondo, superconducting proximity induced on-dot pairing, and ferromagnetic correlations and their influence on electronic tunneling.

The rest of the paper is organized as follows. In Section 2, we introduce our model of the three-terminal hybrid system, and the equivalent slave-boson field Hamiltonian. Then, we present the self-consistent equations of the expectation values of slave-boson operators within the SBMF approach and NGF method. Moreover, the formulas for current and linear conductance, including the local and cross conductances, are given. In Section 3, we present and analyze our numerical calculations for the linear conductance and nonlinear conductance in detail. Finally, a brief summary is given in Section 4.

2. Model and Theoretical Formulation

2.1. Model Hamiltonian

We consider a three-terminal hybrid nanodevice: an interaction QD connected to one superconducting lead and two ferromagnetic leads, as shown in Figure 1. The Hamiltonian of the system can be written as [40]

$$H = H_L + H_R + H_{QD} + H_T, \tag{1}$$

where

$$H_\eta = \sum_{k\sigma} \epsilon_{\eta k\sigma} c^\dagger_{\eta k\sigma} c_{\eta k\sigma}, \tag{2}$$

$$H_{QD} = \sum_\sigma \epsilon_d c^\dagger_{d\sigma} c_{d\sigma} + U n_1 n_2 + \Gamma_s (c^\dagger_{d1} c^\dagger_{d2} + c_{d1} c_{d2}), \tag{3}$$

$$H_T = \sum_{\eta k\sigma} \left(V_{\eta k} c^\dagger_{\eta k\sigma} c_{d\sigma} + \text{H.c.} \right). \tag{4}$$

Here, $\eta = L, R$ denotes the left and right leads, while $\sigma = 1, 2$ represents the spin degree of freedom. In the above equations, $c_{\eta k\sigma}^{\dagger}$ ($c_{\eta k\sigma}$) and $c_{d\sigma}^{\dagger}$ ($c_{d\sigma}$) are creation (annihilation) operators of electrons with spin σ in the η-th ferromagnetic lead and in the QD, respectively. In the dot Hamiltonian H_{QD}, ϵ_d is the energy level of the QD, $n_\sigma = c_{d\sigma}^{\dagger} c_{d\sigma}$, and U is the on-site Coulomb repulsion between opposite spin electrons. H_T depicts the tunneling between the QD and the two ferromagnetic leads, and $V_{\eta k}$ is the corresponding tunneling matrix element. In general, the tunneling amplitude $V_{\eta k}$ is assumed to be independent of spin and energy, and thus the effect of spin-polarized tunneling is captured by the spin-dependent tunneling rates, $\Gamma_{\eta\sigma} = 2\pi \sum_k |V_{\eta k}|^2 \delta(\omega - \epsilon_{\eta k\sigma})$.

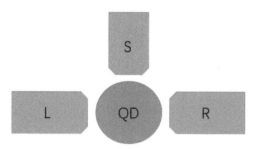

Figure 1. (Color online) Schematic diagram of a quantum dot connected to one superconducting lead and two ferromagnetic leads.

In this paper, since we are only interested in the subgap tunneling, it is natural to consider the limit of an extremely large superconducting gap in the superconducting lead. Therefore, the degree of freedom of the superconducting lead can be integrated out and an effective term can be constructed in the dot Hamiltonian, the third term in Equation (3). The parameter Γ_s plays the role of describing the superconducting proximity effect on the dot. It is evident that this new proximized term mixes the empty state $|0\rangle$ and the doubly occupied state $|\uparrow\downarrow\rangle$ in the dot, and results in two new eigenstates with energies, $E_{\pm} = \varepsilon \pm \sqrt{\varepsilon^2 + \Gamma_s^2}$ (here $\varepsilon = \epsilon_d + U/2$), which are known as the Andreev bound states. What we are interested in this paper is the effect of Andreev reflection on the electron tunneling through an interacting QD in the Kondo regime.

According to the finite-U slave-boson approach, one can introduce four additional auxiliary boson operators, e, p_σ, and d, which are associated with the empty, singly occupied, and doubly occupied electron states, respectively, of the QD, to discuss the above problem without interparticle couplings in an enlarged space with constraints: The completeness relation [46]

$$\sum_\sigma p_\sigma^{\dagger} p_\sigma + e^{\dagger} e + d^{\dagger} d = 1, \tag{5}$$

and the particle number conservation condition

$$c_{d\sigma}^{\dagger} c_{d\sigma} = p_\sigma^{\dagger} p_\sigma + d^{\dagger} d. \tag{6}$$

Within the mean-field scheme, the effective Hamiltonian becomes (please see Appendix A) [46]

$$
\begin{aligned}
H = {} & \sum_\sigma \epsilon_d c_{d\sigma}^{\dagger} c_{d\sigma} + U d^{\dagger} d + \Gamma_s (z_1^* z_2^* c_{d1}^{\dagger} c_{d2}^{\dagger} + z_1 z_2 c_{d1} c_{d2}) + \sum_{\eta k\sigma} (V_{\eta k} c_{\eta k\sigma}^{\dagger} c_{d\sigma} z_\sigma + V_{\eta k}^* c_{d\sigma}^{\dagger} c_{\eta k\sigma} z_\sigma^*) \\
& + \sum_{\eta k\sigma} \epsilon_{\eta k\sigma} c_{\eta k\sigma}^{\dagger} c_{\eta k\sigma} + \lambda^1 (\sum_\sigma p_\sigma^{\dagger} p_\sigma + e^{\dagger} e + d^{\dagger} d - 1) + \sum_\sigma \lambda_\sigma^2 (c_{d\sigma}^{\dagger} c_{d\sigma} - p_\sigma^{\dagger} p_\sigma - d^{\dagger} d),
\end{aligned}
\tag{7}
$$

where three Lagrange multipliers λ^1 and λ_σ^2 are drawn in order to make the constraints valid, and z_σ is the correctional parameters in the hopping term to recover the many-body effect on tunneling with

$$z_\sigma = (1 - d^\dagger d - p_\sigma^\dagger p_\sigma)^{-1/2}(e^\dagger p_\sigma + p_{\bar\sigma}^\dagger d)(1 - e^\dagger e - p_{\bar\sigma}^\dagger p_{\bar\sigma})^{-1/2}. \tag{8}$$

2.2. Self-Consistent Equations

From the effective Hamiltonian Equation (7), one can derive four equations of the motion of slave-boson operators, which serve as the basic equations together with the three constraints. Then, we further apply the mean-field approximation in the statistical expectations of these equations, where all the boson operators are replaced by their respective expectation values. After a lengthy and tedious calculation employing the Langreth technique (please see the Appendix B for the details of derivation), we can obtain the self-consistent equations as follows [49–52]:

$$\Gamma_s \frac{\partial(z_1 z_2)}{\partial e}(R + R^*) + \lambda^1 e + \sum_\sigma \frac{\partial z_\sigma}{\partial e}(Q_\sigma + Q_\sigma^*) = 0, \tag{9}$$

$$\Gamma_s \frac{\partial(z_1 z_2)}{\partial p_1}(R + R^*) + (\lambda^1 - \lambda_1^2)p_1 + \frac{\partial z_1}{\partial p_1}(Q_1 + Q_1^*) + \frac{\partial z_2}{\partial p_1}(Q_2 + Q_2^*) = 0, \tag{10}$$

$$\Gamma_s \frac{\partial(z_1 z_2)}{\partial p_2}(R + R^*) + (\lambda^1 - \lambda_2^2)p_2 + \frac{\partial z_1}{\partial p_2}(Q_1 + Q_1^*) + \frac{\partial z_2}{\partial p_2}(Q_2 + Q_2^*) = 0, \tag{11}$$

$$\Gamma_s \frac{\partial(z_1 z_2)}{\partial d}(R + R^*) + (U + \lambda^1 - \sum_\sigma \lambda_\sigma^2)d + \sum_\sigma \frac{\partial z_\sigma}{\partial d}(Q_\sigma + Q_\sigma^*) = 0, \tag{12}$$

$$\sum_\sigma |p_\sigma|^2 + |e|^2 + |d|^2 - 1 = 0, \tag{13}$$

$$K_\sigma - |p_\sigma|^2 - |d|^2 = 0, \tag{14}$$

where

$$K_1 = \left\langle c_{d1}^\dagger c_{d1} \right\rangle = \int \frac{d\omega}{2\pi i} \left\langle\left\langle c_{d1}; c_{d1}^\dagger \right\rangle\right\rangle^< (\omega) = \frac{1}{2\pi i} \int d\omega G_{d11}^<(\omega), \tag{15}$$

$$K_2 = \left\langle c_{d2}^\dagger c_{d2} \right\rangle = \int \frac{-d\omega}{2\pi i} \left\langle\left\langle c_{d2}^\dagger; c_{d2} \right\rangle\right\rangle^> (\omega) = \frac{-1}{2\pi i} \int d\omega G_{d22}^>(\omega), \tag{16}$$

$$R = \left\langle c_{d1}^\dagger c_{d2}^\dagger \right\rangle = \int \frac{d\omega}{2\pi i} \left\langle\left\langle c_{d2}^\dagger; c_{d1}^\dagger \right\rangle\right\rangle^< (\omega) = \frac{1}{2\pi i} \int d\omega G_{d21}^<(\omega), \tag{17}$$

$$Q_{1\eta} = z_1 \Gamma_{\eta 1} \int \frac{d\omega}{2\pi} \left\{ -\frac{i}{2} \left[\tilde\Gamma_{L1} f_L(\omega) + \tilde\Gamma_{R1} f_R(\omega) \right] |G_{d11}^R(\omega)|^2 \right.$$
$$\left. -\frac{i}{2} \left[\tilde\Gamma_{L2}(1 - f_L(-\omega)) + \tilde\Gamma_{R1}(1 - f_R(-\omega)) \right] |G_{d21}^R(\omega)|^2 + f_\eta(\omega)G_{d11}^A(\omega) \right\}, \tag{18}$$

$$Q_{2\eta} = z_2 \Gamma_{\eta 2} \int \frac{d\omega}{2\pi} \left\{ \frac{i}{2} \left[\tilde\Gamma_{L1}(1 - f_L(\omega)) + \tilde\Gamma_{R1}(1 - f_R(\omega)) \right] |G_{d21}^R(\omega)|^2 \right.$$
$$\left. +\frac{i}{2} \left[\tilde\Gamma_{L2} f_L(-\omega) + \tilde\Gamma_{R1} f_R(-\omega) \right] |G_{d22}^R(\omega)|^2 - f_\eta(-\omega)G_{d22}^A(\omega) \right\}, \tag{19}$$

and

$$Q_\sigma = \sum_\eta Q_{\sigma\eta}. \tag{20}$$

Here, the QD Keldysh NGFs, $G_{d\sigma\sigma'}^{R(A,<,>)}(\omega)$, are the matrix elements of the 2×2 retarded (advanced and correlation) GF matrix $G_d^{R(A,<,>)}(\omega) = \langle\langle \phi; \phi^\dagger \rangle\rangle^{R(A,<,>)}$ defined in the Nambu presentation, in which the mixture Fermion operator, $\phi = (c_{d1}, c_{d2}^\dagger)^T$, has to be introduced to describe the electronic dynamics due to the superconducting proximity effect. For the effective noninteracting Hamiltonian, the retarded and advanced GFs $G_d^{R(A)}$ can be easily written in the frequency domain as

$$\left(G_d^{R(A)}(\omega)\right)^{-1} = \begin{bmatrix} \omega - \epsilon_d - \lambda_1^2 \pm \frac{i}{2}(\tilde{\Gamma}_{L1} + \tilde{\Gamma}_{R1}) & -\Gamma_s z_1 z_2 \\ -\Gamma_s z_1^* z_2^* & \omega + \epsilon_d + \lambda_2^2 \pm \frac{i}{2}(\tilde{\Gamma}_{L2} + \tilde{\Gamma}_{R2}) \end{bmatrix}, \quad (21)$$

with the renormalized parameters, $\tilde{\Gamma}_{\eta\sigma} = |z_\sigma|^2 \Gamma_{\eta\sigma}$. In addition, the correlation GFs $G_d^{<(>)}(\omega)$ can be obtained with the help of the following Keldysh relation typical for a noninteracting system:

$$G_d^{<(>)}(\omega) = G_d^R(\omega) \left[\Sigma_L^{<(>)}(\omega) + \Sigma_R^{<(>)}(\omega)\right] G_d^A(\omega), \quad (22)$$

with the self-energies

$$\Sigma_\eta^<(\omega) = i \begin{bmatrix} \tilde{\Gamma}_{\eta1} f_\eta(\omega) & 0 \\ 0 & \tilde{\Gamma}_{\eta2}[1 - f_\eta(-\omega)] \end{bmatrix}, \quad (23)$$

and

$$\Sigma_\eta^>(\omega) = -i \begin{bmatrix} \tilde{\Gamma}_{\eta1}[1 - f_\eta(\omega)] & 0 \\ 0 & \tilde{\Gamma}_{\eta2} f_\eta(-\omega) \end{bmatrix}, \quad (24)$$

where $f_\eta(\omega) = 1/(e^{\beta(\omega - \mu_\eta)} + 1)$ is the Fermi distribution function of the lead η with the chemical potential μ_η and temperature $1/\beta$.

2.3. The Current and Linear Conductance

The electric current flowing from the lead η into the QD can be obtained from the rate of change of the electron number operator of the left lead:

$$I_\eta = \sum_\sigma I_{\eta\sigma} = -e \sum_\sigma \left\langle \frac{d}{dt} \sum_k c_{\eta k\sigma}^\dagger c_{\eta k\sigma} \right\rangle. \quad (25)$$

After standard calculation, the current for the left lead can be written as [44,45]

$$I_L = I_L^{ET} + I_L^{DAR} + I_L^{CAR}, \quad (26)$$

with

$$I_L^{ET} = \frac{e}{h} \int d\omega \left\{ \tilde{\Gamma}_{L1} \tilde{\Gamma}_{R1} [f_L(\omega) - f_R(\omega)] |G_{d11}^R(\omega)|^2 + \tilde{\Gamma}_{L2} \tilde{\Gamma}_{R2} [f_L(-\omega) - f_R(-\omega)] |G_{d22}^R(\omega)|^2 \right\}, \quad (27)$$

$$I_L^{DAR} = \frac{2e}{h} \int d\omega \tilde{\Gamma}_{L1} \tilde{\Gamma}_{L2} [f_L(\omega) + f_L(-\omega) - 1] \times |G_{d12}^R(\omega)|^2, \quad (28)$$

$$I_L^{CAR} = \frac{e}{h} \int d\omega \left\{ \tilde{\Gamma}_{L1} \tilde{\Gamma}_{R2} [f_L(\omega) + f_R(-\omega) - 1] + \tilde{\Gamma}_{L2} \tilde{\Gamma}_{R1} [f_L(-\omega) + f_R(\omega) - 1] \right\} |G_{d12}^R(\omega)|^2. \quad (29)$$

The corresponding currents for the right lead can be readily obtained by simply exchanging the subscripts L and R in Equations (27)–(29). It is found that the current can be divided into three parts: I_L^{ET} describes the single-particle tunneling current caused by the normal electron transfer (ET) processes from the left lead directly to the right lead; I_L^{DAR} denotes the local Andreev current caused by the direct AR (DAR) processes in which an electron injecting from the left lead forms a Cooper pair in the superconducting lead, and at the same time, is reflected as a hole back into the left lead; and I_L^{CAR} is the nonlocal Andreev current caused by the crossed AR (CAR) processes, which is similar to DAR except that the hole is reflected into another lead, i.e., here, the right lead.

Since we are interested in the interplay between the Andreev bound state and the Kondo effect in the nonlocal subgap tunneling, we choose the bias voltage configuration in this hybrid three-terminal nanodevice as follows: The left lead is biased with the chemical potential V, while the right lead and the superconducting electrode are both in contact with the ground. Therefore, one can define two different linear conductances: The usual local conductance $G_L = \partial I_L / \partial V |_{V=0}$ and the unusual nonlocal (cross) conductance $G_C = \partial I_R / \partial V |_{V=0}$, which is related to the nonlocal current response of the hybrid three-terminal nanodevice to external driving field, i.e., current flowing in the right lead caused by the bias voltage applied to the left lead. From Equations (27)–(29), the local conductance reads

$$G_L = \left. \frac{\partial I_L}{\partial V} \right|_{V=0} = G^{ET} + G^{DAR} + G^{CAR}, \tag{30}$$

and the cross conductance is

$$G_C = \left. \frac{\partial I_R}{\partial V} \right|_{V=0} = G^{ET} - G^{CAR}, \tag{31}$$

where

$$G^{ET} = \frac{e^2}{h} \left(\tilde{\Gamma}_{L1} \tilde{\Gamma}_{R1} |G^R_{d11}(0)|^2 + \tilde{\Gamma}_{L2} \tilde{\Gamma}_{R2} |G^R_{d22}(0)|^2 \right), \tag{32}$$

$$G^{DAR} = \frac{4e^2}{h} \tilde{\Gamma}_{L1} \tilde{\Gamma}_{L2} |G^R_{d12}(0)|^2, \tag{33}$$

$$G^{CAR} = \frac{e^2}{h} \left(\tilde{\Gamma}_{L1} \tilde{\Gamma}_{R2} + \tilde{\Gamma}_{R1} \tilde{\Gamma}_{L2} \right) |G^R_{d12}(0)|^2. \tag{34}$$

It is obvious that all of the three different tunneling processes contribute to the local conductance. Nevertheless, the DAR tunneling process, as expected, has no contribution to the cross conductance. More interestingly, the CAR tunneling process provides a contrary contribution, in comparison with the ET process, to the cross-conductance Equation (31), which is responsible for the negative value of the cross conductance in certain appropriate conditions, as shown in the following section. This opposite role of the CAR can be interpreted in an intuitive way: A hole entering the right lead is physically equivalent to an electron breaking into the QD from the right lead, thus resulting in an opposite current flowing in the right lead. It is important to point out that if the superconducting coupling is switched off ($\Gamma_s = 0$), there are no DAR and CAR processes, and as a result, the cross conductance reduces to the local conductance.

3. Result and Discussion

We suppose that the left and right leads are made from the same material and in the wide band limit, that which is of interest in the present investigation, the ferromagnetism of the leads can be accounted for by the polarization-dependent couplings $\Gamma_{L1} = \Gamma_{R1} = (1+p)\Gamma, \Gamma_{L2} = \Gamma_{R2} = (1-p)\Gamma$ for the parallel (P) alignment, while $\Gamma_{L1} = \Gamma_{R2} = (1+p)\Gamma, \Gamma_{L2} = \Gamma_{R1} = (1-p)\Gamma$ for the anti-parallel (AP) alignment. Here, Γ describes the tunneling coupling between the QD and the nonmagnetic leads, which is taken as the energy unit in the following calculations. In addition, p ($0 \leq p < 1$) denotes the polarization strength of the leads. The Kondo temperature in the case of $p = 0$, given by $T_K = U\sqrt{D}\exp(-\pi/D)/2\pi$ with $D = -2U\Gamma/\epsilon_d(U + \epsilon_d)$, will be set as another dynamical energy scale of the nonlinear conductance.

In the following, we deal with the three-terminal QD system having a fixed finite Coulomb interaction $U = 10$ at zero temperature and consider the effects of changing the bare dot level ϵ_d, the spin polarization p, and the proximity strength Γ_s, respectively.

3.1. Linear Local and Cross Conductances

Firstly, we show the calculated linear conductances in Figure 2, including the local conductance G_L and the nonlocal cross conductance G_C as functions of the bare energy level ϵ_d of the QD at different superconducting coupling strengths, $\Gamma_s = 0, 0.2, 0.5, 1.0, 1.5$, and 2.0, in the case of no spin-polarization $p = 0$. Without the superconducting coupling $\Gamma_s = 0$, $G_L = G_C$ and the linear conductance reaches the unitary limit, G_0 ($G_0 \equiv 2e^2/h$), as expected in the Kondo regime. With increasing the coupling Γ_s, the local conductance G_L raises at the beginning, as seen in Figure 2a, since the AR channel starts to emerge and contribute to the electronic tunneling. A slightly bigger value of conductance, $G_L \simeq 1.1G_0$, than the unitary limit of conductance of single-particle tunneling is reached at the coupling $\Gamma_s = 0.5$ in the Kondo regime. On the other hand, it is known that the resonant AR leads to the unitary limit of conductance, $2G_0$, of the Cooper pair tunneling in the two-terminal hybrid system, e.g., a normal metal-QD-superconductor system [24]. We can therefore deduce that such a larger value of the conductance is a signature indicating that the tunneling event in the present hybrid system is a mixture of the single-particle and Cooper pair tunnelings. Increasing the coupling Γ_s further will, however, cause a decrease in the local conductance G_L. The suppression of G_L can be interpreted as follows: An electron coming from the left lead has much higher probability to form the Cooper pair breaking into the superconducting electrode due to the considerable strength of the coupling $\Gamma_s > 0.5$, and as a result, the ET process is rapidly suppressed. Different from the local conductance, the nonlocal conductance G_C decreases from the beginning and even becomes negative if the proximity-coupling is sufficiently strong. The negative cross conductance means that when the left lead is applied with a voltage which is bigger than the right lead, electrons will, instead of entering into the right lead from the QD, tunnel into the QD out of the right lead. Moreover, we find that when the QD leaves the Kondo regime, the cross conductance becomes positive again.

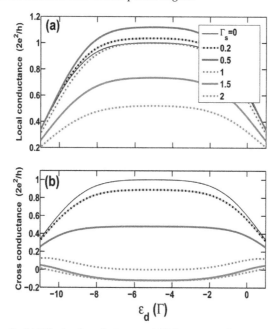

Figure 2. (Color online) (a) The local conductance and (b) the cross conductance vs. the bare dot level ϵ_d at zero temperature for different proximity-coupling strengths Γ_s in the case of normal leads, i.e., $p = 0$.

Such effects of Γ_s are clearly manifested in Figure 3, in which the local and nonlocal conductances, and their three respective parts, G^{ET}, G^{DAR}, and G^{CAR}, are illustrated as functions of the coupling Γ_s for the specific system which has bare dot level, $\epsilon_d = -U/2 = -5$. It is observed that a maximum value of the local conductance, $G_L = 1.125G_0$, is arrived at, $\Gamma_s = 0.58$. After this point of Γ_s, the AR process becomes the predominate tunneling mechanism over the ET process. When the proximity-coupling is equal to the tunnel-coupling, i.e., $\Gamma_s = 1.0$, a new resonance is reached, originating from interplay between the Kondo effect and AR. Consequently, $G^{DAR} = G_0/2$ and $G^{CAR} = G^{ET} = G_0/4$, and the local conductance arrive at the unitary value, $G_L = G_0$ once more. At the same time, the nonlocal conductance completely vanishes, $G_C = 0$, which indicates no current response in the right lead to the bias voltage applied to the left lead.

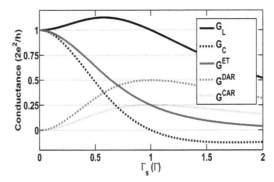

Figure 3. (Color online) The zero temperature local conductance (black-solid line) and the cross conductance (black-dotted line) vs. the proximity coupling Γ_s for the system with a bare dot level at the particle-hole symmetric point, $\epsilon_d = -U/2 = -5$ in the case of normal leads ($p = 0$). The three parts of the conductance are also plotted for illustration purposes.

Secondly, in Figure 4, we investigate the cross conductance G_C as a function of the bare energy level ϵ_d of the QD at different proximity couplings Γ_s in the AP configuration with a large spin polarization $p = 0.5$. In the AP configuration, similar with the case of zero spin polarization $p = 0$, electrons with up-spin and down-spin are equally available in the whole system, favoring the formation of the Kondo-correlated state within a wide dot level range centered at $\epsilon_d = -U/2 = -5$. Meanwhile, since there is no splitting of the renormalized dot levels, $\epsilon_d + \lambda_\sigma^2$, for different spins, the usual tunneling and charging peaks, around $\epsilon_d = 0$ and $-U$, respectively, are relatively narrow. The local conductance G_L vs. ϵ_d curves show a similar behavior as the case of zero spin polarization even in the presence of superconducting coupling Γ_s. Furthermore, since no spin-flip scattering exists in the tunneling processes, in the AP configuration, the majority-spin (e.g., up-spin) states in the left lead increase but the available up-spin (minority-spin) states in the right lead decrease with increasing spin polarization strength, and as a consequence, the transfer of the majority-spin (up-spin) electrons through the QD is suppressed, such that the local conductance goes down and eventually vanishes at $p = 1$ as expected. On the contrary, the available down-spin states in the right lead increase in the AP configuration, which just facilitates the occurrence of the CAR process [30]. Therefore, one can observe that G_C becomes negative in almost the whole region of dot levels, from the mixed-valence regime to the empty orbital regime, even when $\Gamma_s < 1$, and nearly arrives at a considerably bigger negative value, $G_C \simeq -G_0/5$, at the Kondo regime at $p = 0.5$. It is interesting to consider the extreme case of $p = 1$. As mentioned above, in the AP configuration electrons with up-spin and down-spin are identical to each other, preferring the formation of the Kondo-correlated state for all values of p. However, since the up-spin states are almost unavailable in the right lead in the case of large polarization, the ET process for the left lead to the right lead is completely damaged (implying an exactly vanishing conductance in the usual QD system), but the CAR process survives here as a unique tunneling mechanism, exclusively making a contribution to electronic tunneling. It is anticipated that in this case, $G^{ET} = G^{DAR} = 0$ and $G_L = -G_C = G^{CAR} = G_0/2$ (this is the unitary limit of conductance of the single channel).

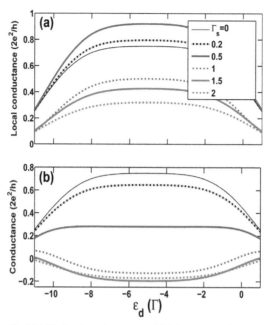

Figure 4. (Colour online) (**a**) The local conductance and (**b**) the cross conductance versus the bare dot level ϵ_d for different proximity-coupling strengths Γ_s in the AP configuration with $p = 0.5$.

The situation is quite different in the case of the P configuration, as demonstrated in Figure 5, in which the two conductances are plotted as functions of bare dot level with spin polarization $p = 0.5$. In the P configuration, finite spin polarization splits the dot level for up- and down-spins and thus broadens the usual resonance peaks around $\epsilon_d = 0$ and $\epsilon_d = -U$ [32–36]. On one hand, since minority-spin electrons are still available in the two electrodes to build the Kondo screening correlation to a certain degree, the central Kondo peak can still be reached at the unitary limit G_0 at the large polarization $p = 0.5$ in the case of $\Gamma_s = 0$. On the other hand, the number of minority-spin electrons is too small to construct the Kondo-correlated state at $p = 0.5$, and thus Kondo-induced conductance enhancement disappears rapidly when the QD moves away from the particle-hole symmetric point $\epsilon_d = -U/2$. These two factors cause the appearance of kinks or splitting peaks in both conductance vs. ϵ_d curves. Besides, it can be observed from Figure 5a that the central Kondo peak in the local conductance is progressively splitting with increasing proximity coupling $\Gamma_s \geq 1.0$ in this P configuration. Furthermore, a decrease in minority-spin states in both leads in the P configuration hinders the emergence of AR processes, which leads to weakly negative cross conductance in the Kondo regime, e.g., $G_C \geq -0.1G_0$, and even causes CAR to totally vanish, thus $G_C \simeq G_L$ at the two usual resonance peaks, as shown in Figure 5b. This states that strong ferromagnetism destroys proximitized superconductivity in this three-terminal hybrid nanosystem.

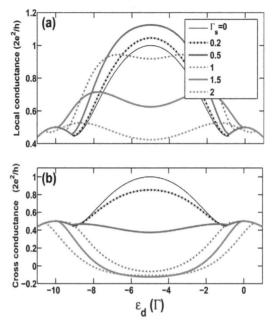

Figure 5. (Color online) (**a**) The local conductance and (**b**) the cross conductance vs. the bare dot level ϵ_d with $U = 10$ at zero temperature for different proximity-coupling strengths Γ_s in the P configuration with $p = 0.5$.

3.2. Nonlinear Local and Cross Conductances

Now, we turn to the investigation of nonlinear tunneling, since the nonlinear differential conductance dI_L/dV is believed to be a very useful tool in experiments aimed at detecting the formation of the Kondo-correlated state due to its proportionality to the transmission spectrum, supposing that the total transmission is unchanged subject to the external bias voltage. In the present three-terminal hybrid device, one can define the local and cross differential conductances, $g_L = \partial I_L/\partial V$ and $g_C = \partial I_R/\partial V$, if the bias voltage V is applied to the left lead and while the superconducting and the right leads are kept grounded. From the Equations (26)–(29), we can obtain that the two diffenertial conductances are both proportional to the normal transmission spectrum $T_N(\omega)$ and the AR spectrum $T_A(\omega)$ at $\omega = V$ at zero temperature, $g_L \propto T_N(V) + aT_A(V)$ and $g_C \propto T_N(V) - bT_A(V)$ (a and b are constants).

Figure 6 shows the local and cross differential conductances as functions of bias voltage at various proximity couplings Γ_s for the system with a single dot level $\epsilon_d = -5$ ($T_K \simeq 0.03$) at the Kondo regime. These curves for weak proximity coupling $\Gamma_s < 1.0$ present a single zero-bias anomaly, which is the signature of the Kondo effect. Nevertheless, there appears non-zero-bias peak with increasing proximity coupling $\Gamma_s \geq 1.0$. It is announced that the Kondo correlation enhances not only the normal ET, but also the AR; nonetheless. the increasing superconducting proximity coupling induces splitting of the Kondo peaks in the normal transmission spectrum as well as the AR spectrum. This peak splitting is the reason that the three parts of the linear conductance are all suppressed when $\Gamma_s > 1.0$, as shown in Figure 3. Finally, one can observe that the negative cross differential conductance becomes positive in the case of large bias voltage. External bias voltage plays a role in dissipation so as to destroy not only the Kondo correlation but the negative nonlocal current response as well.

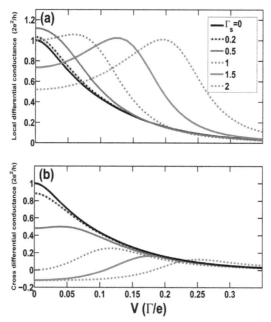

Figure 6. (Color online) The zero-temperature local (**a**) and cross (**b**) differential conductances vs. bias voltage V for various couplings Γ_s for the system with bare dot level $\epsilon_d = -5$ and $U = 10$ in the case of normal leads ($p = 0$).

4. Conclusion

We have theoretically investigated the subgap transport properties of a hybrid nanosystem consisting of an interacting QD connected to one superconducting lead and two ferromagnetic leads. On the basis of the finite-U slave boson mean field approach and the NGF method, we find markedly rich transport features ascribed to the competition among the Kondo correlation, superconducting proximity effect, and spin polarization of electrodes. In the case of weak superconducting proximity coupling, the Kondo-correlated state can still be built, leading to a single zero-bias peak in the voltage-dependent differential conductance. However, the peak height drops down gradually with increasing Γ_s, and when $\Gamma_s \geq 1.0$, a non-zero peak appears. Such strong proximity coupling induces linear cross conductance which is negative in the Kondo region. Spin polarization can further enhance the opposite current response in the right lead (more negative cross conductance) in the AP configuration, because such a configuration is advantageous to the emergence of CAR. In contrast, in the P configuration, the rising spin polarization p blocks the CAR process and also splits the Kondo peak, such that the linear local conductance exhibits four peaks when $\Gamma_s \geq 1.0$, and the linear cross conductance reduces to the normal positive conductance more rapidly.

Author Contributions: Conceptualization, B.D.; Data curation, C.L.; Funding acquisition, B.D.; Investigation, C.L. and B.D.; Supervision, B.D. and X.-L.L; Writing—review & editing, B.D.

Funding: National Natural Science Foundation of China : 11674223.

Acknowledgments: This work was supported by Projects of the National Science Foundation of China under Grant No. 11674223.

Conflicts of Interest: The authors declare no conflict of interest.

Appendix A

In this appendix, we briefly show how to obtain the effective Hamiltonian Equation (7). Within the formulation of the finite-U slave boson approach, the original QD electron operators are replaced by the new varibales as follows: $n_1 n_2 \rightarrow d^\dagger d$, $c_{d\sigma} \rightarrow z_\sigma f_{d\sigma}$, and $c^\dagger_{d\sigma} c_{d\sigma} \rightarrow f^\dagger_{d\sigma} f_{d\sigma}$. The additional degrees of freedom simplify the Coulomb interaction $U n_1 n_2$ as $U d^\dagger d$, but introduce vectors not describing physically real states. Then, two constraint conditions have to be imposed to eliminate the unphysical part of the enlarged Hilbert space $|\Phi\rangle$

$$\left(\sum_\sigma p^\dagger_\sigma p_\sigma + e^\dagger e + d^\dagger d - 1 \right)|\Phi\rangle = 0,$$
$$(f^\dagger_{d\sigma} f_{d\sigma} - p^\dagger_\sigma p_\sigma - d^\dagger d)|\Phi\rangle = 0. \tag{A1}$$

Therefore, the subspace of the enlarged Hilbert space defined by Equation (A1) is equivalent to the original Hilbert space. Applying Dirac's formulation of constrained dynamics, one should then introduce two q-number Lagrange multipliers λ^1 and λ^2_σ, corresponding to the two constraints in Equation (A1), to the Heisenberg equation of motion with respect to the effective Hamiltonian

$$\tilde{H} \equiv H + \lambda^1 \left(\sum_\sigma p^\dagger_\sigma p_\sigma + e^\dagger e + d^\dagger d - 1 \right) + \sum_\sigma \lambda^2_\sigma (f^\dagger_{d\sigma} f_{d\sigma} - p^\dagger_\sigma p_\sigma - d^\dagger d), \tag{A2}$$

and consequently any dynamical observable $\hat{A}$ satisfies the standard equation of motion as a state equation

$$i\hbar \frac{d\hat{A}}{dt}|\Phi\rangle = [\hat{A}, \tilde{H}]|\Phi\rangle. \tag{A3}$$

The next key point of the SBMF approach is to replace all the slave-boson operators and the Lagrange multipliers by their average values according to nonequilibrium steady states (NESS), which can be still expressed in this paper by e, p_σ, d, λ_1, and λ^2_σ. As a result, we can indeed obtain the mean-field expression of the effective Hamiltonian Equation (7) (please note that we still use the notation $c_{d\sigma}$ instead of $f_{d\sigma}$ in the effective Hamiltonian in the main text for the sake of convenience).

Appendix B

It is easily noticed that there are seven averages in total to be determined within the SBMF approach, and the constraints provide three conditions: Equations (13) and (14). Hence, four more conditions are necessary. Originally, Kotliar and Ruckenstein [46] derived them by the saddle-point approximation for the equilibrium free energy, but this approach cannot be applied in this paper, since we are dealing with the NESS. Instead, herein, we can derive those conditions from the equations of motion of the four slave boson fields according to Equation (A3). For instance, the empty boson field e obeys the equation of motion:

$$i\hbar \frac{de}{dt} = \Gamma_s \left([e, z^\dagger_1 z^\dagger_2] c^\dagger_{d1} c^\dagger_{d2} + [e, z_1 z_2] c_{d2} c_{d1} \right) + \sum_{\eta k \sigma} V_{\eta k} \left(c^\dagger_{\eta k \sigma} c_{d\sigma} [e, z_\sigma] + c^\dagger_{d\sigma} c_{\eta k \sigma} [e, z^\dagger_\sigma] \right) + \lambda^1 e. \tag{A4}$$

By replacing the slave-boson operators and Lagrange multipliers to their mean values and evaluating NESS averages of the fermionic operators, one obtains, in the condition of steady state,

$$i\hbar \left\langle \frac{de}{dt} \right\rangle = \Gamma_s \left(\frac{\partial z^*_1 z^*_2}{\partial e^*} \langle c^\dagger_{d1} c^\dagger_{d2} \rangle + \frac{\partial z_1 z_2}{\partial e^*} \langle c_{d2} c_{d1} \rangle \right) + \sum_{\eta k \sigma} V_{\eta k} \left(\frac{\partial z_\sigma}{\partial e^*} \langle c^\dagger_{\eta k \sigma} c_{d\sigma} \rangle + \frac{\partial z^*_\sigma}{\partial e^*} \langle c^\dagger_{d\sigma} c_{\eta k \sigma} \rangle \right) + \lambda^1 e = 0. \tag{A5}$$

Here the commutators of the boson fields are evaluated as $[e, z_\sigma] = \partial z_\sigma / \partial e^\dagger$ and $[e, z_1 z_2] = \partial z_1 z_2 / \partial e^\dagger$, according to the Dirac's quantum algebraic scheme.

Now, we turn to discuss the averages of fermionic operators which are evaluated at a NESS with respect to the fermionic part of the effective Hamiltonian Equation (7). Since we discuss quantum transport through a nanodevice in this paper, a NESS can be defined as being when all electrodes are at equilibrium with their own temperatures $1/\beta_\eta$ and chemical potentials μ_η, and thus the statistical operator can be introduced as

$$\varrho_0 = \frac{1}{\Xi} e^{-\beta_L(H_L - \mu_L N_L)} e^{-\beta_R(H_R - \mu_R N_R)}. \qquad (A6)$$

In the above, Ξ is the normalization constant, H_η and N_η are the Hamiltonian and the total number of particles in the lead η, respectively. According to the nonequilibrium statistical theory, the average of an observable $\hat{A}$ with respect to such a NESS can then be evaluated as $\langle \hat{A} \rangle = Tr(\hat{A}\varrho_0)$. Thereby, we can evaluate the averages of fermionic operators with the help of the NGF technique, for example, the operators of occupation number on the dot, K_1 and K_2 (Equations (15) and (16)), and the order parameter R in Equation (17). Moreover, we can define

$$Q_{1\eta} = \sum_k V_{\eta k} \left\langle c_{d1}^\dagger c_{\eta k 1} \right\rangle = \sum_k V_{\eta k} \int \frac{d\omega}{2\pi i} \left\langle \left\langle c_{\eta k 1}; c_{d1}^\dagger \right\rangle \right\rangle^< (\omega) = \sum_k V_{\eta k} \int \frac{d\omega}{2\pi i} G_{\eta k, d;11}^< (\omega), \qquad (A7)$$

and

$$Q_{2\eta} = \sum_k V_{\eta k} \left\langle c_{\eta k 2}^\dagger c_{d2} \right\rangle = \sum_k V_{\eta k} \int \frac{d\omega}{2\pi i} \left\langle \left\langle c_{\eta k 2}^\dagger; c_{d2} \right\rangle \right\rangle^< (\omega) = \sum_k V_{\eta k} \int \frac{d\omega}{2\pi i} G_{\eta k, d;22}^< (\omega). \qquad (A8)$$

Here, the hybrid NGFs, $G_{\eta k, d}^< (\omega)$ are the matrix elements of the 2×2 hybrid contour-order GF matrix $G_{\eta k, d}^\gamma (\omega) = \langle \langle \mathcal{L}_{\eta k}; \phi^\dagger \rangle \rangle^\gamma$ ($\gamma = R, A, <, >$) with the Nambu representation in the lead $\mathcal{L}_{\eta k} = (c_{\eta k 1}, c_{\eta k 2}^\dagger)^T$. Applying the Langreth theorem, we obtain

$$G_{\eta k, d}^< (\omega) = g_{\eta k}^R (\omega) \hat{V}_{\eta k} G_d^< (\omega) + g_{\eta k}^< (\omega) \hat{V}_{\eta k} G_d^A (\omega), \qquad (A9)$$

where $g_{\eta k}^\gamma$ is the decoupled NGF of the lead η defined as $g_{\eta k}^\gamma (\omega) = \langle \langle \mathcal{L}_{\eta k}; \mathcal{L}_{\eta k}^\dagger \rangle \rangle^\gamma$ and

$$\hat{V}_{\eta k} = \begin{pmatrix} z_1 V_{\eta k} & 0 \\ 0 & -z_2 V_{\eta k}^* \end{pmatrix}. \qquad (A10)$$

Subsequently, we have

$$\sum_k \hat{V}_{\eta k} G_{\eta k, d}^< (\omega) = \sum_k \hat{V}_{\eta k} g_{\eta k}^R (\omega) \hat{V}_{\eta k} G_d^< (\omega) + \hat{V}_{\eta k} g_{\eta k}^< (\omega) \hat{V}_{\eta k} G_d^A (\omega),$$
$$= \Sigma_\eta^R (\omega) G_d^< (\omega) + \Sigma_\eta^< (\omega) G_d^A (\omega), \qquad (A11)$$

with

$$\Sigma_\eta^\gamma (\omega) = \sum_k \hat{V}_{\eta k}^* g_{\eta k}^\gamma (\omega) \hat{V}_{\eta k} = \begin{pmatrix} \sum_k |V_{\eta k}|^2 g_{\eta k;11}^\gamma (\omega) & 0 \\ 0 & \sum_k |V_{\eta k}|^2 g_{\eta k;22}^\gamma (\omega) \end{pmatrix}. \qquad (A12)$$

Noticing the retarded and advanced self energies in the wide band limit,

$$\Sigma_\eta^{R(A)} = \pm \frac{i}{2} \begin{pmatrix} \tilde{\Gamma}_{\eta 1} & 0 \\ 0 & \tilde{\Gamma}_{\eta 2} \end{pmatrix}, \qquad (A13)$$

and Equations (21)–(24), we can further obtain an explicit expression of Equation (A11) and insert this expression into Equations (A7) and (A8) to yield Equations (18) and (19). Finally, we obtain the self-consistent equation Equation (9).

Entropy 2019, 21, 1003

Similarly, from the equations of motion of p_σ and d,

$$i\hbar \frac{dp_\sigma}{dt} = \Gamma_s \left([p_\sigma, z_1^\dagger z_2^\dagger] c_{d1}^\dagger c_{d2}^\dagger + [p_\sigma, z_1 z_2] c_{d2} c_{d1} \right) + \sum_{\eta k \sigma'} V_{\eta k} \left(c_{\eta k \sigma'}^\dagger c_{d\sigma'} [p_\sigma, z_\sigma'] + c_{d\sigma'}^\dagger c_{\eta k \sigma'} [p_\sigma, z_\sigma'^\dagger] \right)$$
$$+ (\lambda^1 - \lambda_\sigma^2) p_\sigma, \tag{A14}$$

$$i\hbar \frac{dd}{dt} = \Gamma_s \left([d, z_1^\dagger z_2^\dagger] c_{d1}^\dagger c_{d2}^\dagger + [d, z_1 z_2] c_{d2} c_{d1} \right) + \sum_{\eta k \sigma} V_{\eta k} \left(c_{\eta k \sigma}^\dagger c_{d\sigma} [d, z_\sigma] + c_{d\sigma}^\dagger c_{\eta k \sigma} [d, z_\sigma^\dagger] \right)$$
$$+ (U + \lambda^1 - \Sigma_\sigma \lambda_\sigma^2) d, \tag{A15}$$

we can get self-consistent equations, Equations (10)–(12).

References

1. Buitelaar, M.R.; Nussbaumer, T.; Schönenberger, C. Quantum Dot in the Kondo Regime Coupled to Superconductors. *Phys. Rev. Lett.* **2002**, *89*, 256801.

2. Russo, S.; Kroug, M.; Klapwijk, T.M.; Morpurgo, A.F. Experimental Observation of Bias-Dependent Nonlocal Andreev Reflection. *Phys. Rev. Lett.* **2005**, *95*, 027002.

3. Cadden-Zimansky, P.; Chandrasekhar, V. Nonlocal Correlations in Normal-Metal Superconducting Systems. *Phys. Rev. Lett.* **2006**, *97*, 237003.

4. Jarillo-Herrero, P.; Van Dam, J.A.; Kouwenhoven, L.P. Quantum supercurrent transistors in carbon nanotubes. *Nature* **2006**, *439*, 953.

5. Van Dam, J.A.; Nazarov, Y.V.; Bakkers, E.P.; De Franceschi, S.; Kouwenhoven, L.P. Supercurrent reversal in quantum dots. *Nature* **2006**, *442*, 667.

6. Sand-Jespersen, T.; Paaske, J.; Andersen, B.M.; Grove-Rasmussen, K.; Jørgensen, H.I.; Aagesen, M.; Sørensen, C.B.; Lindelof, P.E.; Flensberg, K.; Nygård, J. Kondo-Enhanced Andreev Tunneling in InAs Nanowire Quantum Dots. *Phys. Rev. Lett.* **2007**, *99*, 126603.

7. Hofstetter, L.; Csonka, S.; Nygård, J.; Schönenberger, C. Cooper pair splitter realized in a two-quantum-dot Y-junction. *Nature* **2009**, *461*, 960.

8. Grove-Rasmussen, K.; Jørgensen, H.I.; Andersen, B.M.; Paaske, J.; Jespersen, T.S.; Nygård, J.; Flensberg, K.; Lindelof, P.E. Superconductivity-enhanced bias spectroscopy in carbon nanotube quantum dots. *Phys. Rev. B* **2009**, *79*, 134518.

9. Franceschi, S.D.; Kouwenhoven, L.; Schönenberger, C.; Wernsdorfer, W. Hybrid superconductor-quantum dot devices. *Nat. Nanotechnol.* **2010**, *5*, 703.

10. Franke, K.J.; Schulze, G.; Pascual, J.I. Competition of Superconducting Phenomena and Kondo Screening at the Nanoscale. *Science* **2011**, *332*, 940–944. Available online: https://science.sciencemag.org/content/332/6032/940.full.pdf (accessed on 11 October 2019).

11. Dirks, T.; Hughes, T.L.; Lal, S.; Uchoa, B.; Chen, Y.F.; Chialvo, C.; Goldbart, P.M.; Mason, N. Transport through Andreev bound states in a graphene quantum dot. *Nat. Phys.* **2011**, *7*, 386.

12. Schindele, J.; Baumgartner, A.; Schönenberger, C. Near-Unity Cooper Pair Splitting Efficiency. *Phys. Rev. Lett.* **2012**, *109*, 157002.

13. Braunecker, B.; Burset, P.; Levy Yeyati, A. Entanglement Detection from Conductance Measurements in Carbon Nanotube Cooper Pair Splitters. *Phys. Rev. Lett.* **2013**, *111*, 136806.

14. Lee, E.J.H.; Jiang, X.; Houzet, M.; Aguado, R.; Lieber, C.M.; Franceschi, S.D. Spin-resolved Andreev levels and parity crossings in hybrid superconductor Csemiconductor nanostructures. *Nat. Nanotechnol.* **2014**, *9*, 79.

15. Deacon, R.S.; Oiwa, A.; Sailer, J.; Baba, S.; Kanai, Y.; Shibata, K.; Hirakawa, K.; Tarucha, S. Cooper pair splitting in parallel quantum dot Josephson junctions. *Nat. Commun.* **2015**, *6*, 7446.

16. Ruby, M.; Pientka, F.; Peng, Y.; von Oppen, F.; Heinrich, B.W.; Franke, K.J. Tunneling Processes into Localized Subgap States in Superconductors. *Phys. Rev. Lett.* **2015**, *115*, 087001.

17. Fülöp, G.; Domínguez, F.; d'Hollosy, S.; Baumgartner, A.; Makk, P.; Madsen, M.H.; Guzenko, V.A.; Nygård, J.; Schönenberger, C.; Levy Yeyati, A.; Csonka, S. Magnetic Field Tuning and Quantum Interference in a Cooper Pair Splitter. *Phys. Rev. Lett.* **2015**, *115*, 227003.

18. Albrecht, S.M.; Higginbotham, A.P.; Madsen, M.; Kuemmeth, F.; Jespersen, T.S.; Nygård, J.; Krogstrup, P.; Marcus, C.M. Exponential protection of zero modes in Majorana islands. *Nature* **2016**, *531*, 206.

19. Ruby, M.; Peng, Y.; von Oppen, F.; Heinrich, B.W.; Franke, K.J. Orbital Picture of Yu-Shiba-Rusinov Multiplets. *Phys. Rev. Lett.* **2016**, *117*, 186801.

20. Lutchyn, R.M.; Sau, J.D.; Das Sarma, S. Majorana Fermions and a Topological Phase Transition in Semiconductor-Superconductor Heterostructures. *Phys. Rev. Lett.* **2010**, *105*, 077001.

21. Cuevas, J.C.; Levy Yeyati, A.; Martín-Rodero, A. Kondo effect in normal-superconductor quantum dots. *Phys. Rev. B* **2001**, *63*, 094515.

22. Eldridge, J.; Pala, M.G.; Governale, M.; König, J. Superconducting proximity effect in interacting double-dot systems. *Phys. Rev. B* **2010**, *82*, 184507.

23. Golubev, D.S.; Zaikin, A.D. Shot noise and Coulomb effects on nonlocal electron transport in normal-metal/superconductor/normal-metal heterostructures. *Phys. Rev. B* **2010**, *82*, 134508.

24. Martín-Rodero, A.; Yeyati, A.L. Josephson and Andreev transport through quantum dots. *Adv. Phys.* **2011**, *60*, 899–958.

25. Yamada, Y.; Tanaka, Y.; Kawakami, N. Interplay of Kondo and superconducting correlations in the nonequilibrium Andreev transport through a quantum dot. *Phys. Rev. B* **2011**, *84*, 075484.

26. Moghaddam, A.G.; Governale, M.; König, J. Driven superconducting proximity effect in interacting quantum dots. *Phys. Rev. B* **2012**, *85*, 094518.

27. Koga, A. Quantum Monte Carlo study of nonequilibrium transport through a quantum dot coupled to normal and superconducting leads. *Phys. Rev. B* **2013**, *87*, 115409.

28. Kiršanskas, G.; Goldstein, M.; Flensberg, K.; Glazman, L.I.; Paaske, J. Yu-Shiba-Rusinov states in phase-biased superconductor–quantum dot–superconductor junctions. *Phys. Rev. B* **2015**, *92*, 235422.

29. Žitko, R. Spectral properties of Shiba subgap states at finite temperatures. *Phys. Rev. B* **2016**, *93*, 195125.

30. Beckmann, D.; Weber, H.B.; Löhneysen, H.V. Evidence for Crossed Andreev Reflection in Superconductor-Ferromagnet Hybrid Structures. *Phys. Rev. Lett.* **2004**, *93*, 197003.

31. Hofstetter, L.; Geresdi, A.; Aagesen, M.; Nygård, J.; Schönenberger, C.; Csonka, S. Ferromagnetic Proximity Effect in a Ferromagnet–Quantum-Dot–Superconductor Device. *Phys. Rev. Lett.* **2010**, *104*, 246804.

32. Martinek, J.; Utsumi, Y.; Imamura, H.; Barnaś, J.; Maekawa, S.; König, J.; Schön, G. Kondo Effect in Quantum Dots Coupled to Ferromagnetic Leads. *Phys. Rev. Lett.* **2003**, *91*, 127203.

33. Pasupathy, A.N.; Bialczak, R.C.; Martinek, J.; Grose, J.E.; Donev, L.A.K.; McEuen, P.L.; Ralph, D.C. The Kondo Effect in the Presence of Ferromagnetism. *Science* **2004**, *306*, 86–89.

34. Hauptmann, J.R.; Paaske, J.; Lindelof, P.E. Electric-field-controlled spin reversal in a quantum dot with ferromagnetic contacts. *Nat. Phys.* **2008**, *4*, 373.

35. Gaass, M.; Hüttel, A.K.; Kang, K.; Weymann, I.; von Delft, J.; Strunk, C. Universality of the Kondo Effect in Quantum Dots with Ferromagnetic Leads. *Phys. Rev. Lett.* **2011**, *107*, 176808.

36. Dong, B.; Cui, H.L.; Liu, S.Y.; Lei, X.L. Kondo-type transport through an interacting quantum dot coupled to ferromagnetic leads. *J. Phys.-Condes. Matter* **2003**, *15*, 8435–8444.

37. Choi, M.S.; Sánchez, D.; López, R. Kondo Effect in a Quantum Dot Coupled to Ferromagnetic Leads: A Numerical Renormalization Group Analysis. *Phys. Rev. Lett.* **2004**, *92*, 056601.

38. Sánchez, D.; López, R.; Samuelsson, P.; Büttiker, M. Andreev drag effect in ferromagnetic-normal-superconducting systems. *Phys. Rev. B* **2003**, *68*, 214501.

39. Weymann, I.; Wójcik, K.P. Andreev transport in a correlated ferromagnet-quantum-dot-superconductor device. *Phys. Rev. B* **2015**, *92*, 245307.

40. Futterer, D.; Governale, M.; Pala, M.G.; König, J. Nonlocal Andreev transport through an interacting quantum dot. *Phys. Rev. B* **2009**, *79*, 054505.

41. Futterer, D.; Governale, M.; König, J. Generation of pure spin currents by superconducting proximity effect in quantum dots. *EPL (Europhys. Lett.)* **2010**, *91*, 47004.

42. Sothmann, B.; Futterer, D.; Governale, M.; König, J. Probing the exchange field of a quantum-dot spin valve by a superconducting lead. *Phys. Rev. B* **2010**, *82*, 094514.

43. Weymann, I.; Trocha, P. Superconducting proximity effect and zero-bias anomaly in transport through quantum dots weakly attached to ferromagnetic leads. *Phys. Rev. B* **2014**, *89*, 115305.

44. Michałek, G.; Bułka, B.R.; Domański, T.; Wysokiński, K.I. Interplay between direct and crossed Andreev reflections in hybrid nanostructures. *Phys. Rev. B* **2013**, *88*, 155425.

45. Michałek, G.; Domański, T.; Bułka, B.R.; Wysokiński, K.I. Novel non-local effects in three-terminal hybrid devices with quantum dot. *Sci. Rep.* **2015**, *5*, 14572.

46. Kotliar, G.; Ruckenstein, A.E. New Functional Integral Approach to Strongly Correlated Fermi Systems: The Gutzwiller Approximation as a Saddle Point. *Phys. Rev. Lett.* **1986**, *57*, 1362–1365.

47. Schönhammer, K. Variational results as saddle-point approximations: The Anderson impurity model. *Phys. Rev. B* **1990**, *42*, 2591–2593.

48. Aguado, R.; Langreth, D.C. Out-of-Equilibrium Kondo Effect in Double Quantum Dots. *Phys. Rev. Lett.* **2000**, *85*, 1946–1949.

49. Dong, B.; Lei, X.L. Kondo effect and antiferromagnetic correlation in transport through tunneling-coupled double quantum dots. *Phys. Rev. B* **2002**, *65*, 241304.

50. Dong, B.; Lei, X.L. Kondo-type transport through a quantum dot under magnetic fields. *Phys. Rev. B* **2001**, *63*, 235306.

51. Dong, B.; Lei, X.L. Kondo-type transport through a quantum dot: a new finite-Uslave-boson mean-field approach. *J. Phys.-Condes. Matter* **2001**, *13*, 9245–9258.

52. Ma, J.; Dong, B.; Lei, X.L. Spin-Polarized Transport Through a Quantum Dot Coupled to Ferromagnetic Leads: Kondo Correlation Effect. *Commun. Theor. Phys.* **2005**, *43*, 341–348.

53. Bergeret, F.S.; Yeyati, A.L.; Martín-Rodero, A. Interplay between Josephson effect and magnetic interactions in double quantum dots. *Phys. Rev. B* **2006**, *74*, 132505.

Article

Current Correlations in a Quantum Dot Ring: A Role of Quantum Interference

Bogdan R. Bułka * and Jakub Łuczak

Institute of Molecular Physics, Polish Academy of Sciences, ul. M. Smoluchowskiego 17, 60-179 Poznań, Poland; jakub.luczak@ifmpan.poznan.pl
* Correspondence: bogdan.bulka@ifmpan.poznan.pl

Received: 29 April 2019; Accepted: 22 May 2019; Published: 24 May 2019

Abstract: We present studies of the electron transport and circular currents induced by the bias voltage and the magnetic flux threading a ring of three quantum dots coupled with two electrodes. Quantum interference of electron waves passing through the states with opposite chirality plays a relevant role in transport, where one can observe Fano resonance with destructive interference. The quantum interference effect is quantitatively described by local bond currents and their correlation functions. Fluctuations of the transport current are characterized by the Lesovik formula for the shot noise, which is a composition of the bond current correlation functions. In the presence of circular currents, the cross-correlation of the bond currents can be very large, but it is negative and compensates for the large positive auto-correlation functions.

Keywords: quantum interference; shot noise; persistent current

1. Introduction

In 1985, Webb et al. [1] presented their pioneering experiment, showing Aharonov-Bohm oscillations in a nanoscopic metallic ring and a role of quantum interference (QI) in electron transport. Later, Ji et al. [2] demonstrated the electronic analogue of the optical Mach–Zehnder interferometer (MZI), which was based on closed-geometry transport through single edge states in the quantum Hall regime. Theoretical studies [3–7] predicted coherent transport through single molecules with a ring structure, where, due to their small size, one could show constructive or destructive quantum interference effects at room temperatures. From 2011, these predictions have been experimentally verified, using mechanically controllable break junction (MCBJ) and scanning tunneling microscope break junction (STM-BJ) techniques [8,9] in various molecular systems: Single phenyl, polycyclic aromatic, and conjugated heterocyclic blocks, as well as hydrocarbons (for a recent review on QI in molecular junctions, see [10,11] and the references therein).

Our interest is in the internal local currents and their correlations in a ring geometry to see a role of quantum interference. An interesting aspect is the formation of a quantum vortex flow driven by a net current from the source to the drain electrode, which has been studied in many molecular systems [7,10,12–21] (see also [22]). It has also been shown that, under some conditions, a circular thermoelectric current can exceed the transport current [23]. In particular, our studies focus on the role of the states with opposite chirality in the ring and on the QI effect and the circular current. Correlations of the electron currents (shot noise) through edge states in the Mach–Zehnder interferometer have been extensively studied by Buttiker et al. [24–28] (see also [29] and the references therein). However, in a metallic (or molecular) ring, the situation is different than in the MZI, as multiple reflections are relevant to the formation of the circular current. Our studies will show that the transition from laminar to vortex flow is manifested in the shot noise of local currents. In particular, it will be seen in a cross-correlation function for the currents in different branches of the ring, which becomes negative and large in the presence of the circular current.

The paper is organized as follows. In the next chapter, Section 2, we will present the model of three quantum dots in a ring geometry, which is the simplest model showing all aspects of QI and current correlations. The model includes a magnetic flux threading the ring, which changes interference conditions as well as inducing a persistent current. The net transport current and the local bond currents, as well as the persistent current (and their conductances), are derived analytically, by means of the non-equilibrium Keldysh Green function technique. It will be shown that the correlation function for the net transport current can be expressed as a composition of the correlation functions for the local currents inside the ring. We will, also, show all shot noise components; in particular, the one for the net transport current (given by Lesovik's formula [30]). The next chapters, Sections 3–5, present the analyses of the results for the case $\Phi = 0$ (without the magnetic flux), for the case with the persistent current only (without the source-drain bias V), and for the general case (for $V \neq 0$, $\Phi \neq 0$) showing the interplay between the bond currents and the persistent current. Finally, in Section 6, the main results of the paper are summarized.

2. Calculations of Currents and Their Correlations in Triangular Quantum Dot System

2.1. Model

The considered system of three quantum dots (QDs) in an triangular arrangement is presented in Figure 1. This system is described by the Hamiltonian

$$H_{tot} = H_{3QD} + H_{el} + H_{3QD-el}, \tag{1}$$

which consists of parts corresponding to the electrons in the triangular QD system, in the electrodes, and in the coupling between the sub-systems, respectively. The first part is given by

$$H_{3QD} = \sum_{i \in 3QD} \varepsilon_i c_i^\dagger c_i + \sum_{i,j \in 3QD} \left(\tilde{t}_{ij} c_i^\dagger c_j + h.c. \right), \tag{2}$$

where the first term describes the single-level energy, ε_i, at the i-th QD and the second term corresponds to electron hopping between the QDs. Here, the hopping parameters $\tilde{t}_{12} = t_{12}e^{i\phi/3} = \tilde{t}_{21}^*$, $\tilde{t}_{23} = t_{23}e^{i\phi/3} = \tilde{t}_{32}^*$, and $\tilde{t}_{31} = t_{31}e^{i\phi/3} = \tilde{t}_{13}^*$ include the phase shift $\phi = 2\pi\Phi/(hc/e)$, due to presence of the magnetic flux Φ; where hc/e denotes the one-electron flux quantum. The spin of electrons is irrelevant in our studies and so it is omitted. We consider transport in an open system with the left (L) and right (R) electrodes as reservoirs of electrons, each in thermal equilibrium with a given chemical potential μ_α and temperature T_α. The corresponding Hamiltonian is

$$H_{el} = \sum_{k,\alpha \in L,R} \varepsilon_{k,\alpha} c_{k\alpha}^\dagger c_{k\alpha}, \tag{3}$$

where $\varepsilon_{k,\alpha}$ denotes an electron spectrum. The coupling between the 3QD system and the electrodes is given by

$$H_{3QD-el} = \sum_k (t_L c_{kL}^\dagger c_1 + t_R c_{kR}^\dagger c_2 + h.c.), \tag{4}$$

with tunneling from the electrodes given by the hopping parameters t_L and t_R, respectively. The model omits Coulomb interactions and, therefore, one can derive all transport characteristics analytically.

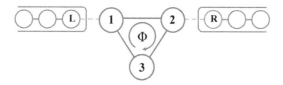

Figure 1. Model of the triangular system of three quantum dots (3QDs) threaded by the magnetic flux Φ and attached to the left (L) and the right (R) electrodes.

2.2. Calculation of Currents

We consider a steady-state current, with the net transport current through the 3QD system, $I^{tr} = I_{12} + I_{13}$, expressed as a sum of the bond currents through the upper and the lower branches

$$I_{ij} = \frac{e}{i\hbar}\left(\tilde{t}_{ij}\langle c_i^\dagger c_j\rangle - \tilde{t}_{ji}\langle c_j^\dagger c_i\rangle\right). \tag{5}$$

We use the non-equilibrium Green function technique (NEGF), which is described in many textbooks (e.g., see [31]). To determine the currents, one calculates the lesser Green functions, $G_{ji}^< \equiv \imath\langle c_i^\dagger c_j\rangle$, by means of the equation of motion method (EOM). The coupling with the electrodes is manifested by the lesser Green functions $g_\alpha^< = 2\pi(g_\alpha^r - g_\alpha^a)f_\alpha$, where $g_\alpha^{r,a}$ denotes the retarded (r) and advanced (a) Green functions in the α electrode, and $f_\alpha = 1/(\exp[(E - \mu_\alpha)/k_B T_\alpha] + 1)$ is the Fermi distribution function for an electron with energy E, with respect to a chemical potential μ_α and at temperature T_α. For any Green function $G_{ji}^<$, we separate contributions from the left and the right electrodes (i.e., we extract the coefficients in front of $g_L^<$ and $g_R^<$) and, after some algebra, the bond current can be expressed as

$$I_{ij} = -\frac{e}{2\pi\hbar}\int_{-\infty}^{\infty} dE\,[\mathcal{G}_{ij}^L(E)f_L - \mathcal{G}_{ij}^R(E)f_R], \tag{6}$$

where the dimensionless conductances for the upper and the lower branches are

$$\mathcal{G}_{12}^L = 2\Gamma_L t_{12}\Im[d_{23,31}d_{23,23}^*]/A, \tag{7}$$
$$\mathcal{G}_{12}^R = 2\Gamma_R t_{12}\Im[d_{23,31}^*d_{31,31}^*]/A, \tag{8}$$
$$\mathcal{G}_{13}^L = 2\Gamma_L t_{31}\Im[d_{12,23}d_{23,23}^*]/A, \text{ and} \tag{9}$$
$$\mathcal{G}_{13}^R = 2\Gamma_R t_{31}\Im[d_{12,31}d_{23,31}]/A. \tag{10}$$

Here, we denote the coefficients: $d_{12,23} = e^{-\imath\phi}t_{12}t_{23} - t_{31}w_2^r$, $d_{12,31} = t_{12}t_{31} - e^{-\imath\phi}t_{23}w_1^r$, $d_{23,31} = e^{\imath\phi}t_{23}t_{31} - t_{12}w_3$, $d_{23,23} = t_{23}^2 - w_2^r w_3$, $d_{31,31} = t_{31}^2 - w_1^r w_3$, and the denominator

$$A = |w_1^r t_{23}^2 + w_2^r t_{31}^2 + w_3 t_{12}^2 - w_1^r w_2^r w_3 - 2t_{12}t_{23}t_{31}\cos\phi|^2, \tag{11}$$

where $w_1^{r,a} = E - \varepsilon_1 - \gamma_L^{r,a}$, $w_2^{r,a} = E - \varepsilon_2 - \gamma_R^{r,a}$, $w_3 = E - \varepsilon_3$, $\Gamma_\alpha = 2\Im[\gamma_\alpha^a]t_\alpha^2$, and $\gamma_\alpha^{r,a} = g_\alpha^{r,a}t_\alpha^2$.

Note that Equation (6) includes the transport current due to the bias voltage applied to the electrodes, as well as the persistent current induced by the magnetic flux [a term proportional to $\sin\phi$], which can be written as $I_{12} = I_{12}^{tr} - I^\phi$ and $I_{13} = I_{13}^{tr} + I^\phi$, respectively. These coefficients are coupled with those in (7)–(10):

$$\mathcal{G}_{12}^L = \mathcal{G}_{12} - \mathcal{G}_\phi^L, \quad \mathcal{G}_{12}^R = \mathcal{G}_{12} + \mathcal{G}_\phi^R, \tag{12}$$
$$\mathcal{G}_{13}^L = \mathcal{G}_{13} + \mathcal{G}_\phi^L, \quad \mathcal{G}_{13}^R = \mathcal{G}_{13} - \mathcal{G}_\phi^R. \tag{13}$$

The first part is

$$I_{ij}^{tr} = -\frac{e}{2\pi\hbar}\int_{-\infty}^{\infty}dE(f_L - f_R)\mathcal{G}_{ij}(E),\tag{14}$$

where the bond conductances are

$$\mathcal{G}_{12} = \Gamma_L\Gamma_R t_{12}w_3[t_{23}t_{31}\cos\phi - t_{12}w_3]/A,\ \text{and}\tag{15}$$

$$\mathcal{G}_{13} = \Gamma_L\Gamma_R t_{23}[t_{12}t_{31}w_3\cos\phi - t_{23}t_{31}^2]/A.\tag{16}$$

The net transport current is $I^{tr} = I_{12}^{tr} + I_{13}^{tr}$ and the transmission is given by

$$\mathcal{T} \equiv \mathcal{G}_{12} + \mathcal{G}_{13} = \Gamma_L\Gamma_R[2t_{12}t_{23}t_{31}w_3\cos\phi - t_{12}^2w_3^2 - t_{23}^2t_{31}^2]/A.\tag{17}$$

The persistent current is expressed as

$$I^{\phi} \equiv -\frac{e}{\pi\hbar}\int_{-\infty}^{\infty}dE\ (\mathcal{G}_{\phi}^L f_L + \mathcal{G}_{\phi}^R f_R),\tag{18}$$

where

$$\mathcal{G}_{\phi}^L = \Gamma_L t_{12}t_{23}t_{31}\sin\phi\ [2t_{23}^2 - (w_2^a + w_2^r)w_3]/A,\ \text{and}\tag{19}$$

$$\mathcal{G}_{\phi}^R = \Gamma_R t_{12}t_{23}t_{31}\sin\phi\ [2t_{31}^2 - (w_1^a + w_1^r)w_3]/A.\tag{20}$$

In the next section, we will show that the voltage bias can induce the circular current, where the bond conductances $\mathcal{G}_{ij}$ are larger than unity or negative.

2.3. Calculation of Current Correlations

Here, we consider a single-particle interference effect which takes place in a Mach–Zehnder or Michelson interferometer, but not in a Hanbury Brown and Twiss situation with a two-particle interference effect. The current fluctuations are described by the operator $\Delta\hat{I}_{ij}(t) - \langle\hat{I}_{ij}(t)\rangle$, and the current–current correlation function is defined as [32]

$$S_{ij,nm}(t,t') \equiv \frac{1}{2}\left[\langle\hat{I}_{ij}(t)\hat{I}_{nm}(t') + \hat{I}_{nm}(t')\hat{I}_{ij}(t)\rangle - 2\langle\hat{I}_{ij}(t)\rangle\langle\hat{I}_{nm}(t')\rangle\right].\tag{21}$$

We consider the steady currents, for which the correlation functions can be represented, in the frequency domain, by their spectral density

$$S_{ij,nm}(\omega) \equiv 2\int_{-\infty}^{\infty}d\tau e^{i\omega\tau}S_{ij,nm}(\tau).\tag{22}$$

In this work, we shall restrict ourselves to studying the current correlations at the zero-frequency limit $\omega = 0$. As the net transport current is $\hat{I}^{tr} = \hat{I}_{12} + \hat{I}_{13}$, its current correlation function can be expressed as a composition of the correlation functions for the bond currents

$$S_{tr,tr} = S_{12,12} + S_{13,13} + 2S_{12,13}.\tag{23}$$

The correlation functions $S_{ij,in}$ can be derived by means of Wick's theorem [31] and are expressed as

$$S_{ij,in} = \frac{e^2}{\pi\hbar}\frac{1}{2}\{t_{ij}t_{in}(\langle c_i^\dagger c_j\rangle\langle c_i c_n^\dagger\rangle + \langle c_i^\dagger c_n\rangle\langle c_i c_j^\dagger\rangle)) - t_{ij}t_{ni}(\langle c_i^\dagger c_i\rangle\langle c_n c_j^\dagger\rangle + \langle c_n^\dagger c_j\rangle\langle c_i c_i^\dagger\rangle))$$
$$+t_{ji}t_{ni}(\langle c_j^\dagger c_i\rangle\langle c_n c_i^\dagger\rangle + \langle c_n^\dagger c_i\rangle\langle c_j c_i^\dagger\rangle)) - t_{ji}t_{in}(\langle c_i^\dagger c_i\rangle\langle c_j c_n^\dagger\rangle + \langle c_j^\dagger c_n\rangle\langle c_i c_i^\dagger\rangle))\}.\tag{24}$$

Once again, we use the NEGF method. As the lesser Green functions, $G^<_{ij} \equiv \imath \langle c^\dagger_i c_j \rangle$, and the greater Green functions, $G^>_{ij} \equiv -\imath \langle c_i c^\dagger_j \rangle$, have the same structure, one should only exchange the Green functions in the electrodes: $g^<_\alpha = 2\pi (g^r_\alpha - g^a_\alpha) f_\alpha \leftrightarrow g^>_\alpha = -2\pi (g^r_\alpha - g^a_\alpha)(1 - f_\alpha)$. Separating coefficients in front of $f_L(1 - f_L)$, $f_R(1 - f_R)$, and $f_L(1 - f_R) + f_R(1 - f_L)$, and after some algebra, one can derive a compact formula for any current–current function. The auto-correlation function for the net transport current is given by the well-known Lesovik formula [30,33,34] (see also [32,35] for a multi-terminal and multi-channel case)

$$S_{tr,tr} = \frac{e^2}{\pi\hbar} \int_{-\infty}^{\infty} dE \left\{ \mathcal{T}^2 \left[f_L(1 - f_L) + f_R(1 - f_R) \right] + \mathcal{T}(1 - \mathcal{T}) \left[f_L(1 - f_R) + f_R(1 - f_L) \right] \right\}, \quad (25)$$

where $\mathcal{T}$ is the transmission through the 3QD system. For a given temperature $T_L = T_R = T$, one has $f_L(1 - f_R) + f_R(1 - f_L) = \coth[(\mu_L - \mu_R)/2k_BT](f_L - f_R)$ and, thus,

$$S_{tr,tr} = 2I^{tr} \coth\left(\frac{eV}{2k_BT} \right) - \frac{e^2}{\pi\hbar} \int_{-\infty}^{\infty} dE\, \mathcal{T}^2 (f_L - f_R)^2. \quad (26)$$

When the scale of the energy dependence ΔE of the transmission $\mathcal{T}$ is much larger than both the temperature and applied voltage (i.e., $\Delta E \gg eV \gg k_BT$), one can obtain the well known explicit relation (see Blanter and Buttiker [32])

$$S_{tr,tr} = \frac{e^2}{\pi\hbar} \left[2k_BT\, \mathcal{T}^2(E_F) + eV \coth\left(\frac{eV}{2k_BT} \right) S^{sh}_{tr,tr} \right]. \quad (27)$$

The first term is the Nyquist-Johnson noise at equilibrium and the second term, with $S^{sh}_{tr,tr} = \mathcal{T}(E_F)(1 - \mathcal{T}(E_F))$, corresponds to the shot noise [30,32,34].

The correlation functions for the bond currents are calculated from Equation (24) and are expressed as

$$S_{ij,ik} = \frac{e^2}{\pi\hbar} \int_{-\infty}^{\infty} dE \left\{ S^{sh}_{ij,ik} \left[f_L(1 - f_R) + f_R(1 - f_L) \right] + [\mathcal{G}^L_{ij}\mathcal{G}^L_{ik} f_L(1 - f_L) + \mathcal{G}^R_{ij}\mathcal{G}^R_{ik} f_R(1 - f_R)] \right\}, \quad (28)$$

where $\mathcal{G}^\alpha_{ij}$ are given by (7)–(10), and the dimensionless spectral functions of the shot noise components are

$$S^{sh}_{12,12} = \Gamma_L \Gamma_R t^2_{12} |d^2_{23,31} - d_{23,23} d^*_{31,31}|^2 / A^2, \quad (29)$$

$$S^{sh}_{13,13} = \Gamma_L \Gamma_R t^2_{31} |d_{12,23} d^*_{23,31} + d_{12,31} d^*_{23,23}|^2 / A^2, \quad (30)$$

$$S^{sh}_{12,13} = \Gamma_L \Gamma_R t_{12} t_{31} \Re[(d^2_{23,31} - d_{23,23} d^*_{31,31})(d^*_{12,23} d^*_{23,31} + d_{12,31} d^*_{23,23})] / A^2, \text{ and} \quad (31)$$

$$S^{sh}_{tr,tr} \equiv S^{sh}_{12,12} + S^{sh}_{13,13} + 2S^{sh}_{12,13} = \Gamma_L \Gamma_R |t_{12}(d^2_{23,31} - d_{23,23} d^*_{31,31}) + t_{31}(d_{12,23} d_{23,31} + d^*_{12,31} d_{23,23})|^2 / A^2. \quad (32)$$

3. Bond Currents and Their Correlations: Driven Circular Current in the Case of $\Phi = 0$

Let us analyse the bond currents in detail; first in the absence of the magnetic flux, $\Phi = 0$, and for a linear response limit $V \to 0$. Using the derivations from the previous section, one can easily calculate the bond conductances and current correlation functions. The results are presented in Figure 2 for an equilateral triangle 3QD system (with all inter-dot hopping parameters $t_{12} = t_{23} = t_{31} = -1$, which is taken as unity in our further calculations) and for various values of the energy level ε_3 at the 3rd QD. The central column corresponds to the case $\varepsilon_1 = \varepsilon_2 = \varepsilon_3 = 0$, when the eigenenergies are given by $E_k = 2t\cos k$, for the wave-vector $k = 0$ and the degenerated state for $k = \pm 2\pi/3$. It can be seen in the transmission (black curve), which is equal to $\mathcal{T} = 1$ at $E = -2$ and $\mathcal{T} = 0$ at $E = 1$, where the Fano resonance takes place, with destructive interference of two electron waves. At low $E < 0$, the incoming wave from the left electrode is split into two branches and the bond conductances are positive, $0 \leq \mathcal{G}_{12}, \mathcal{G}_{13} \leq 1$ (see the blue and green curves in the top panel of Figure 2). The cross-correlation function $S^{sh}_{12,13}$ (for the currents in both branches) is positive (see the red curve in the bottom panel in Figure 2). Note that, at the lowest resonant level, all correlation functions $S^{sh}_{12,12} = S^{sh}_{13,13} = S^{sh}_{12,13} = 0$, which means that the currents in both branches are uncorrelated.

For $E > 0$, the conductances $\mathcal{G}_{12}$ and $\mathcal{G}_{13}$ can be negative and exceed unity (with their maximal absolute values inversely proportional to the coupling Γ_a). This manifests a circular current driven by injected electronic waves to the 3QD system, which can not reach the drain electrode; therefore, they are reflected backwards to the other branch of the ring. The circular current can be characterized by the conductance (see also [16])

$$\mathcal{G}^{dr} \equiv \begin{cases} \mathcal{G}_{12} & \text{for } \mathcal{G}_{12} < 0, \\ -\mathcal{G}_{13} & \text{for } \mathcal{G}_{13} < 0, \end{cases} \tag{33}$$

where the superscript "*dr*" marks the contribution to the circular current driven by the bias voltage, in order to distinguish it from the persistent current induced by the flux (which will be analysed later). There is some ambiguity in definition of the circular current. Our definition (33) is similar to the one given by the condition $\text{sign}[\mathcal{G}_{12}] = -\text{sign}[\mathcal{G}_{13}]$ for the vortex flow, used by Jayannavar and Deo [36] and Stefanucci et al. [16] (see [37]—which refers to [7]).

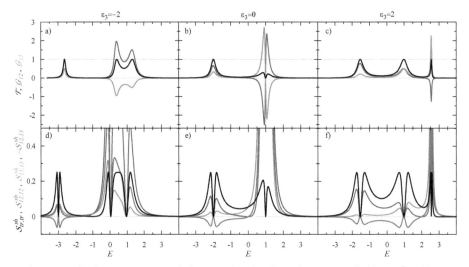

Figure 2. (Top) Transmission and dimensionless bond conductances: $\mathcal{T}$—black, $\mathcal{G}_{12}$—blue, and $\mathcal{G}_{13}$—green. (Bottom) Dimensionless spectral function of the shot noise: $S^{sh}_{tr,tr}$—black, $S^{sh}_{12,12}$—blue, $S^{sh}_{13,13}$—green, and $-S^{sh}_{12,13}$—red; calculated as a function of the electron energy E for the equilateral triangle system of 3QDs (with the inter-dot hopping $t_{12} = t_{23} = t_{31} = -1$, which is taken as unity in this paper) in the linear response limit $V \to 0$. The dot levels are $\varepsilon_1 = \varepsilon_2 = 0$ and $\varepsilon_3 = -2, 0, 2$, for left, center, and right columns, respectively. The coupling with the electrodes is taken to be $\Gamma_L = \Gamma_R = 0.25$. Note that the cross-correlation function $S^{sh}_{12,13}$ (red) is plotted negatively to show the zero crossing more clearly.

For the considered case in Figure 2b, with $\varepsilon_3 = 0$, the circular current is driven counter-clockwise for $0 < E < 1$ and changes its direction to clockwise at the degeneracy point, $E = 1$ (i.e., when $\mathcal{G}_{13}$ becomes negative). All correlation functions are large in the presence of the circular current; their maximum is inversely proportional to Γ_a^2. The cross-correlation $S^{sh}_{12,13}$ is large but negative and, therefore, this component reduces the transport shot noise, $S^{sh}_{tr,tr} = S^{sh}_{12,12} + S^{sh}_{13,13} + 2S^{sh}_{12,13}$ to the Lesovik formula $\mathcal{T}(1 - \mathcal{T})$, which reaches zero at the degeneracy point $E = 1$ (see the black curve in Figure 2e).

This situation is similar to multi-channel current correlations in transport through a quantum dot connected to magnetic electrodes [38], where cross-correlations for currents of different spins usually reduce the total shot noise to a sub-Poissonian noise with Fano factor $F < 1$ (however, in the presence of Coulomb interactions, the cross-correlations can be positive and lead to a super-Poissonian shot noise with $F > 1$).

The plots on the left and right hand sides of Figure 2 give more insight into the circular current effect. They are calculated for the dot level $\varepsilon_3 = \mp 2$ shifted by a gate potential, which breaks the symmetry of the system and removes the degeneracy of the states. Three resonant levels can be observed with $\mathcal{T} = 1$, where two of them are shifted to the left/right for $\varepsilon_3 = \mp 2$; however, the state at $E = 1$ is unaffected. There is still mirror symmetry, for which one gets three eigenstates, where two of them are linear compositions of all local states, but the one at $E = 1$ has the eigenvector $1/\sqrt{2}(c_1^\dagger - c_2^\dagger)|0\rangle$, which is separated for the 3rd QD. Therefore, the bond currents are composed of the currents through all three eigenstates, and their contribution depends on E. From these plots, one can see that the circular current is driven, for $E > \varepsilon_3$, when the cross-correlation $S_{12,13}^{sh}$ becomes negative. The direction of the current depends on the position of the eigenlevels and their current contributions. For $\varepsilon_3 = -2$, the current circulates clockwise, whereas its direction is counter-clockwise for $\varepsilon_3 = 2$.

Here, we assumed a flat band approximation (FBA) for the electronic structure in the electrodes (i.e., the Green functions $g_\alpha^{r,a} = \mp i\pi\rho$, where ρ denotes the density of states). Appendix A presents analytical results for the currents and shot noise in the fully-symmetric 3QD system coupled to a semi-infinite chain of atoms. The results are qualitatively similar. However, the FBA is more convenient for the analysis than the system coupled to atomic chains; in particular, for the cases with $\varepsilon_3 = \mp 2$, when localized states appear at -2.99 and 2.56 (i.e., below/above the energy band of the atomic chain).

The above analysis was performed under the assumption of a smooth energy dependence of the conductance in the small voltage limit $V \to 0$ and at $T = 0$. However, the conductances exhibit sharp resonant characteristics in the energy scale $\Delta E \propto \Gamma_\alpha$ and, therefore, one can expect that these features will be smoothed out with an increase of voltage bias and temperature. Figure 3 presents the Fano factor $F = S_{tr,tr}/2eI^{tr}$, which is the ratio of the current correlation function to the net transport current, which was calculated numerically from Equations (17) and (26). At $E = -2$, one can observe the evolution from the coherent regime, from $F = 0$ to $F = 1/2$ in the sequential regime, for $eV \gg \Gamma_\alpha$ or $k_B T \gg \Gamma_\alpha$. Quantum interference plays a crucial role at $E = 1$, leading to the Fano resonance for which the transmission $\mathcal{T} = 0$ and $F = 1$ in the low voltage/temperature regime. An increase of the voltage/temperature results only in a small reduction of the Fano factor.

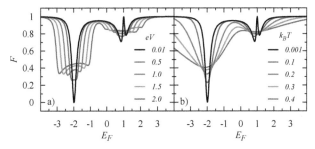

Figure 3. Fano factor as a function of the Fermi energy E_F for the equilateral triangle 3QDs system ($t_{12} = t_{23} = t_{31} = -1$ and $\varepsilon_1 = \varepsilon_2 = \varepsilon_3 = 0$) (**a**) for various bias voltages $eV = 0.01, 0.5, 1.0, 1.5$, and 2.0, at $T = 0$; and (**b**) for various temperatures $k_B T = 0.001, 0.1, 0.2, 0.3$, and 0.4, for $V \to 0$. The coupling to the electrodes is taken as $\Gamma_L = \Gamma_R = 0.25$, and the chemical potentials in the electrodes are $\mu_L = E_F - eV/2$ and $\mu_R = E_F + eV/2$.

4. Persistent Current and Its Noise: The Case $V = 0$

The persistent current and its noise has been studied in many papers (e.g., by Büttiker et al. [39–42], Semenov and Zaikin [43–46], Moskalates [47], and, more recently, by Komnik and Langhanke [48]) using full counting statistics (FCS), as well as in 1D Hubbard rings by exact diagonalization by Saha and Maiti [49] (see, also, the book by Imry [50]).

Here, we briefly present the results for the persistent current and shot noise in the triangle of 3QDs. Notice that, in the considered case, the phase coherence length of electrons is assumed to be larger than the ring circumference, $L_\phi \gg L$ [51]. The circular current is given by Equation (18), which shows that all electrons, up to the chemical potential in the electrodes, are driven by the magnetic flux Φ. Figure 4 exhibits the plots of I^ϕ, derived from Equation (18), for different couplings with the electrodes. In the weak coupling limit, where $\Gamma \to 0$ and the perfect ring is embedded in the reservoir, the persistent current can be simply expressed as

$$I^\phi = e \sum_k v_k f_k = -\frac{e}{\hbar} \sum_k 2t \sin(k + \phi/3) f_k , \tag{34}$$

where $f_k = 1/(\exp[(E_F - E_k)/k_B T] + 1)$ is the Fermi distribution for the electron with wave-vector k, energy $E_k = 2t \cos(k + \phi/3)$, and velocity $v_k = (1/\hbar)\partial E_k/\partial k = (-2t/\hbar) \sin(k + \phi/3)$, and where $\phi = 2\pi\Phi/(hc/e)$ is the phase shift due to the magnetic flux Φ. The sum runs over $k = 2\pi n/(Na)$ for $n = 0, \pm 1$, where $N = 3$ and $a = 1$ is the distance between the sites in the triangle. The current correlator is derived from Equation (24)

$$S_{\phi,\phi} = \frac{e^2}{\hbar} \sum_k 4t^2 \sin^2(k + \phi/3) f_k (1 - f_k) . \tag{35}$$

This result says that fluctuations of the persistent current could occur when the number of electrons in the ring fluctuates (i.e., an electron state moves through the Fermi level and I^ϕ jumps). We show, below, that the coupling with the electrodes (as a dissipative environment) results in current fluctuations [40,41], as well.

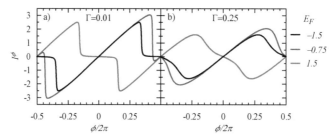

Figure 4. Persistent current I^ϕ versus the flux ϕ threading the equilateral triangle system of 3QDs ($t_{12} = t_{23} = t_{31} = -1$ and $\varepsilon_1 = \varepsilon_2 = \varepsilon_3 = 0$). The coupling is taken as $\Gamma_L = \Gamma_R = \Gamma = 0.01$ and 0.25; the Fermi energies are $E_F = -1.5$ (**black**), -0.75 (**blue**), 1.5 (**red**); and $T = 0$.

At the limit, $V \to 0$, the integrand function of the noise $S_{ij,in}$, Equation (28), is proportional to $f(E)(1 - f(E))$, which becomes the Dirac delta for $T \to 0$ and, therefore, one can analyze the spectral function $S_{\phi,\phi} = S_{12,12} + S_{13,13} - 2S_{12,13}$, where the components are $S_{ij,in} = S_{ij,in}^{sh} + \mathcal{G}_{ij}^L \mathcal{G}_{in}^L + \mathcal{G}_{ij}^R \mathcal{G}_{in}^R$ (see Equations (7)–(10) and (29)–(31)). Figure 5 presents the correlation function $S_{\phi,\phi}$ and its various components for the Fermi energy $E_F = -1.5$ and the strong coupling $\Gamma_L = \Gamma_R = 1$ when fluctuations are large. Notice that the fluctuations of the bond currents $S_{12,12}$ and $S_{13,13}$ (the blue and green curves, respectively) are different, although the average currents are equal. The cross-correlation function $S_{12,13}$ is positive at $\phi = 0$, but it becomes negative for larger ϕ, due to the quantum interference between electron waves passing through different states (as described in the previous section).

Figure 5 also shows $(\mathcal{G}_{12}^L)^2$ (blue-dashed curve) and $(\mathcal{G}_{12}^L)^2$ (blue-dotted curve), which correspond to the local fluctuations of the injected/ejected currents to/from the upper branch on the left and right junctions, respectively (see Equation (28)). The magnetic flux breaks the symmetry, inducing the persistent current and, therefore, the local conductances $\mathcal{G}_{12}^L$ and $\mathcal{G}_{12}^R$ are asymmetric.

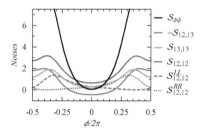

Figure 5. Flux dependence of spectral function of the persistent current correlator $S_{\phi,\phi}$ (**black**) and its components: $S_{12,12}$ (**blue**), $S_{13,13}$ (**green**), $-S_{12,13}$ (**red**), and $S_{12,12}^{LL} = (\mathcal{G}_{12}^L)^2$ (**blue-dashed**), $S_{12,12}^{RR} = (\mathcal{G}_{12}^R)^2$, (**blue-dotted**), respectively. We assume strong coupling: $\Gamma_L = \Gamma_R = 1.0$, $E_F = -1.5$, and $T = 0$.

5. Correlation of Persistent and Transport Currents, $\Phi \neq 0$ and $V \neq 0$

In this section, we analyze the currents and their correlations in the general case, derived from Equations (6), (14), (18), and (28), in the presence of voltage bias and magnetic flux. The results for the conductances and the spectral functions of the shot noise are presented in Figure 6. The magnetic flux splits the degenerated levels at $E = 1$ and destroys the Fano resonance. Figure 6a shows that there is no destructive interference for a small flux $\phi = 2\pi/16$, and the transmission is $\mathcal{T} = 1$ for all resonances. One can observe the driven circular current for $E > 0$, with negative $\mathcal{G}_{12}$ and $\mathcal{G}_{13}$, but their amplitudes are much lower than in the absence of the flux (compare with Figure 2b for $\phi = 0$). For a larger flux, $\phi = 2\pi/4$, there is no driven component of the circular current (see Figure 6b, where $\mathcal{G}_{12}$, $\mathcal{G}_{13} \geq 0$). It can also be seen that, for the state at $E = 0$, the electronic waves pass only through the lower branch of the ring, and the upper branch is blocked (with $\mathcal{G}_{13} = 1$ and $\mathcal{G}_{12} = 0$, respectively).

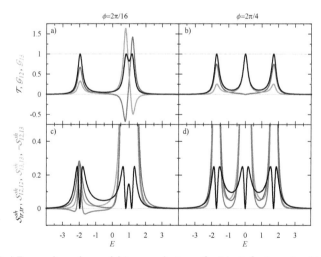

Figure 6. (**Top**) Energy dependence of driven conductance $\mathcal{G}_{12}$ (**blue**), $\mathcal{G}_{13}$ (**green**) and transmission $\mathcal{T}$ (**black**). (**Bottom**) Shot noise $S_{tr,tr}^{sh}$ (**black**) with the components: $S_{12,12}^{sh}$ (**blue**), $S_{13,13}^{sh}$ (**green**), and $-S_{12,13}^{sh}$ (**red**) for the considered triangular 3QD system threaded by the flux $\phi = 2\pi/16$ (**left**) or $\phi = 2\pi/4$ (**right**); the coupling is $\Gamma_L = \Gamma_R = 0.25$, and $T = 0$. Note that we plot $-S_{12,13}^{sh}$.

The lower panel of Figure 6 presents the spectral functions of the shot noise. According the Lesovik formula, $S_{tr,tr}^{sh} = 0$ at the resonant states (as $\mathcal{T} = 1$). This seems to be similar to the case $\phi = 0$ presented in the lower panel in Figure 2. However, there is a great difference in the components of the shot noise $S_{ij,in}^{sh}$, indicating the different nature of transport through these states and the role of quantum interference. Let us focus on the lowest resonant state, at $E = -2$, in Figure 6c, and compare with that in Figure 2e, in the absence of the flux. In the former case, the currents in both branches were uncorrelated, and $S_{12,12}^{sh} = S_{13,13}^{sh} = S_{12,13}^{sh} = 0$. In the presence of the flux, quantum interference becomes relevant, which is seen in the shot noise (Figure 6c). Now, the currents in both branches are correlated; $S_{12,13}^{sh}$ is negative close to resonance and fully compensates for the positive contributions $S_{12,12}^{sh}$ and $S_{13,13}^{sh}$ at resonance. For $\phi = 2\pi/4$ (see Figure 6d), all shot noise components are large, which indicates a strong quantum interference effect.

Figure 7 shows the Fano factor in the presence of the flux $\phi = 2\pi/16$ and for various bias voltages. Compared with the results in Figure 3 for $\phi = 0$, one can see how a small flux can destroy quantum interference and change electron transport. It is particularly seen close to $E = -1$, where the states with opposite chirality are located. In the case $\phi = 0$, one can observe the Fano resonance with a perfect destructive interference, $\mathcal{T} = 0$ and $F = 1$. With an increase of the flux ϕ, the Fano dip disappears, the two states are split, and transmission reaches its maximum value $\mathcal{T} = 1$; the Fano factor $F = 0$ when the splitting $\Delta E > \Gamma_\alpha$. A similar effect was seen in the case of Figure 2, where a change of the position of the local level ε_3 removed the state degeneracy and destroyed the Fano resonance.

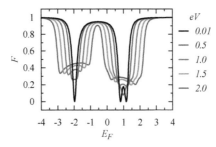

Figure 7. Fano factor as a function of E_F for the considered 3QD system threaded by the flux $\phi = 2\pi/16$ and for various bias voltages $eV = 0.01, 0.5, 1.0, 1.5,$ and 2.0. The coupling is $\Gamma_L = \Gamma_R = 0.25$, the chemical potentials are $\mu_L = E_F - eV/2$ and $\mu_R = E_F + eV/2$, and $T = 0$.

For the strong coupling $\Gamma_L = \Gamma_R = 1$, the intensity of the transport current is comparable to the persistent current and, therefore, one can expect a significant driven circular current. Figure 8 presents the flux dependence of the total circular current I^c and its driven component I^{dr}, as well as the transport current I^{tr}, for various voltages. For the considered case $E_F = 0.9$, the driven current circulates counter-clockwise and deforms the flux dependence of the circular currents, which become asymmetric.

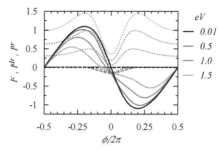

Figure 8. Circular current $I^c = I^{dr} + I^\phi$ (solid curves), its driven component I^{dr} (dashed curves), and the net transport current I^{tr} (dotted curves) versus ϕ for various bias voltages: $eV = 0.01, 0.5,$ 1.0, and 1.5. We assume a strong coupling $\Gamma_L = \Gamma_R = 1$, the chemical potentials are $\mu_L = E_F - eV/2$, $\mu_R = E_F + eV/2$, $E_F = 0.9$, and $T = 0$.

6. Summary

We considered the influence of quantum interference on electron transport and current correlations in a ring of three quantum dots threaded by a magnetic flux. We assumed non-interacting electrons and calculated the bond conductances, the local currents, and the current correlation functions—in particular, the shot noise—by means of the non-equilibrium Keldysh Green function technique, taking into account multiple reflections of the electron wave inside the ring. As we considered elastic scatterings, for which Kirchhoff's current law is fulfilled, the transmission $\mathcal{T} = \mathcal{G}_{12} + \mathcal{G}_{13}$ is a sum of the local bond conductances and the shot noise for the transport current is a composition of the local current correlation functions, $\mathcal{S}_{tr,tr}^{sh} = \mathcal{S}_{12,12}^{sh} + \mathcal{S}_{13,13}^{sh} + 2\mathcal{S}_{12,13}^{sh} = \mathcal{T}(1 - \mathcal{T})$, which gives the Lesovik formula.

In the system, having triangular symmetry, the eigenstates $E_0 = -2$ and $E_\pm = 1$ (with the wavevector $k = 0$ and $k = \pm 2\pi/3$) play a different role in the transport, which is seen in the bond conductances and the shot noise components. An electron wave injected with energy close to E_0 is perfectly split into both branches of the ring and the current cross-correlation function $\mathcal{S}_{12,13}^{sh}$ is positive. At the resonance E_0, the transmission $\mathcal{T} = 1$ and all correlation functions $\mathcal{S}_{12,12}^{sh} = \mathcal{S}_{13,13}^{sh} = \mathcal{S}_{12,13}^{sh} = 0$, which means that the bond currents are uncorrelated. The magnetic flux changes quantum interference conditions and correlates the bond currents; the cross-correlation $\mathcal{S}_{12,13}^{sh}$ becomes negative at the resonance and fully compensates the positive auto-correlation components $\mathcal{S}_{12,12}^{sh}$ and $\mathcal{S}_{13,13}^{sh}$ (with $\mathcal{S}_{tr,tr}^{sh} = 0$).

Quantum interference plays a crucial role in transport through the degenerate states at $E_\pm = 1$, where one can observe Fano resonance with destructive interference. In this region, the circular current I^{dr} can be driven by the bias voltage. The bond conductances have an opposite sign, their maximal value is inversely proportional to the coupling, Γ_α, with the electrodes, and they can be larger than unity. The direction of I^{dr} depends on the bias voltage and the position of the Fermi energy E_F, with respect to the degenerate state $E_\pm$. The auto-correlation functions $\mathcal{S}_{12,12}^{sh}$, $\mathcal{S}_{13,13}^{sh}$ are large (inversely proportional to Γ_a^2) close to the resonance. The cross-correlator $\mathcal{S}_{12,13}^{sh}$ is negative in the presence of the driven circular current. Our calculations show that a small magnetic flux, $\phi = 2\pi/16$, can destroy the Fano resonance, and two resonance peaks (with $\mathcal{T} = 1$) appear. The driven component, I^{dr}, is reduced with an increase of ϕ, and it disappears at $\phi = 2\pi/4$. However, quantum interference still plays a role; the bond currents are strongly correlated (with large $\mathcal{S}_{12,12}^{sh}$ and $\mathcal{S}_{13,13}^{sh}$ and negative $\mathcal{S}_{12,13}^{sh}$). For a large coupling, the driven part I^{dr} can be large and can profoundly modify the total circular current $I^c = I^{dr} + I^\phi$.

We also performed calculations of the bond currents and their correlations for rings with a various number of sites; in particular, for the benzene ring in para-, metha-, and ortho-connection with the electrodes. The results are qualitatively similar to those presented above for the 3QD ring: Quantum interference of the travelling waves with the eigenstates of opposite chirality leads to the driven circular currents, accompanied by large current fluctuations with a negative cross-correlation component. To observe this effect, the two conducting branches should be asymmetric; in particular, in the benzene ring, the driven circular current appears for the metha- and ortho-connections, but is absent in the para-connection, where both conducting branches are equivalent (see also [7]).

An open problem is including interactions between electrons into the calculations of the coherent transport and shot noise. Coulomb interactions can be taken into account in the sequential regime [52], or by using the real-time diagrammatic technique [53–55]; however, in practice, one includes only first- and second-order diagrams with respect to the tunnel coupling and the role of QI is diminished. In principle, one can treat QI on an equal footing with electron interactions in the framework of quantum field theory [56], as was done for the Anderson single impurity model, by means of full counting statistics (FCS), where the average current and all its moments were calculated [57]. However, this is a formidable task, even for the simple 3QD model.

Author Contributions: Both the authors have a similar contribution to the paper in its concept, research, and manuscript preparation.

Funding: The research was financed by National Science Centre, Poland—project number 2016/21/B/ST3/02160.

Conflicts of Interest: The authors declare no conflict of interest.

Appendix A. Coupling to Atomic Chain Electrodes: Analytical Results

The results for the conductances and shot noise may be simplified when we take all hopping integrals equal to t, the same position of the site levels $\varepsilon = 0$, and the symmetric coupling $t_L = t_R = t$, with the electrodes as a semi-infinite atomic chain. In this case, the Green functions in the electrodes are $g^r = e^{ik}/t$ and $g^a = e^{-ik}/t$ and the electron spectrum is $E_k = 2t \cos k$. From Equations (7)–(10), one can calculate the dimensionless bond conductances as

$$\mathcal{G}_{12}^L = 2 \sin k \left[\sin k + \sin 3k - \sin(2k + \phi)\right] / A, \tag{A1}$$

$$\mathcal{G}_{12}^R = 2 \sin k \left[\sin k + \sin 3k - \sin(2k - \phi)\right] / A, \tag{A2}$$

$$\mathcal{G}_{13}^L = 2 \sin k \left[\sin k - \sin(2k - \phi)\right] / A, \text{ and} \tag{A3}$$

$$\mathcal{G}_{13}^R = 2 \sin k \left[\sin k - \sin(2k + \phi)\right] / A, \tag{A4}$$

where the denominator

$$A = 4 + \cos 2\phi - 2 \cos \phi (3 \cos k - \cos 3k) - \cos 4k. \tag{A5}$$

It is seen an asymmetry with respect to the direction of the magnetic flux (to ϕ) for the conductances $\mathcal{G}_{ij}^L$ and $\mathcal{G}_{ij}^R$ from the left and the right electrode. The transmission, $\mathcal{T}$, is expressed as

$$\mathcal{T} \equiv \mathcal{G}_{12}^L + \mathcal{G}_{13}^L = \mathcal{G}_{12}^R + \mathcal{G}_{13}^R = \mathcal{G}_{12} + \mathcal{G}_{13} = 2 \sin^2 k [1 - 4 \cos k (\cos \phi - \cos k)] / A, \tag{A6}$$

where the driven part of the bond conductances are calculated using Equations (15) and (16)

$$\mathcal{G}_{12} = 4 \sin^2 k \cos k (2 \cos k - \cos \phi) / A, \tag{A7}$$

$$\mathcal{G}_{13} = 2 \sin^2 k (1 - 2 \cos \phi \cos k) / A, \text{ and} \tag{A8}$$

$$\mathcal{G}_\phi^L = \mathcal{G}_\phi^R = 2 \sin \phi \sin k \cos 2k / A, \tag{A9}$$

and, from Equations (19) and (20), the part induced by the flux is

$$\mathcal{G}_\phi^L = \mathcal{G}_\phi^R = 2 \sin \phi \sin k \cos 2k / A. \tag{A10}$$

It can be seen that the conductance $\mathcal{G}_{12}$ becomes negative at $k = \pi/2$ (i.e., when the circular current becomes driven).

Entropy **2019**, *21*, 527

The shot noise for the bond currents is expressed as

$$\mathcal{S}_{12,12}^{sh} = 4\sin^2 k \left| e^{\iota\phi}(\cos\phi - 2\cos k) + \cos 2k \right|^2 / A^2, \tag{A11}$$

$$\mathcal{S}_{13,13}^{sh} = 4\sin^2 k \left| \cos k - e^{\iota\phi} \right|^2 / A^2, \tag{A12}$$

$$\mathcal{S}_{12,13}^{sh} = -4\sin^2 k \left[2\cos\phi\cos k(\cos\phi - \cos k)^2 + \sin^2\phi \right] / A^2, \text{ and} \tag{A13}$$

$$\mathcal{S}_{tr,tr}^{sh} = \mathcal{T}(1 - \mathcal{T}) = 4\sin^2 k(\cos\phi - \cos k)^2 [1 - 4\cos k(\cos\phi - \cos k)] / A^2. \tag{A14}$$

Notice that the cross-correlation $\mathcal{S}_{12,13}^{sh}$ can be positive or negative in the laminar or the vortex regime, respectively.

References and Note

1. Webb, R.A.; Washburn, S.; Umbach, C.P.; Laibowitz, R.B. Observation of h/e Aharonov-Bohm Oscillations in Normal-Metal Rings. *Phys. Rev. Lett.* **1985**, *54*, 2696–2699, doi:10.1103/PhysRevLett.54.2696.

2. Ji, Y.; Chung, Y.; Sprinzak, D.; Heiblum, M.; Mahalu, D.; Shtrikman, H. An electronic Mach--Zehnder interferometer. *Nature* **2003**, *422*, 415–418, doi:10.1038/nature01503.

3. Cardamone, D.M.; Stafford, C.A.; Mazumdar, S. Controlling Quantum Transport through a Single Molecule. *Nano Lett.* **2006**, *6*, 2422–2426, doi:10.1021/nl0608442.

4. Solomon, G.C.; Andrews, D.Q.; Hansen, T.; Goldsmith, R.H.; Wasielewski, M.R.; Van Duyne, R.P.; Ratner, M.A. Understanding quantum interference in coherent molecular conduction. *J. Chem. Phys.* **2008**, *129*, 054701, doi:10.1063/1.2958275.

5. Ke, S.H.; Yang, W.; Baranger, H.U. Quantum-Interference-Controlled Molecular Electronics. *Nano Lett.* **2008**, *8*, 3257–3261, doi:10.1021/nl8016175.

6. Donarini, A.; Begemann, G.; Grifoni, M. All-Electric Spin Control in Interference Single Electron Transistors. *Nano Lett.* **2009**, *9*, 2897–2902, doi:10.1021/nl901199p.

7. Rai, D.; Hod, O.; Nitzan, A. Circular currents in molecular wires. *J. Phys. Chem. C* **2010**, *114*, 20583–20594, doi:10.1021/jp105030d.

8. Hong, W.; Valkenier, H.; Mészáros, G.; Manrique, D.Z.; Mishchenko, A.; Putz, A.; García, P.M.; Lambert, C.J.; Hummelen, J.C.; Wandlowski, T. An MCBJ case study: The influence of π-conjugation on the single-molecule conductance at a solid/liquid interface. *Beilstein J. Nanotechnol.* **2011**, *2*, 699–713, doi:10.3762/bjnano.2.76.

9. Guédon, C.M.; Valkenier, H.; Markussen, T.; Thygesen, K.S.; Hummelen, J.C.; van der Molen, S.J. Observation of quantum interference in molecular charge transport. *Nat. Nanotechnol.* **2012**, *7*, 305–309, doi:10.1038/nnano.2012.37.

10. Lambert, C.J. Basic concepts of quantum interference and electron transport in single-molecule electronics. *Chem. Soc. Rev.* **2015**, *44*, 875–888, doi:10.1039/C4CS00203B.

11. Liu, J.; Huang, X.; Wang, F.; Hong, W. Quantum Interference Effects in Charge Transport through Single-Molecule Junctions: Detection, Manipulation, and Application. *Accounts Chem. Res.* **2019**, *52*, 151–160, doi:10.1021/acs.accounts.8b00429.

12. Nakanishi, S.; Tsukada, M. Large Loop Current Induced Inside the Molecular Bridge. *Jpn. J. Appl. Phys.* **1998**, *37*, L1400–L1402, doi:10.1143/JJAP.37.L1400.

13. Nakanishi, S.; Tsukada, M. Quantum Loop Current in a C60 Molecular Bridge. *Phys. Rev. Lett.* **2001**, *87*, 126801, doi:10.1103/PhysRevLett.87.126801.

14. Daizadeh, I.; Guo, J.X.; Stuchebrukhov, A. Vortex structure of the tunneling flow in long-range electron transfer reactions. *J. Chem. Phys.* **1999**, *110*, 8865–8868, doi:10.1063/1.478807.

15. Xue, Y.; Ratner, M.A. Local field effects in current transport through molecular electronic devices: Current density profiles and local nonequilibrium electron distributions. *Phys. Rev. B* **2004**, *70*, 081404, doi:10.1103/PhysRevB.70.081404.

16. Stefanucci, G.; Perfetto, E.; Bellucci, S.; Cini, M. Generalized waveguide approach to tight-binding wires: Understanding large vortex currents in quantum rings. *Phys. Rev. B* **2009**, *79*, 073406, doi:10.1103/PhysRevB.79.073406.

17. Rai, D.; Hod, O.; Nitzan, A. Magnetic field control of the current through molecular ring junctions. *J. Phys. Chem. Lett.* **2011**, *2*, 2118–2124, doi:10.1021/jz200862r.

18. Rai, D.; Hod, O.; Nitzan, A. Magnetic fields effects on the electronic conduction properties of molecular ring structures. *Phys. Rev. B* **2012**, *85*, 155440, doi:10.1103/PhysRevB.85.155440.

19. Yadalam, H.K.; Harbola, U. Controlling local currents in molecular junctions. *Phys. Rev. B* **2016**, *94*, 115424, doi:10.1103/PhysRevB.94.115424.

20. Nozaki, D.; Schmidt, W.G. Current density analysis of electron transport through molecular wires in open quantum systems. *J. Comput. Chem.* **2017**, *38*, 1685–1692, doi:10.1002/jcc.24812.

21. Patra, M.; Maiti, S.K. Modulation of circular current and associated magnetic field in a molecular junction: A new approach. *Sci. Rep.* **2017**, *7*, 43343, doi:10.1038/srep43343.

22. Cabra, G.; Jensen, A.; Galperin, M. On simulation of local fluxes in molecular junctions. *J. Chem. Phys.* **2018**, *148*, 204103, doi:10.1063/1.5029252.

23. Moskalets, M.V. Temperature-induced current in a one-dimensional ballistic ring with contacts. *Europhys. Lett. (EPL)* **1998**, *41*, 189–194, doi:10.1209/epl/i1998-00129-8.

24. Chung, V.S.W.; Samuelsson, P.; Buttiker, M. Visibility of current and shot noise in electrical Mach-Zehnder and Hanbury Brown Twiss interferometers. *Phys. Rev. B* **2005**, *72*, 125320, doi:10.1103/PhysRevB.72.125320.

25. Büttiker, M.; Samuelsson, P. Interference of independently emitted electrons in quantum shot noise. *Ann. Phys.* **2007**, *16*, 751–766, doi:10.1002/andp.200710255.

26. Pilgram, S.; Samuelsson, P.; Förster, H.; Büttiker, M. Full-Counting Statistics for Voltage and Dephasing Probes. *Phys. Rev. Lett.* **2006**, *97*, 066801, doi:10.1103/PhysRevLett.97.066801.

27. Förster, H.; Pilgram, S.; Büttiker, M. Decoherence and full counting statistics in a Mach-Zehnder interferometer. *Phys. Rev. B* **2005**, *72*, 075301, doi:10.1103/PhysRevB.72.075301.

28. Förster, H.; Samuelsson, P.; Pilgram, S.; Büttiker, M. Voltage and dephasing probes in mesoscopic conductors: A study of full-counting statistics. *Phys. Rev. B* **2007**, *75*, 035340, doi:10.1103/PhysRevB.75.035340.

29. Bauerle, C.; Glattli, D.C.; Meunier, T.; Portier, F.; Roche, P.; Roulleau, P.; Takada, S.; Waintal, X. Coherent control of single electrons: A review of current progress. *Rep. Prog. Phys.* **2018**, *81*, 056503, doi:10.1088/1361-6633/aaa98a.

30. Lesovik, G.B. Excess quantum noise in two-dimensional ballistic microcontacts. *Pis'ma Zh. Eksp. Teor. Fiz.* **1989**, *49*, 513–515.

31. Haug, H.; Jauho, A.P. *Quantum Kinetics in Transport and Optics of Semiconductors*; Springer: Berlin/Heidelberg, Germany, 2008, doi:10.1007/978-3-540-73564-9.

32. Blanter, Y.; Büttiker, M. Shot noise in mesoscopic conductors. *Phys. Rep.* **2000**, *336*, 1–166, doi:10.1016/S0370-1573(99)00123-4.

33. Levitov, L.S.; Lesovik, G.B. Charge-transport statistics in quantum conductors. *Pis'ma Zh. Eksp. Teor. Fiz.* **1992**, *55*, 534–537.

34. Lesovik, G.B.; Sadovskyy, I.A. Scattering matrix approach to the description of quantum electron transport. *Physics-Uspekhi* **2011**, *54*, 1007–1059, doi:10.3367/UFNe.0181.201110b.1041.

35. Martin, T.; Landauer, R. Wave-packet approach to noise in multichannel mesoscopic systems. *Phys. Rev. B* **1992**, *45*, 1742–1755, doi:10.1103/PhysRevB.45.1742.

36. Jayannavar, A.M.; Singha Deo, P. Persistent currents in the presence of a transport current. *Phys. Rev. B* **1995**, *51*, 10175–10178, doi:10.1103/PhysRevB.51.10175.

37. One can define the driven current as [7]: $I^{dr} = (I_{12}l_{12} - I_{13}l_{13})/(l_{12} + l_{13})$, where l_{12}, l_{13} denotes the length of the upper and the lower branch. For $I_{13} \rightarrow 0$ one gets a finite circular current, what is an artefact.

38. Bułka, B.R. Current and power spectrum in a magnetic tunnel device with an atomic-size spacer. *Phys. Rev. B* **2000**, *62*, 1186–1192, doi:10.1103/PhysRevB.62.1186.

39. Büttiker, M.; Stafford, C.A. Charge Transfer Induced Persistent Current and Capacitance Oscillations. *Phys. Rev. Lett.* **1996**, *76*, 495–498, doi:10.1103/PhysRevLett.76.495.

40. Cedraschi, P.; Büttiker, M. Suppression of level hybridization due to Coulomb interactions. *J. Phys. Condens. Matter* **1998**, *10*, 3985–4000, doi:10.1088/0953-8984/10/18/009.

41. Cedraschi, P.; Ponomarenko, V.V.; Büttiker, M. Zero-Point Fluctuations and the Quenching of the Persistent Current in Normal Metal Rings. *Phys. Rev. Lett.* **2000**, *84*, 346–349, doi:10.1103/PhysRevLett.84.346.

42. Cedraschi, P.; Buttiker, M. Zero-point fluctuations in the ground state of a mesoscopic normal ring. *Ann. Phys.* **2000**, *289*, 165312, doi:10.1103/PhysRevB.63.165312.

43. Semenov, A.G.; Zaikin, A.D. Fluctuations of persistent current. *J. Phys. Condens. Matter* **2010**, *22*, 485302, doi:10.1088/0953-8984/22/48/485302.

44. Semenov, A.G.; Zaikin, A.D. Persistent current noise and electron-electron interactions. *Phys. Rev. B* **2011**, *84*, 045416, doi:10.1103/PhysRevB.84.045416.

45. Semenov, A.G.; Zaikin, A.D. Persistent currents in quantum phase slip rings. *Phys. Rev. B* **2013**, *88*, 054505, doi:10.1103/PhysRevB.88.054505.

46. Semenov, A.G.; Zaikin, A.D. Quantum phase slip noise. *Phys. Rev. B* **2016**, *94*, 014512, doi:10.1103/PhysRevB.94.014512.

47. Moskalets, M. Persistent currents in ballistic normal-metal rings. *Low Temp. Phys.* **2010**, *36*, 982–989, doi:10.1063/1.3521568.

48. Komnik, A.; Langhanke, G.W. Full counting statistics of persistent current. *Phys. Rev. B* **2014**, *90*, 165107, doi:10.1103/PhysRevB.90.165107.

49. Saha, M.; Maiti, S.K. Circulating current in 1D Hubbard rings with long-range hopping: Comparison between exact diagonalization method and mean-field approach. *Phys. E Low-Dimens. Syst. Nanostruct.* **2016**, *84*, 118–134, doi:10.1016/j.physe.2016.05.042.

50. Imry, Y. *Introduction to Mesoscopic Physics*; Oxford University Press: New York, NY, USA, 1997.

51. Cheung, H.F.; Gefen, Y.; Riedel, E.K.; Shih, W.H. Persistent currents in small one-dimensional metal rings. *Phys. Rev. B* **1988**, *37*, 6050–6062, doi:10.1103/PhysRevB.37.6050.

52. Nazarov, Y.V.; Blanter, Y.M. *Quantum Transport: Introduction to Nanoscience*; Cambridge University Press: Cambridge, UK, 2009, doi:10.1017/CBO9780511626906.

53. Schoeller, H.; Schön, G. Mesoscopic quantum transport: Resonant tunneling in the presence of a strong Coulomb interaction. *Phys. Rev. B* **1994**, *50*, 18436–18452, doi:10.1103/PhysRevB.50.18436.

54. König, J.; Schmid, J.; Schoeller, H.; Schön, G. Resonant tunneling through ultrasmall quantum dots: Zero-bias anomalies, magnetic-field dependence, and boson-assisted transport. *Phys. Rev. B* **1996**, *54*, 16820–16837, doi:10.1103/PhysRevB.54.16820.

55. Thielmann, A.; Hettler, M.H.; König, J.; Schön, G. Cotunneling Current and Shot Noise in Quantum Dots. *Phys. Rev. Lett.* **2005**, *95*, 146806, doi:10.1103/PhysRevLett.95.146806.

56. Kamenev, A. *Field Theory of Non-Equilibrium Systems*; Cambridge University Press: Cambridge, UK, 2011, doi:10.1017/CBO9781139003667.

57. Gogolin, A.O.; Komnik, A. Towards full counting statistics for the Anderson impurity model. *Phys. Rev. B* **2006**, *73*, 195301, doi:10.1103/PhysRevB.73.195301.

Article

Symmetry Properties of Mixed and Heat Photo-Assisted Noise in the Quantum Hall Regime

Flavio Ronetti [1,2], **Matteo Acciai** [1,2,3], **Dario Ferraro** [1,3,*], **Jérôme Rech** [2], **Thibaut Jonckheere** [2], **Thierry Martin** [2] **and Maura Sassetti** [1,3]

[1]   Dipartimento di Fisica, Università di Genova, Via Dodecaneso 33, 16146 Genova, Italy
[2]   Aix Marseille Univ, Université de Toulon, CNRS, CPT, Marseille, France
[3]   SPIN-CNR, Via Dodecaneso 33, 16146 Genova, Italy
*   Correspondence: ferraro@fisica.unige.it

Received: 27 June 2019; Accepted: 23 July 2019; Published: 25 July 2019

Abstract: We investigate the photo-assisted charge-heat mixed noise and the heat noise generated by periodic drives in Quantum Hall states belonging to the Laughlin sequence. Fluctuations of the charge and heat currents are due to weak backscattering induced in a quantum point contact geometry and are evaluated at the lowest order in the tunneling amplitude. Focusing on the cases of a cosine and Lorentzian periodic drive, we show that the different symmetries of the photo-assisted tunneling amplitudes strongly affect the overall profile of these quantities as a function of the AC and DC voltage contributions, which can be tuned independently in experiments.

Keywords: electron quantum optics; photo-assisted noise; charge and heat fluctuations

1. Introduction

The possibility to generate, manipulate and detect single- to few-electron excitations coherently propagating in mesoscopic quantum conductors represents one of the main tasks in condensed matter. Theoretical and experimental studies in this direction culminated with the development of the new field of electron quantum optics (EQO) [1–3]. In this context, a remarkable experimental effort has been devoted to the realization of on-demand sources of electronic wave-packets. With this purpose, two main techniques have been elaborated. The first relies on a driven quantum dot [4,5], which plays the role of a mesoscopic capacitor [6–9], tunnel coupled to a Quantum Hall edge channel. This kind of device allows the periodic injection of an electron and a hole along the edge channels of the Hall bar. An alternative approach involves the generation of purely electronic wave-packets by applying a properly designed train of Lorentzian voltage pulses in time to a quantum conductor [10–13]. The experimental realization of these excitations, dubbed Levitons [14], renewed the interest in the so-called photo-assisted quantum transport, namely the study of the current generated by periodic drives and its fluctuations [15,16].

Even if experimentally challenging, the implementation of EQO in strongly correlated systems, such as Fractional Quantum Hall (FQH) edge channels, could open new and interesting perspectives in the field. In particular, peculiar features like the unexpected robustness of Levitons [17,18] and the crystallization of multiple-electronic wave-packets in the time domain [19,20] have been recently theoretically predicted. The simplest possible set-up to access this new physics requires a periodic train of voltage pulses applied to one of the terminals of a Hall bar in a Quantum Point Contact (QPC) geometry. Here, excitations incoming from the lead are partitioned at the QPC, in an electronic equivalent of the Hanbury–Brown and Twiss optical interferometer [21,22].

Notwithstanding the control of charge transport at the individual excitation level still presenting interesting and fascinating open problems, electric charge is not the only relevant degree of freedom we can look at in the framework of EQO. Indeed, electronic wave-packets also carry energy in a coherent way [23]. This observation is of particular importance in view of the progressive miniaturization of electronic devices, which makes the problem of heat transfer at the nanoscale extremely timely [24], as demonstrated by recent progress in the field of quantum thermodynamics [25]. In this framework, new intriguing challenges are posed by the need of properly transposing concepts such as energy harvesting [26–33], transport [34–40] and exchange [41–43] at the mesoscopic scale.

In this direction, new studies in the EQO domain have also been triggered. For instance, heat current was identified as a useful resource for the full reconstruction of single-electron wave-packets [44]. Intriguingly, also mixed heat-charge correlations [45–47] and heat current noise [48,49] were investigated in the case of single-electron sources with Levitons emerging as robust excitations also for what it concerns heat transport [50]. Despite a direct observation of heat current fluctuations is still lacking, an experimental protocol has been recently proposed in order to extract this quantity from temperature fluctuations [51].

In this paper, we evaluate the photo-assisted mixed and heat noise for FQH states in the Laughlin sequence [52]. Considering the QPC in a weak backscattering regime, we can confine our calculation to the lowest order in the tunneling. The aim will be to investigate the behavior of these quantities by independently tuning the AC and DC components of the voltage drive in the same spirit of what was done for the photo-assisted shot noise generated by electrical current [16,53].

The paper is organized as follows. In Section 2, we illustrate the model used to describe FQH edge channels coupled to a time dependent voltage. The charge current is evaluated at the lowest perturbative order in Section 3, while the corresponding expression for the heat current is derived in Section 4. In Section 5 we consider the charge and heat current fluctuations in terms of mixed and heat noise. Section 6 is devoted to the analysis of the symmetries of the discussed quantities as a function of the dc and ac contribution to the voltages. Finally, in Section 7 we draw our conclusions.

2. Model

We consider a four-terminal FQH bar in the presence of a QPC (see Figure 1). For a quantum Hall state with filling factor ν in the Laughlin sequence $\nu = 1/(2n + 1)$ [52], with $n \in \mathbb{N}$, a single chiral bosonic mode emerges at each edge of the sample [54]. The effective Hamiltonian for right- and left-moving edge states (indicated in the following by R and L respectively) reads

$$H_0 = \sum_{r=R,L} \frac{v}{4\pi} \int_{-\infty}^{+\infty} dx \, [\partial_x \Phi_r(x)]^2,$$ (1)

with $\Phi_{R/L}$ bosonic operators propagating with velocity v, assumed equal for both chiralities, along the edges. Notice that, from now on, we will set $\hbar = 1$ for notational convenience.

The coupling between the electron particle density $\rho_R(x) = \frac{\sqrt{\nu}}{2\pi} \partial_x \Phi_R(x)$ and the coupling with a generic time dependent voltage gate $V(t)$ applied to terminal 1 is encoded by the Hamiltonian

$$H_g = -e \int_{-\infty}^{+\infty} dx \, \Theta(-x - d) V(t) \rho_R(x).$$ (2)

Here, the step function $\Theta(-x - d)$ models a homogeneous contact which is extended with respect to the Hall sample. In the following, we will focus on a periodic voltage drive of the form

$$V(t) = V_{dc} + V_{ac}(t),$$ (3)

with

$$\frac{1}{\mathcal{T}} \int_0^{\mathcal{T}} dt V_{ac}(t) = 0,$$ (4)

$\mathcal{T}$ being the period of the drive.

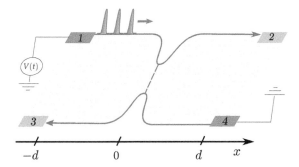

Figure 1. Four-terminal setup for an FQH state in the QPC geometry. Contact 1 is driven by a time dependent voltage $V(t)$ and used as input terminal, contact 4 is grounded, while contacts 2 and 3 are the output terminals where currents and noises are measured. FQH: Fractional Quantum Hall; QPC: Quantum Point Contact.

The time-evolution of the bosonic field Φ_R, according to the Hamiltonian $H = H_0 + H_g$, is given by [17,19]

$$\Phi_R(x,t) = \phi_R\left(t - \frac{x}{v}\right) - e\sqrt{v}\int_{-\infty}^{t-\frac{x}{v}} dt' V(t').$$ (5)

Here, ϕ_R denotes the field which evolves with respect to H_0 only. These characteristic chiral dynamics are a direct consequence of the linear dispersion of edge modes.

Finally, we allow the tunneling of excitations between the two edges by locally approaching them, creating a QPC at $x = 0$. This process can be effectively described by introducing the tunneling Hamiltonian [55]

$$H_T = \Lambda \Psi_R^\dagger(0)\Psi_L(0) + \text{h.c.}$$ (6)

Here, $\Psi_{R/L}$ ($\Psi_{R/L}^\dagger$) are the annihilation (creation) operators for quasi-particles with fractional charge $e^* = -ev$. The Hamiltonian in Equation (6) describes the dominating tunneling process in the weak-backscattering regime [56]. In this regime, the tunneling Hamiltonian H_T can be treated as a small perturbation with respect to H. As a consequence, the time evolution of quantum operators can be constructed in terms of a perturbative series in the tunneling amplitude Λ.

3. Charge Current

Charge current operators for right- and left-moving modes can be defined by means of the continuity equations of densities $\rho_{R/L}(x,t)$, namely

$$- e\partial_t \rho_{R/L}(x,t) + \partial_x J_{R/L}(x,t) = 0.$$ (7)

Due to the fact that the propagation of the modes long the channel is chiral [19], one finds

$$J_{R/L}(x,t) = \mp ev\rho_{R/L}(x - vt),$$ (8)

where $\rho_{R/L}$ are the chiral density operators evolving in time according to the total Hamiltonian $H' = H_0 + H_g + H_T$.

Starting from the definition of the chiral current operator, we can define the operators for charge current entering reservoirs 2 and 3 as

$$J_{2/3}(t) = J_{R/L}(\pm d, t),$$ (9)

where we recall that the interfaces between edge states and contacts 2 and 3 are placed at $x = \pm d$, respectively (see Figure 1). The expansion of $J_{2/3}$ in powers of Λ is given by

$$J_{2/3}(t) = J^{(0)}_{2/3}(t) + J^{(1)}_{2/3}(t) + J^{(2)}_{2/3}(t) + \mathcal{O}(\Lambda^3), \tag{10}$$

with

$$J^{(0)}_{2/3}(t) = \frac{ev\sqrt{v}}{2\pi} (\partial_x \Phi_{R/L}(x,t))_{x=\pm d}, \tag{11}$$

$$J^{(1)}_{2/3}(t) = \pm i\Lambda ev \Psi^\dagger_R\left(0, t - \frac{d}{v}\right)\Psi_L\left(0, t - \frac{d}{v}\right) + \text{h.c.}, \tag{12}$$

$$J^{(2)}_{2/3}(t) = \pm i \int_{-\infty}^{t - \frac{d}{v}} dt'' \left[H_T(t''), +i\Lambda ev \Psi^\dagger_R\left(0, t - \frac{d}{v}\right)\Psi_L\left(0, t - \frac{d}{v}\right) + \text{h.c.} \right]. \tag{13}$$

Tunneling contributions entering reservoirs 2 and 3 are connected by the simple relation

$$J^{(1/2)}_2(t) = -J^{(1/2)}_3(t). \tag{14}$$

The thermal average of current operators will be performed over the initial equilibrium condition, i.e., in the absence of driving voltages (H_g) and tunneling (H_T). It is worth noting that $\langle J^{(1)}_{2/3}(x,t)\rangle$ is zero because it involves a different number of annihilation or creation field operators. Therefore, the average values of charge current operators satisfy

$$\langle J_{2/3}(t)\rangle = \langle J^{(0)}_{2/3}(t)\rangle + \langle J^{(2)}_{2/3}(t)\rangle + \mathcal{O}(\Lambda^3), \tag{15}$$

with

$$\langle J^{(0)}_2(t)\rangle = \frac{e^2 v}{2\pi} V\left(t - \frac{2d}{v}\right), \tag{16}$$

$$\langle J^{(0)}_3(t)\rangle = 0, \tag{17}$$

$$\langle J^{(2)}_{2/3}(t)\rangle = \pm ievv \int_{-\infty}^{t - \frac{d}{v}} dt'' \langle \left[H_T(t''), +i\Lambda \Psi^\dagger_R(d - vt, 0)\Psi_L(d - vt, 0) + \text{h.c.} \right]\rangle. \tag{18}$$

The first term corresponds to the charge current emitted by the reservoir 1, which constitutes the main contribution to the detected currents in reservoir 2. In the absence of tunneling processes ($\Lambda = 0$), these zero-order contributions would correspond to a periodic current generated by $V(t)$. For this reason, the integral over one period $\mathcal{T}$ gives the total charge $\mathcal{C}$ transferred across the edge channel, namely

$$\mathcal{C} = \int_{-\frac{T}{2}}^{\frac{T}{2}} dt \langle J^{(0)}_2(t)\rangle = \frac{e^2 v}{2\pi} \int_{-\frac{T}{2}}^{\frac{T}{2}} dt V\left(t - \frac{2d}{v}\right) = \frac{e^2 v}{\omega} V_{dc} = -eq, \tag{19}$$

where we have introduced the drive frequency $\omega = 2\pi/\mathcal{T}$ of the voltage in Equation (3) and used the property in Equation (4).

Due to the QPC, some of the excitations emitted by contact 1 are backscattered and $\langle J^{(2)}_2(t)\rangle$ ($\langle J^{(2)}_3(t)\rangle$) represents the transmitted (reflected) current [55,57]. Due to the relation in Equation (14), backscattering currents are equal up to a sign, so that we can define

$$J_{BS}(t) = \langle J^{(2)}_3(t)\rangle = -\langle J^{(2)}_2(t)\rangle. \tag{20}$$

According to Equation (18), this backscattering current can be written as

$$J_{BS}(t) = -e\,|\Lambda|^2\,v \int_{-\infty}^{t-\frac{d}{v}} dt'' \left[G_R^< \left(t'', t - \frac{d}{v} \right) G_L^> \left(t'', t - \frac{d}{v} \right) - G_L^< \left(t'', t - \frac{d}{v} \right) G_R^> \left(t'', t - \frac{d}{v} \right) + \right.$$
$$\left. - G_R^< \left(t - \frac{d}{v}, t'' \right) G_L^> \left(t - \frac{d}{v}, t'' \right) + G_L^< \left(t - \frac{d}{v}, t'' \right) G_R^> \left(t - \frac{d}{v}, t'' \right) \right],$$
(21)

where we introduced the quasi-particle Green's functions

$$G_R^< (t', t) = \langle \Psi_R^\dagger(0, t') \Psi_R(0, t) \rangle = e^{-ive \int_t^{t'} d\tau V(\tau)} \langle \psi_R^\dagger(0, t') \psi_R(0, t) \rangle,$$
(22)
$$G_L^< (t', t) = \langle \Psi_L^\dagger(0, t') \Psi_L(0, t) \rangle = \langle \psi_L^\dagger(0, t') \psi_L(0, t) \rangle,$$
(23)

with $\psi_{R/L}$ the quasi-particle field operators evolving with respect to H_0 only. Analogous expressions can be derived for $G_R^>$ and $G_L^>$.

In terms of the bosonized picture [36,58–60], the quasi-particle field operators can be written as coherent states of the bosonic edge modes, namely

$$\psi_R(x, t) = \frac{\mathcal{F}_R}{\sqrt{2\pi a}} e^{ik_F(x - vt)} e^{-i\sqrt{v}\phi_R(x,t)},$$
(24)

$$\psi_L(x, t) = \frac{\mathcal{F}_L}{\sqrt{2\pi a}} e^{ik_F(x + vt)} e^{-i\sqrt{v}\phi_L(x,t)},$$
(25)

with $\mathcal{F}_{R/L}$ Klein factors [59]. Therefore, the expressions for the Green's functions in Equations (22) and (23) become

$$G_R^< (t', t) = \langle \Psi_R^\dagger(0, t') \Psi_R(0, t) \rangle = e^{-ive \int_t^{t'} d\tau V(\tau)} \frac{e^{ik_F v(t' - t)}}{2\pi a} \mathcal{P}_v(t - t'),$$
(26)

$$G_L^< (t', t) = \langle \Psi_L^\dagger(0, t') \Psi_L(0, t) \rangle = \frac{e^{-ik_F v(t' - t)}}{2\pi a} \mathcal{P}_v(t - t'),$$
(27)

where we introduced the function

$$\mathcal{P}_v(\tau) = e^{vW(\tau)},$$
(28)

with

$$W(t) = \left\langle \phi_{R/L}(0, t) \phi_{R/L}(0, 0) - \phi_{R/L}^2(0, 0) \right\rangle =$$

$$= \ln \left[\frac{\left| \Gamma \left(1 + \frac{\theta}{\omega_c} + i\theta t \right) \right|^2}{\left| \Gamma \left(1 + \frac{\theta}{\omega_c} \right) \right|^2 (1 + i\omega_c t)} \right] \simeq \ln \left[\frac{\pi\theta t}{\sinh\left(\pi\theta t \right)(1 + i\omega_c t)} \right].$$
(29)

In the above, Equation θ is the temperature ($k_B = 1$), $\omega_c = v/a$ is the high energy cut-off, with a the finite length cut-off appearing in Equation (25), and $\Gamma(x)$ is the Euler's gamma function. The considered approximation is valid as long as $\omega_c \gg \theta$. Let us notice that the expressions are equal for right- and left-movers and, therefore, we have omitted any label R or L. Moreover, we explicitly used the fact that $W(t)$ is translationally invariant in time.

By inserting these expressions into Equation (21), one finds

$$J_{BS}(t) = 2ive\,|\lambda|^2 \int_0^{+\infty} d\tau \sin \left[ve \int_{t-\tau}^t dt' V(t') \right] (\mathcal{P}_{2v}(\tau) - \mathcal{P}_{2v}(-\tau)),$$
(30)

where we introduced the rescaled tunneling amplitude $\lambda = \Lambda/(2\pi a)$. Notice that the backscattering current satisfies $J_{BS}(t) = J_{BS}(t + \mathcal{T})$ as expected due to the periodicity of the voltage drive $V(t)$. It is thus possible to perform an average over one period of the backscattering current thus finding

$$\overline{J_{BS}(t)} = 2ive\,|\lambda|^2 \int_0^{\mathcal{T}} \frac{dt}{\mathcal{T}} \int d\tau \sin\left[ve \int_{t-\tau}^t dt'\,V(t')\right] \mathcal{P}_{2\nu}(\tau). \tag{31}$$

The presence of a periodic voltage can be conveniently handled by resorting to the photo-assisted coefficients defined by [16,61]

$$p_l(\alpha) = \int_{-\mathcal{T}/2}^{\mathcal{T}/2} \frac{dt}{\mathcal{T}} e^{2i\pi n\frac{t}{\mathcal{T}}} e^{-2i\pi\alpha\varphi(t)}, \tag{32}$$

with

$$\varphi(t) = \int_{-\infty}^t \frac{dt'}{\mathcal{T}} \tilde{V}_{ac}(t'), \tag{33}$$

where $\tilde{V}_{ac}(t)$ is the AC part of $V(t)$ with unitary and dimensionless amplitude. Therefore, the backscattering current assumes the final expression

$$\overline{J_{BS}(t)} = 2ive\,|\lambda|^2 \sum_l |p_l(\alpha)|^2 \int_{-\infty}^{+\infty} d\tau \sin\left[(q+l)\,\omega\tau\right] \mathcal{P}_{2\nu}(\tau). \tag{34}$$

By moving to Fourier space and by introducing the function [58,62–65]

$$\tilde{\mathcal{P}}_\nu(E) = \int_{-\infty}^{+\infty} dt\, e^{iEt} \mathcal{P}_\nu(t) = \left(\frac{2\pi\theta}{\omega_c}\right)^{\nu-1} \frac{e^{E/2\theta}}{\Gamma(\nu)\omega_c} \left|\Gamma\left(\frac{\nu}{2} - i\frac{E}{2\pi\theta}\right)\right|^2, \tag{35}$$

the current in Equation (34) can be recast as

$$\overline{J_{BS}(t)} = ve\,|\lambda|^2 \sum_l |p_l(\alpha)|^2 \left\{\tilde{\mathcal{P}}_{2\nu}[(q+l)\omega] - \tilde{\mathcal{P}}_{2\nu}[-(q+l)\omega]\right\}. \tag{36}$$

This final expression explicitly depends on the rescaled amplitudes of both the AC (α) and DC (q) contribution to the voltage. Even if frequently assumed equal, these two parameters can be tuned independently [16,53].

4. Heat Current

In order to extend the previous analysis to heat transport, we need to properly define the heat current operator. For a system described by a Hamiltonian density $\mathcal{H}$ and with particle number density $\mathcal{N}$, the heat density $\mathcal{Q}$ is given by

$$\mathcal{Q} = \mathcal{H} - \mu\mathcal{N}, \tag{37}$$

with μ the chemical potential of the lead where the heat density is measured. This definition is motivated by the fact that the energy density depends on an arbitrary energy reference and, thus, it is not a well-defined experimental observable. On the contrary, heat is defined as the energy carried by particles with respect to a given chemical potential, thus motivating the expression in Equation (37).

According to this and considering again the chiral propagation of the bosonic modes along the edges [66], in the chiral Luttinger description (see Equation (1)), one defines the heat densities as

$$\mathcal{Q}_{R/L}(x,t) = \frac{v}{4\pi} \left(\partial_x \Phi_{R/L}(x,t)\right)^2, \tag{38}$$

where all the contributions proportional to the chemical potential $\mu = vk_F$ are automatically taken into account [57,67]. Notice that the above equation provides the proper value of the thermal Hall conductance in agreement with the Wiedemann–Franz law [67,68]

The corresponding heat current operators in the terminals 2 and 3 can be expressed in terms of heat density operators [57] as

$$\mathcal{J}_{2/3}(t) = \pm v \, \mathcal{Q}_{R/L}(\pm d, t) \tag{39}$$

due to the chirality of Laughlin edge states.

Proceeding as for the charge current, the heat current operators are represented in powers of the tunneling amplitude Λ,

$$\mathcal{J}_{2/3}(t) = \mathcal{J}_{2/3}^{(0)}(t) + \mathcal{J}_{2/3}^{(1)}(t) + \mathcal{J}_{2/3}^{(2)}(t) + \mathcal{O}\left(|\Lambda|^3\right), \tag{40}$$

with

$$\mathcal{J}_{2/3}^{(0)}(t) = \pm v \mathcal{Q}_{R/L}^{(0)}(\pm d, t), \tag{41}$$

$$\mathcal{J}_{2/3}^{(1)}(t) = \pm i v \int_{-\infty}^{t} dt' \left[H_T(t'), \mathcal{Q}_{R/L}^{(0)}(\pm d, t) \right], \tag{42}$$

$$\mathcal{J}_{2/3}^{(2)}(t) = \pm i^2 v \int_{-\infty}^{t} dt' \int_{-\infty}^{t'} dt'' \left[H_T(t''), \left[H_t(t'), \mathcal{Q}_{R/L}^{(0)}(\pm d, t) \right] \right]. \tag{43}$$

In the above equations, we have denoted with $\mathcal{Q}_{R/L}^{(0)}(x, t)$ the time evolution of heat densities in the absence of tunneling, which can be obtained from the time evolution of bosonic fields in Equation (5) and reads

$$\mathcal{Q}_R^{(0)}(x, t) = \frac{v}{4\pi} \left[(\partial_x \phi_R(x,t))^2 + 2e\sqrt{v}\partial_x\phi_R(x,t)V_{R/L}\left(t \mp \frac{x}{v}\right) + \frac{e^2 v}{v} V_R^2 \left(t \mp \frac{x}{v}\right) \right], \tag{44}$$

$$\mathcal{Q}_L^{(0)}(x, t) = \frac{v}{4\pi} (\partial_x \phi_{R/L}(x,t))^2. \tag{45}$$

The commutators involving $\mathcal{Q}_{R/L}^{(0)}(x, t)$ in Equations (42) and (43) are

$$\mathcal{J}_{2/3}^{(1)}(t) = \pm \dot{Q}_{R/L}(\pm d, t), \tag{46}$$

$$\mathcal{J}_{2/3}^{(2)}(t) = \pm i \int_{-\infty}^{t - \frac{d}{v}} dt'' \left[H_t(t''), \dot{Q}_{R/L}(\pm d, t) \right], \tag{47}$$

where

$$\dot{Q}_R(x, t) = v\Lambda (\partial_x + ik_F) \Psi_R^{\dagger}(x,t)\Psi_L(x,t) + H.c., \tag{48}$$

$$\dot{Q}_L(x, t) = -v\Lambda \Psi_R^{\dagger}(x,t) (\partial_x + ik_F) \Psi_L(x,t) + H.c.. \tag{49}$$

According to this, the average of heat current operators in Equation (40) reads

$$\langle \mathcal{J}_{2/3}(t) \rangle = \langle \mathcal{J}_{2/3}^{(0)} \rangle + \langle \mathcal{J}_{2/3}^{(2)} \rangle. \tag{50}$$

Analogously with the case of the charge current, one has that $\mathcal{J}_{2/3}^{(1)}$ is zero due to the unbalance between annihilation and creation field operators of each chirality. Focusing on the heat current which is backscattered by the QPC into reservoir 3, one can define the backscattering heat current as

$$\mathcal{J}_{BS}(t) = \langle \mathcal{J}_3^{(2)}(t) \rangle. \tag{51}$$

It can be expressed in terms of Green's functions in Equations (26) and (27), by exploiting the explicit expression of $\dot{Q}_L(x,t)$ in Equations (48) and (49), thus finding

$$
\mathcal{J}_{BS}(t) = i|\Lambda|^2 \int_{-\infty}^{t-\frac{d}{v}} d\tau \left[G_R^<\left(t', t - \frac{d}{v}\right)(\partial_t - ik_Fv)\, G_L^>\left(t', t - \frac{d}{v}\right) + \right.
$$
$$
+ G_R^>\left(t', t - \frac{d}{v}\right)(\partial_t + ik_Fv)\, G_L^<\left(t', t - \frac{d}{v}\right) +
$$
$$
- G_R^<\left(t - \frac{d}{v}, t'\right)(\partial_t - ik_Fv)\, G_L^>\left(t - \frac{d}{v}, t'\right) +
$$
$$
\left. - G_R^>\left(t - \frac{d}{v}, t'\right)(\partial_t + ik_Fv)\, G_L^<\left(t - \frac{d}{v}, t'\right) \right].
\tag{52}
$$

By recalling the link between Green's functions and the function $\mathcal{P}_\nu(t)$, the heat backscattering current becomes

$$
\mathcal{J}_{BS}(t) = i|\lambda|^2 \int_0^{+\infty} d\tau \cos\left[ve \int_t^{t-\tau} dt'' V_R(t'')\right] \partial_\tau \left[\mathcal{P}_{2\nu}(\tau) - \mathcal{P}_{2\nu}(-\tau)\right],
\tag{53}
$$

which, averaged over one period, reduces to

$$
\overline{\mathcal{J}_{BS}(t)}(\alpha, q) = |\lambda|^2 \frac{\omega}{2} \sum_l |p_l(\alpha)|^2 (q+l) \left\{ \tilde{\mathcal{P}}_{2\nu}\left[(q+l)\omega\right] - \tilde{\mathcal{P}}_{2\nu}\left[-(q+l)\omega\right] \right\}.
\tag{54}
$$

5. Mixed Noise and Heat Noise

Concerning the fluctuations, together with the conventional current noise [69], it is possible to define both the autocorrelated mixed noise and heat noise [45,46,50]. For the sake of simplicity, we will focus exclusively on the signal detected in reservoir 3 considering the quantities

$$
\mathcal{S}_X = \int_0^T \frac{dt}{T} \int_{-\infty}^{+\infty} dt' \left[\langle \mathcal{I}_3(t') \mathcal{J}_3(t) \rangle - \langle \mathcal{I}_3(t') \rangle \langle \mathcal{J}_3(t) \rangle \right],
\tag{55}
$$

$$
\mathcal{S}_Q = \int_0^T \frac{dt}{T} \int_{-\infty}^{+\infty} dt' \left[\langle \mathcal{J}_3(t') \mathcal{J}_3(t) \rangle - \langle \mathcal{J}_3(t') \rangle \langle \mathcal{J}_3(t) \rangle \right].
\tag{56}
$$

Notice that analogous expressions can be derived focusing on the reservoir 2.

The perturbative expansion of charge and heat current operators in Equations (13) and (40) allows for also expressing these quantities perturbatively in Λ, namely

$$
\mathcal{S}_X = \mathcal{S}_X^{(02)} + \mathcal{S}_X^{(20)} + \mathcal{S}_X^{(11)} + \mathcal{O}\left(|\Lambda|^3\right),
\tag{57}
$$

$$
\mathcal{S}_Q = \mathcal{S}_Q^{(02)} + \mathcal{S}_Q^{(20)} + \mathcal{S}_Q^{(11)} + \mathcal{O}\left(|\Lambda|^3\right),
\tag{58}
$$

where

$$
\mathcal{S}_X^{(ij)} = \int_0^T \frac{dt}{T} \int_{-\infty}^{+\infty} dt' \left\{ \langle \mathcal{I}_3^{(i)}(t') \mathcal{J}_3^{(j)}(t) \rangle - \langle \mathcal{I}_3^{(i)}(t') \rangle \langle \mathcal{J}_3^{(j)}(t) \rangle \right\},
\tag{59}
$$

$$
\mathcal{S}_Q^{(ij)} = \int_0^T \frac{dt}{T} \int_{-\infty}^{+\infty} dt' \left\{ \langle \mathcal{J}_3^{(i)}(t') \mathcal{J}_3^{(j)}(t) \rangle - \langle \mathcal{J}_3^{(i)}(t') \rangle \langle \mathcal{J}_3^{(j)}(t) \rangle \right\}.
\tag{60}
$$

In the perturbative expansions of Equations (57) and (58), the only surviving terms are $S_X^{(11)}$ and $S_Q^{(11)}$, since one can show that [50,66]

$$S_X^{(02)} = S_X^{(20)} = 0, \tag{61}$$

$$S_Q^{(02)} = S_Q^{(20)} = 0. \tag{62}$$

Mixed and heat noises are obtained in terms of Green's functions as

$$S_X = i|\Lambda|^2 \int_0^T \frac{dt}{T} \int_{-\infty}^{+\infty} dt' \left[G_R^< (t',t)(\partial_t - ik_F v)G_L^< (t',t) - G_R^< (t',t)(\partial_t + ik_F v)G_L^< (t',t) \right], \tag{63}$$

$$S_Q = |\Lambda|^2 \int_0^T \frac{dt}{T} \int_{-\infty}^{+\infty} dt' \left[G_R^< (t',t)(\partial_{t'} + ik_F v)(\partial_t - ik_F v)G_L^< (t',t) + \right.$$
$$\left. + G_R^< (t',t)(\partial_{t'} - ik_F v)(\partial_t + ik_F v)G_L^< (t',t) \right] \tag{64}$$

and, in terms of the function $\mathcal{P}_\nu(t)$ in Equation (28), they can be rewritten as

$$S_X = 2\nu e|\lambda|^2 \int_0^T \frac{dt}{T} \int_{-\infty}^{+\infty} dt' \sin\left[\nu e \int_{t'}^t dt'' V(t'') \right] \mathcal{P}_\nu(t'-t) \partial_{t'} \mathcal{P}_\nu(t'-t), \tag{65}$$

$$S_Q = 2|\lambda|^2 \int_0^T \frac{dt}{T} \int_{-\infty}^{+\infty} dt' \cos\left[\nu e \int_{t'}^t dt'' V(t'') \right] \mathcal{P}_\nu(t'-t) \partial_t \partial_{t'} \mathcal{P}_\nu(t'-t). \tag{66}$$

Using again the series expansion of the voltage dependent phase factor and the Fourier transform for $\mathcal{P}_\nu(t'-t)$, one is left with

$$S_X(\alpha,q) = \frac{\nu e\omega}{2}|\lambda|^2 \sum_l |p_l(\alpha)|^2 (q+l) \left\{ \tilde{\mathcal{P}}_{2\nu}\left[(q+l)\omega\right] + \tilde{\mathcal{P}}_{2\nu}\left[-(q+l)\omega\right] \right\}, \tag{67}$$

$$S_Q(\alpha,q) = \frac{|\lambda|^2}{2\pi} \sum_l |p_l(\alpha)|^2 \int_{-\infty}^{+\infty} dE E^2 \tilde{\mathcal{P}}_\nu(E) \left\{ \tilde{\mathcal{P}}_\nu\left[(q+l)\omega - E\right] + \tilde{\mathcal{P}}_\nu\left[-(q+l)\omega - E\right] \right\}, \tag{68}$$

where we have explicitly indicated the dependence on the AC (α) and DC (q) voltage amplitude. To perform the integral in the equation for S_Q, we exploit the identity

$$\int_{-\infty}^{+\infty} \frac{dY}{2\pi} Y^2 \tilde{\mathcal{P}}_{g_1}(Y) \tilde{\mathcal{P}}_{g_2}(X-Y) = \frac{\tilde{\mathcal{P}}_{g_1+g_2}(X)}{1+g_1+g_2} \left[g_1 g_2 \pi^2 \theta^2 + \frac{g_1(1+g_1)}{g_1+g_2} \omega^2 \right] \tag{69}$$

obtaining the expression

$$S_Q(\alpha,q) = |\lambda|^2 \sum_l |p_l(\alpha)|^2 \left[\frac{2\pi^2 \nu^2}{1+2\nu} \theta^2 + \frac{1+\nu}{1+2\nu} (q+l)^2 \omega^2 \right] \left\{ \tilde{\mathcal{P}}_{2\nu}\left[(q+l)\omega\right] + \tilde{\mathcal{P}}_{2\nu}\left[-(q+l)\omega\right] \right\}. \tag{70}$$

At temperature zero, the above expressions reduce to

$$S_X(\alpha,q) = \nu e|\lambda|^2 \frac{\pi}{\Gamma(2\nu)} \left(\frac{\omega}{\omega_c} \right)^{2\nu} \sum_l |p_l(\alpha)|^2 |q+l|^{2\nu} \mathrm{sign}(q+l), \tag{71}$$

$$S_Q(\alpha,q) = \omega|\lambda|^2 \frac{\pi(1+\nu)}{\Gamma(2\nu)(1+2\nu)} \left(\frac{\omega}{\omega_c} \right)^{2\nu} \sum_l |p_l(\alpha)|^2 |q+l|^{2\nu+1}. \tag{72}$$

Here, the chiral free fermion case is recovered directly by inserting $\nu = 1$. It is worth noticing that this peculiar state can be described exactly (to all order in the tunneling amplitude) in terms of the scattering theory [69,70]. Moreover, at this value of the filling factor, the dependence on a (finite length cut-off) disappears.

These noises will be investigated in the following with the aim of carrying out their spectroscopic analysis. For sake of simplicity, we will focused on the zero temperature limit, the finite temperature correction being negligible as far as $\theta \ll \omega$.

6. Results

6.1. Heat Current

The behavior of the photo-assisted charge current as a function of the AC and DC voltage contribution has been already discussed both in the non-interacting [16] and in the strongly interacting regime [17,53], the latter case showing divergencies at zero temperature signature of the limitations of the perturbative approach in this regime. In Figure 2, we report the density plots of its heat counterpart whose functional form has been derived in Equation (54). The first row represents the case of a sinusoidal drive with

$$V_{ac}^{(\cos)}(t) = -\frac{\omega\alpha}{e}\cos(\omega t) \tag{73}$$

both in the free fermion case $\nu = 1$ (top left panel) and at $\nu = 1/3$ (top right panel). Here, all curves are mirror symmetric with respect to $\alpha = 0$ ($\overline{\mathcal{J}_{BS}}(\alpha, q) = \overline{\mathcal{J}_{BS}}(-\alpha, q)$) and $q = 0$ ($\overline{\mathcal{J}_{BS}}(\alpha, q) = \overline{\mathcal{J}_{BS}}(\alpha, -q)$) and consequently satisfy

$$\overline{\mathcal{J}_{BS}}(\alpha, q) = \overline{\mathcal{J}_{BS}}(-\alpha, -q). \tag{74}$$

Both of these remarkable features are a direct consequence of the symmetries of the photo-assisted tunneling amplitudes. Indeed, for this kind of drive, one has [16,61]

$$p_l^{(\cos)}(\alpha) = J_l(-\alpha), \tag{75}$$

with J_l Bessel function of order l. These amplitudes satisfy

$$p_l^{(\cos)}(\alpha) = p_l^{(\cos)}(-\alpha) \tag{76}$$

and

$$p_l^{(\cos)}(\alpha) = p_{-l}^{(\cos)}(\alpha). \tag{77}$$

Considering the case of a periodic train of Lorentzian pulses of the form

$$V_{ac}^{(\mathrm{Lor})}(t) = \frac{\omega}{\pi e}\sum_{l=-\infty}^{+\infty}\frac{\eta}{\eta^2 + \left(\frac{t}{\mathcal{T}} - l\right)^2} - \frac{\omega\alpha}{e}, \tag{78}$$

with width at half height given by η and where the photo-assisted tunneling amplitudes are given by

$$p_l^{(\mathrm{Lor})}(\alpha) = \alpha\sum_{s=0}^{+\infty}\frac{\Gamma(\alpha + l + s)}{\Gamma(\alpha + 1 - s)}\frac{(-1)^s e^{-2\pi\eta(2s+l)}}{\Gamma(l + s + 1)\Gamma(s + 1)}. \tag{79}$$

In this case, one observes that, at $\nu = 1$ (bottom left panel of Figure 2), the heat current is still highly symmetric as in the case of the sinusoidal drive. This is an accidental consequence of the fact that, at this value of the filling factor for every periodic voltage drive, one has [50]

$$\overline{\mathcal{J}_{BS}}(\alpha, q) \propto \sum_{l=-\infty}^{+\infty}|p_l(\alpha)|^2 (q + l)^2 = \frac{e}{\hbar\omega}\int_0^{\mathcal{T}}\frac{dt}{\mathcal{T}}V^2(t) = \frac{e}{\hbar\omega}V_{dc}^2 + \frac{e}{\hbar\omega}\int_0^{\mathcal{T}}\frac{dt}{\mathcal{T}}V_{ac}^2(t), \tag{80}$$

which is manifestly insensitive to the overall sign of both the DC and the AC contribution to the voltage. This is no more true for what concerns the filling factors in the Laughlin sequence (bottom right panel of Figure 2) where

$$\overline{\mathcal{J}_{BS}}(\alpha, q) \neq \overline{\mathcal{J}_{BS}}(-\alpha, q), \tag{81}$$
$$\overline{\mathcal{J}_{BS}}(\alpha, q) \neq \overline{\mathcal{J}_{BS}}(\alpha, -q). \tag{82}$$

However, the condition

$$p_l^{(Lor)}(\alpha) = p_{-l}^{(Lor)}(-\alpha) \tag{83}$$

leads to the residual symmetry

$$\overline{\mathcal{J}_{BS}}(\alpha, q) = \overline{\mathcal{J}_{BS}}(-\alpha, -q). \tag{84}$$

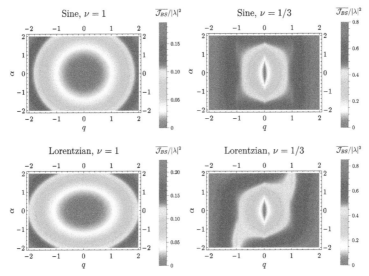

Figure 2. Density plot of the averaged heat current $\overline{\mathcal{J}_{BS}}$ (in units of $|\lambda|^2$) as a function of the DC voltage amplitude q ($\hat{x}$-axis) and the AC voltage amplitude α ($\hat{y}$-axis). Other parameters are: $\eta = 0.1$, $\theta = 0$ and $\omega_c = 10\omega$.

6.2. Mixed Noise

Density plots in Figure 3 show the behavior of the charge-heat mixed noise as a function of the AC voltage amplitude α and of the DC voltage amplitude q. In particular, the first row represents the case of a sinusoidal drive both in the free fermion case $\nu = 1$ (top left panel) and at $\nu = 1/3$ (top right panel). In this case, all curves show the properties $S_X(\alpha, q) = S_X(-\alpha, q)$, $S_X(\alpha, q) = -S_X(\alpha, -q)$ and consequently

$$S_X(\alpha, q) = -S_X(-\alpha, -q). \tag{85}$$

The curves at $\nu = 1$ are increasing (decreasing) for increasing $|\alpha|$ at positive (negative) q, while the opposite is true for $\nu = 1/3$. This overall profile is dictated by the power-law behavior (exponent 2ν in Equation (71)) which is parabolic for $\nu = 1$ and sub-linear ($2\nu = 2/3$, further suppressed by the fast decreasing photo-assisted tunneling amplitude) for $\nu = 1/3$.

While these general considerations about the asymptotic behavior of the curves at increasing $|\alpha|$ still hold in the case of the Lorentzian case (second row of Figure 3), here both plots are manifestly asymmetric. Again, this fact can be directly attributed to the (lack of) symmetries of the photo-assisted tunneling amplitudes $p_l^{(Lor)}(\alpha)$. This asymmetry in α is very evident at small values of $|q|$ and becomes progressively less important by increasing $|q|$. Despite this, it is possible to note that Equation (85) is still satisfied.

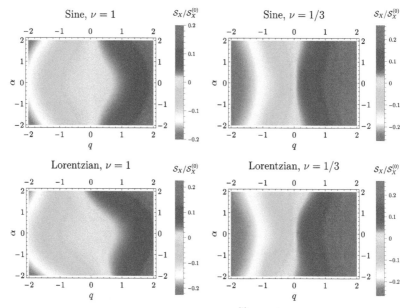

Figure 3. Density plot of the mixed noise $\mathcal{S}_X$ (in units of $\mathcal{S}_X^{(0)} = e|\lambda|^2$) as a function of the DC voltage amplitude q ($\hat{x}$-axis) and the AC voltage amplitude α ($\hat{y}$-axis). Other parameters are: $\eta = 0.1$, $\theta = 0$ and $\omega_c = 10\omega$.

6.3. Heat Noise

The density plots of the heat current fluctuations as a function of the AC voltage amplitude α and of the DC voltage amplitude q are reported in Figure 4. As before, the two upper panels show the sinusoidal drive case both in the free fermion case $\nu = 1$ (top left panel) and at $\nu = 1/3$ (top right panel). Here, according to the conditions in Equations (76) and (77), all curves satisfy $\mathcal{S}_Q(\alpha, q) = \mathcal{S}_Q(-\alpha, q)$. In addition, differently from what happens in the mixed noise case, one has that this quantity is positively defined and that $\mathcal{S}_Q(\alpha, q) = \mathcal{S}_Q(\alpha, -q)$.

Because of the greater power-law in Equation (72) with respect to the one in Equation (71), the curves are always increasing by increasing $|\alpha|$ for both the free and the strongly interacting case. The same is true also for the Lorentzian drive (bottom panels of Figure 4) although, in this case, the asymmetry of the $p_l^{(Lor)}(\alpha)$ directly reflects in the asymmetry of the curves for $\alpha \rightarrow -\alpha$ ($\mathcal{S}_Q(\alpha, q) \neq \mathcal{S}_Q(-\alpha, q)$) and $q \rightarrow -q$ ($\mathcal{S}_Q(\alpha, q) \neq \mathcal{S}_Q(\alpha, -q)$). However, the curves are characterized by the condition

$$\mathcal{S}_Q(\alpha, q) = \mathcal{S}_Q(-\alpha, -q) \tag{86}$$

due to the property in Equation (83).

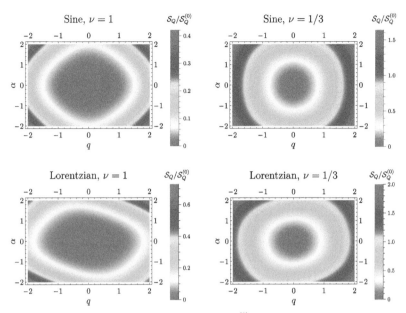

Figure 4. Density plot of the heat noise $\mathcal{S}_Q$ (in units of $\mathcal{S}_Q^{(0)} = \omega |\lambda|^2$) as a function of DC voltage amplitude q ($\hat{x}$-axis) and the AC voltage amplitude α ($\hat{y}$-axis). Other parameters are: $\eta = 0.1$, $\theta = 0$ and $\omega_c = 10\omega$.

7. Conclusions

In this work, we investigated charge-heat mixed noise and heat noise generated by a periodic time-dependent voltage in a quantum point contact geometry implemented in a quantum Hall bar. We focused on the weak backscattering regime of the tunneling barrier and we evaluated fluctuations of charge and heat currents at lowest order in tunneling the amplitude. We provided a spectroscopic analysis of these quantities by tuning independently the AC and DC amplitudes of the voltage drive for integer and fractional filling factors. By focusing on a cosine drive and a periodic train of Lorentzian pulse, we proved that both charge-heat noise and heat noise reveal the symmetry properties of the photo-assisted coefficients of the applied drive. The different power-laws that govern the behavior of mixed and heat noises give rise to two distinct profiles when these quantities are plotted as a function of the AC amplitude. In particular, charge-heat noise displays an opposite monotonicity for the case of integer and fractional filling factors, thus revealing the presence of a sub-linear power-law decay typical of Luttinger liquid physics.

Author Contributions: Conceptualization, F.R., M.A. and D.F.; formal analysis, M.A. and D.F.; writing—original draft preparation, F.R., M.A. and D.F.; writing—review and editing, J.R., T.J., T.M. and M.S.; supervision, T.M. and M.S.

Funding: This work was granted access to the HPC resources of Aix-Marseille Université financed by the project Equip@Meso (Grant No. ANR-10-EQPX29-01). It has been carried out in the framework of project "one shot reloaded" (Grant No. ANR-14-CE32-0017) and benefited from the support of the Labex ARCHIMEDE (Grant No. ANR11-LABX-0033) and the AMIDEX project (Grant No. ANR-11-IDEX-0001-02), funded by the "investissements d'avenir" French Government program managed by the French National Research Agency (ANR).

Acknowledgments: The authors thank Luca Vannucci for useful discussions.

Conflicts of Interest: The authors declare no conflict of interest.

Abbreviations

The following abbreviations are used in this manuscript:

FQH   Fractional Quantum Hall
QPC   Quantum Point Contact
EQO   Electron Quantum Optics

References

1. Grenier, C.; Hervé, R.; Fève, G.; Degiovanni, P. Electron quantum optics in quantum Hall edge channels. *Mod. Phys. Lett. B* **2011**, *25*, 1053.
2. Bocquillon, E.; Freulon, V.; Parmentier, F.D.; Berroir, J.-M.; Plaçais, B.; Wahl, C.; Rech, J.; Jonckheere, T.; Martin, T.; Grenier, C.; et al. Electron quantum optics in ballistic chiral conductors. *Ann. Phys.* **2014**, *526*, 1.
3. Bäuerle, C.; Glattli, D.C.; Meunier, T.; Portier, F.; Roche, P.; Roulleau, P.; Takada, S.; Waintal, X. Coherent control of single electrons: A review of current progress. *Rep. Prog. Phys.* **2018**, *81*, 056503.
4. Fève, G.; Mahé, A.; Berroir, J.-M.; Kontos, T.; Plaçais, B.; Glattli, D.C.; Cavanna, A.; Etienne, B.; Jin, Y. An on-demand coherent single-electron source. *Science* **2007**, *316*, 1169.
5. Ferraro, D.; Rech, J.; Jonckheere, T.; Martin, T. Single quasiparticle and electron emitter in the fractional quantum Hall regime. *Phys. Rev. B* **2015**, *91*, 205409.
6. Büttiker, M.; Thomas, H.; Prêtre, A. Mesoscopic capacitors. *Phys. Lett. A* **1993**, *180*, 364.
7. Moskalets, M.; Samuelsson, P.; Büttiker, M. Quantized Dynamics of a Coherent Capacitor. *Phys. Rev. Lett.* **2008**, *100*, 086601.
8. Ludovico, M.L.; Arrachea, L.; Moskalets, M.; Sanchez, D. Periodic energy transport and entropy production in quantum electronics. *Entropy* **2016**, *18*, 419.
9. Alomar, M.I.; Lim, J.S.; Sanchez, D. Coulomb-blockade effect in nonlinear mesoscopic capacitors. *Phys. Rev. B* **2016**, *94*, 165425.
10. Levitov, L.S.; Lee, H.W.; Lesovik, G.B. Electron counting statistics and coherent states of electric current. *J. Math. Phys.* **1996**, *37*, 4845.
11. Ivanov, D.; Lee, H.W.; Levitov, L.S. Coherent states of alternating current. *Phys. Rev. B* **1997**, *56*, 6839.
12. Keeling, J.; Klich, I.; Levitov, L.S. Minimal Excitation States of Electrons in One-Dimensional Wires. *Phys. Rev. Lett.* **2006**, *97*, 116403.
13. Glattli, D.C.; Roulleau, P. Levitons for electron quantum optics. *Phys. Status Solidi B* **2017**, *254*, 1600650.
14. Dubois, J.; Jullien, T.; Portier, F.; Roche, P.; Cavanna, A.; Jin, Y.; Wegscheider, W.; Roulleau, P.; Glattli, D.C. Minimal excitation states for electron quantum optics using levitons. *Nature* **2013**, *502*, 659.
15. Crépieux, A.; Devillard, P.; Martin, T. Photoassisted current and shot noise in the fractional quantum Hall effect. *Phys. Rev. B* **2004**, *69*, 205302.
16. Dubois, J.; Jullien, T.; Grenier, C.; Degiovanni, P.; Roulleau, P.; Glattli, D.C. Integer and fractional charge Lorentzian voltage pulses analyzed in the framework of photon-assisted shot noise. *Phys. Rev. B* **2013**, *88*, 085301.
17. Rech, J.; Ferraro, D.; Jonckheere, T.; Vannucci, L.; Sassetti, M.; Martin, T. Minimal Excitations in the Fractional Quantum Hall Regime. *Phys. Rev. Lett.* **2017**, *118*, 076801.
18. Acciai, M.; Carrega, M.; Rech, J.; Jonckheere, T.; Martin, T.; Sassetti, M. Probing interactions via non-equilibrium momentum distribution and noise in integer quantum Hall systems at $\nu = 2$. *Phys. Rev. B* **2018**, *98*, 035426.
19. Ronetti, F.; Vannucci, L.; Ferraro, D.; Jonckheere, T.; Rech, J.; Martin, T.; Sassetti, M. Crystallization of levitons in the fractional quantum Hall regime. *Phys. Rev. B* **2018**, *98*, 075401.
20. Ferraro, D.; Ronetti, F.; Vannucci, L.; Acciai, M.; Rech, J.; Jockheere, T.; Martin, T.; Sassetti, M. Hong-Ou-Mandel characterization of multiply charged Levitons. *Eur. Phys. J. Spec. Top.* **2018**, *227*, 1345.
21. Hanbury-Brown, R.; Twiss, R.Q. Correlation between photons in two coherent beams of light. *Nature* **1956**, *177*, 27.
22. Bocquillon, E.; Parmentier, F.D.; Grenier, C.; Berroir, J.-M.; Degiovanni, P.; Glattli, D.C.; Plaçais, B.; Cavanna, A.; Jin, Y.; Fève, G. Electron Quantum Optics: Partitioning Electrons One by One. *Phys. Rev. Lett.* **2012**, *108*, 196803.

23. Calzona, A.; Acciai, M.; Carrega, M.; Cavaliere, F.; Sassetti, M. Time-resolved energy dynamics after single electron injection into an interacting helical liquid. *Phys. Rev. B* **2016**, *94*, 035404.

24. Giazotto, F.; Heikkilä, T.T.; Luukanen, A.; Savin, A.M.; Pekola, J.P. Opportunities for mesoscopics in thermometry and refrigeration: Physics and applications. *Rev. Mod. Phys.* **2006**, *78*, 217.

25. Vinjanampathy, S.; Anders, J. Quantum Thermodynamics. *Contemp. Phys.* **2016**, *57*, 545.

26. Sànchez, R.; Sothmann, B.; Jordan, A.N. Chiral Thermoelectrics with Quantum Hall Edge States. *Phys. Rev. Lett.* **2015**, *114*, 146801.

27. Sànchez, R.; Sothmann, B.; Jordan, A.N. Heat diode and engine based on quantum Hall edge states. *New J. Phys.* **2015**, *17*, 075006.

28. Samuelsson, P.; Kheradsoud, S.; Sothmann, B. Optimal Quantum Interference Thermoelectric Heat Engine with Edge States. *Phys. Rev. Lett.* **2017**, *118*, 256801.

29. Thierschmann, H.; Sànchez, R.; Sothmann, B.; Arnold, F.; Heyn, C.;Hansen, W.; Buhmann, H.; Molenkamp, L.W. Three terminal energy harvester with coupled quantum dots. *Nat. Nanotechnol.* **2015**, *10*, 854.

30. Juergens, S.; Haupt, F.; Moskalets, M.; Splettstoesser, J. Thermoelectric performance of a driven double quantum dot. *Phys. Rev. B* **2013**, *87*, 245423.

31. Erdman, P.A.; Mazza, F.; Bosisio, R.; Benenti, G.; Fazio, R.; Taddei, F. Thermoelectric properties of an interacting quantum dot based heat engine. *Phys. Rev. B* **2017**, *95*, 245432.

32. Mazza, F.; Bosisio, R.; Benenti, G.; Giovannetti, V.; Fazio, R.; Taddei, F. Thermoelectric efficiency of three terminal quantum thermal machines. *New J. Phys.* **2014**, *16*, 085001.

33. Ferraro, D.; Campisi, M.; Andolina, G.M.; Pellegrini, V.; Polini, M. High-Power Collective Charging of a Solid-State Quantum Battery. *Phys. Rev. Lett.* **2018**, *120*, 117702.

34. Vannucci, L.; Ronetti, F.; Dolcetto, G.; Carrega, M.; Sassetti, M. Interference-induced thermoelectric switching and heat rectification in quantum Hall junctions. *Phys. Rev. B* **2015**, *92*, 075446.

35. Ronetti, F.; Vannucci, L.; Dolcetto, G.; Carrega, M.; Sassetti, M. Spin-thermoelectric transport induced by interactions and spin-flip processes in two-dimensional, topological insulators. *Phys. Rev. B* **2016**, *93*, 165414.

36. Ronetti, F.; Carrega, M.; Ferraro, D.; Rech, J.; Jonckheere, T.; Martin, T.; Sassetti, M. Polarized heat current generated by quantum pumping in two-dimensional topological insulators. *Phys. Rev. B* **2017**, *95*, 115412.

37. Ludovico, M.F.; Lim, J.S.; Moskalets, M.; Arrachea, L.; Sànchez, D. Dynamical energy transfer in ac-driven quantum systems. *Phys. Rev. B* **2014**, *89*, 161306(R).

38. Ludovico, M.F.; Moskalets, M.; Sànchez, D.; Arrachea, L. Dynamics of energy transport and entropy production in ac driven quantum electron systems. *Phys. Rev. B* **2016**, *94*, 035436.

39. Ludovico, M.F.; Arrachea, L.; Moskalets, M.; Sànchez, D. Probing the energy reactance with adiabatically driven quantum dots. *Phys. Rev. B* **2018**, *97*, 041416(R).

40. Erlingsson, S.I.; Manolescu, A.; Nemnes, G.A.; Bandarson, J.H.; Sanchez, D. Reversal of Thermoelectric Current in Tubular Nanowires. *Phys. Rev. Lett.* **2017**, *119*, 036804.

41. Mazza, F.; Valentini, S.; Bosisio, R.; Benenti, G.; Giovannetti, V.; Fazio, R.; Taddei, F. Separation of heat and charge currents for boosted thermoelectric conversion. *Phys. Rev. B* **2015**, *91*, 245435.

42. Carrega, M.; Solinas, P.; Braggio, A.; Sassetti, M.; Weiss, U. Functional integral approach to time-dependent heat exchange in open quantum systems: General method and applications. *New J. Phys.* **2015**, *17*, 045030.

43. Carrega, M.; Solinas, P.; Sassetti, M.; Weiss, U. Energy Exchange in Driven Open Quantum Systems at Strong Coupling. *Phys. Rev. Lett.* **2016**, *116*, 240403.

44. Moskalets, M.; Haack, G. Heat and charge transport measurements to access single-electron quantum characteristics. *Phys. Status Solidi B* **2017**, *254*, 1600616.

45. Crépieux, A.; Michelini, F. Mixed, charge and heat noises in thermoelectric nanosystems. *J. Phys. Condens. Matter* **2014**, *27*, 015302.

46. Crépieux, A.; Michelini, F. Heat-charge mixed noise and thermoelectric efficiency fluctuations. *J. Stat. Mech.* **2016**, *2016*, 054015.

47. Battista, F.; Haupt, F.; Splettstoesser, J. Correlations between charge and energy current in ac-driven coherent conductors. *J. Phys. Conf. Ser.* **2014**, *568*, 052008.

48. Battista, F.; Moskalets, M.; Albert, M.; Samuelsson, P. Quantum Heat Fluctuations of Single-Particle Sources. *Phys. Rev. Lett.* **2013**, *110*, 126602.

49. Battista, F.; Haupt, F.; Splettstoesser, J. Energy and power fluctuations in ac-driven coherent conductors. *Phys. Rev. B* **2014**, *90*, 085418.

50. Vannucci, L.; Ronetti, F.; Rech, J.; Ferraro, D.; Jonckheere, T.; Martin, T.; Sassetti, M. Minimal excitation states for heat transport in driven quantum Hall systems. *Phys. Rev. B* **2017**, *95*, 245415.

51. Dashti, N.; Misiorny, M.; Samuelsson, P.; Splettstoesser, J. Probing charge- and heat-current noise by frequency dependent fluctuations in temperature and potential. *Phys. Rev. Appl.* **2018**, *10*, 024007.

52. Laughlin, R.B. Anomalous Quantum Hall Effect: An Incompressible Quantum Fluid with Fractionally Charged Excitations. *Phys. Rev. Lett.* **1983**, *50*, 1395.

53. Vannucci, L.; Ronetti, F.; Ferraro, D.; Rech, J.; Jonckheere, T.; Martin, T.; Sassetti, M. Photoassisted shot noise spectroscopy at fractional filling factor. *J. Phys. Conf. Ser.* **2018**, *969*, 012143.

54. Wen, X.-G. Topological orders and edge excitations in fractional quantum Hall states. *Adv. Phys.* **1995**, *44*, 405.

55. Kane, C.L.; Fisher, M.P.A. Nonequilibrium noise and fractional charge in the quantum Hall effect. *Phys. Rev. Lett.* **1994**, *72*, 724.

56. Kane, C.L.; Fisher, M.P.A. Transport in a one-channel Luttinger liquid. *Phys. Rev. Lett.* **1992**, *68*, 1220.

57. Kane, C.L.; Fisher, M.P.A. Thermal Transport in a Luttinger Liquid. *Phys. Rev. Lett.* **1996**, *76*, 3192.

58. Voit, J. One-dimensional Fermi liquids. *Rep. Prog. Phys.* **1995**, *58*, 977.

59. Von Delft, J.; Schoeller, H. Bosonization for beginners—Refermionization for experts. *Ann. Phys. (Leipzig)* **1998**, *7*, 225.

60. Ferraro, D.; Braggio, A.; Magnoli, N.; Sassetti, M. Neutral modes' edge state dynamics through quantum point contacts. *New J. Phys.* **2010**, *12*, 013012.

61. Ferraro, D.; Ronetti, F.; Rech, J.; Jonckheere, T.; Sassetti, M.; Martin, T. Enhancing photon squeezing one leviton at a time. *Phys. Rev. B* **2018**, *97* 155135.

62. Chamon, C.; Freed, D.E.; Kivelson, S.A.; Sondhi, S.L.; Wen, X.-G. Two point-contact interferometer for quantum Hall systems. *Phys. Rev. B* **1997**, *55*, 2331.

63. Dolcetto, G.; Barbarino, S.; Ferraro, D.; Magnoli, N.; Sassetti, M. Tunneling between helical edge states through extended contacts. *Phys. Rev. B* **2012**, *85*, 195138.

64. Braggio, A.; Sassetti, M.; Kramer, B. Control of spin in quantum dots with non-Fermi liquid correlations. *Phys. Rev. Lett.* **2001**, *87*, 146802.

65. Dolcetto, G.; Sassetti, M.; Schmidt, T.L. Edge physics in two-dimensional topological insulators. *Riv. Nuovo Cimento* **2016**, *39*, 113.

66. Ronetti, F.; Vannucci, L.; Ferraro, D.; Jonckheere, T.; Rech, J.; Martin, T.; Sassetti, M. Hong-Ou-Mandel heat noise in the quantum Hall regime. *Phys. Rev. B* **2019**, *99*, 205406.

67. Kane, C.L.; Fisher, M.P.A. Quantized thermal transport in the fractional quantum hall effect. *Phys. Rev. B* **1997**, *55*, 15832.

68. Slobodeniuk, A.O.; Levkivskyi, I.P.; Sukhorukov, E.V. Equilibration of quantum Hall edge states by an Ohmic contact. *Phys. Rev. B* **2013**, *88*, 165307.

69. Blanter, Y.M.; Büttiker, M. Shot noise in mesoscopic conductors. *Phys. Rep.* **2000**, *336*, 1.

70. Ferraro, D.; Wahl, C.; Rech, J.; Jonckheere, T.; Martin, T. Electronic Hong-Ou-Mandel interferometry in two-dimensional topological insulators. *Phys. Rev. B* **2014**, *89*, 075407.

Article

Electron Traversal Times in Disordered Graphene Nanoribbons

Michael Ridley [1], Michael A. Sentef [2] and Riku Tuovinen [2,*]

[1] The Raymond and Beverley Sackler Center for Computational Molecular and Materials Science, Tel Aviv University, Tel Aviv 6997801, Israel
[2] Max Planck Institute for the Structure and Dynamics of Matter, 22761 Hamburg, Germany
* Correspondence: riku.tuovinen@mpsd.mpg.de

Received: 28 June 2019; Accepted: 25 July 2019; Published: 27 July 2019

Abstract: Using the partition-free time-dependent Landauer–Büttiker formalism for transient current correlations, we study the traversal times taken for electrons to cross graphene nanoribbon (GNR) molecular junctions. We demonstrate electron traversal signatures that vary with disorder and orientation of the GNR. These findings can be related to operational frequencies of GNR-based devices and their consequent rational design.

Keywords: quantum transport; graphene nanoribbons; nonequilibrium Green's function

1. Introduction

A fundamental property limiting the operational frequency of a molecular device is the traversal time τ_{tr} for electronic information to cross between the nanojunction terminals [1]. For instance, in graphene, the cutoff frequency f_{max} is related to the traversal time as $f_{max} = 1/2\pi\tau_{tr}$ [2,3]. For the molecular electrician, this raises the key question of how long it takes for electronic information to propagate across a nanosized device, as this sets a fundamental limit on the speed of the device operation. In quantum mechanics, time does not have the same status as a dynamical variable such as the energy or particle position. In fact, much debate has centred around the correct definition of the traversal time through a generic potential barrier [4–6], as well as the relation of this quantity to the dwell time (time spent in the molecular region) [7], the Larmor clock time (the time taken to move between scattering channels) [8,9], the group delay time (the time delay in the nonlocal propagating wave packet caused by scattering off the potential barrier) [10], or to a generic description of probability distributions via path integrals [11,12]. Crucially, all the aforementioned times are defined in terms of the transmission probability, potential and incident energy of electrons moving in a static scattering theory picture [13], so that a theory which takes strong time-dependence into account is still needed. This is crucial for the understanding of laser-stimulated tunnelling processes and related to the problem of tunnelling times in strong field ionisation experiments [14–16].

Graphene nanoribbon (GNR)-based molecular junctions are excellent candidates for room-temperature transistors, i.e., graphene field-effect transistors (GFETs) [2], GHz–THz frequency modulators [17], and photodetectors [18], due to their high mobility and charge carrier saturation velocities. GNRs can be engineered with band gaps that are tuneable via the nanoribbon symmetry properties and widths [19], and currently, sub-10 nm nanoribbon widths are accessible from chemical fabrication techniques [20–24]. Recent experimental progress shows an inverse scaling of operational frequency with the nanoribbon length [3], or with the square of the nanoribbon length [20] for ribbons whose carrier drift velocity scales inversely with ribbon length. Typically, the maximum operational frequencies of radio frequency (RF) GFETs exceed those of Silicon-based transistors with the same dimensions [3], and can in principle be achieved in the 100 GHz range [25–27]. The cutoff frequencies of GFETs are strongly affected by the presence of defects and flexibility in the nanoribbon, but 100 GHz flexible nanoribbons have also recently been fabricated [28].

Disorder may have a profound impact on the operation of the graphene-based devices. For example, it has been investigated that edge disorder affects the armchair-oriented GNR (AGNR) more than the zigzag-oriented ones (ZGNR) [29,30]. This is because the edge states in ZGNRs are energetically protected against impurity perturbations. This is not the case for AGNRs, in which the edge states are less dominant so that disorder has a much larger effect on the conductance [31,32]. On the other hand, disorder-induced broken chiral symmetry in terms of random bond disorder was considered in Ref. [33].

In this paper, we expand upon a recent proposal [34] to investigate traversal times for electrons moving in disordered GNRs by looking directly at the dynamics of statistical correlations between electronic signals measured in different reservoirs connected to the GNR. We demonstrate that the traversal time has a clearer signature in AGNR than in ZGNR. This is because the charge densities in AGNR structures are more delocalised than in ZGNR, where the formation of standing-edge-state charge waves leads to wave fronts with a spatial orientation lying diagonal across the plane of the nanoribbon [35–37]. We also show that the effect of breaking chiral symmetry (on-site disorder) has less effect on the traversal times compared with the disorder that preserves chirality (hopping disorder). (A random on-site potential breaks the sublattice symmetry and can broaden possible Landau levels, i.e., the chiral symmetry is destroyed. This does not happen with nearest neighbour hopping disorder as hopping between different sublattices is by construction the same in both directions [33,38–40]).

2. Model and Method

Our transport setup (cf. Figure 1) is described by the Hamiltonian $\hat{H} = \hat{H}_{\text{lead}} + \hat{H}_{\text{lead-GNR}} + \hat{H}_{\text{GNR}}$ where the lead environment is given by

$$\hat{H}_{\text{lead}} = \sum_{k\alpha} \epsilon_{k\alpha} \hat{c}_{k\alpha}^\dagger \hat{c}_{k\alpha}. \tag{1}$$

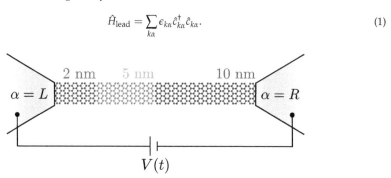

Figure 1. Transport setup for an armchair graphene nanoribbon where the left-most carbon atoms are connected to the $\alpha = L$ lead and the right-most carbon atoms are connected to the $\alpha = R$ lead. For times $t \geq 0$, a voltage bias $V(t)$ is applied in the leads and charge carriers start to flow through the graphene junction. We consider ribbons of varying lengths (2, 5, and 10 nm), and similarly also the zigzag orientation (not shown).

The operators $\hat{c}^{(\dagger)}$ are electron annihilation (creation) operators. At the initial switch-on time t_0, the energy dispersion is shifted by the bias voltage $\epsilon_{k\alpha} \rightarrow \epsilon_{k\alpha} + V_\alpha(t)$. The left-most atoms are connected to the left lead (L), whereas the right-most atoms are connected to the right lead (R). The coupling between the leads ($\alpha = L, R$) and the GNR has the form

$$\hat{H}_{\text{lead-GNR}} = \sum_{mk\alpha} (T_{mk\alpha} \hat{c}_m^\dagger \hat{c}_{k\alpha} + \text{h.c.}), \tag{2}$$

where h.c. stands for the hermitian conjugate. We wish to emphasise that the coupling matrix elements $T_{mk\alpha}$ are kept constant for all times (i.e., we work within the partition-free approach [41–43]). In practice, we choose the lead bandwidth, given by the energy dispersions $\epsilon_{k\alpha}$, to be much larger than the coupling energies so that we may employ the wide-band limit approximation (WBLA), in which the level width matrix $\Gamma_{\alpha,mn}(\omega) = 2\pi \sum_k T_{mk\alpha}\delta(\omega - \epsilon_{k\alpha})T_{k\alpha n}$ is evaluated at the Fermi energy of lead α. The WBLA is justified because we are interested in a regime where the lead-GNR coupling is weaker than the internal hopping within the GNR [44–47], as this also enables us to focus on the effect on traversal time caused by the internal ribbon structure. The GNR is modelled by a single π-orbital tight-binding picture

$$\hat{H}_{\text{GNR}} = \sum_{mn} h_{mn}\hat{c}_m^\dagger \hat{c}_n, \tag{3}$$

where the intramolecular hopping parameters h_{mn} are nonzero for the nearest neighbours only, and set by the typically used carbon–carbon hopping integral in graphene $h_{mn} = t_C = 2.7$ eV [48–51]. Longer-range hoppings could be included in the model similarly but here we wish to preserve the particle–hole symmetry of the undisordered GNR. While our system is completely described above, it would also be feasible to consider hydrogen passivation at the edges of the graphene nanoribbons, as is customary in typical experimental setups [52,53] and in their ab initio modelling [36,54,55], to remove dangling bonds at the edges.

We employ the recently developed time-dependent Landauer–Büttiker (TD-LB) formalism [35,56–64] to compute the two-time current correlation function $C_{\alpha\beta}(t_1, t_2) \equiv \langle \Delta\hat{I}_\alpha(t_1)\Delta\hat{I}_\beta(t_2)\rangle$ between the current deviation operators $\Delta\hat{I}_\alpha(t_1) \equiv \hat{I}_\alpha(t_1) - \langle \hat{I}_\alpha(t_1)\rangle$ between different leads labelled by α and β. The current operator of lead α is related to the particle number there via $\hat{I}_\alpha(t) \equiv q d\hat{N}_\alpha/dt$, where q is the charge of the particle. When there is no variation of current in one of the leads, i.e., $\Delta\hat{I}_\alpha(t_1) = 0$ the correlation between this signal and the current variation in the other leads is trivially zero. The partition-free approach employed here also means that the whole system, described by Equations (1)–(3), is coupled during an initial equilibration period that occurs prior to the switch-on time t_0. This leads to a global initial temperature T and lead-independent initial chemical potential μ.

In a two-terminal junction (cf. Figure 1), the labels α and β can refer to either the left (L) or right (R) terminals, whose energies are shifted symmetrically to create a voltage drop of $V(t)$ across the system $V(t)/2 = V_L(t) = -V_R(t)$. We note here that the driving bias voltage in our setup can be of dc type (sudden quench) or ac type (time-dependent modulation) [35,59,65]. In Ref. [34], it was shown that timescales associated with electron traversal times and internal reflection processes could be seen as resonances in the real part of symmetrised cross-lead correlations $C^{(\times)}(t+\tau, t) \equiv [C_{LR}(t+\tau, t) + C_{RL}(t+\tau, t)]/2$, as a function of the relative time τ. The traversal time τ_{tr} is therefore defined by the following relation:

$$\max \left| \text{Re}\left[C^{(\times)}(t+\tau, t)\right]\right| \equiv \left| \text{Re}\left[C^{(\times)}(t\pm\tau_{\text{tr}}, t)\right]\right|. \tag{4}$$

The motivation behind this definition becomes more apparent when we look at the simulation data of current cross-correlations in Section 3. It is, indeed, observed that the currents in each lead $I_L(t)$ and $I_R(t)$ become most strongly correlated at the time taken for electrons to propagate between the leads following a voltage quench, which is then attributed to electron traversal events.

Within the WBLA, the cross-correlations are evaluated in terms of Keldysh components of the Green's function $G(t_1, t_2)$ (projected onto the GNR subspace) [34]

$$
\begin{aligned}
C^{(\times)}\left(t_{1}, t_{2}\right)=\; & 2 q^{2} \sum_{\alpha \neq \alpha'} \operatorname{Tr}_{\mathrm{GNR}}\left\{\Gamma_{\alpha} G^{>}\left(t_{1}, t_{2}\right) \Gamma_{\alpha'} G^{<}\left(t_{2}, t_{1}\right)\right. \\
& +\; i G^{>}\left(t_{1}, t_{2}\right)\left[\Lambda_{\alpha}^{+}\left(t_{2}, t_{1}\right) \Gamma_{\alpha'}+\Gamma_{\alpha}\left(\Lambda_{\alpha'}^{+}\right)^{\dagger}\left(t_{1}, t_{2}\right)\right] \\
& +\; i\left[\Lambda_{\alpha}^{-}\left(t_{1}, t_{2}\right) \Gamma_{\alpha'}+\Gamma_{\alpha}\left(\Lambda_{\alpha'}^{-}\right)^{\dagger}\left(t_{2}, t_{1}\right)\right] G^{<}\left(t_{2}, t_{1}\right) \\
& -\; \Lambda_{\alpha}^{+}\left(t_{2}, t_{1}\right) \Lambda_{\alpha'}^{-}\left(t_{1}, t_{2}\right)-\left(\Lambda_{\alpha'}^{+}\right)^{\dagger}\left(t_{1}, t_{2}\right)\left(\Lambda_{\alpha}^{-}\right)^{\dagger}\left(t_{2}, t_{1}\right)\Bigg\},
\end{aligned}
\tag{5}
$$

where the $\Lambda_{\alpha}^{\pm}\left(t_{1}, t_{2}\right)$ matrices are defined in terms of convolutions on the real and imaginary branches on the complex time contour [34]. We investigate the dynamics of steady state (the switch-on time is taken to $t_{0} \rightarrow -\infty$) cross-correlations, which are accessible experimentally [66].

We note that there is some spreading in the individual resonant peaks associated with traversal times in the correlator, so that our proposal takes into account the probabilistic nature of electron propagation in accordance with realistic proposals for traversal time distributions [13,14,67]. This is particularly relevant for our approach which enables us to study arbitrary time-dependent biases, e.g., in which the drive is stochastic in time [61]. The Fourier transform of the real part of $C^{(\times)}\left(t+\tau, t\right)$, with respect to the relative time τ, is equivalent to the frequency-dependent power spectrum associated with cross-lead correlations:

$$
\lim_{t_{0} \rightarrow -\infty} \mathcal{F}\left\{\operatorname{Re}\left[C^{(\times)}\left(t+\tau, t\right)\right] ; \omega\right\}=\frac{P_{LR}\left(\omega\right)+P_{RL}\left(\omega\right)}{2}.
\tag{6}
$$

Here, $P_{\alpha\beta}\left(\omega\right)$ is defined as the Fourier transform with respect to τ of the real part of $C_{\alpha\beta}\left(t+\tau, t\right)$ [34]. In practice, the high-frequency component of the current fluctuations can be probed by studying the infrared-to-optical frequency range of light emitted by the junction [66]. It is worth noting that Equation (5) is valid in the transient regime for arbitrary time-dependent bias voltage profiles, whereas in Equation (6) the switch-on time t_{0} is taken to minus infinity ($t_{0} \rightarrow -\infty$) or, equivalently, the observation time t is taken to infinity ($t \rightarrow \infty$). While all the quantities in Equation (5) depend only on the time difference $\tau = t_{1} - t_{2}$ at this "long-time limit" (and the Fourier transform in Equation (6) is taken with respect to this relative time), in the transient regime, it would be required to consider separate frequencies for each time t_{1} and t_{2} in Equation (5).

The central idea of our work is that one should quantify the traversal time for electronic information to cross the system by looking directly at the correlations in the electronic signal itself, rather than trying to build an indirect definition of operational time from the calculation of transmission probabilities. The definition of traversal time here is closely related to the definition of Miller and Pollak, which makes use of flux–flux correlation functions [68]. However, the TD-LB formalism is valid for arbitrary lead temperatures, lead-GNR hybridisation strengths, and time-dependent biases.

3. Results and Discussion

As we consider the WBLA, the detailed electronic structure of the leads is not important for the description of the transport properties of the GNR. We then fix the coupling strength between the GNR and the leads by the frequency-independent resonance width $\Gamma_{\alpha} = t_{C}/10$ corresponding to a weak-coupling regime where the WBLA is a good approximation [44–47]. This is further justified in typical transport setups where the bandwidth of the leads is sufficiently large (e.g., gold electrodes) compared to the applied bias voltage. As we wish to preserve the charge neutrality of the GNR in equilibrium, we set the chemical potential to $\mu = 0$. The global equilibrium temperature is set by $\left(k_{B} T\right)^{-1} = 100 t_{C}^{-1}$ ($T = 313$ K).

3.1. Response to a dc Drive

It is instructive to first study the current correlations in GNRs without disorder. Figure 2 shows the current cross-correlations of undisordered AGNRs and ZGNRs of various lengths and *time-independent* bias voltages. We can make many general observations from the data:

- The signal is more clear for the AGNR than ZGNR. In the AGNR case, the propagating wavefront is coherent [35], so that there is less spread in the resonant traversal time signal than in the corresponding ZGNR case. This relates to the shape of the propagating wavefront, since in AGNR it is flat, whereas in ZGNR it has a triangular shape [35]. The back-and-forth internal reflections of the wavepackets between the electrode interfaces have a fairly regular structure in AGNR which results in a clear signal in the current cross-correlation. This means devices based on ZGNR have a less well-defined operational frequency.
- The current cross-correlations are mostly independent of the strength of the applied voltage. The voltage may affect the shape of the curves slightly, but not the location of the main resonance. This can be related to the group velocity of electrons crossing the GNR, $v_k = d\epsilon_k/dk$, which should not depend on a k-independent shift in the energy dispersion ϵ_k [34].
- Evidently, there is a roughly linear increase of the time-difference between the first maxima with increasing L, due to the time taken for the propagating electron wavefront to cross the structure. The time-difference between the first maxima τ_{max} is related to the traversal time of information through the GNRs via Equation (4).
- Increasing the length in the AGNR does not increase the number of resonant peaks in the cross-correlations, but in ZGNR it leads to a broader range of resonances clustered about a mean traversal time. This dependence on the orientation of the GNR then affects the spread of operational device frequencies.
- The low-frequency regions of the Fourier transforms show resonant frequencies at $\omega = n\Omega_L$ where n is a positive integer and Ω_L is some intrinsic frequency depending on the length of the GNR. In particular, by increasing the length of the GNR more transport channels are opened in the bias window, and therefore more peaks appear in the Fourier spectra.
- From the full frequency ranges of the Fourier transforms (insets), we observe a high-frequency operational cut-off which is smaller for the ZGNR case than for the AGNR case. This is itself an interesting effect, as it sets a limit on switches built with these kind of nanoribbons [2,3].

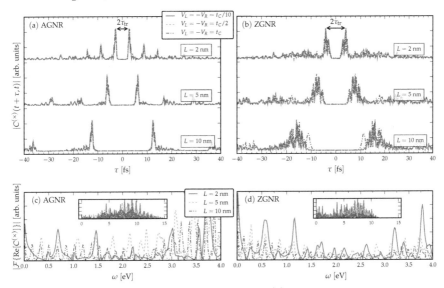

Figure 2. Absolute value of the current cross-correlation $C^{(\times)}(t + \tau, t)$ at long observation times $t \to \infty$ for various undisordered GNR samples with varying bias voltage and the consequent Fourier spectra: (**a**) AGNR of various lengths; (**b**) ZGNR of various lengths; (**c**) the low-frequency region of the Fourier transform of (**a**) for fixed bias voltage $V_L = -V_R = t_C/2$ (the inset shows the full frequency range); and (**d**) the low-frequency region of the Fourier transform of (**b**) for fixed bias voltage $V_L = -V_R = t_C/2$ (the inset shows the full frequency range).

3.2. The Role of Disorder

As we have now established general principles for the traversal times, we concentrate our discussion on disordered GNRs of fixed length and fixed applied (time-independent) voltage. In Figure 3, we introduce disorder into the GNR without breaking chiral symmetry. This is done by drawing a random number from a uniform distribution of width w around the average hopping matrix value t_C. We exclude second and third nearest neighbour hoppings, and we consider randomness only in the hoppings, for we wish to preserve particle–hole symmetry [49,69]. We see that the disorder appears to increase the traversal time and also has the effect of decreasing the quality of the signal, so that multiple side-peaks are visible. These are caused by internal reflections induced by the disorder. In addition, the intrinsic resonant frequencies for the disordered GNRs (shown by the Fourier spectra) are red-shifted due to the hopping disorder. This finding is consistent with the idea of reduced operational frequencies for the GNR devices due to disorder.

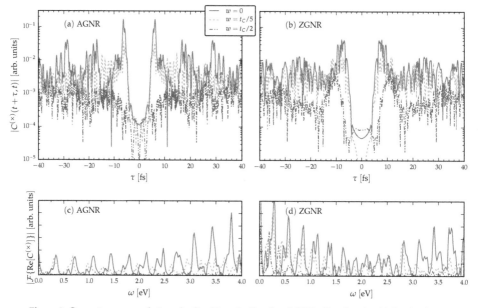

Figure 3. Current cross-correlations for fixed length disordered GNRs ($L = 5$ nm) with fixed voltage $V_L = -V_R = t_C/2$. A uniformly-distributed disorder w is included in the intramolecular hopping energies, which preserves the chiral symmetry.

In Figure 4, we break the chiral symmetry by adding a random term to the on-site energy levels of the GNR. This is also done by drawing a random number from a uniform distribution of width f around the zero on-site energy for the pristine GNR. In Figure 4a, we look at the AGNR case, and we observe in this case the deterioration of a clear traversal time signal as the Anderson localisation increases the average dwell time in the interior of the GNR. In Figure 4b, the average traversal time is once again seen to be larger in the ZGNR case. Interestingly, there is a crossover in both GNR configurations as the disorder destroys the coherence of the propagating wave packet around $f = t_C/2$. In contrast to the case of hopping disorder, here the operational frequencies of the GNR device (shown by the Fourier spectra) remain roughly unchanged for the on-site disorder. This finding could be related to the character of the disorder: While both types of disorder may introduce an effective tunnel barrier around the disordered GNR that the propagating electrons must overcome, the hopping disorder case corresponds to deformation of the lattice geometry, whereas the on-site disorder corresponds to a change in charge neutrality or chemical potential.

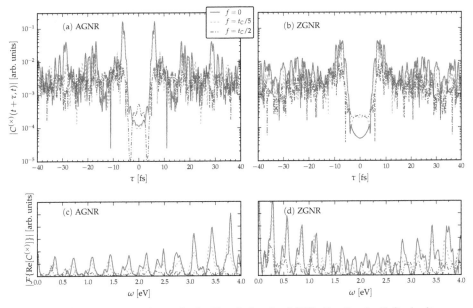

Figure 4. Current cross-correlations for fixed length disordered GNRs ($L = 5$ nm) with fixed voltage $V_L = -V_R = t_C/2$. A uniformly-distributed chiral symmetry-breaking disorder f is included in the on-site energy.

3.3. Response to an ac Drive

We finally address the full two-time character of the cross-correlation in Equation (5). Specifically, we consider the case of ac driving by introducing a monoharmonic driving term to the voltage

$$V(t) = V_0 + A\cos(\Omega t) \tag{7}$$

where the static part is set by $V_0 = t_C/2$ and the amplitude of the ac driving is $A = t_C/2$ with the driving frequency $\Omega = t_C/10$. To reduce the computational effort, we consider only the short nanoribbons in this case ($L = 2$ nm). In Figure 5, we show the propagation of the full two-time cross-correlation from the initial time t_0. AGNR data are shown in Figure 5a–c and ZNGR data are shown in Figure 5d–f. In contrast to the previous steady state results, here we show the initial transient (up to 50 fs), which includes relaxation effects.

We observe that the ac driving does not change the overall picture of traversal time, i.e., the time it takes for the information to traverse through the nanoribbons can be clearly read off from the separation of peaks along the anti-diagonal. Compared to the long-time limit in Figures 2–4, the initial transient only shows some additional oscillations but the main features seen in the steady state data are still visible. The two-time correlations also show the effect of disorder; as in the dc case, the signal gets considerably disturbed for hopping disorder (cf. Figure 3) and for on-site disorder (cf. Figure 4), but in the latter case the signal destruction is less severe. In the disorder energy scale considered in Figure 5 (hopping and on-site disorder strength: $w, f = t_C/2$), it is also observed that the first traversal event peaks are more clearly visible for the case of on-site disorder (Figure 5c,f) compared to the hopping disorder (Figure 5c,e). However, the coherence of subsequent traversal events without disorder (shown in the side-peaks of Figure 5a,d), is strongly suppressed by either kind of disorder. In calculations not shown here, we have checked that other types of ac driving (biharmonic drive, faster/slower modulation, lower/higher intensity) have no effect on the qualitative behaviour of traversal times.

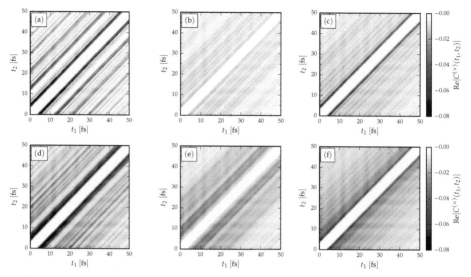

Figure 5. Two-time current cross-correlations for fixed length ($L = 2$ nm) GNRs with ac driving. (a) AGNR without disorder, (b) AGNR with hopping disorder $w = t_C/2$, (c) AGNR with on-site disorder $f = t_C/2$, (d) ZGNR without disorder, (e) ZGNR with hopping disorder $w = t_C/2$, (f) ZGNR with on-site disorder $f = t_C/2$.

4. Conclusions

We have employed the recently developed TD-LB formalism to compute the two-time current correlation functions for disordered GNRs. This methodology is a fast and accurate way of addressing mesoscopic quantum transport phenomena out of equilibrium as it is well-supported by the underlying nonequilibrium Green's function theory [70]. By our analysis, we confirm that the current cross-correlation is a good measure of electron traversal time. We find that the traversal time scales roughly linearly with the length of the GNRs, and that the traversal time also depends strongly on the GNR orientation. In general, the results presented in our work indicate that the traversal time can be most sensibly connected to the group velocity of noninteracting electrons passing through the graphene nanoribbon. The distance between the first maxima in the cross-correlations corresponds to the time for an electron with velocity $v \approx 1$ nm/fs to cross the ribbon. This velocity is consistent with the value of the Fermi velocity in graphene $v_F = 3t_C a/2\hbar$ where $a = 1.42$ Å is the carbon–carbon distance [49]. This can be related to the Büttiker–Landauer time for tunnelling [4] through a rectangular barrier as the time it would take a particle with velocity v_F to traverse the barrier.

We found that disorder in GNRs increases the traversal times, in general, and ultimately destroys the whole picture of coherent information transfer over the GNR junction when the disorder-induced scattering is strong. The "rule-of-thumb" character of our findings is summarised in Figure 6. We considered two types of disorder, one that preserves (hopping disorder) and one that breaks (on-site disorder) the chiral symmetry of the GNR. In Figure 6b, we find that the intrinsic operational frequency of the GNR is redshifted for the hopping disorder while, in Figure 6c, we see that it remains roughly unchanged for the on-site disorder. However, in the latter case, the statistical spread of τ_{tr} is significantly enhanced as the on-site disorder is increased. To measure the current cross-correlation and extract experimental values for τ_{tr}, there exist a range of spectroscopic techniques which relate the field strength of photons emitted from each lead to the current. The zero- and finite-frequency current cross-correlations can then be extracted from these current measurements [66,71].

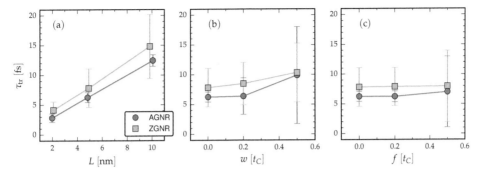

Figure 6. Electron traversal times estimated from the distance between the first maxima in the current cross-correlation (cf. Equation (4) and Figure 2): (**a**) no disorder, fixed voltage $V_L = -V_R = t_C/2$, varying length L; (**b**) fixed length $L = 5$ nm, fixed voltage $V_L = -V_R = t_C/2$, varying hopping disorder w; and (**c**) fixed length $L = 5$ nm, fixed voltage $V_L = -V_R = t_C/2$, varying on-site disorder f. The error bars are empirically estimated from the cross-correlation peaks as the full width at half maximum.

Our noninteracting approach is sufficient for graphene structures since monolayer graphene devices have been experimentally shown to have ballistic transfer lengths in the range from 100 nm at room temperatures to 1 μm at sub-Kelvin temperatures [2,72]. Even though our approach is limited to noninteracting electrons, we expect the current correlations and traversal times to be similarly related even when dealing with, e.g., electron–electron or electron–phonon interactions [47,73–75]. If perturbation theory could be applied, i.e., when the interaction is weak, current correlation or noise simulations are still feasible to perform in terms of the one-particle Green's function [76–81]. Disordered interacting systems have also been studied using mean-field or density-functional theories [82–84]. Here, out-of-equilibrium dynamics only due to voltage bias was considered but also thermal gradients could be included similarly [46,85–88]. At present, for the case of strong interaction, these approaches cannot yet be extended to realistic device structures since considerably more complicated and numerically expensive methods are required [89].

We also confirmed that the overall picture of electron traversal times is not qualitatively changed by introducing an ac driving voltage compared to the response to a dc drive. Possible quantitative differences in the response signals to an ac drive could be related to signatures of photon assisted tunnelling on traversal time [37,65] but, for now, will be left for future work. On the other hand, it would also be interesting to consider, e.g., a short laser pulse for exciting the system out of equilibrium [90–92] instead of the quench of the voltage bias employed in the present work. These topics will also be addressed more thoroughly in a forthcoming paper.

Author Contributions: Conceptualization, M.R. and R.T.; Formal Analysis, M.R., M.A.S. and R.T.; Writing—Original Draft, M.R. and R.T.; Writing—Review & Editing, M.R., M.A.S. and R.T.

Funding: This research was funded by the Raymond and Beverly Sackler Center for Computational Molecular and Materials Science, Tel Aviv University (M.R.) and by the DFG Grant No. SE 2558/2-1 through the Emmy Noether program (M.A.S. and R.T.).

Acknowledgments: We wish to thank Robert van Leeuwen for productive discussions.

Conflicts of Interest: The authors declare no conflict of interest.

References

1. Dragoman, D.; Dragoman, M. Time flow in graphene and its implications on the cutoff frequency of ballistic graphene devices. *J. Appl. Phys.* **2011**, *110*, 014302.
2. Lin, Y.M.; Jenkins, K.A.; Valdes-Garcia, A.; Small, J.P.; Farmer, D.B.; Avouris, P. Operation of graphene transistors at gigahertz frequencies. *Nano Lett.* **2009**, *9*, 422–426.

3. Liao, L.; Bai, J.; Cheng, R.; Lin, Y.C.; Jiang, S.; Qu, Y.; Huang, Y.; Duan, X. Sub-100 nm channel length graphene transistors. *Nano Lett.* **2010**, *10*, 3952–3956.

4. Büttiker, M.; Landauer, R. Traversal time for tunneling. *Phys. Rev. Lett.* **1982**, *49*, 1739.

5. Hauge, E.; Støvneng, J. Tunneling times: A critical review. *Rev. Mod. Phys.* **1989**, *61*, 917.

6. Landauer, R.; Martin, T. Barrier interaction time in tunneling. *Rev. Mod. Phys.* **1994**, *66*, 217.

7. Collins, S.; Lowe, D.; Barker, J. The quantum mechanical tunnelling time problem-revisited. *J. Phys. C Solid State* **1987**, *20*, 6213.

8. Baz', A.I. A quantum mechanical calculation of collision time. *Sov. J. Nucl. Phys.* **1967**, *5*, 161.

9. Rybachenko, V.F. Time of penetration of a particle through a potential barrier. In *In the Intermissions. . . Collected Works on Research into the Essentials of Theoretical Physics in Russian Federal Nuclear Center, Arzamas-16*; Trutnev, Y.A., Ed.; World Scientific: Singapore, 1998.

10. Winful, H.G. Tunneling time, the Hartman effect, and superluminality: A proposed resolution of an old paradox. *Phys. Rep.* **2006**, *436*, 1–69.

11. Yamada, N. Speakable and Unspeakable in the Tunneling Time Problem. *Phys. Rev. Lett.* **1999**, *83*, 3350.

12. Sokolovski, D. Path integral approach to space-time probabilities: A theory without pitfalls but with strict rules. *Phys. Rev. D* **2013**, *87*, 076001.

13. Landsman, A.S.; Keller, U. Attosecond science and the tunnelling time problem. *Phys. Rep.* **2015**, *547*, 1–24.

14. Landsman, A.S.; Weger, M.; Maurer, J.; Boge, R.; Ludwig, A.; Heuser, S.; Cirelli, C.; Gallmann, L.; Keller, U. Ultrafast resolution of tunneling delay time. *Optica* **2014**, *1*, 343–349.

15. Camus, N.; Yakaboylu, E.; Fechner, L.; Klaiber, M.; Laux, M.; Mi, Y.; Hatsagortsyan, K.Z.; Pfeifer, T.; Keitel, C.H.; Moshammer, R. Experimental Evidence for Quantum Tunneling Time. *Phys. Rev. Lett.* **2017**, *119*, 023201.

16. Hofmann, C.; Landsman, A.S.; Keller, U. Attoclock revisited on electron tunnelling time. *J. Mod. Opt.* **2019**, *66*, 1052–1070.

17. Gao, W.; Shu, J.; Reichel, K.; Nickel, D.V.; He, X.; Shi, G.; Vajtai, R.; Ajayan, P.M.; Kono, J.; Mittleman, D.M.; et al. High-contrast terahertz wave modulation by gated graphene enhanced by extraordinary transmission through ring apertures. *Nano Lett.* **2014**, *14*, 1242–1248.

18. Koppens, F.; Mueller, T.; Avouris, P.; Ferrari, A.; Vitiello, M.; Polini, M. Photodetectors based on graphene, other two-dimensional materials and hybrid systems. *Nat. Nanotechnol.* **2014**, *9*, 780.

19. Chen, Y.C.; Cao, T.; Chen, C.; Pedramrazi, Z.; Haberer, D.; De Oteyza, D.G.; Fischer, F.R.; Louie, S.G.; Crommie, M.F. Molecular bandgap engineering of bottom-up synthesized graphene nanoribbon heterojunctions. *Nat. Nanotechnol.* **2015**, *10*, 156.

20. Li, X.; Wang, X.; Zhang, L.; Lee, S.; Dai, H. Chemically derived, ultrasmooth graphene nanoribbon semiconductors. *Science* **2008**, *319*, 1229–1232.

21. Kimouche, A.; Ervasti, M.M.; Drost, R.; Halonen, S.; Harju, A.; Joensuu, P.M.; Sainio, J.; Liljeroth, P. Ultra-narrow metallic armchair graphene nanoribbons. *Nat. Commun.* **2015**, *6*, 10177.

22. Carbonell-Sanromà, E.; Garcia-Lekue, A.; Corso, M.; Vasseur, G.; Brandimarte, P.; Lobo-Checa, J.; de Oteyza, D.G.; Li, J.; Kawai, S.; Saito, S.; et al. Electronic Properties of Substitutionally Boron-Doped Graphene Nanoribbons on a Au(111) Surface. *J. Phys. Chem. C* **2018**, *122*, 16092–16099.

23. Li, J.; Sanz, S.; Corso, M.; Choi, D.J.; Peña, D.; Frederiksen, T.; Pascual, J.I. Single spin localization and manipulation in graphene open-shell nanostructures. *Nat. Commun.* **2019**, *10*, 200.

24. Li, J.; Friedrich, N.; Merino, N.; de Oteyza, D.G.; Peña, D.; Jacob, D.; Pascual, J.I. Electrically Addressing the Spin of a Magnetic Porphyrin through Covalently Connected Graphene Electrodes. *Nano Lett.* **2019**, *19*, 3288–3294.

25. Lin, Y.M.; Dimitrakopoulos, C.; Jenkins, K.A.; Farmer, D.B.; Chiu, H.Y.; Grill, A.; Avouris, P. 100-GHz transistors from wafer-scale epitaxial graphene. *Science* **2010**, *327*, 662.

26. Wu, Y.; Lin, Y.M.; Bol, A.A.; Jenkins, K.A.; Xia, F.; Farmer, D.B.; Zhu, Y.; Avouris, P. High-frequency, scaled graphene transistors on diamond-like carbon. *Nature* **2011**, *472*, 74.

27. Cheng, R.; Bai, J.; Liao, L.; Zhou, H.; Chen, Y.; Liu, L.; Lin, Y.C.; Jiang, S.; Huang, Y.; Duan, X. High-frequency self-aligned graphene transistors with transferred gate stacks. *Proc. Natl. Acad. Sci. USA* **2012**, *109*, 11588–11592.

28. Yu, C.; He, Z.; Song, X.; Liu, Q.; Gao, L.; Yao, B.; Han, T.; Gao, X.; Lv, Y.; Feng, Z.; et al. High-Frequency Flexible Graphene Field-Effect Transistors with Short Gate Length of 50 nm and Record Extrinsic Cut-Off Frequency. *Phys. Status Solidi RRL* **2018**, *12*, 1700435.

29. Areshkin, D.A.; Gunlycke, D.; White, C.T. Ballistic transport in graphene nanostrips in the presence of disorder: Importance of edge effects. *Nano Lett.* **2007**, *7*, 204–210.

30. Dauber, J.; Terrés, B.; Volk, C.; Trellenkamp, S.; Stampfer, C. Reducing disorder in graphene nanoribbons by chemical edge modification. *Appl. Phys. Lett.* **2014**, *104*, 083105.

31. Mucciolo, E.R.; Castro Neto, A.H.; Lewenkopf, C.H. Conductance quantization and transport gaps in disordered graphene nanoribbons. *Phys. Rev. B* **2009**, *79*, 075407.

32. Mucciolo, E.R.; Lewenkopf, C.H. Disorder and electronic transport in graphene. *J. Phys. Condens. Matter* **2010**, *22*, 273201.

33. Zhu, L.; Wang, X. Singularity of density of states induced by random bond disorder in graphene. *Phys. Lett. A* **2016**, *380*, 2233–2236.

34. Ridley, M.; MacKinnon, A.; Kantorovich, L. Partition-free theory of time-dependent current correlations in nanojunctions in response to an arbitrary time-dependent bias. *Phys. Rev. B* **2017**, *95*, 165440.

35. Tuovinen, R.; Perfetto, E.; Stefanucci, G.; van Leeuwen, R. Time-dependent Landauer-Büttiker formula: Application to transient dynamics in graphene nanoribbons. *Phys. Rev. B* **2014**, *89*, 085131.

36. Da Rocha, C.G.; Tuovinen, R.; van Leeuwen, R.; Koskinen, P. Curvature in graphene nanoribbons generates temporally and spatially focused electric currents. *Nanoscale* **2015**, *7*, 8627–8635.

37. Tuovinen, R.; Sentef, M.A.; Gomes da Rocha, C.; Ferreira, M. Time-resolved impurity-invisibility in graphene nanoribbons. *Nanoscale* **2019**, *11*, 12296.

38. Ludwig, A.W.W.; Fisher, M.P.A.; Shankar, R.; Grinstein, G. Integer quantum Hall transition: An alternative approach and exact results. *Phys. Rev. B* **1994**, *50*, 7526.

39. Kawarabayashi, T.; Hatsugai, Y.; Aoki, H. Quantum Hall Plateau Transition in Graphene with Spatially Correlated Random Hopping. *Phys. Rev. Lett.* **2009**, *103*, 156804.

40. Chen, A.; Ilan, R.; de Juan, F.; Pikulin, D.I.; Franz, M. Quantum Holography in a Graphene Flake with an Irregular Boundary. *Phys. Rev. Lett.* **2018**, *121*, 036403.

41. Cini, M. Time-dependent approach to electron transport through junctions: General theory and simple applications. *Phys. Rev. B* **1980**, *22*, 5887.

42. Stefanucci, G.; Almbladh, C.O. Time-dependent partition-free approach in resonant tunneling systems. *Phys. Rev. B* **2004**, *69*, 195318.

43. Ridley, M.; Tuovinen, R. Formal equivalence between partitioned and partition-free quenches in quantum transport. *J. Low Temp. Phys.* **2018**, *191*, 380–392.

44. Zhu, Y.; Maciejko, J.; Ji, T.; Guo, H.; Wang, J. Time-dependent quantum transport: Direct analysis in the time domain. *Phys. Rev. B* **2005**, *71*, 075317.

45. Verzijl, C.J.O.; Seldenthuis, J.S.; Thijssen, J.M. Applicability of the wide-band limit in DFT-based molecular transport calculations. *J. Chem. Phys.* **2013**, *138*, 094102.

46. Covito, F.; Eich, F.G.; Tuovinen, R.; Sentef, M.A.; Rubio, A. Transient Charge and Energy Flow in the Wide-Band Limit. *J. Chem. Theory Comput.* **2018**, *14*, 2495–2504.

47. Ridley, M.; Gull, E.; Cohen, G. Lead Geometry and Transport Statistics in Molecular Junctions. *J. Chem. Phys.* **2019**, *150*, 244107.

48. Reich, S.; Maultzsch, J.; Thomsen, C.; Ordejón, P. Tight-binding description of graphene. *Phys. Rev. B* **2002**, *66*, 035412.

49. Castro Neto, A.H.; Guinea, F.; Peres, N.M.R.; Novoselov, K.S.; Geim, A.K. The electronic properties of graphene. *Rev. Mod. Phys.* **2009**, *81*, 109.

50. Hancock, Y.; Uppstu, A.; Saloriutta, K.; Harju, A.; Puska, M.J. Generalized tight-binding transport model for graphene nanoribbon-based systems. *Phys. Rev. B* **2010**, *81*, 245402.

51. Joost, J.P.; Schlünzen, N.; Bonitz, M. Femtosecond Electron Dynamics in Graphene Nanoribbons— A Nonequilibrium Green Functions Approach Within an Extended Hubbard Model. *Phys. Status Solidi B* **2019**, 1800498.

52. Datta, S.S.; Strachan, D.R.; Khamis, S.M.; Johnson, A.T.C. Crystallographic Etching of Few-Layer Graphene. *Nano Lett.* **2008**, *8*, 1912–1915.

53. Papaefthimiou, V.; Florea, I.; Baaziz, W.; Janowska, I.; Doh, W.H.; Begin, D.; Blume, R.; Knop-Gericke, A.; Ersen, O.; Pham-Huu, C.; et al. Effect of the Specific Surface Sites on the Reducibility of a-Fe2O3/Graphene Composites by Hydrogen. *J. Phys. Chem. C* **2013**, *117*, 20313–20319.

54. Wang, Z.F.; Li, Q.; Zheng, H.; Ren, H.; Su, H.; Shi, Q.W.; Chen, J. Tuning the electronic structure of graphene nanoribbons through chemical edge modification: A theoretical study. *Phys. Rev. B* **2007**, *75*, 113406.

55. Lu, Y.H.; Wu, R.Q.; Shen, L.; Yang, M.; Sha, Z.D.; Cai, Y.Q.; He, P.M.; Feng, Y.P. Effects of edge passivation by hydrogen on electronic structure of armchair graphene nanoribbon and band gap engineering. *Appl. Phys. Lett.* **2009**, *94*, 122111.

56. Landauer, R. Electrical resistance of disordered one-dimensional lattices. *Philos. Mag.* **1970**, *21*, 863.

57. Büttiker, M. Four-terminal phase-coherent conductance. *Phys. Rev. Lett.* **1986**, *57*, 1761.

58. Tuovinen, R.; van Leeuwen, R.; Perfetto, E.; Stefanucci, G. Time-dependent Landauer–Büttiker formula for transient dynamics. *J. Phys. Conf. Ser.* **2013**, *427*, 012014.

59. Ridley, M.; MacKinnon, A.; Kantorovich, L. Current through a multilead nanojunction in response to an arbitrary time-dependent bias. *Phys. Rev. B* **2015**, *91*, 125433.

60. Ridley, M.; MacKinnon, A.; Kantorovich, L. Calculation of the current response in a nanojunction for an arbitrary time-dependent bias: Application to the molecular wire. *J. Phys. Conf. Ser.* **2016**, *696*, 012017.

61. Ridley, M.; MacKinnon, A.; Kantorovich, L. Fluctuating-bias controlled electron transport in molecular junctions. *Phys. Rev. B* **2016**, *93*, 205408.

62. Tuovinen, R.; Säkkinen, N.; Karlsson, D.; Stefanucci, G.; van Leeuwen, R. Phononic heat transport in the transient regime: An analytic solution. *Phys. Rev. B* **2016**, *93*, 214301.

63. Tuovinen, R.; van Leeuwen, R.; Perfetto, E.; Stefanucci, G. Time-dependent Landauer–Büttiker formalism for superconducting junctions at arbitrary temperatures. *J. Phys. Conf. Ser.* **2016**, *696*, 012016.

64. Tuovinen, R.; Perfetto, E.; van Leeuwen, R.; Stefanucci, G.; Sentef, M.A. Distinguishing Majorana Zero Modes from Impurity States through Time-Resolved Transport. *arXiv* **2019**, arXiv:1902.05821.

65. Ridley, M.; Tuovinen, R. Time-dependent Landauer-Büttiker approach to charge pumping in ac-driven graphene nanoribbons. *Phys. Rev. B* **2017**, *96*, 195429.

66. Février, P.; Gabelli, J. Tunneling time probed by quantum shot noise. *Nat. Commun.* **2018**, *9*, 4940.

67. Fertig, H. Traversal-time distribution and the uncertainty principle in quantum tunneling. *Phys. Rev. Lett.* **1990**, *65*, 2321.

68. Pollak, E.; Miller, W.H. New physical interpretation for time in scattering theory. *Phys. Rev. Lett.* **1984**, *53*, 115.

69. Gopar, V.A. Shot noise fluctuations in disordered graphene nanoribbons near the Dirac point. *Phys. E* **2016**, *77*, 23–28.

70. Stefanucci, G.; van Leeuwen, R. *Nonequilibrium Many-Body Theory of Quantum Systems: A Modern Introduction*; Cambridge University Press: Cambridge, UK, 2013.

71. Deng, G.W.; Wei, D.; Li, S.X.; Johansson, J.; Kong, W.C.; Li, H.O.; Cao, G.; Xiao, M.; Guo, G.C.; Nori, F.; et al. Coupling two distant double quantum dots with a microwave resonator. *Nano Lett.* **2015**, *15*, 6620–6625.

72. Miao, F.; Wijeratne, S.; Zhang, Y.; Coskun, U.C.; Bao, W.; Lau, C.N. Phase-Coherent Transport in Graphene Quantum Billiards. *Science* **2007**, *317*, 1530–1533.

73. Galperin, M.; Ratner, M.A.; Nitzan, A. Molecular transport junctions: Vibrational effects. *J. Phys. Condens. Matter* **2007**, *19*, 103201.

74. Swenson, D.W.; Cohen, G.; Rabani, E. A semiclassical model for the non-equilibrium quantum transport of a many-electron Hamiltonian coupled to phonons. *Mol. Phys.* **2012**, *110*, 743–750.

75. Härtle, R.; Cohen, G.; Reichman, D.R.; Millis, A.J. Decoherence and lead-induced interdot coupling in nonequilibrium electron transport through interacting quantum dots: A hierarchical quantum master equation approach. *Phys. Rev. B* **2013**, *88*, 235426.

76. Galperin, M.; Nitzan, A.; Ratner, M.A. Inelastic tunneling effects on noise properties of molecular junctions. *Phys. Rev. B* **2006**, *74*, 075326.

77. Souza, F.M.; Jauho, A.P.; Egues, J.C. Spin-polarized current and shot noise in the presence of spin flip in a quantum dot via nonequilibrium Green's functions. *Phys. Rev. B* **2008**, *78*, 155303.

78. Myöhänen, P.; Stan, A.; Stefanucci, G.; van Leeuwen, R. Kadanoff-Baym approach to quantum transport through interacting nanoscale systems: From the transient to the steady-state regime. *Phys. Rev. B* **2009**, *80*, 115107.

79. Lynn, R.A.; van Leeuwen, R. Development of non-equilibrium Green's functions for use with full interaction in complex systems. *J. Phys. Conf. Ser.* **2016**, *696*, 012020.

80. Miwa, K.; Chen, F.; Galperin, M. Towards Noise Simulation in Interacting Nonequilibrium Systems Strongly Coupled to Baths. *Sci. Rep.* **2017**, *7*, 9735.

81. Cabra, G.; Di Ventra, M.; Galperin, M. Local-noise spectroscopy for nonequilibrium systems. *Phys. Rev. B* **2018**, *98*, 235432.

82. Shepelyansky, D.L. Coherent Propagation of Two Interacting Particles in a Random Potential. *Phys. Rev. Lett.* **1994**, *73*, 2607.

83. Vojta, T.; Epperlein, F.; Schreiber, M. Do Interactions Increase or Reduce the Conductance of Disordered Electrons? It Depends! *Phys. Rev. Lett.* **1998**, *81*, 4212.

84. Karlsson, D.; Hopjan, M.; Verdozzi, C. Disorder and interactions in systems out of equilibrium: The exact independent-particle picture from density functional theory. *Phys. Rev. B* **2018**, *97*, 125151.

85. Eich, F.G.; Di Ventra, M.; Vignale, G. Density-Functional Theory of Thermoelectric Phenomena. *Phys. Rev. Lett.* **2014**, *112*, 196401.

86. Eich, F.G.; Principi, A.; Di Ventra, M.; Vignale, G. Luttinger-field approach to thermoelectric transport in nanoscale conductors. *Phys. Rev. B* **2014**, *90*, 115116.

87. Eich, F.G.; Di Ventra, M.; Vignale, G. Temperature-driven transient charge and heat currents in nanoscale conductors. *Phys. Rev. B* **2016**, *93*, 134309.

88. Eich, F.G.; Ventra, M.D.; Vignale, G. Functional theories of thermoelectric phenomena. *J. Phys. Condens. Matter* **2016**, *29*, 063001.

89. Ridley, M.; Singh, V.N.; Gull, E.; Cohen, G. Numerically exact full counting statistics of the nonequilibrium Anderson impurity model. *Phys. Rev. B* **2018**, *97*, 115109.

90. Kemper, A.F.; Sentef, M.A.; Moritz, B.; Freericks, J.K.; Devereaux, T.P. Direct observation of Higgs mode oscillations in the pump-probe photoemission spectra of electron-phonon mediated superconductors. *Phys. Rev. B* **2015**, *92*, 224517.

91. Sentef, M.A.; Kemper, A.F.; Georges, A.; Kollath, C. Theory of light-enhanced phonon-mediated superconductivity. *Phys. Rev. B* **2016**, *93*, 144506.

92. Kemper, A.F.; Sentef, M.A.; Moritz, B.; Devereaux, T.P.; Freericks, J.K. Review of the Theoretical Description of Time-Resolved Angle-Resolved Photoemission Spectroscopy in Electron-Phonon Mediated Superconductors. *Ann. Phys.* **2017**, *529*, 1600235.

MDPI

St. Alban-Anlage 66

4052 Basel

Switzerland

Tel. +41 61 683 77 34

Fax +41 61 302 89 18

www.mdpi.com

Entropy Editorial Office

E-mail: entropy@mdpi.com

www.mdpi.com/journal/entropy